AIRCRAFT
PROPULSION
and
GAS TURBINE
ENGINES

AIRCRAFT PROPULSION
and
GAS TURBINE ENGINES

Ahmed F. El-Sayed
Zagazig University
Zagazig, Egypt

CRC Press
Taylor & Francis Group
Boca Raton   London   New York

CRC Press is an imprint of the
Taylor & Francis Group, an **Informa** business

CRC Press
Taylor & Francis Group
6000 Broken Sound Parkway NW, Suite 300
Boca Raton, FL 33487-2742

© 2008 by Taylor & Francis Group, LLC
CRC Press is an imprint of Taylor & Francis Group, an Informa business

Library of Congress Cataloging-in-Publication Data

El-Sayed, Ahmed F.
    Aircraft propulsion and gas turbine engines / Ahmed F. El-Sayed.
        p. cm.
    "A CRC title."
    Includes bibliographical references and index.
    ISBN 978-0-8493-9196-5 (hardback : alk. paper)
    1. Airplanes--Turbojet engines. 2. Aircraft gas-turbines. I. Title.

TL709.E42 2006
629.134'353--dc22   2007034182

Visit the Taylor & Francis Web site at
http://www.taylorandfrancis.com

and the CRC Press Web site at
http://www.crcpress.com

Dedication

———

To
My wife Amany, and sons Mohamed, Abdallah, and Khaled

Contents

List of Figures

Preface

Aircraft engine represents the heart of an aircraft exactly as the heart of a human being.

Interest in aircraft engines (aero engines) and gas turbines is increasing day after day. This is due to the escalating use of aircraft as transportation means nationally and internationally on one side and the decisive role of air force in military conflicts from the other side. In addition, gas turbines are nowadays extensively used in electric power generation, gas pipeline compressor drivers as well as in marine propulsion for naval vessels, cargo ships, hydrofoils and other vessels, trains, automobiles, trucks and military tanks. As a consequence, understanding the concepts of aircraft propulsion and performance of aero engines and gas turbines is of great importance for mechanical and aeronautical students, engineers, and researchers in different civil and military fields. Aerodynamics interwoven with thermodynamics, heat transfer, and mechanical design are the foundations of aircraft propulsion and gas turbine design.

This book opens a new window that broadens the vision over the increasingly wide spectrum of propulsive engines and gas turbines. It is divided into two parts. The first part covers the history, classifications, and performance of air-breathing engines, while the second part is concerned with the design and analysis of the rotating and nonrotating modules (or components) of aero engines and gas turbines. The book can be used for two sequential courses in propulsion—an introductory course in jet propulsion and a gas turbine engine components course. The second part can be used as course material for a turbo machinery course by its own.

The first part of the book consists of eight chapters. The first chapter lists chronologically the inventions of aero engines and gas turbines throughout a complete century since the Wright brothers. In 1903, the first heavier-than-air flyer of the Wright brothers shared their flights in the sky with the birds. A piston engine was the source of power in their primitive aircraft and dominated for approximately four decades before the invention of jet engines. Whittle and von Ohain were the coinventors of jet engines. The jet age started in 1939 with turbojet engines and since then the development of aircraft and their power plants has been truly spectacular. Milestones in the history of turbojets and other air-breathing engines including turbofan, turboprop, ramjet, propfan, and the hypersonic engines, namely, turbo ramjet and scramjet, are listed. A unique classification of air-breathing and non–air-breathing engines closes this chapter. Chapter 2 explains how thrust force is generated, together with the types and factors affecting it. The performance parameters influencing both the engine and aircraft are derived and discussed.

Chapter 3 is devoted to pulse and ramjet engines. Both have nonrotating elements and resemble the simplest types of jet engines. Pulsejet was extensively used by the Germans during World War II. Recently, attention is being paid to the new pulse detonation engines (PDE). Both subsonic and supersonic ramjet engines are analyzed. In this and the subsequent chapters treating other air-breathing engines, a performance analysis of a supersonic ramjet operating at different conditions is performed. Two- and three-dimensional illustrations for the results are given. Students are encouraged to use available software to investigate and display the performance of air-breathing engines operating in different flight conditions. Chapter 4 treats turbojet engines, which was the first invented jet engine, in detail. Classifications and analyses of different types are discussed. Detailed analysis of gas cycle for the endeavor of achieving the best performance at different operating conditions is included and illustrated by a case study. Thrust augmentation methods including afterburning (or reheat as frequently used in UK) and water injection are discussed.

In Chapter 5, the currently dominant turbofan engines are analyzed. Enormous innovations of this versatile type are very frequent and have resulted in more than ten different types, and all of them are discussed in detail. Special interest is given to geared double-spool, triple-spool and thrust vectoring engines that power V/STOL aircraft. A parametric study for two case studies, namely, single- and double-spool engines, is given. Turboprop, turboshaft, and propfan engines are discussed in Chapter 6. These types of engines have a different flavor, as power, rather thrust, is the key design factor.

Hypersonic engines, scramjet and turbo ramjet, are analyzed in Chapter 7. Both are candidates for future high-speed planes of the forthcoming decades. Part I ends with Chapter 8, which discusses the various types of industrial gas turbines. The basic Brayton cycle modified by intercoolers, regenerators, and reheaters is analyzed, with available power and fuel consumption as the main issues.

Part II switches from the overall performance of a complete power plant to the performance of its constituents. In Part I, air-breathing engines are looked at as a black box, which is to be opened in Part II. This part is also composed of eight chapters. Chapter 9 examines the intake systems. It starts by defining the different methods of power plant installation in the aircraft, which influences the shape of air intake. Next, both subsonic and supersonic intakes are discussed. Chapter 10 is devoted to combustion chambers, both subsonic and supersonic. Emission, being an environmental constraint, is discussed in some detail. Chapter 11 discusses the exhaust systems. In fact, it is not only a nozzle design but also includes other issues including the thrust reverse and noise suppression systems. Chapters 12 through 15 present a detailed design of turbo machinery (rotating elements) in the engine. Centrifugal compressors are discussed in Chapter 12. A comprehensive classification is given, followed by analysis of its three components and their design and off-design operation. Chapter 13 gives full details of the design procedure for multistage axial compressors. A detailed step-by-step design of axial compressors is proposed. Simplified mechanical design as well as off-design and surge control are also introduced. Practical problems such as erosion and fouling of compressors are discussed outlining their contributions to the performance and lifetime. The next two chapters handle axial and radial inflow turbines. The detailed design procedure for axial turbines is given in Chapter 14. Simplified analyses of turbine cooling, mechanical design, and performance are given. Chapter 15 treats the smaller size radial inflow turbine found in smaller gas turbines and auxiliary power units (APU) of aircraft. Finally, Chapter 16 addresses the matching conditions in both industrial gas turbines and aero engines.

Each chapter includes numerous illustrative examples followed by a Problems Section. A list of important mathematical relations is given at the end of some chapters for clarification. Most of the examples and problems are borrowed from reality or, in other words, handle real aircrafts and engines. However, the given data are not the exact ones. The author used data as close as possible to simulate real behavior.

The book ends with three appendices, including a glossary for technical terms used in the field of aero engines and the gas turbine industry and some data for the available turbofan engines and industrial gas turbines.

In fact, the fruitful discussions with many colleagues (both from industry and academia) and students in several institutions in Egypt, the United States, the United Kingdom, Belgium, Austria, Libya, Japan, and China through the last 25 years were of great help during the preparation of the manuscript.

Ahmed F. El-Sayed

Acknowledgments

It is a great pleasure to acknowledge the continuous help of Professor Darrell Pepper, director, NCACM, University of Nevada, Las Vegas, for his invaluable comments and criticism. He continuously gave very helpful advice and encouragement. I am also grateful to Professor Shaaban Abdallah, Department of Aerospace Engineering, University of Cincinnati, for his support and help. I am deeply indebted to Dr. Hany Moustapha, senior fellow and manager, Pratt & Whitney Canada Technology Programs, for his support and help. I am so thankful to Garry Crook, Rolls-Royce Company; Michel Goulian, Airbus Industries; and Khalid Mahmood, Pratt & Whitney, as well as Jim Stump and Dr. David Wisler both from GE Aviation for their valuable support.

The technical support generously offered by Rolls-Royce Plc, Pratt & Whitney Canada, GE Energy, GE Aviation, Pratt & Whitney, Honeywell, and Airbus Industries is greatly acknowledged. I would like to thank the following agencies and companies for permission to use photos and illustrations: NASA, Boeing Integrated Defense Systems and Boeing Commercial Airplanes, Northrop Grumman Integrated System, EgyptAir, National Archives and Records Administration (NARA), Mus'ee de l'Hydraviation (France), Aviation Image Archive, BERIEF Russian Aircraft, Attonov Company, ALLSTAR website (http://www.allstar.fiu.edu), U.S. military.

I am also thankful to the following publishers: Concepts NREC, AIAA Education series, McGraw-Hill for their permission to use some of their figures. Appreciation is also due to MIT Press and Pearson Co. Honeywell and Airbus Industries are greatly acknowledged.

I owe much to my brother-in-law Riad El-Sharawi, Under Secretary, Ministry of Industry. His great support, help, and encouragement are greatly appreciated. I would like to thank Professor Dr. Khairy Makled, vice president and member of the Board of Trustees for Scientific Affairs, October 6 University, for his continuous support and encouragement.

I cannot forget the support and warm assistance of my former students, Aysam Botros, GE Energy, (USA); Hany Arafa and Hassan Zohier, Zagazig University; Ahmed Hamed, Egypt Air Company; Ahmed Mabrouk, Petroleum Air Service; Mahmoud Khalifa, Ahmed Zaky, and Mohamed Aziz, Air Force; and Ahmed Shousha, Institute of Aviation, Ministry of Aviation, Egypt.

My late parents instilled in me many of the qualities needed for such an undertaking. Their blessings surrounded me everywhere. The long tough hours spent in writing the book passed quickly and happily through the endless support of my family: my wife Amany, who provided a calm and pleasant environment and my sons Mohamed, Abdallah and Khaled for their indulgence, patience and love. Through their cheerful support this book was made possible. In particular, my son Abdallah, a computer specialist, devoted his experience and sacrificed his valuable time in bringing this book into reality. I owe a great deal of gratitude to my grand family members who shared this endeavor with me. My brother Professor Mohamed A. Helal, Cairo University, also had a great role in the advancement of this work. My sisters were the wonderful flowers in my life.

Finally, I would like to thank editor Jonathan W. Plant and publisher Nora Konopka for their endless assistance and backup that helped the manuscript to land safely.

Abstract

The book provides a unified and chronological approach to aero engines and gas turbines from the 1940s to the twenty-first century. Unique classifications for air-breathing and non–air-breathing engines are introduced. Examples are provided for present and past aero engines. All types of aero engines are analyzed including pulse jet, ramjet, turbojet, turbofan, turboprop, and turboshaft engines. Detailed analyses of hypersonic engines, scramjet and turbo ramjets, as well as planned future aircraft are also given. The ever increasing role of gas turbines is discussed in naval, automotive, truck, train, and tank propulsion, electrical generation, and pipeline compression.

The performance of each engine is examined at different operating conditions and the results are given in three-dimensional illustrations combining several variables. Developments and future trends in these engines are highlighted. The aerodynamic and thermodynamic behaviors of nonrotating components or modules including the intake, combustion chamber, and nozzle are given. Comprehensive and complete coverage of the aerothermodynamics of rotating elements, compressors, and turbines is provided. A detailed step-by-step design of axial compressors is proposed. A simplified mechanical design of the rotating components is introduced, along with different cooling techniques and heat transfer in turbines. Practical problems such as erosion and fouling of compressors are discussed, outlining their contributions to off-design conditions. A large number of solved examples are included that cover both aero engine performance and component aerodynamic and mechanical designs. A glossary for the technical terms employed in the aero engine and gas turbine industries is added. Appendices for the technical data of several types of aero engines and industrial gas turbines provide a simple database for the reader.

Part I

Aero Engines and Gas Turbines

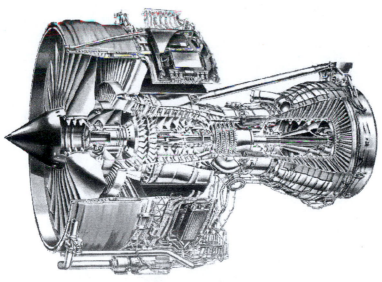

Rolls-Royce Trent 800

1 History and Classifications of Aero Engines

Humans have long dreamt of flying. Various ancient and medieval people who fashioned wings in order to fly met with disastrous consequences in leaping from towers or roofs and flapping their wings vigorously. The Greek myth of Daedalus and his son Icarus who were imprisoned on the island of Crete and tried to escape by fastening wings with wax and flying through the air is also well known. The dream of the humans to fly was achieved only in the twentieth century—at about 10:35 a.m. on Tuesday, December 17, 1903—when Orville Wright made the first successful flight in Kitty Hawk, North Carolina. That flying machine, identified as the Wright Flyer I, was the first heavier-than-air flyer designed and flown by the Wright brothers, Wilbur (1867–1912) and Orville (1871–1948). The Wright brothers, who are the inventors of the first practical airplane, are the premier aeronautical engineers in the history of mankind. A comparison of the Wright Flyer I and the aircraft of the twenty-first century, such as the Boeing 787 and Airbus 350, outlines the miracles that have taken place in the aviation industry. The Wright Flyer I did not have any fuselage and the only position the pilot could fly in was probably by lying prone on the bottom wing. Nevertheless, the effort marked the beginning of human-controlled powered flight. Several years lapsed before the design of the conventional aircraft with a closed fuselage installed to wings, a tail unit, and having an undercarriage or landing gears. Tremendous development in aviation industry now allows passengers in civil aircraft travel in air-conditioned compartments of the fuselage, in comfortable seats, and to eat, and watch video movies up in the air. However, it was a long story how people could be convinced to use aircraft as one of the modes of transportation.

Some milestones in such a long journey spanning nearly one century may be mentioned briefly. For several decades, piston engine coupled with propellers provided the necessary power for early aircraft. The turbojet engines (the first jet engines) invented independently by Sir Frank Whittle in Britain and Dr. von Ohain in Germany powered aircraft from the early 1940s. Such jet engines paved the way to the now, highly sophisticated military and comfortable civil aircraft. During the middle of the twentieth century, airlines relied upon low-speed subsonic aircraft whose flight speeds were less than 250 miles per hour (mph) powered by turbojet and/or turboprop engines. In the late 1960s and early 1970s, the wide-bodied aircraft (Boeing 747, DC-10 and Airbus A 300) powered by turbofan engines flew at transonic speeds, that is, at speeds less than 600 mph. Even now civilian aircraft fly at the same transonic speeds. On the other hand, military airplanes, fighter airplanes, for example, fly at supersonic speeds, that is at speeds less than 1500 mph. Such fighter planes are fitted with turbofan engines that have afterburners. X-planes, which are hypersonic, aircraft fitted with scramjet/rocket engines, fly at speeds less than 6000 mph. Space shuttles, which also have rocket engines, fly at hypersonic speeds of less than 17,500 mph.

It is interesting to compare the flight time between popular destinations such as Los Angeles and Tokyo on different airplanes—9.6 h for Boeing 747, 6.2 h for Concorde, and only 2 h for hypersonic aircraft [1]. It may be stated here that given humankind's endless ambitions, it is difficult to anticipate the shape, speed, and the fuel of the flying machines even for the next few decades.

It is a fact that evolution of aero-vehicles and aero-propulsion are closely linked. Unlike the eternal question of the chicken and the egg, there is no doubt as to which came first. The lightweight and powerful engine enabled humans to design the appropriate vehicle structure for both civil and military aircraft. Owing to the interdependence of the performance characteristics of aero-vehicles (including aircraft, missiles, airships, and balloons) and their aero-propulsion system, the evolution of both aero-propulsion and aero-vehicles will be concurrently reviewed in the next sections. The review of historical inventions will be divided into two phases. The first of these phases is related to the invention of prejet engines, while the second describes the invention and development of jet engines.

1.1 PREJET ENGINES—HISTORY

This section gives a brief description of the long history of flight events. It starts with some activities around the year 250 BC and ends just before the invention of jet engines in 1930s. Activities related to unpowered flight machines and important patents will be described first. Next, powered flights employing internal combustion gasoline engines will be described.

1.1.1 EARLY ACTIVITIES IN EGYPT AND CHINA

Jet propulsion is based on the reaction principle that governs the motion or flight of both aircraft and missiles. Though such a principle was one of the three famous laws of motion stated by Sir Isaac Newton in 1687, ancient Egyptians and Chinese utilized this principle several hundred years before him. The first known reaction engine was built by a noted Egyptian mathematician and inventor, Hero (sometimes called Heron) of Alexandria sometime around the year 250 BC [2]; some references go back to 150 BC [3]. Hero called his device aeolipile; see Figure 1.1. It consisted of a boiler or bowl, which held a supply of water. Two hollow tubes extended up from the boiler and supported a hollow sphere, which was free to turn on these supports. The steam escaped from two bent tubes mounted opposite one another on the surface of a sphere. The force created by the escaping steam transformed the nozzles into jet nozzles caused the sphere to rotate about its axis. It is said that Hero attached a pulley, ropes, and linkages to the axle on which the sphere rotated to use the aeolipile to pull open the temple doors without the aid of any visible power. Further details can be found in the website [4].

The Chinese discovered gunpowder around AD 1000. Some inventive person probably knew that a cylinder filled with gunpowder and open at one end would dart across a surface when the gun powder was ignited. Such a discovery was automatically employed in the Chinese battles by tying tiny cylinders filled with gunpowder to arrows. Thus these arrows would rocket into the air when ignited (Figure 1.2). The records of a battle that took place in China in AD 1232 provides evidence that solid rockets were used as weapons. These early Chinese scientists were the first people to

FIGURE 1.1 Aeolipile of Hero.

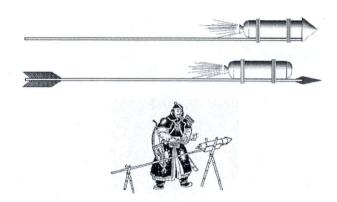

FIGURE 1.2 Chinese fire arrows.

FIGURE 1.3 Wu and his chair.

discover the principle of jet thrust. This same principle is the basis for today's jet engines. A Chinese scholar named Wan Hu planned to use these rockets in flying. A series of rockets lashed to a chair under which a sled had been placed may be seen in Figure 1.3. When the rockets were ignited, both Hu and the chair were obliterated. Hu was thus the first flight martyr.

1.1.2 Leonardo da Vinci

Leonardo da Vinci was possessed by the idea of human flight. He left surviving manuscripts, over 35,000 pages and 500 sketches, that deal with flight between AD 1486 and 1490. One of his famous ornithopter designs illustrates human-powered flight by flapping wings. Although this type of design was always doomed to be a failure, it is again being examined these days in some types of aircraft with hovering wings.

Around AD 1500 da Vinci introduced a sketch for a reaction machine, which was identified as the chimney jack [5]. It is a device for turning roasting spits; see Figure 1.4. The hot air rises up causing several horizontal blades to rotate, which in turn turns the roasting spit through bevel gears and a belt.

1.1.3 Branca's Stamping Mill

In 1629 Giovanni Branca, an Italian engineer invented the first impulse turbine; see Figure 1.5. Pressurized steam exited a boiler through a nozzle and impinged on the blades of a horizontally mounted turbine wheel. The turbine then turned the gear system that operated the stamping mill.

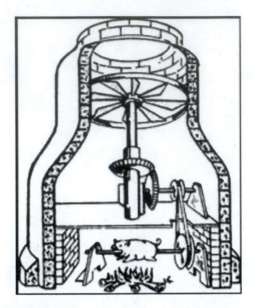

FIGURE 1.4   Chimney jack of Leonardo da Vinci.

FIGURE 1.5   Branca's stamping mill.

1.1.4 NEWTON'S STEAM WAGON

In 1687, Jacob Gravesand, a Dutchman, designed and built a carriage driven by steam power (Figure 1.6). Sir Isaac Newton may have only supplied the idea in an attempt to put his newly formulated laws of motion to the test. The wagon consisted of a boiler fastened to four wheels. The fire beneath the boiler generated the pressurized steam exiting from a nozzle in the opposite direction of the desired movement. A steam cock in the nozzle controlled the speed of the carriage. The proposed motion of such a vehicle relied upon the reaction principle formulated in one of the three famous laws of Newton. However, the steam did not produce enough power to move the carriage.

1.1.5 BARBER'S GAS TURBINE

The first patent for an engine that used the thermodynamic cycle of modern gas turbine (Figure 1.7) was given in 1791 by the Englishman John Barber. This engine included a compressor (of the reciprocating type), a turbine, and a combustion chamber.

FIGURE 1.6 Newton's steam wagon.

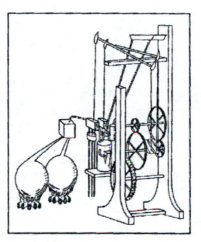

Barber's British Patent—1791

FIGURE 1.7 Barber's gas turbine.

1.1.6 MISCELLANEOUS AERO-VEHICLES' ACTIVITIES IN THE EIGHTEENTH AND NINETEENTH CENTURIES

Several red-letter dates may be listed here:

1. At 1:54 p.m. on November 21, 1783, the first flight of a hot-air balloon designed by Joseph and Etienne Montgolfier carrying Jean Pilatre de Rozier and Marquis d'Arlandes ascended into the air and traveled in a sustained flight for a distance of 5 mi across Paris. Next, the famous French physicist J.A.C. Charles, built and flew a hydrogen-filled balloon (the first use of hydrogen in aeronautics) in its second flight on December 1, 1783. The trip lasted for 2 h and traveled 25 mi.

2. Sir George Cayley in 1799 engraved his concept of an aircraft composed of fuselage, a fixed wing, and a horizontal and vertical tail, a very similar constitution to the now flying aircraft, on a silver disk [6]. In 1807, he also invented the reciprocating hot air engine. This engine operated on the same principle as the modern closed-cycle gas turbine [5]. He also invented a triplane glider (known as the boy carrier) in 1853 and the human carrier glider (1852). Sometime in 1852, he built and flew the world's first human carrying glider. Consequently, he is considered the grandparent of the concept of the modern airplane.

3. Development of internal combustion engines took place largely in the nineteenth century [7]. The first engine was described in 1820 by Reverend W. Cecil. These engines operated in a mixture of hydrogen and air. In 1838, the English inventor William Barnett built a single-cylinder gas engine, which burnt gaseous fuel. The first practical gas engine

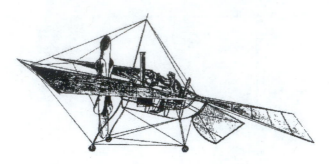

FIGURE 1.8   Du Temple's airplane.

was built in 1860 by the French inventor Jean Lenoir which utilized illuminating gas as a fuel.

The first four-stroke engine was built by the Germans August Otto and Eugen Langen in 1876. As a result four-stroke engines are always called Otto-cycle engines. George Braton in the United States also built a gasoline engine that was exhibited in 1876 in Philadelphia. However, the most successful four-stroke engine was built in Germany in 1885 by Gottlieb Daimler. In the same year a similar engine was also built in Germany by Karl Benz. The Daimler and Benz engines were used in early automobiles. Four-stroke engines were used extensively in the early aircraft.

1. The first airship designed and constructed by the Frenchman, Henri Giffard, was flown on September 24, 1852. This hydrogen-filled airship was powered by a steam engine (his personal design) that drove a propeller. In 1872, a German engineer, Paul Haenein, developed and flew an airship powered by an internal combustion engine fueled also by gaseous hydrogen. The first airship having sufficient control was constructed and flown by Charles Renard and A.C. Kerbs on August 9, 1884.
2. The first powered airplanes, which only hopped off the ground, were constructed by the Frenchman Felix Du Temple in 1874 (Figure 1.8) and the Russian Alexander F. Mozhaiski in 1884. They achieved only hops but not a sustained controlled flight.
3. In 1872, Dr. F. Stoltz designed an engine very similar in concept to the modern gas turbine engines. However, the engine never ran under its own power due to the components' poor efficiencies.
4. The first fully successful gliders in history were designed by Otto Lilienthal during the period 1891–1896 (Figure 1.9). Though he achieved over 2500 successful flights, he was killed in a glider crash in 1896.
5. Samuel Langley achieved the first sustained heavier-than-air unmanned powered flight in history with his small aerodrome in 1896. However, his attempt for manned flight was unsuccessful.

1.1.7   THE WRIGHT BROTHERS

The brothers Wilbur and Orville Wright achieved the first controlled, sustained, powered, heavier-than-air, manned flight in history. Several Internet sites showcase the Wright Brothers' flight; refer for example to the NASA site [8]. Also, many books describe the details of such achievements. One of the most recommended books is listed as Reference 9. Time was ripe for such achievements. This is due to the previous advancements in gliders, aerodynamics, and internal combustion engines. The Wright brothers designed and built three gliders (Gliders I, II, and III); a wind tunnel, propellers, and a light gasoline four-stroke four-cylinder internal combustion engine.

FIGURE 1.9   Glider of Otto Lilienthal.

FIGURE 1.10   Glider I of Wright Brothers. (Courtesy National Archives and Records Administration (NARA).)

At first they designed and built their Glider I in 1900 (Figure 1.10) and Glider II in 1901. However, the lift obtained was only one-third of that obtained from Lilienthal data. Thus, they had to build their own wind tunnel and test different wing models of their design. Through a balance system they measured the lift and drag forces. Moreover, they used the wind tunnel in testing light, long, twisted wooden propellers. Consequently, the Glider III built in 1902 was much better from an aerodynamic point of view. Thus, powered flight was just at their fingertips.

They designed and built their own engine because of the unsuitability of the available commercial engine for their mission. Their engine produced 12 hp and weighed 200 lb. It had 4.375 in. bore, 4 in. stroke, and 240 cubic in. displacement. The cylinders were made of cast iron with sheet aluminum water jackets. The crankcase was made of aluminum alloy. It also had in-head valves with an automatically operated intake valve and mechanically operated exhaust valve. A high-tension magneto is used for ignition.

Finally, they built their Flyer I (Figure 1.11). It had a wing span of 40 ft 4 in. and had a double rudder behind the wing and double elevator in front of the wing. The Wright gasoline engine drove two pusher propellers rotating in opposite directions by means of bicycle-type chains. The wing area was 505 ft^2. Total airplane weight with the pilot was 750 lb. Thus, as mentioned earlier, it achieved the first powered flight under a pilot's control on December 17, 1903. On that day, four flights were made during the morning, with the last covering 852 ft and remaining aloft for 59 s.

FIGURE 1.11   Flyer I of Wright Brothers. (Courtesy National Archives and Records Administration (NARA).)

After that epoch event the Wright brothers did not stop. In 1904 they designed their Flyer II having a more powerful and efficient engine. They made 80 flights during 1904 including the first circular flight. The longest flight lasted 5 min and 4 s traversing 2.75 mi. Further developments led to the Wright Flyer III in June 1905. Both the double rudder and biplane elevator were made larger. They also used a new improved rudder. This Flyer III is considered the first practical airplane in history. It made over 40 flights during 1905.The longest flight covered 24 mi and lasted for 38 min and 3 s.

Between 1905 and 1908, they designed a new flying machine, Wright type A, which allowed two persons to be seated upright between the wings. They also built at least six engines. Thus, their Wright type A was powered by a 40 hp engine. The Wright brothers met the public both in the United States in 1908 with impressive demonstrations by Orville for the army and with a public show by Wilbur at Hunaudieres close to Le Mans in France.

1.1.8   Significant Events up to the 1940s

1.1.8.1   Aero-Vehicle Activities

1. In 1909, the French Louis Bleroit flew his XI monoplane across the English Channel (Figure 1.12). Thus, it was the first time an airplane penetrated natural and political barriers.
2. In 1910, the first seaplane was built and flown by Henri Fabri at Martigues, France. The plane, called a hydravion, powered by a 50 hp Gnome rotary engine, flew 1650 ft on water (Figure 1.13). However, the real and great pioneer of marine flying was Glenn Curtiss. In 1911, he fitted floats to one of his pusher biplanes and flew it off the water. His contribution to marine flying included flying boats and airplanes, which could take off and land on a ship.
3. In 1914, by the time of war, airplanes had to be equipped with guns, bombs, and torpedos. The Vickers Gunbus (Figure 1.14) in England was a biplane airplane, which was considered the first aircraft specially designed as a fighter for the Royal Flying Crops. It was powered by one 100 hp Gnome radial piston engine and had one or two Lewis machine guns.
4. The world's first airline services were the Dirigibles in 1910. With the advances in aircraft design brought about by war, the enclosed cabin airplane became the standard for commercial airline travel by the early 1920s.
5. *Ford Trimotor*, nicknamed *The Tin Goose*, was a three-engine civil transport aircraft first produced in 1926 by Henry Ford and continued until about 1933 (Figure 1.15).

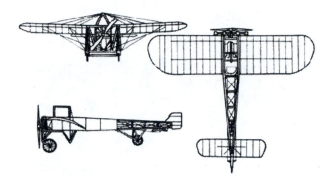

FIGURE 1.12 XI monoplane of Bleroit.

FIGURE 1.13 Henri Fabri. (Courtesy Musée de l'Hydraviation.)

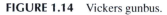

FIGURE 1.14 Vickers gunbus.

This Trimotor was developed from several earlier designs and became America's first successful airliner.

Throughout its lifespan a total of 199 aircraft were produced. It was popular with the military and was sold all over the world. Ford did not make the engines for these aircraft. The original (commercial production) 4-AT had three air-cooled Wright radial engines. The later 5-AT had more powerful Pratt & Whitney engines.

6. The first rigid airship filled with hydrogen called LZ-1, which was designed by Count Ferdinand von Zeppelin made a successful flight in 1900 (Figure 1.16). From 1911 to 1914, five Zeppelin airships were used in commercial air transportation in Germany. In the first

FIGURE 1.15   The Tin Goose.

FIGURE 1.16   Zeppelin LZ-1 airship. (Courtesy Early Aviation Image Archive.)

decade of the twentieth century, airships gained a great development in several countries including Great Britain, France, the United States, Italy, Spain, Poland, Switzerland, and Japan. Successive developments of these airships continued. In 1936, the first commercial air service operated across the North Atlantic between Germany and the United States with the HINDENBERG. At a cruising speed of 78 mph and carrying 50 passengers, it took 65 h for eastbound crossing and 52 h for westbound crossing.

1.1.8.2 Reciprocating Engines

Extensive development and use of airplanes during and after World War I led to a great improvement in engines. Some examples are given below:

1. Rotary-type engines, examples of which are the Gnome-Monosoupape (Figure 1.17) and the Bently.
2. In-line engines, such as the Hispano-Suiza, and such as the inverted in-line engines Menasco Pirate, model C-4.

FIGURE 1.17   The radial piston Gnome.

3. V-type engines such as Rolls-Royce V-12 and the U.S.-made Liberty V-12. Both upright and inverted V-types were available.
4. Radial engines, which could be single-, double-, and multiple-row radial engines. The 28 cylinder Pratt &Whitney R- 4360 engine was used extensively at the end of World War II and after that for both bombers and civil transports.
5. Opposed, flat, or O-type engines.

1.2   JET ENGINES

Before World War II in 1939, jet engines existed only as laboratory items for test. But at the end of the war, in 1945, it was clear that the future of aviation lay with jets. These jet engines gave greater power and thrust, were also compact in size, and simple in their overall layout.

1.2.1   JET ENGINE INVENTORS: Dr. HANS VON OHAIN AND SIR FRANK WHITTLE

Both Dr. Hans von Ohain and Sir Frank Whittle are recognized as the co-inventors of the jet engines [10]. Each worked separately and knew nothing of the other's work. Hans von Ohain, a young German physicist, was in the forefront. He is considered the designer of the first operational turbojet engine. Frank Whittle was the first to register a patent for the turbojet engine in 1930. Hans von Ohain was granted a patent for his turbojet in 1936. However, von Ohain's jet engine was the first to fly in 1939. Frank Whittle's jet first flew in 1941. The detailed histories of both the inventors are given in several books, articles, and internet websites [11–14].

1.2.1.1   Sir Frank Whittle (1907–1996)

Whittle is considered by many to be the father of the jet engine. In January 1930, Frank Whittle submitted his patent application for a jet aircraft engine (Figure 1.18). The patent was granted in 1932. However, he received very little encouragement from the Air Ministry or industry. After receiving support from investment bankers, Power Jets were established in 1936, and Whittle was assigned to the company to work on the design and development of his first jet engine.

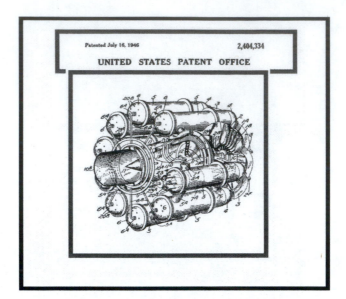

FIGURE 1.18   Patent of Frank Whittle.

FIGURE 1.19   Whittle's W.1 Engine and the Gloster E28/39.

The first run of the experimental engine was in April 1937. This engine had a centrifugal compressor and axial flow turbine. Many problems were encountered. Some were solved during the next year and the experimental engine was reconstructed several times. The resulting engine with 10 combustion chambers performed well. In June 1939, the Air Ministry was finally convinced of the merits of Whittle's invention and decided to have a flight engine. Thus, the first Whittle flight engine was built and was called the Power Jet W.1. The W.1 turbojet engine was designed to produce 1240 lbs at 17,750 rpm. The Air Ministry also decided to have the Gloster Aircraft Company build an experimental airplane called the E28/39. The aircraft was completed in March 1941 and the engine in May 1941. On May 15, 1941, the Gloster E28/39 powered with W.1 Whittle engine (Figure 1.19) took off from Cranwell at 7:40 p.m., flying for 17 min and reaching a maximum speed of around 545 km/h (340 mph). Within days it reached 600 km/h (370 mph) at 7600 m.

1.2.1.2 Dr. Hans von Ohain (1911–1998)

Hans von Ohain started development of the turbojet engine in the early 1930s in the midst of his doctorial studies in Goettinger University in Germany. By 1935, he had developed a test engine to demonstrate his ideas. Ernst Heinkel, an aircraft manufacturer supported him. At the end of February 1937, the He S-1 turbojet engine with hydrogen fuel was tested, which produced a thrust of 250 lb at 10,000 rpm. Next, von Ohain started the development of the He S-3 engine, particularly the liquid

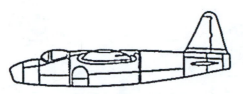

FIGURE 1.20 Heinkel's He-178.

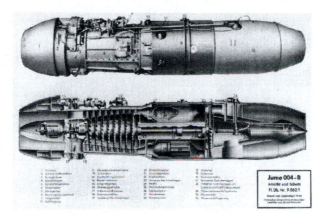

FIGURE 1.21 Jumo 004-B jet engine.

fuel combustor. Detailed design began in early 1938 on the test aircraft, the He-178 (Figure 1.20). In early 1939, both the engine and the airframe were completed. However, the thrust was below requirements. After several adjustments, the engine was ready. On August 27, 1939, Heinkel's test pilot, E. Warsitz, made the first successful flight of a jet-powered aircraft.

1.2.2 Turbojet Engines

1. The Messerschmitt Me 262 jet fighter was the first produced jet aircraft, powered by two Jumo 004-B turbojet engines (Figure 1.21). The Me 262's first flight was on July 18, 1942. The Jumo 004 was designed by Dr. Anslem Franz of the Junkers Engine Company. The Jumo 004 was the most widely used turbojet engine during World War II. The flight speed of the Me 262 was 500 mph (805 km/h) and it was powered by two axial flow jets, each of which developed 1980 pounds of thrust. More than 1600 Me 262 were built. Germany allowed Japan to build its own version of the Me 262 known as the Kikka. The Junker Jumo 004 was also assembled in Czechoslovakia under the name Malsice M-04 and it was used to power the Avia S92/CS92.

2. The British shared Whittle's technology with the United States enabling General Electric company—first engine builder in the United States—to build its first I-A jet engines. On October 1, 1942, America's first jet fighter, Bell XP-59 powered by the General Electric turbojet engine I-16, made its first flight signaling that the jet engine age had arrived in America. General Electric shortly built its J-31 engine, which was the first turbojet engine produced in quantity in the United States.

3. The British continued developing jet engines based on Whittle's designs, with Rolls Royce initiating work on the Nene engine in 1944 (Figure 1.22). The U.S. Navy selected the Nene for a carrier-based jet fighter, the Grumman Panther. Pratt & Whitney firm—another aircraft engine builder—built it under license. Rolls Royce also sold the Nenes to the

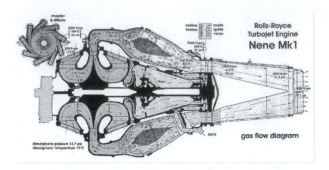

FIGURE 1.22   Nene Turbojet engine. (Reproduced from Rolls-Royce plc, copyright © Rolls-Royce plc 2007. With permission.)

Soviets. A Soviet-built version of the Nene engine subsequently powered the MiG-15 jet Fighter that fought the United States in the Korean War.

4.   Next, Rolls Royce designed and built the Tay engine, which also powered later versions of the Panther fighter.

5.   A.M. Lyulka is the father of the Soviet Jet Engine. In 1938 while he was working for the Kharkov Aviation Institute in a team developing the engine and compressor for the Tupolev Pe-8 heavy bomber, he designed a two-stage centrifugal compressor turbojet engine (RDT-1) with 500 kg (1100 lb) of thrust. Further developments of this engine led to the then USSR's first turbojet engine RDT-1/VDR-2. The second Lyulka turbojet to run was S-18/VDR-3 giving 1250 kg of thrust in August 1945. These two engines did not fly. His third engine TR-1, which is an eight-stage axial compressor developed from S-18 and giving 1300 kg of thrust, was the first engine to fly. It was fitted to Tu-22, Su-11, and Su-10 after clearance in December 1944.

6.   The Lyulka AL-7 represents another famous Russian engine. The prototype was tested in 1952 and the engine was planned to power the Ilyushin Il-54 bomber. The afterburning version AL-7F was created in 1953. In 1957, the Sukhoi Su-9 equipped with this engine exceeded the Mach 2 at 18,000 m. The engine was adopted for the Tupolev Tu-128P in 1960. The Lyulka AL-21 is closely similar to the GE J-79. It entered the service in early 1960s as it powered the interceptor Tu-28P. Later versions such as the AL-21-F3 were used in Sukhoi Su-24 Fencer, Mikoyan Mig-23, and Sukhoi T-10.

7.   Pratt & Whitney in the 1950s put an innovative design for turbojet engines that overcame the drawbacks of earlier engines consuming large quantities of fuel. It was the "dual-spool" concept. The engine had two compressors. Each rotated independently by a separate turbine. Thus, this engine looked as if two engines were combined in one. This approach led to the J-57 engine, which entered the U.S. Air Force in 1953. This engine powered several U.S. Air Force fighters including the F-100, the first airplane to break the sound barrier without going into a dive. Also, eight of the J-57 engines powered the B-52 bomber. It powered also the commercial aircraft Boeing 707 and the Douglas DC-8. However, its military secrets were discovered when the Soviet Union brought down the U-2 spy plane powered by this engine on May 1, 1960.

8.   Rolls Royce also adopted the slender engine designs by the German for its Avon series in 1953. The Avon powered the Hawker Hunter fighter of the 1950s. The Avon-powered Comet was the first turbojet engine to enter the transatlantic service. It also powered the American bomber B-57 and the French jet airliner Caravelle.

9.   The other engine competitor General Electric built its engine J-47, which powered the F-86 fighter and the B-47 bomber. Successive developments of the J-47 led to the J-73 in the early 1950s. This more powerful engine was used in the F-86H aircraft instead of the J-47 engines.

10. After the dual spool concept, another development for the turbojet engine was advanced by General Electric. This was the "variable stator" concept. Through this concept, the turbojet engines can avoid the compressor stall. This phenomenon will be explained later in the compressors chapters. But, it may be mentioned here that stall leads to a loss in thrust and sometimes causes severe damage by breaking off compressor blades.

11. A third advancement was achieved by the thrust augmentation or the afterburner. It is a second combustion chamber fitted between the last (low-pressure) turbine and the nozzle. This combustion chamber utilizes the remaining oxygen in the air in burning more fuel to a much higher temperature than that in the original combustion chamber. Such a thrust augmentation provides nearly 50% or more increase in the thrust force generated by the original engine's configuration.

12. The variable stator design and the afterburner led to the General Electric engine, J-79, which is the first true engine for supersonic flight. The Lockheed F-104 fighter powered by J-79 flew at twice the speed of sound. In May 1958, the F-104 set a record of 1404 mph (2260 km/h) and even at an altitude of 91,249 ft (27,813 m). Other fighters powered by the GE J-79 engines were the B-58 Hustler and the McDonnell Douglas Phantom F-4B.

13. It was interesting to celebrate the fiftieth anniversary of the turbojet engine Viper produced by RR turbojet engine in 2001. It was used to power several military aircraft including the two seat Italian military trainer Aermacchi-MB-326. Numerous versions of Viper were built along the course of its history. Its latest version was produced in Poland to power the Irdya jet.

14. Allison, another builder of aircraft engines in the United States built the J-33 engine, which powered the Lockheed T-33 training version of the F-80 Shooting Star.

15. The Ishikawajima Ne-20 is Japan's first turbojet engine, which was developed during World War II. Few numbers of this engine were manufactured. Two of them were used to power the Kikka airplane on its only flight on August 7, 1945.

16. An example of the Chinese aircraft and its power plants is the Shenyang J-8 fighter airplane powered by the turbojet engine PW-13A (Figure 1.23).

17. The Rolls-Royce/Snecma Olympus 593 turbojet (Figure 1.24) is unique in commercial aviation as the only afterburning turbojet to power a commercial aircraft, the Concorde. The Olympus 593 project was started in 1964, using the Avro Vulcan's Olympus 320 as a basis for development. Bristol Siddeley of the United Kingdom and Snecma Moteurs of France were to share the project. Acquiring Bristol Siddeley in 1966, Rolls-Royce continued as the British partner. Rolls-Royce carried out the development of the original Bristol Siddeley Olympus and engine accessories, while Snecma had the responsibility for variable engine inlet system, the exhaust nozzle/thrust reverser, the afterburner, and the

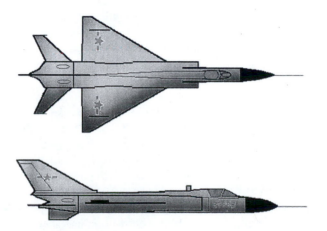

FIGURE 1.23 Shenyang J-8 Chinese fighter.

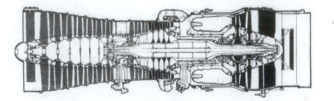

FIGURE 1.24   Rolls-Royce Olympus 593 turbojet engine powering concorde airplane. (Reproduced from Rolls-Royce plc, copyright © Rolls-Royce plc 2007. With permission.)

noise attenuation system. Britain was to have a larger share in production of the Olympus 593 in the same way as France had a larger share in fuselage production. At 15:40 on March 2, 1969, Concorde prototype 001 started its first take-off run with afterburners lit. The four Olympus 593 engines briskly accelerated the aircraft, and after 4700 ft (1.4 km) of runway and at a speed of 205 knots (380 km/h) the aircraft lifted off for the first time.

1.2.3 TURBOPROP AND TURBOSHAFT ENGINES

1. Soon after the turbojets were in the air the turboprop engine was developed. This engine is a gas turbine driving a propeller (previously identified as airscrew). Its early name was "airscrew turbine engine" and it was later given the neater title of turboprop [15]. Since it has mainly two parts, the core engine and the propeller, it produces two thrusts, one with the propeller and the other through the exhaust; a large gearbox makes it possible for the turbine to turn the propeller at a reduced speed to avoid any possible shock waves and flow separation at the propeller tip. There may be an additional turbine (called free or power turbine), which also drives the propeller through the gear box. Turboprop engines are more efficient than other engines when they fly under 30,000 ft and speeds below 400–450 mph.

2. Hungarian Gyorgy Jendrassik, who worked for the Ganz wagon works in Budapest, designed the very first working turboprop engine in 1938, called the "Cs-1." It was produced and flown briefly in Czecho-Slovakia between 1939 and 1942. The engine was fitted to the Varga XG/XH twin-engined reconnaissance bomber but proved very unreliable. Jendrassik had also produced a small-scale 75 kW turboprop in 1937.

3. In the 1940s Rolls-Royce built several turboprop engines for flight speeds of around 400 mph. One of the first engines was the RB50 Trent, which through a reduction gear, turned the five-blade Rotol propeller (7′ 11″ diameter). On September 20, 1945, the Gloster Meteor fitted with two Trents became the world's first turboprop-powered aircraft to fly.

4. Later on Rolls-Royce built the impressive Clyde; although it was one of its earliest turbo-prop engines, it reached 4200 hp plus 830 lb residual jet thrust, equivalent to 4543 equivalent horsepower. Also, DART (Figure 1.25), which is one of the most famous turboprops of RR powered Fairchild Industries' F-27, Grumman Gulfstream, and so on. More than 7100 Darts were sold, flying over 130 million hours. Moreover, another famous Rolls turboprop was the Tyne, which first ran in April 1955. It was a two-spool engine driving through a double-epicyclic gearbox and a four-blade reverse propeller.

5. Another British manufacturer was the Bristol who built Theseus that powered the Avro Lincoln aircraft. Next, Bristol built the Proteus that flew millions of hours and still drives the biggest hovercraft across the British Channel. A third British manufacturer is Armstrong Siddeley, which built Python with 4110 hp that powered the Wyvern carrier strike fighter. It also built Mamba and double Mamba, which powered the Fairey Gannet.

6. The first American turboprop engine was the GE T-31 built by General Electric. The T-31 maiden flight was in the Consolidated Vultee XP-81 on December 21, 1945. The T-31 was mounted in the nose and the J-33 turbojet mounted in the rear fuselage provided added

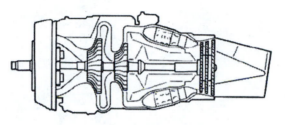

Single-entry two-stage
Centrifugal turbo-propeller

FIGURE 1.25   Rolls-Royce dart turboprop engine. (Reproduced from Rolls-Royce plc, copyright © Rolls-Royce plc 2007. With permission.)

FIGURE 1.26   V-22 Ospery. (Courtesy Boeing.)

thrust. Similar turboprop/turbojet engine combination was also used in the Navy XF2R-1 aircraft. Another General Electric turboprop engine with increasing shaft horsepower (SHP) was the T-56 turboprop powering the Lockheed Martin C-130-H and C-2A Greyhound. Moreover, General Electric built T-64 powering DeHavilland Canada DHC-5 and Buffalo aircraft.

7. Allison (another U.S. aero engine manufacturer) produced several turboprop engines such as GMA 2100 powering L100/C-130J and T56-A-11 powering AC-130A. Also it built the T-406 turboshaft engine that powers Bell Boeing V-22 Ospery aircraft (Figure 1.26).

8. AlliedSignal (previously Garrett AiResearch) is the third US manufacturer for turboprop engines that produced the famous TPF331 turboprop engine, which powered, for example, the Embraer/FAMA CBA-123. It is one of the few pusher turboprops in recent days.

9. The Soviet Union in the 1950s produced the Tu-95 Bear that used TV-12 engines having 4 contra-rotating propellers with supersonic tip speeds to achieve maximum cruise speeds in excess of 575 mph, altitude of 11,300 m, and a range of about 15,000 km. The Bear represents one of the most successful long-range combat and surveillance aircraft and was a symbol of Soviet power projection throughout the end of the twentieth century.

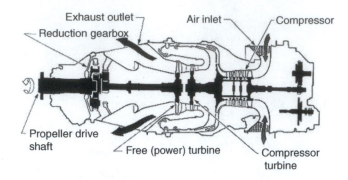

FIGURE 1.27   PT6 Turboprop engine. (Courtesy Pratt & Whitney Canada.)

10. Other turboprops manufactured by the Ukrainian company Motor Sich PJSC are the AI-20D Series (powering the AN-32 aircraft, which also successfully operates in a number of other countries, such as India, Bangladesh, Ethiopia, Peru, Nicaragua, etc.), and the AI-24 installed in the AN-24 aircraft servicing short- and medium-haul routes (both passenger and cargo).

11. Pratt & Whitney of Canada (PWC) built a big family of the popular turboprop/turboshaft engine PT6 and PT6A (Figure 1.27). It is the most popular turboprop engine ever built with 31,000 engines of the PT6A being sold to 5490 customers up to 1997. It powers numerous aircraft including Beechcraft 1900, Dornier DO 128-6, Bell412, and Sikorsky S58T.

12. Turbomeca of France began producing small turboprop engines in 1947. Its first turboprop was the 400 hp Marcadau of 1953, the 650 hp Bastan of 1957. Its 800–1100 shp Astazou was the only turboprop to sell in numbers. However, Turbomeca is recognized as the first company to build a true turboshaft engine. In 1948 they built the first French-designed turbine engine, the 100 shp 782. In 1950 this work was used to develop the larger 280 shp Artouste, which was widely used on the Aerospatiale Alouette II and other helicopters [16].

13. Other worldwide companies include the Italian Fiat that built the turboprop T64-P4D that powered the aircraft G222RM. The Japanese company IHI built T56-IHI-14 and T58-IHI-110 turboprops, which powered the aircraft P-3C and KV-107, respectively.

14. The greatest power turboprop engine is the NK-12 built in 1946 to provide 12,000 hp by the Russian company Kuznetsov and upgraded in 1957 to 15,000 hp, which powers the huge aircraft AN-22. Other famous Russian turboprop engines are the TV3 VK-1500; Isotov company's engines that power the aircraft An-140; medium helicopters; and Canadaair 540C.

15. EPI Europrop International GmbH was created by the four leading European aero engine companies (Industria de Turbo Propulsores, MTU Aero Engines, Rolls-Royce and Snecma) to develop and produce the three shaft 11,000 SHP. TP400-D6 turboprop is used to power Airbus A400M military transport. The A400M will have short takeoffs and landings. It is a two-shaft engine with a free power turbine. Under the original timetable for the A400M project, the first flight should take place at the first quarter of 2008 with 192 deliveries scheduled from October 2009.

16. Turboshaft engines are similar to turboprops. They differ primarily in the function of the turbine. Instead of driving a propeller, the turbine is connected to a transmission system that may drive helicopter rotors. Other turboshaft engines that will be identified as industrial gas turbine engines are used for land and sea applications such as electrical generators, compressors or pumps, and marine propulsion for naval vessels, cargo ships, hydrofoils, and other vessels; see Section 1.2.5.

17. Pratt & Whitney of Canada (PWC) built a big family of the popular turboshaft engine PT6 and PT6A. It powers numerous aircraft including Sikorsky S58T.

18. Also General Electric built several turboshaft engines, including CT58 powering Sikorsky S-61, S-62, and Boeing Vertol CH-47 helicopters. GE T700/CT7 is considered one of its most famous turboshaft engines, which powers Bell AH-1W Super Cobra.
19. Rolls-Royce also built numerous turboshaft engines such as Gem powering Augusta-Westland Lynx helicopter and AE1107C-Liberty turboshaft that powers Bell Boeing V-22 Ospery.
20. One of the earliest Russian turboshaft engines is the D-25V manufactured in 1957 with 4780 kW (5500 hp) that powered the Mi-6, Mi-10, and Mi-12 helicopters.
21. The Ukrainian three-spool turboshaft engine D-136 Series 1 powers the MI-26 transport helicopters that are the largest in the world. Other famous Ukrainian turboshaft engines are the TB3-117 BMA used to power the KA helicopters and the TB3-117 BM powering the MI-17 and MI-8 AMT helicopters.

1.2.4 TURBOFAN ENGINES

1. During 1936–1946 Power Jet and Metropolitan pioneered several kinds of ducted fan jet engines, but it attracted no attention as was the case with patent of Whittle for its jet engine. The Rolls-Royce Conway engine was the first to introduce the "bypass" principle in the 1950s. It featured a large fan located in the front of the engine which resembled a propeller but had many long blades set closely together. The bypass jet engines are frequently denoted as turbofan engines. Turbofan engines had three benefits, namely, a greater thrust, an improved fuel economy, and a quieter operation. The noise level was not a serious problem in those days, but now noise level is a strict condition for engine certification. The Conway powered the four engine Vickers VC-10 jetliner as well as some Boeing 707s and DC-8s.
2. A smaller bypass engine, the Rolls-Royce Spey, was built under license in the United States, which powered the attack plane A-7 flown for both the U.S. Navy and Air Force. One Spey engine powers the Italian/Brazilian AMX airplane while four Spey engines power the BAE SYSTEMS Nimrod aircraft.
3. Both General Electric and Pratt & Whitney built their own turbofan engines after 1965. The first aircraft to use turbofan engines in the United States was the Lockheed C-5A "Galaxy" using the GE TF-39 turbofan engines with a bypass ratio (BPR) of 8. Later on all the U.S. Air Force cargo planes, including the C-17, as well as all the large airliners, including the Boeing 747, used turbofan engines.
4. With the emergence of the wide-bodied airliners in the late 1960s, Rolls-Royce launched the three-spool turbofan engine RB211. This engine is a three-spool engine having three turbines, two compressors, and a fan. After some difficulties, this engine has now established itself at the heart of Rolls-Royce world class family of engines. One of the earliest aircraft powered by RB211 was the Lockheed L-1011 Tri-Star. Later on this engine powered numerous aircraft including Boeing 747, 757, 767, and Tupolev Tu 204-120.
5. In the 1960s, Pratt & Whitney (P&W) built its high bypass ratio turbofan engine JT9D that also powered the wide body Boeing 747 (Figure 1.28). In the 1970s, P&W built its popular engine JT8D-200, which was a quieter, cleaner, and more efficient engine. In the 1990s, P&W built its PW4000 series that powered the Boeing 777.
6. The GE CF6 family of high turbofan engines have gained great popularity since the early 1970s. These engines have powered numerous civil and military aircraft including Airbus A300-600/600R, A310-22, A330, Douglas DC-10, McDonnell Douglas MD-11 Stretch, Boeing 747-200/300/400, and 767-200ER. By the late 1990s, more than 5500 CF6 engines were in service.
7. In 1974, collaboration between General Electric and SNECMA, the leading French aircraft engine manufacturer, resulted in establishing the CFM International, which received the

JT9D-20 Turbofan engine

FIGURE 1.28   JT9D-20 Turbofan engine. (Courtesy Pratt & Whitney.)

first order for CFM56-2 turbofan engine in 1979. CFM56-2 was chosen to re-engine DC-8 Series Super 70s and its military version designated F108, to re-engine the tanker aircraft KC-135. Since then the General Electric and SNECMA CFM56 engines have powered Boeing 737-300/400/500 series, Airbus Industries' A318, A319, A320, A321, and the long-range four-engine Airbus 380. The CFM56-7 power plant was launched in 1993 to power the Boeing 737-600/700/800/900 series. In the second half of the 1990s, more than 3500 CFM engines were delivered worldwide.

8. In the 1980s, a unique five-nation consortium led to the establishment of the International Aero Engines AG (IAE) consisting of the British Rolls-Royce, German MTU, Italian Fiat, American Pratt & Whitney, and the Japanese Aero Engine Corporation (JAEC) and that produced the turbofan engine V2500. Now the IAE is only a four-nation consortium, with the withdrawal of the Italian Fiat company. More than 2920 engines have seen delivered up to April 2006. V2500 powered numerous aircraft including Boeing MD-90 and Airbus corporate jetliners, Airbus A321, 320, and 319.

9. Another alliance between General Electric and P&W was achieved in the 1990s to build the GP7000 series, which powered the super jumbo Airbus A380.

10. Rolls-Royce continued building high technological aero engines—its new family of Trent 500, 700, 800, 900, and 1000. These engines power a big family of Boeing and Airbus airplanes. The Trent 1000 is planned to power the Boeing 7E7 (Figure 1.29). Under construction now is the Trent 1700, which is designed specially to power the Airbus aircraft in production, A350.

11. GE 90 is one of the largest turbofan engines with a BPR of nearly 9.

12. AlliedSignal Garrett is another aero engine manufacturer in the United States. It built the famous engine TFE731 that powered the Lockheed Jetstar II, Dassault Falcon 10, Hawker Siddeley125-700 as well as the three-spool F109 (TFE76) engine.

13. The first Russian turbofan engine was the D-20P, which had advanced thrust, efficiency, and reliabile performance. It is a two-spool engine and is installed in the Tu-124 short-haul airplane for passenger transportation since October, 1962. The two-spool D-30 series, which powered the Tu-134 aircraft, entered service in September 1967. The D-30KP turbofan engine having a takeoff thrust of 12,000 kg$_f$ powers IL-76T, IL-76TD, and IL-76MD versions of the popular long-haul IL-76 airplane for transportation of large-size equipment and cargo. Its commercial use for cargo transportation started since July, 1974.

FIGURE 1.29   Trent 1000 Turbofan engine. (Reproduced from Rolls-Royce plc, copyright © Rolls-Royce plc 2007. With permission.)

14. Several three-spool engines manufactured by the famous Ukrainian company (Motor Sich JSC) are now in flight. Examples are the D-18T Series 1 and 3 powering the AN-124 RUSLAN and AN-225 MRIYA transports. Also the same manufacturer and the other Ukrainian manufacturer Ivchenko Progress ZMKB produce several three-spool turbofans, including the D-436 powering the short-and medium-haul airliners TY-134M, TY-334-100, TY-334-200, D-36 Series 1 and 1E as well as the D-36 Series 2A and 3A installed in AN-72 and AN-74 passenger aircraft. The same two manufacturers produce the low BPR turbofan engines AI-25 TD, which powers the MIG UTS twin-engine trainer airplane.

1.2.5 PROPFAN ENGINE

1. The Metrovick F-2 turbojet engine was rebuilt in 1945 as a turbofan engine and was identified as F.2/4. One of these F.2/4 turbofan engines was fitted with a two-stage open propfan to give an engine identical in layout to the GE UDF of more than 40 years later, but nobody was interested [15].
2. The propfan engines emerged in the early 1970s as the price of fuel began to soar [15]. Propfan is a modified turbofan engine with the fan placed outside of the engine nacelle on the same axis of the compressor. They are also known as ultra-high-bypass (UHBP) engines. It may be a single rotation or a contrarotating engine. Propfan engine combines the best features in turboprop and turbofan engines. The main features of a propfan are more blades but reduced diameter than propeller, swept or scimitar-like profile and thinner blades also compared with blades of turboprops, higher BPR than turbofan, higher propulsive efficiency than both engines and higher flight speed than turboprop engines, and fuel efficiency similar to a turboprop engine.

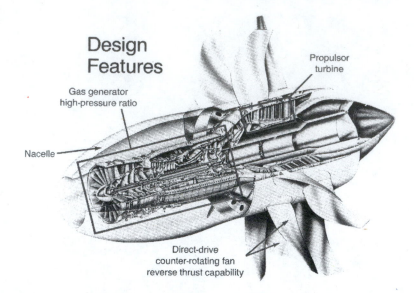

FIGURE 1.30   General electric unducted fan (GE UDF). (Courtesy GE AE.)

3. Hamilton Standard (HamStan) pioneered an idea that NASA took up. In 1974 HamStan tested full-scale propfan having eight blades, assuming cruise altitude of 35,000 ft and Mach number of 0.8 at NASA Ames and Lewis labs. The test resulted in a propfan having a tip sweep of 45° and a thickness/chord ratio of 2% outboard of 0.6 radius.
4. In 1981, NASA–HamStan developed a single rotation propfan driven by the 6000 shp Allison T701 mounted ahead of the swept wing of a Gulfstream II.
5. NASA contracts to General Electric led to the design of the GE UDF (unducted fan) (Figure 1.30). It is based on the aft-fan concept with contrarotating variable pitch propfan blades giving a BPR of about 34, which is 20%–30% more fuel efficient than subsonic turbofan engines.
6. The joint program of PW/Allison led to the 578-DX engine that was demonstrated in Indianapolis in pusher configuration. Similar to the GE engine, it was flown on the same MD-80 from April 1989.
7. Russia and Ukraine designed their NK-93 propfan, which was being tested since 1992 and flown in 1994. It powered D Il-96M and Tu-204 from 1997.
8. The contrarotation D-27 (Figure 1.31) is the most famous propfan developed by Ukraine since 1991. A pusher version of D-27 powered the AN-180, while a tractor version powered the AN-70. Russian and Ukrainians propfans are the only flying propfans until now.

1.2.6   PULSEJET, RAMJET, AND SCRAMJET ENGINES

1.2.6.1   Pulsejet Engine

1. Pulsejet is a simple form of jet propulsion developed from the ramjet. It has spring-loaded inlet valves that slam shut when the fuel ignites. The German Argus As 014 had its first flight at the end of 1942 when fitted to the Fieseler Fi 103/V-1 Flying Bomb (or Buzz Bomb) producing about 750 lb of thrust (Figure 1.32). It was also used to power the Me 328 and versions of the He 280 and He 162A-10.
2. In the United States, Ford manufactured its pulsejet engine PJ-31-1, which was a copy of the Argus As 014. It was fitted to the Republic JB-2 "Thunderburg" and Northrop JB-10. A pair was fitted to the North American P 51 D "Mustang" for tests.

FIGURE 1.31 Ukrainian propfan (or Turbopropfan) D-27. (Courtesy Antonov aircraft.)

FIGURE 1.32 Pulsejet engine.

3. In Japan, a license copy of the Argus As 014 was manufactured and had the name Ka-10.
4. The Russian RD-13/D-10 pulsejet producing around 400 kg of thrust were developed from the German Argus As 014/044 engines.
5. The pulsejet Jumo 226 with a thrust of 500 kg was developed in Dessau, Germany, at the end of World War II. Development continued after the Soviets occupied Dessau. Twelve engines were produced and tested in Germany by fitting them to the Junkers Ju 88G in the period from 1946 to 1947. Next, work was moved to Plant 30 near Moscow.

1.2.6.2 Ramjet and Scramjet Engines

1. The ramjet (sometimes denoted as flying stovepipe) was invented in 1908 by the French scientist Rene Lorin [17]. It consisted of just a tube at one end with the air entering by

the forward motion of the plane. Air is compressed at the inlet and then fuel is added and burnt. The resulting hot gases are expelled at the other end. The ramjet does not contain any turbomachinery (compressor/turbine). Ramjet engine does not start working until the speed of the fight vehicle is around 200 mph, but it gets more efficient the faster it goes. As an airplane increases its speed past Mach 1.0, the air pressure created from the speed of the airflow decreases the need for a compressor. As speeds approach Mach 3.5–4.0, a compressor is not even needed. The ramjet is the most efficient engine because it has few components.

2. Other ramjet pioneers (1913–1947) [18] are Albert Fono (General Electric, United States, patent in 1928), M. Stechkin (Soviet Union, ramjet theory, 1929), and René Leduc (France, patent, 1933).

3. In June 1936, René Leduc demonstrated the practical application of his ramjet engine, which he had been working on since the 1920s. The French government ordered the building of a plane using the new Ramjet engine in 1937. However, work stopped because of the war.

    By November 1946 with the Leduc 010 fitted on top of a specially modified version of the aircraft Sud Est SE-161 "Languedoc" four-engine transport plane "mother ship," the first test flights began on November 16, 1946. The first free gliding flight of the Leduc 010 occurred on October 21, 1947 (Figure 1.33).

4. Other contributors in France were Marcel Warnner, Maurice Roy, and Georges Brun.

5. The development of ramjet started by the Russians in 1933 went on until 1939. The main contributors in the USSR were MM. Bondaryuk (who patented the ramjet VRD-430 with 400 mm diameter, which was manufactured in 1940), Dudakov, Merkulov, Pobedonostsev, Tsander, and Zouyev.

6. In the United States, several contributions were made by those at General Electric (Alexander Lippisch, O. Pabst, M. Trommsdorff, and H. Walter), Roy Marquardt from APL/JHU as well as by the staff at NASA Lewis and Langley research centers. The subsonic ramjets developed by Roy Marquardt in 1945 were later adapted to the Bell Aircraft P-83 and the North American F-82 twin Mustang in 1947. In 1948, a pair of ramjets was fitted onto the wingtip of Air Force's brand new F-80 "Shooting Star" made by Lockheed.

7. The supersonic ramjet developed by Marquardt powered the Rigel missile. The final Rigel flight (before August 1953 when the program was canceled) flew for 31 miles at an altitude of 42,000 feet while reaching a speed of Mach 2.5. Another supersonic ramjet (Scramjet)

FIGURE 1.33   Ramjet Leduc 010.

FIGURE 1.34   X-15. (Courtesy NASA.)

engine powered the X-7 missile, the first flight of which was on April 26, 1951. It achieved a new Mach record of 3.9 at 55,000 ft.

8. Other U.S. ramjets in operational missile programs are those such as the BOMARC missile (1949), Navaho, and Navy's Talos. Target missile drones were made such as XQ-5 "kingfisher" (which afterwards became AQM-60A), "Red Roadrunner" MQM-42A, and the X-7A ramjet test vehicle by Lockheed Skunk Work which later also developed the D-21 drone using a ramjet similar to that first mounted on top of SR-71 and was later used with the B-52. Other missiles powered by ramjets were the Vandal, Triton, super-Talos, and X-15 (Figure 1.34). A ramjet-powered version of cruise missile for underwater launch was achieved and known as IRR Torpedo Tube Vehicle [19].

9. In the United Kingdom, Bristol Aero Engines developed reliable and powerful ramjet engines for Red-Duster Bloodhound II SAM missile. Hawker Siddeley Dynamics developed ramjet engines (Oden) for the sea-dart missiles which were used in the Royal Navy on Type 21 and 22 destroyers in the 1970s.

10. Sub- and supersonic ramjets were developed in other countries. Some of these countries and their engines are listed here: Germany, Eugen Sanger; Sweden, RR2, RRX-1, RRX-5; USSR, SA4 Ganef—SA6 Gainful (SAN3 Goblet); France, ONERA Stataltex, Arsenal (SE 4400, CT41, VEGA), and MATRA, SNECMA (ST 401, ST 402, ST 407).

11. Then there was the development of other ramjets such as "nuclear ramjet," that is, project "Pluto" where the world's first nuclear ramjet engine "Troy-II-A" was operated in 1961—the "first NASP" program Hydrogen Research Engine (HRE) in 1964–1968; the first "dual-mode" scramjet that burned both hydrogen and hydrocarbon fuel (1965–1968); and the ASALM flight test program in the 1970s that went to Mach 7.12. Guinness World Records recognized NASA's X-43A scramjet with a new world speed record for a jet-powered aircraft—Mach 9.6 or nearly 7,000 mph. The X-43A set the new mark and broke its own world record on its third and final flight on November 16, 2004.

1.2.7 INDUSTRIAL GAS TURBINE ENGINES

1. As defined in [20], the term industrial gas turbine is used for any gas turbine other than aircraft gas turbine. Industrial gas turbines have no limitations on their size and weight, and have nearly zero exhaust speeds and a long time between overhaul, which is of the order of 100,000 operating hours. The principal manufacturers for large industrial gas turbines are Alstom, General Electric, and Siemens-Westinghouse, all of whom designed single-shaft engines delivering more than 250 MW per unit. As industrial gas turbine proves itself to be an extremely versatile prime mover, it has been used for a variety of applications, ranging from electric power generation, mechanical drive systems, pump drives for gas or liquid pipelines, chemical industries, and land and sea transport.

2. Gas turbines for electrical power generation since the 1970s were primarily used for peaking and emergency applications. The main advantage of such engines was their ability to produce full power from cold in less than two minutes. Both Rolls-Royce Avon and Olympus engines are used in Great Britain while Pratt & Whitney FT-4 is used in

North America. Countries whose area consists mostly of deserts such as Saudi Arabia extensively use gas turbines in generating electricity since they do not need cooling water.

3. Another application for gas turbine is generating electricity for off-shore platforms. Solar, Ruston turbines, and General Electric LM 500 are used for small power requirements (nearly 5 MW). Rolls-Royce RB211 as well as the Industrial Trent (50, 60) and the General Electric units LM 1600, LM 2500, LM 5000, and LM 6000 are used for larger power requirements (up to 60 MW).

4. Industrial gas turbines have been extensively used in pumping applications for gas and oil transmission pipelines.

5. Great engines in the sky equal great gas turbines aboard ships. Gas turbines have a long history of successful operation in navies. These engines have the advantages of compact size, high power density, and low noise. Several navies such as those of the United Kingdom, the United States, Russia, Canada, Japan, and the Netherlands used gas turbines. Motor Gun Boat in 1947 was the first successful boat to use gas turbine. In 1958, the Rolls-Royce Proteus gas turbine engines were first used in fast patrol boats. The General Electric LM Series was used in the US Navy Bruke Destroyer, AEGIS Gruiser, Italian Lupo Frigate as well as other large ferries, cargo and cruise ships, patrol boats, hydrofoils, and aircraft carriers. The Rolls-Royce Trents were also used in the container ships. In the late 1980s the Canadian DDH-280 (which was the first war ship to use gas turbines in 1970) used both of the P&W FT-4s for boost power and Allison 570s for cruise.

6. The AlliedSignal Lycoming AGT1500 gas turbine was used to power the US tank M1A1 Abram. This tank was used in the Gulf War with great success.

7. Union Pacific used gas turbines for 15–20 years since 1955 to operate freight trains. PT6, the PWC turboshaft engine, was used to operate the CN Turbotrain. High-speed trains powered by turboshaft engines were built by the French and operated in the eastern USA. The Industrial Trent gas turbine allows the starting of large trains.

8. The PT6 turboshaft engine was used to operate the British Columbia Snow Plow. Moreover, the P&W Mobilepac engine, the gas turbine-powered electric generating unit on wheels, was designed as a mobile power trailer.

9. The Rolls-Royce Industrial Trent family is designed to meet the higher power, variable speed demands required by other applications such as natural gas liquefaction, gas transportation, and gas injection for oil recovery.

1.3 CLASSIFICATIONS OF AEROSPACE ENGINES

Aerospace engines [21] are classified into two broad categories, namely, air-breathing or non–air-breathing engines (Figure 1.35). Air-breathing engines (or airbreathers) are engines that use the *air* itself, through which the vehicle is flying, both as an *oxidizer* for the fuel in the combustion chamber and as a *working fluid* for generating thrust [22]. Non–air-breathing engines are the rocket engines wherein the propulsive gas originates onboard the vehicle. Rocket engine is defined as a jet propulsion device that produces thrust by ejecting *stored matter*, called the *propellant* [23].

The air-breathing engines are further divided into reciprocating engines and jet engines. Piston engines were the engines that powered the first successful flight of Wright brothers. These remained the dominant source of power for aircraft propulsion for the next 40 years until jet engines took their place. From 1903 to 1908 piston engines employing water-cooled engines were used. The problem arose that days, was the excess drag and weight caused by the liquid-cooled engines which in turn influenced the airplane performance. By 1908, air-cooled engines replaced the liquid-cooled ones, which led to a substantial savings in weight (30–40%). However, these air-cooled engines did not perform as well as expected. Thus, from about 1915 onward, better designs of liquid-cooled engines

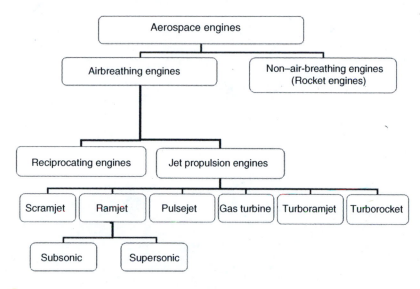

FIGURE 1.35 Classification of aerospace engines.

were able to develop more power than air-cooled ones. Many of the best fighters of World War II were powered by liquid-cooled engines. By the middle of the 1940s the debate between water- and air-cooled engines was over with the air-cooled ones coming out as the winners. NACA developed a cowling enclosing the engine to improve the efficiency of air-cooled aircraft. This enclosure for the engine limited the flow of air over the engine cylinders to the air actually in contact with the cooling fins of the cylinders.

However, piston engines were too heavy to compete with the jet engines for the same power generated. Jet engines have many advantages over reciprocating engines, the most obvious being the capability of higher-altitude and higher-speed performance. Control is simpler because one lever controls both speed and power. With the large airflow, cooling is less complicated. Spark plugs are used only for starting, and the continuous ignition system of the reciprocating engines is not needed. A carburettor and mixture control are not needed.

The dawn of the jet engine was in the 1940s and a tremendous development has been achieved in the next sixty years and until now.

1.4 CLASSIFICATION OF JET ENGINES

Jet engines are further subdivided into numerous categories. Jet engines are subdivided into ramjet, pulsejet, scramjet, gas turbines, turbo-ram, and turbo-rocket engines (Figure 1.35).

A brief description of each type is given here. Further classifications of any of them, if any, are also introduced.

1.4.1 RAMJET

The ramjet engine is an athodyd or aero-thermodynamic-duct to give its full name [24]. It is composed of three elements, namely, an inlet, which is a divergent duct, a combustion zone, and a nozzle, which is either convergent or convergent–divergent. There are no rotating elements, and neither a compressor nor a turbine. Layouts for subsonic and supersonic ramjet are illustrated in Figures 1.36 and 1.37, respectively.

When forward motion is imparted to the engine from an external source, air is forced into the inlet (or air intake). Air entering the inlet loses velocity or kinetic energy and increases its pressure

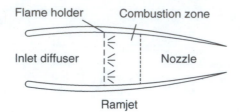

FIGURE 1.36   Subsonic Ramjet engine.

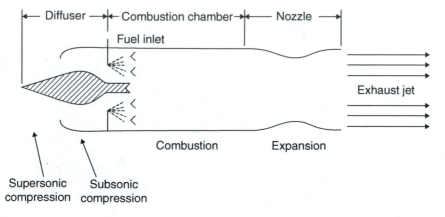

FIGURE 1.37   Supersonic Ramjet engine.

energy since the inlet is a convergent duct. Next, it enters the combustion zone where it is mixed with the fuel and burned. The hot gases are then expelled in the nozzle where the thrust force is generated. The ramjet is inefficient at subsonic flight speeds. As described above, the ramjet has the virtue of maximum simplicity, as no turbomachinery is installed, and maximum tolerance to high-temperature operation and minimum mass-per-unit thrust at suitable flight Mach numbers [25].

1.4.2 PULSEJET

A pulse jet engine (or pulsejet) is a very simple form of internal combustion engine. A pulse jet is similar to a ramjet, except that a series of spring-loaded shutter-type valves (one-way valves) is located ahead of the combustion section. In a pulse jet, combustion is intermittent or pulsing rather than continuous. Air is admitted through the valves, and combustion is initiated, which increases the pressure, closing the valves to prevent backflow through the inlet. The hot gases are expelled through the rear nozzle, producing thrust and lowering the pressure to the point that the valves may open and admit fresh air. Then the cycle is repeated (Figure 1.38).

The most widely known pulse jet was the German V-1 missile, or buzz bomb, used near the end of World War II, which fired at a rate of about 40 Hz. The pulsing effect can also be achieved in a valveless engine, or wave engine, in which the cycling depends on pressure waves traveling back and forth through a properly scaled engine. A pulse-jet engine delivers thrust at zero speed and can be started from rest, but the maximum possible flight speeds are below 960 km/h (600 mph). Poor efficiency, severe vibration, and high noise limited its use to low-cost, pilotless vehicles. Pulse jets use the forward speed of the engine and the inlet shape to compress the incoming air, then shutters at the inlet close while fuel is ignited in the combustion chamber and the pressure of the expanding gases force the jet forward. The shutters then open and the process is repeated.

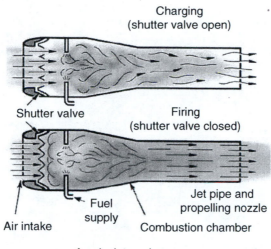

Charging
(shutter valve open)

Shutter valve

Firing
(shutter valve closed)

Air intake

Fuel
supply

Jet pipe and
propelling nozzle

Combustion chamber

A pulsejet engine.

FIGURE 1.38   Pulsejet operation. (Reproduced from Rolls-Royce plc, copyright © Rolls-Royce plc 2007. With permission.)

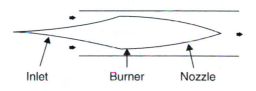

Inlet   Burner   Nozzle

FIGURE 1.39   Scramjet engine.

1.4.3 SCRAMJET

Scramjet is an acronym for Supersonic Combustion Ramjet. The scramjet differs from the ramjet in that combustion takes place at supersonic air velocities through the engine. This allows the scramjet to achieve greater speeds than a conventional ramjet, which slows the incoming air to subsonic speeds before entering the combustion chamber. Projections for the top speed of a scramjet engine (without additional oxidizer input) vary between Mach 12 and Mach 24 (orbital velocity). It is mechanically simple, but vastly more complex aerodynamically than a jet engine. Hydrogen is normally the fuel used. A typical scramjet engine is shown in Figure 1.39.

1.4.4 TURBORAMJET

The turboramjet engine (Figure 1.40) combines the turbojet engine for speeds up to Mach 3 with the ramjet engine, which has good performance at high Mach numbers [24]. The engine is surrounded by a duct that has a variable intake at the front and an afterburning jet pipe with a variable nozzle at the rear. During takeoff and acceleration, the engine functions as a conventional turbojet with the afterburner lit while at other flight conditions up to Mach 3, the afterburner is inoperative. As the aircraft accelerates through Mach 3, the turbojet is shut down and the intake air is diverted from the compressor, by guide vanes, and ducted straight into the afterburning jet pipe, which becomes a ramjet combustion chamber. This engine is suitable for an aircraft requiring high speed and sustained high Mach number cruise conditions where the engine operates in the ramjet mode.

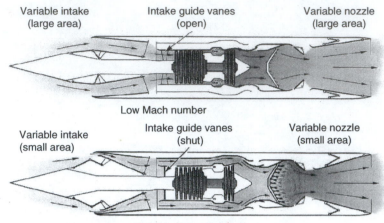

Variable intake
(large area)   Intake guide vanes (open)   Variable nozzle (large area)

Low Mach number

Variable intake
(small area)   Intake guide vanes (shut)   Variable nozzle (small area)

High Mach number

A turbo/ramjet engine.

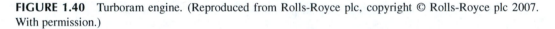

FIGURE 1.40   Turboram engine. (Reproduced from Rolls-Royce plc, copyright © Rolls-Royce plc 2007. With permission.)

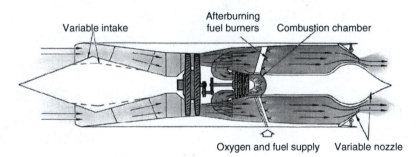

Variable intake   Afterburning fuel burners   Combustion chamber

Oxygen and fuel supply   Variable nozzle

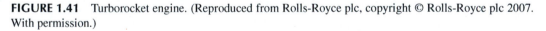

FIGURE 1.41   Turborocket engine. (Reproduced from Rolls-Royce plc, copyright © Rolls-Royce plc 2007. With permission.)

1.4.5 TURBOROCKET

The turborocket engine (Figure 1.41, turborocket) could be considered as an alternative engine to the turbojet/ramjet. However, it has one major difference in that it carries its own oxygen to provide combustion [24]. The engine has a low-pressure compressor driven by a multistage turbine; the power to drive the turbine is derived from combustion of kerosene and liquid oxygen in a rocket-type combustion chamber. Since the gas temperature will be of the order of 3500°C, additional fuel is sprayed into the combustion chamber for cooling purposes before the gas enters the turbine (Figure 1.41). This fuel-rich mixture (gas) is then diluted with air from the compressor and the surplus fuel burnt in a conventional afterburning system. Although the engine is smaller and lighter than the turbojet/ramjet, it has higher fuel consumption. This tends to make it more suitable for an interceptor or space-launcher type of aircraft that requires high speed, high altitude performance, and normally has a flight plan that is entirely accelerative and is of short duration.

1.5   CLASSIFICATION OF GAS TURBINE ENGINES

Gas turbine engines are the power plants for all the flying aircraft/helicopters (and in this case may be denoted as *aero engines* or aero derivative gas turbines) and sources for power in miscellaneous industrial applications in automotives, tanks, marine vessels, and electric power generation. In

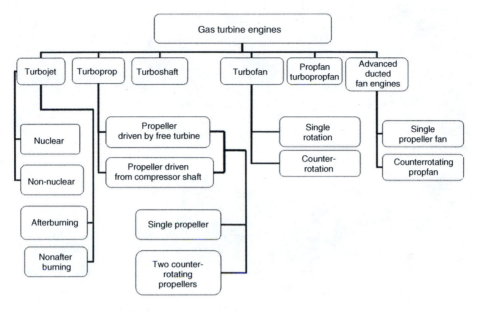

FIGURE 1.42   Classification of gas turbine engines.

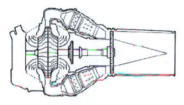

Double-entry single-stage
centrifugal turbojet

FIGURE 1.43   Turbojet single-spool centrifugal compressor. (Reproduced from Rolls-Royce plc, copyright © Rolls-Royce plc 2007. With permission.)

addition to the advantages of the jet engines over the reciprocating engines mentioned in Section 1.3, vibration stresses are relieved as a result of rotating rather than the reciprocating parts. Gas turbines as classified are shown in Figure 1.42.

1.5.1   TURBOJET ENGINES

Turbojet engines were the first type of jet engines used to power aircraft as early as the 1940s. The turbojet engine completely changed air transportation. It greatly reduced the expense of air travel and improved aircraft safety. The turbojet also allowed faster speeds, even supersonic speeds. It had a much higher thrust per unit weight ratio than the piston-driven engines, which led directly to longer ranges, higher payloads, and lower maintenance costs. Military fighters and fast business jets use turbojet engines:

1. Turbojet engines may be classified as single spool or double spool. Each may be further subdivided into centrifugal and axial compressor types. Turbojet single-spool centrifugal compressor in Figure 1.43 and turbojet single-spool axial compressor in Figure 1.44 represent both types of compressors employed in the single spool turbojet. The first turbojet engines of Frank Whittle (W.1) and von Ohain (He S-1) incorporated centrifugal compressors.

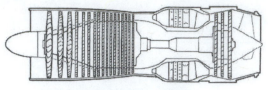

Single-spool axial flow turbojet

FIGURE 1.44 Turbojet single-spool axial compressor. (Reproduced from Rolls-Royce plc, copyright © Rolls-Royce plc 2007. With permission.)

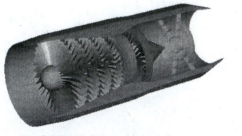

Turbojet with afterburner (single spool)

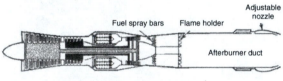

Turbojet with afterburner (double spool)

FIGURE 1.45 Turbojet engine with an afterburner.

2. Turbojets also may be classified as afterburning or nonafterburning. Afterburner is added to get an increased thrust (Figure 1.45). Afterburner is found in fighter aircraft, and is when only absolutely necessary. If a pilot runs too long with the afterburner on, he or she risks running low on fuel before the mission is completed. The only civil transport aircraft that is fitted with afterburner is the Concorde. The turbojet engine was the most popular engine for most high-speed aircraft, in spite of the higher fuel consumption. When high speed and performance are important, the cost of fuel is less important.

3. Turbojet engines may also be classified as nuclear or nonnuclear engines. A typical nuclear turbojet engine is the P-1 reactor/X39 engine complex (reactor P-1 combined to the GE X39 turbojet engine) illustrated in Figure 1.46, where air duct scrolls for connection to the reactor (arrows) with a thrust output of 5000–7000 lb, and is identified as nuclear turbojet nuc x39. It would have been flight tested aboard a highly modified B-36 bomber known as the X-6. The P-1 reactor would have been installed in the X-6's bomb bay for flight and removed shortly after landing.

Extensive studies were performed in the late 1940s and early 1950s. However, all the nuclear engine projects were abandoned in favor of safety.

1.5.2 TURBOPROP

Turboprop engines combine the best features of turbojet and piston engines. The former is more efficient at high speeds and high altitudes, while the latter is more efficient at speeds under 400–450 mph and below 30,000 ft. Consequently, commuter aircraft and military transports tend to feature

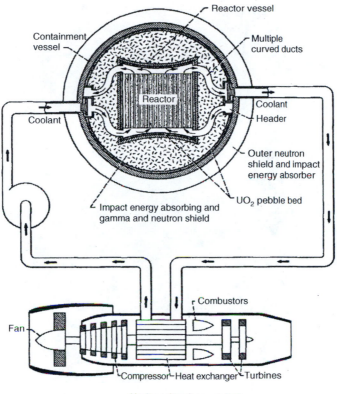

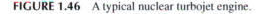

Nuclear aircraft powerplant

FIGURE 1.46   A typical nuclear turbojet engine.

turboprop engines. A turboprop engine differs from a turbojet engine in that the design is optimized to produce rotating shaft power to drive a propeller, instead of thrust from the exhaust gas. The turboprop uses a gas turbine to turn a large propeller. The shaft that connects the propeller to the turbine is also linked to a gearbox that controls the propeller's speed. The propeller is most efficient and quiet, when the tips are spinning at just under supersonic speed. Moreover, no propeller is capable of withstanding the forces generated when it turns at the same speed of the turbine. Turboprop engines may be further classified into two groups, depending on the turbine driving the propeller. In the first group, the propeller is driven by the same gas turbine driving the compressor. In the second group, additional turbine (normally denoted as free power turbine) turns the propeller. Figure 1.47 illustrates a turboprop engine with a single turbine driving the compressor and propeller. Figure 1.48 shows a turboprop engine with a free power turbine.

Turboprop engines can be found on small commuter aircraft.

A nuclear-powered turboprop logistic airplane with a helium-cooled reactor was investigated [39]. A 400,000 lb gross weight airplane was designed for 0.72 flight Mach number at 30,000 ft and powered by 8 turboprop engines (each 4600 hp) (Figure 1.49). The reactor situated at the fuselage had a power of 98.5 MW (Megawatt). Each of the eight engines was coupled to a heat exchanger installed in the wing which receives helium from the reactor at 2250°R and supplies air to the turbine inlet at 1800°R.

1.5.3 TURBOSHAFT

Turboshaft engines are defined here as engines used in powering helicopters. The general layout of a turboshaft is similar to that of a turboprop, the main difference being that the latter produces some residual propulsion thrust to supplement that produced by the shaft-driven propeller (Figure 1.48).

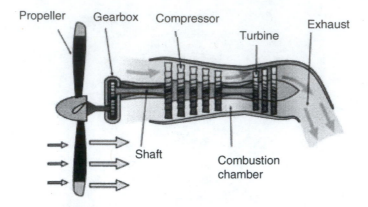

FIGURE 1.47   Turboprop engine with a turbine driving both compressor and propeller.

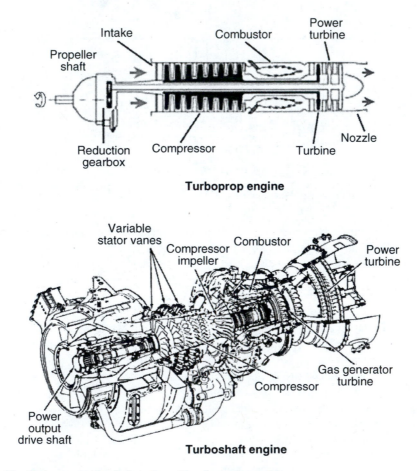

FIGURE 1.48   Turboprop–turboshaft engine with a free power turbine.

Residual thrust on a turboshaft engine is avoided by further expansion in the turbine system and/or truncating and turning the exhaust through 90°. Another difference is that with a turboshaft the main gearbox is part of the vehicle (the helicopter here) while for a turboprop the gearbox is a part of the engine. Virtually all turboshafts have a "free" power turbine, although this is also generally true for modern turboprop engines.

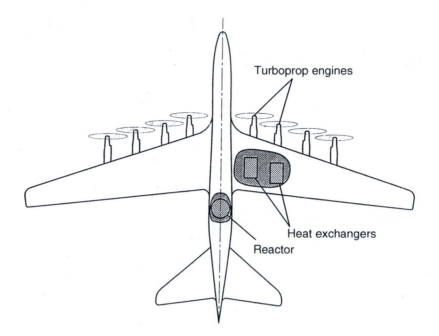

FIGURE 1.49 Nuclear turboprop engines.

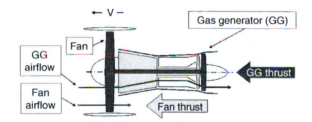

FIGURE 1.50 Turbofan layout.

1.5.4 TURBOFAN ENGINES

Turbofan engines were designed as a compromise between the turboprop and turbojet engines. A turbofan engine includes a large internal propeller (normally denoted a ducted fan) and two streams of air flowing through the engine. The primary stream travels through all the components like a turbojet engine, while the secondary air passes through the fan and is either ducted outside through a second nozzle identified as the cold nozzle or may mix with the hot gases leaving the turbine(s) and both are expelled from a single nozzle (Figure 1.50). Turbofan engines have a better performance, greater fuel economy than turbojet at low power setting, low speed, and low altitude. Figure 1.51 represents the numerous types of turbofan engines.

Turbofan engines may be classified as

1. Forward fan or aft fan configuration (Figures 1.52 and 1.53)
2. Low or high (Figures 1.54 and 1.55)
   High BPR turbofan jet engines like JT9, CF6, CFM56, PW4000, and GE90 are found in large commercial airliners. The BPR is now 5:1 or greater. At subsonic speed, high-bypass turbofans are more fuel efficient and quieter than other types of jet engines, making them

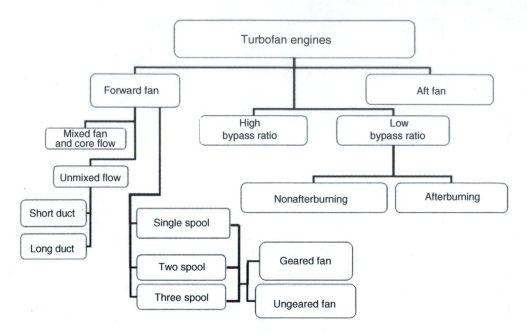

FIGURE 1.51   Classification of turbofan engines.

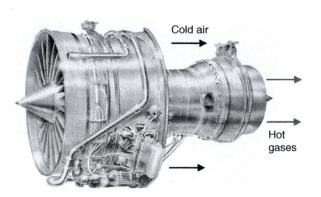

FIGURE 1.52   Forward fan. (Reproduced from Rolls-Royce plc, copyright © Rolls-Royce plc 2007. With permission.)

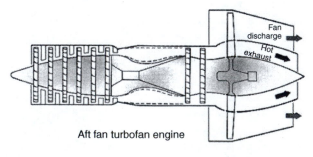

FIGURE 1.53   Aft fan.

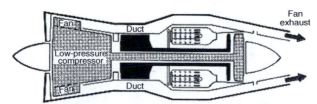

Forward-fan engine with long duct and unmixed exhaust

FIGURE 1.54 Low bypass turbofan.

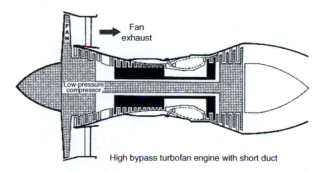

High bypass turbofan engine with short duct

FIGURE 1.55 High bypass turbofan.

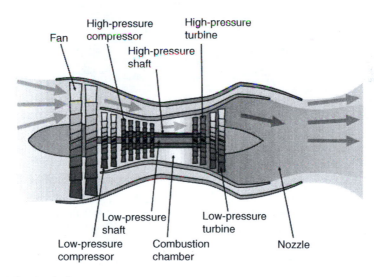

FIGURE 1.56 Mixed turbofan.

ideally suited for commercial aircraft. Low BPR turbofan engines are found mainly in military airplanes.

3. The forward fan may be of the mixed or unmixed types. In the former, the fan (cold) air stream passes through the fan where it is compressed and ejected from the cold nozzle, while the hot stream passes through the gas generator and is ejected through another hot nozzle. In the latter types, the cold stream leaving the fan mixes with the hot stream leaving the last turbine before leaving the engine from one nozzle. Figure 1.56 illustrates a mixed engine. The famous mixed engine is the Pratt & Whitney JT8 engine. The unmixed turbofan is already illustrated in Figure 1.54.

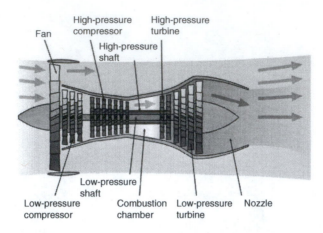

FIGURE 1.57   Unmixed turbofan long duct.

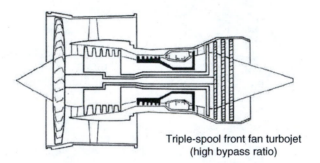

Triple-spool front fan turbojet
(high bypass ratio)

FIGURE 1.58   Typical three-spool turbofan forward fan. (Reproduced from Rolls-Royce plc, copyright ©
Rolls-Royce plc 2007. With permission.)

4. The unmixed turbofan may be of the short or long ducts. Figures 1.54 and 1.57 illustrate
   the general layout of both types. Mixed flow engines are mostly military ones.
5. Forward fan may also be classified as single-, double- (two), and triple- (three) spool
   engines. Nowadays, the single-spool turbofan engines are rare, while the majority of
   turbofan engines are of the two- or three-spool types. Figure 1.55 illustrates a two-spool
   unmixed turbofan engine having a high BPR and a short duct. Numerous high BPR engines
   are two-spool engines, such as the GE90, the CF6, and the PW4000. Figure 1.56 represents
   a typical turbofan engine employed in military applications. Triple- or three-spool engines
   are also prevalent these days. It includes many RR engines including the RB211 and Trent
   series. Garrett AFT3 represents a medium bypass ratio of 2.81 turbofan engine. A typical
   layout for a three-spool engine is depicted in Figure 1.58, an example for which is the
   Rolls-Royce Trent 800 engine.
6. Forward fan may be also classified as either geared or ungeared fan. Lycoming ALF502R
   (Figure 1.59) and PW8000 are examples of such geared fans.
7. Low BPR turbofan engines may also be classified as afterburning or nonafterburning
   engines. The former resemble all the recent military airplanes. These engines are also of
   the mixed low BPR type. An example of these engines is the F100 (Figure 1.60). Because
   of their high fuel consumption, afterburners are not used for extended periods (a notable
   exception is the Pratt & Whitney J58 engine used in the SR-71 Blackbird). Thus, they are
   only used when it is important to have as much thrust as possible. This includes takeoffs
   from short runways (as in aircraft carriers) and in combat situations.

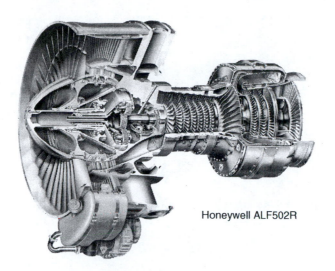

Honeywell ALF502R

FIGURE 1.59   Honeywell ALF502R. (Courtesy Honeywell Aerospace.)

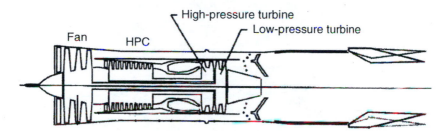

FIGURE 1.60   F-100 two-spool afterburning turbofan engine.

8. Turbofan engines may also be classified as nuclear and nonnuclear engines. Though nuclear engines were abandoned in the early 1960s, extensive studies were performed from 1946 to 1961. The lines of thought were that for aircraft with gross weight of 0.45 million kilograms (1 million pounds) such as Boeing 747 and Lockheed C-5, it was practical to use nuclear fuel for their propulsion. Moreover, long operation (of the order of 10,000 h) between refueling and long range have an economic attraction. However, complete shielding of the flight and ground crew as well as the passengers was a dragging factor. Moreover, public safety in case of aircraft accidents is another major concern. Though some studies made by NASA in the 1970s were in favor of using nuclear engines [26], these engines are not adopted until now.

9. It is not common to find an airplane like Maritime patrol ASW (A-40) and SAR (Be-42) amphibian manufactured in Russia, which is powered by two 11 7.7 kN (26,455 lb) Aviadvigatel D-30KPV turbofan engines together with two 24.5 kN (5510 lb) RKBM (formerly Klimov) RD-60K booster turbojet engines as booster engines (Figure 1.61).

1.5.5   PROPFAN ENGINES

Propfan engines are sometimes identified in former Soviet Union countries as turbopropfan (like D-27 engine), unducted fan or ultra high BPR. It may be of one of the following two categories:

1. Single rotation
2. Counterrotating

FIGURE 1.61   A-40 aircraft powered by two turbofans and two booster turbojets. (Courtesy Albatros Russian Aircraft.)

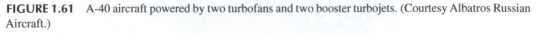

(a)

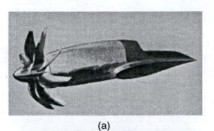

(b)

FIGURE 1.62   Single rotation propfan: (a) single rotation and (b) single rotation with swirl recovery vanes. (Courtesy NASA.)

Single rotation has a single forward unducted fan that combines the advantages of the propeller of a turboprop and the fan of a turbofan engine (Figure 1.62). The second type has two propellers, one behind the other, each of which rotates in an opposite direction (CR). Counterrotating designs offer the best performance [2]. This has some similarities with the aft-fan type but it is composed of two counterrotating fans (Figure 1.63) here. Though this design is the most efficient option, it is prohibitively expensive, complex or heavy for certain applications. The single rotation propfan with a swirl recovery vane (Figure 1.62b) is a third propfan configuration that fills a middle ground between single and counter rotation types. In this configuration, a set of stationary vanes are mounted a distance aft of the single rotation propfan blades. These vanes, much like the rotating aft blade row of a counterrotating design, straighten the propeller swirl resulting in increased efficiency [40].

1.5.6 ADVANCED DUCTED FAN

These designs are essentially turbofans with large, swept fan blades that have pitch control and reduction gearing similar to propfans, but the fans are enclosed in ducts like turbofan. The BPR

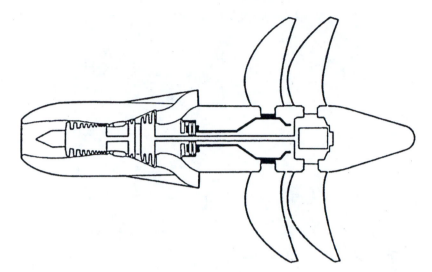

FIGURE 1.63 Counterrotating propfan. (Reproduced from Rolls-Royce plc, copyright © Rolls-Royce plc 2007. With permission.)

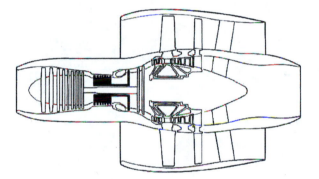

Counterrotating fan-concept (high bypass ratio)

FIGURE 1.64 Counterrotating advanced ducted fan. (Reproduced from Rolls-Royce plc, copyright © Rolls-Royce plc 2007. With permission.)

for advanced ducted fan is from 15:1 to 25:1 [2]. There are two basic designs: one with geared, variable pitch, single propeller fan, and the other with counterrotating blades with pitch control (Figure 1.64). The need for such engines has been spurred by increasing airplane traffic, which raises noise, environmental, and fuel consumption issues. Extensive work has been done in some aero engine manufacturing companies like Pratt &Whitney MTU and Fiat Avio for the design of this type of engines. A thin-lip, slim-line nacelle is required to give such a high bypass ratio.

1.6 INDUSTRIAL GAS TURBINES

The words "industrial gas turbine," refer to all turboshaft engines that are not installed in helicopters. Turboshaft engines are extremely versatile and are used in electric powerplants, offshore oil drilling, locomotives, tanks, trucks, hovercrafts, and marine vessels (Figure 1.65). In large powerplants used in generating electricity, it may be used as a combined cycle with steam powerplants (Figure 1.66).

FIGURE 1.65   GE-7B Gas turbine. (Courtesy GE Power.)

FIGURE 1.66   GE-7H Combined cycle gas turbine. (Courtesy GE Power.)

1.7   NON–AIR-BREATHING ENGINES

The different types of non–air-breathing engines or rocket engines are illustrated in Figure 1.67. However, these types of engines are beyond the scope of this book.

1.8   FUTURE OF AIRCRAFT AND POWER PLANT INDUSTRIES

A. The twenty-first century is characterized by a revolutionary development in both aircraft and power plant industry. On the *military side*, achieving air superiority is the main target of research and development (R&D) of giant companies. Toward that end continuous developments in the aircraft structure, materials, navigation and radar systems, exterior aerodynamic shape, other systems, computers, armaments, and jet engines take place. There are two distinct workers in the United States developing the latest technological achievements in this regard, namely Boeing and Lockheed Martin.

The *Phantom Works* division is the main R&D arm of Boeing Company. Recent Phantom projects include some unmanned aerial vehicles (UAVs are remotely piloted or self-piloted aircraft that can carry cameras, sensors, communications equipment, or other payloads) and a large capacity transport aircraft. Examples of these UAVs are as follows:

- A160 Hummingbird UAV helicopter.

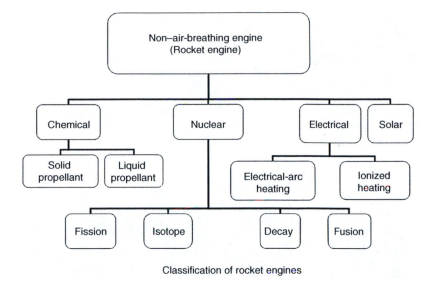

Classification of rocket engines

FIGURE 1.67 Classifications of non–air-breathing engines.

- X-45 series Unmanned Combat Air vehicle (UCAV) powered by GE F404 engine. The first of the three planned X-45C aircraft was originally scheduled to be completed in 2006, with capability demonstrations scheduled for early 2007. By 2010 Boeing hoped to complete an autonomous aerial refueling of the X-45C by a KC-135 Stratotanker. As of March 2006, the US Air Force has decided not to continue with the X-45 project and next, the US Navy started its own UCAS program. The contract was awarded to Northrop Grumman's proposed naval X-47, thus ending the X-45 program. Northrop by that point had already been responsible for the first autonomous carrier landing of a UAV.

The large-capacity transport craft Pelican ULTRA (Figure 1.68), intended for military or civilian use, would have a wingspan of 500 ft, a cargo capacity of 1400 tons, and a range of 10,000 mi. Powered by four turboprop engines, its main mode would be to fly 20–50 ft over water, though it would also be capable of overland flight at a higher altitude but lesser range. The craft would be capable of carrying 17 M-1 battle tanks on a single sortie at speeds 10 times those of current container ships.

Lockheed Martin has several programs including the F-35 Lightning, X-35, F-22, and X-31.

The F-35 Lightning II—descended from the X-35 of the Joint Strike Fighter (JSF) program—is a single-seat, single-engine military strike fighter. Demonstrator aircraft flew in 2000; a production model first took flight on 15 December 2006.

The X-35 [28] refers to a JSF having a short takeoff (400 ft runway at a speed of 80 mph) and a vertical landing by swiveling down the rear portion of the engine. It will enter the service in 2008. It has a top speed of 1.6 Mach number but can fly supersonically at a Mach number 1.4 without afterburner which gives the fighter a greater operating range and allows for stealthier flight operation. It is powered by two JSF119-611 afterburning turbofan engines (each of 35,000 lb thrust) manufactured by Pratt & Whitney and a shaft driven lift fan by Rolls-Royce.

The F-22 Raptor [29] is an air superiority fighter having a top speed of 1.7 Mach number and ceiling of 50,000 ft. Its first flight was in 1997 but will enter service in 2010. It will fly at supersonic speeds without the afterburner. It is a stealthy aircraft so it carries its weapons in internal bays. It is powered by two Pratt & Whitney F119-PW-100 turbofan engines, each of which has a 35,000 lb thrust. Thrust vectoring helps the pilot in both dog fighting maneuverability and landing.

The X-31 is the first international experimental airplane [28]. It represents a joint project between the United States and Germany. It enhances super maneuvers with a top speed of 1.3 Mach number and extremely short takeoff and landing. Thus, it can operate from a small airfield close to the enemy

FIGURE 1.68   Boeing Pelican ULTRA.

front line. It lifts the nose at final approach to reduce the speed at a steep angle with only two feet clearance from ground. It is powered by a single General Electric P404-GE-400 turbofan engine, producing 16,000 pounds thrust.

The X-31 has three paddles mounted around the engine nozzles to redirect the flow of exhaust gases. The moveable nozzles provide more control and maneuverability at high angles of attack when normal flight control surfaces are less effective. Information gained from the X-31 program could be used for future application to highly maneuverable aircraft.

Despite the scheduled retirement of the F117 Nighthawk [30] between 2017 and 2025, its retirement was surprisingly announced in January 2006. It is the world's number one stealth aircraft. It can penetrate deep into the enemy's airspace and deliver its cargo, precision laser-guarded bombs, and head back home. Its flat surface designed in sharp angles deflects the incoming radar signals away from the aircraft. Moreover, its two GE F404-F1D2 turbofan engines are almost silent so it cannot be heard.

Another future field is employing the solar and fuel cell system for powering light airplanes like UAVs for extended flight time. Helios Prototype is a solar and fuel cell system-powered UAV that NASA tested. On August 13, 2001, it sustained flight at above 96,000 ft (29,250 m) for 40 min and at one time it flew as high as 96,863 ft (29,524 m). The solar electric Helios Prototype flying wing is shown in Figure 1.69 over the Pacific Ocean during its first test flight on solar power from the U.S. Navy's Pacific Missile Range Facility at Kauai, Hawaii, on July 14, 2001. The 18-h flight was a functional checkout of the aircraft's systems and performance in preparation for an attempt to reach sustained flight at 100,000 ft altitude.

B. On the other hand, in civil transports, the two competitors Boeing and Airbus will provide the following airplanes: Boeing 787 Dreamliner [31] as well as the Airbus A350 and A380 [32] nearly by the end of the first decade of the twenty-first century. Comfort of passengers and fuel economy are the common features of both aircraft.

Boeing 787 Dreamliner, which is on track for early 2008, is a superefficient airplane with unmatched fuel efficiency and exceptional environmental performance. Passengers will have increased comfort and convenience. It has three different families, namely, 787-3, 787-8, and 787-9. The first family will accommodate 290–330 passengers and be optimized for routes of

FIGURE 1.69   Solar electric Helios. (Courtesy NASA.)

300–3500 nautical miles. Boeing 787-8 Dreamliner will carry 210–250 passengers on routes of 8000–8500 nautical miles, while the 787-9 Dreamliner will carry 260–290 passengers for routes of 8600–8800 nautical miles. Boeing 787 will have as much as 50% of its structure made of composite materials. This will reduce the risk of corrosion and lower the overall maintenance costs and maximize airline revenue by keeping airplanes flying as much as possible. Another innovative application is the move from hydraulically actuated brakes to electric [39]. Moreover, its designer aims at incorporating a health monitoring system that will allow the airplane to self-monitor and report maintenance requirements to ground-based computer systems. This will aid in troubleshooting the 787. Airplane systems' information and fully integrated support will help maintenance and engineering organizations quickly isolate failed components and reduce return-to-service times. Boeing selected both General Electric and Rolls-Royce to develop engines for the 787. Advances in engine technology will contribute as much as 8% of the increased efficiency of the new airplane.

The airbus A380 [32] is the most ambitious airplane now having several configurations; namely A380-800, A380-800F, A380-700 and A380-900. The first is the 525-seat double deck configuration (upper deck extends along the entire length of the fuselage) in standard three-class configuration or up to 853 people in full economy class configuration. The second is the freight version having a listed payload capacity exceeded only by the Antonov An-225. The other two are the shortened 480 seat version and the stretched 656 seat version. It had its first commercial flight on October 25, 2007.

Various Internet sites that provide interesting information on aircraft are found in References 33–36.

The work on *hypersonic airplanes* is vigorously carried out in several countries. A hypersonic vehicle means an aerospace plane that would take off from a normal runway as a conventional airplane and then accelerate to near-orbital speeds within the atmosphere using scramjet engines (Figure 1.70). Hypersonic vehicles have long been touted as potential high-speed commercial transports to replace the current sub- and supersonic airplanes.

Within the first decade of the twenty-first century, it is expected that the *hyper soar* will reach a flight Mach number of 10 (Figure 1.71) *hyper soar aircraft*. It is interesting to say that some aerospace companies, airlines, and government officials have proposed vehicles cruising at Mach 7–12 capable of carrying passengers from New York to Tokyo in less than 2 h.

Within the next decade (2010–2020), aurora scramjet-powered airplane will reach a Mach number of 20 (Figure 1.72).

C. Now turning to the air-breathing engines, the technological developments assure that the ideal power plant would be the most fuel efficient at its thrust, the most reliable, the lightest, the quietest, and the cleanest engine, all made at the lowest cost as thoroughly discussed by Rolls-Royce [37]. The fan in turbofan engines features hollow swept blades with

FIGURE 1.70   Hypersonic transport.

FIGURE 1.71   Hyper soar aircraft.

FIGURE 1.72   Hypersonic aurora aircraft.

advanced aerodynamics giving the lightest, strongest, quietest, and most efficient fans. Moreover, future engines will use smarter ribbed steel and titanium casings. All recent RR engines employ three-shaft (spool) configurations. Future advancement in material and manufacturing technology will allow compressor weight savings up to 70% with the introduction of blings. Blings are lightweight replacements of traditional compressor disks featuring integral blades and using advanced silicon carbide-reinforced metal matrix to withstand high levels of stress.

The combustor is designed to have a stable flame over the different operating conditions. Advanced cooling methods are adopted to ensure integrity and long life as the flame temperature is nearly one thousand above the melting temperature of the combustor material. The lean burn combustion systems currently studied will bring the NOx level of the RR engines by 2010 at a level of half that required by legislation. By 2020 80% NOx reduction relative to those specified by CAEP 2 regulation is planned.

Advanced single crystal material and thermal barrier coatings are used on static and rotating turbine component. Advanced cooling techniques allow increasing the turbine temperature to improve the engine efficiency. Aerodynamic advancements including counterrotating systems, 3-D design, and end wall profiling will lead to lighter blades with no efficiency penalty.

Engines are controlled by the Full Authority Digital Electronic Control (FADEC) system, which is becoming smaller, lighter, and more reliable.

A noise reduction target by 2020 to be one-half its level at 2001 is set as a strict goal.

A 10% reduction in fuel consumption of all newly designed engines by 2010 compared with the equivalent in1998 engines is also planned.

Power plant developments by Pratt & Whitney [26] cover the following aspects:

- Integrally bladed rotors: in most stages disks and blades are made of a single piece of metal for better performance and less air leakage.
- Long chord and shroudless fan blades: wider and stronger fan blades with a ring of metal around most jet engines.
- Low-aspect, high-stage-loading compressor blades that have wider chord blades for greater strength.
- Alloy C high-strength, burn-resistant titanium compressor stator.
- Alloy C in augmentor (if available) and nozzle: the same heat resistant titanium alloy to protect the aft components.
- Float wall combustor: thermally isolated panels of oxidation-resistant high cobalt material to make combustion chamber more durable.
- Fourth generation FADEC.
- Improved supportability: easy access for inspection, maintenance, and replacement of components/modules.

D. Concerning the supersonic transports (SST), two groups are performing studies in this regard.

1. Manufacturers of aircraft (Boeing) and aero engines (P&W, GE, RR, and Snecma): Improvement of engine-specific fuel consumption (SFC) for high-speed civil transport (HSCT) at its subsonic flight conditions is one of the main concerns over the past 20 years. The original U.S. SST had very poor subsonic SFC. High subsonic SFC penalizes the mission performance by reducing the efficiency during subsonic mission legs and by requiring larger amounts of reserve fuel. The key to good subsonic and supersonic SFC is a variable-cycle engine (VCE). The major objective for a future HSCT application is to provide some degree of engine cycle variability that will not significantly increase the cost, the maintenance requirements, or the overall complexity of the engine. The VCE must have a good economic payoff for the airline while still providing more mission flexibility and reducing the reserve fuel requirements so that more payload can be carried. In the past, VCEs were designed with large variations in BPR to provide jet noise reduction. However, these types were complicated and did not perform well. Today, the trend is toward

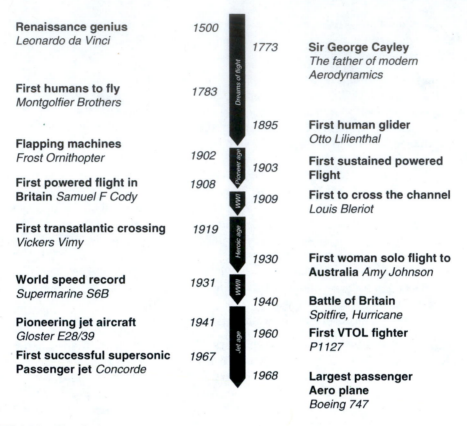

FIGURE 1.73   Historical events.

turbojets or low-bypass engines that have the ability to improve off-design performance by adjustment of compressor bleed or by a relatively small variation in BPR. The current engine offerings from Pratt & Whitney and General Electric fall into this category. Both of these engines will require an effective jet noise suppressor. Rolls-Royce SNECMA favors other approaches. One is a tandem fan that operates as a turbojet cycle for cruise but opens a bypass inlet and nozzle for higher flow at subsonic speeds. A second approach is to increase the BPR by incorporating an additional fan and turbine stream into the flow path at subsonic speeds.

2. NASA Lewis: Following are five of the most promising engine concepts studied:

   a. Turbine bypass engine (TBE) is a single-spool turbojet engine that possesses turbofan-like subsonic performance but produces the largest jet velocity of all the concepts. Hence, it needs a very advanced technology mixer–ejector exhaust nozzle with about 18 decibels (dB) suppression ability to attain FAR 36 Stage III noise requirements without oversizing the engine and reducing power during takeoff. This level of suppression could be reached if the ejector airflow equals 120% of the primary flow.

   b. The VCE alters its BPR during flight to better match varying requirements. However, although its original version defined in the 1970s relied on an inverted velocity profile exhaust system to meet less stringent FAR 36 Stage II noise goals, the revised version needs a more powerful 15 dB suppression solution. A 60% mass flow-augmented mixer–ejector nozzle together with modest engine oversizing would satisfy this requirement. It should not be inferred from the above that the TBE

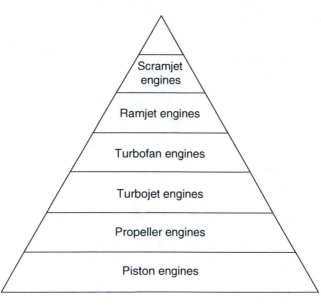

FIGURE 1.74 Evolution of aero engines.

Propulsion system	Jet velocity (m/s)
Helicopter lifting rotor	up to 30
Propeller	30–200
Remote ducted fan, unducted fan or propfan	100–300
Turbofan	200–600
Turbojet (sea-level,static)	350–600
Turbojet ($M = 2.0$ at 36,000 ft, ~ 600 m/s)	900–1200
Ramjet ($M = 2.0$ at 36,000 ft, ~ 600 m/s)	900–1200
Ramjet ($M = 4.0$ at 36,000 ft, ~ 1200 m/s)	1800–2400
Solid-Propellant rocket	1500–2600
Liquid-Propellant rocket	2000–3500
↓impractical-------practical↑	
Nuclear rocket (solid core reactor)	9000–10,000
Nuclear rocket (fusion reactor)	25,000
Solar	20,000
Arc	6000
Plasma and Ion	$90–30 \times 10^6$
Photon	300×10^6

FIGURE 1.75 Classification of aero engines based on their speeds.

needs a 120% mass flow-augmented mixer–ejector nozzle, while the VCE only needs one that is 60%. There is uncertainty concerning the best combination of mass flow augmentation, acoustic lining, and engine oversizing for both engines.

c. A relative newcomer, the fan-on-blade ("Flade") engine is a variation of the VCE. It has an auxiliary third flow stream deployed during takeoff by opening a set of inlet guide vanes located in an external annular duct surrounding the VCE. The auxiliary annular duct is pressurized by extension to the fan blades and is scrolled into the lower half of the engine before exhausting to provide a fluid acoustic shield. It also requires a relatively modest mixer–ejector exhaust nozzle of approximately 30% flow augmentation.

d. The fourth concept is the mixed flow turbofan (MFTF) with a mixer–ejector nozzle.

e. The final engine concept is a TBE with an inlet flow valve (TBE/IFV). The IFV is activated during takeoff to permit auxiliary inlets to feed supplementary air to the rear compressor stages while the main inlet air is compressed by just the front compressor stages. While a single-spool TBE/IFV still needs a mixer–ejector exhaust nozzle, it seems possible to avoid that complexity with a two-spool version because of greater flow handling ability in the takeoff mode.

E. In numbers, the forecast for the second decade (2007–2017) lists the following points [38]:

1. More than 59,800 turbofan engines will be produced.
2. Over 22,850 turboshaft engines will be produced.
3. Sales of turboprop engines will be rebounded in face of escalating airline fuel cost.
4. Some 23,000 APU/GPUs will be produced.
5. Nearly 27,700 engines for missiles, drones, and UAVs will be produced.
6. Some 11,000 micro turbines for electrical generation (30–500 kW) will be produced.
7. Engines for marine power for surface vessels of all sizes, commercial and military that are worth $4.7 billion will be manufactured.
8. Concerning airplanes, 7900 large commercial jets and 3865 fighter/attack/jet trainer aircraft will be produced. Some 700 business jets will be produced annually. Concerning military transport aircraft, Airbus will garner the lion's share (24%), Lockheed Martin will rank second (18% share) and Boeing's share will fall to 12.3%.

CLOSURE

The evolution of jet engines since the dream of mankind many centuries ago until the twenty-first century together with forecasting for the near future, say within one or two of the coming decades was discussed. A summary of some historical events in the period from 1500 to 1968 is given in Figure 1.73.

A summary for the evolution of aero engines is given in Figure 1.74. Moreover, a concise table for the different propulsion systems and the flight speeds for their powered vehicles is given in Figure 1.75.

PROBLEMS

1.1 Since the first flight of Wright brothers in 1903 and until now, endless developments in aircraft and engine industries have been achieved. It is required to identify some milestones in such a long journey by listing the FIRST engines of the following categories:
turbojet engine, turbojet engine with afterburner, turbofan engine, supersonic turbofan engine, high BPR turbofan engine, turboprop engine, and propfan engine.

1.2   Do you think the hen and the egg eternal question applies to the aircraft and engine?

1.3   Describe the British and German patents for the first jet engines.

1.4   Compare between the first engine invented by Frank Whittle and von Ohain.

1.5   Complete the following table:

Name	Contribution
Hero	
Wan Hu	
Leonardo da Vinci	
Giovanni Branca	
Isaac Newton	
Felix du Temple	
John Barber	
Samuel Langley	
Wright Brothers	
Frank Whittle	

1.6   Which engines have no moving parts?

1.7   What is the type of engine that powers most of today's airliners?

1.8   What is the type of engine that would be used in a helicopter?

1.9   How is a turboprop different from a turbojet?

1.10 List the names of some Chinese aircraft and engines.

1.11 Identify engines available for the following aircraft B787, B777, B767, B757, B747-200, B747-400, MD80, B737-800/500, A310, A320, A330, A340, B 787, A380, A350, the Sukhoi Su-9, Tupolev Tu-128P, DC-8, U-2 spy, Shenyang J-8, Rafale, Mirage 5, Swedish Ja-37 Viggen, Mig 23, Mig 29, Tornado ADV, Eurofighter V-22 Ospery, AN-225, Convair 580, Fairchild F-27, Lockheed L100 Hercules, Cessna Citation II, Gulfstream Aerospace (IAI) G100, Bombardier Canadair 415, Bell 412, Sikorsky S-58, Lockheed Martin F-117A Nighthawk, Lockheed Martin P-3 Orion, Mi-12, ATR 42-500, ATR 72, Panavia Tornado, AH-64 A Apache, SAAB 340B, BOMBADIER Q400 DASH 8, MiG-AT, Rutan Voyager.

1.12 State the type of each of the following engines and mention the names of the some aircraft they are used to power:
      F-100, F-101, F-110, RB-I99, RB 211, Nene, JT 8D, RR Pegasus T56, V2500, CFM56, GE90, Lyulka AL-7, JT9, CF6, TFE 731, TPE331, TV3 VK-1500, AL-31 F, Trent 700, Olympus 593, T700-GE, PW100, SM146, DR-1700, CF34-8, Ha 102

1.13 Why does the RR Pegasus have counterrotating shafts?

1.14 Write down the advantages of the following engines: RB-211 and the JT9D or CF6.

1.15 What are the key differences between the PT-6 and the Garrett 331?

1.16 Compare between turboprop, turbofan, and propfan engines (draw figures to illustrate their configurations).

1.17 Compare between ramjet and scramjet engines.

1.18 Will physical fitness of passengers influence the design of supersonic and hypersonic transports?

1.19 What engines are suitable for Airborne Warning and Control System (AWACS) aircraft?

1.20 Compare between turboprop, turboshaft, and industrial gas turbines.

1.21 Write the names of two VTOL and V/STOL aircraft and discuss the requirements in their power plants.

1.22 What is the possible engine type for UAV aircraft?

1.23 What is the largest fan diameter, highest BPR, and largest overall pressure ratio of turbofan engines up to the year 2007?

1.24   Which companies are associated with the following engines?
       CFM 56, CF6, JT-3D, GP 7000, V 2500, RB 211
1.25   Compare between:
       a.   Piston engines powering aircraft and turboprop engines
       b.   Turbo/ramjet and turbo-rocket engines
1.26   Write short notes on the shown Curtiss flying boat.
1.27   Classify the following engines.

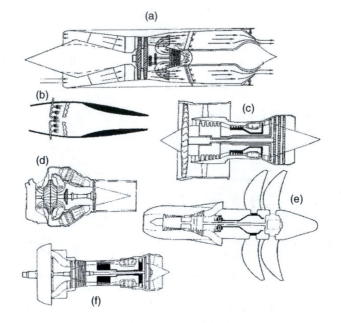

(a)

(b)

(c)

(d)

(e)

(f)

PROBLEM 1

REFERENCES

1.   G. Daniel Brewer, *Hydrogen Aircraft Technology*, CRC Press Inc., 1991, p. 248.
2.   *The Aircraft Gas Turbine Engine and Its Operation*, Pratt & Whitney Operating Instructions 200, 1988, Section 1.
3.   J.V. Casamassa and R.D. Bent, *Jet Aircraft Power Systems*, 3rd edn., McGraw-Hill Book Co., 1965, chap. 1.
4.   http://www.aircav.com/histturb.html
5.   I.E. Treager, *Aircraft Gas Turbine Engine Technology*, 3rd edn., McGraw-Hill, Inc., 1999.
6.   J.D. Anderson, Jr., *Introduction to Flight*, 3rd edn., McGraw-Hill, Inc., 1989.
7.   M.J. Kroes and T.W. Wild, *aircraft Powerplants*, 7th edn., Glencoe/McGraw-Hill, 2002.
8.   http://wright.nasa.gov/airplane/eng.html
9.   T.D. Crouch, *The Bishop's Boys: a Life of Wilbur and Orville Wright*, Amazon.com, April 2003.
10. F.J. Malina, R.C. Truax, and A.D. Baxter, Historical Development of Jet Propulsion, Part A of Jet Propulsion Engines, O.E. Lancaster (Ed.), *Vol. XII of High Speed Aerodynamics and Jet Propulsion*, C.D. Donaldson (Ed.), Princeton University Press, Princeton, NJ, 1959.
11. H. von Ohain, The evolution and future of aeropropulsion systems, In *The Jet Age*, W.J. Boyne and D.S. Lopez (Eds.), Smithsonian Institution Press, Washington, DC, 1979, pp. 25–46.
12. H. von Ohain, *Face to Face*, AIAA Aerospace America, July 1990, pp. 8–9.
13. F.W. Whittle, *The Early History of the Whittle Jet Propulsion Gas Turbine*, IMechE Clayton Lecture, October 5, 1945.

14. F.W. Whittle, The birth of jet engine in Britain, In *The Jet Age*, W.J. Boyne and D.S. Lopez (Eds.), Smithsonian Institution Press, Washington, DC, 1979, pp. 1–24.
15. B. Gunston, *The Development of Jet and Turbine Aero Engines*, Patrick Stephens Limited, imprint of Haynes Publishing, 1997.
16. http://en.wikipedia.org/wiki/Turboshaft
17. M.J. Zucrow, *Aircraft and Missile Propulsion*, Vols I and II, John Wiley & Sons, Inc., 1958.
18. P.J. McMahon, *Aircraft Propulsion*, Harpert & Row, 1971.
19. P.J. Walterup, M.E. White, F. Zarlingo, and S. Garvlin, *History of Ramjet and Scramjet Propulsion Development for U.S. Navy Missiles*, John Hoipkins APL Digest, volume 18, Number 2, 1997.
20. H.I.H. Saravanamuttoo, G.F.C. Rogers, and H. Cohen, *Gas Turbine Theory*, Prentice Hall, 5th edn., 2001.
21. http://www.geae.com/education/engines101
22. R.D. Archer and M. Saarlas, *An Introduction to Aerospace Propulsion*, Prentice-Hall, Inc., 1996, p. 1.
23. G.P. Sutton, *Rocket Propulsion Elements—An Introduction to the Engineering of Rockets*, 5th ed., John Wiley & Sons, 1986, p. 1.
24. *The Jet Engine*, Rolls-Royce plc., p. 3, 5th edn., Reprinted 1996 with revisions.
25. J.P. Hill and C. Peterson, *Mechanics and Thermodynamics of Propulsion*, 2nd edn., 1992, Addison Wesley Publication Company, Inc., p. 155.
26. F.E. Rom, *Airbreathing Nuclear Propulsion-A New Look*, NASA TM X-2425, December 1971.
27. http://www.globalaircraft.org/planes/x-35_jsf.pl
28. http://www.f22fighter.com/
29. http://en.wikipedia.org/wiki/F-117A_Nighthawk
30. http://www.boeing.com/commercial/787family/
31. http://www.airliners.net/info/stats.main?id=29
32. http://en.wikipedia.org/wiki/Airbus_A380
33. http://www.jetplanes.co.uk/
34. http://www.pw.utc.com/
35. http://www.dfrc.nasa.gov/Gallery/Photo/index.html
36. "*The future Now*" Rolls Royce vcom 9362, April 2004.
37. Forecast International, Aerospace, Defense, Electronics and Power Systems, Forecast International Inc., Newtown, CT 06470, USA, 2007.
38. J. Hale, *Boeing 787 From the Ground Up*, Boeing, Aero magazine, Issue 24 _Quarter 04, 2006.
39. R.H. Cavicchi, H.H. Ellerbrock, E.W. Hall, H.J. Happler, J.N.B. Livingood, and F.C. Schwenk, *Design Analysis of a Subsonic Nuclear-Powered Logistic Airplane with Helium-Cooled Reactor*, NASA TM X-28, 1959.
40. R.S. Ciszek, Propfan Propulsion Concepts: Technology Review, Design Methodology, State-of-the-Art Design and Future Outlook, Senior Thesis, University of Virginia, Department of Mechanical and Aerospace Engineering, March 2002.

2 Performance Parameters of Jet Engines

2.1 INTRODUCTION

The designer of an aircraft engine must recognize the differing requirements for takeoff, climb, cruise and maneuvering, the relative importance of these being different for civil and military applications and for long- and short-haul aircraft. In the early aircraft, it was common practice to focus on the take-off thrust. This is no longer adequate for later and present day aircraft. For long-range civil aircraft such as Boeing 747, 777, and Airbus A340 as well as the future Boeing 787, Airbus A380 (the world's truly double-deck airliner), and A350 XWB (extra wide body expected in 2012), the fuel consumption through some ten or more flight hours is the dominant parameter. Military aircraft have numerous criteria such as the rate of climb, maneuverability for fighters, short takeoff distance for aircraft operating from air carriers and maximum ceilings for high-altitude reconnaissance aircraft such as SR-71 Blackbird aircraft. For civil and military freighter airplanes the maximum payload is the main requirement.

In all types of aircraft, the engines are expected to provide efficiently the thrust force necessary for their propelling during different flight phases and at different operating conditions, including hottest or coldest ambient temperatures and rainy, windy, or snowing weather.

This chapter resembles a first window for air-breathing engines. It starts by a derivation for the thrust force or the propelling force generated in the direction opposite to the flow of air entering the engine in accordance with Sir Isaac Newton's laws of motion. Consequently, all jet engines including rocket motors belong to the class of power plants called *reaction engines*. It is the internal imbalance of forces within the gas turbine engines that gives all reaction engines their names. The propulsive force developed by a jet engine is the result of a complex series of actions and reactions that occur within the engine. The thrust constituents and the different factors affecting the thrust are next explained. Some of these factors are related to the engine; others are related to the medium in which the engine operates.

The performance of jet engines is evaluated through the propulsive, thermal, and overall efficiencies. The propeller efficiency of turboprop engines is also evaluated. Fuel consumption is properly evaluated through a parameter identified as the thrust-specific fuel consumption (TSFC), which is the ratio of fuel flow rate into the engine to the generated thrust. Thus, different jet engines may be compared. The range of aircraft is a combined engine/aircraft parameter where the fuel consumption through the engine is coupled to the aircraft's lift and drag forces.

2.2 THRUST FORCE

Thrust force is the force responsible for propelling the aircraft in its different flight regimes.

It is in addition to the lift, drag, and weight that together represent the four forces that govern the aircraft motion. During the cruise phase of flight, where the aircraft is flying steadily at a constant speed and altitude, the four forces are in equilibrium in pairs—lift and weight as well as thrust and drag. During landing, thrust force is either fully or partially used in braking the aircraft through a thrust-reversing mechanism. The basic conservation laws of mass and momentum are used in their integral forms to derive an expression for thrust force.

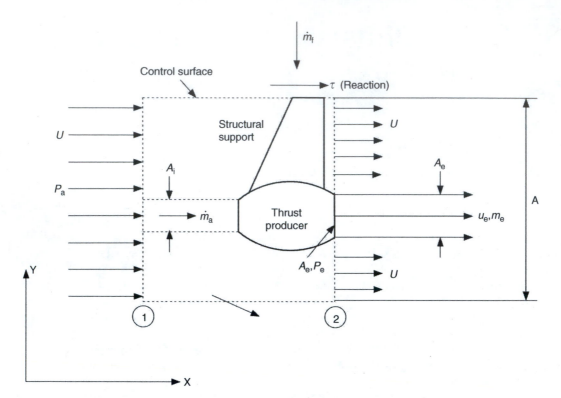

FIGURE 2.1 Thrust generated in an aero engine.

Consider a schematic diagram (Figure 2.1) for an engine with a part of its pod installation (i.e., a structural support for hanging the engine to the wing). Next define a control volume which control surface passes through the engine outlet (exhaust) plane (2) and extends far upstream at (1). The two side faces of the control volume are parallel to the flight velocity u. The upper surface side cuts the structural support while the lower one is far from the engine. The surface area at planes (1) and (2) are equal and denoted by A. The stream tube of air entering the engine has an area A_i at plane (1), while the exhaust area for gases leaving the engine is A_e. The velocity and pressure over plane (1) are u (which is the flight speed) and P_a (ambient pressure at this altitude). Over plane (2) the velocity and pressure are still u and P_a except over the exhaust area of the engine A_e where the values will be u_e and P_e. The x and y directions employed here are chosen parallel and normal to the centerline of the engine.

The following assumptions are made:

1. The flow is steady within the control volume; thus all the properties within the control do not change with time.
2. The external flow is reversible; thus the pressures and velocities are constant over the control surface except over the exhaust area P_e of the engine.

Conservation of mass across the engine gives

$$\dot{m}_a + \dot{m}_f = \dot{m}_e$$

where $\dot{m}_a$ and $\dot{m}_e$ are the rate of mass flow of air entering the engine and exhaust gases leaving the engine respectively, which are expressed as:

$$\dot{m}_a = \rho u A_i, \quad \dot{m}_e = \rho_e u_e A_e$$

thus:

$$\dot{m}_f = \rho_e u_e A_e - \rho u A_i \tag{2.1}$$

The fuel-to-air ratio is defined here as

$$f = \frac{\dot{m}_f}{\dot{m}_a}$$

thus $\dot{m}_e = \dot{m}_a(1 + f)$.

Apply the continuity equation over the control volume

$$\frac{\partial}{\partial t} \iiint_{CV} \rho d\upsilon + \oiint_{CS} \rho \bar{u} \cdot d\bar{A} = 0$$

For a steady flow, $\frac{\partial}{\partial t} \iiint_{CV} \rho d\upsilon = 0$, then

$$\dot{m}_e + \dot{m}_s + \rho u (A - A_e) - \dot{m}_a - \dot{m}_f - \rho u (A - A_i) = 0$$

where $\dot{m}_s$ is the side air leaving the control volume.

Rearranging and applying Equation 2.1 we get the side mass flow rate as

$$\dot{m}_s = \rho u (A_e - A_i) \tag{2.2}$$

According to the momentum equation

$$\sum \bar{F} = \frac{\partial}{\partial t} \iiint_{CV} \rho \bar{u} d\upsilon + \oiint_{CS} \bar{u} \left(\rho \bar{u} \cdot d\bar{A} \right)$$

where $\sum \bar{F}$ is the vector sum of all forces acting on the material within the control volume, which are surface forces (pressure force as well as the reaction to thrust force through the structural support denoted by τ) and the body force (which is the gravitational force here).

For steady flow

$$\sum \bar{F} = \oiint_{CS} \bar{u} \left(\rho \bar{u} \cdot d\bar{A} \right)$$

The x-component of the momentum equation

$$\sum F_x = \oiint_{CS} u_x \left(\rho \bar{u} \cdot d\bar{A} \right) = (P_a - P_e) A_e + \tau \tag{a}$$

If the sides of the control volume are assumed to be sufficiently distant from the engine, then the side mass flow rate leaves the control volume nearly in the x-direction.

Now $\oiint u_x \left(\rho \bar{u} \cdot d\bar{A} \right) = \dot{m}_e u_e + u [\rho u (A - A_e)] + \dot{m}_s u - \dot{m}_a u - u [\rho u (A - A_i)]$

$$\therefore \oiint u_x \left(\rho \bar{u} \cdot d\bar{A} \right) = \dot{m}_e u_e - \dot{m}_a u - \rho u^2 (A_e - A_i) + \dot{m}_s u \tag{b}$$

with $\dot{m}_s = \rho u\,(A_e - A_i)$ from Equation 2.2

$$\therefore \oiint u_x\,(\rho \bar{u} \cdot d\bar{A}) = \dot{m}_e u_e - \dot{m}_a u$$

From Equations a and b then:

$$\dot{m}_e u_e - \dot{m}_a u = -\,(P_e - P_a)\,A_e + \tau$$

$$\therefore \tau = \dot{m}_a\,[(1+f)\,u_e - u] + (P_e - P_a)\,A_e \qquad (2.3a)$$

where τ is the net thrust, $\dot{m}_a\,[(1+f)\,u_e]$ the momentum thrust, $(P_e - P_a)\,A_e$ the pressure thrust, $\dot{m}_a\,[(1+f)\,u_e] + (P_e - P_a)\,A_e$ the gross thrust, and $\dot{m}_a u$ is the momentum drag

Equation 2.3a can then be rewritten employing the above definitions as

Net thrust = Gross thrust − Momentum drag

or

Net Thrust = Momentum thrust + Pressure thrust − Momentum drag

If the nozzle is unchoked, then $P_e = P_a$, the pressure thrust cancels in Equation 2.3a. The thrust is then expressed as:

$$\therefore \tau = \dot{m}_a\,[(1+f)\,u_e - u] \qquad (2.3b)$$

In many cases the fuel-to-air ratio is neglected in the above equation and the thrust force equation is reduced to the simple form:

$$\therefore \tau = \dot{m}_a (u_e - u) \qquad (2.3c)$$

The thrust force in turbojet engine attains high values as the exhaust speed is high and much greater than the air (flight) speed; $u_e/u \gg 1$. For turboprop engines, the high value of thrust is achieved by the very large quantity of the airflow rate, though the exhaust and flight speeds are very close.

In a similar way, the thrust force for two stream engines such as *turbofan* and *propfan* engines can be derived. It will be expressed as

$$\tau = \dot{m}_h\,[(1+f)\,u_{eh} - u] + \dot{m}_c\,(u_{ec} - u) + A_{eh}\,(P_{eh} - P_a) + A_{ec}\,(P_{ec} - P_a) \qquad (2.4)$$

where $f = \dot{m}_f/\dot{m}_h$ is the fuel-to-air ratio, $\dot{m}_h$ the air mass flow passing through the hot section of engine (combustion chamber and turbine(s)), $\dot{m}_c$ the air mass flow passing through the fan, u_{eh} the velocity of hot gases leaving the turbine nozzle, u_{ec} the velocity of cold air leaving the fan nozzle, P_{eh} exhaust pressure of the hot stream, P_{ec} the exhaust pressure of the cold stream, A_{eh} the exit area for the hot stream, and A_{ec} is the exit area for the cold stream.

The specific thrust is defined as the thrust per unit air mass flow rate $(T/\dot{m}_a)$, which can be obtained from Equations 2.3 and 2.4. It has the dimensions of velocity (say m/s).

Example 1 Air flows through a jet engine at the rate of 30 kg/s and the fuel flow rate is 1 kg/s. The exhaust gases leave the jet nozzle with a relative velocity of 610 m/s. Pressure equilibrium exists over the exit plane. Compute the velocity of the airplane if the thrust power is 1.12×10^6 W.

Solution: The thrust force is expressed as

$$\tau = \left(\dot{m}_a + \dot{m}_f \right) u_e - \dot{m}_a u$$

$$\text{Thrust power} = \tau \times u$$

$$\text{Thrust power} = \left(\dot{m}_a + \dot{m}_f \right) u_e u - \dot{m}_a u^2$$

$$1.12 \times 10^6 = (31)(610)u - 30u^2$$

$$30u^2 - 18,910u + 1.12 \times 10^6 = 0$$

or

$$u = \frac{18,910 \pm 10^3 \sqrt{357.588 - 134.4}}{60}$$

thus, either $u = 564.15$ m/s or $u = 66.17$ m/s.

Example 2   The idling turbojet engines of a landing airplane produce forward thrust when operating in a normal manner, but they can produce reverse thrust if the jet is property deflected. Suppose that, while the aircraft rolls down the runway at 100 mph, the idling engine consumes air at 100 lb_m/s and produces an exhaust velocity of 450 ft/s.

(a) What is the forward thrust of the engine?
(b) What is the magnitude and direction (forward or reverse) if the exhaust is deflected 90° and if the mass flow is kept constant?

Solution: Forward thrust has positive values and reverse thrust has negative values:

(a) The flight speed is $U = 100 \times 1.4667 = 146.67$ ft/s. The thrust force represents the horizontal or the x-component of the momentum equation:

$$T = \dot{m}_a(U_e - U)$$

$$T = \left(100 \, \frac{\text{lb}_m}{\text{s}} \right) \times (450 - 146.67) \, \frac{\text{ft}}{\text{s}} \times \frac{1}{32.2}$$

$$T = 942 \, \text{lb}_f$$

(b) Since the exhaust velocity is now vertical due to thrust reverse application, it has a zero horizontal component; thus the thrust equation is

$$T = \dot{m}_a(U_e - U)$$

$$T = \frac{100}{32.2} \times (0 - 146.67)$$

$$T = -455.5 \, \text{lb}_f \quad \text{(reverse)}$$

Example 3 It is required to calculate and plot the momentum drag as well as momentum, pressure, gross and net thrusts versus the flight speed for a turbojet engine powering an aircraft flying at 9 km (ambient temperature and pressure are 229.74 K and 30.8 kPa) and having the following characteristics: $A_i = 0.235 \ m^2$, $A_e = 0.25 \ m^2$, $f = 0.02$, $P_e = 200$ kPa, $U_e = 600$ m/s.

The flight speed varies from 500 to 6000 km/h. Two cases are to be considered:

1. The air mass flow rate is constant and equal to 40 kg/s irrespective of the variation of flight speed.
2. The mass flow rate varies with the flight speed.

Solution:
Case 1   The mass flow rate is constant and equal to 40 kg/s.

The momentum thrust (T_{momentum}) is constant and given by the relation

$$T_{\text{momentum}} = \dot{m}_a(1 + f)U_e = 40 \times 1.02 \times 600 = 24480 \ N$$

The pressure thrust (T_{pressure}) is also constant and calculated as

$$T_{\text{pressure}} = A_e \times (P_e - P_a) = 0.25 \times (200 - 30.8) \times 10^3 = 42300 \ N$$

The gross thrust (T_{gross}) is constant and equal to the sum of momentum and pressure thrusts

$$T_{\text{gross}} = T_{\text{momentum}} + T_{\text{pressure}} = 66780 \ N$$

The momentum drag for flight speed varying from 500 to 6000 km/h is given by the relation:

$$D_{\text{momentum}} = \dot{m}_a U = 40 \, (\text{kg/s}) \times \frac{U \, (\text{km/h})}{3.6} = 11.11 \times U \ (N)$$

which is a linear relation in the flight speed U.

The net thrust is then:

$$T_{\text{net}} = T_{\text{gross}} - D_{\text{momentum}} = 66780 - 11.11 \times U \ (N)$$

The net thrust varies linearly with the flight speed. The results are plotted in Figure 2.2. The net thrust must be greater than the total aircraft drag force during acceleration and equal to the drag at steady cruise flight. Zero net thrust results from the intersection of the gross thrust and ram drag.

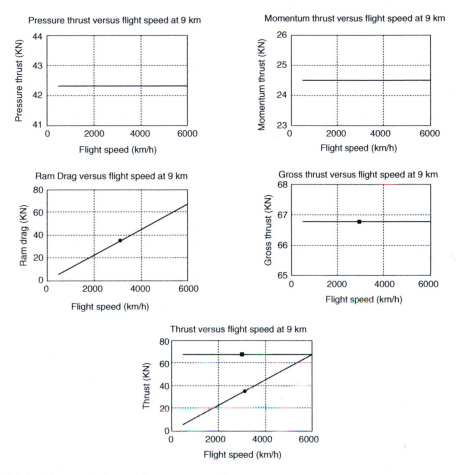

FIGURE 2.2 Thrust variations with constant mass flow rate.

Case 2 The mass flow rate varies linearly with the flight speed according to the relation:

$$\dot{m}_a = \rho_a U A_i = \frac{P_a}{RT_a} U A_i = \frac{30.8 \times 10^3}{287 \times 229.74} \times \frac{U}{3.6} \times 0.235 \text{ (kg/s)}$$

$$\dot{m}_a = 0.03049 \times U \text{ (kg/s)}$$

The momentum thrust varies linearly with the flight speed as per the relation

$$T_{momentum} = \dot{m}_a(1+f)U_e = 0.03049 \times U \times 1.02 \times 600 = 18.66 \times U \text{ (N)}$$

The pressure thrust is constant and has the same value as in Case 1

$$T_{pressure} = A_e \times (P_e - P_a) = 0.25 \times (200 - 30.8) \times 10^3 = 42300 \, N$$

The gross thrust varies linearly with the flight speed

$$T_{gross} = T_{momentum} + T_{pressure} = 18.66 \times U + 42300 \text{ (N)}$$

The momentum drag for flight speed varying from 500 to 6000 km/h is given by the relation:

$$D_{\text{momentum}} = \dot{m}_a U = 0.03049 \times U \text{ (kg/s)} \times \frac{U \text{ (km/h)}}{3.6} = 8.47 \times 10^{-3} \times U^2 \text{ (N)}$$

which is a quadratic relation in the flight speed U.

The net thrust is then:

$$T_{\text{net}} = T_{\text{gross}} - D_{\text{momentum}} = 42300 + 18.66 \times U - 8.47 \times 10^{-3} \times U^2 \text{ (N)}$$

The above relations are plotted in Figure 2.3.

Example 4   Figure 2.4(a) illustrates a typical single-spool axial-flow turbojet engine. It is required to calculate the distribution of the thrust force for each component (compressor, diffuser, combustion chamber, turbine, jet pipe, and nozzle) during ground run.

Prove that the sum of these thrust forces is equal to the thrust force developed by the engine using the usual thrust force equation.

Solution: This example illustrates how the thrust is generated from the unbalanced forces and momentum created within the engine itself. For simple one-dimensional steady flow through an

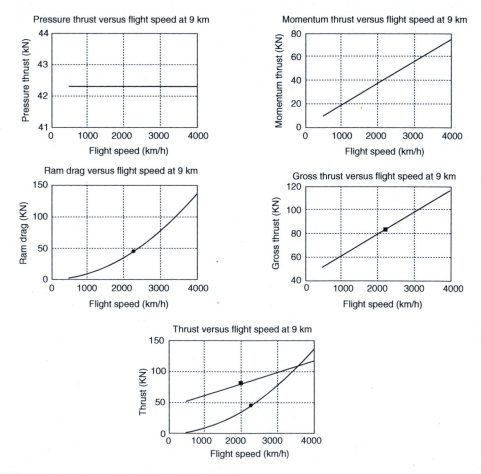

FIGURE 2.3   Thrust variations with variable mass flow rate.

	Compressor	Diffuser	Combustion Chamber	Turbine	Jet Pipe	Nozzle
Outlet area (m^2)	0.117	0.132	0.374	0.310	0.420	0.214
Outlet velocity (m/s)	124	112	94	271	196	584
Outlet gage pressure (kPa)	648	655	641	145	145	41
Mass flow rate (kg/s)	69.4	69.4	69.4	69.4	69.4	69.4

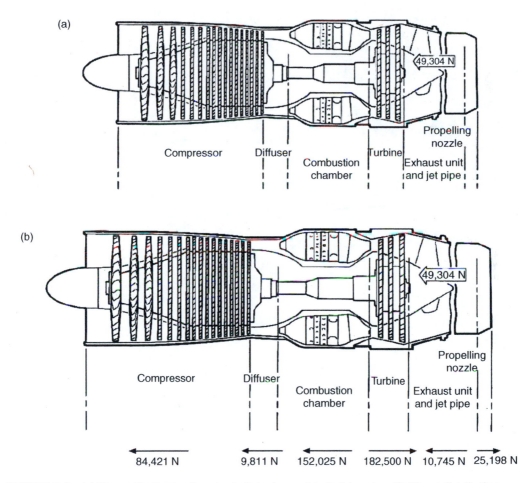

FIGURE 2.4 (a) Thrust distribution for a typical single-spool turbojet engine. (b) Thrust distribution across different modules of the engine. (Reproduced from Rolls-Royce plc, copyright © Rolls-Royce plc 2007. With permission.)

engine or any of its modules [1], the thrust force can be expressed as:

$$\text{Thrust} = (PA)_e - (PA)_i + (\dot{m}\,U)_e - (\dot{m}\,U)_i$$

Neglecting the fuel mass flow rate, the above relation is reduced to the following formula:

$$\text{Thrust} = P_e A_e - P_i A_i + \dot{m}\,(U_e - U_i)$$

where U_e and P_e are the velocity and pressure at the exit of each component while U_i and P_i are the velocity and pressure at the inlet to this component. Now this equation is applied to different components.

Compressor:

$$T_{comp} = 648 \times 10^3 \times 0.117 + 69.4\,(124 - 0)$$

$$= 75816 + 8605$$

$$T_{comp} = 84421\,N$$

Since the value is positive, its direction is forward.

$$T_{diffuser} = 655 \times 10^3 \times 0.132 - 648 \times 10^3 \times 0.117 + 69.4 \times 112 - 69.4 \times 124$$

$$= 86460 + 7773 - 84421$$

$$T_{diffuser} = 9812\,N$$

Since the value is positive, its direction is forward.

$$T_{CC} = 641 \times 10^3 \times 0.374 + 69.4 \times 94 - 94233$$

$$= 246257 - 94233$$

$$T_{CC} = 152025\,N$$

Since the value is positive, its direction is forward.

Turbine assembly:

$$T_{Turbine} = 145 \times 10^3 \times 0.31 + 69.4 \times 271 - 246257$$

$$= 63757 - 246257$$

$$T_{Turbine} = -182500\,N$$

Since the value is negative, its direction is rearward.

$$T_{Tailpipe} = 145 \times 10^3 \times 0.42 + 69.4 \times 196 - 63757$$

$$= 74502 - 63757$$

$$T_{Tailpipe} = 10745\,N$$

Since the value is positive, its direction is forward.

Nozzle:

$$T_{Nozzle} = 41 \times 10^3 \times 0.214 + 69.4 \times 584 - 74502$$

$$= 49303 - 74502$$

$$T_{Nozzle} = -25198\,N$$

Since the value is negative, its direction is rearward.

The magnitude and direction of the thrust forces generated at different components are plotted on the modules as shown Figure 2.4(b).

	Forward	Rearward
	84,421	−182500
Force	9,811	−25198
	152,025	
	10,745	
Total	257,002	−207,698

Total thrust $= 49{,}304\ N$

Engine thrust equation:

$$T = (P_e - P_a)\,A_e + \dot{m}\,U_e$$

$$T = P_{e_{gage}} \times A_e + \dot{m}\,U_e$$

$$T = 41 \times 10^3 \times 0.214 + 69.4 \times 584$$

$$T = 49{,}304\,N$$

Equal values for the thrust force are obtained from the summation of the thrust forces generated at different elements and the general formula for the engine as a whole.

2.3 FACTORS AFFECTING THRUST

As seen from Equation 2.3 for the thrust of a single stream aero engine (ramjet or turbojet engine), the thrust force depends on the inlet and outlet mass flow rates, fuel-to-air ratio, flight speed, exhaust speed, and exhaust pressure. Though it looked like a simple task to identify the factors listed above, each of them is dependent on several parameters. For example, the inlet air mass flow rate influencing both the momentum thrust and momentum drag is dependent on several variables, including the flight speed, ambient temperature and pressure, humidity, altitude, and rotational speed of the compressor. The outlet gas mass flow rate is dependent on the fuel added, air bleed, as well as water injection. The pressure thrust term depends on the turbine inlet temperature, flight altitude, and the nozzle outlet area and pressure. The momentum thrust is also dependent on the jet nozzle velocity. These parameters [2] can be further explained as given below.

2.3.1 JET NOZZLE

The outlet area and pressure of the exhaust nozzle affect the net thrust. The nozzle is either of the convergent or convergent–divergent type. Convergent nozzles may be choked or unchoked. For a choked convergent nozzle, the speed of the exhaust gases is equal to the sonic speed, which is mainly influenced by the exhaust gas temperature. The exhaust pressure for a choked nozzle is greater than the ambient pressure and thus the pressure thrust has a nonzero value. The pressure thrust depends on both the area of the exhaust nozzle and on the difference between the exit and ambient pressures. If the

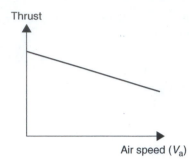

FIGURE 2.5   Variation of thrust force with air speed.

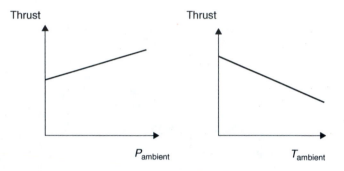

FIGURE 2.6   Variation of the thrust force with air temperature and pressure.

nozzle is unchoked then the jet velocity varies with the atmospheric pressure. The exhaust pressure is equal to the ambient pressure and the pressure thrust is zero.

2.3.2 AIR SPEED

The air speed, sometimes denoted as the approach speed, is equal to the flight speed in the thrust force; see Equation 2.3, derived from the control volume in Figure 2.1. Such a parameter has a direct effect on the net thrust. If the exhaust gas velocity is constant and the air velocity is increased, then the difference between both velocities $[(1 + f) u_e - u]$ is decreased, leading to a decrease also in the net thrust. If the air mass flow and the fuel-to-air ratio are assumed constants, then a linear decrease in the net thrust is enhanced (Figure 2.5).

2.3.3 MASS AIR FLOW

The mass airflow $\dot{m}_a$ is the most significant parameter in the thrust equation. It depends on the air temperature and pressure as both together determine the density of the air entering the engine. In free air, a rise in temperature will decrease the density. Thus, as the temperature increases, the thrust decreases. On the contrary, an increase in the pressure of free air increases its density and, consequently, its thrust increases. The effect of both air temperature and pressure is illustrated in Figure 2.6. In brief, the density affects the inlet air mass flow and it directly affects thrust. A 10,000 lb thrust engine for instance, might generate only about 8000 lb of thrust in a hot day, while on a cold day this same engine might produce as much as 12,000 lb of thrust.

2.3.4 ALTITUDE

As outlined above, air temperature and pressure have significant effects on the thrust. Thus, we need to know how the ambient temperature and pressure vary with height above the sea level. The variation depends to some extent on the season and latitude. However, it is usual to work with the "standard" atmosphere. The International Standard Atmosphere (ISA) corresponds to average values at middling latitudes and yields a temperature decrease by about 3.2 K per 500 m of altitude up to nearly 11,000 m (36,089 ft). The variations of ambient temperature and pressure are given by the following relations:

$$T(K) = 288 - 0.0065 \times z$$

$$P(\text{bar}) = 1.01325 - 0.000112 \times z + 3.8e^{-9} \times z$$

After 11,000 m the temperature stops falling, but the pressure continues to drop steadily with increasing altitude. Consequently, above 11,000 m (36,089 ft) the thrust will drop off more rapidly (Figure 2.7). This makes 11,000 m the optimum altitude for long-range cruising at nominal speed, just before the rapidly increased effect of altitude on thrust. It may be concluded that the effect of altitude on thrust is really a function of density.

2.3.5 RAM EFFECT

The movement of the aircraft relative to the outside air causes air to be rammed into the engine inlet duct. Ram effect increases the airflow to the engine, which in turn, means more gross thrust. However, it is not so easy; ram effects combine two factors, namely, the air speed increase and, at the same time, increase in the pressure of the air and the airflow into the engine. As described earlier, the increase of air speed reduces the thrust, which is sketched in Figure 2.8 as the curve "A". Moreover, the increase of the airflow will increase the thrust, which is sketched by the curve "B" in the same

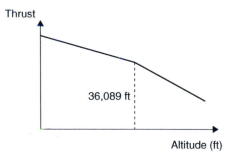

FIGURE 2.7   Variation of the thrust force with altitude.

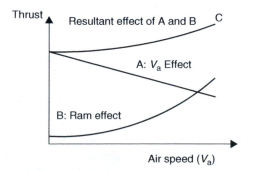

FIGURE 2.8   Effect of ram pressure on thrust.

figure. The curve "C" is the result of combining curves "A" and "B." The increase of thrust due to ram becomes significant as the air speed increases, which will compensate for the loss in thrust due to the reduced pressure at high altitude. Ram effect is thus important in high-speed fighter aircraft. Also modern subsonic jet-powered aircraft fly at high subsonic speeds and higher altitudes to make use of the ram effect.

Finally, it is fruitful to classify the factors affecting thrust into two groups [3]:

1. Factors related to the engine
2. Factors related to the surrounding medium

The first group includes

1. The rotational speed (rpm), which influences both the compressor pressure ratio and the turbine work
2. Exhaust nozzle size that influences the jet velocity
3. Fuel flow rate and turbine inlet temperature, which affect the combustor heat generation
4. Compressor bleed flow, water injection, and components performance, which lead to increase of the specific work

The first group thus contributes to both the air mass flow rate and jet velocity.
The second group includes

1. Forward (air) speed that leads to thrust decrease and more fuel consumption
2. Air density (influenced by the flight altitude, humidity, hot/cold day) that influences the airflow leading to a thrust increase if the airflow is increased and vice versa

2.4 ENGINE PERFORMANCE PARAMETERS

The performance of an aircraft engine may be criticized by its ability to provide the necessary thrust force needed in propelling an aircraft efficiently in its different flight regimes.

The engine performance parameters are identified as

1. Propulsive efficiency
2. Thermal efficiency
3. Propeller efficiency
4. Overall efficiency
5. Takeoff thrust
6. Specific fuel consumption
7. Aircraft range

Military aircraft are powered by engines that fulfill their mission requirements. For this reason the takeoff thrust and maneuverability are the critical issues with some sacrifice of fuel consumption in some types of aircraft such as fighters and interceptors. For civil transport specific fuel consumption and the aircraft range are the critical design issues. In both types several efficiencies related to the conversion of heat generated by fuel burning into thrust force are important.

2.4.1 PROPULSIVE EFFICIENCY

Propulsive efficiency is the efficiency of the conversion of the kinetic energy of air when it passes through the engine into a propulsive power. It is identified by manufacturers of aero engine as an external efficiency. It is influenced by the amount of the energy wasted in the propelling nozzle(s) and is denoted by (η_p).

It is defined as

$$\eta_p = \frac{\text{Thrust power}}{\text{Power imparted to engine airflow}} \qquad (2.5a)$$

or simply

$$\eta_p = \frac{\text{Thrust power}}{\text{Thrust power} + \text{power wasted in the exhaust}} \qquad (2.5b)$$

which will be denoted here as the *first* expression for the propulsive efficiency.

Another definition used in many texts is

$$\eta_p = \frac{\text{Thrust power}}{\text{Rate of kinetic energy added to engine airflow}} \qquad (2.6)$$

This will be referred to as the *second* expression.

For *ramjet*, *scramjet*, and *turbojet* engines, the first expression has the form:

$$\eta_p = \frac{uT}{uT + (1/2)\, \dot{m}_e\, (u_e - u)^2}$$

$$= \frac{u\left\{ \dot{m}_a\, [(1+f)\, u_e - u] + A_e\, (P_e - P_a) \right\}}{u\left\{ \dot{m}_a\, [(1+f)\, u_e - u] + A_e\, (P_e - P_a) \right\} + (1/2)\, \dot{m}_a\, (1+f)\, (u_e - u)^2} \qquad (2.7)$$

This expression is normally used by manufacturers of aero engines [2–5] as well as some texts [6,7]. The second expression adopted by other authors [8,9], is expressed as

$$\eta_p = \frac{2uT}{\dot{m}_a\left[(1+f)\, u_e^2 - u^2\right]} \qquad (2.8a)$$

$$\eta_p = \frac{2u\left\{ \dot{m}_a\, [(1+f)\, u_e - u] + A_e\, (P_e - P_a) \right\}}{\dot{m}_a\left[(1+f)\, u_e^2 - u^2\right]} \qquad (2.8b)$$

The two expressions are identical if the nozzle is unchoked ($P_e = P_a$) and the fuel-to-air ratio (f) is negligible. Under such conditions the following *third* expression is obtained:

$$\eta_p = \frac{2u}{u + u_e} = \frac{2}{1 + (u_e/u)} \qquad (2.9)$$

Considering Equations 2.3c and 2.9 it is important to notice that

1. If the exhaust speed is much greater than the air (flight) speed, $u_e \gg u$, then the thrust force $T \to$ maximum, $\eta_p \to 0$. This case represents the takeoff condition where $u = 0$.
2. If the exhaust speed is nearly equal to the flight speed, $u_e/u \approx 1$, then the thrust force $T \to 0$, $\eta_p \to$ maximum (100%).

For this reason turboprop engines have higher propulsive efficiency compared with turbojet engines, as in the former the exhaust speed is close to the flight speed, while for turbojet engines, the exhaust speed is much higher than the flight speed.

For bypass engines (turbofan and propfan), the air coming into the engine is spitted into two streams, the first passes through the fan/propfan, and is known as the cold stream, while the other passes through the engine core, compressor, combustion chamber and subsequent modules, and is known as the hot stream. Applying the same principle and employing the first expression we get the following form:

$$\eta_p = \frac{u\,(T_h + T_c)}{u\,(T_h + T_c) + W_h + W_c} \tag{2.10a}$$

where T_h and T_c are the thrust force generated by the hot and cold streams respectively, while W_h and W_c are the wake losses of the hot and cold streams respectively. Their values are expressed as

$$T_h = \dot{m}_h\,[(1+f)\,u_{eh} - u] + A_{eh}\,(P_{eh} - P_a)$$

$$T_c = \dot{m}_c\,[u_{ec} - u] + A_{ec}\,(P_{ec} - P_a)$$

$$W_h = \tfrac{1}{2}\dot{m}_h\,(1+f)\,(u_{eh} - u)^2 = \tfrac{1}{2}\dot{m}_{eh}(u_{eh} - u)^2$$

$$W_c = \tfrac{1}{2}\dot{m}_c\,(u_{ec} - u)^2$$

where $\dot{m}_{eh} = (1+f)\dot{m}_h$.

This relation can be written also as

$$\eta_p = \frac{Tu}{Tu + 0.5\left\{\dot{m}_{eh}\,(u_{eh} - u)^2 + \dot{m}_c\,(u_{ec} - u)^2\right\}} \tag{2.10b}$$

this is the first expression for the propulsive efficiency in turbofan engine.

The *second* expression (2.6), can be also written using the two hot and cold streams.

$$\eta_p = \frac{2uT}{\dot{m}_h\left\{(1+f)\,u_{eh}^2 + \beta u_{ec}^2 - (1+\beta)\,u^2\right\}} \tag{2.11a}$$

When the two nozzles of the hot and cold streams are unchoked and the fuel-to-air ratio (f) is negligible, both expressions yield the following expression:

$$\eta_p = \frac{2u\,[u_{eh} + \beta u_{ec} - (1+\beta)\,u]}{u_{eh}^2 + \beta u_{ec}^2 - (1+\beta)\,u^2} \tag{2.11b}$$

where β is the BPR, which is the ratio between the mass flow rates of the cold air and hot air, or $\beta = \dot{m}_c/\dot{m}_h$.

For turbofan engines, the simple relation expressed in Equation 2.9 cannot be applied.

The propulsive efficiency for turboprop, turbojet, and turbofan engines is illustrated in Figure 2.9. Figure 2.10 illustrates the propulsive efficiency for single and contrarotating propfan engines.

Example 5

A turbojet engine is powering a fighter airplane. Its cruise altitude and Mach number are 10 km and 0.8, respectively. The exhaust gases leave the nozzle at a speed of 570 m/s and a pressure of 0.67 bar. The exhaust nozzle is characterized by the ratio $A_e/\dot{m}_a = 0.006\,\mathrm{m}^2 \cdot \mathrm{s/kg}$. The fuel-to-air ratio is 0.02.

It is required to calculate

(a) The specific thrust $(T/\dot{m}_a)$.
(b) The propulsive efficiency using the different expressions defined above.

Solution:

(a) At altitude 10 km, the ambient temperature and pressure are
     $T_a = 223.3\ \mathrm{K}$ and $P_a = 0.265\ \mathrm{bar}$

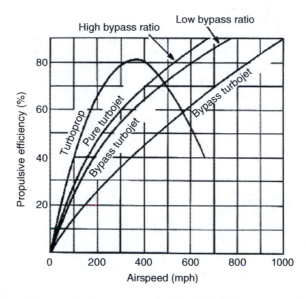

FIGURE 2.9 Propulsive efficiency for turboprop, turbojet, and turbofan engines. (Reproduced from Rolls-Royce plc, copyright © Rolls-Royce plc 2007. With permission.)

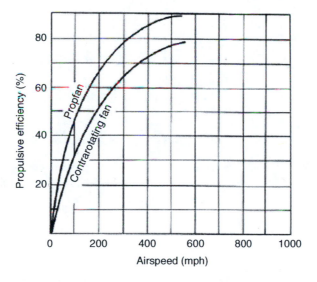

FIGURE 2.10 Propulsive efficiency for single and contrarotating propfan engines. (Reproduced from Rolls-Royce plc, copyright © Rolls-Royce plc 2007. With permission.)

The flight speed $u = M\sqrt{\gamma R T_a} = 239.6$ m/s

From Equation 2.3, the specific thrust is

$$\frac{T}{\dot{m}_a} = [(1+f)\,u_e - u] + (P_e - P_a)\frac{A_e}{\dot{m}_a}$$

$$\frac{T}{\dot{m}_a} = (1.02 \times 570 - 239.6) + 0.006 \times (0.67 - 0.265) \times 10^5$$

Then $T/\dot{m}_a = 584.77$ N · s/kg.

(b) The propulsive efficiency is calculated using the different expressions.

The first expression given by Equation 2.7 can be rewritten as

$$\eta_p = \frac{2(T/\dot{m}_a)u}{2(T/\dot{m}_a)u + (1+f)(u_e - u)^2}$$

Substitution will give

$$\eta_p = \frac{2 \times 584.77 \times 239.6}{2 \times 584.77 \times 239.6 + 1.02 \times (570 - 239.6)^2} = 0.7156 = 71.56\%$$

Next, the second expression given by Equation 2.8 can be rewritten as

$$\eta_p = \frac{2\left(T/\dot{m}_a\right)u}{(1+f)\,u_e^2 - u^2}$$

which, by substitution, gives

$$\eta_p = \frac{2 \times 584.77 \times 239.6}{(1.02) \times (570)^2 - (239.6)^2} = 1.022 = 102.2\%$$

This value confirms that this expression must not be used if the nozzle is choked.
Finally, using the third expression given by Equation 2.9 gives

$$\eta_p = \frac{2u}{u_e + u} = 0.5918 = 59.18\%$$

Though this expression is simple, yet it gives a very rough estimate of the propulsive efficiency.

Example 6   A turbofan engine is powering an aircraft flying at Mach number 0.9, at an altitude of 33,000 ft, where the ambient temperature and pressure are −50.4° and 26.2 kPa. The engine bypass ratio is 3, and the hot airflow passing through the engine core is 22.7 kg/s. Preliminary analysis provided the following results:

$$\text{Fuel-to-air ratio } f = 0.015$$

$$P_{ec} = 55.26 \text{ kPa} \quad u_{ec} = 339.7 \text{ m/s} \quad A_{ec} = 0.299 \text{ m}^2$$
$$P_{eh} = 32.56 \text{ kPa} \quad u_{eh} = 452 \text{ m/s} \quad A_{eh} = 0.229 \text{ m}^2$$

Calculate the thrust force and the propulsive efficiency using the first and second expressions.

Solution: From the given data, it is clear that both nozzles are choked since their exit pressures are higher than the ambient pressure of the flight altitude.
The flight speed is

$$u = M\sqrt{\gamma R T_a} = 269.2 \text{ m/s}$$

The cold airflow $\dot{m}_c = 3 \times 22.7 = 68.1$ m/s
Fuel flow rate $\dot{m}_f = f \times \dot{m}_h = 0.015 \times 22.7 = 0.3405$ kg/s
The hot exhaust flow rate $= \dot{m}_h + \dot{m}_f = 23.04$ kg/s

The thrust force is calculated from the relation:

$$T = \dot{m}_h \left[(1+f)\,u_{eh} - u\right] + \dot{m}_c\,(u_{ec} - u) + A_{eh}\,(P_{eh} - P_a) + A_{ec}\,(P_{ec} - P_a)$$

$$= (22.7)\,[(1.015)\,(452.1 - 269.2)] + (68.1)\,[339.72 - 269.2]$$

$$+ (0.229)\,[32.56 - 26.2] \times 10^3 + (0.299)\,[55.267 - 26.2] \times 10^3$$

$$= 19.257\ \text{kN}$$

The first expression for the propulsive efficiency, Equation 2.10b is

$$\eta_p = \frac{Tu}{Tu + 0.5\left\{\dot{m}_{eh}(u_{eh} - u)^2 + \dot{m}_c(u_{ec} - u)^2\right\}} = 0.9032 = 90.32\%$$

The second expression for choked nozzle is given by Equation 2.11a

$$\eta_p = \frac{2uT}{\dot{m}_h\left\{(1+f)\,u_{eh}^2 + \beta u_{ec}^2 - (1+\beta)\,u^2\right\}} = 1.7304 = 170.04\%$$

Comment: As depicted in Examples 5 and 6, the second expression for the propulsive efficiency provides efficiencies greater than unity. Such an astonishing value of a propulsive efficiency greater than unity can be overcome if the effective jet velocity proposed in References 6 and 10 is employed. The proposed effective jet velocity is expressed as:

$$u_e' = u_e + (P_e - A_a)\,\frac{P_e}{\dot{m}_e}$$

Such an effective jet velocity means that a full expansion of the gases to the ambient pressure is assumed. In such a case the second expression for the propulsive efficiency will be exactly equal to the first expression. This effective velocity was also employed by Rolls-Royce [4]. The effective jet velocity is identified by aero engine manufacturers as the fully expanded jet velocity (in the exhaust plume).

2.4.2 THERMAL EFFICIENCY

It is the efficiency of energy conversion within the power plant itself. So, it is considered as an internal efficiency while the propulsive efficiency resembles an external efficiency as stated earlier. Thermal efficiency is denoted by (η_{th}). Two forms of such efficiency are defined for the following two groups of engines:

1. *Ramjet, Scramjet, turbojet, and turbofan engines*

$$\eta_{th} = \frac{\text{Power imparted to engine airflow}}{\text{Rate of energy supplied in the fuel}} \qquad (2.12)$$

It is easy to note that the denominator in Equation 1.5a is equal to the numerator in Equation 2.12.

For a *ramjet* and *turbojet* engine:

Using the first expression for the propulsive efficiency, the following expression for (η_{th}) is obtained:

$$\eta_{th} = \frac{Tu + \frac{1}{2}\dot{m}_a\,(1+f)\,(u_e - u)^2}{\dot{m}_f Q_R} \qquad (2.13)$$

where Q_R is the heat of reaction of the fuel used. Other names for Q_R are the calorific value of fuel [8] or the lower heating value (LHV) [3]. It is also alternatively written as Q_{HV}.

Assuming an unchoked nozzle, then

$$\dot{W}_{out} = \tfrac{1}{2}\left[\left(\dot{m}_a + \dot{m}_f\right)u_e^2 - \dot{m}_a u^2\right]$$

$$\eta_{th} = \frac{\tfrac{1}{2}\left[\left(\dot{m}_a + \dot{m}_f\right)u_e^2 - \dot{m}_a u^2\right]}{\dot{m}_f Q_R}$$

$$\eta_{th} = \frac{\left[(1+f)u_e^2 - u^2\right]}{2fQ_R} \tag{2.14a}$$

For an unchoked nozzle and negligible (f), then

$$\eta_{th} = \frac{u_e^2 - u^2}{2fQ_R} \tag{2.14b}$$

For a two-stream engine (turbofan and prop fan), the following expression is employed:

$$\eta_{th} = \frac{Tu + \tfrac{1}{2}\dot{m}_h(1+f)(u_{eh} - u)^2 + \tfrac{1}{2}\dot{m}_c(u_{ec} - u)^2}{\dot{m}_f Q_R} \tag{2.15}$$

Similarly, assuming unchoked nozzles and negligible (f), we get

$$\eta_{th} = \frac{u_{eh}^2 + \beta u_{ec}^2 - (1+\beta)u^2}{2fQ_R} \tag{2.16}$$

2. *Turboprop and turboshaft engines*

   The output of turboprop or turboshaft engines is largely a shaft power. In this case thermal efficiency is defined as:

   $$\eta_{th} = \frac{SP}{\dot{m}_f Q_R} \tag{2.17}$$

   where SP is the shaft power.

2.4.3 Propeller Efficiency

Propellers are used in two types of aero engines: piston and turboprop. In both cases, shaft power is converted to thrust power. Propeller efficiency (η_{pr}) is defined as the ratio of the thrust power generated by the propeller ($TP \equiv uT_{pr}$) to the shaft power (SP):

$$\eta_{pr} = \frac{TP}{SP} = \frac{uT_{pr}}{SP} \tag{2.18}$$

when an appreciable amount of thrust is obtained from the exhaust gases T_e in a turboprop engine, an equivalent thrust power, ETP is expressed in terms of the shaft power, propeller efficiency, and exhaust thrust power; thus:

$$ETP = (SP) \times \eta_{pr} + T_e u \tag{2.19a}$$

Alternatively, an equivalent shaft power (ESP) is used instead of the equivalent thrust power. They are related to each other by the relation:

$$ESP = ETP/\eta_{pr} = SP + \frac{T_e u}{\eta_{pr}} \tag{2.19b}$$

Moreover, the exhaust thrust is expressed as

$$T_e = \dot{m}_h \left[(1 + f) u_e - u \right] \tag{2.20a}$$

The total thrust force is given by the relation:

$$T = T_{pr} + T_e \tag{2.20b}$$

The propeller efficiency is then expressed as

$$\eta_{pr} = \frac{uT}{ESP} \tag{2.21}$$

Such propeller efficiency replaces the propulsive efficiency of other types of aero engines discussed above. More details will be given in Chapter 6.

2.4.4 OVERALL EFFICIENCY

The product of the propulsive and thermal efficiencies ($\eta_p \times \eta_{th}$) or ($\eta_{pr} \times \eta_{th}$) as appropriate is called the overall efficiency.

In all cases

$$\eta_o = \frac{Tu}{\dot{m}_f Q_R} \tag{2.22}$$

For a turbojet with unchoked nozzle and negligible fuel-to-air ratio (f), then from Equation 2.9:

$$\eta_o = \eta_{th} \frac{2u}{u + u_e} \tag{2.23}$$

Example 7   Boeing 747 aircraft is powered by four CF-6 turbofan engines manufactured by General Electric Company. Each engine has the following data:

Thrust force	24.0 kN
Air mass flow rate	125 kg/s
Bypass ratio	5.0
Fuel mass flow rate	0.75 kg/s
Operating Mach number	0.8
Altitude	10 km
Ambient temperature	223.2 K
Ambient pressure	26.4 kPa
Fuel heating value	42,800 kJ/kg

If the thrust generated from the fan is 75% of the total thrust, determine

(a)   The jet velocities of the cold air and hot gases
(b)   The specific thrust
(c)   The thrust specific fuel consumption (TSFC)
(d)   The propulsive efficiency
(e)   The thermal efficiency
(f)   The overall efficiency

(Assume that the exit pressures of the cold and hot streams are equal to the ambient pressure).

Solution:

(a)   Since the total air mass flow rate through the engine is $\dot{m}_a = 125$ kg/s, then the flow rates are calculated from the bypass ratio as follows:

$$\dot{m}_c \equiv \dot{m}_{Fan} = \frac{5}{6} \times 125 = 104.2 \text{ kg/s}$$

$$\dot{m}_h \equiv \dot{m}_{core} = (1/6) \times 125 = 20.8 \text{ kg/s}$$

Subscript c, which stands for the cold section of the engine and fan, is used alternatively. Also, subscript h, which stands for the hot section and core, is used alternatively.

Fuel-to-air ratio $f = \dot{m}_{fuel}/\dot{m}_h = \frac{0.75}{20.8} = 0.036$

$$T_{fan} = 0.75T = 18 \text{ kN}$$

$$T_{core} = 0.25T = 6 \text{ kN}$$

Flight speed $U = M\sqrt{\gamma R T_a} = 0.8\sqrt{1.4 \times 287 \times 223.2}$

$$U = 240 \text{ m/s}$$

Exit velocity from fan (cold air)

$$\text{since } T_{Fan} = \dot{m}_{Fan}\left(U_{e\,Fan} - U_{Flight}\right)$$

$$\therefore U_{e\,Fan} = \frac{T_{Fan}}{\dot{m}_{Fan}} + U_{Flight} = \frac{18000}{104.2} + 240 = 173 + 240$$

$$U_{e\,Fan} = 413 \text{ m/s}$$

Exhaust velocity from engine core (hot gases)

Since $T_{Core} = \dot{m}_{Core}\left[(1+f)\,U_{e\,C} - U\right]$

$$\therefore U_{e\,C} = \frac{\left[T_{Core}/\dot{m}_{Core} + U\right]}{1+f} = \frac{1}{1.036}[6000/20.8 + 240]$$

$$U_{e_{Core}} = 510 \text{ m/s}$$

(b)   Specific thrust $= T/\dot{m}_a = \frac{24}{125} = 0.192$ kN · s/kg $= 192$ m/s

(c)   TSFC $= \dot{m}_{Fuel}/T = \frac{0.75}{24} = 0.03125$ kg · Fuel/kN/s

(d)   The propulsive efficiency

Since both cold (fan) and hot (core) nozzles are unchoked, then, the propulsive efficiency can be expressed as

$$\eta_P = \frac{T \times U}{(1/2)\,\dot{m}_h\left[(1+f)\,U_{eh}^2 - U^2\right] + (1/2)\,\dot{m}_c\left[U_c^2 - U^2\right]}$$

$$= \frac{2 \times 24000 \times 240}{20.8\left[1.036 \times (510)^2 - (240)^2\right] + (104.2)\left[(413)^2 - (240)^2\right]}$$

$$= \frac{11.52 \times 10^6}{4.409 \times 10^6 + 11.77 \times 10^6} = 0.712 = 71.2\%$$

Here also the propulsive efficiency can be written as

$$\eta_p = \frac{2U\,\{[(1+f)\,U_{e\,Core} - U] + \beta\,(U_{eFan} - U)\}}{\left[(1+f)\,U_{ecore}^2 - U^2\right] + \beta\,(U_{fan}^2 - U^2)}$$

(e) *The thermal efficiency*

$$\eta_{th} = \frac{\dot{m}_{Core}\left[(1+f)\,U_{e\,Core}^2 - U^2\right] + \dot{m}_{Fan}\left(U_{eFan}^2 - U^2\right)}{2\dot{m}_{fuel}Q_{HV}}$$

$$= \frac{16.179 \times 10^6}{2 \times 0.75 \times 42.8 \times 10^6} = 0.2546$$

$$= 25.46\%$$

(f) *The overall efficiency*

$$\eta_o = \eta_P\eta_{th} = 17.93\%$$

Example 8 For a turbofan engine with unchoked nozzles, prove that the value of the maximum overall efficiency is given by the relation:

$$\eta_{o\,max} = \frac{(1+\beta)\,u^2}{fQ_R}$$

where the flight speed u is given by the relation:

$$u = \frac{(1+f)\,u_h + \beta u_c}{2\,(1+\beta)}$$

Deduce the appropriate relations corresponding to a single-stream engine (ramjet, scramjet, and turbojet engines).

Solution: From Equation 2.4 with the cold mass flow rate equal to the bypass ratio times the hot mass flow rate, the following relation is obtained

$$T = \dot{m}_h\,(1+f)\,u_{eh} + \beta\dot{m}_h u_{ec} - \dot{m}_h\,(1+\beta)\,u$$

Substituting in the overall efficiency expression to we get

$$\eta_o = \frac{Tu}{\dot{m}_f Q_R} = \frac{u\left[(1+f)\,u_{eh} + \beta u_{ec} - (1+\beta)\,u\right]}{fQ_R} \tag{a}$$

Differentiate with respect to the flight speed:

$$\frac{\partial \eta_o}{\partial u} = \frac{(1+f)\,u_{eh} + \beta u_{ec} - 2\,(1+\beta)\,u}{\text{constant}}$$

Setting $\partial \eta_o / \partial u = 0$ to obtain the maximum value of η_o

$$\therefore u = \frac{(1+f)\,u_h + \beta u_c}{2\,(1+\beta)} \tag{b}$$

Substitute to get the maximum value of the overall efficiency:

$$\eta_{max} = \frac{1}{fQ_R}\left\{\frac{(1+f)\,u_h + \beta u_c}{2\,(1+\beta)}\right\}\left[(1+f)\,u_h + \beta u_c - (1+\beta)\left\{\frac{(1+f)\,u_h + \beta u_c}{2\,(1+\beta)}\right\}\right]$$

$$= \frac{(1+f)\,u_h + \beta u_c}{2\,(1+\beta)fQ_R}\left[(1+f)\,u_h + \beta u_c - \frac{(1+f)\,u_h + \beta u_c}{2}\right]$$

$$= \frac{\{(1+f)\,u_h + \beta u_c\}^2}{4\,(1+\beta)fQ_R}$$

The above expression defines the maximum overall efficiency in terms of the exhaust speeds of both the hot and cold stream.

To obtain an expression for the maximum overall efficiency in terms of the flight speed, substitute Equation (b) in (a) to get the requested relation:

$$\eta_{o\,mx} = \frac{(1+\beta)\,u^2}{fQ_R}$$

From the above relation, the maximum overall efficiency for a single-stream engine (ramjet, scramjet, and turbojet engines) is obtained by setting the bypass ratio equal to zero; thus its value will be

$$\eta_{o\,mx} = \frac{u^2}{fQ_R}$$

The corresponding value for the flight speed will be

$$u = \frac{(1+f)\,u_e}{2}$$

2.4.5 TAKEOFF THRUST

This is an important parameter that defines the ability of an aero engine to provide a static and low speed thrust, which enables the aircraft to take off under its own power. Both ramjet and scramjet engines are not self-accelerating propulsion systems from static conditions; they require acceleration to an appreciable velocity by a boost system before they are capable of providing net positive thrust. Thus, they are excluded from this discussion.

From Equation 2.3, the static thrust of a turbojet engine with an unchoked nozzle is expressed by the relation:

$$T_{\text{takeoff}} = \dot{m}_a \, (1 + f) \, u_e \tag{a}$$

From Equation 2.13 when $u = 0$,

$$(\eta_{\text{th}})_{\text{static}} = \dot{m}_a \, (1 + f) \, \frac{u_e^2}{2 \dot{m}_f Q_R} \tag{b}$$

From Equations (a) and (b), the static or takeoff thrust is given by the relation:

$$T_{\text{takeoff}} = \frac{2 \eta_{\text{th}} \dot{m}_f Q_R}{u_e} \tag{2.24}$$

For a given flow rate and thermal efficiency,

$$T_{\text{takeoff}} \propto \frac{1}{u_e}$$

The above relation outlines one of the advantages of turboprop engines over turbojet and turbofan engines. Turboprop engines accelerate a large mass flow rate of air to a small exhaust velocity that in turn increases the take-off thrust. Consequently, aircraft powered by turboprop engines can take off from a very short runway.

2.4.6 Specific Fuel Consumption

This performance parameter of the engine has a direct influence on the costs of aircraft trip and flight economics. Fuel consumption is either defined per unit thrust force (for ramjet, turbojet, and turbofan engines) or per horsepower (e.g., for turboprop and piston-propeller engines).

1. *Ramjet, turbojet, and turbofan engines:*
    The TSFC is defined as:

$$\text{TSFC} = \frac{\dot{m}_f}{T} \tag{2.25a}$$

where the thrust force (T) is expressed by Equation 2.3 for a turbojet engine and Equation 2.4 for a turbofan engine; values of TSFC strongly depend on the flight speed. So its typical values for both turbojet and turbofan are defined for static condition. Typical values [9] for turbojet engines are 0.075–0.11 kg/N · h (0.75–1.1 ib/ib$_f$ · h), while turbofans are more economic and have the following range 0.03–0.05 kg/N · h (0.3–0.5 ib/ib$_f$ · h). However, for ramjet reference values for TSFC are defined at flight Mach number of 2. Typical values are 0.17–0.26 kg/N · h (1.7–2.6 ib/ib$_f$ · h). Some empirical formulae are presented in Reference 11 for TSFC. These formulae have the form:

$$\text{TSFC} = (a + b M_0) \sqrt{\theta} \tag{2.25b}$$

where a and b are constants that vary from one engine to another, M_0 is the flight Mach number, and θ is the dimensionless ratio (T_a / T_{ref}), which is the ratio between the ambient

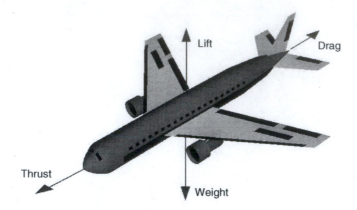

FIGURE 2.11 Forces on airplane during cruise.

temperature at the flight altitude and the standard temperature at sea level (288.2 K). For a high bypass ratio turbofan, the values of these constants are $a = 0.4$ and $b = 0.45$.

2. *Turboprop engines*
   For engines that produce shaft power, fuel consumption is identified by brake-specific fuel consumption (BSFC) or simply SFC, and defined as

$$SFC \equiv BSFC = \frac{\dot{m}_f}{SP} \tag{2.26}$$

When appreciable thrust is produced by the hot gases, the fuel consumption is identified by the equivalent EBSFC or simply equivalent specific fuel consumption (ESFC) and is defined by

$$ESFC = \frac{\dot{m}_f}{ESP} \tag{2.27}$$

Typical values [9] for ESFC are 0.45–0.60 ((ib/h)/hp) or 0.27–0.36 (kg/kW · h). The values of the corresponding constants in the empirical relation (2.25b) are $a = 0.2$ and $b = 0.9$ [11].

2.4.7 AIRCRAFT RANGE

Aircraft range is a design parameter of the aircraft based on which, the number, location, and capacity of fuel tanks in the aircraft are determined. Aircraft weight is composed of the structural weight, the payload (including the crew members, food, and service supply), and the fuel weight. During any trip only the fuel weight is changeable.

Now consider the cruise phase in a flight trip where the aircraft flies at a constant speed. Ignoring the climb and descent phases, it may be assumed that the cruise distance is equal to the trip distance of the aircraft. During cruise, the aircraft is subjected to two vertical forces [lift (L) and weight (mg), which are equal in magnitude] and two horizontal forces [thrust (T) and drag (D), which are also equal in magnitude]; see Figure 2.11.

Then

$$T = D = mg \left(\frac{D}{L} \right)$$

where m is the instantaneous aircraft mass and g is the gravitational acceleration.

Since, the rate of fuel consumption is equal to the rate of decrease of the mass of aircraft,

$$\overset{\bullet}{m}_f = -\frac{dm}{dt} = \left(-\frac{dm}{ds}\right) \times \left(\frac{ds}{dt}\right) = -u\frac{dm}{ds}$$

where s is the distance along the flight path.

$$\therefore \overset{\bullet}{m}_f = -u\left(\frac{T}{D}\right)\frac{dm}{ds}$$

$$= -\frac{uT}{D}\frac{L}{mg}\frac{dm}{ds}$$

$$\therefore ds = -\frac{u}{g}\frac{T}{\overset{\bullet}{m}_f}\frac{L}{D}\frac{dm}{m}$$

Integration of the above equation yields the range of aircraft (S) as follows:

$$S = \frac{u}{g}\frac{T}{\overset{\bullet}{m}_f}\frac{L}{D}\ln\left(\frac{m_1}{m_2}\right) \tag{2.28}$$

where m_1 and m_2 are the initial and final mass of aircraft.

From Equation 2.25, Equation 2.28 is reformulated as

$$S = \frac{u}{g}\frac{1}{\text{TSFC}}\frac{L}{D}\ln\left(\frac{m_1}{m_2}\right) \tag{2.29}$$

The above equation is known as *Breguet's* equation, which was derived in 1920. Another expression for the range is obtained from the propulsive efficiency (Equations 2.22 and 2.28) as follows:

$$S = \frac{\eta_o Q_R}{g}\frac{L}{D}\ln\left(\frac{m_1}{m_2}\right) \tag{2.30}$$

In Equations 2.29 and 2.30, the ratio between the lift and drag forces may be alternatively written as, C_L/C_D, where the lift force coefficient is calculated as usual from the relation:

$$C_L = \frac{W}{\frac{1}{2}\rho U^2 S_W}$$

where W and S_W are the aircraft weight and wing area.

The drag coefficient is calculated from the relation:

$$C_D = K_1 C_L^2 + K_2 C_L + C_{D_o}$$

where the constants K_1, K_2, and C_{D_o} are typically functions of the flight Mach number and wing configuration (flap position, etc.). The C_{D_o} term is the zero lift drag coefficient, which accounts for both frictional and pressure drag in the subsonic flight and wave drag in supersonic flight [11]. The K_1 and K_2 terms account for the drag due to lift. Normally K_2 is very small and approximately equal to zero in most fighter aircraft.

It is clear from Equations 2.29 and 2.30 that the following should be done in order to maximize the range of an aircraft flying at a specified speed [12]:

1. Fly at the maximum lift to drag ratio.
2. Have the highest possible overall efficiency.

3. Have the lowest specific fuel consumption.
4. Have the highest possible ratio between the aircraft weights at the start and end of cruise.

For a propeller or reciprocating or turboprop engines a similar analysis may be followed. From Equations 2.21 and 2.27, then

$$\therefore \frac{uT}{\overset{\bullet}{m_f}} = \frac{\eta_{pr}}{ESFC}$$

Substituting the above equation in Equation 2.28, the following equation for the range is obtained:

$$S = \frac{\eta_{pr}}{g} \frac{1}{ESFC} \frac{L}{D} \ln\left(\frac{m_1}{m_2}\right) \qquad (2.31)$$

Similarly, to maximize the range, the lift-to-drag ratio, propeller efficiency, and the ratio of the aircraft weight at the start to that at the end of cruise have to be chosen as maximums while the fuel consumption (ESFC) has to be minimum.

Airlines and aircraft manufacturers normally use the following forms for Breguet's relation:

For *turbojet/turbofan engines*:

$$S = \frac{u}{c} \frac{L}{D} \ln \frac{m_1}{m_2} \qquad (2.32)$$

and for *turboprop engines:*

$$S = \frac{\eta}{c} \frac{L}{D} \ln \frac{m_1}{m_2} \qquad (2.33)$$

where c is the specific fuel consumption [13] expressed in (lbs × fuel/lb × thrust/h) or (lbs × fuel/ESHP/h).

Another definition is also employed, namely, the *specific range* [14], which is given in miles per pound of fuel (mi/lb).

(a) *For a turbojet/turbofan*

$$\frac{mi}{lb} = \frac{u}{[c\,(D/L)\,W]} = \frac{u}{c} \frac{L}{D} \frac{1}{W} \qquad (2.34a)$$

If u is in knots, mi/lb will be in nautical miles per pound of fuel.

(b) *For a turboprop engine*

$$\frac{mi}{lb} = 325 \left(\frac{\eta_{prop}}{c}\right) \left(\frac{L}{D}\right) \left(\frac{1}{W}\right) \qquad (2.34b)$$

Example 9   A Boeing 747 aircraft has a lift-to-drag ratio of 17. The fuel-to-air ratio is 0.02 and the fuel heating value is 45,000 kJ/kg. The ratio between the weight of the aircraft at the end and start of cruise is 0.673. The overall efficiency is 0.35.

1. Calculate the range of aircraft.
2. What will be the fuel consumed in the cruise if the takeoff mass of aircraft is 385,560 kg?
3. If the fuel consumed during the engine start, warming, and climb is 4.4% of the initial aircraft weight and the fuel consumed during descent, landing, and engine stop is 3.8% of the aircraft weight at the end of cruise, calculate the fuel consumed in the whole trip.

Solution: From Equation 2.30 and the given data the aircraft range (S) is

$$S = \frac{0.35 \times 45 \times 10^6}{9.8} \times 17 \times \ln\left(\frac{1}{0.673}\right) \text{ m}$$

$$= \frac{0.35 \times 45 \times 10^3}{9.8} \times 17 \times 0.396 \text{ km}$$

$$= 10{,}819 \text{ km}$$

Fuel consumed in the trip $= (1 - 0.673) \times 385{,}560 = 126{,}078 \text{ kg}$

$$\text{Weight at landing} = \text{Weight at engine start} \times \frac{\text{Weight at the start of cruise}}{\text{Weight at engine start}}$$

$$\times \frac{\text{Weight at the end of cruise}}{\text{Weight at the start of cruise}} \times \frac{\text{Weight after landing and engine stop}}{\text{Weight at the end of cruise}}$$

$$= 385{,}560 \times (1 - 0.044) \times 0.673 \times (1 - 0.038)$$

$$= 385{,}560 \times 0.956 \times 0.673 \times 0.962$$

$$= 238{,}638 \text{ kg}$$

Fuel consumed in the whole trip $= 146{,}921$ kg.

2.4.8 RANGE FACTOR

The range factor (RF) is defined as

$$\text{RF} = \frac{1}{g} \frac{u}{\text{TSFC}} \frac{L}{D} = \frac{1}{g} \frac{u}{\text{TSFC}} \frac{C_L}{C_D} \tag{2.35}$$

The minimum fuel consumption for a distance(s) occurs at the condition where the RF is maximum.

2.4.9 ENDURANCE FACTOR

The endurance factor is defined as

$$\text{EF} = \frac{1}{g} \frac{1}{\text{TSFC}} \frac{L}{D} = \frac{1}{g} \frac{1}{\text{TSFC}} \frac{C_L}{C_D} \tag{2.36}$$

The minimum fuel consumption as seen from Equation 2.36 for a time, flight time (t) occurs when the endurance factor is maximum.

Example 10   *DASSAULT MIRAGE G* is a two seat Strike and Reconnaissance fighter powered by one *SNECMA TF-306C* turbofan engine. It has the following characteristics:

Flight Mach number	0.8
Altitude	65,000 ft
Ambient temperature	216.7 K
Ambient pressure	5.5 kPa
Fuel heating value	42,700 kJ/kg

Thrust force	53.4 kN
Air mass flow rate	45 kg/s
Fuel mass flow rate	2.5 kg/s
Aircraft gross weight (65,000 ft)	156 kN
Aircraft takeoff weight	173.3 kN
Wing area	26.4 m^2
Fuel weight	5. kN
Maximum lift coefficient	$C_{L\,max=1.8}$

$C_{D0} = 0.012$ $K_1 = 0.2$ $K_2 = 0.0$

Air density at take-off	1.225 kg/m^3
Air density at 650,000 ft	0.88 kg/m^3

Calculate:

1. The specific thrust
2. TSFC
3. The exit velocity
4. The thermal efficiency
5. The propulsive efficiency
6. The overall efficiency
7. The range factor

Solution:

$$U = M_o\sqrt{\gamma R T_o} = 0.8\sqrt{1.4 \times 287 \times 216.7} = 236 \text{ m/s}$$

1. The specific thrust $= \dfrac{T}{\dot{m}_a} = \dfrac{53.4 \times 10^3}{45} = 1186.7$ m/s

2. TSFC $= \dfrac{\dot{m}_f}{T} = \dfrac{2.65}{53.4} = 0.046 \ \dfrac{\text{kg/s}}{\text{kN}} = 46 \ \dfrac{\text{mg}}{\text{N.s}}$

3. The thrust force is $T = (\dot{m}_a + \dot{m}_f)U_e - \dot{m}_a U$

$$U_e = \frac{T + \dot{m}_a U}{\dot{m}_a + \dot{m}_f} = \frac{53,000 + 45 \times 236}{45 + 2.5} = 1339.4 \text{ m/s}$$

$$U_e = \frac{T + \dot{m}_a U}{\dot{m}_a + \dot{m}_f} = \frac{53,000 + 45 \times 236}{45 + 2.5} = 1339.4 \text{ m/s}$$

4. The thermal efficiency is given by the relation

$$\eta_{th} = \frac{\dot{W}_{out}}{\dot{Q}_{in}}$$

$$\dot{W}_{out} = \tfrac{1}{2}\left[\left(\dot{m}_a + \dot{m}_f\right)U_e^2 - \dot{m}_a U^2\right] = 0.5\left[(45 + 2.5)(1339.4)^2 - 45 \times (236)^2\right]$$

$$\dot{W}_{out} = 41.35 \times 10^6 \text{ W}$$

$$\dot{Q}_{in} = \dot{m}_f Q_{HV} = 107 \times 10^6 \text{ W}$$

Substituting the above values to get $\eta_{th} = 0.386 = 38.6\%$.

5. The propulsive efficiency $\eta_p = \dfrac{TU}{\dot{W}_{out}} = \dfrac{53.4 \times 10^3 \times 236}{41.35 \times 10^6} = 0.3047 = 30.47\%$

$$\eta_o = \eta_p \eta_{th} = 0.1179 = 11.79\%$$

$$C_L = \frac{W}{\frac{1}{2}\rho U^2 S_W} = \frac{156,000}{0.5 \times 0.088 \times (236)^2 \times 66.4} = 0.959$$

$$C_D = K_1 C_L^2 + K_2 C_L + C_{D_o}$$

$$C_D = 0.2 \times 0.919 + 0.012 = 0.195$$

$$RF = \frac{C_L}{C_D} \frac{U}{TSFC} \frac{1}{g} = \frac{0.959}{0.195} \times \frac{236}{0.053 * 10^{-3}} \times \frac{1}{9.81} = 2232 \text{ km}$$

2.4.10 Specific Impulse

The specific impulse (I_{sp}) is defined as the thrust per unit fuel flow rate, or

$$I_{sp} = \frac{T}{\dot{m}_f g} \qquad (2.37)$$

This quantity enters directly into the calculation of the fractional weight change of aircraft (or rocket) during flight. Thus, from Equation 2.28, the range can be expressed in terms of the specific impulse as follows:

$$S = u I_{sp} \frac{L}{D} \ln\left(\frac{m_1}{m_2}\right) \qquad (2.38)$$

The specific impulse is equally applied for both rockets and aircraft. The unit of specific impulse is time (s).

Example 11 The maximum range of an aircraft is given by the relation:

$$S_{max} = \left(\frac{V_g}{TSFC}\right)\left(\frac{1}{g}\right)\left(\frac{L}{D}\right)\ln\left(\frac{m_1}{m_2}\right)$$

where V_g is the air relative speed including the effect of wind as shown in Figure 2.12 for either head wind or tail wind conditions:

1. Calculate the mass of fuel consumed during a trip where its range is 4000 km, flight speed is 250 m/s, $L/D = 10$, TSFC $= 0.08$ kg/N/h, and $m_1 = 50,000$ kg in the following two cases:
   (i) Head wind $= 50$ m/s
   (ii) Tail wind $= 50$ m/s
2. Calculate the time for such a trip in the above two cases.

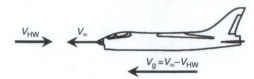 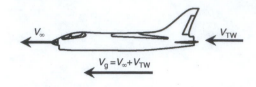

FIGURE 2.12   Head and tail wind conditions.

Solution: This example illustrates the effect of wind on both fuel consumption and the time of a trip. First, consider the case:

Tail wind

1. Fuel consumption: The equation for the range is now rewritten in terms of the relative speed rather than the airplane speed.

$$S = \frac{V_g}{\text{TSFC}} \frac{1}{g} \frac{L}{D} \ln \frac{m_1}{m_2}$$

$$\therefore \ln \frac{m_1}{m_2} = \frac{(S)\,(g)\,(\text{TSFC})}{V_g\,(L/D)}$$

When $V_g = V_\infty + V_{TW} = 250 + 50 = 300$ m/s

$$\ln \frac{m_1}{m_2} = \frac{4000 \times 10^3 \times 9.8 \times 0.08}{300 \times 10 \times 3600} = 0.2903$$

$$\frac{m_1}{m_2} = 1.3369$$

Fuel consumed $= m_1 - m_2 = m_1 \left(1 - \dfrac{1}{1.3369}\right) = 50,000 \times 0.252 = 12,600$ kg.

Fuel consumed for the case of tail wind (i.e., supporting wind) $= 12,600$ kg.

$\therefore m_2 = 37,400$ kg.

   (i)   *Head wind case:*

$$V_g = V_\infty - V_{HW} = 250 - 50 = 200 \text{ m/s}$$

$$\ln \frac{m_1}{m_2} = 0.43555$$

$$\frac{m_1}{m_2} = 1.54582$$

Fuel consumed $= m_1 \left(1 - \dfrac{1}{1.154582}\right)$

Fuel consumed for the case of head wind $= 17,654$ kg

$$\therefore m_2 = 50,000 - 17,654 = 32,345 \text{ kg}$$

2. Trip time (t):

$$\text{Trip time } (t) = \frac{S}{V_g} = \frac{L/D}{(g)\,(\text{TSFC})} \ln \frac{m_1}{m_2}$$

For tail wind case:

$$\text{Time} = \frac{10}{(9.8)\,(0.08)} \ln \frac{50,000}{37,400}$$

$$\text{Time} = 3.70\,\text{h}$$

For the head wind case:

$$\text{Time} = \frac{10}{(9.8)\,(0.08)} \ln \frac{50,000}{32,345}$$

$$\text{Time} = 5.556\,\text{h}$$

From the above calculation tail wind—as expected—helps in shortening the trip time and saves fuel. On the contrary, for head wind, the trip time and fuel consumed are increased. What are the corresponding values for zero wind speed?

Example 12 An airplane is powered by four engines and has the following data:

Drag coefficient	$C_d = 0.097$
Wing area	$A = 260\,\text{m}^2$
Flight speed	$u = 300\,\text{m/s}$
Air density	$\rho = 0.414\,\text{kg/m}^3$
Inlet area of each engine	$A_i = 3.14\,\text{m}^2$

It is required to calculate the jet speed assuming that the nozzles are unchoked and the fuel-to-air ratio is negligible.

Now, due to some trouble, one engine is shut down. It is required to prove that the new flight speed u will be given by the relation:

$$u = \frac{6A_i u_e}{C_d A + 6A_i}$$

where u_e is the speed of the exhaust gases, which is constant irrespective of other engine failures.

Calculate this flight speed based on the new available thrust forces.

Solution: For steady flight conditions, the drag force is equal to the sum of thrust forces generated by the four engines or

$$D = \tfrac{1}{2}\rho u^2 A C_d$$

$$D = \tfrac{1}{2} \times 0.414 \times (300)^2 \times 260 \times 0.097 \, \text{N}$$

$$D = 469.85 \, \text{KN}$$

$$T = \frac{D}{4} = 117.46 \, \text{KN}$$

$$T = \dot{m}(u_e - u) = \rho u A_i (u_e - u)$$

$$u_e = u + \frac{T}{\rho u A_i} = 300 + \frac{117,460}{0.414 \times 300 \times 3.14}$$

$$u_e = 601.2 \, \text{m/s}$$

Since the air mass flow rate into each engine is expressed as

$$\dot{m} = \rho u A_i$$

Thrust force for one engine is

$$\therefore T = \dot{m}(u_e - u) = \rho u A_i (u_e - u)$$
$$\text{Drag force} = D = \tfrac{1}{2}\rho u^2 A C_d$$

For *n* operative engines

$$D = nT$$

$$\tfrac{1}{2}\rho u^2 A C_d = n\rho u A_i (u_e - u)$$

$$u A C_d = 2n A_i (u_e - u)$$

$$u(C_d A + 2n A_i) = 2n A_i u_e$$

$$\therefore u = \frac{2n A_i u_e}{C_d A + 2n A_i}$$

$\therefore$ For three operative engines, $n = 3$

$$\therefore u = \frac{6 A_i u_e}{C_d A + 6 A_i}$$

Keeping the exhaust speed same as obtained above, then, the new value for flight speed after one engine shutdown is calculated from the relation:

$$u = \frac{6 A_f u_e}{C_d A + 6 A_f}$$

$$= \frac{6 \times 3.14 \times 600}{0.0966 \times 260 + 6 \times 3.14}$$

$$= \frac{11304}{43.956} = 257.16 \text{ m/s}$$

Thus the flight speed is reduced from 300 to 257.16 m/s.

Now, if a second engine were shut down or, $n = 2$

$$\therefore u = \frac{4A_f u_e}{C_d A + 4A_f} = \frac{4 \times 3.14 \times 600}{0.0966 \times 260 + 4 \times 3.14}$$

$$= \frac{7536}{34.536} = 218.2 \text{ m/s}$$

Again, the flight speed is reduced to only 218.2 m/s.

PROBLEMS

2.1 A turbojet engine powers an aircraft that is flying at a speed of 240 m/s, has an exhaust speed of 560 m/s, and a specific thrust of 525 N · s/kg. Using the three different formulae, calculate the propulsive efficiency. What are your comments?
(Neglect the fuel-to-air ratio).

2.2 Starting from the thrust force equation for an unchoked nozzle engine, prove that:

$$\eta_{0\text{max}} = \frac{(1 + f)^2 u_e^2}{4fQ_R}$$

If λ is the ratio of the maximum overall efficiency to the overall efficiency ($\lambda = \eta_{0\text{max}}/\eta_0$), then prove that the flight speed u may be expressed by the relation:

$$u = \frac{(1 + f)u_e}{2} \left[1 \pm \sqrt{\frac{\lambda - 1}{\lambda}} \right]$$

Now, consider an aircraft fitted with a single engine, which has the following data:
   Air mass flow rate $\dot{m}_a = 100$ kg/s
   Exhaust speed $u_e = 770$ m/s
   Overall efficiency $\eta_0 = 0.16$
   Maximum overall efficiency $\eta_{0\text{max}} = 0.18$
   Fuel-to-air ratio $f = 0.02$
   Lift/drag ratio $L/D = 10$
   Aircraft mass ratio at takeoff and landing $m_1/m_2 = 1.2$
Calculate:
   (a) Possible value(s) for flight speed
   (b) Heating value (Q_R)
   (c) Possible range(s) of aircraft
   (d) Trip time(s)
   (e) The aircraft masses m_1 and m_2
   (f) Thrust force(s)

2.3 For a turbojet engine, prove that the thrust force (T) and propulsive efficiency (η_p) are given by the following relations:

$$T = \dot{m}_a u (B + C)$$

and

$$\eta_p = \frac{B + C}{D + C - 1}$$

where

$$C = \frac{A_e P_a}{\dot{m}_a u}\left[\left(\frac{P_e}{P_a}\right) - 1\right]$$

$$D = \frac{1}{2}(1 + f)\left[\left(\frac{u_e}{u}\right)^2 + 1\right]$$

Next, if the nozzle is unchoked, then:

$$\eta_p = \frac{B}{D - 1}$$

Moreover, if the nozzle is unchoked and (f) is negligible, then:

$$\eta_p = \frac{2u}{u + u_e}$$

2.4 For a turbojet engine having the following data:

$$\dot{m}_a = 50 \text{ kg/s} \quad f = 0.015 \quad u_e = 500 \text{ m/s}$$

$$P_a = 60 \text{ kPa} \quad P_e = 120 \text{ kPa} \quad A_e = 0.25 \text{ m}^2$$

(a) Plot the relation between the propulsive efficiency η_p and the speed ratio (u/u_e).
(b) Plot the relation between the thrust force and the speed ratio (u/u_e).
    The flight speed u varies from 0 to 500 m/s.
2.5 To examine the effect of fuel-to-air ratio (f), plot the above two relations for the following values of $f, f = 0.0, 0.01, 0.015, 0.02$.
2.6 To examine the effect of exhaust pressure, plot the above relations for the case in problem 5, with variable P_e. Take $P_e = 60, 90, 120,$ and 150 kPa.
2.7 Apply Reynolds Transport equation

$$\frac{DN}{Dt} = \oiint_{cs} \eta \rho \bar{u} \cdot d\bar{A} + \frac{\partial}{\partial t}\iiint_{cv} \eta \rho d\Psi$$

to a turbofan engine to derive the thrust equation

$$T = \dot{m}_h (1 + f) u_h + \dot{m}_c u_c - \dot{m}_a u + A_h (P_h - P_a) + A_c (P_c - P_a)$$

where dV is the element of volume,

$$\dot{m}_a = \dot{m}_c + \dot{m}_h,$$

u_h, u_c the exhaust speeds of the hot and cold streams, respectively,
A_h, A_c the exhaust areas for the hot and cold streams, respectively,
P_a the ambient pressure,
u the flight speed,
f the fuel-to-air ratio.

2.8   Show that the specific thrust and the TSFC for a turbofan engine can be expressed as:

$$\frac{T}{\dot{m}_a} = \frac{1+f}{1+\beta} u_{eh} + \frac{\beta}{1+\beta} u_{ec} - u$$

and

$$\text{TSFC} = \frac{f}{(1+\beta)\,(T/\dot{m}_a)}$$

2.9   The JT9D high bypass ratio turbofan engine at maximum static power ($V_0 = 0$) on a sea level, standard day ($P_0 = 14.696$ psia, $T_0 = 518.7°$R) has the following data:
Air mass flow rate through the core is 247 lb$_m$/s, the air mass flow rate through the fan bypass duct is 1248 lb$_m$/s, the exit velocity from the core is 1190 ft/s, the exit velocity from the bypass duct is 885 ft/s and the fuel flow rate into the combustor is 15,570 lbm/h. For the case of exhaust pressures equal to ambient pressure ($P_0 = P_e$), estimate the following:
  (a)   The thrust of the engine
  (b)   The thermal efficiency of the engine (heating value of jet fuel is about $18,400$ Btu/lbm)
  (c)   The propulsive efficiency and TSFC of the engine

2.10   An aircraft is powered by two turbojet engines having the following characteristics:
Exhaust velocity $u_e = 2300$ km/h
Maximum overall efficiency $\eta_{0\,max} = 0.15$
Fuel-to-air ratio $f = 0.0175$
Lift-to-drag ratio $L/D = 10$
Mass ratio at the beginning and end of the trip $m_1/m_2 = 1.2$
The maximum range is expressed by the relation:

$$S_{max} = \frac{\eta_{0\,max} Q_R}{g} \frac{L}{D} \ln\left(\frac{m_1}{m_2}\right)$$

where the maximum overall efficiency ($\eta_{0\,max}$) is stated in problem (2) and the corresponding flight speed is given by the relation $u = \frac{(1+f)}{2} u_e$.
Calculate:
  (a)   Flight speed (u)
  (b)   Heating value (Q_R)
  (c)   Maximum range of aircraft
  (d)   Trip time
  (e)   The value of m_1 and m_2

2.11   A fighter aircraft flies at speed of 250 m/s. It is powered by a single turbojet engine having an inlet area of 0.35 m^2. Use the ISA standard tables for altitudes from 3 to 18 km to
  (a)   Plot the air mass flow rate entering the engine versus altitude.
  (b)   Plot the thrust force against the altitude assuming the exhaust speed is constant and equal to 600 m/s, the fuel-to-air ratio $f = 0.025$, and the nozzle is unchoked.

2.12 The range factor of aircraft (RF) is defined as

$$\text{RF} = \frac{C_L}{C_D} \frac{u}{\text{TSFC}}$$

with

$$u = M\sqrt{\gamma R T_a}$$

and

$$\text{TSFC} = (1.7 + 0.26M)\sqrt{\theta}$$

Plot the RF versus the Mach number for altitudes of 0, 5, 10, and 20 km.
where $\theta = T/T_{SL}$, T is the temperature at any altitude, and T_{SL} is the sea level temperature—both are in Kelvin degrees. The aircraft weight is 150 kN, the drag coefficient is given by the relation,

$$C_D = K_1 C_L^2 + K_2 C_L + C_{D0},$$

where $C_{D0} = 0.012$, $C_L = 0.3$, $K_1 = 0.2$, and $K_2 = 0.0$.

2.13 What is the ESFC of a turboprop engine that consumes 1500 lb of fuel per hour and produces 700 lb of exhaust thrust and 3300 shaft horsepower (shp) during flight at 250 mph. The propeller efficiency is 0.9.

2.14 For a turbofan engine with unchoked nozzles, prove that if the fuel-to-air ratio is negligible ($f \approx 0$), the propulsive efficiency is expressed as:

$$\eta_p = \frac{2u\left[u_{e_h} + \beta u_{e_c} - (1+\beta)u\right]}{u_{e_h}^2 + \beta u_{e_c}^2 - (1+\beta)u^2}$$

Next, plot the relation η_p versus u for the following cases:

β	u (m/s)				u_{ec} (m/s)	u_{eh} (m/s)
0.5	0	200	450	700	750	1000
2.0	0	200	400	600	700	950
5.0	0	100	200	300	500	800
8.0	0	100	200	320	450	750

REFERENCES

1. R. Crawford and R. Schulz, *Fundamentals of Air Breathing Propulsion*, University of Tennessee Space Institute, Aero propulsion Short Course, PS-749.
2. United Technologies Pratt Whitney, *The Aircraft Gas Turbine Engine and Its Operation*, P&W Operation Instruction 200, 1988, pp. 1–10.
3. H. Moustapha, *Turbomachinery & Propulsion*, Pratt & Whitney Canada Inc., Course Handbook, p. 61.
4. *The Jet Engine*, Rolls-Royce Plc.
5. J. Kurzke, *Aero-Engine Design: A State of the Art, Preliminary Design*, Von Karman Institute Lecture Series, 2002–2003, April 2003, p. 14.
6. M.J. Zucrow, *Aircraft and Missile Propulsion, Volume I: Thermodynamics and Fluid Flow and Application to Propulsion Engines*, John Wiley & Sons, 1958, p. 121.
7. H. Ashly, *Engineering Analysis of Flight Vehicles*, Addison-Wesley Publication Co, 1974, p. 140.

8. H.I.H. Saravanamuttoo, G.F.C. Rogers, and H. Cohen, *Gas Turbine Theory*, Prentice-Hall, 5th edn., 2001.

9. P. Hill and C. Peterson, *Mechanics and Thermodynamics of Propulsion*, 2nd edn., 1992, Addison Wesley Publication Company, Inc., p. 155.

10. D.G. Shepherd, *Aerospace Propulsion*, American Elsevier Publishing Company, Inc., 1972, p. 28.

11. J.D. Mattingly, *Elements of Gas Turbine Propulsion*, McGraw-Hill International Edition, 1996, p. 44.

12. J.D. Anderson, Jr., *Aircraft Performance and Design*, WCB/McGraw-Hill, 1999, p. 293.

13. G. Corning, *Supersonic And Subsonic CTOL and VTOL Airplane Design*, Box No.14, College Park, Maryland, 4th edn., 1979, p. 2.38.

14. R. Shevell, *Fundamentals of Flight*, Prentice-Hall, Inc., 1983, p. 265.

3 Pulsejet and Ramjet Engines

3.1 INTRODUCTION

The two types of engines treated in this chapter are air-breathing engines of the athodyd type. This abbreviation stands for aero-thermodynamic duct. Thus, these engines have no major rotating parts (fan/compressor or turbine) and consist of an entry duct, a combustion chamber, and a nozzle [1].

The first engine that will be described and analyzed is the pulsejet engine. Pulsejet engine operates intermittently and has found limited applications [2]. The reasons are the difficulty of its integration into manned aircraft as well as its high fuel consumption, poor efficiency, severe vibration, and high noise. The second engine described here will be the ramjet engine. The ramjet engine is appropriate for supersonic flight speeds [3], where the ram compression of the air becomes sufficient to overcome the need for mechanical compression achieved normally by compressors or fans in other jet engines. If the flight speed is so high that fuel combustion must occur supersonically, then this ramjet is called a scramjet [4].

3.2 PULSEJET ENGINES

3.2.1 INTRODUCTION

Pulsejet engine is a very simple jet engine, which comprises an air intake, a combustion chamber, and an acoustically resonant exhaust pipe. It is a jet engine in which the intake of air is intermittent; thus, combustion occurs in pulses, resulting in a pulsating thrust rather than a continuous one.

Presently, there are two types of pulsejets, namely, *valved and valveless* jet engines. A third future type known as pulse detonation engine (PDE) is in the research and testing phase of production. The valved type is fitted with a one-way valve while a valveless engine has no mechanical valves at its intake. The PDE is a concept currently in active development to create a jet engine that operates on the supersonic detonation of fuel.

Historically, Martin Wiberg (1826–1905) developed the first pulsejet in Sweden. The first type—valved or traditional pulsejet—was used to power a German cruise missile called the Vergeltungswaffe 1 (Vengeance 1), or V1, and normally referred to as the V-1 flying bomb (refer to Figure 1.38). It was extensively used in bombing England and Belgium during World War II. The engines made a distinctive sound, leading the English to call them "buzz bombs."

The pulsejet engine is extremely simple, cheap, and easy to construct. However, it has low reliability, poor fuel economy, and very high noise levels. The high noise levels make them impractical for applications other than the military and similar restricted applications. Pulsejets have been used to power experimental helicopters, the engines being attached to the extreme ends of the rotor blades. They have also been used in both tethered and radio-control model aircraft. The speed record for tethered model aircraft is 186 miles per hour (299 km/h), set in the early 1950s.

The valved-type pulsejet engine has a set of one-way valves (or check valves) through which the incoming air passes. The valving is accomplished through the use of reed valves that consist of thin flexible metal or fiberglass strips fixed on one end that open and close upon changing pressures across opposite sides of the valve much like heart valves do. Fuel in the form of a gas or liquid aerosol is either mixed with the air in the intake or injected into the combustion chamber. When the air–fuel mixture is ignited, these valves slam shut, which means that the hot gases can only leave through the engine's tailpipe, thus creating forward thrust. The cycle frequency is primarily dependent on the length of the engine. For a small model-type engine the frequency may be typically around 250 pulses per second, whereas for a larger engine such as the one used on the German V1 flying bomb, the frequency was closer to 45 pulses per second. Once the engine is running it requires only an input of fuel, but it usually requires forced air and an ignition method for the fuel–air mix. Once running, the engine is self-sustaining. The main drawback of valved type is that the valves require regular replacement and represent a weak link in the engine's reliability. The pulsejet powered V1 "flying bombs" were only good for about 20–30 min of continuous operation. So, by eliminating the valves, it should be possible to make an engine that is as simple as a ramjet having absolutely no moving parts. Thus, a *valveless* type is employed.

The name valveless is really a misnomer. These engines have no mechanical valves, but they do have aerodynamic valves, which, for the most part, restrict the flow of gases to a single direction just as their mechanical counterparts. Thus, this type of pulsejet has no mechanically moving parts at all and in that respect is similar to a ramjet.

3.2.2 Valved Pulsejet

Figure 3.1 illustrates the simple construction of valved type together with its cycle on the temperature–entropy (T–S) diagram.

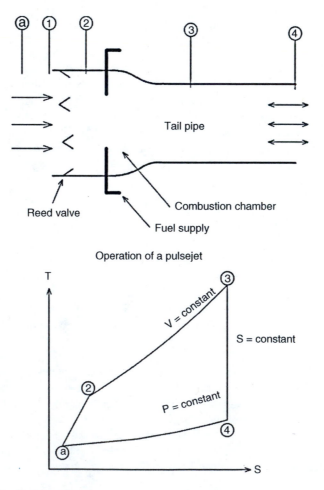

FIGURE 3.1 Valved pulsejet engine.

The cycle can be described briefly as follows:

1. The air is sucked into the combustion chamber through a bank of spring-loaded check (reed) valves by the vacuum created by the exhaust of the previous cycle. These valves are normally closed, but if a predetermined pressure differential exists, they will open to permit high-pressure air from the diffusing section to pass into the combustion chamber. They never permit flow from the combustion chamber back into the diffuser.
2. A spark plug initiates the combustion process, which occurs at something approaching a constant volume process [5]. Such combustion occurs in the form of an explosion, which raises the pressure in the combustion chamber to a high level and closes the spring valve at the intake.
3. The resultant high pressure and temperature force the gases to flow out of the tail pipe at high velocity.

At the end of the discharge, the inertia of the gas creates a vacuum in the combustion chamber that together with the ram pressure developed in the diffuser causes sufficient pressure differential to open the check valves. A new charge of air enters the chamber and a new cycle starts. The frequency of the cycle depends on the size of the engine and the dynamic characteristics of the valves must be matched carefully to this frequency. Small-size engines operate at frequencies as high as 300–400 cycles per second, and large engines operate at frequencies as low as 40 cycles per second.

To start the engine, a carefully adjusted amount of fuel is sprayed into the cold combustion chamber in order to create a mixture and the resultant strong explosion starts the cycle. The fuel flow, injected directly into the combustion chamber, is continuous throughout the cycle with some variation in fuel flow resulting from the pressure in the combustion chamber. This fuel flow fluctuation can be neglected.

1. *Intake or diffuser*

   This process occurs from state (a) to state (2) as shown in Figure 3.1. State (a) is far upstream, while state (1) is at the pulsejet inlet and state (2) is just aft of the check valve upstream of the combustion chamber. Owing to ram effect, both the pressure and temperature rise from the ambient conditions to the values to be calculated hereafter. The total conditions at state (1) are given by the relations:

$$P_{01} = P_{0a} = P_a \left(1 + \frac{\gamma_c - 1}{2} M^2 \right)^{\frac{\gamma_c}{\gamma_c - 1}} \tag{3.1}$$

   Here M is the flight Mach number and γ_c is the specific heat ratio for cold air.
   The stagnation temperatures for states (a), (1), (2) are also equal

$$T_{01} = T_{0a} = T_a \left(1 + \frac{\gamma_c - 1}{2} M^2 \right) \tag{3.2}$$

   If it is assumed that the diffuser is an ideal one, then the total pressure and temperature at state (2) is equal to those at state (1), or

$$P_{02} = P_{01} \tag{3.3}$$

   and

$$T_{02} = T_{01} \tag{3.4}$$

The total temperature at state (2) is always equal to its value at state (1). However, due to losses in the diffuser and check valves, the pressure at state (2) is less than that at state (1). It may even reach half its value at state (1). In general, if the diffuser efficiency (η_d) is specified, then the total pressure will be calculated from the relation:

$$P_{02} = P_a \left(1 + \eta_d \frac{\gamma_c - 1}{2} M_a^2 \right)^{\frac{\gamma_c}{\gamma_c - 1}} \tag{3.5}$$

2. *Combustion chamber*

   Combustion takes place at nearly a constant volume process, thus

$$P_{03} = P_{02} \left(\frac{T_{03}}{T_{02}} \right)$$

Fuel is added to the combustion chamber and burnt. The mass flow rate of the burnt fuel is calculated from the energy balance of the combustion chamber.

$$(\dot{m}_a + \dot{m}_f) Cp_h T_{03} = \dot{m}_a Cp_c T_{02} + \eta_b \dot{m}_f Q_R$$

With $f = \dot{m}_f / \dot{m}_a$, η_b is the burner's (or combustion chamber) efficiency, and C_{pc}, C_{ph} are the specific heats for the cold air and hot gases.

The fuel-to-air ratio is determined from the relation:

$$f = \frac{Cp_h T_{03} - Cp_c T_{02}}{\eta_b Q_R - Cp_h T_{03}} \tag{3.6}$$

Tail Pipe

The gases expand isentropically in the tail pipe to the ambient pressure. Thus, $p_7 = p_a$ and the temperature of the exhaust gases is determined from the relation:

$$\left(\frac{T_{03}}{T_4} \right) = \left(\frac{P_{03}}{P_a} \right)^{\frac{\gamma_h - 1}{\gamma_h}} \tag{3.7}$$

The exhaust velocity is now calculated from the relation:

$$U_e = \sqrt{2 Cp_h T_{03} \left[1 - \left(\frac{P_a}{P_{03}} \right)^{\frac{\gamma_h - 1}{\gamma_h}} \right]} \tag{3.8}$$

The thrust force is now calculated from the relation:

$$T = \dot{m}_a [(1 + f) U_e - U] \tag{3.9}$$

The specific thrust is ($T/\dot{m}_a$) and the thrust-specific fuel consumption (TSFC) is given by the relation

$$\text{TSFC} = \frac{\dot{m}_f}{T} = \frac{f}{(T/\dot{m}_a)} \tag{3.10}$$

Example 1   A pulsejet engine is employed in powering a vehicle flying at a Mach number of 2 at an altitude of 40,000 ft. The engine has an inlet area 0.084 m^2. The pressure ratio at combustion chamber is $P_{03}/P_{02} = 9$, fuel heating value is 43,000 kJ/kg, and combustion efficiency is 0.96. Assuming ideal diffuser ($P_{02} = P_{0a}$), it is required to calculate

1. The air mass flow rate
2. The maximum temperature
3. The fuel-to-air ratio
4. The exhaust velocity
5. The thrust force
6. The thrust-specific fuel consumption (TSFC)

Solution: At altitude 40,000 ft, the ambient temperature and pressure are 216.65 K and 18.75 kPa, respectively.

The flight speed is

$$U = M\sqrt{\gamma R T_a} = 2\sqrt{1.4 \times 287 \times 216.65}$$
$$= 590 \text{ m/s}$$

The air mass *f* low rate is

$$\dot{m}_a = \rho_a U A_i = \frac{P_a}{R T_a} U A_i = 14.99 \text{ kg/s}$$

Diffuser: The total temperature at the intake is equal to the total upstream temperature, or

$$T_{02} = T_{0a} = T_a \left(1 + \frac{\gamma_c - 1}{2} M^2 \right)$$
$$= 216.65 \left(1 + \frac{1.4 - 1}{2} \times 2^2 \right)$$
$$= 390 \text{ K}$$

The total pressure at the diffuser outlet also equals the free stream total pressure, as the flow through the diffuser is assumed ideal.

$$P_{02} = P_{0a} = P_a \left(\frac{T_{0a}}{T_a} \right)^{\frac{\gamma}{\gamma - 1}}$$
$$P_{02} = 146.7 \text{ kPa}$$

Combustion Chamber: A constant volume process is assumed; then

$$T_{03} = T_{02} \left(\frac{P_{03}}{P_{02}} \right)$$
$$= 9 T_{02} = 3510 \text{ K}$$

The fuel-to-air ratio is calculated from Equation 3.6 as

$$f = \frac{1.148 \times 3510 - 1.005 \times 390}{0.96 \times 43,000 - 1.148 \times 3510}$$

$$= 0.0976$$

Tail pipe: The exhaust speed is calculated from Equation 3.8 as

$$U_e = \sqrt{2 \times 1148 \times 3510 \left[1 - \left(\frac{18.75_a}{1320.3}\right)^{0.25}\right]}$$

$$= 2297 \text{ m/s}$$

The specific thrust is

$$\frac{T}{\dot{m}_a} = (1 + f)\, U_e - U$$

$$\frac{T}{\dot{m}_a} = 1931.4 \, \frac{\text{N} \cdot \text{s}}{\text{kg}}$$

The thrust force is then $T = \dot{m}_a (T/\dot{m}_a) = 28955$ N
The TSFC is calculated from Equation 3.10

$$\text{TSFC} = \frac{0.0976}{1931.4} = 5.06 \times 10^{-5} \, \frac{\text{kg}}{\text{N} \cdot \text{s}}$$

$$\text{TSFC} = 0.182 \, \frac{\text{kg}}{\text{N} \cdot \text{h}}$$

3.2.3 Valveless Pulsejet

Valveless pulsejet engine is sometimes identified as acoustic jet engine [6]. This idea was the brainchild of the French propulsion research group SNECMA. They developed these valveless pulsejets in the late forties for use in drones. One application was the Dutch AT-21 target drone built by Aviolanda Aircraft from 1954 to 1958. The main design difficulties encountered for valved pulse-jets are difficult-to-resolve wear issues. Thus, a valveless pulsejet (or pulsejet) is designed to replace valved type. Valveless pulsejets are low in cost, lightweight, powerful, and easy to operate. They have all the advantages and most of the disadvantages of conventional valved pulsejets. Fuel consumption is excessively high and the noise level is unacceptable as per recent standards. However, they do not have the troublesome reed valves that need frequent replacement. They can operate for their entire useful life with practically zero maintenance. They have been used to power model aircraft, experi-mental go-karts, and even some unmanned military aircraft such as cruise missiles and target drones. These engines have no mechanical valves, but they do have aerodynamic valves that, for the most part, restrict the flow of gases to a single direction just as their mechanical counterparts. Thus, the intake and exhaust pipes usually face the same direction. This necessitates bending the engine into a "U" shape as shown in Figure 3.2 or placing a 180° bend in the intake tube. The Lockwood–Hiller is an example for the "U" shape design. When the air–fuel mixture inside the engine ignites, hot gases will rush out through both the intake tube and the exhaust tube, since the aerodynamic valves "leak." If both tubes were not facing in the same direction, less thrust would be generated because the reactions from the intake and exhaust gas flows would partially cancel each other.

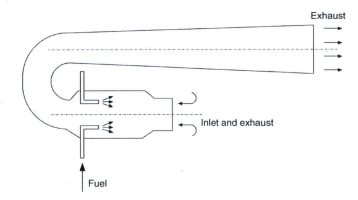

FIGURE 3.2   Valveless pulsejet engine.

In this type of pulsejet, combustion process generates two shock wave fronts, and one travels down each tube. By properly "tuning" the system, a stable, resonating combustion process can be achieved, which yields a considerable thrust.

Successful valveless pulsejets have been built from a few centimeters in length to huge sizes, though the largest and smallest have not been used for propulsion. The smallest ones are only successful when extremely fast-burning fuels such as acetylene or hydrogen, for example, are employed. Medium- and larger-sized engines can be made to burn almost any flammable material that can be delivered uniformly to the combustion zone, though of course volatile flammable liquids (gasoline, kerosene, various alcohols) and standard fuel gases (propane, butane, MAPP gas) are easiest to use. Because of the deflagrating nature of pulsejet combustion, these engines are extremely efficient combustors, producing practically no hazardous pollutants, even when using hydrocarbon fuels. With modern high temperature metals for the main structure, engine weight can be kept extremely low.

Untill now, the physical size of successful valveless designs has always been somewhat larger than valved engines for the same thrust value, though this is theoretically not a requirement. An ignition system of some sort is required for engine startup. In the smallest sizes, forced air at the intake is also typically needed for startup. There is still much room for improvement in the development of really efficient, fully practical designs for propulsion uses.

3.2.4 PULSE DETONATION ENGINE

The pulse detonation engine (PDE) marks a new approach toward noncontinuous combustion in jet engines and promises higher fuel efficiency compared even to turbofan jet engines. With the aid of the latest design techniques and high pulse frequencies the drawbacks of the early designs can be overcome. To date no practical PDE engine has been put into production, but several test bed engines have been built by Pratt & Whitney and General Electric, which have proven the basic concept. Extensive research work is also carried out in different NASA centers. In theory, the design can produce an engine with the efficiency far surpassing gas turbine with almost no moving parts. These systems should be put to use in the near future.

All regular jet engines operate on the *deflagration* of fuel, that is, the rapid but subsonic combustion of fuel. The PDE is a concept currently in active development to create a jet engine that operates on the supersonic *detonation* of fuel.

PDEs are an extension of pulsejet engines. They share many similarities. However, there is one important difference between them: PDEs detonate, rather than deflagrate, their fuel. Detonation of fuel is a supersonic combustion of fuel that results in immense pressure, which in turn is used as thrust. These combustors have an advantage over traditional near-constant pressure combustors in being more thermodynamically efficient by approximating constant volume combustion. However,

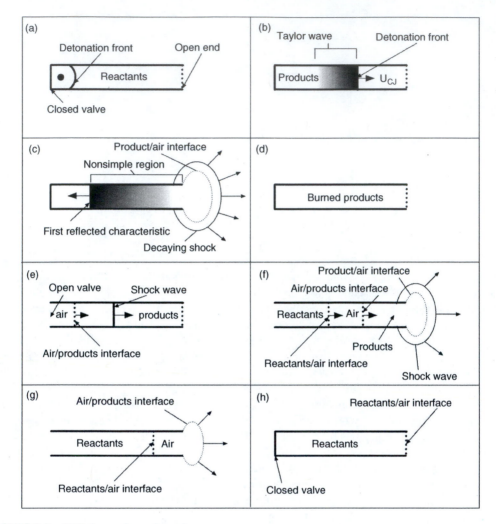

FIGURE 3.3 PDE thermodynamic cycle.

some analysis [7] suggests that detonation is neither a constant volume nor constant pressure combustion, but rather a distinctive change in the properties of the thermodynamic state specific to the detonation process [7].

The main objective of PDE is to provide an efficient engine that is primarily used for high-speed (about Mach 5) civilian transport as well as many military applications including supersonic vehicles, cruise missiles, afterburners, UAVs, SSTO Launchers, and rockets.

PDEs use intermittent detonation waves to generate thrust. The cycle of PDE is illustrated in Figure 3.3. Unlike the pulsejet, combustion in PDE is supersonic, effectively an explosion instead of burning, and the shock wave of the combustion front inside the fuel serves the purpose of shutters of valved pulsejet. A detonation propagating from the closed end of the tube is followed by an unsteady expansion wave (called Taylor wave) whose role is to bring the flow to rest near the closed end of the tube. When the shock wave reaches the rear of the engine and exits, the combustion products are ejected in "one go," the pressure inside the engine suddenly drops, and the air is pulled in front of the engine to start the next cycle.

PDE operation is not determined by the acoustics of the system and can be directly controlled. PDEs typically operate at a frequency of 50–100 Hz, which means that a typical cycle time is of the order of 10–20 ms. Since PDE produces a higher specific thrust than a comparable ramjet engine at low supersonic speeds, it is suitable for use as part of a multistage propulsion system. The specific

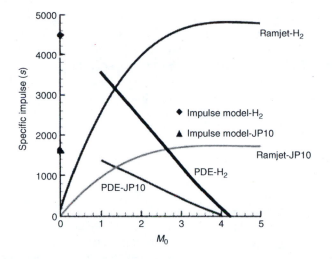

FIGURE 3.4 Specific impulse for PDE compared to other engines.

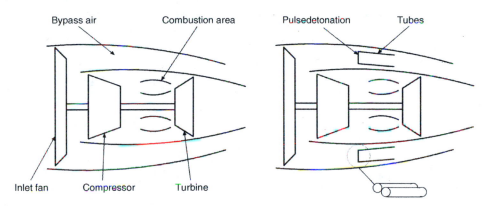

FIGURE 3.5 Hybrid turbofan-PDE.

impulse for PDE is plotted versus several other pulse and ramjet engines for comparison as shown in Figure 3.4. Single-tube supersonic impulse PDE with straight detonation has higher performance than ideal ramjet engines for flight Mach number up to 1.35. The PDE can provide static thrust for a ramjet or scramjet engine, or operate in combination with turbofan systems.

PDE can be classified as pure (standalone), combined cycles, and hybrid turbomachinery cycles [8]. Pure PDE, as the name implies, consists of an array of detonation tubes, an inlet, and a nozzle. The applications of pure PDEs are mainly military, as they are light, easy to manufacture, and have higher performance around Mach 1 than current engine technologies. This makes them an ideal form of propulsion for missiles, unmanned vehicles, and other small-scale applications.

However, their noise and the drop in efficiency at higher Mach number imply that pure PDEs are not likely to be used often for large-scale applications.

Combined-cycle PDEs may provide the most exciting possibilities for aviation. Adding a PDE to the flowpath of a ramjet or scramjet would make an engine capable of operating efficiently at as high as Mach 5.5. These engines would seem initially suitable for high altitude, high-speed aircraft [9].

Hybrid PDEs make use of detonative combustion in place of constant pressure combustion, usually in combination with turbomachinery. For example, a hybrid turbofan PDE would combine both systems: the central core engine would still turn the large fan in front, but the bypass would flow into a ring of PDEs.

The bypass enters pulse detonation tubes that surround the standard combustion chamber (Figure 3.5). The tubes are then cyclically detonated; one detonates while the others fill with air or are primed with fuel. This combination promises to require simpler engine mechanisms and yield higher thrust with lower fuel consumption as examined by General Electric (GE).

Hybrid PDEs will allow commercial aircraft powered by subsonic gas turbines to be faster, more efficient, and more environmentally friendly. NASA is projecting that the intercity travel will reduce significantly by the year 2007, and intercontinental travel time will also reduce by 2022. Similarly, hybrid supersonic gas turbines can also be used in military applications. Generally hybrid PDEs will deliver the same thrust of a turbofan engine but with less fuel consumption. Moreover, they would produce significantly more thrust without requiring additional fuel.

3.3 RAMJET ENGINES

A ramjet has no moving parts, much like a valveless pulsejet but they operate with continuous combustion rather than the series of explosions that give a pulsejet its characteristic noise. Figure 3.6 illustrates a supersonic ramjet. Ramjet engine may be of the subsonic or supersonic type. Although ramjet can operate at subsonic flight speed, the increasing pressure rise accompanying higher flight speeds renders the ramjet most suitable for supersonic flight [10]. For this reason, subsonic ramjets are seldom (if ever) used these days. The ramjet has been called a flying stovepipe, because it is open at both ends and has only fuel nozzles in the middle. A straight stovepipe would not work, however; a ramjet must have a properly shaped inlet-diffusion section to produce low-velocity, high-pressure air at the combustion section and it must also have a properly shaped exhaust nozzle. High-speed air enters the inlet, is compressed as it is slowed down, is mixed with fuel and burned in the combustion chamber, and is finally expanded and ejected through the nozzle. The air rushing toward the inlet of an engine flying at high speeds is partially compressed by the so-called ram effect. If the flight speed is supersonic, part of this compression actually occurs across a shock system that precedes the inlet. When the flight speed is high enough, this compression can be sufficient to operate an engine without a compressor. Once the compressor is eliminated, the turbine is no longer required and it can also be omitted. Ramjets can operate at speeds above 320 km/h (about 200 mph) or at as low Mach number as 0.2. However, the fuel consumption is horrendous at these low velocities. The operation of ramjet does not become competitive with that of a turbojet until speeds of about $M_a = 2.5$ or even more are reached. Ramjets become practical for military applications only at very high or supersonic speeds. The combustion region is generally a large single combustor, similar to the afterburner [5]. For the combustion process to be efficient, the air must be compressed sufficiently. This is possible only when the free stream Mach number exceeds about 3, and therefore ramjets have been practical for only a few missile applications. A *hybrid* engine, part turbojet, part ramjet (identified as turbo-ram engine), was also used on the SR-71 high-speed reconnaissance aircraft and is a topic of current research interest for several possible hypersonic applications. In brief, because the ramjet depends

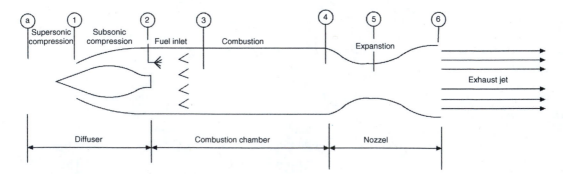

FIGURE 3.6 Ramjet engine and its different states.

on the compression of the inrushing air for its operation, or in other words depends on the flight speed, a vehicle powered by a ramjet cannot develop static thrust and therefore cannot accelerate a vehicle from the stationary position. The vehicle must first be accelerated by other means to a sufficiently high speed before using a ramjet as a propulsive device. It may be dropped from a plane or launched with rocket assistance.

3.3.1 IDEAL RAMJET

Ramjet engine as described above is composed of three modules (same as pulsejet engine), namely, an intake or diffuser, a combustion chamber, and a nozzle. To analyze ramjet performance, the different states of its three modules are defined here. State (a), standing for ambient, is far upstream of the engine. Diffusion process takes place partly outside the engine and partly inside the engine. The first part of diffusion takes place from the far upstream, state (a), to the engine entrance, state (1). The second part of diffusion takes place inside the engine, from state (1) to state (2), where state (2) is at the end of the diffusion section. Thus, the first module of the engine, the diffuser, is located between states (1) and (2). Air then enters the combustion chamber at subsonic speeds. This is achieved by either a normal shock wave or multiple oblique shock waves followed by a normal shock wave. Fuel is next, injected as fine droplets, which mixes rapidly with the mixture and is next ignited by a spark. The mixture then passes through a "flame holder" at state (3) to stabilize the flame and facilitate a good combustion process. The flame holder may be of grill type, which provides a type of barrier to the burning mixture while allowing hot, expanding gases to escape through the exhaust nozzle. The high-pressure air coming into the combustion chamber keeps the burning mixture from effectively reacting toward the intake end of the engine.

The end of the combustion chamber is identified by state (4). Thus the second module of the engine, combustion chamber is located between states (2) and (4). Combustion in general raises the temperature of the mixture to approximately 3000 K before the products of combustion expand to high velocities in the nozzle. Although the walls of combustion chambers cannot tolerate temperature much above 1200 K, they can be kept much cooler than the main fluid stream by a fuel injection pattern that leaves a shielding layer of relatively cool air next to the walls [10]. The expansion process takes place at the third module of the engine, namely the nozzle. Expansion starts at the convergent section between state (4) and the nozzle throat, state (5). The nozzle has next a diverging section that ends at state (6). Sometimes state (6) is denoted by (e) resembling the exhaust or exit condition. Thus, the nozzle is situated between states (4) and (6). The nozzle is shaped to accelerate the flow. Thus, the exit velocity is greater than the free stream velocity, and thrust is created.

The cycle of events within the engine (Figure 3.7) is described below. Here all the processes within the engine are assumed ideal with no losses.

For isentropic (reversible adiabatic) flow inside the engine, no pressure drop will be encountered in the three modules of the engine, thus

$$P_{0a} = P_{02} = P_{04} = P_{06} \tag{3.11}$$

Moreover, since neither work nor heat addition or rejection takes place in the intake and nozzle, from the first law of thermodynamics, equal total enthalpy (and thus total temperature) is presumed. Thus

$$\left.\begin{array}{l} T_{0a} = T_{02} \\ T_{04} = T_{06} \end{array}\right\} \tag{3.12}$$

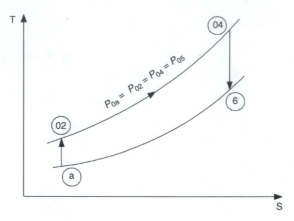

FIGURE 3.7   Ideal thermodynamic cycle for ramjet engine.

Full expansion of the hot gases within the nozzle is assumed, thus

$$P_a = P_6 = P_e \tag{3.13}$$

The relation between total and static conditions (temperature and pressure) at the inlet and outlet of the engine, states (a) and (6 or e) are

$$\left. \begin{array}{l} \dfrac{T_{0a}}{T_a} = 1 + \dfrac{\gamma_a - 1}{2} M^2 = \dfrac{T_{02}}{T_a} \\[3mm] \dfrac{T_{0e}}{T_e} = \dfrac{T_{06}}{T_6} = 1 + \dfrac{\gamma_6 - 1}{2} M_e^2 = \dfrac{T_{04}}{T_e} \end{array} \right\} \tag{3.14}$$

$$\left. \begin{array}{l} \dfrac{P_{0a}}{P_a} = \left(1 + \dfrac{\gamma_a - 1}{2} M^2 \right)^{\frac{\gamma_a}{\gamma_a - 1}} \\[4mm] \dfrac{P_{06}}{P_e} = \left(1 + \dfrac{\gamma_6 - 1}{2} M_e^2 \right)^{\frac{\gamma_6}{\gamma_6 - 1}} \end{array} \right\} \tag{3.15}$$

In Equations 3.14 and 3.15 (γ_a, γ_6) are the specific heat ratios for air and exhaust gases respectively. If we ignore the variations in fluid properties, (R, γ), then from Equations 3.11 and 3.13, we get

$$\frac{P_{06}}{P_e} = \frac{P_{0a}}{P_a} \tag{3.16}$$

Then from Equation 3.15, we have

$$M_e = M_a \tag{3.17}$$

or

$$u_e = \frac{a_e}{a} u = \sqrt{\frac{\gamma_6 R T_e}{\gamma_a R T_a}} u \tag{3.18a}$$

The flight and exhaust Mach numbers are equal, but the flight and exhaust speeds are not. This difference generates the thrust force.

Again, assuming constant (γ, R) within the engine, then,

$$u_e = \sqrt{\frac{T_e}{T_a}} u$$

From Equation 3.14

$$\therefore u_e = u\sqrt{\frac{T_{04}}{T_{0a}}} \equiv u\sqrt{\frac{T_{04}}{T_{02}}} \tag{3.18b}$$

The fuel-to-air ratio (f): To get the fuel-to-air ratio, apply the energy equation to the combustion process

$$\dot{m}_a h_{02} + \dot{m}_f Q_R = (\dot{m}_a + \dot{m}_f) h_{04}$$

Q_R = heating value of the fuel

$$h_{02} + f Q_R = (1 + f) h_{04}$$

$$C_{p2} T_{02} + f Q_R = (1 + f) C_{p4} T_{04}$$

but $T_{02} = T_{0a}$

$$\therefore f = \frac{(C_{p4} T_{04}/C_{p2} T_{0a}) - 1}{(Q_R/C_{p2} T_{0a}) - (C_{p4} T_{04}/C_{p2} T_{0a})} \tag{3.19}$$

The thrust force (T): Since full expansion in the nozzle is assumed (i.e., $P_e = P_a$), then the thrust force is expressed as

$$T = \dot{m}_a (1 + f) u_e - \dot{m}_a u$$

$$\frac{T}{\dot{m}_a} = (1 + f) u_e - u = u\left[(1 + f)\frac{u_e}{u} - 1\right]$$

Now, the flight speed and exhaust speed can be written as

$$u = Ma = M\sqrt{\gamma R T_a}$$

$$u_e = u\sqrt{\frac{T_{04}}{T_{0a}}} = u\sqrt{\frac{T_{04}}{T_a}\frac{T_a}{T_{0a}}} = u\sqrt{\frac{T_{04}}{T_a}}\sqrt{\frac{1}{1 + \frac{\gamma-1}{2}M^2}} \tag{3.20}$$

From Equation 3.19, a final expression for the thrust force will be expressed as

$$\frac{T}{\dot{m}_a} = M\sqrt{\gamma R T_a}\left[(1 + f)\sqrt{T_{04}/T_a}\left(1 + \frac{\gamma-1}{2}M^2\right)^{-1/2} - 1\right] \tag{3.21}$$

The TSFC is

$$\therefore \text{TSFC} = \frac{\dot{m}_f}{T} = \frac{f}{T/\dot{m}_a} \tag{3.22}$$

3.3.2 REAL CYCLE

Now, consider the performance of a real ramjet engine. Aerodynamic losses in the intake, combustion chamber, and nozzle lead to a stagnation pressure drop in every element. Moreover, losses occur during combustion, which are accounted for by combustion or burner efficiency. Figure 3.8 illustrates the effect of irreversibilties on the different processes in the cycle. Neither the compression process in the diffuser nor the expansion process in the nozzle is any more isentropic. Moreover, combustion will not take place at a constant pressure.

Performance of diffuser is characterized by a stagnation pressure ratio

$$r_d = \frac{P_{02}}{P_{0a}}$$

Stagnation pressure ratio in combustion chamber is

$$r_c = \frac{P_{04}}{P_{02}}$$

Stagnation pressure ratio in nozzle is

$$r_n = \frac{P_{06}}{P_{04}}$$

The overall stagnation pressure ratio is

$$\frac{P_{06}}{P_{0a}} = r_d r_c r_n \tag{3.23}$$

It will be shown later that irreversibilities will lead to the inequality of the flight and exhaust Mach numbers. The properties (γ, R) of the fluid through the different parts of the engine are assumed constant.

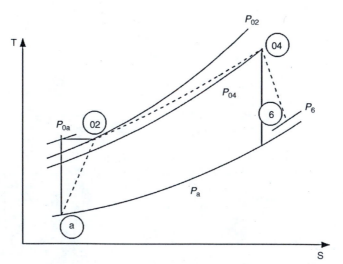

FIGURE 3.8   Real cycle of ramjet engine.

Now, since the stagnation pressure of the free stream is expressed as

$$\frac{P_{0a}}{P_a} = \left(1 + \frac{\gamma - 1}{2}M^2\right)^{\frac{\gamma}{\gamma - 1}}$$

$$\therefore \left(\frac{P_a}{P_{0a}}\right)^{\frac{\gamma - 1}{\gamma}} \left(1 + \frac{\gamma - 1}{2}M^2\right) = 1$$

and

$$\frac{P_{06}}{P_6} = \left(1 + \frac{\gamma - 1}{2}M_e^2\right)^{\frac{\gamma}{\gamma - 1}}$$

$$\therefore M_e^2 = \frac{2}{\gamma - 1}\left[\left(\frac{P_{06}}{P_6}\right)^{\frac{\gamma - 1}{\gamma}} - 1\right]$$

$$= \left(\frac{2}{\gamma - 1}\right)\left[\left(1 + \frac{\gamma - 1}{2}M^2\right)\left(\frac{P_a}{P_{0a}}\right)^{\frac{\gamma - 1}{\gamma}}\left(\frac{P_{06}}{P_6}\right)^{\frac{\gamma - 1}{\gamma}} - 1\right]$$

$$\therefore M_e^2 = \left(\frac{2}{\gamma_e - 1}\right)\left[\left(1 + \frac{\gamma - 1}{2}M^2\right)\left(\frac{P_{06}}{P_6}\frac{P_a}{P_{0a}}\right)^{\frac{\gamma - 1}{\gamma}} - 1\right]$$

Then from Equation 3.23, the exhaust Mach number is

$$M_e^2 = \left(\frac{2}{\gamma - 1}\right)\left[\left(1 + \frac{\gamma - 1}{2}M^2\right)\left(r_d r_c r_n \frac{P_a}{P_e}\right)^{\frac{\gamma - 1}{\gamma}} - 1\right] \quad (3.24a)$$

Now, define (m) as

$$m = \left(1 + \frac{\gamma - 1}{2}M^2\right)\left(r_d r_c r_n \frac{P_a}{P_e}\right)^{\frac{\gamma - 1}{\gamma}} \quad (3.25a)$$

$$\therefore M_e^2 = \left(\frac{2}{\gamma - 1}\right)(m - 1) \quad (3.24b)$$

From Equation 3.24, another expression for (m) is

$$m = \left(1 + \frac{\gamma - 1}{2}M_e^2\right) \quad (3.25b)$$

If $r_d = r_c = r_n = 1$, and $P_e = P_a$ then $M_e = M$.

If heat transfer from the engine is assumed negligible (per unit mass of fluid) then the total exhaust temperature $T_{06} = T_{04}$.

$$\frac{T_{06}}{T_6} = \frac{T_{04}}{T_e} = \left(1 + \frac{\gamma - 1}{2} M_e^2\right)$$

$$u_e = M_e \sqrt{\gamma R T_e} = M_e \sqrt{\gamma R T_{04} \frac{T_e}{T_{04}}}$$

$$= M_e \sqrt{\frac{\gamma R T_{04}}{\left(1 + \frac{\gamma-1}{2} M_e^2\right)}}$$

$$= M_e \sqrt{\frac{\gamma R T_{04}}{m}}$$

Substituting for M_e from Equation 3.24a, then

$$u_e = \sqrt{\frac{2\gamma R T_{04} (m - 1)}{(\gamma - 1) m}} \tag{3.26}$$

Since irreversibilities have no effect on total temperatures throughout the engine, then the fuel-to-air ratio can be given by Equation 3.19, but modified for burner efficiency (η_b)

$$f = \frac{\left(C_{p4} T_{04}/C_{p2} T_{0a}\right) - 1}{\left(\eta_b Q_R/C_{p2} T_{0a}\right) - \left(C_{p4} T_{04}/C_{p2} T_{0a}\right)} \tag{3.27}$$

The specific thrust is expressed as usual by the relation:

$$\frac{T}{\dot{m}_a} = [(1 + f) u_e - u_a] + \frac{1}{\dot{m}_a} (P_e - P_a) A_e \tag{3.28}$$

Then since, $u = M \sqrt{\gamma R T_a}$, specific thrust will be expressed as

$$\frac{T}{\dot{m}_a} = (1 + f) \sqrt{\frac{2\gamma_e R T_{04} (m - 1)}{(\gamma - 1) m}} - M \sqrt{\gamma R T_a} + \frac{P_e A_e}{\dot{m}_a} \left(1 - \frac{P_a}{P_e}\right) \tag{3.29}$$

The specific thrust and TSFC are plotted in Figure 3.9 for ideal and real engines. The ratio between specific heats is assumed constant, $\gamma = 1.4$. The maximum temperatures investigated are $T_{04} = 2000$ K or 3000 K. The pressure ratios in the diffuser, combustion chamber, and nozzle are $r_d = 0.7$, $r_c = 0.97$, and $r_n = 0.96$, respectively. The burner is assumed ideal with $\eta_b = 1$ and fuel heating value is $Q_R = 43,000$ kJ/kg.

For any given peak temperature, the thrust per unit mass flow rate $\dot{m}_a$, and for real ramjet it is less than the ideal. However, the TSFC for the real ramjet is higher than, for the ideal. Also for real ramjet there is a reasonably well-defined minimum TSFC.

Example 2 A ramjet has a flight speed of $M_a = 2.0$ at an altitude of 16,200 m where the temperature is 216.6° K and the pressure is 10.01 kPa. An axisymmetric inlet fitted with a spike has a deflection angle of 12°. Neglect frictional losses in the diffuser and combustion chamber. The inlet area is $A_1 = 0.2\,\mathrm{m}^2$. The maximum total temperature in the combustion chamber is 2500 K. Heating value of fuel is 45,000 kJ/kg. The burner efficiency is $\eta_b = 0.96$. The nozzle expands to atmospheric

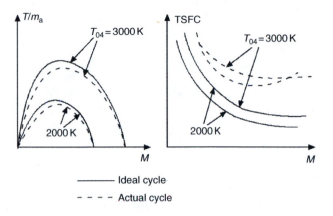

FIGURE 3.9   Specific thrust and TSFC for ideal and real ramjet engines.

pressure for maximum thrust with $\eta_n = 0.96$. The velocity entering the combustion chamber is to be kept as large as possible but the Mach number is not greater than 0.25.

Assuming the fluid to be air ($\gamma = 1.4$), compute

1. The stagnation pressure ratio of the diffuser (r_d)
2. The inlet Mach number to the combustion chamber
3. The stagnation pressure ratio in the combustion chamber (r_c)
4. The stagnation pressure ratio in the nozzle (r_n)
5. The flight and exhaust speeds
6. The thrust force
7. The specific thrust (TSFC)

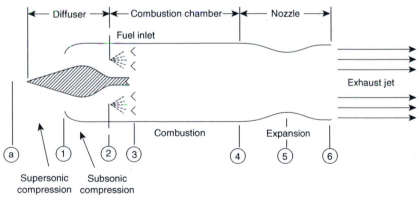

Schematic diagram of a ramjet engine.

Solution: The total conditions for air before entering the engine are

$$\frac{T_{0a}}{T_a} = 1 + \frac{\gamma - 1}{2}M_a^2 = 1.8$$

$$\frac{P_{0a}}{P_a} = \left(1 + \frac{\gamma - 1}{2}M_a^2\right)^{\frac{\gamma}{\gamma - 1}} = 7.824$$

Thus, $T_{0a} = 390\ K$ and $P_{0a} = 7.824 P_a$.

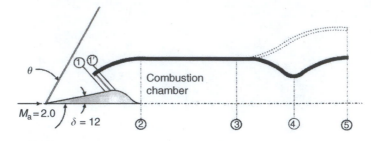

FIGURE 3.10   Shock wave pattern in ramjet engine.

The total temperature is constant in the intake section and so

$$T_{0a} = T_{01} = T'_{01} = T_{02} = 390 \text{ K}$$

The flight speed is $U = M\sqrt{\gamma R T_a} = 590$ m/s.

Now, let us analyze the intake. Since the flight speed is supersonic two shock waves must be generated in the intake to deliver air at a subsonic speed to the combustion chamber as shown in Figure 3.10. The first shock is an oblique shock wave attached to the apex of the spike, which is to be followed by a normal shock wave.

1. *Oblique shock wave*

    The complete governing equations for oblique and normal shock waves are found in Section 3.5; some are rewritten here.

    With a free stream Mach number $M_a = 2.0$ and a deflection angle $\delta = 12°$, the shock wave angle will be either determined from gas dynamics tables or iteratively from the following relation:

    $$\tan \delta = \frac{2 \cot \theta \left(M_1^2 \sin^2 \theta - 1 \right)}{(\gamma + 1) M_1^2 - 2 \left(M_1^2 \sin^2 \theta - 1 \right)}$$

    Both result in the value $\theta = 41.56°$. As shown in Figure 3.11, the Mach number upstream the oblique shock wave can be decomposed into normal and tangential components. The normal component of the free stream Mach number is

    $$M_{an} = M_a \sin \theta = 2.0 \sin 41.56° = 1.326$$

    The normal component of the Mach number downstream the normal shock wave will also be determined either from normal shock wave tables or from the relation

    $$M_{1n}^2 = \frac{(\gamma - 1) M_{an}^2 + 2}{2\gamma M_{an}^2 - (\gamma - 1)} \tag{1}$$

    The result is $M_{1n} = 0.773$.

    The total pressure ratio also may be determined from normal shock wave tables or from the relation

    $$\frac{P_{01}}{P_{0a}} = \left[\frac{(\gamma + 1) M_{an}^2}{2 + (\gamma - 1) M_{an}^2} \right]^{\frac{\gamma}{\gamma - 1}} \left[\left(\frac{2\gamma}{\gamma + 1} \right) M_{an}^2 - \left(\frac{\gamma - 1}{\gamma + 1} \right) \right]^{-\left(\frac{1}{\gamma - 1} \right)} \tag{2}$$

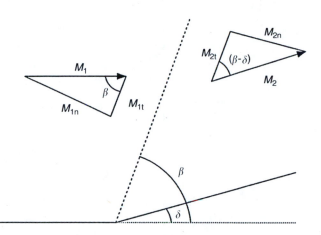

FIGURE 3.11   Oblique shock wave pattern.

Thus the pressure ratio is $P_{01}/P_{0a} = 0.9746$.

The Mach number for flow leaving the oblique shock wave is

$$M_1 = \frac{M_{1n}}{\sin(\theta - \delta)} = \frac{0.773}{\sin(41.56 - 12)} = 1.567$$

2. *Normal shock wave*

    The second (normal) shock wave originates at the engine cowl (nose lips). With $M_1 = 1.567$, from normal shock wave tables, or Equation 1, $M_1' = 0.678$, and also the total pressure ratio either from tables or Equation 2 is

$$\frac{P_{01}'}{P_{01}} = 0.90719$$

The pressure ratio is then

$$\frac{P_{01}'}{P_{0a}} = \frac{P_{01}'}{P_{01}} \frac{P_{01}}{P_{0a}} = 0.90719 \times 0.9746 = 0.884$$

Since there are no losses in the diffuser, then

$$P_{02} = P_{01}'$$

The stagnation pressure ratio in the intake is

$$r_d = \frac{P_{02}}{P_{0a}} = 0.884$$

Combustion Chamber

The flow in the combustion chamber is a flow with heat transfer or a Rayleigh flow. Now, if $M_2 = 0.25$, then from Rayleigh flow tables or the governing equations in Section 3.5

$$\frac{T_{02}}{T_0^*} = 0.2568$$

Then

$$T_0^* = T_{02} \frac{T_0^*}{T_{02}} = \frac{390}{0.2568} = 1518 \text{ K}$$

$$\frac{T_{03}}{T_0^*} = \frac{2500}{1518} = 1.64615 > 1.0$$

which is impossible.

Thus, adding fuel to make $T_{03} = 2500$ K means that the flow will be choked ($M_3 = 1.0$) and $M_2 < 0.25$. Thus to find M_2

$$\frac{T_{02}}{T_0^*} = \frac{T_{02}}{T_{03}} \frac{T_{03}}{T_0^*} = \frac{390}{2500}(1) = 0.156$$

From Rayleigh tables or equations, then $M_2 = 0.189$, and $P_{02}/P_0^* = 1.2378$. Mach number at state 3 is unity, thus $P_{03} = P_0^* = P_0^*/P_{02} \, P_{02}$

$$P_{03} = \frac{1}{1.2378}(0.884 \, P_{0a}) = 0.714 \, P_{0a} = 5.586 P_a$$

The stagnation pressure ratio in the combustion chamber is

$$r_c = \frac{P_{03}}{P_{02}} = \frac{1}{1.2378} = 0.808$$

The fuel-to-air ratio

$$f = \frac{C_P (T_{03} - T_{0a})}{\eta_b Q_R - C_P T_{0a}}$$

$$f = \frac{1.005(2500 - 390)}{(0.98)(45,000) - (1.005)(2500)} = 0.05099$$

Nozzle

Since $M_3 = 1.0$ the nozzle diverges immediately. The gases expand in the nozzle and leave at the same ambient pressure. The nozzle efficiency is

$$\eta_n = \frac{T_{03} - T_5}{T_{03} - T_{5s}}$$

$$T_5 = T_{03} \left[1 - \eta_n \left\{ 1 - \frac{T_{5s}}{T_{03}} \right\} \right]$$

$$T_5 = T_{03} \left[1 - \eta_n \left\{ 1 - \left(\frac{P_a}{P_{03}} \right)^{\frac{\gamma-1}{\gamma}} \right\} \right]$$

With $P_{03}/P_a = 5.586$

$$T_5 = 2500 \left[1 - 0.96 \left\{ 1 - (5.58)^{-0.2857} \right\} \right] = 1567 \text{ K}$$

To calculate the stagnation pressure ratio in the nozzle, the isentropic total pressure at the nozzle outlet is then

$$\frac{P_{05}}{P_5} = \left(\frac{T_{05}}{T_5}\right)^{\frac{\gamma}{\gamma-1}} = \left(\frac{2500}{1567}\right)^{3.5} = 5.129$$

$$\therefore P_{05} = 5.129 P_a$$

The nozzle total pressure ratio is $r_n = P_{05}/P_{03} = 0.918$.
The engine overall pressure ratio is

$$\frac{P_{05}}{P_{0a}} = r_d r_c r_n = 0.655$$

Since $T_5/T_{05} = 1567/2500 = 0.6268$, then from isentropic flow tables, the exhaust Mach number $M_5 = 1.725$.

The exhaust speed is then $V_5 = M_5 a_5 = 1.725\sqrt{(1.4)(287)(1567)} = 1368$ m/s.

The mass flow rate is calculated from the known area at state (1), then with $M_1 = 1.567$, the static temperature and pressure at state (1) are given from normal shock tables as

$$T_1 = \frac{T_1}{T_a} T_a = 1.207 \times 216.66 = 261.5 \text{ K}$$

$$P_1 = \frac{P_1}{P_a} P_a = 1.8847 \times 10.01 = 18.8659 \text{ kPa}$$

The air density at state (1) is then $\rho_1 = P_1/RT_1 = 0.2513$ kg/m^3.
Moreover, the air speed is given by $V_1 = M_1\sqrt{\gamma RT_1} = 429.8$ m/s.

$$\dot{m} = \rho_1 V_1 A_1$$

$$\dot{m} = \rho_1 V_1 A_1 = (0.2513)(0.2)(429.8) = 21.6 \text{ kg/s}$$

The thrust force is given by the relation

$$T = \dot{m}_a\left[(1+f)V_5 - V_a\right] = 21.6\left[1.05099 \times 1368 - 590\right]$$

$$= 19897 \text{ N}$$

$$\text{TSFC} = \frac{\dot{m}_f}{T} = \frac{f\dot{m}_a}{T}$$

$$= \frac{(0.05099)(21.6)}{19897} = 0.00005535 \text{ kg/N} \cdot \text{s}$$

$$= 0.199 \text{ kg/N.h}$$

Example 3   For an ideal ramjet, it is required to prove that the thermal efficiency is expressed by

$$\eta_{th} = 1 - \left(\frac{1}{\tau_r}\right)$$

where $\tau_r = T_{02}/T_a$.
(Hint: use the following approximations: $f = Cp\,(T_{04} - T_{02})\,/Q_R$ and $\dot{m}_f \ll \dot{m}_a$)

As derived in Chapter 2, the thermal efficiency is expressed by the relation

$$\eta_{th} = \frac{(\dot{m}_a + \dot{m}_f)\,V_6^2 - \dot{m}_a V^2}{2\dot{m}_f Q_R}$$

Now, with $\dot{m}_a \ll \dot{m}_f$, then

$$\eta_{th} = \frac{\dot{m}_a\left(V_6^2 - V^2\right)}{2\dot{m}_f Q_R}$$

with $f = \dot{m}_f/\dot{m}_a$ and $f = Cp\left(T_{04} - T_{02}\right)/Q_R$

$$\therefore \eta_{th} = \frac{\left(V_6^2 - V^2\right)}{2Cp\left(T_{04} - T_{02}\right)}$$

The flight speed $V = M\sqrt{\gamma R T_a}$
The exhaust velocity $V_6 = M_e\sqrt{\gamma R T_6}$
For an ideal ramjet $M_e = M$

$$\therefore \eta_{th} = \frac{M^2\left(\gamma R T_6 - \gamma R T_a\right)}{2Cp\left(T_{04} - T_{02}\right)}$$

$$= \frac{\gamma R T_a M^2 \left(\frac{T_6}{T_a} - 1\right)}{2Cp T_{02}\left(\frac{T_{04}}{T_{02}} - 1\right)} \tag{3}$$

Since

$$\frac{P_{04}}{P_6} = \frac{P_{02}}{P_a}$$

$$\left(\frac{T_{04}}{T_6}\right)^{\left(\frac{\gamma-1}{\gamma}\right)} = \left(\frac{T_{02}}{T_a}\right)^{\left(\frac{\gamma-1}{\gamma}\right)}$$

or $\quad \frac{T_{04}}{T_6} = \frac{T_{02}}{T_a} \quad$ as $\gamma =$ constant

$$\therefore \frac{T_{04}}{T_{02}} = \frac{T_6}{T_a} \tag{4}$$

From Equations 3 and 4

$$\therefore \eta_{th} = \frac{\gamma R T_a M^2}{2Cp T_{02}} = \frac{\gamma R T_a M^2}{2\left(\frac{\gamma R}{\gamma-1}\right)} T_{02}$$

But

$$\tau_r = \frac{T_{02}}{T_a} = \left(1 + \frac{\gamma-1}{2}M^2\right)$$

Then

$$\eta_{th} = \frac{\left(\frac{\gamma-1}{2}\right)M^2}{T_{02}/T_a} = \frac{\left(\frac{\gamma-1}{2}\right)M^2}{\left(1+\frac{\gamma-1}{2}M^2\right)} = 1 - \frac{1}{\left(1+\frac{\gamma-1}{2}M^2\right)}$$

$$\therefore \eta_{th} = 1 - \frac{1}{\tau_r}$$

Example 4 For an ideal ramjet engine, prove that the thrust is expressed by the relation

$$T = mu\left(\sqrt{\tau_b} - 1\right), \quad \tau_b = \frac{T_{04}}{T_{02}}$$

For a flow at stratosphere region of the standard atmosphere, where T_a = constant, and if the maximum temperature T_{04} = constant (full-throttle operation), prove that the thrust attains a maximum value when the flight Mach number M_0 is given by:

$$M_0 = \sqrt{\frac{2}{\gamma-1}\left(\tau_\lambda^{\frac{1}{3}} - 1\right)}, \quad \tau_\lambda = \frac{T_{04}}{T_a}$$

Plot the relation

$$\left(\frac{T}{\dot{m}_a\, a_0}\right) \text{ versus } M_0 \quad \text{for} \quad \tau_\lambda = 9.0$$

where a_0 is the sonic speed at the flight altitude.

Solution: The specific thrust for an ideal ramjet engine with negligible fuel-to-air ratio f and unchoked nozzle is given by the relation:

$$\frac{T}{\dot{m}} = (u_e - u) = u\left(\frac{u_e}{u} - 1\right)$$

From Equation 3.18b, the ratio between the exhaust and flight speeds is equal to the ratio between the maximum temperature and the total free stream temperature

$$\frac{u_e}{u} = \sqrt{\frac{T_{04}}{T_{0a}}}$$

Define this temperature ratio as

$$\tau_b = \frac{T_{04}}{T_{02}}$$

$$\therefore \frac{u_e}{u} = \sqrt{\tau_b}$$

$$\frac{T}{\dot{m}_u} = \left(\frac{u_e}{u} - 1\right) = \left(\sqrt{\tau_b} - 1\right) \tag{a}$$

Since $\tau_\lambda = \dfrac{T_{04}}{T_a}$ and $\tau_b = \dfrac{T_{04}}{T_{02}}$

$$\therefore \tau_\lambda = \frac{T_{04}}{T_{02}} \frac{T_{02}}{T_a} = \tau_b \left(1 + \frac{\gamma - 1}{2} M_0^2\right)$$

$$\tau_b = \frac{\tau_\lambda}{\left(1 + \frac{\gamma-1}{2} M_0^2\right)}$$ (b)

Since the flight speed (u) is equal to $M_0 a_0$, then from Equations a and b,

$$\therefore \frac{T}{\dot{m}} = M_0 a_0 \left(\sqrt{\frac{\tau_\lambda}{\left(1 + \frac{\gamma-1}{2} M_0^2\right)}} - 1\right)$$

$$\frac{\partial (T/\dot{m})}{\partial M_0} = a_0 \left(\sqrt{\frac{\tau_\lambda}{\left(1 + \frac{\gamma-1}{2} M_0^2\right)}} - 1\right) + M_0 a_0 \left((-0.5)\left(\frac{\gamma-1}{2}\right)(2M_0)\frac{\sqrt{\tau_\lambda}}{\left(1 + \frac{\gamma-1}{2} M_0^2\right)^{3/2}}\right)$$

The specific thrust is a maximum when

$$\frac{\partial (T/\dot{m})}{\partial M_0} = 0$$

$$\sqrt{\tau_\lambda}\left(1 + \frac{\gamma - 1}{2} M_0^2\right) - \left(1 + \frac{\gamma - 1}{2} M_0^2\right)^{3/2} - \left(\frac{\gamma - 1}{2} M_0^2\right)\sqrt{\tau_\lambda} = 0$$

$$\sqrt{\tau_\lambda} = \left(1 + \frac{\gamma - 1}{2} M_0^2\right)^{3/2}$$

$$\tau_\lambda^{1/3} = \left(1 + \frac{\gamma - 1}{2} M_0^2\right)$$

$$M_o^2 = \frac{2}{\gamma-1}\left(\tau_\lambda^{1/3} - 1\right)$$

$$(M_0)\,\underset{\text{thrust}}{\text{max.}} = \sqrt{\frac{2}{\gamma-1}\left(\tau_\lambda^{1/3} - 1\right)}$$

Now to plot the relation ($T/\dot{m}a_0$) against M_0, since

$$\frac{T}{\dot{m}a_0} = M_0\left(\sqrt{\tau_b} - 1\right)$$

From the above relation, the thrust, or $T/\dot{m}a_0 = 0$, when (1) $M_0 = 0$, and (2) $\tau_b = 1$
Since $\tau_\lambda = \tau_b \left(1 + \frac{\gamma-1}{2} M_o^2\right)$, then at $\tau_b = 1$,

$$\tau_\lambda = \left(1 + \frac{\gamma - 1}{2} M_0^2\right)$$

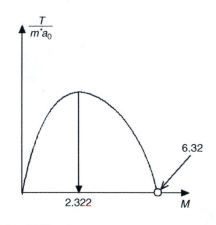

FIGURE 3.12 Plot for the relation $(T/\dot{m}_a a_0)$.

and

$$M_0 = \sqrt{\frac{2}{\gamma - 1}(\tau_\lambda - 1)} = \sqrt{\frac{2}{0.4}(9 - 1)} = 6.32$$

Thus the thrust is also equal to zero when the flight Mach number $M_0 = 6.32$. Since the thrust is equal to zero at $M_0 = 0$ and $M_0 = 6.32$, it will attain a maximum value at an intermediate value calculated from the relation

$$M_0 = \sqrt{\frac{2}{\gamma - 1}\left(\tau_\lambda^{1/3} - 1\right)} = \sqrt{\frac{2}{0.4}\left(9^{1/3} - 1\right)} = 2.322$$

Thus the thrust is maximum at $M_0 = 2.322$

A plot for the relation $(T/\dot{m}_a a_0)$ versus M_0 for $\tau_\lambda = 9.0$ is illustrated in Figure 3.12.

Comment

The above example illustrates the range of operations for an ideal ramjet engine. The engine can develop a thrust force when the flight Mach number is some value close to zero. However, this is normally not the case due to the different sources of irreversibilities in the engine. Example 5 clarifies that a real ramjet engine can develop thrust force only at higher values of Mach number that is close to or greater than 0.5 depending on the total pressure ratios within its three modules.

Example 5   A ramjet has low-speed stagnation pressure losses given by $r_d = 0.90$, $r_c = 0.95$, $r_n = 0.96$. The ambient and maximum temperatures are 222 K and 1950 K, respectively. The fuel-to-air ratio is assumed to be very small such that $f \ll 1$. What is the minimum flight Mach number at which the engine develops positive thrust? Assume the nozzle is unchoked.

Solution: The thrust equation is expressed by Equation 3.29 as follows:

$$\frac{T}{\dot{m}_a} = (1 + f)\sqrt{\frac{2\gamma_e R_e T_{04}(m - 1)}{(\gamma - 1)m}} - M\sqrt{\gamma R T_a} + \frac{P_e A_e}{\dot{m}_a}\left(1 - \frac{P_a}{P_e}\right)$$

where

$$m = \left(1 + \frac{\gamma - 1}{2}M^2\right)\left(r_d r_c r_n \frac{P_a}{P_e}\right)^{\frac{\gamma - 1}{\gamma}}$$

for an unchoked nozzle $P_a = P_e$, also assuming small fuel to air ratio; $f \ll 1$, then

$$\frac{T}{\dot{m}_a} = \sqrt{\frac{2\gamma R T_{04}\,(m-1)}{(\gamma-1)\,m}} - M\sqrt{\gamma R T_a}$$

Rewrite m as

$$m = \left(1 + \alpha_1 M^2\right)\alpha_2$$

where $\alpha_1 = \frac{\gamma-1}{2}$ and $\alpha_2 = (r_d r_c r_n)^{\frac{\gamma-1}{\gamma}}$ also define

$$\alpha_3 = \sqrt{\gamma R T_a}$$

and

$$\alpha_4 = \frac{2\gamma R T_{04}}{\gamma - 1}$$

$$\therefore \frac{T}{\dot{m}_a} = \sqrt{\frac{\alpha_4\left[\alpha_2\left(1 + \alpha_1 M^2\right) - 1\right]}{\alpha_2\left(1 + \alpha_1 M^2\right)}} - \alpha_3 M$$

When $T/\dot{m}_a = 0$

$$\therefore \frac{\alpha_4\left[\alpha_2\left(1 + \alpha_1 M^2\right) - 1\right]}{\alpha_2\left(1 + \alpha_1 M^2\right)} = \alpha_3^2 M^2$$

$$\therefore \alpha_2\alpha_4\left(1 + \alpha_1 M^2\right) - \alpha_4 = \alpha_2\alpha_3^2 M^2\left(1 + \alpha_1 M^2\right)$$

$$\alpha_1\alpha_2\alpha_3^2 M^4 + \alpha_2\alpha_3^2 M^2 - \alpha_1\alpha_2\alpha_4 M^2 - \alpha_2\alpha_4 + \alpha_4 = 0$$

Writing the above equation in the general form

$$AM^4 + BM^2 + C = 0$$

where $A = \alpha_1\alpha_2\alpha_3^2$, $B = \alpha_2\alpha_3^2 - \alpha_1\alpha_2\alpha_4$, $C = \alpha_4 - \alpha_2\alpha_4 = (1 - \alpha_2)\alpha_4$, then: $M^2 = -B \pm \sqrt{B^2 - 4AC}/2A$,

Now substitute the numerical values:

$$\alpha_1 = 0.2, \quad \alpha_2 = 0.945, \quad \alpha_3 = 298.7, \quad \alpha_4 = 3917 \times 10^3$$

then $A = 16.86 \times 10^3$, $B = -655.98 \times 10^3$, and $C = 215 \times 10^3$

$$M^2 = \frac{656 \times 10^3 \pm 10^3\sqrt{430336 - 14499}}{2 \times 16.86 \times 10^3}$$

$M^2 = 0.33$, which gives $M = 0.574$ or $M^2 = 38.55$, which yields $M = 6.2088$.

Thrust starts to build up when the flight Mach number $M = 0.574$ and not zero as depicted in Example 4. A plot for the specific thrust versus the flight Mach number is given in Figure 3.13.

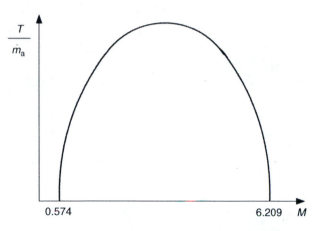

FIGURE 3.13   Specific thrust versus flight Mach number for a real ramjet engine.

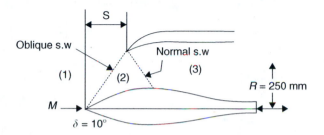

FIGURE 3.14   Variable geometry intake of a ramjet engine.

Spike motion

Supersonic ramjet engine is fitted with a spike that moves axially in and out of the engine. This motion is necessary to adjust the oblique shock wave originating from the spike to the intake lips and avoid supersonic spillage and also penetrating shocks into the intake. This is the case for the turboram engine powering the SR-71 aircraft.

The motion of spike is automatically adjusted by the engine control unit (ECU). The following example illustrates this motion.

Example 6   A ramjet engine is fitted with a variable geometry intake where a center body moves forward or backward to adjust the oblique shock wave to the intake lips (Figure 3.14). The half angle of the spike is 10°. The flight Mach number M_1 in the supersonic regime varies from $M_{minimum}$ to 3.5. The inlet radius of the intake is 250 mm.

(a)   Find the minimum Mach number $M_{min} > 1.0$ that corresponds to an attached shock wave to the apex.
(b)   Find the distance (S) wave to the apex of the center body.
(c)   Does the center body move forward or rearward when the flight Mach number increases?
(d)   What would be the value of (S) if the Mach number $M_1 = 2.5$? Calculate the distance moved by the spike.
(e)   Calculate the total pressure ratio P_{03}/P_{01} in the case of $M_1 = 3.0$.

Solution: It is known that for a certain deflection angle, there is a minimum Mach number that provides an attached shock wave. If the Mach number is reduced a detached shock wave exists.

(a)   From shock wave graphs (Figure 3.15), for a semi-vertex angle of the spike of 10°, then the minimum Mach number that corresponds to an attached shock wave is $M_{min} = 1.42$ and the corresponding wave angle $[\sigma]_{min} = 68°$.

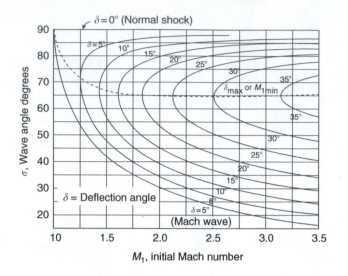

FIGURE 3.15 Shock angle versus inlet Mach number and turning angle [11].

(b) The distance S corresponding to this minimum Mach number is

$$\tan 68 = \frac{R}{S}$$

$$S = \frac{R}{\tan 68} = \frac{250}{2.4751} = 101.0 \text{ mm}$$

(c) Figure 3.15, illustrating the shock wave angle versus the deflection angle (σ versus δ), defines two types of shock waves, namely, strong and weak shock waves. The strong shock wave is close to the normal one and has a large angle, where the flow downstream of the shock is subsonic. The weak shock wave occurs at small angles. The flow downstream of the weak shock wave is supersonic. The angle of the weak shock wave decreases as the Mach number increases. Thus, the spike (or the center body) moves forward to keep the shock wave attached to the intake lips.

(d) At $M_1 = 2.5 \Rightarrow \sigma = 33°$

$$S = \frac{R}{\tan 33} = \frac{250}{0.6494} = 384.07 \text{ mm}$$

Thus, the spike moves forward by a distance (Δ) as shown in Figure 3.16.

$$\Delta = 384.07 - 101.0 = 283.07 \text{ mm}$$

(e) Following the same procedure of Example 2, then for flight Mach number $M_1 = 3.0$, the shock angle is $\sigma = 27°$.

The corresponding total pressure ratio is $P_{02}/P_{01} = 0.97$.

Now, to evaluate the pressure ratio between states (3) and (1), the Mach number at state (2) is to be first determined. For the oblique shock wave then with $M_1 = 3.0$ and $\delta = 10°$, $M_2 = 2.5$.

A normal shock wave is developed downstream the oblique shock wave.

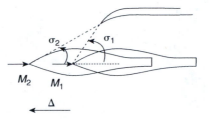

FIGURE 3.16   Spike motion due to increase of flight Mach number.

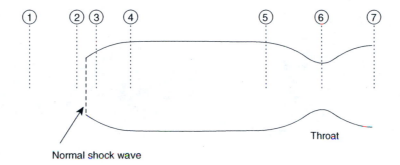

FIGURE 3.17   Ramjet engine with normal shock wave at inlet.

The corresponding pressure ratio (from tables or equations in Section 3.5) is $P_{03}/P_{02} = 0.499$. The overall pressure ratio is

$$\frac{P_{03}}{P_{01}} = \frac{P_{03}}{P_{02}}\frac{P_{02}}{P_{01}} = 0.484$$

Example 7:   Figure 3.17 illustrates a ramjet traveling in level flight at an altitude of 55,000 ft with a constant speed of 536 m/s. The intake conditions have been simplified to represent the formation of a normal shock wave immediately at the inlet. It may be assumed that the pressure external to the duct is everywhere that of the ambient atmosphere. The inlet process 2–3 is an isentropic diffusion, and the combustion process 3–4 occurs through the addition of fuel at constant area frictionless duct, the stagnation temperature at 4 being 1280 K. Process 4–5 is an isentropic expansion. Assume that the air–fuel ratio is 40/1. Take the working fluid as air with $\gamma = 1.4$ for all processes. The areas of the intake and combustion chamber are $A_2 = 0.0929$ m^2 and $A_3 = 0.1858$ m^2.

(a) Calculate the mass flow rate.
(b) Calculate the throat area at (5).
(c) Calculate the pressure drop in the combustion chamber.
(d) Calculate the thrust developed by the ramjet if the nozzle expands the gas down to the ambient pressure.
(e) Repeat (c) for a convergent nozzle.
(f) Calculate the propulsive efficiency for cases (c) and (d) above.
(g) Calculate the TSFC in both cases.
(h) Draw the cycle on T–S diagram for both cases of convergent–divergent and convergent nozzles.

Solution: At an altitude of 55,000 ft, the ambient conditions are

$$T_a = -56.5°C = 216.5 \text{ K}, \quad P_a = 9.122 \text{ KPa}$$

The sonic speed is $a = \sqrt{\gamma R T_a} = 295$ m/s

The flight Mach number is $M_1 = \frac{V_f}{a} = 1.818$

$$\frac{T_{0a}}{T_a} = 1 + \frac{\gamma - 1}{2} M^2 = 1.66$$

$$\frac{P_{0a}}{P_a} = \left(1 + \frac{\gamma + 1}{2} M^2\right)^{\frac{\gamma}{\gamma - 1}} = 5.9$$

The ambient total conditions are $T_{0a} = 360$ K

$$P_{0a} = 53.86 \text{ kPa}$$

Just upstream the engine, the static and total conditions are equal to the ambient conditions or

$$P_{01} = P_{0a}, \quad P_1 = P_a, \quad T_1 = T_a, \quad T_{01} = T_{0a}$$

The density is $\rho_1 = \frac{P_1}{RT_1} = \frac{9122}{287 \times 216.5} = 0.147$ kg/m³.

Normal shock wave
Normal shock wave exists between states (1) and (2). From normal shock wave tables or normal shock relations with $M_1 = 1.82$, then

$$\therefore \ M_2 = 0.6121, \quad \frac{T_{02}}{T_2} = 1.0773, \quad \frac{P_{02}}{P_{01}} = 0.803, \quad \frac{T_2}{T_1} = 1.5465$$

$$P_{02} = 43.25 \text{ kPa}, \quad T_2 = 334.8 \text{ K}, \quad T_{02} = 360 \text{ K} \equiv T_{01}$$

The mass flow rate $m° = \rho_1 V_1 A_1$

$$m° = 0.147 \times 536 \times 0.0929 = 7.31 \text{ kg/s}$$

Intake
From state (2) to state (3), the flow is isentropic, thus

$$P_{03} = P_{02} = 43.25 \text{ kPa} \quad \text{and} \quad T_{03} = T_{02} = 360 \text{ K}$$

From isentropic flow tables or equations in Section 3.5, with $M_2 = 0.6121$

$$\frac{A_2}{A^*} = 1.16565$$

since $\dfrac{A_3}{A^*} = \dfrac{A_3}{A^*} \dfrac{A^*}{A_2} = \dfrac{A_3}{A_2} \dfrac{A_2}{A^*} = 2 \times 1.16565$

$$\frac{A_3}{A^*} = 2.3313$$

From isentropic flow table or equation in Section 3.5, the Mach number at the inlet of combustion chamber is

$$M_3 = 0.26$$

This subsonic flow is appropriate for combustion.

Combustion chamber
From state (3) to state (4), a frictionless, heat addition constant area duct exists; thus the process represents a *Rayleigh* flow. From Rayleigh flow tables or equations in Section 3.5, with $M_3 = 0.26$

$$\therefore \quad \frac{T_{03}}{T_0^*} = 0.27446, \quad \frac{P_{03}}{P_0^*} = 1.214$$

Next, with $T_{04} = 1280$ K

$$\therefore \quad \frac{T_{04}}{T_0^*} = \frac{T_{04}}{T_{03}} \frac{T_{03}}{T_0^*} = \frac{1280}{360} \times 0.27446 = 0.9759$$

$\therefore$ From Rayleigh flow tables (equations) $M_4 = 0.83$ and $\frac{P_{04}}{P_0^*} = 1.014$

$$\therefore \quad \frac{P_{04}}{P_{03}} = \frac{P_{04}}{P_0^*} \frac{P_0^*}{P_{03}} = \frac{1.014}{1.214} = 0.8352$$

$$P_{04} = 36.12 \text{ kPa}$$

The pressure drop in the combustion chamber is

$$\Delta P_{CC} = P_{03} - P_{04} = 43.25 - 36.12 = 7.13 \text{ kPa}$$

This is a rather large pressure drop.

Nozzle
From state (4) to state (5), the flow in nozzle is isentropic flow:

At $M_4 = 0.83$, then $\dfrac{A_4}{A^*} = 1.02696$

The throat area is A^* and calculated as

$$A_5 \equiv A^* = \frac{A_4}{(A_4/A^*)} = \frac{0.1858}{1.02696} = 0.1809 \text{ m}^2$$

Since $\dfrac{P_6}{P_{06}} = \dfrac{P_6}{P_{04}} = \dfrac{P_1}{P_{04}} = \dfrac{9.122}{36.12} = 0.2525$

From isentropic tables or the governing equations in Section 3.5, $M_6 = 1.555$.

Then $A_6/A^* = 1.211$ and $T_6/T_{06} = 0.67545$.

Thus $A_6 = 1.211 \times 0.1809 = 0.21906 \text{ m}^2$.

Since $T_{06} = T_{04} = 1280$ K, then $T_6 = 864.6$ K.

The jet speed is now calculated as $V_6 = M_6\sqrt{\gamma R T_6} = 916.5$ m/s.

Thrust Force $= m^\circ \left[(1 + f)V_6 - V_f\right] + P_1\,(A_6 - A_1)$.

Since $f = \dfrac{1}{40} = 0.025$, then the thrust force is

$$T = (7.31)\,[(1.025)\,(916.5) - 536] + 9.122 \times 10^3\,(0.219 - 0.0929) = 4099\ \text{N}$$

If the nozzle is convergent (from state (4) to state (6))
Check for choking

$$\frac{P_{04}}{P_a} = \frac{36.12}{9.122} = 3.959$$

$$\frac{P_{04}}{P_c} = \left(\frac{\gamma + 1}{2}\right)^{\frac{\gamma}{\gamma - 1}} = 1.8929$$

Since $\dfrac{p_{04}}{p_a} > \dfrac{p_{04}}{p_c}$, then the nozzle, is choked

Thus $p_6 = p_c = p_{04} \times \dfrac{P_c}{P_{04}} = \dfrac{36.12}{1.8929} = 19.08\ \text{kPa}$

$$T_6 = T_c = \frac{T_{04}}{(\gamma + 1)/2} = \frac{1280}{1.2} = 1067\ \text{K}$$

$$V_6 = \text{Sonic speed} = \sqrt{\gamma R T_6} = 654\ \text{m/s}$$

From isentropic flow tables or equation, with $M_4 = 0.83$, then $\dfrac{A_4}{A^*} = 1.02708$

$$A_6 = A^* = \frac{0.1858}{1.02708} = 0.1809\ \text{m}^2$$

The thrust force equation is now expressed as $T = m^\circ\,[(1 + f)\,V_6 - V_1] + P_6 A_6 - P_1 A_1$

$$T = 7.31\,[(1.025)\,(654) - 536] + 19.08 \times 10^3 \times 0.1809 - 9.122 \times 10^3 \times 0.0929$$
$$= 3587\ \text{N}$$

Propulsive efficiency
The general expression for the propulsive efficiency is Equation 2.7, which is rewritten here

$$\eta_p = \frac{T V_f}{T V_f + \frac{1}{2}\dot{m}_a\,(V_e - V_f)^2}$$

The exhaust mass flow rate $\dot{m}_e = \dot{m}_a\,(1 + f) = 7.31 \times 1.025 = 7.493\ \text{kg/s}$.

Case (1)   Convergent–divergent nozzle ($V_e = 916.5\ \text{m/s}$)

$$\eta_p = \frac{4099 \times 536}{4090 \times 536 + \frac{1}{2}7.493\,(916.5 - 536)^2}$$
$$= 80.19\%$$

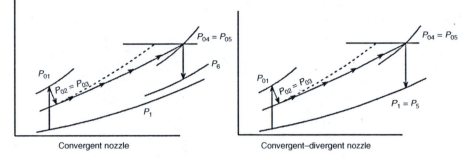

FIGURE 3.18   T–S diagram for convergent and convergent–divergent nozzles.

Case (2)   Convergent nozzle ($V_e = 654$ m/s)

$$\eta_p = \frac{3587 \times 536}{3587 \times 536 + \frac{1}{2} \times 7.493 \, (654 - 536)^2}$$

$$= 97.35\%$$

The TSFC is expressed as $\text{TSFC} = \dfrac{m_f^\circ}{T} = \dfrac{f}{T/m_a^\circ}$

Case (1)   With $T = 4090$N, then $T/m_a^\circ = 559.5$ N $\cdot$ s/kg

$$\text{TSFC} = \frac{0.025}{559.5} \frac{\text{kg}}{\text{N} \cdot \text{s}} = 0.16 \frac{\text{kg}}{\text{N} \cdot \text{h}}$$

Case (2)   Now $T = 3587$ N

$$\text{TSFC} = \frac{0.025 \times 3600}{(3587/7.31)} = 0.1834 \frac{\text{kg}}{\text{N} \cdot \text{h}}$$

Comments

1. Convergent nozzle results in less thrust and consequently higher TSFC compared with a convergent–divergent nozzle
2. The propulsive efficiency for the case of convergent nozzle is higher than that of a convergent–divergent nozzle as the exhaust speed is sonic in the former case and supersonic in the latter.

Figure 3.18 illustrates the cycle on the T–S diagram for both convergent and convergent–divergent nozzles.

3.4   CASE STUDY

The performance of a ramjet engine is examined here. The variables in ramjet engines are

1. Flight Mach number
2. Flight altitude
3. Maximum temperature of the engine
4. Fuel heating value
5. Pressure ratio of three modules (intake, combustion chamber, and nozzle)

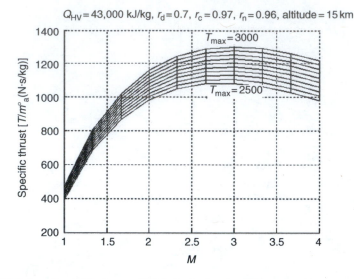

FIGURE 3.19   Variation of specific thrust with Mach number for different maximum temperatures.

These parameters influence the engine performance. Measures for performance are

1. Specific thrust
2. Fuel-to-air ratio
3. Efficiencies: propulsive, thermal, and overall

The following ranges for variables are examined:

1. The flight Mach number is successively increased from 1.0 to 4.0
2. Both sea level and 15 km altitudes
3. Maximum temperature from 2500 to 3000 K
4. Two fuels types were considered:
    (a) Hydrocarbon with heating value ranging from 44,786 to 50,010 kJ/kg
    (b) Hydrogen fuel having heating value of 119,954 kJ/kg
5. Intake pressure ratio $r_d = 0.7$, combustion chamber pressure ratio of $r_c = 0.97$, nozzle pressure ratio $r_n = 0.96$

Figure 3.19 illustrates the specific thrust versus Mach number for different maximum temperatures. As the maximum temperature increases, the specific thrust increases. At the same time increasing the maximum temperature due to burning more fuel leads to increase in the TSFC (Figure 3.20) and increase of the fuel-to-air ratio (Figure 3.21).

Alternatively, the above variables can be drawn versus the maximum temperature for different Mach numbers. As an example, Figure 3.22 illustrates the variation of specific thrust versus maximum temperature for different Mach numbers. It is clarified from Figure 3.22 that a nearly linear relation persists between the specific thrust and maximum temperature. Same conclusion is found for the fuel-to-air ratio variation with maximum temperature as shown in Figure 3.23.

The contribution of the fuel heating value on both TSFC and fuel-to-air ratio is plotted in Figures 3.24 and 3.25, respectively. As the fuel heating value increases the fuel consumption decreases. Thus, the TSFC and fuel-to-air ratio decrease with increase of the fuel heating value.

The variation of propulsive efficiency with Mach number for different maximum temperatures is plotted in Figure 3.26. This figure is distinct as the efficiency drops at first and then increases

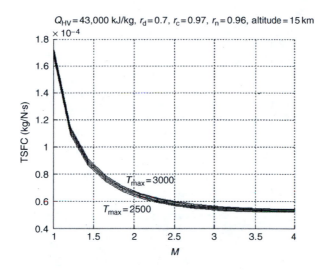

FIGURE 3.20   Variation of TSFC with Mach number for different maximum temperatures.

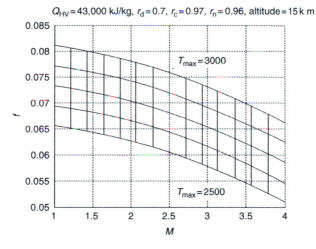

FIGURE 3.21   Variation of fuel-to-air ratio with Mach number for different maximum temperatures.

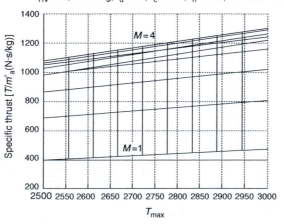

FIGURE 3.22   Variation of specific thrust versus the maximum temperature for different Mach number.

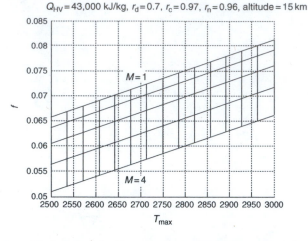

FIGURE 3.23   Variation of fuel-to-air ratio (f) versus the maximum temperature for different Mach number.

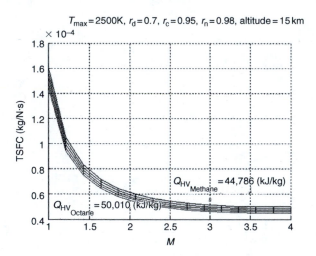

FIGURE 3.24   Variation of TSFC versus the Mach number for different fuel heating values.

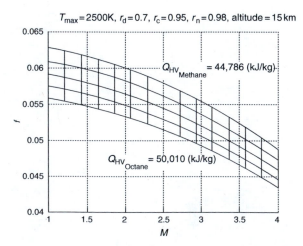

FIGURE 3.25   Variation of fuel-to-air ratio (f) versus the Mach number for different fuel heating values.

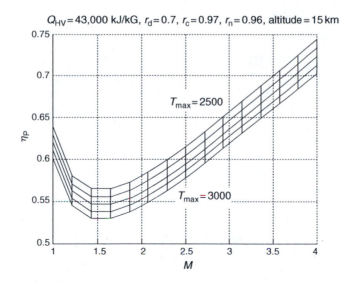

FIGURE 3.26   Variation of propulsive efficiency with Mach number for different maximum temperatures.

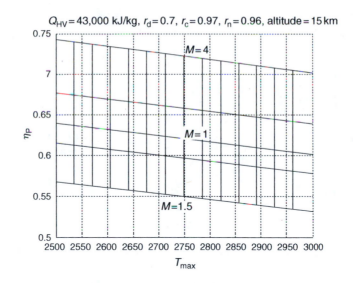

FIGURE 3.27   Variation of propulsive efficiency with maximum temperature for different Mach number.

with Mach number. The increase of maximum cycle temperature leads to a drop in the propulsive efficiency for the same Mach number.

The variation of propulsive efficiency with maximum temperature for different Mach number is plotted in Figure 3.27. From both Figures 3.26 and 3.27 the minimum propulsive efficiency is seen at Mach number close to 1.5.

The thermal efficiency increases with Mach number. Figure 3.28 illustrates this relation for a maximum temperature of 2500 K.

The overall efficiency is plotted in Figure 3.29 versus Mach number for different maximum temperatures. It is interesting to notice that both the thermal and overall efficiencies increase continuously with Mach number and the kink found in the propulsive efficiency is no longer seen.

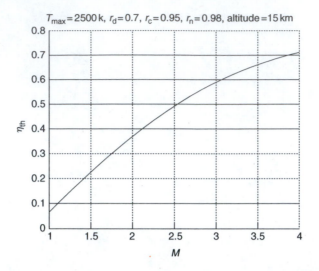

FIGURE 3.28 Variation of thermal efficiency with Mach number for a maximum temperature of 2500 K.

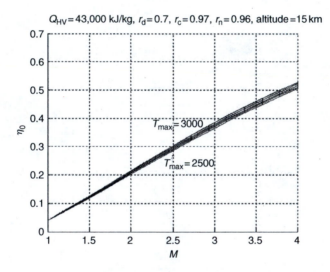

FIGURE 3.29 Variation of overall efficiency with Mach number for different maximum temperatures.

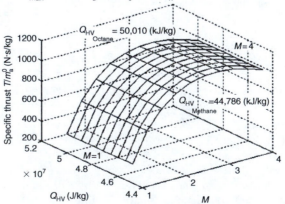

FIGURE 3.30 Variation of specific thrust with Mach number and heating value.

$Q_{HV} = 43{,}000$ kJ/kg, $r_d = 0.7$, $r_c = 0.97$, $r_n = 0.96$, altitude = 15 km

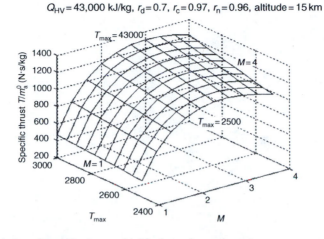

FIGURE 3.31   Variation of specific thrust with Mach number and maximum temperature.

$T_{max} = 2500$ K, $r_d = 0.7$, $r_c = 0.95$, $r_n = 0.98$, altitude = 15 km

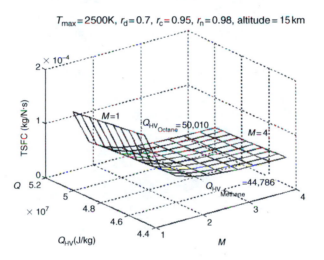

FIGURE 3.32   Variation of TSFC with Mach number and heating value.

$Q_{HV} = 43{,}000$ kJ/kg, $r_d = 0.7$, $r_c = 0.97$, $r_n = 0.96$, altitude = 15 km

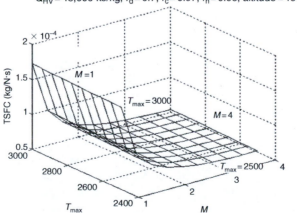

FIGURE 3.33   Variation of TSFC with Mach number and maximum temperature.

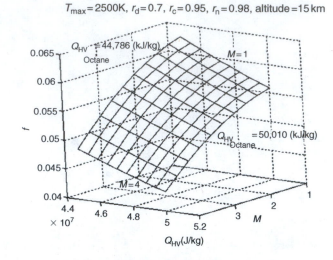

FIGURE 3.34   Variation of fuel-to-air ratio with Mach number and heating value.

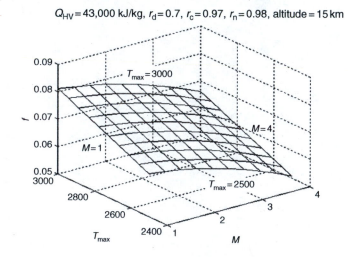

FIGURE 3.35   Variation of fuel-to-air ratio with Mach number and maximum temperature.

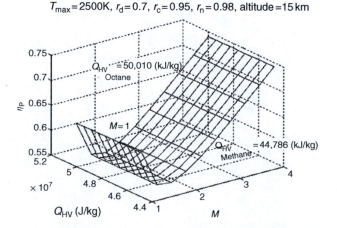

FIGURE 3.36   Variation of propulsive efficiency with Mach number and heating value.

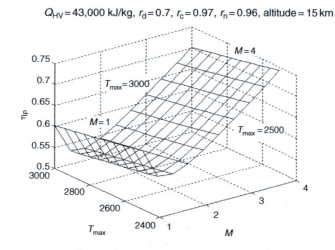

FIGURE 3.37 Variation of propulsive efficiency with Mach number and maximum temperature.

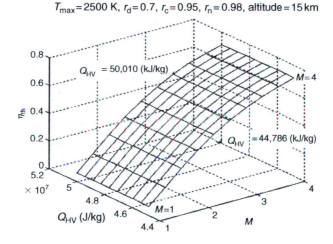

FIGURE 3.38 Variation of thermal efficiency with Mach number and heating value.

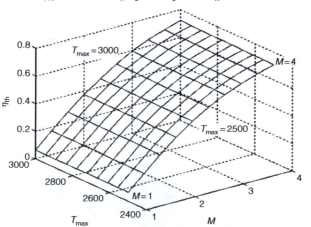

FIGURE 3.39 Variation of thermal efficiency with Mach number and maximum temperature.

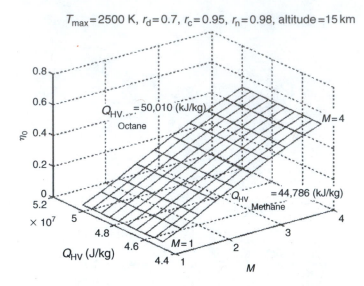

FIGURE 3.40 Variation of overall efficiency with Mach number and heating value.

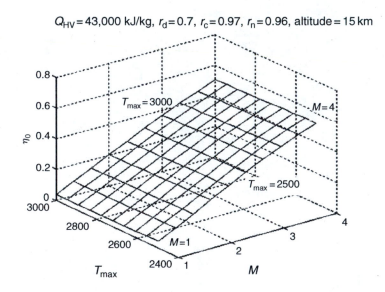

FIGURE 3.41 Variation of overall efficiency with Mach number and maximum temperature.

It is interesting to have three-dimensional plots that illustrate the effect of two variables simultaneously on one of the performance parameters. Figures 3.30 through 3.41 illustrate these 3D plots.

Since hydrogen fuel is expected for future ramjet engines where its heating value is approximately 240% of that of present hydrocarbon fuels, it was examined in this study also. The TSFC and the fuel-to-air ratio of hydrogen-fueled engines are very much less than those using hydrocarbon fuels. Their values for hydrogen fuel are nearly 40% those for hydrocarbon fuels. Figures 3.42 and 3.43 are assembled figures for sea level and 15 km altitude. Each of them contains six figures representing the specific thrust, TSFC, fuel-to-air ratio, propulsive, thermal, and overall efficiencies.

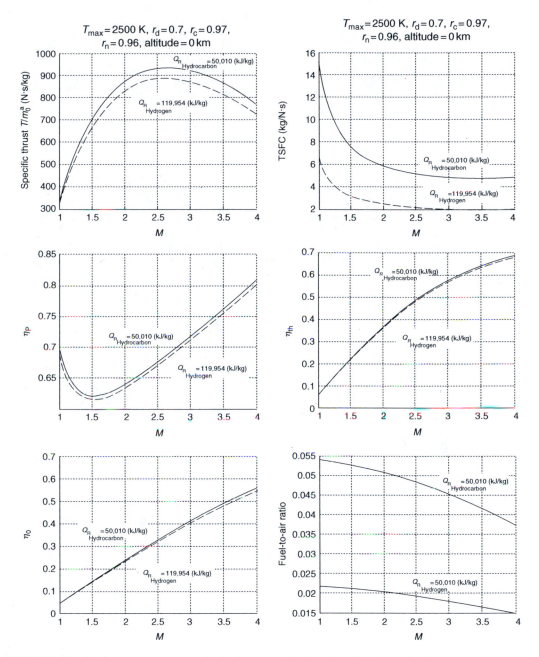

FIGURE 3.42 Performance parameters for hydrogen and hydrocarbon fuels at sea level.

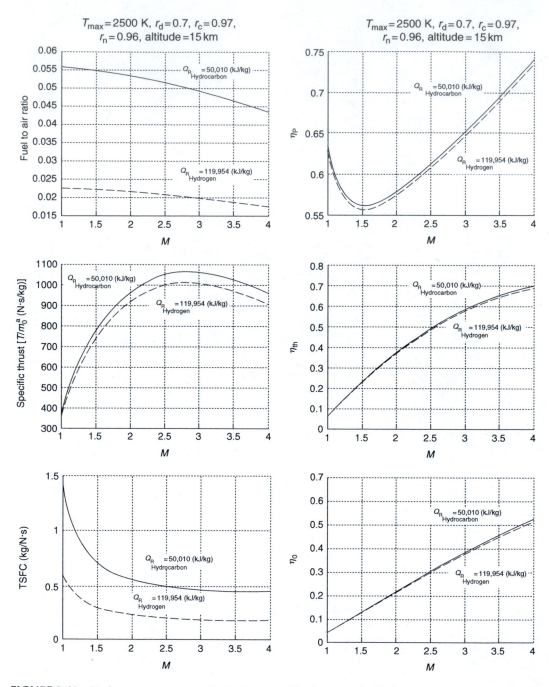

FIGURE 3.43   Performance parameters for hydrogen and hydrocarbon fuels at 15 km.

3.5 SUMMARY AND GOVERNING EQUATIONS FOR SHOCK WAVES AND ISENTROPIC FLOW

3.5.1 SUMMARY

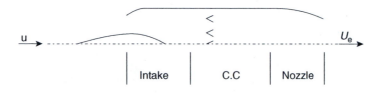

$$T = \dot{m}_a \left[(1+f) u_e - u \right]$$

$$\frac{T}{\dot{m}_a} = \left[(1+f) u_e - u \right]$$

$$\eta_p = \frac{2u \left((1+f) u_e - u \right)}{(1+f) u_e^2 - u^2}$$

$$\eta_{th} = \frac{(1+f) u_e^2 - u^2}{2f Q_{HV}}$$

$$\eta_o = \frac{u \left((1+f) u_e - u \right)}{f Q_{HV}}$$

$$\text{TSFC} = \frac{f}{u \left((1+f) \frac{u_e}{u} - 1 \right)}$$

3.5.2 NORMAL SHOCK WAVE RELATIONS

$$M_2^2 = \frac{(\gamma - 1) M_1^2 + 2}{2\gamma M_1^2 - (\gamma - 1)}$$

$$\frac{T_2}{T_1} = \left[\frac{2}{(\gamma + 1) M_1^2} + \left(\frac{\gamma - 1}{\gamma + 1} \right) \right] \left[\left(\frac{2\gamma}{\gamma + 1} \right) M_1^2 - \left(\frac{\gamma - 1}{\gamma + 1} \right) \right]$$

$$\frac{P_2}{P_1} = \left[1 + \left(\frac{2\gamma}{\gamma + 1} \right) \left(M_1^2 - 1 \right) \right]$$

$$\frac{\rho_2}{\rho_1} = \left[\frac{(\gamma + 1) M_1^2}{2 + (\gamma - 1) M_1^2} \right]$$

$$\frac{P_{02}}{P_{01}} = \left[\frac{(\gamma + 1) M_1^2}{2 + (\gamma - 1) M_1^2} \right]^{\frac{\gamma}{\gamma - 1}} \left[\left(\frac{2\gamma}{\gamma + 1} \right) M_1^2 - \left(\frac{\gamma - 1}{\gamma + 1} \right) \right]^{-\left(\frac{1}{\gamma - 1} \right)}$$

3.5.3 Oblique Shock Wave Relations

$$\tan \delta = \frac{2 \cot \beta \left(M_1^2 \sin^2 \beta - 1 \right)}{(\gamma + 1) M_1^2 - 2 \left(M_1^2 \sin^2 \beta - 1 \right)}$$

$$M_2^2 \sin^2 (\beta - \delta) = \frac{(\gamma - 1) M_1^2 \sin^2 \beta + 2}{2\gamma M_1^2 \sin^2 \beta - (\gamma - 1)}$$

$$\frac{P_2}{P_1} = \left[1 + \left(\frac{2\gamma}{\gamma + 1} \right) \left(M_1^2 \sin^2 \beta - 1 \right) \right]$$

$$\frac{T_2}{T_1} = \left[\frac{2}{(\gamma + 1) M_1^2 \sin^2 \beta} + \left(\frac{\gamma - 1}{\gamma + 1} \right) \right] \left[\left(\frac{2\gamma}{\gamma + 1} \right) M_1^2 \sin^2 \beta - \left(\frac{\gamma - 1}{\gamma + 1} \right) \right]$$

$$\frac{\rho_2}{\rho_1} = \left[\frac{(\gamma + 1) M_1^2 \sin^2 \beta}{2 + (\gamma - 1) M_1^2 \sin^2 \beta} \right]$$

$$\frac{P_{02}}{P_{01}} = \left[\frac{(\gamma + 1) M_1^2 \sin^2 \beta}{2 + (\gamma - 1) M_1^2 \sin^2 \beta} \right]^{\frac{\gamma}{\gamma - 1}} \left[\left(\frac{2\gamma}{\gamma + 1} \right) M_1^2 \sin^2 \beta - \left(\frac{\gamma - 1}{\gamma + 1} \right) \right]^{-\left(\frac{1}{\gamma - 1} \right)}$$

3.5.4 Rayleigh-Flow Equations

$$\frac{P}{P^*} = \left[\frac{\gamma + 1}{1 + \gamma M^2} \right]$$

$$\frac{T}{T^*} = \frac{M^2 (\gamma + 1)^2}{(1 + \gamma M^2)^2}$$

$$\frac{T_0}{T_0^*} = \frac{2 (1 + \gamma) M^2}{(1 + \gamma M^2)^2} \left[1 + \frac{\gamma - 1}{2} M^2 \right]$$

$$\frac{P_0}{P_0^*} = \frac{(1 + \gamma)}{(1 + \gamma M^2)} \left(\frac{1 + \frac{\gamma - 1}{2} M^2}{\frac{\gamma + 1}{2}} \right)^{\frac{\gamma}{\gamma - 1}}$$

3.5.5 Isentropic Relation

$$\frac{A}{A^*} = \frac{1}{M} \left(\frac{1 + \frac{\gamma - 1}{2} M^2}{\frac{\gamma + 1}{2}} \right)^{\frac{\gamma + 1}{2(\gamma - 1)}}$$

PROBLEMS

3.1 A pulsejet similar to the German engine has the following data:

Net thrust   1400 lb
Flight altitude   10,000 ft
Speed   400 mph
Inlet diameter   1.34 ft
Total pressure drop in the inlet   20%
TSFC   10 lb/lb$_f$h
Fuel heating value   14,000 Btu/lb$_m$

Assume $\gamma = 1.4$ and $Cp = 0.24\ Btu/lb_m^{OR}$

Calculate
(a) Fuel-to-air ratio
(b) Maximum temperature in the combustion chamber
(c) Exhaust velocity
(d) Propulsive efficiency
(e) Thermal efficiency

3.2 Why is mechanical efficiency not an issue with ramjets?

3.3 How is thrust created in ramjets?

3.4 A small vehicle is powered by a pulsejet. The available net thrust is 6000 N and the traveling speed is 200 km/h. The gases leave the engine with an average velocity of 360 m/s. Assume pressure equilibrium exists at the outlet plane and the fuel-to-air ratio is 0.06.
(a) Compute the mass flow rate required.
(b) Calculate the inlet area (assume 16°C and one atmosphere).
(c) Calculate the thrust power.
(d) Calculate the propulsive efficiency.

3.5 (a) Prove that for an ideal ramjet having a choked nozzle, the specific thrust is expressed by the relation:

$$\frac{T}{\dot{m}_a} = \sqrt{\gamma_c RT_a}\left[(1+f)\sqrt{\frac{2\gamma_h T_{04}}{\gamma_c\,(\gamma_h+1)\,T_a}} - M\right] + \frac{A_e P_a}{\dot{m}_a}\left[\frac{\left(1+\frac{\gamma_c-1}{2}M^2\right)^{\frac{\gamma_c}{\gamma_c-1}}}{\left(\frac{\gamma_h+1}{2}\right)^{\frac{\gamma_h}{\gamma_h-1}}} - 1\right]$$

(b) Plot figures to illustrate the TSFC and $T/\dot{m}_a$ versus the flight Mach number M for different values of the maximum cycle temperatures in both cases of ideal and real cycles.

3.6 A ramjet has low-speed stagnation pressure losses given by $r_d = 0.80$, $r_c = 0.97$, $r_n = 0.96$. The ambient and maximum temperatures are 216 K and 2050 K respectively. The fuel-to-air ratio is assumed very small such that $f \ll 1$. Using the thrust force in Problem 3.5, determine the minimum flight Mach number at which the engine develops positive thrust? Assume the nozzle is unchoked and $\gamma = 1.4$.

3.7 Following the same procedure of Example 4, prove that the specific impulse is expressed as

$$I_{sp} = \frac{uQ_R}{CpT_a g\left(1 + \frac{\gamma-1}{2}M^2\right)}\left(\frac{\sqrt{\tau_b}-1}{\tau_b-1}\right)$$

where $\tau_b = \dfrac{T_{03}}{T_{02}}$

3.8 A ramjet engine is powering an aircraft flying at a Mach number M. The thermodynamic cycle is considered ideal. Comparison is to be made between the two cases of choked and unchoked nozzle. Prove that:

$$\frac{V_j}{V_c} = M \sqrt{\frac{\gamma + 1}{2\left(1 + \frac{\gamma-1}{2}M^2\right)}}$$

$$\frac{P_c}{P_a} = \left(\frac{2}{\gamma + 1} + \frac{\gamma - 1}{\gamma + 1}M^2\right)^{\frac{\gamma}{\gamma-1}}$$

where V_j is the exhaust gas velocity when the nozzle is unchoked, while V_c and P_c are the exhaust velocity and pressure when the nozzle is choked.

Next, if $M = 1$,

(a) Plot the thermodynamic cycle on the T–S diagram. Write also the thrust equation.
(b) Prove that the thrust force is given by the relation:

$$T = \rho_a A_i \left(\gamma R T_a\right)\left[\sqrt{\frac{2}{\gamma + 1}\frac{T_{04}}{T_a}} - 1\right]$$

3.9 For the shown ramjet engine, if $T_{02} = T_6$ prove that:

$$\frac{T_{04}}{T_a} = \left(1 + \frac{\gamma - 1}{2}M^2\right)^2$$

$$V_6 = V\left(1 + \frac{\gamma - 1}{2}M^2\right)^{1/2}$$

where T_{04} is the maximum temperature and V_6 is the exhaust velocity.

3.10 For an ideal ramjet engine, prove that if θ_b is constant, then the thrust parameter $(T/\dot{m}\,a_a)$ is maximum when $\theta_a = \theta_b^{\frac{1}{3}}$, where $\theta_a = T_{0a}/T_a$, $\theta_b = T_{0max}/T_a$, $T = $ thrust, $a_a = \sqrt{\gamma R T_a}$.

3.11 A ramjet engine is powering an airplane flying with Mach number $M = 4$, at an altitude of 50,000 ft, where the ambient temperature and pressure are 205 K and 11.6 kPa, respectively. The fuel heating value is 45,000 kJ/kg and the peak temperature is 3000 K. The air mass flow rate is 50 kg/s. The ramjet diffuser losses are given by:

$$\frac{P_{02}}{P_{0a}} = 1 - 0.1\,(M - 1)^{1.5}$$

where M is the flight Mach number. The stagnation pressure ratio across the flame holders (from state 2 to 3) is 0.97 and the stagnation pressure loss in the combustor (from state 3 to 4) is 5%. The nozzle and combustion efficiencies are 0.95 and 0.98, respectively. Assume the propellant is a perfect gas having $R = 0.287$ kJ/(kg.K) throughout, but having $\gamma = 1.4$ from state a (ambient) to state 3 and $\gamma = 1.3$ from states 3 to 6. It is required to calculate:

(a) The fuel-to-air ratio
(b) The nozzle exit area for the cases of choked (convergent nozzle) and unchocked (de Laval nozzle)
(c) The thrust force in the two cases stated in item b
(d) TSFC

REFERENCES

1. *The Jet Engine*, Rolls-Royce plc., p. 3, 5th edn., Reprinted 1996 with revisions.
2. J.V. Foa, *Element of Flight Propulsion*, Wiley, New York, 1960.
3. R.D. Archer and M. Saarlas, *An Introduction to Aerospace Propulsion*, Prentice-Hall, Inc., 1996, p. 2.
4. F.S. Billing, Research on supersonic combustion, *AIAA Journal of Propulsion and Power*, 9, 499–514, 1993.
5. R.D. Zucker and O. Biblarz, *Fundamental Gas Dynamics*, 2nd edn., 2002, p. 366.
6. J. Roskam and C-T.E. Lan, *Airplane Aerodynamics and Design*, DAR Corporation, 1997, pp. 258.
7. W. Heiser and D. Pratt, Thermodynamic cycle analysis of pulse detonation engines, *Journal of Propulsion and Power*, 18, 68–76, 2002.
8. D.E. Paxson, *A Performance Map for Ideal Airbreathing Pulse Detonation Engine*, AIAA-2001–3465, 2001.
9. L.A. Povinelli, *Pulse Detonation Engine for High Speed Flight*, NASA/TM 2002-211908.
10. P. Hill, and C. Peterson, *Mechanics and Thermodynamics of Propulsion*, 2nd edn., Addison Wesley Publication Company, Inc., 1992, p. 155.
11. A.H. Shapiro, *The Dynamics and Thermodynamics of Compressible Fluid Flow*, Vol. I, Ronald Press, New York, p. 536.

4 Turbojet Engine

4.1 INTRODUCTION

Turbojet is the basic engine of the jet age. It is a reaction engine invented by the two fathers of jet engines, Frank Whittle from Britain and von Ohain from Germany. The first flight of an airplane powered by a turbojet engine was the He178 German aircraft powered by the He S-3 engine on August 27, 1939. On the basis of von Ohain's work, the German engine designer Anslem Franz developed his turbojet engine that powered the jet fighter Me 262 built by the firm Messerschmitt. Frank Whittle in England, having no knowledge of Ohain's engine, built his W.1 turbojet engine, which powered the Gloster E28/39 aircraft. As described in Chapter 1, the future of aviation lay with jets.

The disadvantage of ramjet engine is that its pressure ratio depends on the flight Mach number. It cannot develop takeoff thrust and does not perform well unless the flight Mach number is much greater than 1 [(refer to the figure $(T/\dot{m}_a)$ versus M); Example 6, Ramjet]. To overcome this disadvantage, a compressor is installed in the inlet duct so that even at zero flight speed air could be drawn into the engine. This compressor is driven by the turbine installed downstream of the combustion chamber and connected to the compressor by a central shaft. Addition of the two rotating parts or modules (compressor and turbine) converts the ramjet into a turbojet. The compressor, combustion chamber, and turbine constitute what is called *gas generator*. Air is squeezed into the compressor many times its normal atmospheric pressure and then forced into the combustion chamber. Fuel is sprayed into the compressed air, then ignited and burned continuously in the combustion chamber. This raises the temperature of the fluid mixture from about 1100°F to 1300°F. The resulting burning gases expand rapidly rearward and pass through the turbine, which drives the compressor. The turbine extracts energy from the expanding gases to drive the compressor, which intakes more air. If the turbine and compressor are efficient, the pressure at the turbine discharge will be nearly twice the atmospheric pressure, and this excess pressure is sent to the nozzle to produce a high-velocity stream of gases. These gases bounce back and shoot out of the rear of the exhaust, thus producing a thrust pushing the plane forward. Substantial increases in thrust can be obtained by employing an afterburner or augmenter. It is a second combustion chamber positioned after the turbine and before the nozzle. The afterburner increases the temperature of the gas ahead of the nozzle. The result of this increase in temperature is higher jet velocity and more push. Thus, there is an increase of about 40% in thrust at takeoff and a much larger percentage at high speeds once the plane is in the air.

Classification of turbojet engines is shown in Figure 4.1. Turbojet engines may be either a nuclear or a nonnuclear engine. In the late 1940s and through the 1950s [1,2], the preliminary works were carried out for developing the nuclear-powered rockets and jet engines. Both the United States and the Soviet Union conducted rigorous research and development (R&D) programs in this field, but ultimately canceled their respective programs owing to technical difficulties and growing safety concerns, such as catastrophic atomic radiation, that might be encountered in the case of crashes/accidents of such nuclear airplanes. For the past 40 years, little research related to nuclear-powered jet engine has been carried out. However, recent discoveries, in the field of controlled or triggered nuclear decay [3–5] together with the advancement of material, airframe, and engine design, have reinvigorated the possibility of running aircraft on nuclear power.

Another classification for turbojet engines may be emphasized as follows. Turbojet engines may be classified as either afterburning or nonafterburning. As described earlier, the afterburner is a second chamber where afterburning (or reheat) is a method of augmenting the basic thrust of

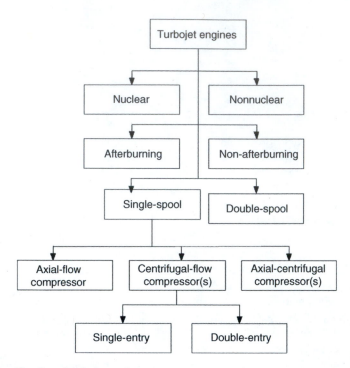

FIGURE 4.1 Classification of turbojet engines.

an engine to improve the aircraft takeoff, climb, and (for military aircraft) combat performance. Afterburning consists of the introduction and burning of raw fuel between the engine turbine and the jet pipe propelling nozzle, utilizing the unburned oxygen in the exhaust gas to support combustion. The increase in the temperature of the exhaust gas increases the velocity of the jet leaving the propelling nozzle and therefore increases the engine thrust by 50% or more. This increased thrust could be obtained by the use of a larger engine, but this would increase the weight and overall fuel consumption.

A third classification of the turbojet engines is based on their *number of spools*. Turbojets may have one or two (sometimes identified as double or dual) spool engines. A single-spool turbojet has one shaft with a compressor and a turbine mounted at either shaft ends whereas a two-spool engine has two shafts, compressors, and turbines. The first compressor next to the intake is identified as the low-pressure compressor (LPC), which is driven by the low-pressure turbine (LPT). Both, together with their shaft, are denoted as the low-pressure spool. The high-pressure compressor (HPC) is situated downstream of the LPC. In addition, this compressor is driven by the high-pressure turbine (HPT). Both are coupled by the high-pressure shaft (which is concentric with the low-pressure shaft) and constitute the modules of the high-pressure spool. Figure 4.2 illustrates the layout and modules of the single- and double-spool turbojet engines.

The single-spool engine may be further classified based on the *type of compressor* employed. The compression section may be composed of either a single or the double compressors. Moreover, the compressor may be of the axial or the radial (centrifugal) type. Thus, numerous cases may exist. A single axial or centrifugal compressor is seen in some turbojets. In other cases, two compressors assembled in series are found. These compressors may be either two centrifugal compressors or an axial compressor followed by a centrifugal compressor.

Finally, this classification closes with the *type of entry* associated with the two centrifugal compressors case. Centrifugal compressors may have either single or double entries if they are assembled in a back-to-back configuration.

Single-rotor turbojet JT12, J60

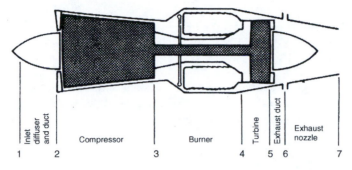

Dual-rotor turbojet JT3, JT4, J52

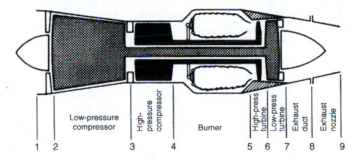

FIGURE 4.2   Layout of single- and dual-spool turbojet engine.

4.2   SINGLE SPOOL

4.2.1   EXAMPLES OF ENGINES

1. *Rolls-Royce Nene* (powered Vickers_Viking_1a-G-Agrn aircraft)

    Nonafterburning engine
    Length: 97 in. (2464 mm)
    Diameter: 49.5 in. (1257 mm)
    Compressor: dual-entry centrifugal compressor with two-sided impeller
    Combustors: 9 cans combustion chambers
    Turbine: single-stage axial flow
    Fuel type: aviation kerosene with 1% lubricating oil
    Thrust: 5000 lbf (22.2 kN) at 12,400 rpm for takeoff
    Specific fuel consumption 1.04 lb/(h· lbf)
    Thrust-to-weight ratio 1:1

2. *Klimov_Vk-1* (powered MiG17 and Lim-R_B1 aircraft)

    Nonafterburning engine
    Length: 102 in. (2600 mm)
    Diameter: 51.0 in. (1300 mm)
    Compressor: centrifugal
    Dry weight: 1395 lbf
    Thrust: 5955 lbf (26.5 kN)

Specific fuel consumption: 1.07 lb/(h · lbf) [109.1 kg/(kN · h)]

Thrust-to-weight ratio: 4.27:1

VK-1F, an improved version with a simple afterburner and variable nozzle, was developed for the main production version, the MiG-17F

3. *General Electric J47*

The engine was produced in at least 17 different series and was used to power Air Force aircraft such as the F-86, XF-91, B-36, B-45, B-47, and XB-51

Compressor: 12-stage axial compressor

Turbine: single-stage axial turbine

Weight: 2707 lbs

Thrust: 5670 lbs

Maximum rpm: 7950

Maximum operating altitude: 50,000 ft

4.2.2 THERMODYNAMIC ANALYSIS

As described earlier, single-spool turbojet may have either one or two compressors as well as a single driving turbine. It may or may not have an afterburner. Figure 4.3 illustrates a single-spool turbojet engine having a single compressor and an afterburner together with designations for each state. The different processes that are encountered within the engine are described here.

(a)–(1):   The air flows from far upstream, where the velocity of air relative to engine is the flight velocity up to the intake, usually with some deceleration during cruise and acceleration during takeoff.

(2)–(3):   The air flows through the inlet diffuser and ducting system, where the air velocity is decreased as the air is carried to the compressor inlet.

(2)–(3):   The air is compressed in a dynamic compressor.

(3)–(4):   The air is "heated" by mixing and burning of fuel in the air.

(4)–(5):   The air is expanded through a turbine to obtain power to drive the compressor.

(5)–(6):   The air may or may not be further heated by the addition and burning of fuel in an afterburner.

(6)–(7):   The air is accelerated and exhausted through the exhaust nozzle.
            If the engine is not fitted with an afterburner, then states 5 and 6 are coincident. It will be simpler not to discard point 6 in the following analysis. Thus, the flow in the nozzle remains from points 6 to 7. The amount of mass flow is usually set by flow choking in the nozzle throat.

4.2.3 IDEAL CASE

The components except the burners are assumed to be reversible adiabatic or isentropic. Moreover, the burners are replaced by frictionless heaters; thus, the velocities at stations 2 through 6 are negligible.

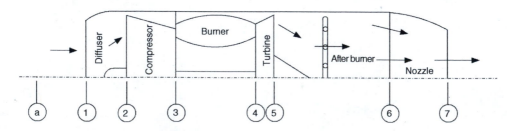

FIGURE 4.3   Single-spool afterburning turbojet.

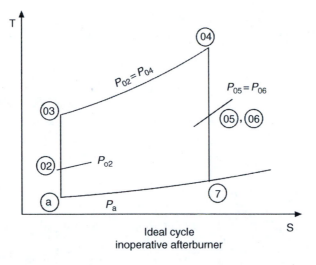

FIGURE 4.4   Temperature–entropy diagram for ideal turbojet, inoperative afterburner.

An analysis of the different components of the engine will be followed (Figure 4.4).

1. *Intake or inlet:*   During cruise, the static pressure rises from (a) to (1) outside the intake and from (1) to (2) inside the intake. Air is decelerated relative to the engine. Since the velocity at (2) is assumed to be zero and the deceleration is isentropic, the total or stagnation pressure at states (a), (1), and (2) are constant and equal:

$$P_{02} = P_{01} = P_{0a} = P_a \left(1 + \frac{\gamma - 1}{2} M^2 \right)^{\gamma/(\gamma-1)} \tag{4.1}$$

The stagnation temperatures for states (a), (1), and (2) are also equal:

$$T_{02} = T_{01} = T_{0a} = T_a \left(1 + \frac{\gamma - 1}{2} M^2 \right) \tag{4.2}$$

2. *Compressor:*   The pressure ratio of the compressor, π_C, is assumed to be known; thus, the pressure and temperature at the outlet of the compressor are evaluated from the corresponding values at the inlet of the compressor (outlet of the intake) from the relations

$$P_{03} = (P_{02})(\pi_c) \tag{4.3a}$$

and

$$T_{03} = T_{02} \left[\frac{P_{03}}{P_{02}} \right]^{(\gamma-1)/\gamma} \tag{4.3b}$$

3. *Combustion chamber:*   The turbojet is assumed to be with a single-spool engine. Thus, the outlet conditions of the compressor will be those at the inlet to the combustion chamber. Combustion process takes place from (3) to (4). In this process, fuel is injected in an atomized form, which is then evaporated and mixed with air. Spark plugs initiate the

combustion process. Since no pressure drop is assumed for the ideal case, the pressures at the inlet and outlet of the combustion chamber are equal; or

$$P_{03} = P_{04} \tag{4.4}$$

The temperature at the outlet of the turbine is determined from metallurgical limits set by the turbine blade material and is known as the turbine inlet temperature (TIT). Fuel is added to the combustion chamber and burnt. The mass flow rate of the burnt fuel is calculated from the energy balance of the combustion chamber.

$$(\dot{m}_a + \dot{m}_f)Cp_h\, T_{04} = \dot{m}_a Cp_c T_{03} + \dot{m}_f Q_R$$

With $f = \dot{m}_f/\dot{m}_a$, then the fuel-to-air ratio is determined using the following relation:

$$f = \frac{Cp_h T_{04} - Cp_c T_{03}}{Q_R - Cp_h T_{04}} \tag{4.5}$$

4. *Turbine:* The power consumed in the compression from (2) to (3) must be supplied through the turbine in expansion from (4) to (5). If the ratio of the power needed to drive the compressor to the power available in the turbine is (λ), then the energy balance for the compressor turbine shaft is

$$W_c = \lambda W_t$$

Here (λ) is of the range 75%–85%. Thus, in terms of the temperature differences:

$$Cp_c(T_{03} - T_{02}) = \lambda(1 + f)Cp_h(T_{04} - T_{05})$$

$$\left(\frac{T_{05}}{T_{04}}\right) = 1 - \frac{(Cp_c/Cp_h)T_{02}}{\lambda(1 + f)T_{04}}\left[\left(\frac{T_{03}}{T_{02}}\right) - 1\right] \tag{4.6}$$

Then, the turbine and compressor pressure ratios are related by

$$\left(\frac{P_{05}}{P_{04}}\right) = \left\{1 - \frac{(Cp_c/Cp_h)T_{02}}{\lambda(1 + f)T_{04}}\left[\left(\frac{P_{03}}{P_{02}}\right)^{(\gamma_c - 1/\gamma_c)} - 1\right]\right\}^{\gamma_h/(\gamma_h - 1)} \tag{4.7a}$$

From Equation 4.2

$$\left(\frac{P_{05}}{P_{04}}\right) = \left\{1 - \frac{(Cp_c/Cp_h)T_a}{\lambda(1 + f)T_{04}}\left(1 + \frac{\gamma - 1}{2}M^2\right)\left[\left(\frac{P_{03}}{P_{02}}\right)^{(\gamma_c - 1/\gamma_c)} - 1\right]\right\}^{\gamma_h/(\gamma_h - 1)} \tag{4.7b}$$

If the pressure drop in the turbine is calculated before temperature drop calculation, then the temperature at the turbine outlet T_{05} may be calculated from the isentropic relation:

$$\frac{T_{05}}{T_{04}} = \left(\frac{P_{05}}{P_{04}}\right)^{(\gamma_h - 1)/\gamma_h} \tag{4.8}$$

5. *Afterburner:* If the jet engine is without an afterburner, then no work or heat transfer occurs downstream of station (5). The stagnation enthalpy remains constant throughout the rest of the engine. However, if there is an afterburner, we have two cases: is the afterburner either operative or inoperative. If the afterburner is inoperative, then states (5) and (6) are coincident. Both the temperatures and pressures are equal:

$$T_{06} = T_{05}$$

$$P_{06} = P_{05}$$

For an operative afterburner, a subscript (A) is added to the symbols of the temperature and the pressure to denote operative afterburner. In this case, an additional amount of fuel is burnt that raises the temperature to (T_{06A}), which is much higher than the TIT. The reason for such a high temperature is that the downstream element is the nozzle, which is a nonrotating part. Thus, the walls are subjected only to thermal stresses rather than the combined thermal and mechanical stresses as in the turbine(s). For an ideal cycle, then

$$P_{06A} = P_{05}$$

and

$$T_{06A} = T_{MAX}$$

There is an additional fuel quantity ($\dot{m}_{fab}$), which is added and burnt in the afterburner. The conservation of mass and energy within the afterburner yields the following equations:

$$\dot{m}_6 = \dot{m}_5 + \dot{m}_{fab}$$

$$\dot{m}_5 h_{05} + \dot{m}_{fab} Q_R = \dot{m}_6 h_{06A}$$

$$\therefore \dot{m}_5 Cp_5 T_{05} + \dot{m}_{fab} Q_R = (\dot{m}_5 + \dot{m}_{fab}) Cp_6 T_{06A}$$

$$\text{with } f_{ab} = \frac{\dot{m}_{fab}}{\dot{m}_a}$$

$$\therefore (1+f) Cp_5 T_{05} + f_{ab} Q_R = (1 + f + f_{ab}) Cp_6 T_{06A}$$

The corresponding fuel-to-air ratio within the afterburner (f_{ab}) is evaluated from the relation

$$f_{ab} = \frac{(1+f)(Cp_{6A} T_{06A} - Cp_5 T_{05})}{Q_R - Cp_{6A} T_{06A}} \tag{4.9}$$

In Equation 4.9, the variation of the specific heat with temperature is taken into account.

6. *Nozzle:* The exhaust velocity is obtained from the conservation of energy in the nozzle. The hot gases expand in the nozzle from (6) or (6A) to (7) depending on whether the afterburner is operative or inoperative. Hereafter, the two cases will be considered.

(a) *Inoperative afterburner*

The nozzle is to be checked first to define whether it is choked or not. Thus, the critical pressure is calculated from the relation:

$$\left(\frac{P_{06}}{P_c}\right) = \left(\frac{\gamma_h + 1}{2}\right)^{\gamma_h/(\gamma_h-1)} \quad (4.10a)$$

The critical pressure is then compared with the ambient pressure. If it is greater than or equal to the ambient pressure, then the nozzle is choked. This means that $P_7 = P_c$ and the nozzle outlet temperature (T_7) is then calculated from the relation:

$$\left(\frac{T_{06}}{T_7}\right) = \left(\frac{\gamma_h + 1}{2}\right) \quad (4.11a)$$

The exhaust velocity is then calculated as

$$V_7 = \sqrt{\gamma_h R T_7} \quad (4.12a)$$

On the other hand if the ambient pressure is greater than the critical pressure, then the nozzle is unchoked and the exhaust pressure is equal to the ambient pressure, $P_7 = P_a$, and the temperature of the exhaust gases is determined from the relation:

$$\left(\frac{T_{06}}{T_7}\right) = \left(\frac{P_{06}}{P_a}\right)^{(\gamma_h-1)/\gamma_h}. \quad (4.13a)$$

The exhaust velocity is now calculated from the relation:

$$V_7 = \sqrt{2C_{p_n}(T_{06} - T_7)} = \sqrt{2C_{p_n}T_{06}\left[1 - \left(\frac{P_a}{P_{06}}\right)^{(\gamma_h-1)/\gamma_h}\right]}. \quad (4.14a)$$

(b) *Operative afterburner*

In this case, the expansion process in the nozzle starts from the points (6A) to (7); refer to Figure 4.5. Again a check for the state of the nozzle is performed by calculating the critical pressure from Equation 4.10, or

$$\left(\frac{P_{06A}}{P_c}\right) = \left(\frac{\gamma_h+1}{2}\right)^{\gamma_h/(\gamma_h-1)} \quad (4.10b)$$

where $P_{06A} = P_{05}$. Again, the critical pressure is compared with the ambient pressure. If $P_c > P_a$, then $P_{7A} = P_c$ and the nozzle is choked. The outlet temperature (T_{7A}) is much greater than that in the case of inoperative afterburner and calculated from the relation:

$$\left(\frac{T_{06A}}{T_{7A}}\right) = \left(\frac{\gamma_h + 1}{2}\right) \quad (4.11b)$$

If the nozzle is choked, then the exhaust velocity may be calculated from the following equation:

$$V_{7ab} = \sqrt{\gamma_h R T_{7A}} \quad (4.12b)$$

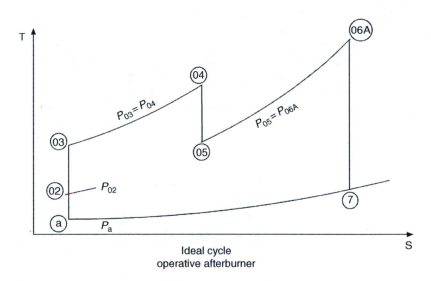

Ideal cycle
operative afterburner

FIGURE 4.5 Temperature–entropy diagram for ideal turbojet, operative afterburner.

If the nozzle is unchoked, then the exhaust pressure is again an ambient one. The velocity of the exhaust gases is calculated from the relation

$$V_{7ab} = \sqrt{2Cp_n T_{06A}\left[1 - \left(\frac{P_a}{P_{06A}}\right)^{(\gamma_h-1)/\gamma_h}\right]} \qquad (4.14b)$$

The thrust force is expressed by the following equation:

$$T = \dot{m}_a[(1+f)V_e - V] + (P_e - P_a)A_e \qquad (2.3a)$$

V_e is the exhaust jet speed, which may be either V_7 or V_{7ab} depending on the case of operative or inoperative afterburner.

Similarly, the propulsive, thermal, and overall efficiencies may be calculated from appropriate relations as described in Chapter 2.

Now, let us compare the velocity of the jet leaving the turbojet engine in the case of operative and inoperative afterburner. Since the shape of the constant pressure lines on the T–S diagram is divergent, the enthalpy drop in the nozzle for operative afterburner is greater than in the case of the inoperative one. The jet velocity is proportional to the square of the enthalpy drop

$$h_{06} - h_7 = \frac{V_7^2}{2}$$

Hence, the exhaust velocity V_7 in the case of afterburning will be greater than in the nonafterburning case.

$$(V_7)_{afterburning} \approx 1.5(V_7)_{nonafterburning} \qquad (4.15)$$

This is due to the absence of highly stressed material in the afterburner, which allows T_{06A} being much higher than TIT; T_{04}.

The big advantage of an afterburner is that the jet speed increases, refer to Equation 4.15, which in turn significantly increases the thrust of the engine without adding much weight or complexity to the engine. Generally, an afterburner is nothing but a set of fuel injectors, a tube and a flame holder that the fuel burns in, and an adjustable nozzle. A jet engine with an afterburner needs an adjustable nozzle so that it can work with the afterburners both on and off.

The disadvantage of an afterburner is that it uses a lot of fuel, refer to Equation 4.9, for the power it generates. Therefore, most planes use afterburners sparingly. For example, a military jet would use its afterburners when taking off from the short runway on an aircraft carrier or during a high-speed maneuver in a dogfight.

Example 1   A single-spool turbojet engine (without an afterburner) is running on ground ($M = 0$). Prove that for an ideal process and unchoked nozzle, the pressure ratio of the turbine (π_t) is related to the pressure ratio of the compressor (π_c) by the relation:

$$\pi_t = \frac{1}{\left[1 - (C_{P_c}/C_{P_h})\,(T_a/T_{04})\,\left(\pi_c^{(\gamma_c-1)/\gamma_c} - 1\right)\right]^{\gamma_t/(\gamma_t-1)}}$$

and the jet speed is expressed as

$$V_j = \sqrt{2C_{P_h}\left[T_{04} - \frac{C_{P_c}}{C_{P_h}}T_a\left(\pi_c^{(\gamma_c-1)/\gamma_c} - 1\right)\right]\left\{1 - \left(\frac{\pi_t}{\pi_c}\right)^{(\gamma_h-1)/\gamma_h}\right\}}$$

where $\pi_t = P_{04}/P_{05}$, T_a and T_{04} are the ambient and maximum temperatures, respectively. Assume that the fuel-to-air ratio is negligible and the work consumed by the compressor is equal to the work developed by the turbine.

Solution: The successive elements will be analyzed as follows:

1. *Diffuser:*   For a ground run of the engines ($M = 0$) and an ideal diffusion process, we obtain the following from Equations 4.1 and 4.2

$$T_a = T_{0a} = T_{02} = T_7$$
$$P_{02} = P_{0a} = P_a$$

2. *Compressor:*   For an ideal compression process,

$$\frac{T_{03}}{T_{02}} = \pi_c^{(\gamma_c-1/\gamma_c)} \tag{a}$$

where $\pi_c = P_{03}/P_{02}$.

3. *Combustion chamber:*   For no pressure loss in the combustion chamber and negligible fuel-to-air ratio,

$$P_{04} = P_{03}$$
$$f \ll 1$$

4. *Turbine:*   The pressure and temperature ratios at the turbine are expressed as follows:

$$\frac{T_{04}}{T_{05}} = \pi_t^{(\gamma_t - 1)/\gamma_t}$$

where $\pi_t = P_{04}/P_{05}$.

Assuming that the work developed by the turbine is equal to the work consumed by the compressor, then

$$W_t = W_c$$

For negligible fuel-to-air ratio

$$C_{P_t} (T_{04} - T_{05}) = C_{P_c} (T_{03} - T_{02}) \tag{b}$$

$$C_{P_t} T_{04} \left(1 - \frac{1}{\pi_t^{(\gamma_t - 1/\gamma_t)}} \right) = C_{P_c} T_{02} \left(\pi_c^{(\gamma_c - 1)/\gamma_c} - 1 \right)$$

With $T_{02} = T_a$

$$\frac{1}{\pi_t^{(\gamma_t - 1/\gamma_t)}} = 1 - \frac{C_{P_c}}{C_{P_h}} \frac{T_a}{T_{04}} \left[\pi_c^{(\gamma_c - 1)/\gamma_c} - 1 \right]$$

$$\pi_t = \frac{1}{\left[1 - (C_{P_c}/C_{P_h}) (T_a/T_{04}) \left(\pi_c^{(\gamma_c - 1)/\gamma_c} - 1 \right) \right]^{(\gamma_t/\gamma_t - 1)}}. \tag{c}$$

5. *Nozzle:*   For an unchoked nozzle and ideal expansion, the jet speed is

$$V_j = \sqrt{2 C_{P_h} T_{06} \left[1 - \left(\frac{P_7}{P_{06}} \right)^{(\gamma_h - 1)/\gamma_h} \right]}.$$

From (a) and (b), then

$$T_{05} = T_{04} - \frac{C_{P_c}}{C_{P_h}} T_a \left[\pi_c^{(\gamma_c - 1)/\gamma_c} - 1 \right] \equiv T_{06}.$$

Since

$$\frac{P_7}{P_{06}} = \frac{P_7}{P_{02}} \frac{P_{02}}{P_{03}} \frac{P_{03}}{P_{04}} \frac{P_{04}}{P_{05}} \frac{P_{05}}{P_{06}}$$

Then with $P_6 = p_a = P_{02}$, $P_{03} = P_{04}$ and $P_{05} = P_{06}$

$$\therefore \frac{P_7}{P_{06}} = \frac{\pi_t}{\pi_c}$$

$$V_j = \sqrt{2 C_{P_h} \left[T_{04} - \frac{C_{P_c}}{C_{P_h}} T_a \left(\pi_c^{(\gamma_c - 1)/\gamma_c} - 1 \right) \right] \left\{ 1 - \left(\frac{\pi_t}{\pi_c} \right)^{(\gamma_h - 1)/\gamma_h} \right\}} \tag{d}$$

Example 2   The airplane *FIAT G91Y* is a single-seat Strike and Reconnaissance fighter powered by two *GENERAL ELECTRIC J85-GT-13A* turbojets each rated at 12.12 KN at an altitude of 9150 m, where the ambient conditions are 32 kPa and 240 K. The pressure ratio across the compressor is 12 and temperature at the turbine inlet is 1400 K. The aircraft speed is 310 m/s. Assume ideal operation for all components; assume unchoked nozzle and constant specific heat in all processes, $C_p = 1005\,\text{J/kgK}$. The heating value of the fuel is 42,700 kJ/kg.

Determine

(a)   The fuel-to-air ratio
(b)   The velocity of the exhaust gases
(c)   The air mass flow rate
(d)   The propulsive efficiency
(e)   The thermal efficiency
(f)   The overall efficiency

Solution: The given data are as follows:
Thrust $T = 12.12\,\text{kN}$
The flight altitude is 9150 m, at which the ambient pressure and temperature are

$$P_a = 32\,\text{kPa} \quad \text{and} \quad T_a = 240\,\text{K}$$

The compressor pressure ratio $\pi_c = 12$
The TIT and also the maximum temperature in the engine is

$$T_{max} = 1400\,\text{K}$$

Fuel heating value $Q_R = 42{,}700\,\text{kJ/kg}$
Flight speed $V_f = 310\,\text{m/s}$
Specific heat $C_p = 1005\,\text{J/kg} \cdot \text{K}$

Diffuser: Since all the processes are ideal, then the total temperature is

$$T_{02} = T_{0a} = T_a + \frac{V_f^2}{2C_p} = 240 + \frac{(310)^2}{2 \times 1005} = 287.8\,\text{K}$$

The total pressure is calculated from the relation

$$\frac{P_{02}}{P_a} = \left(\frac{T_{02}}{T_a}\right)^{\gamma/(\gamma-1)} = \left(\frac{287.8}{240}\right)^{3.5}$$

$$P_{02} = 60.4\,\text{kPa}$$

Compressor: The compressor outlet pressure is obtained from the relation

$$P_{03} = \pi_c \times P_{02} = 725.1\,\text{kPa} \equiv P_{04}$$

The temperature at the outlet from the compressor is obtained from the relation

$$\frac{T_{03}}{T_{02}} = \pi_c^{(\gamma-1)/\gamma} = 12^{0.286} = 2.035$$

$$T_{03} = T_{02} \times 2.035 = 587.5\,\text{K}$$

Combustion chamber: From Equation 4.5, the fuel-to-air ratio is

$$f = \frac{C_P(T_{04} - T_{03})}{Q_R - C_P T_{04}} = \frac{1.005(1400 - 586)}{42700 - 1.005 \times 1400} = 0.0198$$

Assuming equal work in both the compressor and turbine, and equal specific heats, then from Equation 4.6,

$$T_{05} = T_{04} - \frac{T_{03} - T_{02}}{1 + f} = 1106.1 \quad K = T_{06}$$

The pressure at the outlet of the turbine is obtained as follows:

$$\frac{P_{04}}{P_{05}} = \left(\frac{T_{04}}{T_{05}}\right)^{\gamma/(\gamma-1)} = \left(\frac{1400}{1106.1}\right)^{3.5} = 2.281$$

$$P_{05} = 317.86\,\text{kPa} = P_{06}.$$

Nozzle: For an isentropic flow in the nozzle, the temperature ratio within the nozzle is

$$\frac{T_{06}}{T_7} = \left(\frac{P_{06}}{P_7}\right)^{(\gamma-1)/\gamma} = \left(\frac{317.86}{32}\right)^{0.286} = 1.9283$$

$$T_7 = 563.6\,\text{K}$$

The exhaust speed is then

$$V_j = \sqrt{2C_P(T_{06} - T_7)} = 1044.34\,\text{m/s}$$

Since the thrust force for unchoked nozzle is expressed as $T = \dot{m}_a[(1 + f)V_j - V_f]$ the air mass flow rate is

$$\dot{m}_a = \frac{T}{(1 + f)V_j - V_f} = \frac{12.12 \times 1000}{1.0198 \times 1044.2 - 310} = \frac{12120}{754.9}$$

$$\dot{m}_a = 16.06\,\text{kg/s}$$

The fuel mass flow rate is $\dot{m}_a = f \times \dot{m}_a = 0.317\,\text{kg/s}$.

The propulsive efficiency is calculated using two of the expressions discussed in Chapter 2 as follows:

(a) *Expression 1*

$$\eta_p = \frac{T \times V}{T \times V + 0.5\dot{m}_j \left(V_j - V\right)^2} = 45.97\%$$

(b) *Expression 2*

$$\eta_p = \frac{2V}{V + V_j} = 45.79\%$$

As shown, the results are very close as the nozzle is unchoked. The difference is due to the fuel addition in the jet mass flow in Expression 1.

The thermal efficiency

$$\eta_{th} = \frac{T \times V + 0.5\dot{m}_j \left(V_j - V\right)^2}{\dot{m}_f Q_R} = 60.36\%$$

The overall efficiency

$$\eta_o = \eta_p \times \eta_{th} = 0.4597 \times 0.6036 = 0.2775 = 27.75\%$$

Example 3   A single-spool turbojet engine has the following data:

Flying Mach number	0.6
Ambient conditions	$T_a = 222\,K,\ P_a = 25kPa$
Compressor pressure ratio	6
Maximum temperature	1300 K
Fuel heating value	45,000 kJ/kg
Exhaust nozzle area	0.2 m^2

All the processes are assumed ideal and no pressure is lost either in the combustion chamber or in the inoperative afterburner.

1. Deduce that the nozzle is choked.
2. Calculate the thrust force.
3. Calculate the thrust force if the nozzle is unchoked.

Solution: The given data are rearranged as follows:

$$M = 0.6,\ T_a = 222\,K,\ P_a = 25\,kPa,\ \pi_c = 6$$

$$T_{o\,max} = 1300\,K,\ Q_R = 45,000\,kJ/kg,\ A_e = 0.2\,m^2$$

The flight speed

$$V = M\sqrt{\gamma R T_a} = 0.6\sqrt{1.4 \times 287 \times 222} = 179.2\,m/s$$

Diffuser

$$T_{02} = T_a \left(1 + \frac{\gamma - 1}{2}M^2\right) = 238 \text{ K}$$

$$P_{02} = P_a \left(1 + \frac{\gamma - 1}{2}M^2\right)^{\gamma/(\gamma-1)} = 31.88 \text{ kPa}$$

Compressor

$$T_{03} = T_{02}\pi_c^{(\gamma-1)/\gamma} = 397.3 \text{ K}$$

$$P_{03} = \pi_c P_{02} = 191.33 \text{ kPa}$$

Combustion chamber

$$T_{04} = 1300 \text{ K}$$

$$P_{04} = P_{03} = 191.33 \text{ kPa}$$

$$f = \frac{Cp(T_{04} - T_{03})}{Q_R - CpT_{04}} = \frac{1.01(1300 - 397.3)}{45000 - 1.01 \times 1300} = 0.02087$$

Turbine: From Equation 4.6, with constant specific heats and $\lambda = 0.8$, then

$$T_{05} = T_{04} - \frac{(T_{03} - T_{02})}{0.8(1+f)} = 1105 \text{ K}$$

$$P_{05} = P_{04}\left(\frac{T_{05}}{T_{04}}\right)^{\gamma/(\gamma-1)} = 182.63 \text{ kPa}$$

Tail pipe: Ideal flow is assumed in the tail pipe. Thus there is no pressure or temperature drop, or

$$T_{06} = T_{05} \text{ and } P_{06} = P_{05}$$

Nozzle
Case 1 Check if the nozzle is choked.

$$P_c = \frac{P_{06}}{((\gamma + 1)/2)^{\gamma/(\gamma-1)}} = 96.48 \text{ kPa}$$

Since $P_a = 25$ kPa, then $P_c > P_a$, thus the nozzle is choked.

$$\therefore P_7 = P_C = 96.48 \text{ kPa}$$

$$\therefore T_7 = T_c = \frac{T_{06}}{(\gamma + 1)/2} = 920.8 \text{ K}$$

$$V_7 = \text{sonic speed} = \sqrt{\gamma RT_7} = 608.26 \text{ K}$$

The mass flow rate is calculated as follows:

$$\dot{m} = \rho_7 V_7 A_7 = \frac{P_7}{RT_7} V_7 A_7 = \frac{96.48 \times 10^3}{287 \times 920.8} \times 608.26 \times 0.2 = 44.41 \, \text{kg/s}$$

The thrust force T is expressed by the relation

$$T = \dot{m}_a[(1+f)V_7 - V] + A_e(P_e - P_a)$$
$$= 44.41 \times (1.02087 \times 608.26 - 179.2) + 0.2 \times 10^3 \times (96.48 - 25)$$
$$= 33914 \, N = 33.914 \, \text{kN}$$

Case 2   The nozzle is unchoked, then

$$P_7 = P_a = 25 \, \text{kPa}$$

$$\therefore T_7 = T_{05} \left(\frac{P_7}{P_{05}} \right)^{(\gamma-1/\gamma)} = 625.7 \, K$$

The exhaust speed is calculated as follows: $V_7 = \sqrt{2Cp(T_{05} - T_7)} = 983.96 \, \text{m/s}$

This means that the exhaust Mach number, $M_7 = \dfrac{V_7}{\sqrt{\gamma RT_7}} = 1.962$

The thrust force is calculated as follows:

$$T = \dot{m}_a[(1+f)V_7 - V] = 44.41 \times (1.02087 \times 983.96 - 179.2) = 36651 \, N$$
$$= 36.65 \, \text{kN}$$

Thus, the thrust force for choked nozzle is less than its value for unchoked nozzle.

Example 4   Compare between the thrust-specific fuel consumption (TSFC) of a ramjet engine and an afterburning turbojet engine, when both have flight Mach number of 2 at 45,000 ft altitude, where $T_a = -69.7°C, P_a = 14.75 \, \text{kPa}$.

Both engines have the same maximum temperature of 2225 K and same exhaust speed. For the turbojet engine the compressor pressure ratio is 8 and the turbine outlet temperature is 700 K.

Conventional hydrocarbon fuel is used with $Q_R = 43,000 \, \text{kJ/kg}$. For simplicity, assume constant properties; or $\gamma = 1.4$, $C_P = 1.005 \, \text{kJ/kg/k}$, no aerodynamic loss in both engines, all the processes are ideal, and no pressure losses in either the combustion chamber or afterburner. Assume also complete expansion of the gases in the nozzles to the ambient pressure.

Next, for the turbojet engine, what will be the ratio of the power consumed in the compressor to that available in the turbine?

Solution:

1. Ramjet engine

The gases in the nozzle expand to the ambient condition. With

$$T_a = -69.7 + 273.15 = 203.5 \, K$$
$$P_a = 14.75 \, \text{kPa} \quad \text{and} \quad M = 2$$

The flight speed is $u_a = M\sqrt{\gamma R T_a} = 2\sqrt{(1.4)(287)(203.5)}$

$$\therefore u_a = 571.8\,\text{m/s}$$

$$\frac{T_{0a}}{T_a} = 1 + \frac{\gamma - 1}{2}M^2 = 1 + 0.2 \times 4 = 1.8$$

$$\therefore T_{0a} = 366.3\,\text{K}$$

$$\frac{P_{0a}}{P_a} = \left(1 + \frac{\gamma - 1}{2}M^2\right)^{\gamma/(\gamma-1)} = (1.8)^{3.5} = 7.824$$

$$P_{0a} = 115\,\text{kPa}$$

Diffuser: For ideal diffuser, then

$$T_{02} = T_{0a} = 366.3\,\text{K}$$
$$P_{0a} = P_{02} = P_{04} = 115\,\text{kPa}$$

Combustion chamber: The fuel-to-air ratio is

$$f = \frac{C_P(T_{04} - T_{0a})}{Q_R - C_P T_{04}}$$
$$f = 0.04582$$

Nozzle: Since all the processes are ideal, then the velocity ratio is obtained from Equation 3.18b as

$$\frac{u_e}{u_a} = \sqrt{\frac{T_{04}}{T_{0a}}} = \sqrt{\frac{2225}{366.3}} = 2.465$$

$$\therefore u_e = 1410\,\text{m/s}$$

The specific thrust is

$$\therefore \frac{T}{\dot{m}_a} = u_a\left[(1+f)\frac{u_e}{u_a} - 1\right] = 572[1.04582 \times 2.465 - 1]$$

$$\frac{T}{\dot{m}_a} = 902.6\,\frac{\text{N}}{(\text{kg/s})}$$

$$\text{TSFC} = \frac{f}{(T/\dot{m}_a)} = 0.1828\,\frac{\text{kg}}{\text{N}\cdot\text{h}}$$

2. Turbojet

An ideal cycle is assumed and the nozzle is unchoked. The outlet conditions of the diffuser are the same as in the ramjet case,

$$\therefore T_{02} = 366.3\,\text{K} \quad \text{and} \quad P_{02} = 115\,\text{kPa}$$

Moreover, there is no pressure drop in the combustion chamber, thus $P_{04} = P_{03}$.

There is no pressure drop in the afterburner also, thus $P_{05} = P_{06A}$.

Nozzle: Since the nozzle is unchoked, and the exhaust speed is the same as in the case of ramjet, then

$$T_{06A} = T_7 + \frac{V_7^2}{2C_P}$$

$$T_7 = T_{06A} - \frac{V_7^2}{2C_P} = 2225 - \frac{(1410)^2}{2 \times 1005}$$

$$T_7 = 1236 \, \text{K}$$

$$\frac{P_{06A}}{P_7} \equiv \frac{P_{06A}}{P_a} = \left(\frac{T_{06A}}{T_7}\right)^{\gamma/(\gamma-1)} = \left(\frac{2225}{1236}\right)^{1.4/0.4} = 7.829$$

$$P_{06A} = P_{05} = 115.5 \, \text{kPa}$$

Compressor

$$\frac{P_{03}}{P_{02}} = \pi_c = 8$$

$$\therefore P_{03} = 920 \, \text{kPa} \equiv P_{04}$$

$$\frac{T_{03}}{T_{02}} = (8)^{(\gamma-1)/\gamma} = 8^{(0.4/1.4)} = 8^{0.286} = 1.811$$

$$T_{03} = 663 \, \text{K}$$

Turbine

$$\frac{T_{04}}{T_{05}} = \left(\frac{P_{04}}{P_{05}}\right)^{(\gamma-1)/\gamma} = \left(\frac{P_{03}}{P_{06A}}\right)^{0.4/1.4}$$

$$= \left(\frac{920}{115.5}\right)^{0.286} = 1.8103$$

$$\because T_{05} = 700 \, \text{K}$$

$$\therefore T_{04} = 1267.2 \, \text{K}$$

Combustion chamber

$$f = \frac{C_P(T_{04} - T_{03})}{Q_R - C_P T_{04}} = \frac{1.005 \times (1267.2 - 663)}{43,000 - 1.005 \times 1267.2}$$

$$f = 0.0146$$

To determine the fuel-to-air ratio of afterburner f_{ab}, consider the energy balance of afterburner.

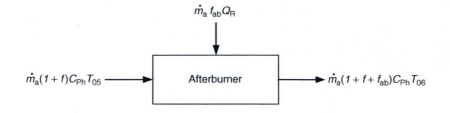

From Equation 4.9, the afterburner fuel-to-air ratio is

$$f_{ab} = \frac{(1+f)c_P(T_{06} - T_{05})}{Q_R - c_P T_{06}} = \frac{(1.0171)(1.005)(2225 - 700)}{43,000 - 1.005 \times 2225}$$

$$f_{ab} = 0.03815$$

The total fuel-to-air ratio of the turbojet engine is $f_{total} = f + f_{ab}$

$$f_{total} = 0.05275$$

$$\frac{T}{\dot{m}_a} = (1 + f_{tot})u_e - u = 1.05275 \times 1410 - 572$$

$$\frac{T}{\dot{m}_a} = 912.4 \frac{N}{(kg/s)}$$

$$\text{TSFC} = 0.20813 \frac{kg}{N \cdot h}$$

SUMMARY

	Ramjet	Turbojet
$T/\dot{m}_a(N \cdot s/kg)$	902.6	912.4
TSFC $\left(\dfrac{kg}{N \cdot h}\right)$	0.1828	0.20813

The afterburning turbojet engine produces more thrust at the expense of more fuel consumption compared to the ramjet.

The ratio between the power consumed in driving the compressor out of the total turbine power [i.e., (λ)] is to be calculated from the relation

$$Cp(T_{03} - T_{02}) = \lambda Cp(T_{04} - T_{05})$$

$$\lambda = \frac{663 - 336}{1267.2 - 700} = 0.57652$$

4.2.4 Actual Case

The temperature entropy diagram for the real case is shown in Figure 4.6 for inoperative afterburner and Figure 4.7 for operative afterburner.

1. All components are irreversible but they are adiabatic (except burners); thus, isentropic efficiencies for the intake, compressor, turbine, and nozzle are employed.

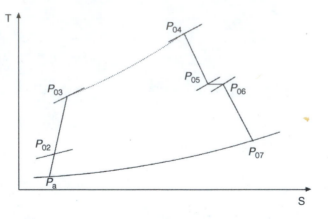

FIGURE 4.6   T–S diagram for real cycle of single-spool inoperative afterburner.

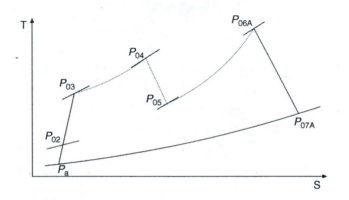

FIGURE 4.7   T–S diagram for real cycle of single-spool operative afterburner.

2. Friction in the air intake (or diffuser) reduces the total pressure from its free stream value and increases its entropy. The total temperature at the outlet of the intake is higher than in the isentropic case, which depends on the intake or diffuser efficiency (η_d).

3. The compression of the air in the compressor is accompanied by losses due to friction, turbulence, separation, shocks, and so on. Consequently, the entropy of the air also increases during its flow through the compressor. Moreover, at the outlet of the compressor, the temperature is higher than the corresponding isentropic temperature. Such increase in temperature depends on the compressor efficiency (η_c).

4. A portion of the compressed air is utilized in cooling the turbine discs, blades, and the supporting bearings through an air bleed system. Thus, the air mass flow rate in the succeeding modules is somewhat smaller than that entering the compressor.

5. The burners are not simple heaters and the chemical composition of the working fluid changes during combustion process. The larger the fuel-to-air ratio (f) the greater the deviation in the chemical composition of the products of combustion from that of the air. Losses in the combustion process are encountered owing to many factors including imperfect combustion, physical characteristics of the fuel, as well as thermal losses due to conduction and radiation. Such losses are handled by introducing the burner efficiency (η_b). Pressure drop due to skin friction and pressure drag in the combustors (normally 3–6% of the total pressure of the entering air) must also be taken into account.

6. The expansion process in turbine is very nearly adiabatic. However, because of friction an increase in the entropy is encountered. Moreover, the outlet temperature is higher

than that of the isentropic case. Thus, the available power from the turbine is less than that in the isentropic case. The expansion process is associated with the turbine efficiency (η_t).

7. Finally, the expansion process in the nozzle is similar to that in the turbine and influenced by skin friction. It is also governed by the adiabatic efficiency (η_n).

8. It is worth mentioning here that air/gas velocities within the gas generator are ignored. The velocities at the inlet to intake and outlet of nozzle are only calculated.

The engine performance is estimated by defining the adiabatic efficiencies as follows:

Diffuser efficiency: $0.7 < \eta_d < 0.9$ (depending strongly on flight Mach number)
Compressor efficiency: $0.85 < \eta_c < 0.90$
Burner efficiency: $0.97 < \eta_b \leq 0.99$ same value applies for the efficiency of afterburner
Turbine efficiency: $0.90 < \eta_t < 0.95$
Nozzle efficiency: $0.95 < \eta_n < 0.98$

The different processes through the engine modules are described again here.

1. *Intake:*   The pressure ratio within the inlet may be given. The outlet pressure is obtained from the relation:

$$r_d = \frac{P_{02}}{P_{0a}} \tag{4.16a}$$

Alternatively, the efficiency (η_d) of the inlet (sometimes denoted as intake or diffuser) is given. The outlet pressure will be given by

$$P_{02} = P_a \left(1 + \eta_d \frac{\gamma_c - 1}{2} M_a^2 \right)^{\gamma_c/(\gamma_c - 1)} \tag{4.16b}$$

$$T_{02} = T_{0a} = T_a \left(1 + \frac{\gamma_c - 1}{2} M_a^2 \right) \tag{4.2}$$

Outside the engine, the total pressure remains constant, thus

$$P_{01} = P_{0a} = P_a \left(1 + \frac{\gamma - 1}{2} M^2 \right)^{\gamma/(\gamma - 1)}$$

The outlet temperature of the diffuser is independent from the losses and so will be calculated from Equation 4.2 as in the ideal case.

2. *Compressor:*   From state (2) to (3) an irreversible adiabatic compression process takes place, which is associated with the isentropic efficiency of the compressor, η_c. The outlet conditions are now

$$P_{03} = (P_{02})(\pi_c) \tag{4.17}$$

$$T_{03} = T_{02} \left[1 + \frac{\pi_c^{(\gamma_c - 1)/\gamma_c} - 1}{\eta_c} \right] \tag{4.18}$$

3. *Combustion chamber:* The stagnation pressure at the outlet of combustion chamber, state (4) is less than its value at inlet, state (3), because of fluid friction. The pressure drop is either given as a definite value or as a percentage. Thus, the outlet pressure from the combustion chamber is expressed either as

$$P_{04} = P_{03} - \Delta P_{cc} \qquad (4.19a)$$

or

$$P_{04} = P_{03}(1 - \Delta P_{cc}\%) \qquad (4.19b)$$

The outlet temperature of the combustion chamber is defined as high as the turbine material limitations will allow. The fuel-to-air ratio is now calculated taking into consideration the efficiency of burners η_b from the relation:

$$f = \frac{Cp_h T_{04} - Cp_c T_{03}}{\eta_b Q_R - Cp_h T_{04}} \qquad (4.20)$$

4. *Turbine:* From states (4) to (5) the fluid expands through the turbine, providing shaft power input to the compressor plus any mechanical losses or accessory power. Equation 4.6 is also valid here and the outlet temperature to the turbine T_{05} is calculated from this equation. However, the outlet pressure is calculated considering the adiabatic efficiency of the turbine (η_t), thus

$$\frac{P_{05}}{P_{04}} = \left[1 - \frac{1}{\eta_t}\left(1 - \frac{T_{05}}{T_{04}}\right)\right]^{\gamma_h/(\gamma_h-1)} \qquad (4.21a)$$

Other relations (similar to those of the ideal case) for evaluating the pressure ratio are

$$\left(\frac{P_{05}}{P_{04}}\right) = \left\{1 - \frac{(Cp_c/Cp_h)T_{02}}{\lambda(1+f)\eta_c\eta_t T_{04}}\left[\left(\frac{P_{03}}{P_{02}}\right)^{(\gamma_c-1)/\gamma_c} - 1\right]\right\}^{\gamma_h/(\gamma_h-1)} . \qquad (4.21b)$$

From Equation 4.2

$$\left(\frac{P_{05}}{P_{04}}\right) = \left\{1 - \frac{(Cp_c/Cp_h)T_a}{\lambda(1+f)\eta_c\eta_t T_{04}}\left(1 + \frac{\gamma-1}{2}M^2\right)\left[\left(\frac{P_{03}}{P_{02}}\right)^{(\gamma_c-1)/\gamma_c} - 1\right]\right\}^{\gamma_h/(\gamma_h-1)} \qquad (4.21c)$$

5. *Afterburner:* State (6) depends on the geometry of the engine. The pressure at the outlet of the afterburner will be less than its value at the inlet, whether the afterburner is operative or inoperative.

(a) *Inoperative afterburner*

A treatment similar to the combustion chamber is considered. Thus, based on the value of the pressure drop within the afterburner due to the skin friction and the drag from the flame holders,

$$P_{06} = P_{05} - \Delta P_{ab} \qquad (4.22a)$$

or

$$P_{06} = P_{05}(1 - \Delta P_{ab}\%) \tag{4.22b}$$

No further fuel is burnt as the afterburner is inoperative.
The temperature is constant in the afterburner duct:

$$T_{06} = T_{05}$$

(b) *Operative afterburner*
Additional amount of fuel is burnt in the afterburner, which leads to a steep rise in temperature to the maximum temperature in the cycle ($T_{06A} = T_{MAX}$). The pressure drop is obtained using Equations 4.22a and b but P_{06A} replaces P_{06}. The afterburner fuel-to-air ratio is calculated from the relation

$$f_{ab} = \frac{(1+f)(Cp_{6A}T_{06A} - Cp_5 T_{05})}{\eta_{ab} Q_R - Cp_{6A} T_{06A}} \tag{4.23}$$

6. *Nozzle:* A check for nozzle choking is also performed first by calculating the critical pressure. The two cases of operative and inoperative are again considered.
   a. *Inoperative afterburner*
      The critical pressure is obtained from the relation

$$\frac{P_{06}}{P_c} = \frac{1}{[1 - (1/\eta_n)(\gamma_h - 1)/(\gamma_h + 1)]^{\gamma_h/(\gamma_h - 1)}} \tag{4.24}$$

Here η_n is the efficiency of the nozzle. If the nozzle is unchoked, then the outlet pressure is equal to the ambient pressure. The jet speed is now evaluated from the relation

$$V_7 = \sqrt{2Cp_h(T_{06} - T_7)} = \sqrt{2Cp_h \eta_n T_{06}\left[1 - \left(\frac{P_a}{P_{06}}\right)^{(\gamma_h - 1)/\gamma_h}\right]} \tag{4.25a}$$

or

$$V_7 = \sqrt{\frac{2\gamma_h \eta_n RT_{06}}{(\gamma_h - 1)}\left[1 - \left(\frac{P_a}{P_{06}}\right)^{(\gamma_h - 1)/\gamma_h}\right]} \tag{4.25b}$$

If the nozzle is choked, then the outlet temperature (T_7) is calculated from the relation

$$\left(\frac{T_{06}}{T_7}\right) = \left(\frac{\gamma_h + 1}{2}\right) \tag{4.11a}$$

The jet speed is calculated from relation

$$V_7 = \sqrt{\gamma_h RT_7} \tag{4.12a}$$

(b) *Operative afterburner*
The expansion process in the nozzle starts from state (06A) to state (7)

The critical pressure is calculated from the relation

$$\frac{P_{06A}}{P_c} = \frac{1}{[1 - (1/\eta_n)(\gamma_h - 1)/(\gamma_h + 1)]^{\gamma_h/(\gamma_h-1)}} \qquad (4.26)$$

If the nozzle is unchoked the jet speed will be

$$V_{7ab} = \sqrt{2C_{ph}\eta_n T_{06A}\left[1 - \left(\frac{P_a}{P_{06}}\right)^{(\gamma_h-1)/\gamma_h}\right]} \qquad (4.27a)$$

If the nozzle is choked, then

$$V_{7ab} = \sqrt{\gamma_h R T_{7A}} \qquad (4.27b)$$

A general relation for the exhaust speed (whether choked or unchoked) may be obtained from Equation 4.25 by replacing P_a by P_7, thus

$$V_{7ab} = \sqrt{2C_{ph}\eta_n T_{06A}\left[1 - \left(\frac{P_7}{P_{06A}}\right)^{(\gamma_h-1)/\gamma_h}\right]} \qquad (4.28)$$

The two engine parameters defining the performance of engine are the specific thrust and the specific fuel consumption.
The specific thrust is expressed by the relation

$$\frac{T}{\dot{m}_a} = [(1 + f + f_{ab})V_7 - V] + \frac{A_7}{\dot{m}_a}(P_7 - P_a) \qquad (4.29)$$

The TSFC is given by

$$TSFC = \frac{\dot{m}_f + \dot{m}_{fab}}{T} \qquad (4.30a)$$

Substituting from Equation 4.26 to get

$$TSFC = \frac{f + f_{ab}}{(1 + f + f_{ab})V_7 - V + (A_7/\dot{m})(P_7 - P_a)} \qquad (4.30b)$$

For inoperative afterburner, the same Equations 4.29 and 4.30 are used but the afterburner fuel-to-air ratio f_{ab} is set equal to zero.

Example 5   A single-spool turbojet engine has the following data:

Compressor pressure ratio	8
TIT	1200 K
Air mass flow rate	15 kg/s
Aircraft flight speed	260 m/s
Flight altitude	7000 m
Isentropic efficiency of intake	0.9
Isentropic efficiency of compressor and turbine	0.9

Combustion chamber pressure loss $= 6\%$ of delivery pressure
Combustion chamber efficiency   0.95
Isentropic efficiency of propelling nozzle   0.9

Calculate the propelling nozzle area, the net thrust developed, and the TSFC.
If the gases in the jet pipe are reheated to 2000 K where a pressure loss of 3% is encountered, calculate

1. The percentage increase in nozzle area
2. The percentage increase in net thrust

$$(T_a = 242.7 \text{ K}, P_a = 41.06 \text{ kPa}, Q_R = 43{,}000 \text{ kJ/kg})$$

Solution:

Case 1   Inoperative afterburner

Diffuser

$$T_{02} = T_a + \frac{V^2}{2\,Cp} = 242.7 + \frac{(260)^2}{2 \times 1005} = 276.3 \text{ K}$$

$$P_{02} = P_a \left[1 + \eta_d \left(\frac{T_{02} - T_a}{T_a} \right) \right]^{\gamma/(\gamma-1)} = 1.5089\,P_a = 61.95 \text{ kPa}$$

Compressor

$$P_{03} = \pi_c P_{02} = 8 \times 61.95 = 495.6 \text{ kPa}$$

$$T_{03} = T_{02} \left[1 + \frac{\pi_C^{(\gamma c-1)/\gamma c} - 1}{\eta_C} \right] = 525.7 \text{ K}$$

Combustion chamber

$$P_{04} = (1 - \Delta P_{cc})\,P_{03} = 0.94\,P_{03} = 475.8 \text{ kPa}$$

$$T_{04} = 1200 \text{ K}$$

$$f = \frac{Cp_t\,T_{04} - Cp_c T_{03}}{\eta_b\,Q_R - Cp_t\,T_{04}} = 0.023.$$

Turbine: Energy balance for compressor and turbine:

$$T_{05} = T_{04} - \frac{Cp_c\,(T_{03} - T_{02})}{Cp_t\,(1 + f)} = 1200 - \frac{1.005 \times (525.7 - 276.3)}{1.157 \times 1.023} = 988.2 \text{ K}$$

$$P_{05} = P_{04} \left[1 - \frac{T_{04} - T_{05}}{\eta_t\,T_{04}} \right]^{\gamma/(\gamma-1)} = 476 \left[1 - \frac{1200 - 988.2}{0.9 \times 1200} \right]^4 = 198.8 \text{ kPa}$$

Tail pipe: No losses is assumed if the afterburner is inoperative, then

$$T_{06} = T_{05}$$

$$P_{06} = P_{05}$$

Nozzle: Check whether the nozzle is choked or not.

$$\frac{P_{06}}{P_c} = \frac{1}{[1 - (1/\eta_n)(\gamma_n - 1)/(\gamma_n + 1)]^{\gamma_n/(\gamma_n-1)}} = 1.907$$

$$P_c = 104.2 \text{ kPa}$$

Since $P_c > P_a$, then the nozzle is choked.

$$\therefore P_7 = P_c = 104.2 \text{ kPa}$$

$$T_7 = \frac{T_{06}}{(\gamma_n + 1)/2} = 848.2 \text{ K}$$

$$V_7 = \sqrt{\gamma_n R T_7} = 569 \text{ m/s}$$

To get the nozzle area

$$\dot{m} = \dot{m}_a (1 + f) = 15 \times 1.023 = 15.345 \text{ kg/s}$$

Since $\dot{m} = \rho_7 V_7 A_7$ and $\rho_7 = \dfrac{P_7}{RT_7} = 0.428 \text{ kg/m}^3$

$$\therefore A_7 = \frac{\dot{m}}{\rho_7 V_7} = 0.063 \text{ m}^2$$

$$T = \dot{m}_a [(1 + f) V_7 - V] + A_7 (P_7 - P_a)$$

$$= 15 [1.023 \times 569 - 260] + 0.063 \times 10^3 (104.2 - 41.06)$$

$$= 8809 \text{ N}$$

$$\text{TSFC} = \frac{\dot{m}_f}{T} = \frac{f \dot{m}_a}{T} = \frac{0.023 \times 15}{8809} = 3.916 \times 10^{-5} \frac{\text{kg}}{\text{N} \cdot \text{s}}$$

Case 2   Operative afterburner

A pressure drop of 3% is encountered in the afterburner, thus

$$P_{06A} = 0.97 P_{05} = 192.836 \text{ kPa}$$

The maximum temperature is now $T_{06A} = 2000$ K

$$f_{ab} = \frac{Cp_t (T_{06A} - T_{05})}{\eta_{ab} Q_R - Cp_t T_{06A}} = \frac{1.157 \times (2000 - 988.2)}{0.9 \times 43,000 - 1.157 \times 2000} = 0.0322$$

Check also nozzle choking with

$$\frac{P_{06A}}{P_c} = \frac{1}{(1 - (1/\eta_n)(\gamma_n - 1)/(\gamma_n + 1)))^{\gamma_n/(\gamma_n-1)}} = 1.907$$

$$\therefore P_c = 101.12 \text{ kPa}$$

$\therefore P_c > P_a \therefore$ Nozzle is choked also.

$$P_{7A} = P_c = 101.12 \text{ kPa},$$

$$T_{7A} = \frac{T_{06A}}{(\gamma_n + 1)/2} = 1716.7 \text{ K}$$

$$V_{7A} = a_{7A} = \sqrt{\gamma_n R T_{7A}} = 809.5 \text{ m/s}$$

The nozzle area is calculated from the relation

$$A_{7A} = \frac{\dot{m}_a \, (1 + f + f_{ab})}{\rho_{7A} \, V_{7A}} = \frac{R \, T_{7A} \, \dot{m}_a \, (1 + f + f_{ab})}{P_{7A} \, V_{7A}}$$

$$= \frac{287 \times (1716.7) \times (15) \times (1 + 0.023 + 0.0322)}{101.12 \times 10^3 \times (809.5)}$$

$$A_{7A} = 0.0952 \, \text{m}^2$$

The thrust force for operative afterburner is

$$T_{ab} = \dot{m}_a \, [(1 + f + f_{ab}) \, V_{7A} - V] + A_{7A} \, (P_{7A} - P_a)$$

$$= 15 \, [1.0552 \times 809.5 - 260] + 0.0952 \times 10^3 \, (101.12 - 41.06) = 14,634.6 \, \text{N}$$

$$= 14.635 \, \text{kN}.$$

The percentage of thrust increase when the afterburner is operative is

$$\Delta T\% = \frac{T_{ab} - T}{T}\% = \frac{14.635 - 8.809}{8.809} = 66\%$$

The percentage increase in exhaust nozzle area is $\Delta A\%$.

$$\Delta A\% = \frac{A_{7A} - A_7}{A_7}\% = 51.1\%$$

Example 6 A single-spool turbojet engine is powering an aircraft flying at Mach number of 0.85 at altitude where the ambient conditions are $T_a = 233$ K and $P_a = 26.4$ kPa. The nozzle is choked and the gases are leaving the nozzle with velocity of 600 m/s; the nozzle area is 0.2 m^2. The maximum temperature is 1200 K and fuel heating value is 43,000 kJ/kg. It is required to calculate

(a) Compressor pressure ratio
(b) The fuel-to-air ratio
(c) Thrust force
(e) Overall efficiency

Solution:

(a) Compressor inlet temperature is

$$T_{02} = T_a \left(1 + \frac{\gamma - 1}{2} M^2\right) = 255 \, \text{K}$$

Since no losses are assumed, then states (5) and (6) are coincident, or

$$T_{06} = T_{05}, \quad P_{06} = P_{05}$$

Since the nozzle is choked

$$\frac{T_{05}}{T_C} = \frac{\gamma + 1}{2}$$

$$V_j = \sqrt{\gamma R T_C} = \sqrt{\frac{2\gamma}{\gamma + 1} R T_{05}}$$

$$T_{05} = \frac{\gamma + 1}{2\gamma R} V_j^2 = 1075 \text{ K}$$

The temperature drop in the turbine is

$$(T_{04} - T_{05}) = 1200 - 1075 = 125 \text{ K}$$

The fuel-to-air ratio is

$$f = \frac{C_{ph} T_{04} - C_{pc} T_{03}}{Q_R - C_{ph} T_{04}}$$

Since $W_c = W_t$

$$C_{pc}(T_{03} - T_{02}) = (1 + f) C_{ph}(T_{04} - T_{05})$$

$$\therefore (T_{03} - T_{02}) = \frac{C_{ph}}{C_{pc}} \left(1 + \frac{C_{ph} T_{04} - C_{pc} T_{03}}{Q_R - C_{ph} T_{04}}\right)(T_{04} - T_{05})$$

$$T_{03} = \frac{T_{02} + (C_{ph}/C_{pc})(1 + (C_{ph} T_{04}/(Q_R - C_{ph} T_{04}))(T_{04} - T_{05})}{1 + (C_{ph}(T_{04} - T_{05})/(Q_R - C_{ph} T_{04}))} = 398 \text{ K}$$

$$\pi_c = \left(\frac{T_{03}}{T_{02}}\right)^{\gamma/(\gamma-1)} = \left(\frac{398}{255}\right)^{3.5} = 4.75$$

(b) The fuel-to-air ratio is then

$$f = \frac{C_{ph} T_{04} - C_{pc} T_{03}}{Q_R - C_{ph} T_{04}} = 0.02374$$

(c) To calculate the thrust, the pressure at the exit of the engine must be defined first. It will be calculated from the ambient pressure as follows:

$$P_{02} = P_a \left(1 + \frac{\gamma - 1}{2} M^2\right)^{\gamma_c/(\gamma_c - 1)} = 42.34 \text{ kPa}$$

$$P_{03} = \pi_c P_{02} = 201.1 \text{ kPa}$$

$$P_{03} = P_{04}$$

$$P_{05} = P_{04} \left(\frac{T_{05}}{T_{04}}\right)^{\gamma_c/(\gamma_c-1)} = 129.5 \text{ kPa}$$

$$P_6 = P_C = \frac{P_{05}}{((\gamma_h + 1)/2)^{\gamma_h/(\gamma_h-1)}} = 70.31 \text{ kPa}$$

The mass flow rate is calculated from the outlet conditions

$$T_6 = T_C = \frac{T_{05}}{(\gamma_h + 1)/2} = 922.7 \text{ K}$$

$$\rho_6 = \rho_c = \frac{P_c}{RT_c} = 0.266 \text{ kg/m}^3$$

$$\dot{m}_e = \rho_6 V_6 A_6 = 31.86 \text{ kg/s}$$

$$\dot{m}_a = \frac{\dot{m}_e}{1+f} = 31.12 \text{ kg/s}$$

$$\dot{m}_f = \dot{m}_e - \dot{m}_a = 0.74 \text{ kg/s}$$

$$T = \dot{m}_e V_j - \dot{m}_a V_f + A_e (P_e - P_a) = 19.98 \text{ kN}$$

(d) The overall efficiency

$$\eta_o = \frac{TV_f}{\dot{m}_f Q_R} = 0.1597 = 15.97\%$$

4.2.5 COMPARISON OF OPERATIVE AND INOPERATIVE AFTERBURNERS

For an inoperative afterburner both the specific thrust $(T/\dot{m}_a)$ and the TSFC are dependent on

1. Compressor pressure ratio (π_c)
2. Flight Mach number
3. Maximum temperature or TIT (T_{04})

For an operative afterburner or in the case of augmented thrust, an additional parameter is considered, namely, the maximum temperature in the engine T_{06A}.

Now consider the cases of unchoked nozzle with and without thrust augmentation. If more precise analysis is assumed by considering the exact values for the specific thrust

$$\frac{T_{ab}}{\dot{m}_a} = [(1 + f + f_{ab}) \, V_{7ab} - V] + \frac{A_{7ab}}{\dot{m}_a} (P_{7ab} - P_a) \qquad (4.31a)$$

$$\frac{T}{\dot{m}_a} = [(1 + f) \, V_7 - V] + \frac{A_7}{\dot{m}_a} (P_7 - P_a) \qquad (4.31b)$$

where

$$V_{7ab} = \sqrt{2Cp_{n_{ab}}T_{06A}\eta_{n_{ab}}\left\{1 - \left(\frac{P_{7A}}{P_{06A}}\right)^{((\gamma-1)/\gamma)_{n_{ab}}}\right\}}$$

(4.32a)

$$V_7 = \sqrt{2Cp_n T_{06}\eta_n\left\{1 - \left(\frac{P_7}{P_{06}}\right)^{((\gamma-1)/\gamma)_n}\right\}}$$

(4.32b)

$$\upsilon = \frac{V}{V_7}$$

Consider the case of unchoked nozzle.

$$\frac{T_{ab}}{T} = \frac{(1+f+f_{ab})/(1+f))(V_{7ab}/V_7) - (1/(1+f))\upsilon}{1 - (\upsilon/(1+f))}$$

$$\frac{T_{ab}}{T} = \frac{\left\{\left(\frac{Cp_{n_{ab}}T_{06A}\eta_{n_{ab}}}{Cp_n T_{06}\eta_n}\left([1 - (P_{7A}/P_{06A})^{((\gamma-1)/\gamma)_{n_{ab}}}]/[1 - (P_7/P_{06})^{(\gamma-1)/\gamma)_n}]\right)\right)^{1/2}(1+f+f_a)/(1+f)\right\} - (\upsilon/(1+f))}{\{1 - (\upsilon/(1+f))\}}$$

(4.33)

Consider the following simplifying assumptions:

$$P_7 = P_{7A}, \quad P_{06} = P_{06A}, \quad \gamma_n = \gamma_{n_{ab}}$$
$$1 + f + f_a \approx 1 + f \approx 1$$

Then the ratio between the jet (exhaust) speeds for operative and inoperative afterburner is

$$\frac{V_{7ab}}{V_7} = \sqrt{\frac{T_{06A}}{T_{06}}}$$

(4.34)

The ratio between the thrust force for operative and inoperative afterburner is given as

$$\frac{T_{ab}}{T} = \frac{(T_{06A}/T_{06})^{1/2} - \upsilon}{1 - \upsilon}$$

(4.35)

Since $T_{06} = T_{05}$ the thrust ratio in fact is dependent on the temperature ratio between the temperature after and before the afterburner. When Equation 4.35 is plotted, as in Figure 4.8, the curves give an optimistic value of (T_{ab}/T) because of the assumptions made.

The curves show that

1. For any value of (T_{06A}/T_{06}), the increase in (υ) leads to an increase in (T_{ab}/T).
2. At any value of υ $(\equiv V/V_7)$, the increase of (T_{06A}/T_{06}) will lead to an increase in the augmented thrust ratio.
3. Since the afterburning engine can deliver large thrusts at larger values of υ $(\equiv V/V_7)$ than are practical for a normal engine, the afterburning engine can operate with larger values of υ_{ab} $(\equiv V/V_{7ab})$ and still develop larger values of thrust per unit area.

4. Since the propulsive efficiency for unchoked nozzle and negligible fuel burnt with respect to air mass flow rate is expressed as

$$\eta_p = \frac{2V}{V + V_{7ab}} = \frac{2}{1 + (1/v_{ab})}$$

then, as v_{ab} increases, the propulsive efficiency increases.

During takeoff ($V = 0$), Equation 4.35a is reduced to

$$\frac{T_{ab}}{T} = \left(\frac{T_{06A}}{T_{06}}\right)^{1/2} \equiv \frac{V_{7ab}}{V_7} \tag{4.36}$$

Equation 4.36 is plotted in Figure 4.9 to illustrate the effect of temperature ratio.

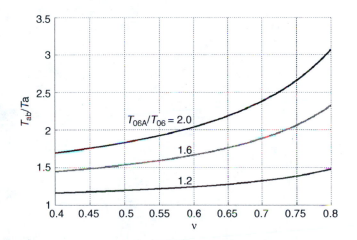

FIGURE 4.8 Ratio between thrust force for operative and inoperative afterburner.

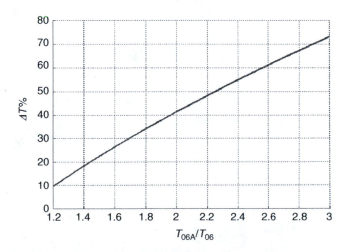

FIGURE 4.9 Percentage of thrust increase and temperature ratio.

Consider the following case for static operation with the following temperature values:

$$T_{06} = T_{05} = 959 \text{ K}$$

$$T_{06A} = 2000 \text{ K}$$

$$T_{04} = 1200 \text{ K}$$

$$T_{03} = 650 \text{ K}$$

From Equation 4.36, the increase in momentum thrust amounts to $\sqrt{\dfrac{2000}{959}} = 1.44$.

As an approximation the increase in fuel-to-air ratio is obtained from the ratio of temperature differences for both cases of operative and inoperative afterburner, or

$$\frac{(2000 - 959) + (1200 - 650)}{(1200 - 650)} = 2.89$$

An increase in thrust of 44% is obtained at the expense of 189% increase in fuel consumption, which represents the thrust augmentation during takeoff conditions. However, at high forward speed, the gain is much greater and is often more than 100%. This is due to the increase in momentum thrust for a fixed ram drag (momentum drag).

It is interesting also to prove that for choked nozzles, relation 4.34 is also correct. The exhaust temperature for inoperative afterburner (or no afterburner at all) is

$$T_7 = \left(\frac{2}{\gamma + 1}\right) T_{06} = \frac{(V_7^2)}{\gamma R}$$

For operative afterburner, the exhaust temperature is

$$T_{7A} = \left(\frac{2}{\gamma + 1}\right) T_{06A} = \frac{V_{7ab}^2}{\gamma R}$$

Again the ratio between the jet speeds for operative and inoperative afterburner is expressed as

$$\therefore \frac{V_{7ab}}{V_7} = \sqrt{\frac{T_{06A}}{T_{06}}}$$

However, the thrust ratio is no longer expressed by Equation 4.36 as the pressure thrust must be included.

The increase in jet speed and thrust force when the afterburner is lit provide improved takeoff and climb characteristics as shown in Figure 4.10. Thus for military applications, the time saved in takeoff and reaching the needed altitude is most important. Moreover, the short takeoff distance is needed if an airplane takes off from an air carrier. Operation of afterburner is accompanied by an increase in fuel consumption as extra fuel is burnt, which is calculated from Equation 4.23. However, it was found that the amount of fuel required to reduce the time taken to reach the operation height in Figure 4.10 is not excessive [6].

4.3 TWO-SPOOL ENGINE

In this section, the thermodynamic analysis of a two-spool turbojet engine is given. There are two types, namely, turbojet engine with no afterburner and turbojet engine with an afterburner (Figure 4.11). Mostly, the second type of turbojet engines are fitted to military airplanes. There is

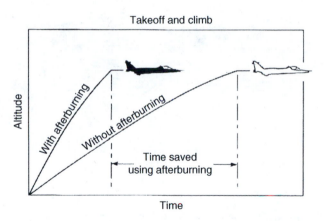

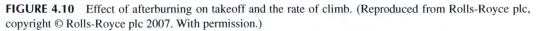

FIGURE 4.10 Effect of afterburning on takeoff and the rate of climb. (Reproduced from Rolls-Royce plc, copyright © Rolls-Royce plc 2007. With permission.)

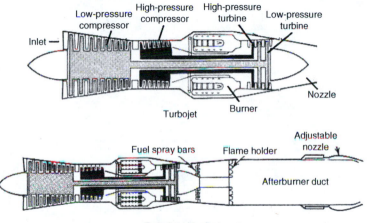

FIGURE 4.11 Layout of turbojet with and without afterburner.

only one engine of this type, namely, the Rolls-Royce/SNECMA Olympus 593 Mrk 610 that powers the commercial aircraft Concorde. The two-spool engines are composed of two compressors coupled to two turbines through two spools or shafts. As in the case of single spool with afterburner turbojet, the afterburning engines have two combustors, namely, the conventional combustion chamber and the afterburner installed downstream the LPT. Both variable area intake and nozzle are necessary here.

4.3.1 Nonafterburning Engine

4.3.1.1 Examples of Engines

GE J57-P-43 WB (Turbojet with water injection for thrust augmentation)

> Length: 169 in. (4300 mm)
> Diameter: 35 in. (900 mm)
> Dry weight: 3875 lb (1757 kg)

Two axial compressors of total 16 stages

> Overall pressure ratio: 12:1
> TIT: 1600°F (870°C)

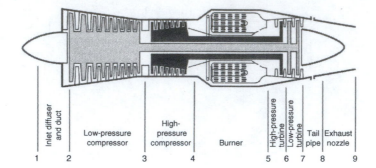

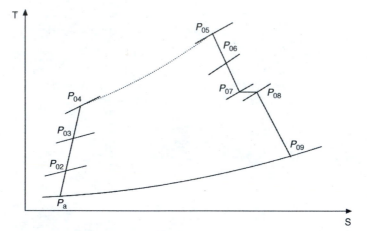

FIGURE 4.12   Stations of a turbojet engine.

FIGURE 4.13   Cycle for inoperative two-spool turbojet.

Thrust: 11,200 lbf (49.9 kN) with water–methanol injection
Specific fuel consumption: 0.755 lb/(h.lbf) [77.0 kg/(h.kN)] at maximum military power
Thrust-to-weight ratio: 2.9:1 (28.4 N/kg)

4.3.2.2 Thermodynamic Analysis

A simplified layout of turbojet engine is shown in Figure 4.12. Moreover, the cycle or the successive processes are shown in Figure 4.13. Real case will only be considered here, as the ideal performance can be considered as a special case where the efficiencies are set equal to unity with no pressure losses.

Now the successive elements will be examined.

1. *Intake:*   The inlet conditions of the air entering the inlet are the ambient pressure and temperature (P_a and T_a), respectively. The intake has an isentropic efficiency (η_d). For a flight Mach number (M_a), the temperature and pressure at the outlet of the intake are T_{02} and P_{02} given by the same relations as for the single spool in the following equations:

$$P_{02} = P_a \left(1 + \eta_d \frac{\gamma_c - 1}{2} M_a^2 \right)^{\gamma_c/(\gamma_c - 1)} \tag{4.16b}$$

$$T_{02} = T_{0a} = T_a \left(1 + \frac{\gamma_c - 1}{2} M_a^2 \right) \tag{4.2}$$

2. *Low-pressure compressor (LPC):*   For a known compressor pressure ratio (π_{c1}) and isentropic efficiency (η_{c1}), the temperature and pressure at the outlet of the LPC are given by the following relations:

$$P_{03} = (P_{02})\,(\pi_{c1}) \tag{4.37}$$

$$T_{03} = T_{02}\left[1 + \frac{\pi_{c1}^{(\gamma_c - 1)/\gamma_c} - 1}{\eta_{c1}}\right] \tag{4.38}$$

3. *High-pressure compressor (HPC):*   Similarly, both the pressure ratio (π_{c2}) and its isentropic efficiency (η_{c2}) are known. Thus, the temperature and pressure at the outlet of the HPC are given by the following relations:

$$P_{04} = (P_{03})\,(\pi_{c2}) \tag{4.39}$$

$$T_{04} = T_{03}\left[1 + \frac{\pi_{c2}^{(\gamma_c - 1)/\gamma_c} - 1}{\eta_{c2}}\right] \tag{4.40}$$

4. *Combustion chamber:*   The temperature at the end of combustion process T_{05} is generally known. It is the maximum temperature in the cycle if the afterburner is inoperative. The pressure at the end of the combustion depends on the pressure drop in the combustion process itself. Thus, depending on the known expression for the pressure drop in the combustion chamber it may be expressed as

$$P_{05} = P_{04} - \Delta P_{cc} \tag{4.41}$$

or

$$P_{05} = P_{04}\,(1 - \Delta P_{cc}\%) \tag{4.42}$$

The energy balance for the combustion chamber yields the following relation for the fuel-to-air ratio (f) defined as the ratio between the mass of fuel burnt to the air mass flow rate through the gas generator

$$\dot{m}_a(1 + f)Cp_h\,T_{05} = \dot{m}_a Cp_c T_{04} + \eta_b \dot{m}_f Q_R$$

with

$$f = \frac{\dot{m}_f}{\dot{m}_a}$$

$$f = \frac{(Cp_h / Cp_c)\,(T_{05}/T_{04}) - 1}{\eta_b\,(Q_R / Cp_c T_{04}) - (Cp_h / Cp_c)\,(T_{05}/T_{04})} \tag{4.43}$$

5. *High-pressure turbine (HPT):*   The power generated in the HPT is used in driving the HPC in addition to some of the accessories. If the ratio of the power needed to drive the HPC to the power available from the HPT is (λ_1) then, the energy balance for the compressor turbine shaft is

$$W_{HPC} = \lambda_1 W_{HPT}$$

Here (λ_1) is of the range 75%–85%. Thus, in terms of the temperature differences

$$Cp_c\,(T_{04} - T_{03}) = \lambda_1\,(1+f)\,Cp_h\,(T_{05} - T_{06})$$

$$\left(\frac{T_{06}}{T_{05}}\right) = 1 - \frac{(Cp_c/Cp_h)\,T_{03}}{\lambda_1\,(1+f)\,T_{05}}\left[\left(\frac{T_{04}}{T_{03}}\right) - 1\right] \tag{4.44}$$

Then the pressure ratios of the HPT and HPC are related by

$$\left(\frac{P_{06}}{P_{05}}\right) = \left\{1 - \frac{(Cp_c/Cp_h)\,T_{03}}{\lambda_1\,(1+f)\,\eta_{c2}\eta_{t2}T_{05}}\left[\left(\frac{P_{04}}{P_{03}}\right)^{(\gamma_c-1/\gamma_c)} - 1\right]\right\}^{\gamma_h/(\gamma_h-1)} \tag{4.45}$$

where η_{t2} is the isentropic efficiency of the HPT.

6. *Low-pressure turbine (LPT):*   The power consumed in the compression from (2) to (3) must be supplied through the LPT in expansion from (6) to (7). If the ratio of the power needed to drive the compressor to the power available in the turbine is (λ_2), then the energy balance for the low-pressure spool is

$$W_{\text{LPC}} = \lambda_2 W_{\text{LPT}}$$

Here (λ_2) is of the range 75%–85%. Thus, in terms of the temperature differences

$$Cp_c\,(T_{03} - T_{02}) = \lambda_2\,(1+f)\,Cp_h\,(T_{06} - T_{07})$$

$$\left(\frac{T_{07}}{T_{06}}\right) = 1 - \frac{(Cp_c/Cp_h)\,T_{02}}{\lambda_2\,(1+f)\,T_{06}}\left[\left(\frac{T_{03}}{T_{02}}\right) - 1\right] \tag{4.46}$$

Then the turbine and compressor pressure ratios are related by

$$\left(\frac{P_{07}}{P_{06}}\right) = \left\{1 - \frac{(Cp_c/Cp_h)\,T_{02}}{\lambda_2\,(1+f)\,\eta_{c1}\eta_{t1}T_{06}}\left[\left(\frac{P_{03}}{P_{02}}\right)^{(\gamma_c-1)/\gamma_c} - 1\right]\right\}^{\gamma_h/(\gamma_h-1)} \tag{4.47}$$

From the diffuser part (Equation 4.2)

$$\left(\frac{P_{07}}{P_{06}}\right) = \left\{1 - \frac{(Cp_c/Cp_h)\,T_a}{\lambda_2\,(1+f)\,\eta_{c1}\eta_{t1}T_{06}}\left(1 + \frac{\gamma-1}{2}M^2\right)\left[\left(\frac{P_{03}}{P_{02}}\right)^{(\gamma_c-1)/\gamma_c} - 1\right]\right\}^{\gamma_h/(\gamma_h-1)} \tag{4.48}$$

7. *Jet pipe:*   The jet pipe following the LPT and preceding the nozzle is associated with a slight pressure drop, while the total temperature remains unchanged.

$$P_{08} = P_{07} - \Delta P_{\text{jet pipe}} \tag{4.49a}$$

$$T_{08} = T_{07} \tag{4.49b}$$

8. *Nozzle:* First of all a check for nozzle choking is performed. Thus, for an isentropic efficiency of the nozzle, η_n, the critical pressure is obtained from the relation

$$\frac{P_{08}}{P_c} = \frac{1}{[1 - (1/\eta_n)((\gamma_h - 1)/(\gamma_h + 1))]^{\gamma_h/(\gamma_h - 1)}} \tag{4.50}$$

If the nozzle is unchoked, then the outlet pressure is equal to the ambient pressure. The jet speed is now evaluated from the relation

$$V_9 = \sqrt{2C_{p_h}\eta_n T_{08}\left[1 - \left(\frac{P_a}{P_{08}}\right)^{(\gamma_h - 1)/\gamma_h}\right]} \tag{4.51a}$$

If the nozzle is choked, then the outlet temperature (T_8) is calculated from the relation

$$\left(\frac{T_{08}}{T_9}\right) = \left(\frac{\gamma_h + 1}{2}\right)$$

The jet speed is expressed as

$$V_9 = \sqrt{\gamma_h R T_9} \tag{4.51b}$$

4.3.3 Afterburning Engine

Most military engines together with the Rolls-Royce SNECMA Olympus 593 Mrk 610 that powers the Concorde are fitted with an afterburner.

4.3.3.1 Examples of Two-Spool Afterburning Turbojet Engines

1. *Pratt & Whitney J57-P-23:*
   It powered the following military aircraft: Boeing B-52, Boeing C-135, Convair F-102 Delta Dagger, Convair XB-60, Douglas A3D Skywarrior, Lockheed U-2, McDonnell F-101, and North America F-100
   Length: 244 in. (6200 mm)
   Diameter: 39 in. (1000 mm)
   Dry weight: 5175 lb (2347 kg)
   Two-spool axial compressor of total 16 stages
   Overall pressure ratio: 11.5:1
   TIT: 1600°F (870°C)
   Thrust: 11,700 lbf (52.0 kN) dry
   17,200 lbf (76.5 kN) with afterburner
   Specific fuel consumption: 2.1 lb/(h · lbf) (214.2 kg/h · kN)

2. *General Electric J79-GE-17:*
   It powered the following military aircraft: Lockheed F-104, B-58 Hustler, and the McDonnell Douglas Phantom F-4
   Length: 17.4 ft (5.3 m)
   Diameter: 3.2 ft (1.0 m)
   Dry weight: 3850 lb (1750 kg)
   Two-spool axial compressor of total 17 stages with variable stator vanes

Overall pressure ratio: 13.5:1
TIT: 1210°F (655°C)
Thrust: 11,905 lbf (52.9 kN) dry
17,835 lbf (79.3 kN) with afterburner
Specific fuel consumption: 1.965 lb/ (h · lbf) (200 kg/h · kN) with afterburner

3. *Rolls-Royce/SNECMA Olympus 593 Mrk 610:*
   It powered the Concorde commercial airplane
   Dry weight: 7000 lb
   Air mass flow rate: 410 lbs/s
   Two-spool axial compressor of total 14 stages: 7 stages for both of the LPC and HPC
   Overall pressure ratio: 15.5:1
   Burner: Annular with 16 fuel nozzles
   Turbine: Single stage for each of the HPT and LPT
   Thrust: 32,000 lbf without afterburner at full power
   38,050 lbf with afterburner at full power
   Exhaust: Constant diameter afterburning jet pipe, convergent primary nozzle, variable
   area divergent secondary nozzle with thrust reverser
   Specific fuel consumption: 0.59 lb/(h · lbf) dry
   1.13 lb/(h · lbf) with afterburner
   Thrust/weight: 5.4:1

4. *4- Al-21:*
   It powered the following Russian military aircraft: MiG-19S and MiG-19PF
   Two spool having axial compressors
   Overall pressure ratio: 14.75:1
   TIT: 2000°F (1100°C)
   Thrust: 17,175 lbf (76.4 kN) dry
   24,675 lbf (109.8 kN) with afterburner
   Specific fuel consumption: 0.76 lb/(h · lbf) (77.5 kg/h · kN) at idle
   0.86 lb/(h · lbf) (87.7 kg/h · kN) at maximum military power
   1.86 lb/(h · lbf) (189.7 kg/h · kN) with afterburner
   Thrust/weight: 6.6:1

4.3.2.2 Thermodynamic Analysis

A simplified diagrammatic sketch for the engine and cycle is plotted on the T–S diagram, which is shown in Figures 4.14 and 4.15. The LPC is driven by the LPT while the HPC is driven by the HPT. Concerning the Olympus 593 engine, the HPT also drives the accessory gearbox via a tower shaft and bevel gear in front of the HPC.

The same treatment for all the modules upstream to the afterburner is applied here. Now, consider the afterburner. Owing to fuel addition in the afterburner, a steep rise in temperature to a value much greater than that obtained in the combustion chamber is visible. As expected, because of the fuel lines, igniters, and flame holder a pressure drop is encountered in the afterburner. Thus, the pressure at the outlet to the afterburner is determined from one of the following relations:

$$P_{08_A} = P_{07} - \Delta P_{ab} \tag{4.52a}$$

or

$$P_{08_A} = P_{07}(1 - \Delta P_{ab}\%) \tag{4.52b}$$

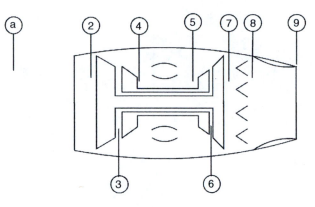

FIGURE 4.14   Stations for a two-spool turbojet engine with operative afterburner.

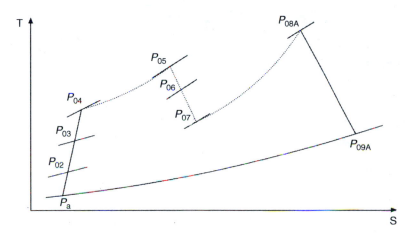

FIGURE 4.15   Cycle for a two-spool turbojet engine with operative afterburner.

The maximum temperature in the cycle is obtained at the end of the combustion process:

$$T_{08A} = T_{MAX}$$

The afterburner fuel-to-air ratio is calculated from the energy balance in the afterburner. The final relation is

$$f_{ab} = \frac{(1+f)\,(Cp_{8A}\,T_{08A} - Cp_7 T_{07})}{\eta_{ab} Q_R - Cp_{8A} T_{08A}} \tag{4.53}$$

Nozzle
The very hot gases leaving the afterburner expand in the nozzle from the state (08A) to state (9). As usual a check for nozzle choking is performed as follows:

$$\frac{P_{08A}}{P_c} = \frac{1}{\left[1 - (1/\eta_n)\left((\gamma_h - 1)/(\gamma_h + 1)\right)\right]^{\gamma_h/(\gamma_h - 1)}} \tag{4.54}$$

If the nozzle is unchoked the jet speed will be

$$V_{9ab} = \sqrt{2C_{ph}\eta_n T_{08A}\left[1 - \left(\frac{P_a}{P_{08}}\right)^{(\gamma_h - 1)/\gamma_h}\right]} \qquad (4.55a)$$

If the nozzle is choked, then the exhaust gases leave the nozzle with a temperature of T_{9A}, which is calculated from the relation

$$\left(\frac{T_{08}}{T_{9A}}\right) = \left(\frac{\gamma_h + 1}{2}\right)$$

The exhaust speed is then calculated from the relation

$$V_{9ab} = \sqrt{\gamma_h R T_{9A}} \qquad (4.55b)$$

The performance parameters are the specific thrust and the TSFC as well as the three efficiencies. The specific thrust is expressed by the relation

$$\frac{T}{\dot{m}_a} = [(1 + f + f_{ab}) V_9 - V] + \frac{A_9}{\dot{m}_a}(P_9 - P_a) \qquad (4.56)$$

The TSFC is given by

$$\text{TSFC} = \frac{\dot{m}_f + \dot{m}_{fab}}{T}$$

Substituting from Equation 4.56, the TSFC is

$$\text{TSFC} = \frac{f + f_{ab}}{(1 + f + f_{ab}) V_9 - V + (A_9/\dot{m}_a)(P_9 - P_a)} \qquad (4.57)$$

For inoperative afterburner, the same Equations 4.56 and 4.57 are used but the afterburner fuel-to-air ratio f_{ab} is set equal to zero.

Turbojet engines resemble a one stream flow engine, and then from Chapter 2, the efficiencies are calculated as follows. The propulsive efficiency is obtained from the relation

$$\eta_p = \frac{TV}{TV + (1/2)\dot{m}_e (V_e - V)^2}$$

where the mass and velocity of the gases leaving the nozzle are expressed in general as $\dot{m}_e$ and V_e.

The thermal efficiency is expressed as

$$\eta_{th} = \frac{TV + (1/2)\dot{m}_a (1 + f)(V_e - V)^2}{\dot{m}_f Q_R}$$

The overall efficiency is then

$$\eta_0 = \eta_p \times \eta_{th}$$

Example 7 A two-spool turbojet engine has the following data:

η_d	η_{c1} η_{c2}	η_{cc}	η_{t1} η_{t2}	η_{ab}	η_n	π_{c_1}	π_{c_2}	ΔP_{cc}	ΔP_{ab}	TIT (K)	T_{max} (K)	$A_i(m^2)$	$\dot{m}$ (kg/s)
0.9	0.88	0.95	0.94	0.95	0.96	3.35	4.65	0.03	0.05	1200	1500	0.9519	186

Fuel heating value is 45,000 kJ/kg.

A. It is required to calculate
   1. The temperature and pressure at the inlet and outlet of each module of the engine
   2. Fuel-to-air ratio for both the combustion chamber and afterburner
   3. The exit area for both operative and inoperative afterburner
   4. The specific thrust and the TSFC
   5. The propulsive, thermal, and overall efficiency
B. Recalculate as above but when the airplane powered by the above engine is flying at 8 km altitude with a Mach number 2.04. The maximum temperature at the afterburner is increased to 2000 K and the air mass flow rate is increased to 314.64 kg/s.

In both cases the nozzle is unchoked.

Solution: Using the previous equations, the results are arranged in the following table:

Properties	Sea Level	Altitude 8 km (26,264 ft)
T_a (K)	288.16	236.23
P_a (Pa)	1.013125×10^5	3.5651×10^4
M	0	2.04
V (m/s)	0	628.5893
T_{02}(K)	288.16	432.9772
P_{02}(Pa)	1.013125×10^5	4.862×10^4
P_{03}	3.3944×10^5	1.6288×10^5
T_{03}	442.8617	635.3745
P_{04}	1.5784×10^6	7.5739×10^5
T_{04}	686.9657	1032.2
P_{05}	1.531×10^6	7.3467×10^5
T_{05} (TIT)	1200	1200
P_{06}	6.0126×10^5	1.6603×10^5
T_{06}	964.9536	849.7425
P_{07}	3.4105×10^5	6.0315×10^4
T_{07}	854.0723	671.0998
P_{08}	3.2399×10^5	6.0315×10^4
T_{08} (T_{max})	Operative $A/B = 1500$ Inoperative $A/B = T_{07} = 854.0723$	Operative $A/B = 2000$ Inoperative $A/B = T_{07} = 671.0998$
$P_9 = P_a$	1.013125×10^5	3.5651×10^4
T_9	1136.9	1785.2
V_e (m/s)	913.1156	702.2306

Continued

Properties	Sea Level	Altitude 8 km (26,264 ft)
T (N)	Operative $A/B = 1.7582 \times 10^5$ Inoperative $A/B = 1.7266 \times 10^5$	Operative $A/B = 3.3388 \times 10^4$ Inoperative $A/B = 2.4987 \times 10^4$
η_p	0	0.9222
η_{th}	0.5977	0.1922
η_0	0	0.1802
f	0.0166	0.0082
f_{ab}	0.0186	0.038
A_e (m^2)	Operative $A/B = 0.4381$ Inoperative $A/B = 0.4032$	Operative $A/B = 1.0424$ Inoperative $A/B = 1.0045$

Continued

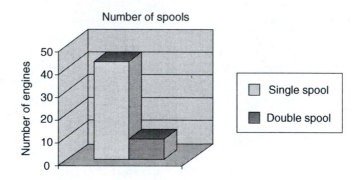

FIGURE 4.16   Survey for number of spools.

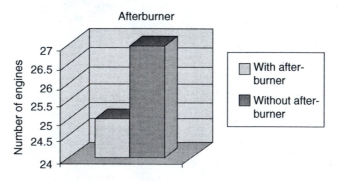

FIGURE 4.17   Survey for afterburning engines.

4.4 STATISTICAL ANALYSIS

A survey for grand sector of the turbojet engines outlined that

1. The single-spool engine is nearly 12% of the turbojet engine and the rest (88%) comprises two-spool engines (Figure 4.16).
2. The turbojet engines fitted with afterburner comprise 48% of the total turbojet engines (Figure 4.17).

4.5 THRUST AUGMENTATION

If the thrust of an engine has to be increased above the original design value, one of the following methods may be employed:

1. Increase of TIT, which will increase the specific thrust and hence the thrust for a given engine size
2. Increase of the mass flow rate through the engine without altering the cycle parameter

Both techniques imply some redesign of the engine and either/or both methods may be used to update the existing engine. For example, the inlet area may be increased, but this would be obtained at the expense of increased drag and weight. However, in some circumstances, there will be a requirement for temporary increase in thrust for a short time, for example, during takeoff, climb, and acceleration from subsonic to supersonic speed, or during combat maneuvers. For these circumstances, the thrust must be augmented for the fixed engine size. Numerous schemes have been proposed, but the two methods most widely used are

1. Water injection
2. Reheat or afterburning

4.5.1 WATER INJECTION

Water injection is an old aviation technology that was previously used to enable an increase in engine thrust or power during takeoff. The thrust force depends to a large extent on the mass, the density, or the airflow passing through the engine. There is, therefore, a reduction in thrust as the atmospheric pressure decreases with altitude and/or the ambient air temperature increases (on very hot days). Thus, it is frequently necessary to provide some means of thrust augmentation for nonafterburning turbojet or turbofan engines [7]. The thrust (or power for turboprop or turboshaft engines and stationary gas turbine) may be restored or even boosted as much as 10%–30% for takeoff by cooling the airflow with *water* or *water–methanol* (methyl alcohol) injection. The technique is to inject a finely atomized spray of water or W/M (a mixture of water and methanol) into either the inlet of compressor or into the combustion chamber inlet [7]. The compressor inlet water injection system is either all water or

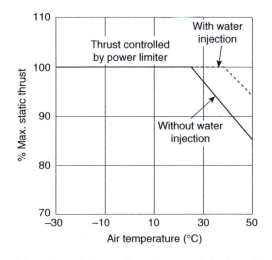

FIGURE 4.18 Thrust restoration of a turbojet engine using water injection. (Reproduced from Rolls-Royce plc, copyright © Rolls-Royce plc 2007. With permission.)

a mixture of water and methanol, but the combustion chamber water injection is always water and methanol. When methanol is added to water it gives antifreezing properties and also provides an additional source of fuel. The maximum thrust of an engine using water or W/M injection is called the "wet rating." Figure 4.18 illustrates a typical turbojet engine thrust restoration. Water has to be very pure (so-called demineralized) since any impurities may cause rapid buildup of hard deposits on vanes and rotor blades. It vaporizes rapidly causing intense cooling [8]. Turbojet engines usually employ water injection into the combustion chamber inlet, while in turboprop engines water or W/M is injected into the compressor inlet. Once injected into the compressor inlet, the water increases the air density and cools the turbine gas temperature, enabling extra fuel to be burned, which further adds to the power. Injection into the combustion chamber gives better distribution and makes higher water flow rates possible. It increases the mass flow through the turbine, relative to that through the compressor. The pressure and temperature drop across the turbine is reduced as depicted from the following compressor—turbine energy balance

$$\dot{m}_a C_{pc} (T_{03} - T_{02}) = (\dot{m}_a + \dot{m}_f + \dot{m}_{\text{waterinjection}}) C_{ph} (T_{04} - T_{05})$$

This leads to an increased jet pipe pressure, which in turn results in additional thrust. Injection of water only reduces the TIT, which enables an increase in fuel flow rate and an increase in the maximum rotational speed of the engine and thus provides additional thrust. When methanol is used with water, the TIT is restored partially or fully by the burning of the methanol in the combustion chamber [6]. As aircraft engines have matured and become capable of generating ever more thrust, water injection for new engines has been abandoned, although there are still a few aircraft in service that continue to use water injection. This is due to the extra weight, cost, and extra maintenance arising from the additional system of tanks, pumps, and control system [8]. Moreover, for turbojets, "wet" takeoffs mean excruciating noise and usually a lot of black smoke [8].

Although water injection was first used over 45 year ago on commercial transport aircraft Boeing's 707-120 with Pratt & Whitney JT3C-6 engines and later the Boeing 747-100 and 200 aircraft with Pratt & Whitney JT9D-3AW and -7AW engines for thrust augmentation, it has not been used on aircraft to reduce emissions. Recently, water injection was revisited again to evaluate the feasibility of its use in commercial aircraft, but this time for emission reduction [9,10]. Moreover, newer high bypass ratio engines experience higher thrust lapse rates with altitude, which results in higher core temperature. Again water injection systems are evaluated for possible use all the way to top of climb [11]. These recent studies are in favor of water injection in the combustion chamber inlet rather than injection into the inlet of the compressor. Moreover, it assured that water injection may reduce takeoff NOx emission more than 50%. When using water injection throughout climb, turbine life would further improve but the water weight penalties would present unacceptable payload and airplane performance penalties.

4.5.2 Afterburning

Afterburning (or reheat) is a method of augmenting the basic thrust of an engine to improve the takeoff, climb, and combat performance (for military aircraft). Although the increased thrust can be obtained by a larger engine, it would increase the weight, frontal area, and overall fuel consumption [6]. Afterburning provides the best method for thrust augmentation for short periods. Afterburning is another combustion chamber located between the LPT (or the turbine if only one turbine exists) and the jet pipe propelling nozzle. Fuel is burnt in this second combustion chamber utilizing the unburned oxygen in the exhaust gas. As described earlier, this will increase the temperature and velocity of the exhaust gases leaving the propelling nozzle and therefore increases the engine thrust.

The burners of the afterburner are arranged so that the flame is concentrated around the axis of the jet pipe. Thus, a portion of the discharge gases flow along the walls of the jet pipe and protect these walls from the afterburner flame, the temperature of which is in excess of 1700°C. The area of

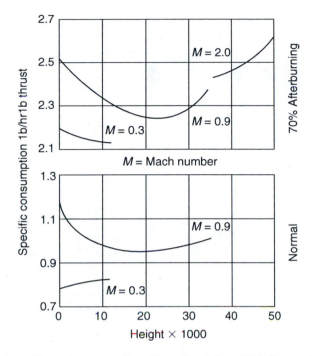

FIGURE 4.19   Thrust-specific fuel consumption. (Reproduced from Rolls-Royce plc, copyright © Rolls-Royce plc 2007. With permission.)

the afterburning jet is larger than that of a normal jet pipe would be for the same engine, to reduce the velocity of the gas stream. The afterburning jet pipe is fitted with either a two-position or a variable-area propelling nozzle.

The increase of thrust in afterburning engine is also accompanied by an increase in specific fuel consumption. Therefore, afterburning is limited to periods of short duration. Fuel is not burnt in the afterburner as efficiently as in the engine combustion chamber, since the pressure in the afterburner is not the peak pressure in the cycle. Figure 4.19 illustrates the TSFC for operative and inoperative afterburners at different flight Mach numbers.

4.5.3   Pressure Loss in Afterburning Engine

In the afterburner diffuser the velocity of the gases leaving the LPT (or turbine) in conventional turbojet engines ranges from 180 to 360 m/s, which is far too high for a stable flame to be maintained. Thus, the flow is diffused before it enters the afterburner combustion zone thereby decreasing the velocity and increasing the static pressure. A flame stabilizer is located downstream the fuel burner to provide a region in which turbulent eddies are formed to assist combustion. This reduction in speed also helps in reducing the total pressure losses due to momentum and friction. The conditions that must be satisfied to obtain an efficient diffusion make a compromise between small diffusion angle to avoid separation losses and length to avoid excessive skin friction losses. Since low velocities at the exit section that tends to increase the frontal area of the engine while an efficient diffusion requires a small diffusion angle which in turn tends to make the diffusion section too long, the design is usually dictated by limiting the afterburner diameter to the maximum diameter of the engine.

An additional state is defined now (5A) between the outlet of the turbine (5) and the outlet of the afterburner (6A) for a single-spool afterburning turbojet (Figure 4.20), while for a two-spool engine, the point (7A) is added between the points (8) and (8A) (see Figure 4.21). The total pressure loss in

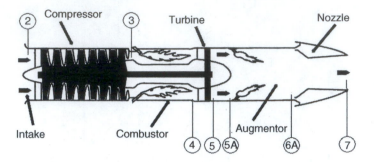

FIGURE 4.20   Stations for single-spool afterburning engine.

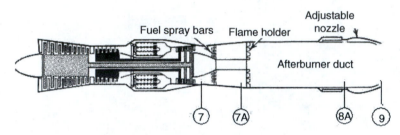

FIGURE 4.21   Stations in the afterburner for two-spool afterburning engine.

the afterburner arises from three sources:

1. Loss in the afterburning diffuser
2. Momentum pressure loss due to heat addition in the afterburner
3. Pressure loss due to drag of flame holders and wall friction known as afterburner friction pressure drop

Hereafter, the case of a single spool will be considered. Analogy between the two engines would yield relevant relations.

1. *Loss in the afterburning diffuser:*   The isentropic efficiency of the diffuser that will be employed here is the typical total-to-static efficiency which is normally employed for diffusers or in general for non-rotating elements. It is expressed as follows:

$$\eta_{d_{ab}} = \frac{(T_{05A})_s - T_5}{T_{05A} - T_5} \quad \text{with } T_{05A} = T_{05}, \qquad \text{then}$$

$$\eta_{d_{ab}} = \frac{[(T_{05A})_s / T_{05}] - (T_5 / T_{05})}{1 - (T_5 / T_{05})}$$

with

$$\frac{(T_{05A})_s}{T_{05}} = \left(\frac{P_{05A}}{P_{05}} \right)^{(\gamma - 1)/\gamma}$$

and A defined as

$$A = \frac{T_5}{T_{05}} = \left(1 + \frac{\gamma - 1}{2} M_5^2\right)^{-1} \tag{4.58a}$$

$$\therefore \eta_{d_{ab}} = \frac{(P_{05A}/P_{05})^{(\gamma-1)/\gamma} - A}{1 - A}$$

Thus

$$\left(\frac{P_{05A}}{P_{05}}\right) = [A + \eta_{d_{ab}}(1 - A)]^{\gamma/(\gamma-1)} \tag{4.58b}$$

The diffuser efficiency becomes increasingly important as V_5 is increased. The velocity of the gases discharged by the afterburner diffuser (V_{5A}) is determined primarily by the cross-sectional area of the tailpipe and it is desirable to keep it small without the frontal area of the afterburning engine exceeding that for the normal engine.

2. *Momentum total pressure loss in afterburner:* The pressure loss due to the rate of change in the momentum of gases flowing through the afterburner corresponds to the differential pressure required for accelerating the gases (due to heat addition) from V_{5A} to V_{6A}. The momentum total pressure loss ($P_{05A} - P_{06A}$) depends on the afterburner temperature rise ratio (T_{06A}/T_{05A}), the afterburner inlet Mach number M_{5A}, and the outlet Mach number M_{6A}.

Assuming the duct between stations 5A and 6A has a constant cross-sectional area and is frictionless, then, for such a simple adiabatic flow, the governing equations will be

Continuity $\rho_{5A} V_{5A} = \rho_{6A} V_{6A}$

State $P_{5A} = \rho_{5A} R_{5A} T_{5A}$ and $P_{6A} = \rho_{6A} R_{6A} T_{6A}$

Momentum $P_{5A} + \rho_{5A} V_{5A}^2 = P_{6A} + \rho_{6A} V_{6A}^2$

Energy $Cp_{5A} T_{05A} = Cp_{5A} T_{5A} + \frac{V_{5A}^2}{2}$

$Cp_{6A} T_{06A} = Cp_{6A} T_{6A} + \frac{V_{6A}^2}{2}$

From these equations we get

$$\frac{T_{06A}}{T_{05A}} = \frac{Cp_{5A}(\gamma_{5A} - 1)}{Cp_{6A}(\gamma_{6A} - 1)} \frac{\gamma_{6A}^2 M_{6A}^2}{\gamma_{5A}^2 M_{5A}^2} \frac{\left[1 + \gamma_{5A} M_{5A}^2\right]^2}{\left[1 + \gamma_{6A} M_{6A}^2\right]^2} \frac{\left[1 + ((\gamma_{6A} - 1)/2) M_{6A}^2\right]}{\left[1 + ((\gamma_{5A} - 1)/2) M_{5A}^2\right]} \tag{4.59}$$

This equation gives the outlet Mach number M_{6A}. The total pressure ratio after some manipulation is thus

$$\frac{P_{06A}}{P_{05A}} = \frac{\left[1 + \gamma_{5A} M_{5A}^2\right]}{\left[1 + \gamma_{6A} M_{6A}^2\right]} \frac{\left[1 + ((\gamma_{6A} - 1)/2) M_{6A}^2\right]^{\gamma_{6A}/(\gamma_{6A}-1)}}{\left[1 + ((\gamma_{5A} - 1)/2) M_{5A}^2\right]^{\gamma_{5A}/(\gamma_{5A}-1)}} \tag{4.60}$$

From the above equation, it is noticed that raising M_{5A} not only increases the loss in total pressure but also reduces the operating range of the afterburner.

3. *Friction pressure loss:* The friction pressure loss is determined experimentally with the afterburner inoperative. The friction pressure loss due to the drag of the flame holders and

wall friction of the tailpipe is expressed in terms of drag coefficient C_D and inlet dynamic pressure of the gas

$$\Delta P_F = C_D \cdot \tfrac{1}{2} \rho_{5A} V_{5A}^2$$

$$C_D \le 1.0$$

By careful design, the decrease in thrust of the normal engine due to friction pressure loss in the afterburner can be kept as small as 1% of the normal thrust.

Example 8   In an afterburning jet engine, the gases enter the afterburner at a temperature $T_0 = 900$ K whether the afterburner is operating or not. Then the gases' temperature rises to 2000 K when the burner is operating and the stagnation pressure is about 5% less than its value without burning. If the average specific heat in the nozzle rises from 1.1 to 1.315 kJ/kg · K because of burning, how much relative change in exhaust nozzle area is required as the burner is ignited? It may be assumed that the nozzle is choked in both cases and that the mass flow is unchanged.

Solution:

Since $R = 287$ J/kg · K, $Cp = \dfrac{\gamma R}{\gamma - 1}$, then

1. For $(C_p)_{\text{inoperative}\atop ab} = 1100$ J/kg · K, then $\gamma_1 = 1.353$

2. For $(C_p)_{\text{operative}\atop ab} = 1315$ J/kg · K, then $\gamma_2 = 1.279$

The process is plotted in Figure 4.22 for operative and inoperative afterburner.

Since the mass flow rate $\dot{m} = P \times A \times M\sqrt{\gamma/RT}$, assuming no losses in the nozzle, then the mass flow for a choked nozzle is

$$\dot{m}_{\text{choked}} = P^* \times A^* \sqrt{\dfrac{\gamma}{RT^*}}$$

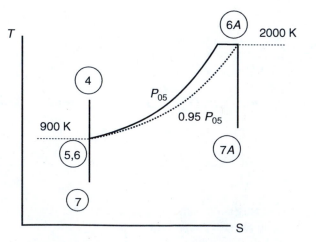

FIGURE 4.22   Temperature–entropy diagram for operative and inoperative afterburner.

This equation can be expressed in terms of the total conditions between states (6) and (7) as

$$\dot{m}_{\text{choked}} = \frac{A^* P_{06}}{\sqrt{RT_{06}}} \sqrt{\gamma} \left(\frac{2}{\gamma + 1} \right)^{(\gamma+1)/2(\gamma-1)}$$

where A^* is the throat area, which is here the outlet area of the convergent nozzle.
Since in both cases of operative and inoperative afterburner, the nozzle is choked and the mass flow is constant

$$\therefore \frac{A_1^* P_0}{\sqrt{R \times 900}} \sqrt{\gamma_1} \left(\frac{2}{\gamma_1 + 1} \right)^{(\gamma_1+1)/2(\gamma_1-1)} = \frac{A_2^* \times 0.95 P_0}{\sqrt{R \times 2000}} \sqrt{\gamma_2} \left(\frac{2}{\gamma_2 + 1} \right)^{(\gamma_2+1)/2(\gamma_2-1)}$$

$$\therefore \frac{A_2^*}{A_1^*} = \sqrt{\frac{2000}{900}} \frac{1}{0.95} \sqrt{\frac{1.353}{1.279}} \frac{(0.85)^{3.3329}}{(0.8776)^{4.084}}$$

The area ratio is

$$A_2^*/A_1^* = 1.4907 \times 1.0526 \times 1.0285 \times \frac{0.5817}{0.5867} = 1.6.$$

4.6 SUPERSONIC TURBOJET

Military airplanes fitted with turbojet engines are fitted with supersonic intakes. Figure 4.23 illustrates an example for such engines. When the afterburner is lit, the airplane will fly at a supersonic speed. Thus, shock waves are generated in the intake as described in Chapter 3. A single normal shock wave is avoided due to the large losses encountered. Normally, at least one oblique shock wave followed by a normal shock wave is seen in the diffuser or intake, upstream of the compressor. From the shock wave tables or graphs, the pressure ratio in the diffuser ($r_d \equiv P_{02}/P_{0a}$) is determined. Then the intake efficiency (η_d), which is dependent on both the flight Mach number and the pressure ratio in the intake (r_d), is calculated from the relation

$$\eta_d = \frac{\left(1 + ((\gamma - 1)/2)M^2\right)(r_d)^{(\gamma-1)/\gamma} - 1}{((\gamma - 1)/2)M^2} \tag{4.61}$$

Other modules are treated in the same procedure as described previously.

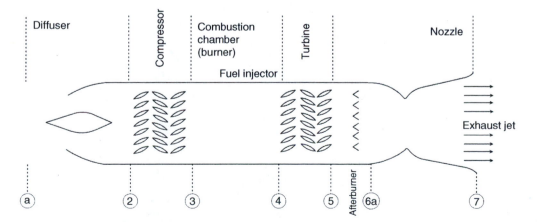

FIGURE 4.23 Layout for a supersonic afterburning turbojet.

Example 9   A single-spool afterburning turbojet engine is powering a fighter airplane flying at Mach number $M_a = 2.0$ at an altitude of 16,200 m where the temperature is 216.6°K and the pressure is 10.01 kPa. The inlet is of the axisymmetric type and is fitted with a spike having a deflection angle 12°. The air mass flow rate is 15 kg/s. The compressor has a pressure ratio of 5 and isentropic efficiency of 0.85. The pressure loss in the combustion chamber is 6% and the heating value of fuel is 45,000 kJ/kg. The burner efficiency is $\eta_b = 0.96$. The TIT is 1200 K and its isentropic efficiency is 0.9. The maximum temperature in the afterburner is 2000 K. The pressure drop in the afterburner is 3% and the afterburner efficiency is 0.9. The nozzle efficiency is $\eta_n = 0.96$. Calculate

1. The stagnation pressure ratio of the diffuser (r_d) and its isentropic efficiency
2. The thrust force

Take $C_{pt} = 1.157$, $C_{pc} = 1.005$ kJ/kg · K.

Solution:

1. *Intake:*   The intake here is identical to that in Example 3, Chapter 3. Two shock waves are developed in the intake; the first is an oblique one, while the second is normal. The important results from this example are

     The total conditions for air upstream of the engine are $T_{0a} = 390$ K and $P_{0a} = 7.824 P_a$.

     The flight speed is $V = M\sqrt{\gamma R T_a} = 590$ m/s.

     The stagnation pressure ratio in the intake is $r_d = \dfrac{P_{02}}{P_{0a}} = 0.884$.

     From Equation 4.61, the diffuser efficiency is

$$\eta_d = \frac{(1 + 0.2 \times 4)(0.884)^{0.286} - 1}{0.2 \times 4} = 0.922$$

     The stagnation temperature of air leaving the diffuser is $T_{02} = T_{0a} = 390$ K. The stagnation pressure of air leaving the intake is

$$P_{02} = 0.884\, P_{0a} = 0.884 \times 7.824\, P_a = 0.884 \times 7.824 \times 10.01 = 69.23 \ \text{kPa}$$

2. *Compressor:*

$$P_{03} = \pi_c P_{02} = 5 \times 69.23 = 346.15 \ \text{kPa}$$

$$T_{03} = T_{02}\left[1 + \frac{\pi_C^{(\gamma_C - 1)/\gamma_C} - 1}{\eta_C}\right] = 658.2 \ \text{K}$$

3. *Combustion chamber:*

$$P_{04} = (1 - \Delta P_{cc})\,, \ P_{03} = 0.94\, P_{03} = 325.38 \ \text{kPa}$$

$$T_{04} = 1200 \ \text{K, then}$$

$$f = \frac{C_{pt}\, T_{04} - C_{pc}\, T_{03}}{\eta_b\, Q_R - C_{pt}\, T_{04}} = 0.0174$$

4. *Turbine:*   Energy balance for compressor and turbine

$$Cp_t (1+f)(T_{04} - T_{05}) = Cp_c (T_{03} - T_{02})$$

$$T_{05} = T_{04} - \frac{Cp_c (T_{03} - T_{02})}{Cp_t (1+f)} = 1200 - \frac{1.005 \times (658.2 - 390)}{1.157 \times 1.0174} = 970.9 \text{ K}$$

$$P_{05} = P_{04} \left[1 - \frac{T_{04} - T_{05}}{\eta_t \, T_{04}} \right]^{(\gamma_t/\gamma_t - 1)} = 325.38 \left[1 - \frac{1200 - 970.9}{0.9 \times 1200} \right]^4 = 125.37 \text{ kPa}$$

5. *Tail pipe:*   A pressure drop of 3% is encountered in the afterburner; thus

$$P_{06A} = 0.97 \, P_{05} = 121.6 \text{ kPa}$$

The maximum temperature is now $T_{06A} = 2000$ K

$$f_{ab} = \frac{Cp_t (T_{06A} - T_{05})}{\eta_{ab} \, Q_R - Cp_t \, T_{06A}} = \frac{1.157 \times (2000 - 970.9)}{0.9 \times 45,000 - 1.157 \times 2000} = 0.0312$$

Check nozzle choking with

$$\frac{P_{06A}}{P_c} = \frac{1}{(1 - (1/\eta_n)(\gamma_n - 1)/(\gamma_n + 1))^{\gamma_n/(\gamma_n - 1)}} = 1.907$$

$$\therefore P_c = 63.77 \text{ kPa}$$
$$\because P_c > P_a$$

$\therefore$ Nozzle is choked also.

$$P_{7A} = P_c = 63.77 \text{ kPa}$$

$$T_{7A} = \frac{T_{06A}}{(\gamma_n + 1)/2} = 1716.9 \text{ K}$$

$$V_{7A} = a_{7A} = \sqrt{\gamma_n R T_{7A}} = 809.5 \text{ m/s}$$

The nozzle area is calculated from the relation

$$A_{7A} = \frac{\dot{m}_a (1 + f + f_{ab})}{\rho_{7A} \, V_{7A}} = \frac{R \, T_{7A} \, \dot{m}_a (1 + f + f_{ab})}{P_{7A} \, V_{7A}}$$

$$= \frac{287 \times (1716.9) \times (15) \times (1 + 0.017 + 0.0312)}{63.77 \times 10^3 \times (809.5)}$$

$$= 0.15 \text{ m}^2$$

The thrust force for operative afterburner is

$$T_{ab} = \dot{m}_a \left[(1 + f + f_{ab}) V_{7A} - V \right] + A_{7A} (P_{7A} - P_a)$$

$$= 15 \left[1.0486 \times 809.5 - 590 \right] + 0.15 \times 10^3 (63.77 - 10.01) = 11,946 \text{ N}$$

$$= 11.946 \text{ kN}$$

4.7 OPTIMIZATION OF THE TURBOJET CYCLE

Performance parameters for a jet engine are the specific thrust, TSFC, fuel-to-air ratio, afterburner fuel-to-air ratio (for afterburning engines), as well as the propulsive, thermal, and overall efficiencies. The factors influencing these parameters are as follows:

1. Compressor(s) pressure ratio(s) (π_c for a single spool and π_{c1}, π_{c2} for dual spool)
2. Outlet temperature from the combustion chamber, which is the inlet temperature to the turbine (if single-spool engine) or the HPT (if two-spool engine). It is normally denoted as TIT or TET (turbine entry temperature)
3. Efficiencies (diffuser, compressor(s), combustion chamber, turbine(s), and nozzle)
4. Flight speed
5. Flight altitude
6. Maximum temperature in the afterburner, if available

Specific thrust

$$F_S = \frac{T}{\dot{m}_a} = (1+f)\, U_e - U_a + \frac{A_e}{\dot{m}} \, (P_e - P_a) \quad (\text{N} \cdot \text{s/kg air})$$

The TSFC

$$\text{TSFC} = \frac{\dot{m}_f}{T} = \frac{\dot{m}_f}{\dot{m}_a F_s}$$

$$\text{TSFC} = \frac{f}{F_s} \quad (\text{kg fuel/N} \cdot \text{s})$$

The following case study is considered for this parametric study:

   Two altitudes are considered: sea level and 8 km
   The compressor pressure ratio varies up to 50
   The TIT varies from 1200 to 1500 K
   The flight Mach number varies from 0.1 to 1.5 for a nonafterburning engine and up to 2.7
      for an afterburning engine
   The maximum temperature in the afterburner varies from 2000 to 3000 K

The following values are considered for efficiencies: diffuser (0.9), compressor (0.88), combustion chamber (90.95), turbine (0.94), afterburner (0.95), and nozzle (0.96). For two-spool engine, the efficiencies of both compressors are equal and also both turbines have the same efficiency. Fuel heating value is 45,000 kJ/kg. The pressure drop in the combustion chamber and afterburner is 3% and 5%, respectively. The results of this analysis are discussed here.

Figures 4.24 and 4.25 illustrate the operation for a nonafterburning single-spool turbojet at sea level. Each figure combines four figures: Figure 4.24 represents the variations in specific thrust or the TSFC due to changes in Mach number, compressor pressure ratio, and TIT.

At a constant compressor pressure ratio π_c, an increase in TIT increases both the specific thrust and TSFC. However, the increase in specific thrust is more important than the penalty in increased TSFC, as shown from the two left-side figures.

- For a fixed TIT, the increase of overall pressure ratio reduces the TSFC. The specific thrust increases at first and then reduces. Thus, for each TIT there is an optimum (π_c). The optimum (π_c) for maximum specific thrust increases as the TIT increases.
- For a constant pressure ratio and TIT, the increase in flight Mach number leads to an increase in TSFC and decrease in specific thrust. This is due to the increase in inlet

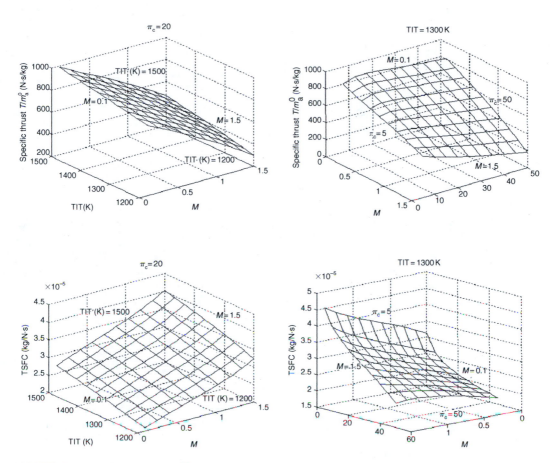

FIGURE 4.24 Variations of specific thrust and TSFC for a single-spool turbojet at sea level for inoperative afterburner.

momentum drag and the increase in compressor outlet temperature and the resultant decrease in fuel consumption.

Figure 4.25 illustrates the variations in the fuel-to-air ratio and propulsive efficiencies due to changes in flight Mach number, compressor pressure ratio, and TIT.

- For a constant TIT, the fuel-to-air ratio decreases with the increase in flight Mach number and increase in the compressor pressure ratio.
- For a constant TIT also, the propulsive efficiency increases with the increase of flight Mach number and remains nearly unchanged with the increase in compressor pressure ratio.
- For a constant compressor pressure ratio the fuel-to-air ratio increases with the increase in the TIT. On the contrary, the fuel-to-air ratio decreases as the flight Mach number increases.
- For a constant compressor pressure ratio the propulsive efficiency increases with the increase of flight Mach number and remains nearly unchanged with the increase in TIT.

  Figures 4.26 and 4.27 illustrate the operation of a single-spool turbojet with operative afterburner at sea level. Figure 4.26 represents the variations in specific thrust or the TSFC due to changes in Mach number, compressor pressure ratio, and afterburner temperature, maintaining a constant TIT of 1300 K.
- At a constant compressor pressure ratio π_c, an increase in maximum temperature in the afterburner increases both the specific thrust and TSFC. Moreover, the increase in Mach number for a constant compressor pressure ratio decreases the specific thrust and increases TSFC.

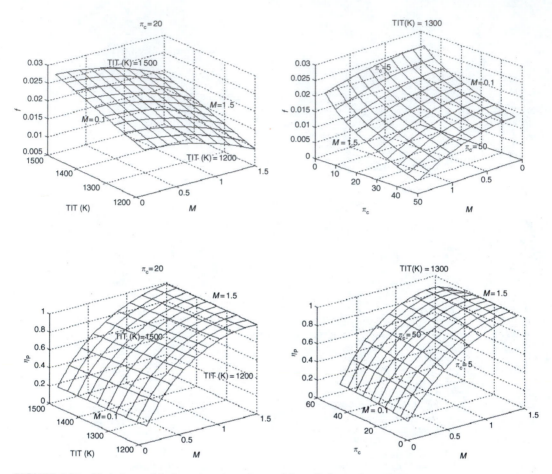

FIGURE 4.25 Variations of fuel-to-air ratio and propulsive efficiency for a single-spool turbojet at sea level for inoperative afterburner.

- For a fixed maximum temperature in the afterburner, the increase in overall pressure ratio increases the specific thrust and slightly influences the TSFC. The increase in Mach number for a constant maximum temperature decreases the specific thrust and increases TSFC.

  Figure 4.27 illustrates the variation of fuel-to-air ratio (f), afterburner fuel-to-air ratio (f_{ab}), and propulsive efficiency at sea level for operative afterburner.

- For a constant compressor pressure ratio, TIT, and maximum temperature, the fuel-to-air ratio decreases as the flight Mach number increases. It is important to notice that there is an upper limit for Mach number at takeoff. If the Mach number increases above a certain value, then the temperature rise in the intake and compressor yields greater temperature at the inlet to combustion chamber than its design maximum temperature. This explains the negative values of fuel-to-air ratio at Mach numbers close or equal to 2.7.

- For a constant Mach number, TIT, and maximum temperature, the fuel-to-air ratio decreases with the increase of the compressor pressure ratio.

- For a constant Mach number, TIT, and maximum temperature, the afterburner fuel-to-air ratio increases with the increase in the compressor pressure ratio.

- For a constant compressor pressure ratio, TIT, and maximum temperature, the afterburner fuel-to-air ratio increases as the flight Mach number increases.

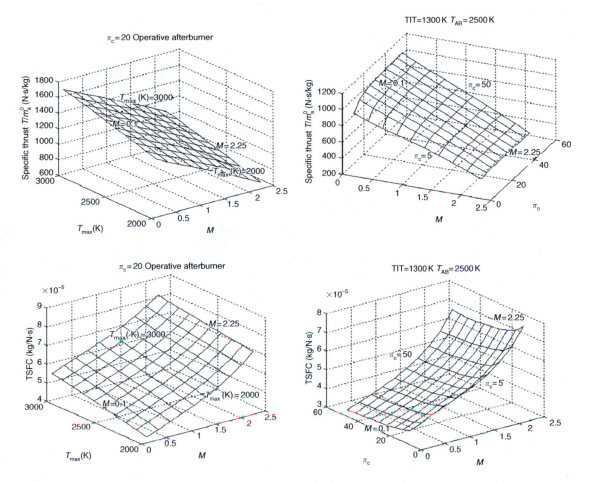

FIGURE 4.26 Variations of specific thrust and TSFC for a single-spool turbojet at sea level with operative afterburner.

- For a constant compressor pressure ratio, constant TIT, and maximum temperature, the propulsive efficiency increases with the increase in flight Mach number and is nearly constant for different compressor pressure ratios.

Figures 4.28 and 4.29 illustrate the operation of a single-spool turbojet with inoperative afterburner at an altitude of 8 km. A comparison of Figures 4.24 and 4.28 outlines that the performance is quite similar in both altitudes. TSFC is somewhat higher at 8 km than at sea level for the same compressor pressure ratio, Mach number, and TIT. The reason is that the ambient temperature and pressure are less than the corresponding values at sea level particularly at low Mach numbers. This is quite clear from the fuel-to-air ratios (see Figures 4.25 and 4.29). However, since the air mass flow rate at sea level is much higher than its value at 8 km, the thrust at sea level will be much greater than the altitude thrust. Moreover, the fuel consumption is also higher at sea level.

The propulsive efficiency at sea level is slightly higher than at 8 km for the same Mach number. This is because the values of flight speed for the same Mach number at sea level are higher than at 8 km since the ambient temperature at sea level is higher than its value at 8 km.

The effect of operating afterburner on the engine performance at an altitude of 8 km is illustrated in Figures 4.30 through 4.32. Figure 4.30 illustrates the specific thrust and TSFC variations due to Mach number, compressor pressure ratio, and maximum temperature variations. This figure has the

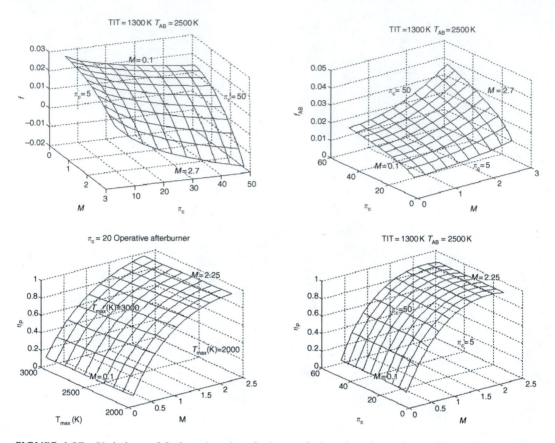

FIGURE 4.27 Variations of fuel-to-air ratio, afterburner fuel-to-air ratio, and propulsive efficiency for a single-spool turbojet at sea level.

same trends as sea level operation; see Figure 4.26. Also TSFC is higher at 8 km when compared with that at the sea level.

The fuel-to-air ratio and afterburner fuel-to-air ratio are plotted in Figure 4.31 for an 8 km altitude. Comparing Figures 4.27 and 4.31 yields a higher fuel-to-air ratio at 8 km as compared with that at the sea level and a lower afterburner fuel-to-air ratio at 8 km with respect to sea level values.

The propulsive efficiency for afterburning engine at 8 km is plotted in Figure 4.32. It has the same trend as in Figure 4.27.

The variations of TSFC are plotted versus specific thrust for different compressor pressure ratio and TIT in Figure 4.33.

Example 10 Figure 4.33 illustrates the performance map for a turbojet engine having the following operating conditions and some other data given below:

Ambient temperature	247.9 K	Ambient pressure	46.0 kPa
Compressor pressure ratio	10	Turbine inlet temperature	1600 K
Diffuser efficiency	0.95	Compressor efficiency	0.8
Burner efficiency	0.98	Turbine efficiency	0.9
Nozzle efficiency	0.9	Pressure drop in combustion chamber	1%
Exhaust area	0.3 m^2		

$$C_{p_h} = 1.148 \, \text{kJ/kg} \cdot \text{K}, \quad C_{p_c} = 1.005 \, \text{kJ/kg} \cdot \text{K}, \quad Q_R = 45{,}0000 \, \text{kJ/kg}$$

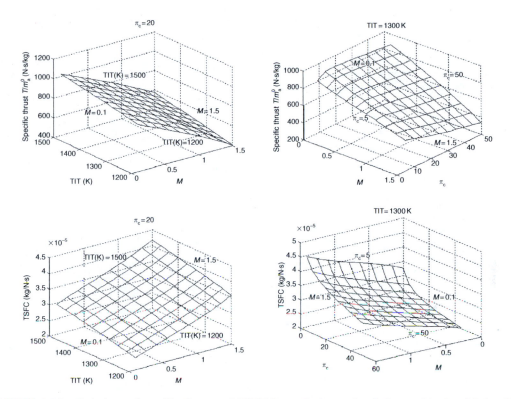

FIGURE 4.28 Variations of specific thrust and TSFC for a single-spool turbojet at altitude of 8 km for inoperative afterburner.

It is required to calculate

1. The fuel-to-air ratio
2. The compressor outlet temperature
3. The flight Mach number
4. The turbine outlet temperature
5. The air mass flow rate
6. The inlet area

Solution: From the performance map with known TIT (1600 K) and compressor pressure ratio (10):

1. The specific thrust $(T/\dot{m}_a)$ is 815 N · s/kg
2. The specific fuel consumption is.3.5×10^{-5} kg/N · s

Since

$$\text{TSFC} = \frac{\dot{m}_f}{T}$$

$$\therefore f = \frac{\dot{m}_f}{\dot{m}_a} = \frac{\dot{m}_f}{T}\frac{T}{\dot{m}_a} = \left(3.5 \times 10^{-5}\frac{\text{kg}}{\text{N} \cdot \text{s}}\right)\left(815\frac{\text{N} \cdot \text{s}}{\text{kg}}\right)$$

$$\therefore f = 0.02853$$

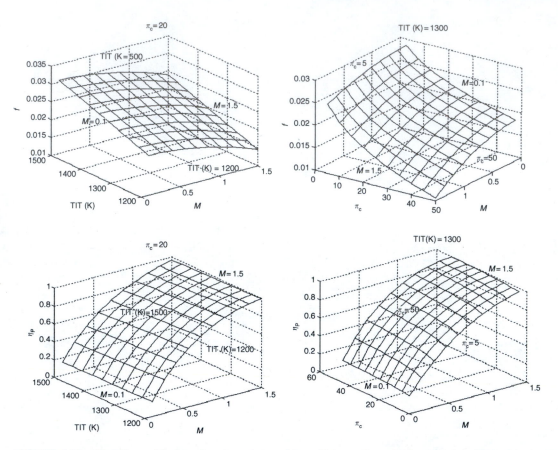

FIGURE 4.29 Variations of fuel-to-air ratio and propulsive efficiency for a single-spool turbojet at altitude of 8 km for inoperative afterburner.

From the energy balance in the combustion chamber:

$$f = \frac{Cp_{cc}T_{04} - Cp_{c}T_{03}}{\eta_{b}Q_{R} - Cp_{cc}T_{04}}$$

$$\therefore T_{03} = \frac{1}{Cp_{c}}\left[Cp_{cc}T_{04} - f\left(\eta_{b}Q_{R} - Cp_{cc}T_{04}\right)\right]$$

$$= \frac{1}{1.005}\left[1.148 \times 1600 - 0.02853\left(0.98 \times 45{,}000 - 1.148 \times 1600\right)\right]$$

$$= 631 \text{ K}$$

The compressor outlet temperature is given by the relation

$$T_{02} = \frac{T_{03}}{\left[1 + \left(\left(\pi_{c}^{(\gamma_{c}-1)/\gamma_{c}} - 1\right)/\eta_{c}\right)\right]} = 291.5 \text{ K}$$

The temperature ratio within the diffuser is

$$\frac{T_{02}}{T_{a}} = \frac{T_{0a}}{T_{a}} = 1 + \frac{\gamma - 1}{2}M^{2}$$

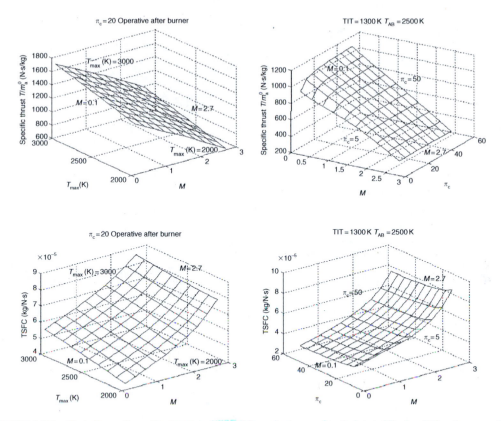

FIGURE 4.30   Variations of specific thrust and TSFC for a single-spool turbojet at altitude of 8 km for operative afterburner.

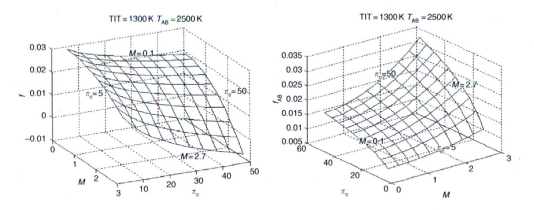

FIGURE 4.31   Variations of fuel-to-air ratio and afterburner fuel-to-air ratio for a single-spool turbojet at 8 km.

$$\therefore M = \sqrt{\frac{2}{\gamma - 1}\left(\frac{T_{02}}{T_a} - 1\right)} = \sqrt{\frac{2}{0.4}\left(\frac{291.5}{247.9} - 1\right)}$$

$$M = 0.938$$

The flight speed is then

$$V_f = M\sqrt{\gamma R T_a} = 296 \text{ m/s}$$

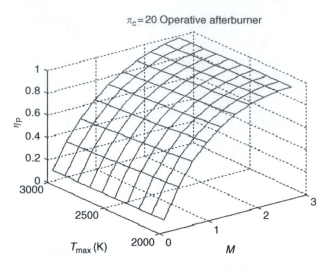

FIGURE 4.32   Variations of propulsive efficiency for a single-spool turbojet at 8 km.

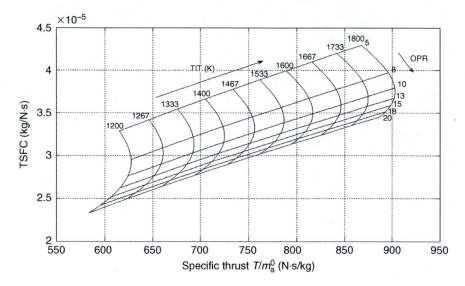

FIGURE 4.33   TSFC versus specific thrust.

The diffuser pressure ratio is

$$P_{02} = P_a \left(1 + \eta_d \frac{\gamma_c - 1}{2} M_a^2 \right)^{\gamma_c/(\gamma_c - 1)} = 79 \text{ kPa}$$

The compressor outlet pressure is

$$P_{03} = (P_{02})(\pi_c) = 790 \text{ kPa}$$

Owing to the pressure drop in the combustion chamber, the outlet pressure for the gases leaving the combustion chamber is

$$P_{04} = (1 - \Delta P) P_{03} = 0.99 \times 790 = 782.1 \text{ kPa}$$

From the energy balance between the compressor and turbine, the turbine outlet temperature is

$$T_{05} = T_{04} - \frac{Cp_c\,(T_{03} - T_{02})}{(1+f)\,Cp_h} = 1600 - \frac{1.005\,(631 - 291.5)}{1.02853 \times 1.148}$$

$$= 1311\ \text{K}$$

The pressure ratio in the turbine is

$$\frac{P_{05}}{P_{04}} = \left[1 - \frac{1}{\eta_t}\left(1 - \frac{T_{05}}{T_{04}}\right)\right]^{(\gamma_h/\gamma_h - 1)} = \left[1 - \frac{1}{0.9}\left(1 - \frac{1311}{1600}\right)\right]^4 = 0.4082$$

Since $P_{03} = P_{04} = 782.1\ \text{kPa}$
Then $P_{05} = 319.2\ \text{kPa}$
Now, the nozzle is to be checked out for choking:

$$\frac{P_{05}}{P_c} = \frac{1}{[1 - (1/\eta_n)(\gamma_h - 1)/(\gamma_h + 1)]^{\gamma_h/(\gamma_h - 1)}} = \frac{1}{\left(1 - (1/0.95)\,(0.33/2.33)\right)^4} = 1.907$$

$$P_c = 167.4\ \text{kPa}$$

Since P_c is greater than P_a, then the nozzles are choked. The gases leave the nozzle at the temperature

$$T_7 = T_c = \frac{T_{05}}{(\gamma + 1)/2} = 1123.7\ \text{K}$$

The jet speed is then

$$V_7 = V_c = \sqrt{\gamma R T_7} = \sqrt{1.333 \times 287 \times 1123.7} = 655.75\ \text{m/s}$$

The specific thrust is given by the relation

$$\frac{T}{\dot{m}_a} = (1+f)\,V_7 - V_f + \frac{A_e}{\dot{m}_a}\,(P_7 - P_a)$$

$$815 = 1.02853 \times 655.75 - 296 + \frac{0.3}{\dot{m}_a}\,(167.4 - 46) \times 10^3$$

The mass flow rate is $\dot{m}_a = 83.43\ \text{kg/s}$.
Finally, since the mass flow rate is given by the relation:

$$\dot{m}_a = \rho_i V_f A_i = \frac{P_i}{RT_i} V_f A_i$$

The inlet area is then $A_i = 0.436\ \text{m}^2$.

Hint: From the performance map, it is seen that for a constant TIT, the specific thrust increases with the increase in compression pressure ratio at first and then reduces. Thus, for each TIT there is an optimum (π_c) that yields a maximum specific thrust. This optimum (π_c) increases as the TIT increases.

Summary of mathematical relations

Ideal and Actual Cycles for a Turbojet Engine

Element	Ideal cycle	Actual cycle
Diffuser	$\eta_d = 1$	$T_{02} = T_a \left(1 + \dfrac{\gamma_c - 1}{2} M^2 \right)$
		$P_{02} = P_a \left(1 + \eta_d \dfrac{\gamma_c - 1}{2} M^2 \right)^{\gamma_c/(\gamma_c - 1)}$
Compressor	$\eta_c = 1$	$T_{03} = T_{02} \left(1 + \dfrac{\pi_c^{(\gamma_c - 1/\gamma_c)} - 1}{\eta_c} \right)$
		$P_{03} = \pi_c P_{02}$
Combustion chamber	$\eta_b = 1$ $\Delta P_0 \% = 0$	$T_{04} = TIT$
		$P_{04} = (1 - \Delta P\%) P_{03}$
		$f = \dfrac{C_{p_h} T_{04} - C_{p_c} T_{03}}{\eta_b Q_R - C_{p_h} T_{04}}$
		$\dot{m}_f = f \dot{m}_a$
Turbine	$\eta_m = 1$ $\eta_t = 1$	$T_{05} = T_{04} - \dfrac{C_{p_c}}{\eta_b \lambda (1 + f) C_{p_h}} (T_{03} - T_{02})$
		$P_{05} = P_{04} \left(1 - \dfrac{T_{04} - T_{05}}{\eta_t T_{04}} \right)^{\gamma_h/(\gamma_h - 1)}$
Afterburner	$\Delta P_{ab} = 0$	$P_{06} = (1 - \Delta P_{ab} \%) P_{05}$
		For inoperative afterburner $T_{06} = T_{05}$
		For operative afterburner $T_{06} = T_{0_{max}}$
	$\eta_{ab} = 1$	$f_{ab} = \dfrac{(1 + f) C_{p_h} (T_{06} - T_{05})}{\eta_{ab} Q_R - C_{p_h} T_{06}}$
		$\dot{m}_{fab} = f_{ab} \dot{m}_a$
Nozzle	$\eta_n = 1$	$P_c = P_{06} \left(1 - \dfrac{1}{\eta_n} \dfrac{\gamma_h - 1}{\gamma_h + 1} \right)^{\gamma_h/(\gamma_h - 1)}$
		The nozzle is choked if $P_c \geq P_a$
		$P_7 = P_c$
		$T_7 = T_c = \dfrac{T_{06}}{(\gamma_h + 1)/2}$
		$V_7 = \sqrt{\gamma_h R T_7}$
		The nozzle is unchoked if $P_c < P_a$
		$P_7 = P_a$
		$T_7 = T_{06} \left[1 - \eta_n \left\{ 1 - \left(\dfrac{P_7}{P_{06}} \right)^{(\gamma_h - 1)/\gamma_h} \right\} \right]$
		$V_7 = \sqrt{2 C_{p_h} (T_{06} - T_7)}$

$$T = \dot{m}_a [(1 + f + f_{ab}) V_7 - V_f] + A_7 (P_7 - P_a)$$

$$\text{TSFC} = \frac{\dot{m}_f + \dot{m}_{ab}}{T} = \frac{f + f_{ab}}{T/\dot{m}_a}$$

$$\eta_p = \frac{TV}{TV + 0.5 \dot{m}_a (1 + f + f_{ab}) (V_7 - V)^2}$$

$$\eta_{th} = \frac{TV + 0.5\dot{m}_a\left(1 + f + f_{ab}\right)\left(V_7 - V\right)^2}{\dot{m}_a\left(f + f_{ab}\right)Q_{HV}}$$

$$\eta_o = \frac{TV}{\dot{m}_a\left(f + f_{ab}\right)Q_{HV}}$$

C_{P_c} : Specific heat at constant pressure for cold air
C_{P_h} : Specific heat at constant pressure for hot gases
$\dot{m}_a$: Air mass flow rate
$\dot{m}_f$: Fuel mass flow rate
f : Fuel-to-air ratio
$\dot{m}_{f_{ab}}$: Fuel burnt in afterburner
f_{ab} : Afterburner fuel-to-air ratio
Q_R : Fuel heating value
γ_c : Ratio of specific heats for cold air
γ_h : Ratio of specific heats for hot gases

PROBLEMS

4.1 The following data refer to the operating conditions for an aircraft turbojet engine:

Barometric pressure	101 kPa
Aircraft speed	500 km/h
Compressor air flow	45 kg/s
Exhaust nozzle area	0.2 m^2
Exhaust nozzle pressure	200 kPa
Exhaust gas velocity	450 m/s
Fuel flow	0.8 kg/s

(a) Calculate the ram drag, momentum thrust, pressure thrust, gross thrust, net thrust, and specific thrust.
(b) If the aircraft speed is increased to 600, 700, 800, and 900 km/h, while all other data are maintained constant, plot the ram drag, gross thrust, and net thrust versus the aircraft speed.
(c) If the mass flow rate is proportional to the aircraft speed and the speed also changes to be 600, 700, 800, and 900 km/h, then calculate the new value of the air mass flow and plot again the ram drag, gross thrust, and net thrust versus the aircraft speed.

4.2 Calculate the specific thrust and the TSFC of a single-spool turbojet engine having the following peculiarities:

Cruise velocity of 280 m/s at altitude of 7000 m
Intake efficiency of 93%
Compressor pressure ratio of 8:1 and efficiency of 87%
Burner efficiency of 98%
Pressure drop in the combustion chamber of 4% of the delivery pressure of the compressor
TIT of 1200 K and efficiency of 90%
Mechanical efficiency of 99%
Nozzle efficiency 95%
Fuel heating value 44,000 kJ/kg

4.3 Derive an expression for the ratio between the thrust forces of a turbojet engine fitted with an afterburner when the afterburner is operative (T_{ab}) and inoperative (T) with choked nozzle similar to that in Equation 4.33.

4.4 A single-spool turbojet engine powers an aircraft flying at Mach number of 0.5, at an altitude where $T_a = 226.7$ k and $P_a = 28.7$ kPa.

It has the following data:
Air mass flow rate 20 kg/s
Compressor pressure ratio 8
Maximum cycle temperature 1200 K
Fuel heating value Q_R 43,000 kJ/kg
All the components of engine are ideal.
Calculate the thrust force, fuel-to-air ratio, and TSFC.

4.5 Consider a simple turbojet engine with a compressor pressure ratio of 8, a TIT 1200 K, and a mass flow of 15 kg/s, when the aircraft is flying at 260 m/s at an altitude of 7000 m. Assuming the following components, efficiencies, and I.S.A conditions, calculate the propelling nozzle area required, the net thrust developed, and the TSFC.
Isentropic efficiency of intake 0.9
Isentropic efficiency of compressor 0.88
Isentropic efficiency of turbine 0.92
Isentropic efficiency of propelling nozzle 0.95
Combustion chamber pressure loss is 6% of compressor delivery pressure
Combustion chamber efficiency 0.98
If the gases in the jet pipe are reheated to 2000 K and the combustion chamber pressure loss is 3% of the pressure at outlet from turbine, calculate
The percentage increase in nozzle area required if the air mass flow is to be unchanged.
The percentage increase in net thrust.

$$(T_a = 242.7 \text{ K}, \ P_a = 41.06 \text{ kPa}, \ Q_R = 43,000 \text{ kJ/kg})$$

4.6 A turbojet engine is installed in a fighter airplane flying at Mach number of 1.3. The ambient conditions are 236.21 K and 35.64 kPa. The exhaust conditions are 1800 K and 100 kPa. The compressor pressure ratio is 10. The engine's inlet and exhaust areas are 0.3 m² and 0.25 m², respectively. The afterburner is operating at this case. Assuming ideal processes, calculate:

(i)   The maximum temperature and pressure at the afterburner
(ii)   The outlet temperature and pressure of the compressor
(iii)   The turbine inlet and outlet temperature (neglect fuel-to-air ratio -f- in power balance)

4.7 A turbojet engine is fitted with an afterburner. During a ground run, it has the following characteristics:

$$P_a=101 \text{ kPa}, \ T_a=288 \text{ K}, \ \pi_c=6, \ T_{04}=1600 \text{ K}, \ T_{06A}=2600 \text{ K}, \ Q_R=45,000 \text{ kJ/kg}$$

Assuming that all the engine modules behave ideally and the afterburner is operative, calculate
(i) The total fuel-to-air ratio of both the combustion chamber and afterburner.
(ii) The exhaust speed.
(iii) The specific thrust.
Now, if another turbojet engine is theoretically investigated where no afterburner is available and the gases in the combustion chamber attain a maximum temperature of 2600 K, it is required to calculate the quantities calculated in items (i), (ii), and (iii) above. Next, compare both corresponding values. Write a brief note on the disadvantage of the second engine.

4.8 A fighter aircraft is powered by a single turbojet engine. The aircraft is flying at an altitude of 10 km where the ambient conditions are 223.23 K and 26.48 kPa with a Mach number of 1.2. The exhaust gases leave the nozzle at a speed of 900 m/s. The compressor has a pressure ratio of 8. Neglecting the fuel-to-air ratio, it is required to calculate the TIT. Assume all efficiencies are equal to unity and the specific heat is constant.

4.9 A turbojet engine powers a fighter aircraft flying at an altitude of 13 km at a speed of 500 m/s. If the maximum temperature in the combustion chamber is 1600 K, the compressor pressure ratio is 6, and fuel heating value is 45,000 kJ/kg, assume the highest values of efficiency for each module. Calculate the exhaust speed.

If it is needed to increase the turbine power by 30% for accessories requirement and the exhaust speed is maintained the same, what will be the increase necessary in the TIT?

4.10 A turbojet engine is installed in a fighter airplane flying at Mach number of 1.3. The ambient conditions are 236.21 K and 35.64 kPa. The exhaust conditions are 1800 K and 100 kPa. The compressor pressure ratio is 10. The areas of the intake and exit are $0.3\,\mathrm{m}^2$ and $0.25\,\mathrm{m}^2$, respectively. The afterburner is operating at this case. Assume ideal processes and variable properties (C_p and γ) within the engine, plot the cycle on the T–S diagram. Next, calculate

(a) The maximum temperature and pressure at the afterburner.
(b) The outlet temperature and pressure of the compressor.
(c) The turbine inlet and outlet temperature (neglect f in power balance).
(d) The thrust force (calculate f and the new turbine outlet temperature assuming TIT is kept the same and $Q_R = 42{,}000$ kJ/kg).

Next, if a convergent–divergent nozzle is assumed to replace the above convergent nozzle, what will be the new value of the thrust force? Write your comments.

4.11 A turbojet engine is operating under the following conditions:

Flight speed at sea level, standard day	0
Air flow entering compressor	1 lb/s (1 kg/s)
Compressor pressure ratio (total-to-total)	12
Efficiencies	
Diffuser	100%
Compressor	87%
Turbine	89%
Jet nozzle	100%
TIT (stagnation)	2530°R (1400 K)

Assuming mass of fuel added is negligible:

(a) Calculate the thrust developed by and the heat added to this engine assuming a converging nozzle.
(b) Calculate the thrust developed by, the heat added to, and the thermal efficiency of this engine assuming a converging-diverging nozzle, that is, expansion in the nozzle back to ambient pressure.

4.12 Consider the turbojet in Example 1. Prove that if $\pi_t = \pi_c = \pi$, $\gamma_t = \gamma_c = \gamma$ and $C_{pc} = C_{pt}$, then the derived relations for the pressure ratio and the jet velocity are reduced to

$$\pi = \left(\frac{T_{04}}{T_a}\right)^{(\gamma/\gamma - 1)} \quad \text{and} \quad V_j = 0$$

Explain why the exhaust jet velocity attained a zero value.

4.13 The following data refer to a typical turbojet engine fitted with an afterburner; it is required to compare the afterburning and nonafterburning engine specific thrust, TSFC, and nozzle exit area.

The turbojet engine has the following characteristics and operating conditions:

Altitude	40,000 ft (12,200 m)
Ambient temperature	216.7°K
Ambient pressure	18.75 kPa
Flight Mach number	2.0
Mass flow rate	250 kg/s
Compressor stagnation pressure ratio	8.0

Pressure loss in burner and afterburner:

- Dry operation   1%
- Wet operation   3%

Turbine inlet temperature 1400°K

Efficiencies:

- Diffuser   98%
- Compressor   85%
- Burner   98%
- Turbine   90%
- Nozzle   97%
- Mechanical   99%

Specific heat ratios:

- Cold stream   1.4
- Hot stream   1.33

Fuel heating value 45,000 kJ/kg

The maximum temperature with operative afterburner is 1900 K.

4.14 Prove that for ideal turbojet engine

$$\frac{T}{\dot{m}_a a_a} = \left\{ \sqrt{\left\{1 + \frac{2}{(\gamma - 1) M_a^2}\right\} (\tau_b - 1)(\tau_c - 1) + \tau_b} - 1 \right\}$$

where $a_a = \sqrt{\gamma R T_a}$, $\tau_c = T_{03}/T_{02}$ and $\tau_b = T_{04}/T_{03}$ and M_a is the flight Mach number.

4.15 An airplane is weighing 182.24 KN and is powered by two turbojet engines. During flight at constant speed the following data was recorded.
Flight speed = 188 m/s, exhaust speed = 456 m/s
Ambient conditions $T_a = 288$ K, $P_a = 101.3$ kPa
Ratio of airplane drag to lift coefficients $(C_D/C_L) = 0.135$, fuel-to-air ratio $f = 0.00913$.
Calculate the intake area of the engine. Next, when airplane flew at altitude H = 9 Km (where $T_a = 229.5$ K, $P_a = 20.7$ kPa), the drag decreased by 40% and fuel-to-air ratio (f) decreased by 20%, calculate:

- New air mass flow rate
- Flight speed and Mach number
- Exhaust speed

(Assume the nozzle is unchoked, C_D = constant.)

4.16 Compare between TSFC of a turbojet and a ramjet which are being considered for flight at $M = 1.5$ and 50,000 ft altitude (ambient condition of pressure and temperature 11 kPa and 216 K, respectively). The turbojet pressure ratio is 8 and maximum temperature is 1130 K, for the ramjet the maximum temperature is 2240 K. For simplicity, ignore aerodynamics losses in both engines. Conventional hydrocarbon fuels having a heating value of 45,000 kJ/kg are used. Assume $\gamma = 1.4$ and $C_p = 1.005$ kJ/kg · K.

REFERENCES

1. R.G. Perel'Man, *Soviet Nuclear Propulsion (Yadernyye Dviagateli)*, Triumph Publishing Co., Washington, D.C., 1959.

2. D.J. Keirn, The USAF nuclear propulsion programs, In *Nuclear Flight: The United States Air Force Programs for Atomic Jets, Missiles, and Rockets*, K.F. Gantz, (Ed.), Duell, Sloan, and Pearce, New York, 1960.

3. C.B. Collins, M.C. Iosif, R. Dussart, J.M. Hicks, S.A. Karamian, C.A. Ur, I.I. Popsescu, V.I. Kirischuk, J.J. Carroll, H.E. Roberts, P. McDaniel, and C.E. Crist, Accelerated emission of gamma rays from the 31-years isomer of 178Hf induced by X-ray irradiation, *Physical Review Letters*, 82-695-698, January 1999.

4. P. McDaniel, *Triggered Isomer Research Program: Propulsion Aspects*, Memorandum from AFRL/DEPA, undated.

5. C.E. Hamilton, *Design Study of Triggered Isomer Heat Exchanger—Combustion Hybrid Jet Engine For High Altitude Flight*, M.Sc. thesis, AFIT, WPAFB, OH 45433-7765, 2002.

6. *The Jet Engine*, Rolls-Royce plc, 5th edn., 1996, p. 178.

7. W.W. Bathie, *Fundamentals of Gas Turbines*, 2nd edn., 1996, p. 227.

8. B. Gunston, *The Development of Jet and Turbine Aero Engines*, Patrick Stephens Limited, imprint of Haynes Publishing, 1997, pp. 79.

9. D.L. Daggett, S. Ortanderl, D. Eames, J.J. Berton, and C.A. Snyder, *Revisiting Water Injection for Commercial Aircraft*, SAE Paper No. 2004-01-3108.

10. D.L. Daggett, *Water Misting and Injection of Commercial Aircraft Engines to Reduce Airport*, NOx, NASA CR 2004-212957, March 2004.

11. D.J.H. Eames, *Short Haul Civil Tiltrotor Contingency Power System Preliminary Design*, NASA CR-NAS3-970209, Jan. 2006.

5 Turbofan Engines

5.1 INTRODUCTION

Turbofan engines were first termed by Rolls-Royce as bypass turbojet. Boeing Company sometimes identifies them as fanjets [1]. Turbofan engines are the most reliable engines ever developed. Fundamentally, turbofan engines are fuel-efficient and quiet turbine engines. They feature continuous combustion and smooth rotation, unlike the internal combustion engine of an automotive. Similar to the turbojet engines, the gas generator in turbofan engines has three sections:

- Fan unit and compressor section
- Combustion chamber
- Turbine section

The compressors pressurize air and feed it aft. Most of the air goes around the engine core through a nozzle-shaped chamber. The rest goes through the engine core where it mixes with fuel and ignites. The hot expanding combustion efflux passes through the turbine section, spinning the turbine as it exits the engine.

The spinning turbine turns the engine shaft. The rotating shaft spins the fan on the front of the engine. The fan compresses more air and keeps this continuous cycle going.

Although during 1936–1946 the two companies Power Jet and Metropolitan-Vickers pioneered various kinds of ducted-fan jet engines, nobody had an interest in these engines and looked at them as rather complicated turbojet engines. Ten years later Rolls-Royce introduced such bypass turbojet engines having a bypass ratio (BPR) of three and more [2]. General Electric (GE) led the way in 1965 with the TF39 (BPR of 8) and still leads with GE90 (BPR nearly 9).

The turbofan engine has several advantages over both the turboprop and the turbojet engines. The fan is not as large as the propeller, so the increase of speeds along the blade is less. Thus, a turbofan engine can power a civil transport flying at transonic speeds up to Mach 0.9. In addition, by enclosing the fan inside a duct or cowling, the aerodynamics is better controlled. There is less flow separation at higher speeds and less trouble with shock developing. The turbofan may suck in more airflow than a turbojet, thus generating more thrust. Like the turboprop engine, the turbofan has low fuel consumption compared with the turbojet. The turbofan engine is the best choice for high-speed, subsonic commercial airplanes. Advantages of turbofan engines are as follows:

- The fan is not as large as the propeller; therefore, higher aircraft velocities can be reached before vibrations occur. The aircraft is able to reach transonic speeds of Mach 0.9.
- The fan is more stable than a single propeller and therefore, if the vibration velocity is reached, vibrations are less apparent and do not disrupt the airflow as significantly.
- The fan is encased in a duct or cowling; therefore, the aerodynamics of the airflow is controlled a lot better providing greater efficiency.
- For geared turbofan engines, the gearbox required to translate the energy from the compressor/fan to the turbine is relatively small and less complex as the fan is smaller. This reduces the weight and aerodynamic drag loss that are present on the turboprop design.

- The smaller fan is more efficient and takes in air at a greater rate than the propeller allowing the engine to produce greater thrust.
- The turbofan engine is a much more fuel-efficient design than the turbojet and is able to equal some of the high performance velocities.

As described in the classification section of Chapter 1, numerous types of turbofan exist. A detailed analysis of these types will be given in the following sections.

5.2 FORWARD FAN UNMIXED SINGLE-SPOOL CONFIGURATION

The main components here are the intake, fan, fan nozzle, compressor, combustion chamber, turbine, and turbine nozzle. The turbine drives both the fan and compressor. A schematic diagram of the fan as well as the cycle (successive processes) is illustrated in Figures 5.1 and 5.2. Here the BPR is defined as

$$\beta = \frac{\text{Bypass airflow}}{\text{Primary airflow}} = \frac{\dot{m}_{\text{cold}}}{\dot{m}_{\text{hot}}} \equiv \frac{\dot{m}_{\text{fan}}}{\dot{m}_{\text{core}}}$$

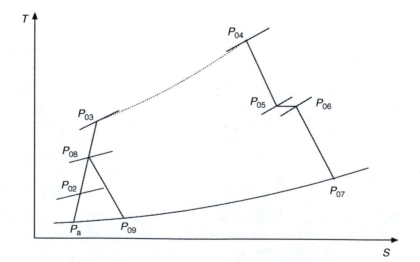

FIGURE 5.1   Schematic diagram of a single-spool turbofan engine.

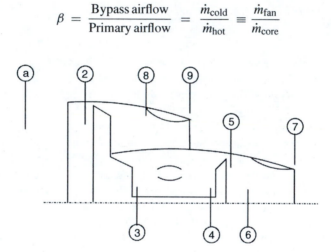

FIGURE 5.2   Temperature–entropy (T–S) diagram for the single-spool turbofan engine.

Thus if the air mass through the core (compressor) is ($\dot{m}_a$), then the bypass air mass flow rate is ($\beta \dot{m}_a$). Sometimes the core mass flow rate is identified as the gas generator mass flow rate ($\dot{m}_{GG}$). However, this designation will not be used in this text.

Now the successive elements will be examined.

1. *Intake:*   The inlet conditions of the air entering the inlet are the ambient temperature and pressure (P_a and T_a respectively). The intake has an isentropic efficiency (η_d). For a flight Mach number of (M_a), then the temperature and pressure at the outlet of the intake are T_{02} and P_{02} given by the relations

$$P_{02} = P_a \left(1 + \eta_d \frac{\gamma_c - 1}{2} M_a^2 \right)^{\gamma_c/(\gamma_c - 1)} \tag{5.1}$$

$$T_{02} = T_{0a} = T_a \left(1 + \frac{\gamma_c - 1}{2} M_a^2 \right) \tag{5.2}$$

where η_d is the diffuser isentropic efficiency and γ_c is the specific heat ratio for air (cold stream).

2. *Forward fan:* For a known fan pressure ratio (π_f) and isentropic efficiency (η_f), then the temperature and pressure at the outlet of the fan are given by the following relations:

$$P_{08} = (P_{02}) (\pi_f) \tag{5.3}$$

$$T_{08} = T_{02} \left[1 + \frac{1}{\eta_f} \left(\pi_f^{(\gamma_c - 1)/\gamma_c} - 1 \right) \right] \tag{5.4}$$

3. *Compressor:* Similarly, both the compressor pressure ratio (π_c) and its isentropic efficiency (η_c) are known. Thus, the temperature and pressure at the outlet of the compressor are given by the following relations:

$$P_{03} = (P_{08}) (\pi_c) \tag{5.5}$$

$$T_{03} = T_{08} \left[1 + \frac{1}{\eta_c} \left(\pi_c^{(\gamma_c - 1)/\gamma_c} - 1 \right) \right] \tag{5.6}$$

4. *Combustion chamber:* The temperature at the end of combustion process T_{04} is generally known. The maximum temperature in the cycle, which is frequently identified as the turbine inlet temperature (TIT) occurs here. The pressure at the end of combustion depends on the pressure drop in the combustion process itself. Thus, depending on the known expression for the pressure drop in the combustion chamber it may be expressed as

$$P_{04} = P_{03} - \Delta P_{cc} \tag{5.7}$$

or

$$P_{04} = P_{03} (1 - \Delta P_{cc}\%) \tag{5.8}$$

The energy balance for the combustion chamber yields the following relation for the fuel-to-air ratio (f), defined as the ratio between the mass of fuel burnt and the air mass flow rate through the core (hot)

$$f = \dot{m}_f / \dot{m}_a$$

$$\dot{m}_a (1+f) Cp_h T_{04} = \dot{m}_a Cp_c T_{03} + \eta_b \dot{m}_f Q_R$$

$$f = \frac{(Cp_h/Cp_c)(T_{04}/T_{03}) - 1}{\eta_b (Q_R/Cp_c T_{03}) - (Cp_h/Cp_c)(T_{04}/T_{03})} \tag{5.9}$$

5. *Turbine:* Since the turbine drives both the compressor and turbine, then the energy balance for this spool per unit air mass flow rate is given by the following relation:

$$W_t = W_c + W_f$$

From the above relation, the temperature ratio across the turbine is deduced as follows:

$$\dot{m}_a (1+f) Cp_h (T_{04} - T_{05}) = \dot{m}_a Cp_c (T_{03} - T_{08}) + (1+\beta) \dot{m}_a Cp_c (T_{08} - T_{02})$$

$$T_{04} - T_{05} = \frac{Cp_c \{[(T_{03} - T_{08})] + (1+\beta)[(T_{08} - T_{02})]\}}{Cp_h (1+f)}$$

$$\therefore \frac{T_{05}}{T_{04}} = 1 - \frac{[(T_{03} - T_{08})] + (1+\beta)[(T_{08} - T_{02})]}{(T_{04})(Cp_h/Cp_c)(1+f)} \tag{5.10a}$$

The above relation can be rewritten as

$$\frac{T_{05}}{T_{04}} = 1 - \frac{[(T_{03} - T_{02})] + \beta[(T_{08} - T_{02})]}{(T_{04})(Cp_h/Cp_c)(1+f)} \tag{5.10b}$$

This temperature ratio can be further expressed as follows:

$$\therefore \frac{T_{05}}{T_{04}} = \left\{ 1 - \frac{(1/\eta_c)\left[(P_{03}/P_{02})^{(\gamma_c-1)/\gamma_c} - 1\right] + (\beta/\eta_f)\left[(P_{08}/P_{02})^{(\gamma_c-1)/\gamma_c} - 1\right]}{(1+f)(Cp_h/Cp_c)(T_{04}/T_{02})} \right\} \tag{5.10c}$$

The expansion ratio through the turbine can now be evaluated as long as its adiabatic efficiency (η_t) and the ratio of specific heats (γ_h) of the turbine working fluid are known.

$$\frac{P_{05}}{P_{04}} = \left[1 - \frac{1}{\eta_t}\left(1 - \frac{T_{05}}{T_{04}}\right) \right]^{\gamma_h/(\gamma_h-1)} \tag{5.11a}$$

$$\frac{P_{05}}{P_{04}} = \left[1 - \frac{1}{\eta_t}\left\{ \frac{(1/\eta_c)\left[(P_{03}/P_{02})^{(\gamma_c-1)/\gamma_c} - 1\right] + (\beta/\eta_f)\left[(P_{08}/P_{02})^{(\gamma_c-1)/\gamma_c} - 1\right]}{(1+f)(Cp_h/Cp_c)(T_{04}/T_{02})} \right\} \right]^{\gamma_h/(\gamma_h-1)} \tag{5.11b}$$

6. *Turbine nozzle:* Assume that there are no changes in the total pressure and total temperature in the jet pipe between the turbine and nozzle. Thus $P_{05} = P_{06}$ and $T_{05} = T_{06}$. Next, a check for nozzle choking is performed. Thus for a nozzle isentropic efficiency of η_{n_t}, the critical pressure is calculated from the relation

$$\frac{P_{06}}{P_c} = \frac{1}{[1 - (1/\eta_{n_t})(\gamma_h - 1)/(\gamma_h + 1)]^{\gamma_h/(\gamma_h - 1)}}$$

Now if the nozzle is an ideal, then $\eta_{n_t} = 1$, the above equation is reduced to

$$\left(\frac{P_{06}}{P_c}\right) = \left(\frac{\gamma_h + 1}{2}\right)^{\gamma_h/(\gamma_h - 1)}$$

If $P_c \geq P_a$ then the nozzle is choked. The temperature of the gases leaving the nozzle is obtained from the relation

$$\left(\frac{T_{06}}{T_7}\right) = \left(\frac{\gamma_h + 1}{2}\right)$$

In this case, the gases leave the nozzle at a speed equal to the sonic speed or

$$V_7 = \sqrt{\gamma_h R T_7} \tag{5.12a}$$

If the nozzle is unchoked ($P_7 = P_a$), then the speed of the gases leaving the nozzle is now given by

$$V_7 = \sqrt{2C_{p_h} T_{06} \eta_{n_t} \left[1 - (P_a/P_{06})^{(\gamma_h - 1)/\gamma_h}\right]} \tag{5.12b}$$

The pressure ratio in the nozzle is obtained from the relation

$$\frac{P_{06}}{P_a} = \frac{P_{06}}{P_{05}} \frac{P_{05}}{P_{04}} \frac{P_{04}}{P_{03}} \frac{P_{03}}{P_{02}} \frac{P_{02}}{P_{0a}} \frac{P_{0a}}{P_a}$$

7. *Fan nozzle:* The fan nozzle is also checked to determine whether choked or unchoked. Thus, the critical pressure is calculated from the relation

$$\frac{P_{08}}{P_c} = \frac{1}{[1 - (1/\eta_{fn})(\gamma_c - 1)/(\gamma_c + 1)]^{\gamma_c/(\gamma_c - 1)}}$$

Now if the nozzle is an ideal, then $\eta_{fn} = 1$ and the above equation will be reduced to

$$\left(\frac{P_{08}}{P_c}\right) = \left(\frac{\gamma_c + 1}{2}\right)^{\gamma_c/(\gamma_c - 1)}$$

If $P_c \geq P_a$ then the fan nozzle is choked. The temperature of the gases leaving the nozzle is obtained from the relation

$$\left(\frac{T_{08}}{T_9}\right) = \left(\frac{\gamma_c + 1}{2}\right)$$

Then the gases leave the nozzle at a speed equal to the sonic speed or

$$V_9 = \sqrt{\gamma_c R T_9} \tag{5.13a}$$

If the nozzle is unchoked ($P_9 = P_a$), then the speed of the gases leaving the nozzle is now given by

$$V_9 = \sqrt{\frac{2\gamma_c R T_{08} \eta_{fn}}{(\gamma_c - 1)} \left[1 - (P_a/P_{08})^{(\gamma_c - 1)/\gamma_c}\right]} \tag{5.13b}$$

$$\frac{P_{08}}{P_a} = \frac{P_{08}}{P_{02}} \frac{P_{02}}{P_a}$$

The thrust force is now obtained from the general relation

$$\frac{T}{\dot{m}_a} = (1 + f) V_7 + \beta V_9 - U (1 + \beta) + \frac{1}{\dot{m}_a} [A_7 (P_7 - P_a) + A_9 (P_9 - P_a)]$$

$$= (1 + f) V_7 + \beta (V_9 - U) - U + \frac{1}{\dot{m}_a} [A_7 (P_7 - P_a) + A_9 (P_9 - P_a)] \tag{5.14a}$$

The specific thrust, the thrust force per total air mass flow rate ($\dot{m}_{at}$), is further evaluated from the relation

$$\frac{T}{\dot{m}_{at}} = \frac{T}{\dot{m}_h + \dot{m}_c} = \frac{T}{\dot{m}_a (1 + \beta)} = \frac{(1 + f)}{(1 + \beta)} V_7 + \frac{\beta}{(1 + \beta)} V_9 - U$$

$$+ \frac{1}{\dot{m}_a (1 + \beta)} [A_7 (P_7 - P_a) + A_9 (P_9 - P_a)] \tag{5.14b}$$

The thrust-specific fuel consumption (TSFC) is

$$\text{TSFC} = \frac{\dot{m}_f}{T} = \frac{\dot{m}_f}{\dot{m}_a} \frac{\dot{m}_a}{T} = \frac{f}{T/\dot{m}_a} \tag{5.15}$$

Two important questions for the design of turbofan engines are as follows:

1. What is the best value for the BPR (β)?
2. What is the best pressure ratio of the secondary (bypass) stream (π_f)?

The answers to these questions depend on a number of factors including

- Components efficiencies
- Structural weights
- Nature of application

The increase in thrust ΔT gained by adding a fan to a turbojet engine will be of the form shown in Figure 5.3. Associated with the increase in mass flow an increase in drag ΔD is achieved due to the larger engine installation. Thus, for the desired flight conditions the best BPR would be that

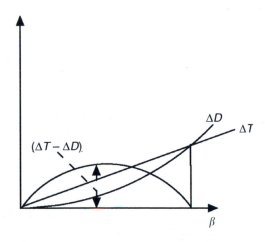

FIGURE 5.3   Thrust and drag variation of a turbofan engine with BPR (β).

which corresponds to the maximum difference between the increased thrust and the increased drag ($\Delta T - \Delta D$), the net thrust gain.

If, on the other hand, takeoff thrust is a very important consideration, then, it might be worthwhile to increase the BPR beyond this value, since, at takeoff, the drag would be small.

Moreover, both the jet velocities of the cold air from the fan nozzle and the hot gases from the turbine nozzle are less than the jet velocity of turbojet engines. This represents an additional advantage of turbofan engine over simple turbojet engines as the jet noise is correlated to the jet velocity. The intensity of jet noise has been shown to be proportional to the eighth power of the velocity of the jet relative to the ambient air, thus small reductions in velocity means significant reductions in noise. It is well known that, the noise of the jet engine, especially those noise components with frequencies disagreeable to humans, appear to come from the compressor.

Another possible advantage of turbofan engines over the turbojet ones is associated with the lower temperature of the hot exhausts before the nozzle. Thus, the exhaust is then much less susceptible to infrared detection.

5.3 FORWARD FAN UNMIXED TWO-SPOOL ENGINES

5.3.1 THE FAN AND LOW-PRESSURE COMPRESSOR ON ONE SHAFT

The famous engine of this configuration is the General Electric CF6 engine series. This engine is composed of a low-pressure spool with a single-stage fan and a three-stage low-pressure compressor (LPC) (sometimes denoted as booster by aero engines manufacturers). Both are driven by a five-stage low-pressure turbine (LPT). The high-pressure spool is composed of a 14-stage high-pressure compressor (HPC) driven by a two-stage high-pressure turbine (HPT).

Another famous example is Pratt and Whitney PW4000 series.

A typical turbofan of this configuration together with the corresponding T–S diagram is shown in Figures 5.4 and 5.5.

The different modules are analyzed here.

1. *Intake:* The same governing equations for the intake in the single-spool configuration, that is, Equations 5.1 and 5.2 are applied here.

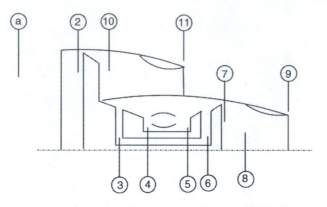

FIGURE 5.4 Layout of a two-spool turbofan engine (fan and compressor driven by LPT).

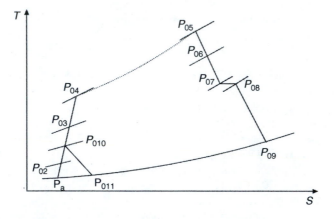

FIGURE 5.5 Temperature–entropy diagram of a two-spool turbofan engine.

2. *Fan:* The following equations are applied:

$$P_{010} = (P_{02})(\pi_f) \tag{5.16}$$

$$T_{010} = T_{02}\left[1 + \frac{1}{\eta_f}\left(\pi_f^{(\gamma-1)/\gamma} - 1\right)\right] \tag{5.17}$$

3. *Low-pressure compressor:* The pressure ratio achieved in this compressor is less than 2.

$$P_{03} = (P_{010})(\pi_{LPC}) \tag{5.18}$$

$$T_{03} = T_{010}\left[1 + \frac{1}{\eta_{LPC}}\left(\pi_{LPC}^{(\gamma-1)/\gamma} - 1\right)\right] \tag{5.19}$$

4. *High-pressure compressor:*

$$P_{04} = (P_{03})(\pi_{HPC}) \tag{5.20}$$

$$T_{04} = T_{03}\left[1 + \frac{1}{\eta_{HPC}}\left(\pi_{HPC}^{(\gamma-1)/\gamma} - 1\right)\right] \tag{5.21}$$

In Equations 5.17 through 5.21, $\gamma \equiv \gamma_c$.

5. *Combustion chamber:* The pressure at the outlet of the combustion chamber is obtained from the pressure drop in the combustion chamber

$$P_{05} = P_{04} - \Delta P_{cc} \tag{5.22a}$$

or

$$P_{05} = P_{04}(1 - \Delta P_{cc}\%) \tag{5.22b}$$

The temperature at the outlet of the combustion chamber is also the maximum temperature in the engine and known in advance. Thus, the fuel-to-air ratio is calculated from the relation

$$f = \frac{(Cp_h/Cp_c)\,(T_{05}/T_{04}) - 1}{\eta_b\,(Q_R/Cp_c T_{04}) - (Cp_h/Cp_c)\,(T_{05}/T_{04})} \tag{5.23}$$

6. *High-pressure turbine:* To calculate the temperature and pressure at the outlet of the HPT, an energy balance between the HPC and HPT is expressed by the relation

$$\dot{m}_a\,Cp_c\,(T_{04} - T_{03}) = \dot{m}_a\,(1 + f)\,Cp_h\,(T_{05} - T_{06}) \tag{5.24}$$

From the above relation, the temperature at the outlet of the turbine T_{06} is defined. Moreover, from the known isentropic efficiency of HPT (η_{HPT}) then the outlet pressure from the HPT (P_{06}) may be obtained as explained above in the single-spool turbofan (Equations 5.11a and 5.11b).

7. *Low-pressure turbine:* An energy balance between the fan and LPC from one side and the LPT on the other side is expressed by the relation

$$\beta \dot{m}_a Cp_c\,(T_{010} - T_{02}) + \dot{m}_a\,Cp_c\,(T_{03} - T_{02}) = \dot{m}_a\,(1 + f)\,Cp_h\,(T_{06} - T_{07}) \tag{5.25a}$$

or

$$(1 + \beta)\,\dot{m}_a Cp_c\,(T_{010} - T_{02}) + \dot{m}_a\,Cp_c\,(T_{03} - T_{010}) = \dot{m}_a\,(1 + f)\,Cp_h\,(T_{06} - T_{07}) \tag{5.25b}$$

$$\therefore T_{07} = T_{06} - \left(\frac{1+\beta}{1+f}\right)\left(\frac{Cp_c}{Cp_h}\right)(T_{010} - T_{02}) - \left(\frac{Cp_c}{Cp_h}\right)\left(\frac{T_{03} - T_{010}}{1+f}\right)$$

The pressure at the outlet is obtained from the relation

$$P_{07} = P_{06}\left(1 - \frac{T_{06} - T_{07}}{\eta_{t2} \times T_{06}}\right)^{\gamma_h/(\gamma_h - 1)}$$

Now, if there is an *air bleed* from the HPC at a station where the pressure is P_{03b}, then the energy balance with the HPT gives

$$\dot{m}_a Cp_c (T_{03b} - T_{03}) + \dot{m}_a (1 - b)(T_{04} - T_{03b}) = \dot{m}_a (1 + f - b) Cp_h (T_{05} - T_{06})$$

(5.26)

where $b = \dot{m}_b / \dot{m}_a$ is the air bleed ratio defined as the ratio between the air bled from the HPC and the core air flow rate.

Moreover, such a bleed has its impact on the energy balance of the low-pressure spool as the air passing through the LPT is now reduced.

$$(1 + \beta) \dot{m}_a Cp_c (T_{010} - T_{02}) + \dot{m}_a Cp_c (T_{03} - T_{010}) = \dot{m}_a (1 + f - b) Cp_h (T_{06} - T_{07})$$

(5.27)

The flow in the jet pipe is frequently associated with a pressure drop mainly due to skin friction.

Thus the pressure upstream of the turbine nozzle is slightly less than the outlet pressure from the turbine. The temperature however, is the same. Thus

$$P_{08} = P_{07} \left(1 - \Delta P_{\text{jet pipe}} \right)$$

$$T_{08} = T_{07}$$

Concerning the other modules or components, a similar procedure is followed. The exhaust velocities of both the cold air from the fan nozzle and the hot gases from the turbine nozzles are obtained after checks for choking. The thrust force, specific thrust, and the TSFC are calculated.

Example 1   A two-spool turbofan engine is to be examined here. The low-pressure spool is composed of a turbine driving the fan and the LPC. The high-pressure spool is composed of a HPC and a HPT. Air is bled from an intermediate state in the HPC.

The total pressure (in psia) and total temperature (in °C) during a ground test ($M = 0.0$) are recorded and shown in the following table:

Station	Fan Inlet	Fan Outlet	LPC Outlet	Air Bleed Location	HPC Outlet	CC Outlet	HPT Outlet	LPT Outlet
P_0 (psia)	14.7	23.2	35.6	110	347	332	86.4	21.4
T_0 (°C)	15	62.7	109.3	275	501	1286	906	512

The fan and turbine nozzles have isentropic efficiency of 0.9.
It is required to calculate the following:

(a) The fan isentropic efficiency (stations 2–F2.5)
(b) The high-pressure ratio isentropic efficiency (stations 3–4)
(c) The fuel-to-air ratio (f)

(d) The air bleed ratio (b)

(e) The BPR (β)

(f) The thrust force if the total air mass flow rate is 280 kg/s

Assume the following values for the different variables

$$\eta_b = 0.96$$

$$Q_{HV} = 45{,}000 \text{ kJ/kg}$$

$$\gamma_{air} = 1.4$$

$$\gamma_{gases} = 1.33$$

Solution: The modules of the engine and its different processes plotted on the T–S diagram are shown in Figures 5.4 and 5.5.

(a) Fan module (from stations 2 to 10)

$$T_{02} = 15 + 273 = 288 \text{ K}$$

$$T_{010} - T_{02} = 62.7 - 15 = 47.7 \text{ K}$$

The fan isentropic efficiency may be expressed as in Figure 5.6.

$$\eta_f = \frac{T_{010S} - T_{02}}{T_{010} - T_{02}} = \frac{T_{02}\left[(P_{010}/P_{02})^{(\gamma-1)/\gamma} - 1\right]}{T_{010} - T_{02}} = \frac{288\left[(23.2/14.7)^{0.286} - 1\right]}{47.7}$$

$$= \frac{288 \times 0.1394}{47.7}$$

$$= 0.8416$$

(b) High-pressure compressor

With $T_{03} = 109.3 + 273 = 382.3 \text{ K}$, then the isentropic efficiency of the high-pressure compressor is given from the relation

$$\eta_{HPC} = \frac{T_{04S} - T_{03}}{T_{04} - T_{03}} = \frac{T_{03}\left[(P_{04}/P_{03})^{(\gamma-1)/\gamma} - 1\right]}{T_{04} - T_{03}} = \frac{382.3\,[0.9178]}{391.7}$$

$$\eta_{HPC} = 0.895$$

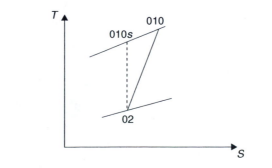

FIGURE 5.6   Isentropic efficiency.

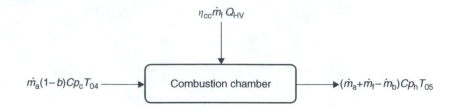

(c) Fuel-to-air ratio

$$f = \frac{\dot{m}_f}{\dot{m}_a} = \frac{(1-b)\,(Cp_h T_{05} - Cp_c T_{04})}{\eta_b Q_{HV} - Cp_h T_{05}} = (1-b)\,\frac{1.147 \times 1{,}559 - 1.005 \times 774}{0.96 \times 45{,}000 - 1.147 \times 1{,}559}$$

$$f = 0.0244\,(1-b) \tag{1}$$

(d) Energy balance for high-pressure spool

$$Cp_h\,(1 + f - b)\,(T_{05} - T_{06}) = Cp_c\,(T_{03b} - T_{03}) + Cp_c\,(1-b)\,(T_{04} - T_{03b})$$

Substituting from (1)

$$1.147\,[1 + 0.0244\,(1-b) - b]\,(380) = 1.005\,[165.7 + 226 - 226b]$$
$$1.147\,[1.0244 - 1.0244b]\,(380) = 393.65 - 227.13b$$
$$b = 0.2406$$
$$\text{Bleed} = 24.06\%$$

Comment
This bleed ratio looks like a great percentage; however as will be described in the turbine cooling sections, each blade row may need some 2% for cooling, and the two turbines here have a total of five stages and so 10 blade rows. In addition, bleeding has its other applications for cabin air conditioning as well as other anti-icing applications.

$$\text{Fuel-to-air ratio}\ (f) = 0.0185$$

This ratio for fuel-to-air ratio is a reasonable figure, which assures the value previously obtained for the bleed ratio.

(e) Energy balance for low-pressure spool

$$(1 + f - b)\,Cp_h\,(T_{06} - T_{07}) = \beta Cp_c\,(T_{010} - T_{02}) + Cp_c\,(T_{03} - T_{02})$$
$$0.7779 \times 1.147 \times 394 = \beta \times 1.005 \times 47 + 1.005 \times 94$$
$$\beta = 5.4425$$

Thus the BPR of this engine is 5.4425.

Fan nozzle

The first step in nozzle analysis is to check whether it is choked or not by calculating the pressure ratio.

$$\frac{P_{010}}{P_c} = \frac{1}{\left(1 - \left((1/\eta_{fn})\left((\gamma_c - 1)/(\gamma_c + 1)\right)\right)\right)^{\gamma_c/(\gamma_c-1)}}$$

$$= \frac{1}{\left(1 - (1/0.9)(1.4 - 1)/(1.4 + 1)\right)^{1.4/(1.4-1)}} = 2.0478$$

$$P_c = 11.32 \text{ psi}$$

Since $P_c < P_a$, then the fan nozzle is unchoked.

$$V_{e_{fan}} = \sqrt{2Cp_c\eta_n T_{010}\left\{1 - \left(\frac{P_a}{P_{010}}\right)^{(\gamma_c-1)/\gamma_c}\right\}} = \sqrt{2 \times 1005 \times 0.9 \times 335.7 \times 0.1223} = 272.7 \text{ m/s}$$

Turbine nozzle

Here also a check for the choking of turbine nozzle is followed.

$$\frac{P_{08}}{P_c} = \frac{1}{\left(1 - \left((1/\eta_{tn})\left((\gamma_h - 1)/(\gamma_h + 1)\right)\right)\right)^{\gamma_h/(\gamma_h-1)}}$$

$$= \frac{1}{\left(1 - (1/0.9)(1.33 - 1)/(1.3 + 1)\right)^{1.33/(1.33-1)}} = 1.9835$$

$$P_c = 11.32 \text{ psi}$$

Since $P_c < P_a$, then the turbine nozzle is also unchoked.

$$V_{e_h} = \sqrt{2Cp_h\eta_n T_{08}\left\{1 - \left(\frac{P_a}{P_{08}}\right)^{(\gamma_h-1)/\gamma_h}\right\}} = \sqrt{2 \times 1147 \times 0.9 \times 785\left\{1 - \left(\frac{14.7}{21.4}\right)^{0.25}\right\}}$$

$$V_{e_h} = 381 \text{ m/s}$$

Since the total air mass flow rate is

$$\dot{m}_a = 280 \text{ kg/s}$$

the fan (cold) air mass flow rate is

$$\dot{m}_{fan} = \frac{\beta}{1 + \beta}\dot{m}_a = \frac{5.4425}{6.4425} \times 280 = 236.5 \text{ kg/s}$$

And the core (hot) mass flow rate is

$$\dot{m}_{core} = 43.5 \, kg/s$$

Thrust force (T)

$$T = \dot{m}_{fan} V_{e_{fan}} + (1 + f - b) \dot{m}_{core} V_{e_{core}}$$
$$T = 236.5 \times 272.7 + 0.7779 \times 43.5 \times 381 = 77375 \, N$$
$$\therefore T = 77.4 \, kN$$

Example 2   One of the techniques used in short takeoff and landing (STOL) fighters is the flap blowing turbofan engines. Air is bled from the LPC and ducted to the flap system in the wing trailing edge (Figure 5.7). Such accelerated air flow creates greater lift than would be possible by the wings themselves. Here are the data for one of theses engines: During takeoff at sea level ($T_a = 288$ K and $P_a = 101$ kPa) the engine has a BPR of 2.0 and the core engine air flow is 22.7 kg/s; 15% of this air is bled at the exit of the LPC. The components efficiencies are

$$\eta_f = 0.9, \quad \eta_c = 0.88, \quad \eta_b = 0.96, \quad \eta_t = 0.94, \quad \eta_n = 0.95$$

The total pressure recovery factor for the intake is $r_d = P_{02}/P_{0a} = 0.98$. The pressure ratio of the fan and both the LPC and HPC are 2.5, 3.0, and 5.0, respectively. The maximum allowable turbine total inlet temperature is 1500 K. The total pressure loss in the combustion chamber is 3%. The fuel heating value is 43,000 kJ/kg. It is required to calculate

(a) The properties of the working fluid in all the states
(b) The thrust force
(c) The TSFC
(d) The overall efficiency

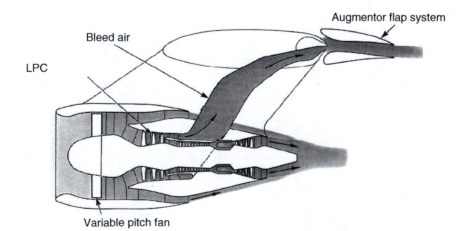

FIGURE 5.7   Flap blowing engine. (Reproduced from Rolls-Royce plc, copyright © Rolls-Royce plc 2007. With permission.)

Solution: The given data are

$$T_a = 288 \text{ K}, \quad P_a = 101 \text{ kPa}$$

$$\beta = 2.0, \quad \dot{m}_{core} = 22.7 \text{ kg/s}, \quad b = 15\%$$

$$\eta_c = 0.88, \quad \eta_f = 0.9, \quad \eta_b = 0.94, \quad \eta_n = 0.95$$

$$r_d = \frac{P_{02}}{P_{0a}} = 0.98$$

$$\pi_{ctotal} = 15, \quad \pi_{LPC} = 3.0, \quad \pi_{HPC} = 5.0, \quad \pi_f = 2.5,$$

$$T_{0\,max} = 1500 \text{ K}, \quad \Delta P_{cc} = 3\%, \quad \text{and} \quad Q_{HV} = 43{,}000 \text{ kJ}$$

Figure 5.8a illustrates the core and bleed mass flow rates, (b) illustrates the core and fan mass flow rates, and (c) and (d) illustrate the modules and stations of the engine and the T–S diagram. For takeoff operation

$$P_{02} = P_a = 101 \text{ kPa}$$

$$T_{02} = T_a = 288 \text{ K}$$

$$P_{03} = P_{02} \times \pi_{fan} = 101 \times 2.5 = 252.5 \text{ kPa}$$

The isentropic efficiency of the fan will be calculated as seen from Figure 5.6.

$$\eta_f = \frac{T_{03s} - T_{02}}{T_{03} - T_{02}} = \frac{T_{02}\left[(\pi_{fan})^{(\gamma-1)/\gamma} - 1\right]}{T_{03} - T_{02}}$$

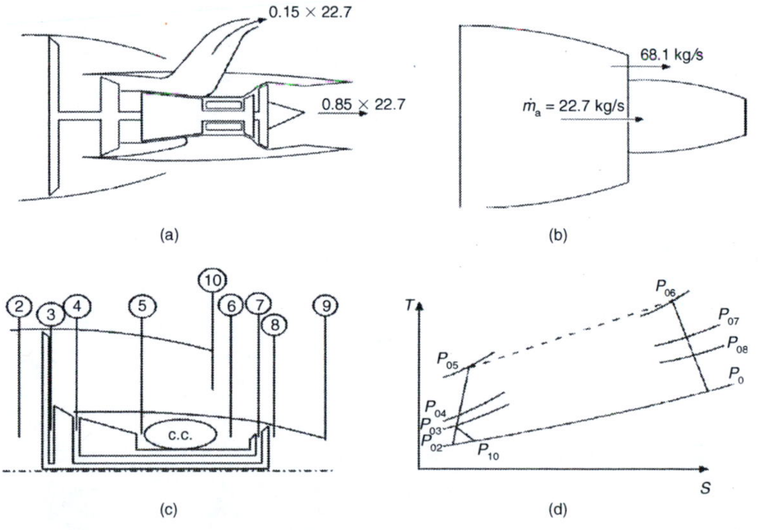

FIGURE 5.8 Flap blowing engine (a) and (b) air flow through the engine (c) station numbering (d) T–S diagram.

$$T_{03} = 383.9 \text{ K}$$

$$P_{04} = P_{03} \times \pi_{\text{LPC}} = 757.5 \text{ kPa}$$

$$T_{04} = T_{03}\left[1 + \frac{1}{\eta_c}\left(\pi_{\text{LPC}}^{(\gamma-1)/\gamma} - 1\right)\right] = 383.9\left[1 + \frac{1}{0.88}(1.369 - 1)\right] = 544.9 \text{ K}$$

$$P_{05} = P_{04} \times \pi_{\text{HPC}} = 3787.5 \text{ kPa}$$

$$T_{05} = T_{04}\left[1 + \frac{1}{\eta_c}\left(\pi_{\text{HPC}}^{(\gamma-1)/\gamma} - 1\right)\right] = 544.9\,[1.6643] = 906.9 \text{ K}$$

$$P_{06} = P_{05} \times [1 - \Delta P_{56}\%] = P_{05} \times 0.97 = 3673.9 \text{ kPa}$$

$$(\dot{m}_a)_{\text{LPC}} = 22.7 \text{ kg/s}$$

$$(\dot{m}_a)_{\text{HPC}} = (\dot{m}_a)_{\text{LPC}} \times (1 - b) = 19.3 \text{ kg/s}$$

$$\dot{m}_f = \frac{(\dot{m}_a - \dot{m}_b)\,[Cp_h T_{06} - Cp_c T_{05}]}{\eta_{cc} Q_{\text{HV}} - Cp_h T_{06}}$$

$$\dot{m}_a - \dot{m}_b = 0.85\dot{m}_{\text{core}} = 19.3 \text{ kg/s}$$

$$f = \frac{\dot{m}_f}{\dot{m}_a - \dot{m}_b} = \frac{[1.147 \times 1.500 - 1.005 \times 906.9]}{0.96 \times 43{,}000 - 1.147 \times 1{,}500} = \frac{809}{39560} = 0.0205$$

$$\eta_P \dot{m}_f\, Q_{\text{HV}}$$

$$\dot{m}_a(1-b)Cp_c T_{05} \longrightarrow \boxed{\text{Combustion chamber}} \longrightarrow \dot{m}_a(1+f-b)Cp_h T_{06}$$

Energy balance for high-pressure spool

From Figure 5.8c, the energy balance assuming nearly no mechanical losses in power transmission from the turbine to the compressor, is then

$$\dot{m}_a (1 - b)\, Cp_c\, (T_{05} - T_{04}) = \dot{m}_a (1 + f - b)\, Cp_h\, (T_{06} - T_{07})$$

$$T_{07} = T_{06} - \left(\frac{1 - b}{1 + f - b}\right)\frac{Cp_c}{Cp_h}(T_{05} - T_{04}) = 1400 - \frac{0.85}{0.87088}\frac{1.005}{1.147}(906.9 - 544.9)$$

$$T_{07} = 1089.2 \text{ K}$$

$$\eta_t = \frac{T_{06} - T_{07}}{T_{06} - T_{07s}} = \frac{T_{06} - T_{07}}{T_{06}\left[1 - (P_{07}/P_{06})^{(\gamma-1)/\gamma}\right]}$$

$$\frac{P_{07}}{P_{06}} = \left[1 - \frac{T_{06} - T_{07}}{\eta_t T_{06}}\right]^{\gamma/(\gamma-1)} = \left(1 - \frac{310.8}{0.94 \times 1400}\right)^4 = 0.34$$

$$P_{07} = 1249.126 \text{ kPa}$$

Energy balance for low-pressure spool

From Figure 5.9, the energy balance

$$\beta \dot{m}_a Cp_c\,(T_{03} - T_{02}) + \dot{m}_a Cp_c\,(T_{04} - T_{02}) = \dot{m}_a (1 + f - b)\, Cp_h\,(T_{07} - T_{08})$$

$$T_{08} = 1204.6 - \frac{1.005}{0.87008 \times 1.147}\{3 \times (383.9 - 288) + (544.9 - 288)\}$$

$$T_{08} = 656.6 \text{ K}$$

$$\frac{P_{08}}{P_{07}} = \left[1 - \frac{T_{07} - T_{08}}{\eta_t T_{07}}\right]^{\gamma/(\gamma-1)} = 0.0709$$

$$P_{08} = 137.04 \text{ kPa}$$

Check hot nozzle choking.

$$\frac{P_{08}}{P_c} = \frac{1}{(1 - (1/\eta_{tn})(\gamma - 1)/(\gamma + 1))^{(\gamma/\gamma-1)}} = \frac{1}{(1 - (1/0.9)(1.33 - 1)/(1.33 + 1))^4} = 1.907$$

$$P_c = 71.86 \text{ kPa}$$

Since $P_c < P_a$, then the hot nozzle is unchoked

$$P_9 = P_a = 101 \text{ kPa}$$

$$\eta_{tn} = \frac{T_{08} - T_9}{T_{08} - T_{9s}} = \frac{T_{08} - T_9}{T_{08}\left[1 - (P_a/P_{08})^{(\gamma-1)/\gamma}\right]}$$

$$T_9 = 610.78 \text{ K}$$

$$V_9 = \sqrt{2C_P(T_{08} - T_9)} = \sqrt{1.147 \times 287 \times (656.6 - 610.78)} = 324.2 \text{ m/s}$$

Check cold (fan) nozzle

$$\frac{P_{03}}{P_c} = \frac{1}{(1 - (1/\eta_n)(\gamma - 1)/(\gamma + 1))^{\gamma/(\gamma-1)}} = \frac{1}{((1 - (1/0.95))((1.4 - 1)/(1.4 + 1)))^{3.5}} = 1.9643$$

$$P_c = 128.54 \text{ kPa}$$

Since $P_c > P_a$, then the fan nozzle is choked.

$$P_{10} = P_c = 128.54 \text{ kPa}$$

$$T_{10} = T_{03}\left[1 - \eta_{fn}\left\{1 - \left(\frac{P_c}{P_{03}}\right)^{(\gamma-1)/\gamma}\right\}\right]$$

$$= 383.9\left[1 - 0.95\left\{1 - \left(\frac{1}{1.964}\right)^{0.286}\right\}\right] = 319.8 \text{ m/s}$$

$$V_{10} = \sqrt{\gamma R T_{10}} = \sqrt{1.4 \times 287 \times 319.8}s$$

$$V_{10} = 358.5 \text{ m/s}$$

The thrust force

$$T = \beta \dot{m}_a (V_{10} - V_{\text{Flight}}) + \dot{m}_a \left[(1 + f - b) V_9 - V_{\text{Flight}}\right] + A_{10}(P_{10} - P_a)$$

$$V_{\text{Flight}} = 0$$

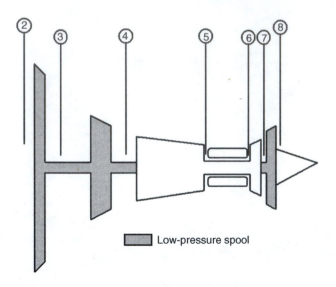

FIGURE 5.9   The two spools of the flap blowing engine.

The exhaust area of the fan nozzle must be first calculated.

$$A_{10} = \frac{\beta \dot{m}_a R T_{10}}{P_{10} V_{10}} = \frac{3 \times 22.7 \times 287 \times 319.8}{128{,}540 \times 358.5} = 0.136 \text{ m}^2$$

$$T = 3 \times 22.7 \times 358.5 + 22.7 \times 0.87088 \times 324.2 + 0.136 \times 128.54 \times 1000$$

$$T = 24{,}414 + 6{,}409 + 17{,}481 = 48{,}304 \text{ N}$$

$$T = 48.304 \text{ kN}$$

$$\text{TSFC} = \frac{\dot{m}_f}{T} = \frac{f \times \dot{m}_{cc}}{T} = \frac{0.02088 \times 19.3}{48.304} = 0.00834 \text{ kg/kN} \cdot \text{S}$$

The overall efficiency is

$$\eta_o = \frac{TV}{\dot{m}_f Q_{HV}} = 0$$

Since the flight speed is zero as the aircraft is still on ground then both the propulsive and overall efficiencies are zeros.

5.3.2   Fan Driven by the LPT and the Compressor Driven by the HPT

The famous example for such an engine is the GE Rolls-Royce (RR) F136 engine. It is a 40,000 lb thrust engine manufactured by Rolls-Royce and General Electric companies. This engine has three-stage fan (RR), five-stage compressor (GE), advanced annular combustor (RR), one-stage HPT (GE) and a three-stage LPT, first stage (GE), and the second and third stages (RR).

Here each spool is composed of a single compressing element; either a fan or a compressor. A schematic drawing for such a configuration together with its T–S diagram is shown in Figures 5.10 and 5.11.

Only the changes in the analysis between this case and that in Section 5.3.1 will be discussed here.

1. Energy balance between the fan and the LPT taking a possible bleed from the HPC

$$(1 + \beta) \dot{m}_a C p_c (T_{03} - T_{02}) = \dot{m}_a (1 + f - b) C p_h (T_{06} - T_{07}) \qquad (5.28)$$

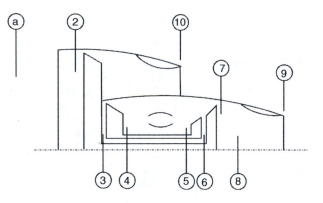

FIGURE 5.10 Layout for a two-spool turbofan (fan driven by LPT and compressor driven by HPT).

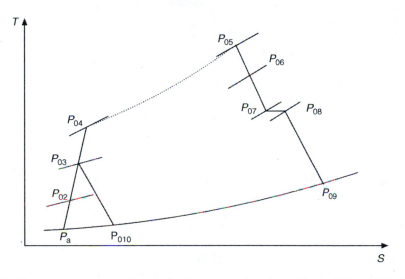

FIGURE 5.11 Temperature–entropy diagram for the two-spool engine in Figure 5.10.

2. Energy balance between the HPC and the HPT:

Assuming that there is an air bleed from station 3b, then

$$\dot{m}_a Cp_c (T_{03b} - T_{03}) + \dot{m}_a (1 - b) Cp_c (T_{04} - T_{03b}) = \dot{m}_a (1 + f - b) Cp_h (T_{05} - T_{06}) \quad (5.29)$$

If there is no bleed from the compressor then put $b = 0$ in Equations 5.28 and 5.29.

5.3.3 A GEARED FAN DRIVEN BY THE LPT AND THE COMPRESSOR DRIVEN BY THE HPT

Geared turbofans are the next big step in engine performance, efficiency, and economy. Conventional turbofans consist of a high-pressure spool and a low-pressure spool. Since the fan is normally part of the low spool, both are turning at the same speed. This speed, however, is often a compromise. The fan really operates more efficiently at low rotational speeds (rpm) while the rest of the low spool is more efficient at higher speeds. Putting a reduction gear in between these components makes it possible for the fan and the low spool to run at their optimum speed. This in turn will minimize the possibility of formation of shock waves in the fan leading to a higher efficiency. The other advantages are higher fuel efficiency, reduced noise levels, and fewer engine parts (fewer compressors and turbine stages).

EXAMPLES FOR THIS CONFIGURATION

An example for this new design is the aero engine PW8000 [3]. Pratt & Whitney launched its program, Advanced Technology Fan Integrator (AFTI) and combined Hartford-based engineering resources with those of Pratt & Whitney Canada and making use of the ADP project (Advanced Ducted Propulsion) of the German manufacturer Motoren- und Turbinen-Union (MTU) together with its partner Fiat for several key components, particularly their gearbox.

Pratt & Whitney says the AFTI demonstrator combines Hartford-based engineering resources with those of Pratt & Whitney Canada, while also enlisting the help of MTU and Fiat for several key components.

The fan rotates at nearly one-third of the speed of the low-pressure spool (3200/9000 rpm). Moreover, the fan BPR is either 10 or 11 for a fan diameter of 1.93 or 1.83 m.

The fan is a single-stage axial configuration (Figure 5.12) driven by a planetary gearbox. The LPC is a three-stage axial type. The HPC is also five-stage axial type. The HPC is a single-stage axial type while the low pressure turbine is a three-stage axial type also. Thus the total number of stages is 13 while similar turbofan engines have 20 or more stages. The fuel saving is 9%. The reduction in noise is 30 dB (cumulative) below current Stage 3 limitations.

The PW8000 will share the core with the PW6000, which is Pratt's other new engine but is designed as a conventional turbofan. While the PW6000 is designed for the thrust class of 70–101 kN (15,750–22,950 lb), the more powerful PW8000 geared fan is aimed at the 110–155 kN (24,750–34,875 lb) category. Expected year of flight is 2008 after completion of tests in 2007.

Another geared fan is the Pratt & Whitney Canada PW800 [4].

PW800 was shown for the first time in its stand in the 2001 Paris Air show, Le Bourget. Its thrust is in the 10,000–20,000 lb range. It represents a new family of environmentally friendly "green" engines that will burn 10% less fuel, be quieter, and cleaner. The German partner MTU will provide both the HPC and the LPT. Moreover, their partner Fiat Avio will provide the fan drive gearbox assembly and the accessory gearbox.

A third example is the Honeywell ALF502R (Figure 5.13).

Now, to examine this engine, there are no changes in the thermodynamic analysis from the above case except in the energy balance for the low spool.

If the mechanical efficiency of the gearbox is η_{gb} and the ratio of the bled air in the compressor is (b) then the energy balance of the low-pressure spool is expressed by the relation

$$(1 + \beta)\, \dot{m}_a Cp_c\, (T_{03} - T_{02}) = \eta_{gb} \dot{m}_a\, (1 + f - b)\, Cp_h\, (T_{06} - T_{07}) \tag{5.30}$$

The energy balance relation for the high-pressure spool is given by the relation 5.29.

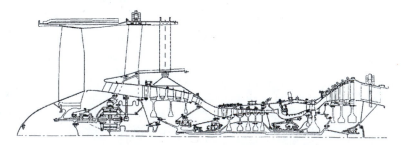

FIGURE 5.12   Geared fan (SAE International Aerospace Engineering online).

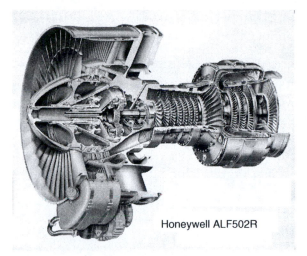

Honeywell ALF502R

FIGURE 5.13 Honeywell ALF502R Geared Turbofan.

5.4 FORWARD FAN UNMIXED THREE-SPOOL ENGINE

The three-spool engine is composed of a low-, intermediate-, and high-pressure spools running at different speeds. The fan and the LPT compose the low-pressure spool. The intermediate spool is composed of an intermediate pressure compressor and intermediate pressure turbine. The high-pressure spool is also composed of a HPC and HPT. RR was the first aero engine manufacturer to design, develop, and produce the three-spool turbofan engine. The RR RB211 was the first three-spool engine to enter service. Later on several manufacturers developed and manufactured this type of engines.

The main advantages of three-spool arrangement are as follows:

1. Shorter modules and shafts that result in a shorter engine

   A. Single-stage fan with no booster stages
   B. Fewer overall compressor stages and fewer variable stages
   C. Shorter HPC
2. Higher efficiencies as each spool is running at its "right speed"
3. Greater engine rigidity
4. Lighter weight
5. Fan and core compressors can run at their individual optimum speeds.

The main drawbacks of this three-spool category are that they are more complex to build and maintain [5].

EXAMPLES FOR THREE-SPOOL ENGINES

1. Rolls-Royce RB211

RB211-22 Series [5]: This is the first of the whole series, which first saw service in 1972. Its thrust rating is 42,000 lbf (169 kN). Being the pioneer three-shaft engine it underwent difficult gestation. However, it improved during service and matured into a reliable engine.

RB211-524 Series: A development of the -22, it featured a very mature design. Its thrust rating is 50,000–60,600 lbf (222–270 kN). It was first fitted into Boeing 747 in 1977. Its excellent service record led it to be fitted to the improved Lockheed L-1011 in 1981.

An improved version, -524G rated at 58,000 lbf (258 kN) and -524H rated at 60,600 lbf (270 kN), featuring FADEC, was offered with Boeing 747-400 and Boeing 767. The -524G and H is the first to feature the wide-chord fan, which increases efficiency, reduces noise, and gives added protection against foreign object damage. This was later adopted by GE and Pratt & Whitney for their engines.

RB211-535 Series: This is essentially a scaled down version of the -524. Its thrust range spans from 37,000 to 43,100 lbf (165–192 kN). It powers Boeing 757 and the Russian Tupolev TU-204 airliner. The latter series shares common features with the later series -524 such as wide-chord fan and Full Authority Digital Engine Control (FADEC).

2. Rolls-Royce Trent

It is a family of high bypass developed from RB211 with thrust ratings spanning between 53,000 and 95,000 lbf (236–423 kN). The Trent's advanced layout provides lighter weight and better performance compared to the original RB211 and other comparable competing engines. It features the wide-chord fan and single crystal HPT blades inherited from later generations of the RB211, but with improved performance and durability.

The core turbomachinery is brand new, giving better performance, noise and pollution levels.

The main points different from two-spool engine are also introduced. The layout of this engine and the T–S diagram are shown in Figures 5.14 and 5.15. Air bleed is also extracted from the intermediate pressure compressor at station (3b).

1. Energy balance of the first spool (Fan and LPT)

$$(1 + \beta) \dot{m}_a Cp_c (T_{03} - T_{02}) = \dot{m}_a (1 + f - b) Cp_h (T_{08} - T_{09}) \qquad (5.31)$$

2. Energy balance for the intermediate spool (IPC and IPT)

$$\dot{m}_a Cp_c (T_{03b} - T_{03}) + (1 - b) \dot{m}_a Cp_c (T_{04} - T_{03b}) = \dot{m}_a (1 + f - b) Cp_h (T_{07} - T_{08}) \qquad (5.32)$$

3. Energy balance for the high-pressure spool (HPC and HPT)

$$\dot{m}_a (1 - b) Cp_c (T_{05} - T_{04}) = \dot{m}_a (1 + f - b) Cp_h (T_{06} - T_{07}) \qquad (5.33)$$

The same procedure discussed earlier in evaluating the fuel-to-air ratio and jet velocities of the cold air and hot gases from the fan and turbine nozzles may be followed here. The propulsive, thermal, and overall efficiencies are obtained following the appropriate equations from Chapter 2.

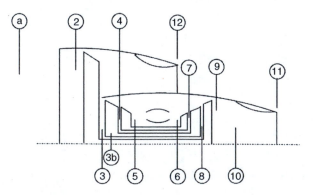

FIGURE 5.14   Layout of a three-spool engine.

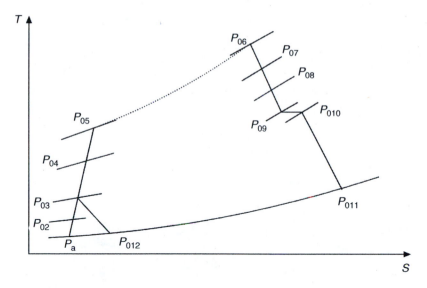

FIGURE 5.15   T–S diagram for a three-spool turbofan.

Example 3   A three-spool turbofan (forward fan) is shown in Figure 5.14. It has the following data:

Fan pressure ratio	1.42:1
Overall compressor pressure ratio	25
HPC pressure ratio	5
BPR ($\dot{m}_{cold}/\dot{m}_{hot}$)	5:1
Fan air mass flow rate	500 kg/s
Fuel-to-air ratio (at takeoff thrust)	0.0177
Ambient pressure	1.0 bar
Ambient temperature	288 K
Total pressure recovery at inlet	98%
Efficiencies	
Fan	99%
Intermediate compressor	89%
HPC	90%
HPT	90%
Intermediate pressure turbine	90%
LPT	90%
Combustion chamber	98%
Mechanical efficiency	99%
Percentage bleeding air of core mass flow	2%
Fuel heating value	45,000 kJ/kg
Specific heat at constant pressure	$Cp_{air} = 1.005$ kJ/kg/K
	$Cp_{gases} = 1.147$ kJ/kg/K
Hot gases nozzle and Fan nozzle efficiency	87%
Percentage total pressure drop in combustion chamber relative to HPC	2%
Percentage pressure losses in the jet pipe relative to LPT	2%

$\gamma_{air}=1.4$, $\gamma_{gases}=1.33$

Engine weight	28.8 kN
Maximum diameter	2172 mm
Maximum frontal area	2.79 m^2

Calculate the take-off thrust at $M = 0.2$

For all elements up to the combustion chamber $\gamma = 1.4$, $Cp = 1.005$ kJ/kg K.

Solution: The same states designation of Figure 5.14 is used here. The T–S will be as shown in Figure 5.15. However, bleed is alternatively assumed from a station downstream of the HPC and upstream of the combustion chamber.

1. *Air inlet*

$$U_a = M\sqrt{\gamma RT_a} = 68 \text{ m/s}$$

$$T_{02} = T_{0a} = T_a\left(1 + \frac{\gamma - 1}{2}M_a^2\right) = 290.3\text{K}$$

$$P_{0a} = P_a\left(1 + \frac{\gamma - 1}{2}M_a^2\right)^{\gamma/(\gamma - 1)} = 1.0283 \text{ bar}$$

$$P_{02} = 0.98\, P_{0a} = 1.008 \text{ bar}$$

2. *Fan*

$$P_{03} = \pi_f \times P_{02} = 1.42 \times 1.008 = 1.431 \text{ bar}$$

$$T_{03} = T_{02}\left(1 + \frac{\pi_f^{(\gamma-1)/\gamma} - 1}{\eta_f}\right) = 290.3\left(1 + \frac{1.42^{0.286} - 1}{0.99}\right) = 321 \text{ K}$$

3. *Intermediate-pressure compressor*

$$P_{04} = \pi_{LPC} \times P_{03} = 5 \times 1.431 = 7.155 \text{ bar}$$

$$T_{04} = T_{03}\left(1 + \frac{\pi_{IPC}^{(\gamma-1)/\gamma} - 1}{\eta_{IPC}}\right) = 321 \times 1.656\text{K} = 532 \text{ K}$$

4. *High-pressure compressor*

$$P_{05} = \pi_{HPC} \times P_{04} = 5 \times 7.155 = 35.775 \text{ bar}$$

$$T_{05} = T_{04}\left\{1 + \left[\frac{\pi_{HPC}^{(\gamma-1)/\gamma} - 1}{\eta_{HPC}}\right]\right\} = 532\left\{1 + \left[\frac{(5)^{0.286} - 1}{0.90}\right]\right\} = 532 \times 1.6489 = 877.2 \text{ K}$$

5. *Combustion chamber (CC):* For all the hot elements including CC, turbines, and nozzles

$$\gamma = 1.33, \quad Cp = 1.147 \text{ kJ/kg} \cdot \text{K}$$

Energy balance of combustion chamber
A bleed from the outlet of the HPC is considered here.

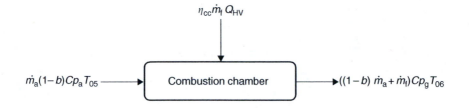

$$\text{with } b = \dot{m}_b/\dot{m}_a \quad \text{and} \quad f = \dot{m}_f/\dot{m}_a$$

$$\dot{m}_a \left[(1 - b)\, Cp_a T_{05} \right] + \eta_{cc}\dot{m}_f Q = \left[(1 - b)\, \dot{m}_a + \dot{m}_f \right] Cp_g T_{06}$$

$$(1 - b)\, Cp_a T_{05} + \eta_{cc} fQ = (1 + f - b)\, Cp_g T_{06}$$

$$T_{06} = \frac{(1 - b)\, Cp_a T_{05} + \eta_{cc} fQ}{(1 + f - b)\, Cp_g} = 1437 \text{ K}$$

$$P_{06} = 0.98\, P_{05} = 0.98 \times 35.775 = 35.06 \text{ bar}$$

6. High-pressure turbine

$$\eta_m W_{HPT} = W_{HPC}$$

$$\eta_m \dot{m}_{hot} Cp_g\, (T_{06} - T_{07}) = \dot{m}_{a_{core}} Cp_a\, (T_{05} - T_{04})$$

$$\dot{m}_{hot} = (1 + f - b)\, \dot{m}_{a_{core}} = 0.9977\, \dot{m}_{a_{core}}$$

$$\Delta T_{067} = T_{06} - T_{07} = \frac{Cp_a\, (T_{05} - T_{04})}{0.9977 \times \eta_m \times Cp_g} = \frac{1.005\, (877.2 - 532)}{0.9977 \times 0.99 \times 1.147} = 306.22 \text{ K}$$

$$T_{07} = T_{06} - \Delta T_{067} = 1130 \text{ K}$$

$$P_{07} = P_{06} \left[1 - \frac{\left[1 - (T_{07}/T_{06}) \right]}{\eta_{HPT}} \right]^{\gamma_g/(\gamma_g - 1)}$$

$$= 35.06 \left[1 - \frac{\left[1 - (1130/1437) \right]}{0.9} \right]^{1.33/0.33} = 11.859 \text{ bar}$$

7. Intermediate-pressure turbine: From energy balance

$$\eta_m W_{IPT} = W_{IPC}$$

$$\eta_m \dot{m}_{hot} Cp_g\, (T_{07} - T_{08}) = \dot{m}_{a_{core}} Cp_a\, (T_{04} - T_{03})$$

$$\Delta T_{078} = \frac{Cp_a\, (T_{04} - T_{03})}{0.9977 \times \eta_m \times Cp_g} = \frac{1.005\, (532 - 321)}{0.9977 \times 0.99 \times 1.147} = 187.2 \text{ K}$$

$$T_{08} = T_{07} - \Delta T_{078} = 1130 - 187.2 = 942.8 \text{ K}$$

$$P_{08} = P_{07} \left[1 - \frac{\left[1 - (T_{08}/T_{07}) \right]}{\eta_{IPT}} \right]^{\gamma_g/(\gamma_g - 1)} = 0.44$$

$$P_{08} = 5.24 \text{ bar}$$

8. *Low-pressure turbine*: From energy balance

$$\eta_m W_{LPT} = W_{fan}$$

$$\eta_m Cp_g \dot{m}_{hot} (T_{08} - T_{09}) = Cp_a \left(\dot{m}_{a_{core}} + \dot{m}_{fan} \right) (T_{03} - T_{02})$$

$$\eta_m Cp_g \times 0.9977 \dot{m}_{a_{core}} (T_{08} - T_{09}) = Cp_a \left(\dot{m}_{a_{core}} + \beta \dot{m}_{a_{core}} \right) (T_{03} - T_{02})$$

$$\Delta T_{089} = \frac{Cp_a (1 + \beta)}{\eta_m Cp_g \times 0.9977} (T_{03} - T_{02}) = 165 \text{ K}$$

$$T_{09} = T_{08} - \Delta T_{089} = 943 - 165 = 778 \text{ K}$$

$$P_{09} = P_{08} \left[1 - \frac{\left(1 - (T_{09}/T_{08})\right)}{\eta_{LPT}} \right]^{\gamma_g/(\gamma_g - 1)} ,$$

$$\left[1 - \frac{\left(1 - (778/943)\right)}{0.9} \right]^{(1.33/0.33)} = 5.24 \times 0421 = 2.207 \text{ bar}$$

9. *Hot nozzle*: The inlet temperature and pressure to the hot nozzle are

$$T_{010} = T_{09} = 778 \text{ K}$$

$$P_{010} = P_{09} \times \left(1 - \Delta P_{jet\,pipe} \right) = 0.98 P_{09} = 2.163 \text{ bar}$$

To check whether the hot nozzle is choked or not

$$\frac{P_{010}}{P_C} = \frac{1}{(1 - (1/\eta_n)(\gamma - 1)/(\gamma + 1))^{\gamma/(\gamma-1)}} = 1.8892$$

But $P_{010}/P_a = 2.163$, since $P_{010}/P_a > P_{09}/P_C$ then the nozzle is choked

$$\therefore P_{11} = P_C = \frac{P_{010}}{P_{010}/P_C} = \frac{2.163}{1.8892} = 1.1449 \text{ bar}$$

$$\frac{T_{11}}{T_{010}} = \frac{2}{\gamma + 1} = \frac{2}{1.33 + 1} = 0.858$$

$$T_{11} = 667.5 \text{ K}$$

$$U_{11} = a_{11} = \sqrt{\gamma R T_{11}} = \sqrt{1.33 \times 287 \times 667.5} = 505 \text{ m/s}$$

$$\rho_{11} = \frac{P_{11}}{R T_{11}} = \frac{1.1449 \times 10^5}{287 \times 667.5} = 0.597 \text{ kg/m}^3$$

$$\dot{m}_{core} = 0.9977 \dot{m}_{hot}$$

But $\dot{m}_{\text{hot}} = \dot{m}_{\text{fan}}/\beta = 500/5 = 100$ kg/s

$$\dot{m}_{a_{\text{core}}} = 99.77 \text{ kg/s}$$

$$\dot{m}_{a_{\text{core}}} = \rho_{11} U_{11} A_{11}$$

$$A_{11} = \frac{\dot{m}_{a_{\text{core}}}}{\rho_{11} U_{11}} = \frac{99.77}{0.59 \times 505} = \frac{99.77}{301.7} = 0.3307 \text{ m}^2$$

Thrust due to hot gases

$$T_{\text{hot}} = \dot{m}_{\text{core}} U_{11} - \dot{m}_{\text{hot}} U_a + A_{11} (P_{11} - P_a)$$
$$= 99.77 \times 505 - 100 \times 68 + 0.3307 (1.145 - 1.0) \times 10^5$$
$$= 48.39 \text{ kN}$$

10. *Fan nozzle*: Next the fan nozzle is also checked for choking

$$\frac{P_{03}}{P_a} = 1.431$$

$$\frac{P_{03}}{P_C} = \frac{1}{(1 - ((1/\eta_{\text{fn}})(\gamma - 1)/(\gamma + 1)))^{\gamma/(\gamma-1)}} = \frac{1}{(1 - ((1/0.97)(1.4 - 1)/(1.4 + 1)))^{1.4/(1.4-1)}}$$

$$\frac{P_{03}}{P_C} = 1.899$$

$$\frac{P_{03}}{P_a} < \frac{P_{03}}{P_C}$$

$\therefore$ The fan nozzle is unchoked.

$$P_{12} = P_a$$

$$\frac{T_{12}}{T_{03}} = \left[1 - \eta_n \left\{ 1 - \left(\frac{P_{12}}{P_{03}} \right)^{(\gamma-1)/\gamma} \right\} \right]$$

$$\frac{T_{12}}{T_{03}} = \left[1 - 0.97 \left[1 - \left(\frac{1}{1.431} \right)^{(0.4/1.4)} \right] \right] = 0.9054$$

$$T_{12} = 290.65 \text{ K}$$

$$U_{12} = \sqrt{2 C p_a (T_{03} - T_{12})} = \sqrt{2 \times 1005 \times (321 - 290.65)} = 247 \text{ m/s}$$

Thrust generated by the fan is T_{fan}.

$$T_{\text{fan}} = \dot{m}_{\text{fan}} (U_{12} - U_a) = 500 (247 - 68)$$

$$T_{\text{fan}} = 89.49 \text{ kN}$$

$$\text{Total thrust} = T_{\text{hot gases}} + T_{\text{fan}} = 137.88 \text{ kN}$$

$$\dot{m}_f = f \dot{m}_a = 0.017 \times 100 \times 3600 = 6120 \text{ kg/h}$$

$$\text{TSFC} = \frac{\dot{m}_f}{T} = \frac{6120}{137.88 \times 10^3} = 0.04438 \text{ kg/N} \cdot \text{h}$$

5.5 FORWARD FAN MIXED-FLOW ENGINE

Mixed turbofan engines are always found in either single- or two-spool engines. It was used in the past for military applications only. Nowadays it is used in both civil and military aircraft. An example for a two-spool engine is the CFM56 series. The cold compressed air leaving the fan will not be directly exhausted as previously described but it flows in a long duct surrounding the engine core and then mixes with the hot gases leaving the LPT. Thus the cold air is heated while the hot gases are cooled. Only one mixed exhaust is found.

A layout of a single-spool mixed turbofan and its T–S diagram is shown in Figures 5.16 and 5.17.

5.5.1 MIXED-FLOW TWO-SPOOL ENGINE

Most of the mixed turbofan engines now are two-spool ones. If mixed turbofan engines are analyzed versus unmixed turbofan engines, the following points are found [6]:

1. The optimum fan pressure ratio for a mixed-flow engine is generally lower than that for a separate flow for a given BPR.
2. At a given fan pressure ratio, the mixed-flow engine has a lower BPR and therefore a higher specific thrust.
3. The amount of power that the LPT supplies to drive the fan will be smaller.
4. Possibly one LPT stage less is sufficient.
5. Other features to be considered are
   - Thrust gain due to mixing
   - Noise
   - Weight
   - Reverse thrust.

Hereafter, a detailed analysis of this category will be given. Figures 5.18 and 5.19 present the engine layout and its T–S diagram.

The requirements for the mixing process are equal static pressures and also equal velocities. Thus, from the layout designation these two conditions are specified as $P'_3 = P_7$ and $V'_3 = V_7$, which means that if no pressure losses in the bypass duct connecting the cold and hot streams and no pressure loss in the mixing process occur, then

$$P_{03} = P'_{03} = P_{07} = P_{08} \qquad (5.34)$$

If losses occur in the fan bypass duct, then

$$P'_{03} = P_{03} - \Delta P_{\text{fan/duct}}$$
$$P_{07} = P_{08} = P_{03} - \Delta P_{\text{fan/duct}}$$

In the above equation, no pressure drop is considered during the mixing process.

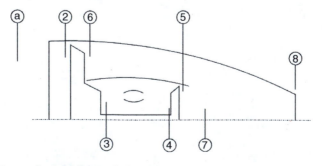

FIGURE 5.16   Single-spool mixed-flow turbofan.

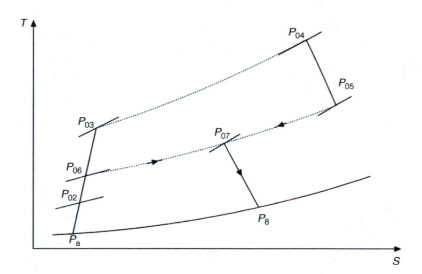

FIGURE 5.17   T–S diagram of single-spool mixed-flow turbofan.

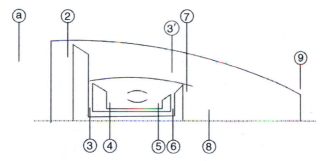

FIGURE 5.18   Layout of a mixed two-spool turbofan.

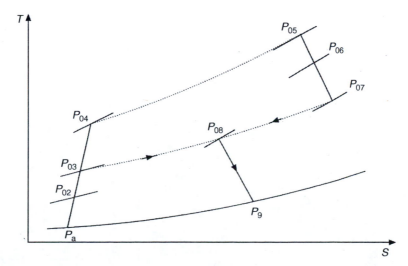

FIGURE 5.19   T–S diagram for two-spool mixed turbofan.

In some engines like the CFM56 series the mixing process takes place in a mixer preceding the nozzle. It results in a quieter engine than if the mixer was not present. Figure 5.20 illustrates a typical mixed two-spool engine.

1. *Energy balance for the low-pressure spool*: Considering a mechanical efficiency for the low-pressure spool of (η_{m1}), then

$$W_{fan} = \eta_{m1} W_{LPT}$$

or

$$\dot{m}_a (1 + \beta) Cp_c (T_{03} - T_{02}) = \eta_{m1} \dot{m}_a (1 + f) Cp_h (T_{06} - T_{07}) \quad (5.35)$$

2. *Energy balance for the high-pressure spool*:
   Also a second mechanical efficiency for the high-pressure spool is (η_{m2}) assumed.

$$W_{HPC} = \eta_{m2} (1 + f) \dot{m}_a Cp_h (T_{05} - T_{06})$$

Thus

$$\dot{m}_a Cp_c (T_{04} - T_{03}) = \eta_{m2} \dot{m}_a (1 + f) Cp_h (T_{05} - T_{06}) \quad (5.36)$$

3. *Mixing process*: The hot gases leaving the LPT and the cold air leaving the fan bypass duct are mixed and give new properties at state (8). Thus such a process is governed by the first law of thermodynamics as follows:

$$H_{03} + H_{07} = H_{08}$$
$$\dot{m}_c Cp_c T_{03} + (1 + f) \dot{m}_h Cp_h T_{07} = [\dot{m}_c + (1 + f) \dot{m}_h] Cp_h T_{08}$$

which is reduced to

$$\beta Cp_c T_{03} + (1 + f) Cp_h T_{07} = (1 + \beta + f) Cp_h T_{08} \quad (5.37)$$

Now for a better evaluation of the gas properties after mixing we can use mass-weighted average properties of the gases at state (8) as follows:

$$Cp_8 = \frac{Cp_7 + \beta Cp_3}{1 + \beta}$$

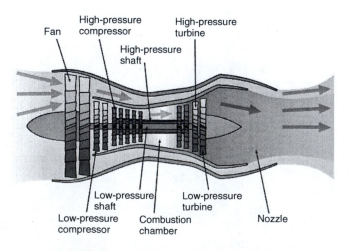

FIGURE 5.20   Two-spool mixed turbofan engine with a mixer.

$$R_8 = \frac{R_7 + \beta R_3}{1 + \beta}$$

$$\gamma_8 = \frac{Cp_8}{Cp_8 - R_8}$$

Consider the real case of mixing where normally losses are encountered, where pressure drop is associated with the mixing process. Such a pressure loss is either given as the value $\Delta P_{\mathrm{mixing}}$ or as a ratio r_m in the mixing process

$$P_{08} = P_{07} - \Delta P_{\mathrm{mixing}} \quad \text{or} \quad P_{08} = r_m P_{07} \quad \text{where} r_m < 1 \approx 0.98$$

The thrust force for unchoked nozzle is then given by the relation

$$T = \dot{m}_a \left[(1 + f + \beta) U_e - (1 + \beta) U \right] \tag{5.38}$$

Example 4 Figure 5.21 illustrates the mixed high BPR turbofan engine CFM56-5C. It has the following data:

BPR	6.6
Typical TSFC	0.04 kg/(N·h)
Cruise thrust	29,360 N
Total mass flow rate at cruise	100 kg/s
During takeoff operation	
Takeoff thrust	140,000 N
Total air mass flow rate at takeoff	474.0 kg/s
Overall pressure ratio	37.5
Total temperature at inlet to the LPT	1251 K
Exhaust gas temperature	349 K
Ambient temperature	288.0 K
Ambient pressure	101 kPa

Assuming that all the processes are ideal and there are no losses in both the combustion chamber and mixing process

A. Prove that the thrust force is given by the relation

$$T = \frac{\dot{m}_a (1 + \beta)(u_e - u)}{1 - \mathrm{TSFC} \times u_e}$$

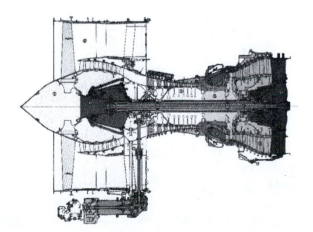

FIGURE 5.21 CFM56-5C mixed high BPR turbofan engine. (Courtesy GE Aviation.)

B. Calculate

    1.   The exhaust velocity at both takeoff and cruise (use the above relation)
    2.   The fuel-to-air ratios at both takeoff and cruise conditions

C. At takeoff condition, calculate

    1.   The total temperature at the outlet of the HPC
    2.   The total temperature at the inlet to the HPT
    3.   The pressure ratio of the HPC
    4.   The fan pressure ratio
    5.   The temperature of the hot gases before mixing
    6.   The pressure ratio of the LPC (booster)

Assume that the nozzle is unchoked.

Solution:
(A) For a mixed-flow turbofan engine, with an unchoked nozzle, the thrust force is expressed as

$$T = \dot{m}_a \left[(1 + f + \beta) u_e - (1 + \beta) u \right] \qquad \text{(a)}$$

The TSFC

$$\text{TSFC} = \frac{\dot{m}_f}{T} = \frac{f \dot{m}_a}{T} \qquad \text{(b)}$$

From Equations (a) and (b)

$$\therefore\ T = \dot{m}_a (1 + \beta) u_e + \text{TSFC} \times T u_e - \dot{m}_a (1 + \beta) u$$
$$T (1 - \text{TSFC} \times u_e) = \dot{m}_a (1 + \beta) (u_e - u)$$

or

$$T = \dot{m}_a (1 + \beta)(u_e - u)/1 - \text{TSFC} \times u_e$$

(B)   (1) Rearranging the above relation to get

$$u_e = \frac{T + \dot{m}_a (1 + \beta) u}{T\,(\text{TSFC}) + \dot{m}_a (1 + \beta)} \qquad \text{(c)}$$

At cruise, the flight speed is

$$u = M\sqrt{\gamma R T_a} = 0.8\sqrt{1.4 \times 287 \times 218.7} = 237.3 \text{ m/s}$$
$$\therefore u_e = \frac{29{,}360 + 100 \times 237.3}{29{,}360 \times (0.04/3{,}600) + 100} = 529.17 \text{ m/s}$$

From Equation (c), at takeoff $u = 0$

$$(u_e)_{\text{takeoff}} = \frac{T}{T\,(\text{TSFC}) + \dot{m}_a (1 + \beta)}$$
$$(u_e)_{\text{takeoff}} = \frac{140{,}000}{140{,}000 \times (0.04/3{,}600) + 474} = 294.4 \text{ m/s}$$

(2) Since $f = \dot{m}_f / \dot{m}_h$ and $\dot{m}_h = \dot{m}_t/(1 + \beta)$,
where $\dot{m}_h$ is the air passing through the hot section of the engine (sometimes called the core air) and $\dot{m}_t$ is the total mass flow rate through the engine

$$\therefore f = \frac{\dot{m}_f (1 + \beta)}{\dot{m}_t}$$

Moreover,

$$\text{TSFC} = \frac{\dot{m}_f}{T}$$

$$\therefore f = \frac{T\,(\text{TSFC})\,(1 + \beta)}{\dot{m}_t}$$

$$(f)_{\text{cruise}} = \frac{29{,}360 \times 0.04 \times 7.6}{100} = 0.02479$$

$$(\dot{m}_f)_{\text{cruise}} = \frac{0.02479 \times 100}{7.6} = 0.326 \text{ kg/s}$$

Moreover, $(f)_{\text{T/O}} = 140{,}000 \times 0.04 \times 7.6/474 = 0.02494$

$$(\dot{m}_f)_{\text{T/O}} = \frac{0.02494 \times 474}{7.6} = 1.555 \text{ kg/s}$$

(C)   (1) At the end of compression process, the delivery total temperature of the HPC (state 5) is

$$T_{05} = T_a\,(\text{OPR})^{(\gamma_c - 1)/\gamma_c} = 288(37.5)^{0.286} = 812 \text{ K}$$

(2) The energy balance at the combustor (from states 5 to 6)

$$\dot{m}_f h_{05} + \dot{m}_f Q_R = (\dot{m}_a + \dot{m}_f)\,h_{o6}$$

Then

$$\dot{m}_f C p_c T_{05} + \dot{m}_f Q_R = (\dot{m}_a + \dot{m}_f)\,C p_h T_{06}$$

$$T_{06} = \frac{f Q_R + C p_c T_{05}}{(1 + f)\,C p_h}$$

Then substituting the value of (f) at takeoff

$$T_{06} = \frac{0.02494 \times 45{,}000 + 1.005 \times 812}{1.02494 \times 1.148} = 1647.4 \text{ K}$$

(3) The energy balance of the high-pressure spool yields

$$C p_c\,(T_{05} - T_{04}) = (1 + f)\,C p_h\,(T_{06} - T_{07})$$

$$C p_c T_{05} \left[1 - \left(\frac{P_{04}}{P_{05}} \right)^{(\gamma_c - 1)/\gamma_c} \right] = (1 + f)\,C p_h\,(T_{06} - T_{07})$$

The pressure ratio P_{04}/P_{05} may be calculated as follows:

$$\frac{P_{04}}{P_{05}} = \left[1 - \frac{(1+f)\,Cp_h\,(T_{06} - T_{07})}{Cp_c\,T_{05}}\right]^{\gamma_c/(\gamma_c-1)}$$

$$\frac{P_{04}}{P_{05}} = \left[1 - \frac{1.02494 \times 1.148 \times (1647.4 - 1251)}{1.005 \times 812}\right]^{3.5} = 0.05148$$

The pressure ratio of the high-pressure compressor is then

$$\pi_{HPC} = \frac{P_{05}}{P_{04}} = 19.423$$

(4) The exhaust velocity is obtained from the relation

$$u_e^2 = 2Cp\,(T_{09} - T_{10}) = 2CpT_{10}\left[\left(\frac{P_{09}}{P_{10}}\right)^{(\gamma_c-1)/\gamma_c} - 1\right]$$

$$\frac{P_{09}}{P_{10}} = \left[1 + \frac{u_e^2}{2CpT_{10}}\right]^{\gamma_c/(\gamma_c-1)} = \left[1 + \frac{(294.4)^2}{2 \times 1005 \times 349}\right]^{3.5}$$

$$\frac{P_{09}}{P_{10}} = 1.503$$

The conditions to be satisfied for proper mixing yield equal total pressure at the fan outlet, LPT, and the end of mixing process; that is

$$P_{03} = P_{08} = P_{09}$$

Also, since a complete expansion is assumed at the nozzle, thus $P_{10} = P_a$

$$\therefore \frac{P_{09}}{P_{10}} = \frac{P_{03}}{P_a} = \pi_f$$

$\therefore$ Fan pressure ratio

$$\pi_f = 1.503$$

(5) Since the fan pressure ratio is now calculated, then

$$T_{03} = T_{02}\pi_f^{(\gamma_c-1)/\gamma_c} = (288)\,(1.503)^{0.286} = 323.6 \text{ K}$$

Also

$$\frac{T_{09}}{T_{10}} = \pi_f^{(\gamma_c-1)/\gamma_c}$$

$$T_{09} = (349)(1.503)^{0.286} = 398.6 \text{ K}$$

Now, from the energy balance within the mixing process

$$\beta \dot{m}_a C p_c T_{03} + (1+f)\dot{m}_a C p_h T_{08} = (1+f+\beta)\dot{m}_a C p_c T_{09}$$

$$T_{08} = \left(\frac{1+f+\beta}{1+f}\right)\frac{C p_c}{C p_h}T_{09} - \left(\frac{\beta}{1+f}\right)\frac{C p_c}{C p_h}T_{03}$$

$$T_{08} = \left(\frac{1+0.02494+6.6}{1.024949}\right)\frac{1.005}{1.148}\times 398.6 - \frac{6.6}{1.02494}\frac{1.005}{1.148}\times 323.6$$

$$= 775.15\,\text{K}$$

(6) The booster pressure ratio (π_{LPC})

$$\pi_{LPC} = \frac{\text{OPR}}{\pi_f \times \pi_{HPC}} = \frac{37.5}{19.423\times 1.503}$$

$$\pi_{LPC} = 1.2846$$

Example 5 A two-spool turbofan engine is in its preliminary design stages. Two configurations for the engine are to be compared; namely the mixed and the unmixed types (Figure 5.22a and 5.22b, respectively). The following characteristics are to be considered: $M = 0.9$, altitude $= 33,000$ ft (where $T_a = 222$ K and $P_a = 26.2$ kPa), fan pressure ratio ($\pi_f = 2$), compressor pressure ratio ($\pi_c = 8$), maximum temperature $= 1400$ K, pressure drop in combustion chamber $\Delta P_{cc} = 3\%$, fuel-to-air ratio $f = 0.017$, fuel heating value $Q_R = 43,000$ kJ/kg, efficiencies for diffuser, compressor, fan, combustion chamber, turbines, and nozzles are respectively, $\eta_d = 0.98$, $\eta_c = 0.88$, $\eta_f = 0.9$, $\eta_{cc} = 0.96$, $\eta_t = 0.94$, $\eta_n = 0.95$. Variations in the specific heats are considered within the engine for cold and hot streams as follows: $Cp_d = Cp_c = Cp_f = 1.005$ kJ/kg/K, $\gamma_c = 1.4$, $Cp_t = Cp_n = 1.2$ kJ/kg/K, $\gamma_h = 1.33$.

Both cases have the same BPR. For the mixing type, the outlet streams of fan and LPT have same total pressure just before mixing. Calculate

(i) The BPR (β);
(ii) TSFC

Solution: The given data may be expressed as:

$M = 0.9$

$H = 33,000\,\text{ft}, \quad P_a = 26.2\,\text{kPa}, \quad T_a = 222\,\text{K}, \quad f = 0.017, \quad Q_R = 43,000\,\text{kJ/kg}$

$\pi_f = 2, \quad \pi_c = 8, \quad T_{max} = 1400\,\text{K}, \quad \Delta P_{cc} = 3\%$

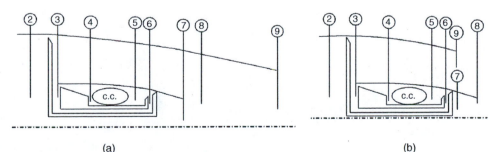

(a) (b)

FIGURE 5.22 (a) Mixed and (b) unmixed turbofan engines.

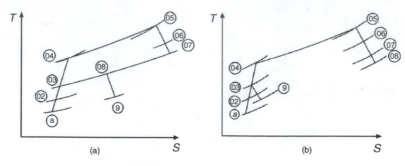

FIGURE 5.23 T–S diagram of mixed (a) and unmixed (b) turbofan engines.

$$\eta_d = 0.98, \quad \eta_c = 0.88, \quad \eta_f = 0.9, \quad \eta_{cc} = 0.96, \quad \eta_t = 0.94, \quad \eta_n = 0.95, \quad \eta_{nf} = 0.95$$
$$Cp_d = Cp_c = Cp_f = 1.005 \text{ kJ/kg} \cdot \text{K}, \quad \gamma_c = 1.4$$
$$Cp_t = Cp_n = 1.2 \text{ kJ/kg} \cdot \text{K}, \quad \gamma_h = 1.33$$

The cycle for both mixed and unmixed cases is plotted in Figure 5.23.

(A) *Mixed turbofan*

$$u = 0.9\sqrt{\gamma R T_a} == 0.9\sqrt{(1.4)\,(287)\,(222)} = 268.8 \text{ m/s}$$

$$T_{02} = T_a \left[1 + \frac{\gamma - 1}{2} M^2\right] = 222\left[1 + 0.2\,(0.9)^2\right] = 258.8 \text{ K}$$

$$T_{02} = 258.8 \text{ K}$$

$$\frac{P_{02}}{P_a} = \left(1 + \eta_d \frac{\gamma - 1}{2} M^2\right)^{\gamma/(\gamma - 1)} = \left[1 + (0.98)\,(0.2)\,(0.9)^2\right]^{3.5} = (1.15876)^{3.5} = 1.676$$

$$P_{02} = 43.9 \text{ kPa}$$

$$P_{03} = P_{02}\,(2) = 87.8 \text{ kPa}$$

$$T_{03} = T_{02}\left[1 + \frac{1}{\eta_f}\left(\pi_f^{(\gamma-1)/\gamma} - 1\right)\right] = T_{02}\left[1 + \frac{1.219 - 1}{0.9}\right]$$

$$T_{03} = 322 \text{ K}$$

$$\frac{P_{04}}{P_{03}} = r_c = 8$$

$$P_{04} = 702.4 \text{ kPa}$$

$$T_{04} = T_{03}\left[1 + \frac{1}{\eta_c}\left\{\pi_c^{(\gamma-1)/\gamma} - 1\right\}\right]$$

$$T_{04} = 344\left[1 + \frac{1}{0.88}\left(8^{0.286} - 1\right)\right] = 612 \text{ K}$$

$$P_{05} = 0.97 \quad P_{04} = 681.3 \text{ kPa}$$

$$T_{05} = 1400 \text{ K}$$

With $\eta_m = 1.0$, then $W_{HPT} = W_{HPC}$

$$\dot{m}_a (1+f) Cp_t (T_{05} - T_{06}) = \dot{m}_a Cp_c (T_{04} - T_{03})$$

With $f = 0.017$

$$1.017 \times 1200 (1400 - T_{06}) = 1005 (612 - 322)$$

$$T_{06} = 1161 \text{ K}$$

$$\frac{P_{06}}{P_{05}} = \left[1 - \frac{1}{\eta_t} \left(1 - \frac{T_{06}}{T_{05}} \right) \right]^{\gamma/(\gamma-1)}$$

$$\frac{P_{06}}{P_{05}} = \left[1 - \frac{1}{0.94} \left(1 - \frac{1161}{1400} \right) \right]^4 = 0.4485$$

$$P_{06} = 305.6 \text{ kPa}$$

Mixing condition

$$P_{07} = P_{03} = P_{08}$$

$$\therefore P_{07} = 87.8 \text{ kPa}$$

$$\frac{T_{07}}{T_{06}} = 1 - \eta_t \left\{ 1 - \left(\frac{P_{07}}{P_{06}} \right)^{(\gamma-1)/\gamma} \right\}$$

$$T_{07} = 1 - 0.94 \left\{ 1 - \left(\frac{87.8}{305.6} \right)^{0.25} \right\} = 869 \text{ K}$$

Power balance between LPT and fan

$$W_F = W_{LPT}$$

$$\therefore (\dot{m}_h + \dot{m}_c) Cp_f (T_{03} - T_{02}) = (1+f) \dot{m}_h Cp_t (T_{06} - T_{07})$$

with $\beta = \dot{m}_c / \dot{m}_h$, then

$$\therefore (1 + \beta) (1005) (322 - 259) = (1.017) (1200) (1161 - 869)$$

$$1 + \beta = 5.635$$

$$\beta = 4.635$$

The process from 7 to 8 is a mixing process.

$$\therefore H_{03} + H_{07} = H_{08}$$

$$\dot{m}_c Cp_c T_{03} + (1+f) \dot{m}_h Cp_h T_{07} = [\dot{m}_c + \dot{m}_h (1+f)] Cp_h T_{08}$$

$$\beta Cp_c T_{03} + (1+f) Cp_h T_{07} = [(1+f) + \beta] Cp_h T_{08}$$

$$T_{08} = \frac{4.635 \times 1005 \times 322 + 1200 \times 869 \times 1.017}{(1 + 0.017 + 4.635) \times 1200} = 377.5 \text{ K}$$

Propelling nozzle (Process 8–9)

$$\frac{P_{08}}{P_a} = \frac{87.8}{26.2} = 3.35$$

But the critical pressure is calculated from the relation;

$$\frac{P_{08}}{P_c} = \left(\frac{1}{1 - (1/\eta_n)(\gamma - 1)/(\gamma + 1)}\right)^{\gamma/(\gamma-1)}$$

$$\frac{P_{08}}{P_c} = \left(\frac{1}{1 - (1/0.95)(0.33/2.33)}\right)^4 = (1.1752)^4 = 1.9076$$

$$\frac{P_{08}}{P_a} > \frac{P_{08}}{P_c}$$

∴ Nozzle is choked.

$$P_9 = P_c = \frac{P_{08}}{1.9076} = 46.026\,\text{kPa}$$

$$T_9 = T_c = \frac{T_{08}}{(\gamma + 1)/2} = \frac{T_{08}}{1.167} = 323.6\,\text{K}$$

$$\rho_9 = \frac{P_9}{RT_9} = \frac{46.026 \times 10^3}{287 \times 323.6}$$

$$\rho_9 = 0.496\,\text{kg/m}^3$$

$$u_9 = C_9 = \sqrt{\gamma RT_9} = \sqrt{1.33 \times 287 \times 323.6}$$

$$u_9 = 351.5\,\text{m/s}$$

$$\dot{m}_9 = \dot{m}_a + \dot{m}_f = \dot{m}_a(1 + f) = \rho_9 u_9 A_9$$

$$\therefore \frac{A_9}{\dot{m}_a} = \frac{1 + f}{\rho_9 u_9} = \frac{1.017}{0.496 \times 351.5}$$

$$\frac{A_9}{\dot{m}_a} = 5.83 \times 10^{-3} \left(\frac{\text{m}^2}{\text{kg/s}}\right)$$

Specific thrust

$$\frac{T}{\dot{m}_a} = [(1 + f)u_9 - u] + \frac{A_9}{\dot{m}_a}(P_9 - P_a)$$

$$\frac{T}{\dot{m}_a} = [(1.017)(351.5) - 268.8] + 5.83 \times 10^{-3} \times (46.026 - 26.2) \times 10^3$$

$$\frac{T}{\dot{m}_a} = 88.68 + 115.59 = 204.27 \left(\frac{\text{N} \cdot \text{s}}{\text{kg}}\right)$$

$$\text{TSFC} = \frac{f}{(T/\dot{m}_a)} = 0.2996 \frac{\text{kg fuel}}{\text{N} \cdot \text{h}}$$

(B)   *Unmixed turbofan*

$$\frac{P_{03}}{P_a} = \frac{87.8}{26.2} = 3.35$$

$$\frac{P_{03}}{P_c} = \left(\frac{1}{1 - ((1/\eta_{nf})(\gamma - 1))/(\gamma + 1)}\right)^{\gamma/(\gamma-1)}$$

$$= \left(\frac{1}{1 - (1/0.95)\,(0.4/2.4)}\right)^{3.5}$$

$$= (1.2127)^{3.5} = 1.964$$

$$\frac{P_{03}}{P_a} > \frac{P_{03}}{P_c}$$

$\therefore$ Fan nozzle is choked.

$$P_9 = P_c$$

$$P_9 = \frac{P_{03}}{1.964} = \frac{87.8}{1.964} = 44.7\,\text{kPa}$$

$$T_9 = T_c = \frac{T_{03}}{1.2127} = \frac{322}{1.2127} = 265.5\,\text{K}$$

$$\rho_9 = \frac{P_9}{RT_9} = \frac{44.7 \times 10^3}{287 \times 265.5} = 0.587\,\text{kg/m}^3$$

$$u_9 = \sqrt{\gamma RT_9} = \sqrt{1.4 \times 287 \times 265.5} = 326.6\,\text{m/s}$$

As previously defined

$$T_{03} = 322\,\text{K}$$

$$T_{04} = 612\,\text{K}$$

Since $T_{05} = 1400\,\text{K}$, then the energy balance between HPC and HPT yields

$$W_{\text{HPC}} = W_{\text{HPT}}$$

$$Cp_c\,(T_{04} - T_{03}) = (1 + f)\,Cp_h\,(T_{05} - T_{06})$$

$$1005 \times (612 - 322) = (1.017) \times 1200\,(1400 - T_{06})$$

$$T_{06} = 1161\,\text{K}$$

Since $P_{06}/P_{05} = \{1 - (1/\eta_t)\,(1 - (T_{06}/T_{05}))\}^{\gamma/(\gamma-1)} = \{1 - (1/0.94)\,(1 - (1161/1400))\}^4 = 0.4486$ with $P_{05} = 681.3\,\text{kPa}$. Then $P_{06} = 305.6\,\text{kPa}$.

Now, from the energy balance between fan and LPT,

since $W_f = W_{LPT}$

$$\therefore (1 + \beta) Cp_c (T_{03} - T_{02}) = (1 + f) Cp_h (T_{06} - T_{07})$$

With same β

$$5.635 \times 1005 \times (322 - 259) = 1.017 \times 1200 \times (1161 - T_{07})$$

$$T_{07} = 869\,K$$

$$\frac{P_{07}}{P_{06}} = \left\{ 1 - \frac{1}{\eta_t} \left(1 - \frac{T_{07}}{T_{06}} \right) \right\}^{\gamma/(\gamma-1)} = \left\{ 1 - \frac{1}{0.94} \left(1 - \frac{869}{1161} \right) \right\}^4 = 0.288$$

$$P_{07} = 88.02\,kPa$$

$$\frac{P_{07}}{P_a} = \frac{88.02}{26.2} = 3.359$$

$$\frac{P_{07}}{P_c} = \left(\frac{1}{1 - (1/\eta_n)(\gamma - 1)/(\gamma + 1)} \right)^{\gamma/(\gamma-1)} = \left(\frac{1}{1 - (1/0.95)(0.33/2.33)} \right)^4 = (1.1752)^4$$

$$\frac{P_{07}}{P_c} = 1.9076$$

$$\therefore \frac{P_{07}}{P_{06}} > \frac{P_{07}}{P_c}$$

$$\therefore P_c > P_a$$

The hot nozzle is choked.

$$P_8 = P_c = \frac{P_{07}}{1.9076} = 46.14\,kPa$$

$$T_8 = T_c = \frac{T_{07}}{1.167} = 744.8\,K$$

$$\rho_8 = \frac{P_8}{RT_8} = \frac{46.14 \times 10^3}{287 \times 744.8}$$

$$\rho_8 = 0.216\,kg/m^3$$

$$u_8 = \sqrt{\gamma RT_8} = \sqrt{1.33 \times 287 \times 744.8s}$$

$$u_8 = 533\,m/s$$

$$\rho_8 u_8 A_8 = (1 + f)\,\dot{m}_h$$

$$\frac{A_8}{\dot{m}_h} = \frac{1 + f}{\rho_8 u_8} = \frac{1.017}{0.216 \times 533}$$

$$\frac{A_8}{\dot{m}_h} = 8.83 \times 10^{-3} \frac{m^2}{kg/s}$$

Define the total mass flow rate $\dot{M} = \dot{m}_h + \dot{m}_c = (1 + \beta)\,\dot{m}_h$.

Since the thrust force is defined as

$$T = \beta \dot{m}_h (u_9 - u) + A_9 (P_9 - P_a) + \dot{m}_h [(1 + f) u_8 - u] + A_8 (P_8 - P_a)$$

$$\frac{T}{\dot{M}} \equiv \frac{T}{(1 + \beta) \dot{m}_h} = \frac{\beta}{1 + \beta} (u_9 - u) + \frac{A_9}{(1 + \beta) \dot{m}} (P_9 - P_a) + \left(\frac{1}{1 + \beta}\right) \{(1 + f) u_8 - u\}$$

$$+ \frac{A_8}{(1 + \beta) \dot{m}_h} (P_8 - P_a)$$

Now since $\rho_9 u_9 A_9 = \beta \dot{m}_h$

$$\therefore \frac{A_9}{\beta \dot{m}_h} = \frac{1}{\rho_9 u_9} = \frac{1}{0.58 \times 328.3} = 5.2517 \times 10^{-3} \frac{m^2}{kg/s}$$

$$\therefore \frac{A_9}{(1 + \beta) \dot{m}_h} = \frac{A_9}{\beta \dot{m}_h} \frac{\beta}{1 + \beta} = 5.2517 \times 10^{-3} \times \frac{4.635}{5.635}$$

$$\frac{A_9}{(1 + \beta) \dot{m}_h} = 4.319 \times 10^{-3} \frac{m^2}{kg/s}$$

$$\frac{T}{\dot{M}} = \frac{4.635}{5.635} (328.3 - 268.8) + 4.319 \times 10^{-3} (44.7 - 26.2) \times 10^3$$

$$+ \frac{1}{5.635} \{(1.017)(533) - 268.8\} + \frac{8.83 \times 10^{-3}}{5.635} (46.14 - 26.2) \times 10^3$$

$$= 48.95 + 79.9 + 48.49 + 31.245$$

$$\frac{T}{\dot{M}} = 208.58 \frac{N}{kg/s}$$

$$\text{TSFC} = \frac{f}{(T/\dot{M})} = \frac{0.017}{208.58} \times 3600 = 0.2934 \left(\frac{kg\ fuel}{Nh}\right)$$

From the above results it may be deduced that for the same BPR, fan, and compressor pressure ratios, the unmixed flow ensures a slight higher specific thrust and better fuel economy.

5.6 MIXED TURBOFAN WITH AFTERBURNER

5.6.1 INTRODUCTION

Since 1970 most recent military aircraft are powered by a low BPR turbofan fitted with an afterburner [7]. The first afterburning turbofan engine was the Pratt & Whitney TF30. Afterburning gives a significant thrust boost for takeoff particularly from short runways like air carriers, transonic acceleration, and combat maneuvers, but is very fuel intensive. This engine is a mixed low BPR forward fan one. The mixed flow still has a sufficient quantity of oxygen for another combustion process. Thus an afterburner is installed downstream of the LPT and upstream of the nozzle. Prodigious amounts of fuel are burnt in the afterburner when it is lit. This rises the temperature of exhaust gases by a significant amount, which results in a higher exhaust velocity/engine specific thrust. For a turbofan engine afterburning (or reheat) offers greater gains because of the relatively low temperature after mixing of the hot and cold streams and the large quantity of the excess air available for combustion.

This nozzle is normally of the variable area type to furnish a suitable media for different operating conditions. The variable geometry nozzle must open to a larger throat area to accommodate the extra volume flow when the afterburner is lit. Unlike the main combustor, where the integrity of the

Modern afterburning turbofan engine

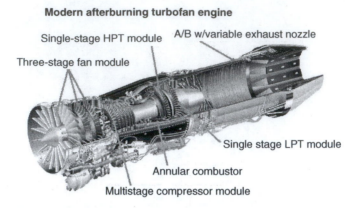

FIGURE 5.24   Afterburning two-spool turbofan engine (GE). (Courtesy GE Aviation.)

downstream turbine blades must be preserved, an afterburner can operate at the ideal maximum (stoichiometric) temperature (i.e., about 2100 K (3780 R)). Now, at a fixed total applied fuel-to-air ratio, the total fuel flow for a given fan airflow will be the same, regardless of the dry specific thrust of the engine. However, a high specific thrust turbofan will, by definition, have a higher nozzle pressure ratio, resulting in a higher afterburning net thrust and, therefore, lower afterburning specific fuel consumption (SFC). However, high specific thrust engines have a high dry SFC. A typical photo for such an engine is shown in Figure 5.24.

Examples of this type of engines and its maximum thrust are Pratt & Whitney series F100-220 (40K), F100-229 (48K), F100-232 (28 K), F101 (51 K), F119 (65 K), Pratt & Whitney F119, Eurojet EJ200, and the GE F110.

Some details on the afterburning engine **Adour Mk.104** are given here.

This engine is installed to Sepecat Jaguar aircraft. It has the following characteristics:

- Length 117 in. (with afterburner removed 77 in.) and inlet diameter 22 in.
- Two-stage axial fan, five-stage axial compressor
- Annular straight-through flow combustion chamber
- One stage for the HPT, one stage for the LPT
- Fully modulated afterburner
- Takeoff thrust (dry) 5260 lbs and with afterburner 8000 lbs

A Russian engine is the RD-133 AFTERBURNING TURBOFAN TVC ENGINE.

- Length 4230 mm, maximum diameter 1040 mm
- Weight 1055 kg
- Full reheat power setting ($H = 0$, $V = 0$, ISA), 89 kN
- Maximum dry power setting ($H = 0$, $V = 0$, ISA), 55 kN
- BPR 0.437
- Maximum TIT, 1720 K

5.6.2 Ideal Cycle

A typical layout of such an engine together with its T–S diagram for an ideal cycle is shown in Figures 5.25 and 5.26.

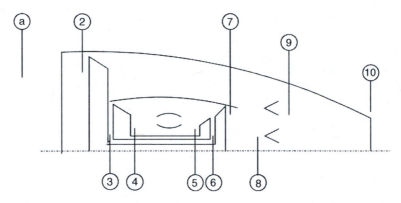

FIGURE 5.25 Layout of a typical two-spool mixed afterburning engine.

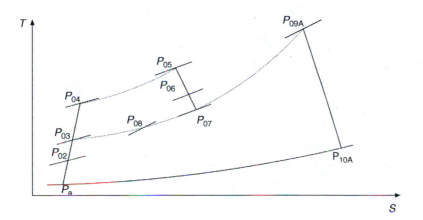

FIGURE 5.26 T–S diagram for two-spool mixed afterburning engine.

The same conditions necessary for mixing described above still hold. Thus the following relation for the total pressure may be stated:

$$P_{03} = P_{07} = P_{08} = P_{09A}$$

where P_{09A} is the total pressure at the outlet of the afterburner, when it is operative. Moreover, the outlet temperature of the afterburner T_{09A} is the maximum temperature within the engine, which is known in advance. It is limited by the maximum temperature that the material of afterburner can withstand. However, it is much higher than the total TIT due to the absence of centrifugal stresses created by the rotation of the turbine blades.

Now the energy balance in the afterburner gives the following afterburner fuel-to-air ratio (f_{ab}). With (f_{ab}) defined as

$$f_{ab} = \frac{\dot{m}_{f_{ab}}}{\dot{m}_c + \dot{m}_h} \tag{5.39a}$$

Then the balance of the afterburner yields the relation

$$f_{ab} = \frac{C_{p9A} T_{09A} - C_{p8} T_{08}}{Q_{HV} - C_{p9A} T_{09A}} \tag{5.39b}$$

Now for a convergent–divergent nozzle, a full expansion to the ambient pressure occurs. Thus the jet speed will be given by the relation

$$\therefore V_{10A} = \sqrt{2Cp_{10A}T_{09A}\left[1 - \left(\frac{P_a}{P_{09A}}\right)^{(\gamma_{10}-1)/\gamma_{10}}\right]} \tag{5.40}$$

The thrust force is now

$$T = \dot{m}_e V_{10A} - \dot{m}_a V_\infty$$

With the exhaust mass flow rate defined as

$$\dot{m}_e = m_c + \dot{m}_h\,(1+f) + \dot{m}_{f_{ab}}$$

$$\dot{m}_e = \{\dot{m}_c + \dot{m}_h\,(1+f)\}\,(1+f_{ab})$$

Then the thrust is now expressed as

$$T = \{\dot{m}_c + \dot{m}_h\,(1+f)\}\,(1+f_{ab})\,V_{10A} - (\dot{m}_c + \dot{m}_h)\,V_\infty \tag{5.41}$$

5.6.3 Real Cycle

The same procedure described in the real mixed turbofans previously mentioned will be followed.

The losses in the mixing process are also governed by either one of the following equations for losses:

$$P_{08} = P_{07} - \Delta P_{mixing}$$

Or $P_{08} = r_m P_{07}$ where $r_m < 1 \approx 0.98$.

The losses in the afterburner are also governed by the relation

$$P_{09A} = P_{08} - \Delta P_{ab}$$

The afterburning combustion is no longer ideal but controlled by the efficiency of burners. Thus the afterburning fuel-to-air ratio is now

$$f_{ab} = \frac{Cp_{9A}T_{09A} - Cp_8 T_{08}}{\eta_{ab}Q_{HV} - Cp_{9A}T_{09A}} \tag{5.39c}$$

5.7 AFT FAN

There are few turbofan engines of the aft fan types. The most famous is the GE CF700 engine.

CF700

The CF700 turbofan engine was developed as an aft fan engine to power the Rockwell Sabre 75 version of the Sabreliner aircraft as well as the dassult Falcon. There are over 400 CF700 aircraft in operation around the world, with an experience of 10 million service hours. The CF700 was the first small turbofan to be certified by the Federal Aviation Administration (FAA). Moreover, it was also used to train Moon-mounted astronauts in Apollo program as the power plant for the Lunar Landing Training vehicle.

Specifications

- Fan/compressor stages: 1/8, core/power turbine stages: 2/1
- Maximum diameter: 33 in, length: 75.5 in
- Dry weight (lbs): 725–737

Application examples: Falcon 20 and Sabreliner 75A/80A.

Figures 5.27 and 5.28 show the layout of a typical aft fan together with its T–S diagram.

The engine is composed of a single spool (compressor driven by turbine as normal) and another turbine/fan configuration. This LPT has blades extending outside the engine core to form the fan section. The blading of the turbine fan must be designed to give turbine blade sections for the hot stream and compressor blade section for the cold stream. Thus a blading of high cost arises, because the entire blade must be made from the turbine material of high weight.

Moreover, manufacturing of these blades are so complicated where the inner part is a turbine blade while the outer part is a fan blade. Thus the convex shape of the inner part of the blade becomes concave in the outer part of the blade and vice versa. This is a necessity as the direction of rotation of the compressor blades must be from the pressure side to the suction side of the opposite blade while the turbine is from the suction side to the pressure side. In addition, if the turbine blades have

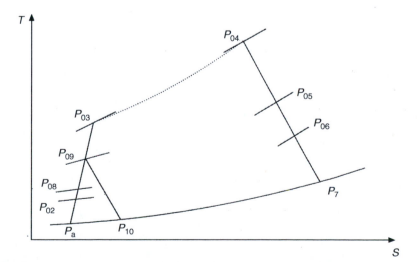

Aft fan turbofan engine

FIGURE 5.27 Layout of aft fan configuration.

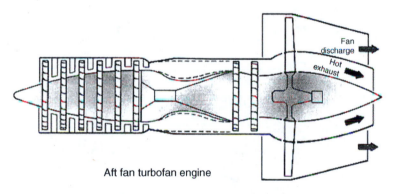

FIGURE 5.28 T–S diagram for aft fan.

to be cooled, then a highly sophisticated manufacturing process is needed. Finally, sealing between the two hot and cold streams is needed.

1. *Core or gas generator intake*: The intake for the core or the gas generator has the same governing equations, that is, Equations 5.1 and 5.2
2. *Fan intake*: Similar governing equations are used with the efficiency of fan diffuser rather than the efficiency of the core diffuser

$$P_{08} = P_a \left(1 + \eta_{fd}\frac{\gamma_c - 1}{2}M_a^2\right)^{\gamma_c/(\gamma_c-1)} \tag{5.42}$$

$$T_{08} = T_{0a} = T_a \left(1 + \frac{\gamma_c - 1}{2}M_a^2\right) \tag{5.43}$$

Note that

$$T_{02} = T_{08} \quad \text{while} \quad P_{02} \neq P_{08} \quad \text{as } \eta_{fd} \neq \eta_d$$

3. *Compressor*: For known values of the pressure ratio and efficiency of the compressor, then the outlet pressure and temperature of the compressor are obtained as

$$P_{03} = \pi_c P_{02}$$

$$T_{03} = T_{02}\left[1 + \frac{1}{\eta_c}\left(\pi_c^{(\gamma_c-1)/\gamma_c} - 1\right)\right]$$

4. *Fan*: The outlet pressure and temperature of the fan section are determined as follows:

$$P_{09} = \pi_f P_{08}$$

$$T_{09} = T_{08}\left[1 + \frac{1}{\eta_f}\left(\pi_f^{(\gamma_c-1)/\gamma_c} - 1\right)\right]$$

5. *Combustion chamber*: The pressure at the outlet of the combustion chamber is defined as

$$P_{04} = P_{03} - \Delta P_{cc}$$

$$P_{04} = P_{03}(1 - \Delta P_{cc}\%)$$

While the maximum cycle temperature T_{04} is known, the fuel-to-air ratio is defined as usual from the typical relation

$$f = \frac{Cp_h T_{04} - Cp_c T_{03}}{\eta_b Q_{HV} - Cp_h T_{04}}$$

6. *Compressor turbine or the HPT*: The temperature at the outlet from the turbine driving the compressor is determined from the energy balance between the compressor and turbine

incorporating a mechanical efficiency η_{m1}

$$W_c = \eta_{m1} W_{HPT}$$

$$Cp_c (T_{03} - T_{02}) = \eta_{m1} (1 + f) Cp_h (T_{04} - T_{05}) \tag{5.44}$$

Thus the temperature at the outlet of the HPT T_{05} is now known. From this temperature and the isentropic efficiency of the HPT, then P_{05} can be calculated as follows:

$$P_{05} = P_{04} \left(1 - \frac{T_{04} - T_{05}}{\eta_{t1} \times T_{04}} \right)^{\gamma_h/(\gamma_h - 1)} \tag{5.44a}$$

7. *Fan turbine*: Also from the energy balance between the fan and the LPT incorporating a second mechanical efficiency η_{m2},

$$W_f = W_{LPT}$$

$$\beta \dot{m}_a Cp_c (T_{09} - T_{08}) = \eta_{m2} \dot{m}_a (1 + f) Cp_h (T_{05} - T_{06}) \tag{5.45}$$

8. *Nozzles*: The aft fan is also of the unmixed type. Thus, it also has two outlet streams; one exits from the cold (fan) section through the fan nozzle. The other is the hot stream and exits from the turbine nozzle. For both nozzles, a check for their choking is needed. Thus the outlet pressures of the hot and cold streams (P_7, P_{10}) are calculated if different from the ambient pressure. The outlet velocities (V_7, V_{10}) are also calculated.

The general thrust equation is now

$$T = \dot{m}_a [(1 + f) V_7 - V] + \beta \dot{m}_a (V_{10} - V) + A_7 (P_7 - P_a) + A_{10} (P_{10} - P_a) \tag{5.46}$$

5.8 V/STOL

Vertical takeoff and landing (VTOL) or STOL are desirable characteristics for any type of aircraft, provided that its normal performance characteristics are fulfilled [8]. This type of engine requires a takeoff thrust of about 120% of the takeoff weight, because the lift has to be produced by the propulsion system [9]. Many approaches were tried since the 1950s for both civil and military applications. However, it became clear that VTOL was not economical for civil applications and there were major problems with engine noise in urban areas. Development then focused on military applications, primarily for fighters; the VTOL capability offered the chance to operate from unprepared areas, air carriers, or from battle-damaged airfields. This is achieved by what is known as the lift/propulsion engines. This type of engine is capable of providing thrust for both normal wing-prone flight and for lift. It is achieved by changing the direction of thrust either by a deflector system consisting of one, two, or four swiveling nozzles or by a device known as a switch-in deflector, which redirects the exhaust gases from a rearward-facing propulsion nozzle to one or two downward-facing lift nozzles.

5.8.1 SWIVELING NOZZLES

The only really successful system to date is the single-engined vectored thrust installation of the RR Pegasus in the Harrier (Figure 5.29). The Pegasus is basically a conventional unmixed turbofan engine having separate exhausts, with vectoring nozzles for both of hot and cold streams. The bypass cold air is exhausted into a pair of swiveling nozzles, one on each side of the aircraft. Moreover, the hot gases are also exhausted via two swiveling nozzles, one on each side of the airplane. In this way,

FIGURE 5.29   Harrier aircraft. (Reproduced from Rolls-Royce plc, copyright © Rolls-Royce plc 2007. With permission.)

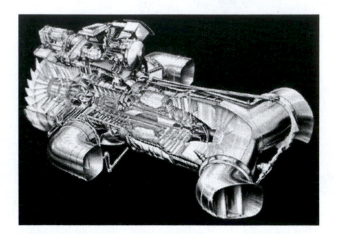

FIGURE 5.30   Harrier engine (Pegasus). (Reproduced from Rolls-Royce plc, copyright © Rolls-Royce plc 2007. With permission.)

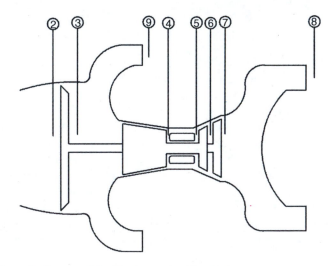

FIGURE 5.31   Schematic diagram of the Harrier engine (Pegasus).

the first ducted fan lift/propulsion engine (the Pegasus) (Figure 5.30) evolved. The Pegasus engine develops a vertical takeoff thrust up to 98 kN (17,000 lbf). The exhaust nozzles may be rotated to provide any combination of horizontal and vertical thrust.

The Pegasus engine is of the two-spool unmixed type and has a three-stage fan, while the compressor has eight stages. Each is driven by a two-stage turbine. A schematic diagram for the Pegasus engine and its T–S diagram are shown in Figures 5.31 and 5.32.

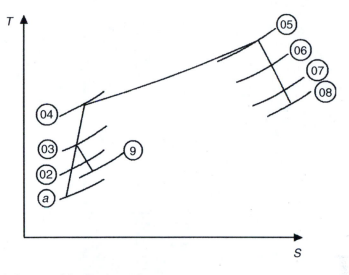

FIGURE 5.32 T–S diagram of the Harrier engine.

Example 6   Hawker Siddeley Harrier G.R.MK.1 illustrated in Figure 5.29 is a single seat V/STOL ground attack fighter. It is powered by a single turbofan engine. The fan and turbine exhausts pass through two pairs of swiveling nozzles as shown. During takeoff the plane moves vertically. The nozzles are rotated 90° such that the thrust force is now a lift force. The turbofan engine has the following characteristics:

Static take-off thrust	98.0 kN
Fan pressure ratio	1.7
Compressor pressure ratio	6.0
Maximum temperature	1,200 K
Fuel heating value	43,000 kJ/kg
BPR	1.0

The fan is driven by the HPT. For sea level conditions ($T_a = 288$ K, $P_a = 101$ kPa), assume that

1. All the processes are ideal
2. Both fan and turbine nozzles are unchoked
3. Constant properties equivalent to the cold air ($Cp = 1.005$ kJ/kg/K and $\gamma = 1.4$)
   Calculate
   (a) The total mass flow rate
   (b) The thermal efficiency

Solution: Since the BPR $\beta = 1$

$$\therefore \dot{m}_c = \dot{m}_h = \tfrac{1}{2}\dot{m}_a$$

Based on the simplifying assumptions listed above, the T–S diagram of Figure 5.32 is replaced by that illustrated in Figure 5.33.
Diffuser: For static operation
$P_{02} = P_a = 101$ kPa and $T_{02} = T_a = 288$ K

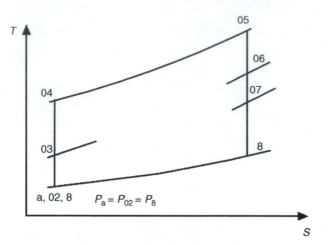

FIGURE 5.33   T–S diagram for ideal processes of the Pegasus at takeoff.

Fan

$$P_{03} = \pi_f \, P_{02} = 1.7 \times 101$$
$$P_{03} = 171.7 \text{ kPa}$$
$$T_{02} = T_a = 288 \text{ K}$$
$$T_{03} = T_{02} \, \pi_f^{(\gamma_c-1)/\gamma_c} = 335.2 \text{ K}$$

Compressor

$$P_{04} = \pi_C P_{03} = 1030.2 \text{ kPa}$$
$$T_{04} = (T_{03}) \, \pi^{(\gamma_c-1)/\gamma_c} = 559.6 \text{ K}$$

Combustion chamber

$$f = \frac{Cp\,(T_{05} - T_{04})}{Q_R - Cp\,T_{04}} = \frac{1.005 \times (1,200 - 559.6)}{43,000 - 1.005 \times 1,200} = \frac{643.6}{41,794} = 0.0154$$
$$P_{05} = P_{04} = 1030.2\,\text{kPa}$$

Energy balance for the high-pressure spool

$$W_C = W_t$$
$$\dot{m}_h\,Cp\,(T_{04} - T_{03}) = \dot{m}_h\,(1+f)\,Cp\,(T_{05} - T_{06})$$
$$T_{04} - T_{03} = (1+f)(T_{05} - T_{06})$$
$$T_{06} = T_{05} - \frac{(T_{04} - T_{06})}{(1+f)} = 1200 - \frac{(559.6 - 335.2)}{(1 + 0.0154)} = 979 \text{ K}$$
$$P_{06} = P_{05} \left(\frac{T_{06}}{T_{05}}\right)^{\gamma/(\gamma-1)} = 1030.2 \left(\frac{979}{1200}\right)^{3.5} = 505 \text{ kPa}$$

Energy balance for the low-pressure spool (fan and free-power turbine)

$$W_f = W_{ft}$$

$$\dot{m}_a Cp \ (T_{03} - T_a) = \dot{m}_h \ (1 + f) \ Cp \ (T_{06} - T_{07}) = \left(\frac{\dot{m}_a}{1 + \beta}\right) \ (1 + f) \ Cp \ (T_{06} - T_{07})$$

$$T_{07} = T_{06} - \left(\frac{1 + \beta}{1 + f}\right) \ (T_{03} - T_a) = 979 - \frac{2}{1.0154} \ (335.2 - 288) = 886 \text{ K}$$

$$P_{07} = P_{06} \left(\frac{T_{07}}{T_{06}}\right)^{\gamma/(\gamma-1)} = 505 \left(\frac{886}{979}\right)^{3.5} = 356.2 \text{ kPa}$$

$$T_8 = T_{07} \left(\frac{P_8}{P_{07}}\right)^{(\gamma-1)/\gamma} = 886 \left(\frac{101}{356.2}\right)^{0.286} = 617.9 \text{ K}$$

$$V_8 = \sqrt{2 \ Cp \ (T_{07} - T_8)} = \sqrt{2 \times 1005 \times (886 - 617.9)} = 734 \text{ m/s}$$

$$V_9 = \sqrt{2 \ Cp \ (T_{03} - T_a)} = \sqrt{2 \times 1005 \times (335.2 - 288)} = 308 \text{ m/s}$$

Thrust force will be

$$\frac{T}{\dot{m}_a} = \frac{\beta}{1 + \beta} V_9 + \frac{(1 + f)}{(1 + \beta)} \ V_8 = \left[\frac{308}{2} + \frac{1.0154}{2} \times 734\right] = 525.65 \text{ N.s/kg}$$

Since the thrust force is 98 kN then

$$98 \times 10^3 = 526.65 \ \dot{m}_a$$

$$\therefore \dot{m}_a = 186.07 \text{ kg/s}$$

Thermal efficiency: The general equation for η_{th} of a turbofan engine (unchoked case) is

$$\eta_{th} = \frac{(\beta/(1 + \beta)) \ V_{ec}^2 + (1 + f)/(1 + \beta)V_{eh}^2 - V_f^2}{(2f/(1 + \beta)) \ Q_s}$$

For takeoff conditions, then

$$\eta_{th} = \frac{\dot{m}_a \left[(\beta/(1 + \beta))V_9^2 + (1 + f)/(1 + \beta)V_8^2\right] - 0}{2 \ \dot{m}_f \ Q_R}$$

with $\beta = 1, f = \dot{m}_f/\dot{m}_h = 2\dot{m}_f/\dot{m}_a$

$$\therefore 2 \ \dot{m}_f = f\dot{m}_a$$

$$\eta_{th} = \frac{V_9^2 + (1 + f) \ V_8^2}{2fQ_R} = \frac{(308)^2 + (1.0154) \ (734)^2}{2 \times 0.0154 \times 43 \times 10^6} = 0.484 = 48.4\%$$

5.8.2 Switch-In Deflector System

Another suggested technique for thrust vectoring is the switch-in deflector. It is not available in reality until now. However, it is visible. One of the switch-in-deflector systems is used in the tandem fan or hybrid fan vectored thrust engine [8] (Figure 5.34).

The turbofan shown is of the mixed-flow afterburning turbofan type. A deflector is positioned in between the tandem fans. During cruise the engine operates as a mixed-flow turbofan engine with the valve switched in the off position. During takeoff or lift thrust, the valve is switched so that the exhaust flow from the front part of the fan exhausts through a downward-facing lift nozzles and a secondary inlet is opened to provide the required airflow to the rear part of the fan and the main engine. In fact, it cannot be said that it is a typical unmixed turbofan engine. The only difference is that the air passing through the fan is not split into two streams, cold and hot ones. However, two exhaust streams exist as in the unmixed turbofan engines.

Figure 5.35 illustrates a schematic diagram for the engine during cruise and takeoff conditions. A detailed thermodynamic analysis of the above shown engine is given here.

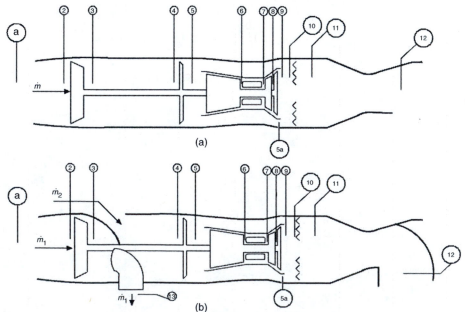

FIGURE 5.34 Vectored thrust engine. (Reproduced from Rolls-Royce plc, copyright © Rolls-Royce plc, 2007. With permission.)

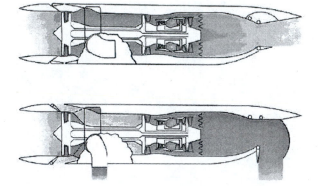

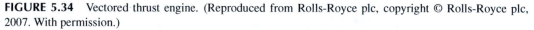

FIGURE 5.35 A schematic diagram for the engine during (a) cruise and (b) takeoff conditions.

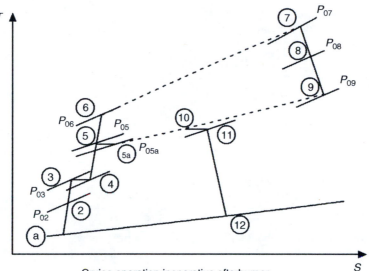

Cruise operation inoperative afterburner

FIGURE 5.36 T–S diagram for cruise conditions.

(a) Cruise
The different processes are presented in Figure 5.36.

Intake: Air is sucked into the engine core through the intake. The ambient conditions depend on the cruise altitude. The compression process through the intake; from far upstream; state (a) to state (2); just upstream of the fan, is governed by the flight speed (V) or the free stream Mach number (M) and the isentropic efficiency of the diffuser (η_d). The outlet pressure and temperature are

$$P_{02} = P_a \left(1 + \eta_d \frac{\gamma_c - 1}{2} M_a^2 \right)^{\gamma_c/(\gamma_c-1)} \tag{5.47}$$

$$T_{02} = T_{0a} = T_a \left(1 + \frac{\gamma_c - 1}{2} M_a^2 \right) \tag{5.48}$$

Fan:

First part: The first part (stage) of the fan has a pressure ratio and isentropic efficiency of (π_{f1}, η_{f1}). The inlet conditions are (T_{02}, P_{02}). The outlet pressure and temperature are calculated as follows:

$$P_{03} = (P_{02})(\pi_{f1}) \tag{5.49}$$

$$T_{03} = T_{02} \left[1 + \frac{1}{\eta_{f1}} \left(\pi_{f1}^{(\gamma_c-1)/\gamma_c} - 1 \right) \right] \tag{5.50}$$

Second part (stage): A pressure drop in the long duct separating the two parts of the fan is ΔP_f, then

$$P_{04} = P_{03} - \Delta P_f$$

The total temperature remains the same.

$$T_{04} = T_{03}$$

Concerning the second part (stage) of the fan with a pressure ratio and isentropic efficiency of (π_{f2}, η_{f2}), the pressure and temperature at the fan outlet are

$$P_{05} = (P_{04})(\pi_{f2}) \tag{5.51}$$

$$T_{05} = T_{04}\left[1 + \frac{1}{\eta_{f2}}\left(\pi_{f2}^{(\gamma_c-1)/\gamma_c} - 1\right)\right] \tag{5.52}$$

Compressor: The compressed air leaving the fan is divided into two streams. The cold stream passes in an annulus duct surrounding the engine (from state 5 to state 5a), while the hot stream proceeds forward toward the compressor and the subsequent hot section. This division occurs in a proportionality of the BPR (β), where

$$\beta = \frac{\dot{m}_c}{\dot{m}_h}$$

Only one compressor exists here and the compression process within the compressor depends on its pressure ratio and isentropic efficiency (π_c, η_c). The outlet conditions are

$$P_{06} = (P_{05})(\pi_c) \tag{5.53}$$

$$T_{06} = T_{05}\left[1 + \frac{1}{\eta_c}\left(\pi_c^{(\gamma_c-1)/\gamma_c} - 1\right)\right] \tag{5.54}$$

A quantity of air is bled from a position downstream the compressor for several reasons including turbine cooling.

This bled air is $\dot{m}_b$, where the bleed ratio is expressed as

$$b = \frac{\dot{m}_b}{\dot{m}_h}$$

Combustion chamber: The maximum temperature in the outlet of the combustion chamber is known (T_{07}) and the pressure within the combustion chamber enhances a pressure drop; thus

$$P_{07} = P_{06}(1 - \Delta P_{cc}) \tag{5.55}$$

The fuel burnt in the combustion chamber is determined from the energy balance

$$f = \frac{(1-b)(Cp_h T_{07} - Cp_c T_{06})}{\eta_b Q_R - Cp_h T_{07}} \tag{5.56}$$

High-pressure turbine: The HPT drives the HPC. The mechanical efficiency of the high-pressure spool is (η_{m1}). The outlet pressure and temperature are obtained from the energy balance between the compressor and turbine.

$$\eta_{m1} W_{HPT} = W_c$$

$$\eta_{m1}\dot{m}_h(1+f-b)Cp_h(T_{07}-T_{08}) = \dot{m}_c Cp_c(T_{06}-T_{05})$$

The outlet temperature is then obtained from the above relation as

$$T_{08} = T_{07} - \frac{Cp_c(T_{06}-T_{05})}{\eta_{m1} Cp_h(1+f-b)} \tag{5.57}$$

The outlet pressure depends on the isentropic efficiency of the HPT (η_{t1}) and is given by the relation

$$\frac{P_{08}}{P_{07}} = \left[1 - \frac{1 - (T_{08}/T_{07})}{\eta_{t1}}\right]^{\gamma_h/(\gamma_h - 1)} \tag{5.58}$$

Low-pressure turbine: The LPT drives the tandem fan, considering a mechanical efficiency for this low-pressure spool of (η_{m2}), then the energy balance for the turbine and fan is

$$\eta_{m2} W_{LPT} = W_{f1} + W_{f2}$$

$$\eta_{m2} \dot{m}_h (1 + f - b) Cp_h (T_{08} - T_{09}) = \dot{m}_a Cp_c (T_{05} - T_{02})$$

Since

$$\dot{m}_h = \frac{\dot{m}_a}{1 + \beta}$$

Then the outlet temperature from the LPT will be

$$T_{09} = T_{08} - \frac{(1 + \beta)}{\eta_{m2}(1 + f - b)}(T_{05} - T_{02}) \tag{5.59}$$

Mixing process: The cold air passes through the cold duct (denoted here as fan duct) surrounding the hot section experiences a pressure drop (ΔP_{fd}), where

$$P_{05a} = P_{05} - \Delta P_{fd} \tag{5.60a}$$

The air leaves the fan duct at the same temperature as it enters; thus

$$T_{05a} = T_{05} \tag{5.61}$$

This cold air mixes with the hot gases leaving the LPT. An energy balance is necessary to determine its state before entering the subsequent elements. The conditions for this mixing process are equal static pressure and speeds, which prove equal total pressures for air and gases at states (5a) and (9) respectively, or

$$P_{05a} = P_{09} \tag{5.60b}$$

The energy balance for the mixing process is

$$\dot{m}_c Cp_c T_{05a} + (1 + f - b) \dot{m}_h Cp_h T_{09} = [\dot{m}_c + (1 + f - b) \dot{m}_h] Cp_h T_{010}$$

Thus the total temperature after mixing process will be

$$T_{010} = \frac{\beta Cp_c T_{05a} + (1 + f - b) Cp_h T_{09}}{(1 + \beta + f - b) Cp_h} \tag{5.62}$$

This mixing process also experiences a slight pressure drop of (ΔP_m) where

$$P_{010} = (1 - \Delta P_m) P_{09}$$

Afterburner: During cruise, the afterburner is inoperative. Thus the hot gases pass through it. Again the presence of flame holders, igniters, and other elements in the afterburner results in a pressure drop (ΔP_{ab}); thus

$$P_{11} = (1 - \Delta P_{ab}) P_{010} \tag{5.63}$$

The total temperature remains the same as in the inlet to afterburner; thus

$$T_{011} = T_{010} \tag{5.64}$$

Nozzle: The gases expend through the nozzle to the ambient pressure; thus the outlet temperature is

$$T_{12} = T_{011} \left[1 - \eta_n \left\{ 1 - \left(\frac{P_{12}}{P_{011}} \right)^{(\gamma_h - 1)/\gamma_h} \right\} \right] \tag{5.65}$$

The jet or exhaust speed is

$$V_j = \sqrt{2 C p_h (T_{011} - T_{12})} \tag{5.66}$$

Thrust force: The thrust force for such unchoked nozzle is given by the relation

$$T = \dot{m}_a \left[\left(1 + \frac{f - b}{1 + \beta} \right) V_j - V \right] \tag{5.67}$$

(B) *Takeoff or lift thrust*: During takeoff or lift thrust, the valve is switched off so that the air mass flow rate ($\dot{m}_1$) passing through the front part of the fan exhausts through a downward-facing lift nozzles at the cold exhaust speed (V_{ef}). A secondary inlet is opened to provide the required airflow rate ($\dot{m}_2$) to the rear part of the fan and the main engine. The resulting hot gases also leave the engine vertically downward to generate a second lift thrust. Figure 5.37 illustrates the T–S diagram for takeoff conditions.

Now, the different processes are discussed.

Intake: The outlet conditions are same as the ambient ones

$$T_{02} = T_a, \quad P_{02} = P_a$$

Fan: The first part of the fan compresses the air to the same outlet conditions as expressed above in the cruise case. However, this compresses air flows through the downward nozzles to the ambient pressure. The exhaust speed is then

$$V_{ef} = V_{13} = \sqrt{2 C p_c \eta_{fn} T_{03} \left\{ 1 - \left(\frac{P_a}{P_{03}} \right)^{(\gamma_c - 1)/\gamma_c} \right\}} \tag{5.68}$$

The second part of the fan receives the secondary airflow rate ($\dot{m}_2$), which enters the engine through the secondary port. This air enters at the ambient conditions, thus

$$T_{04} = T_a, \quad P_{04} = P_a \tag{5.69}$$

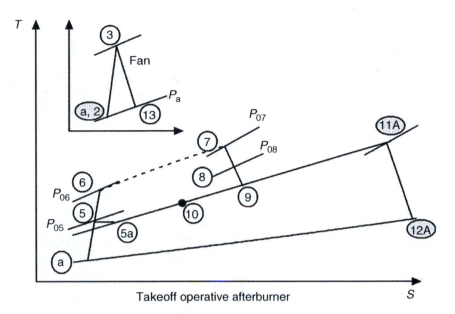

FIGURE 5.37   T–S diagram for takeoff condition.

The outlet conditions will be calculated by the same Equations 5.51 and 5.52.

Compressor: The compressed air leaving the fan ($\dot{m}_2$) is divided into two streams. The cold stream passes in an annulus duct surrounding the engine, while the hot stream proceeds toward the compressor and the subsequent hot section. This division occurs in a proportionality of the BPR (β).

The outlet conditions are determined from the same Equations 5.53 and 5.54.

A quantity of air is bled from a position downstream the compressor for several reasons including turbine cooling.

Also an amount of air $\dot{m}_b$ is bled just downstream of the compressor.

Combustion chamber: Combustion process raises the temperature to (T_{07}), which is known in advance. The pressure within the combustion chamber enhances a pressure drop; thus Equation 5.55 is used to determine the outlet pressure. The fuel burnt in the combustion chamber is determined from Equation 5.56.

High-pressure turbine: The HPT drives the HPC. The mechanical efficiency of the high-pressure spool is (η_{m1}). The outlet pressure and temperature are obtained from Equations 5.57 and 5.58.

Low-pressure turbine: The LPT drives the tandem fan, considering a mechanical efficiency for this low-pressure spool of (η_{m2}); then the energy balance for the turbine and fan is

$$\eta_{m2} W_{LPT} = W_{f1} + W_{f2}$$

$$\eta_{m2} \dot{m}_h \left(1 + f - b\right) Cp_h \left(T_{08} - T_{09}\right) = \dot{m}_1 Cp_c \left(T_{03} - T_{02}\right) + \dot{m}_2 Cp_c \left(T_{05} - T_{04}\right)$$

The outlet temperature from the LPT will be

$$T_{09} = T_{08} - \frac{Cp_c}{\eta_{m2}(1+f-b)}[\dot{m}_1(T_{03} - T_{02}) + \dot{m}_2(T_{05} - T_{04})] \tag{5.70}$$

The outlet temperature is determined from the relation

$$\frac{P_{09}}{P_{08}} = \left[1 - \frac{1 - (T_{09}/T_{08})}{\eta_{t2}}\right]^{\gamma_h/(\gamma_h - 1)}$$

Mixing process: The cold air passes through the cold duct (denoted here as fan duct) surrounding the hot section. Equations 5.60 and 5.61 are again used to determine the properties of the cold air just before mixing.

This cold air mixes with the hot gases leaving the LPT. Thus the total temperature after mixing process will be also calculated from Equation 5.62. The total pressure after the mixing process is determined from Equation 5.63.

Afterburner: During takeoff the afterburner is operative. Thus another combustion process takes place. The temperature of these hot gases is raised to the maximum temperature in the engine, which is (T_{011A}). This combustion process is associated with a pressure loss $(\overline{\Delta P_{ab}})$. A bar is added to this pressure drop to discriminate between the dry and wet operation of the afterburner; thus

$$P_{11} = (1 - \overline{\Delta P_{ab}})P_{010} \tag{5.71}$$

The total temperature remains the same as in the inlet to afterburner; thus

$$T_{011A} = T_{max} \tag{5.72}$$

A large quantity of additional fuel is burned during the operation of afterburner. The energy balance for the afterburner yields the following afterburner fuel-to-air ratio.

$$f_{ab} = \frac{(1 + \beta + f - b)Cp_h(T_{max} - T_{010})}{\eta_{ab}Q_R - Cp_hT_{max}} \tag{5.73}$$

where $f_{ab} = \dot{m}_{fab}/\dot{m}_h$.

Nozzle: The gases expend through the nozzle to the ambient pressure; thus the outlet temperature is

$$T_{12A} = T_{011A}\left[1 - \eta_n\left\{1 - \left(\frac{P_a}{P_{011A}}\right)^{(\gamma_h - 1)/\gamma_h}\right\}\right] \tag{5.74}$$

The jet or exhaust speed is

$$V_{12A} = \sqrt{2Cp_h(T_{011A} - T_{12A})} \tag{5.75}$$

Thrust force: The thrust force for such unchoked nozzle is given by the relation

$$T = \dot{m}_1 V_{13} + \frac{\dot{m}_2}{(1+\beta)}(1 + \beta + f + f_{ab} - b)V_{12A} \tag{5.76}$$

5.9 PERFORMANCE ANALYSIS

The turbofan engine GE90 represents the investment of GE in the future of wide-body aircraft. Over the past two decades, GE's CF6 and CFM56* engines have been chosen to power more than 50% of all new aircraft ordered with a capacity of 100 passengers or more.

Example 7   The data of a high BPR (similar to GE90) during cruise operation are listed below. It is required to calculate thrust, fuel mass flow rate, fuel consumption, TSFC, specific thrust, propulsive efficiency, thermal efficiency, and overall efficiency; sketch the T–S diagram and plot the engine map for TIT ranging from 1380 to 1600 and OPR in the range from 15 to 43 (Figure 5.38).

GE90 Engine Data
   Area of hot nozzle $= 1.0111 \text{ m}^2$
   Area of cold nozzle $= 3.5935 \text{ m}^2$
   Main engine parameters

Component	Real Case
η_d (Intake efficiency)	0.98
η_{fn} (Isentropic fan nozzle efficiency)	0.95
η_f (Fan polytropic efficiency)	0.93
$\eta_{C_{LPC}}$ (LPC efficiency)	0.91
$\eta_{C_{HPC}}$ (HPC efficiency)	0.91
η_{CC} (Combustion champers efficiency)	0.99
ΔP_{cc} Combustion pressure loss (ratio)	0.02
Q_{HV} (J/kg)	4.80E + 07
η_m (Mechanical efficiency)	0.99
$\eta_{t_{HPT}}$ (HPT efficiency)	0.93
$\eta_{t_{LPT}}$ (LPT efficiency)	0.93
η_n (Isentropic nozzle efficiency)	0.95
β (BPR)	8.1
π_f (fan pressure ratio)	1.7
$\pi_{C_{LPC}}$ (LPC pressure ratio)	1.14
$\pi_{C_{HPC}}$ (HPC pressure ratio)	21.0

The engine operation conditions

M (Mach number)	0.85
Altitude (m)	10668
T_a (K)	218.934
P_a (kPa)	23.9
TIT (K)	1380

Solution:

Diffuser:

$$P_{02} = P_a \left[1 + \eta_d \left(\frac{\gamma_c - 1}{2} \right) M^2 \right]^{\gamma_c/(\gamma_c - 1)} = 37.994 \text{ kPa}$$

$$T_{02} = T_a \left[1 + \left(\frac{\gamma_c - 1}{2} \right) M^2 \right] = 250.44 \text{ K}$$

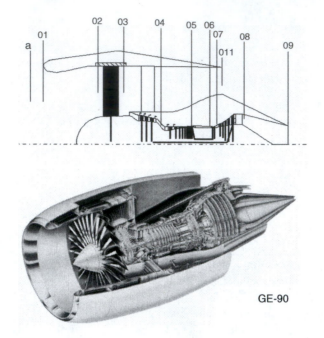

GE-90

FIGURE 5.38   High bypass ratio GE90 turbofan engine. (Courtesy GE Aviation.)

Fan:

$$P_{03} = \pi_f \times P_{02} = 64.589 \text{ kPa}$$

$$T_{03} = T_{02} \left[1 + \frac{\left(\pi_f^{(\gamma_c - 1)/\gamma_c} - 1 \right)}{\eta_{fn}} \right] = 294.52 \text{ K}$$

Low-pressure compressor:

$$P_{04} = \pi_{C_{LPC}} \times P_{03} = 73.632 \text{ kPa}$$

$$T_{04} = T_{03} \left(1 + \frac{\left(\pi_{C_{LPC}}^{(\gamma_c - 1)/\gamma_c} - 1 \right)}{\eta_{C_{LPC}}} \right) = 306.87 \text{ K}$$

High-pressure compressor:

$$P_{05} = \pi_{C_{HPC}} \times P_{04} = 1546.0 \text{ kPa}$$

$$T_{05} = T_{04} \left(1 + \frac{\left(\pi_{C_{HPC}}^{(\gamma_c - 1)/\gamma_c} - 1 \right)}{\eta_{C_{HPC}}} \right) = 774.45 \text{ K}$$

Combustion chamber:

$$T_{06} = 1380 \text{ K}$$

$$P_{06} = 0.98 \times P_{05} = 1515.347 \text{ kPa}$$

$$f = \frac{(Cp_h \times T_{06} - Cp_c \times T_{05})}{(\eta_b \times Q_{HV} - Cp_h \times T_{06})} = 0.01737$$

High-pressure turbine: From energy balance

$$\eta_m W_{HPT} = W_{HPC}$$

$$T_{07} = T_{06} - \left(\frac{Cp_c \times (T_{05} - T_{04})}{\eta_m Cp_h (1 + f)} \right) = 973.79 \text{ K}$$

$$P_{07} = P_{06} \left(1 - \frac{T_{06} - T_{07}}{\eta_{tHPT} \times T_{06}} \right)^{\gamma_h/(\gamma_h - 1)} = 330.7 \text{ kPa}$$

Low-pressure turbine: From the energy balance

$$\eta_m W_{LPT} = W_{fan} + W_{LPC}$$

$$(1 + \beta) \dot{m}_a Cp_c (T_{03} - T_{02}) + \dot{m}_a Cp_c (T_{04} - T_{03}) = \eta_m \dot{m}_a (1 + f) Cp_h (T_{07}' - T_{08})$$

$$T_{08} = T_{07} - \left(\frac{1 + \beta}{1 + f} \right) \left(\frac{Cp_c}{\eta_m Cp_h} \right) (T_{03} - T_{02}) - \left(\frac{Cp_c}{\eta_m Cp_h (1 + f)} \right) (T_{04} - T_{03}) = 614.6 \text{ K}$$

$$P_{08} = P_{07} \left(1 - \left(\frac{T_{06} - T_{07}}{\eta_{tLPT} \times T_{06}} \right) \right)^{\gamma_h/(\gamma_h - 1)} = 43.83 \text{ kPa}$$

Hot nozzle:

$$P_{c1} = P_{08} \left(1 - \frac{1}{\eta_n} \left(\frac{\gamma_h - 1}{\gamma_h + 1} \right) \right)^{\gamma_h/(\gamma_h - 1)} = 22.84 \text{ kPa}$$

$$\because P_{c1} \langle P_a$$

$\therefore$ The nozzle is unchoked and $P_9 = P_{c1}$

$$P_9 = P_{c1} = 28.34 \text{ kPa}$$

$$T_{c1} = \frac{T_{08}}{(\gamma_h + 1)/2} = 560.27 \text{ K}$$

$$T_9 = T_{c1} = 560.27 \text{ K}$$

$$V_9 = \sqrt{2Cp_h \eta_h T_{08} \left[1 - (\frac{p_a}{p_{08}})^{(\gamma_h - 1)/\gamma_h} \right]} = 434.26 \text{ m/s}$$

Cold nozzle:

$$P_{c2} = P_{03} \left(1 - \frac{1}{\eta_{fn}} \left(\frac{\gamma_c - 1}{\gamma_c + 1} \right) \right)^{\gamma_c/(\gamma_c - 1)} = 32.88 \text{ kPa}$$

$$\because P_{c2} \rangle P_a$$

$\therefore$ the nozzle is choked $M_e = 1$ and $P_{11} = P_{c2}$

$$P_{11} = P_{c2} = 3.288 \times 10^4 \text{ Pa}$$

$$T_{c2} = \frac{T_{03}}{(\gamma_c + 1)/2} = 243.22 \text{ K}$$

$$T_{11} = T_{c2} = 243.22 \, \text{K}$$

$$V_{11} = \sqrt{(\gamma_c R T_{11})} = 314 \, \text{m/s}$$

Specific thrust:

$$\frac{T}{\dot{m}_a} = \frac{\beta}{1+\beta} V_{11} + \frac{1}{1+\beta} (1+f) V_9 - V_f + \left(\frac{1+f}{1+\beta}\right)\left(\frac{RT_9}{P_9 V_9}\right)(P_9 - P_a)$$

$$+ \frac{\beta}{1+\beta} \frac{RT_{11}}{P_{11} V_{11}} (P_{11} - P_a) = 130.49 \, \text{N.s/kg}$$

Thrust:

$$\dot{m}_a = \rho_a \times A_{\text{inlet}} \times V_{\text{inlet}}$$

$$= \frac{P_a}{R \times T_a} \times 6.006 \times M \times \sqrt{\gamma_c \times R \times T_a} = 576.1 \, \text{kg/s}$$

$$T = \left(\frac{T}{\dot{m}_a}\right) \times \dot{m}_a = 75.15 \, \text{kN}$$

$$\text{TSFC} = \frac{f}{(T/\dot{m}_a)} = 1.331 \times 10^{-4} \, \text{kg/N/s}$$

Mass flow rate:

$$\dot{m}_f = f \times \dot{m}_{\text{hot}} = 1.0996 \, \text{kg/s}$$

Fuel consumption:

$$\text{fuel consumpation} = \dot{m}_f \times \frac{3600}{1000} = 3.959 \, \text{ton/h}$$

Propulsive efficiency:

$$\eta_p = \frac{(T/\dot{m}_a) V_f}{(T/\dot{m}_a) V_f + (0.5 \times ((1+f)/(1+\beta))) (V_9 - V_f)^2 + 0.5 \times (\beta/1+\beta)(V_{11} - V_f)^2} = 0.9023$$

Thermal efficiency:

Since the nozzles are choked, the expression for thermal efficiency will be given by the following relation:

$$\eta_{th} = \frac{(T/\dot{m}_a) V_f + (0.5 \times ((1+f)/(1+\beta))) (V_9 - V_f)^2 + 0.5 \times (\beta/1+\beta)(V_{11} - V_f)^2}{(f/1+\beta) \times Q_{HV}}$$

Substituting the values in the above equation gives the following result:

$$\eta_{th} = 0.3979$$

If the nozzles are unchoked, then the above relation is reduced to the well-known form

$$\eta_{th} = \frac{0.5\left[(\beta/1+\beta)(V_{11})^2 + (1+f/1+\beta)(V_9)^2 - (V_f)^2\right]}{(f/1+\beta) \times Q_{HV}}$$

Overhaul efficiency:

$$\eta_o = \eta_{th} \times \eta_p = 0.359$$

Figure 5.39 illustrates the air cycle for the turbofan engine plotted on the T–S diagram after calculating the temperature and pressure at each module of the engine. A special program is designed and executed using Mat Lab to evaluate the performance of the engine through successive changes in the overall pressure ratio from 15 to 43 and also TIT variation from 1380 to 1600. The results are plotted in Figure 5.40.

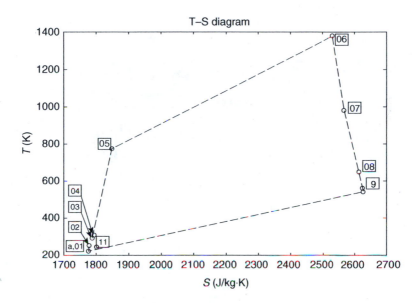

FIGURE 5.39 T–S diagram for high bypass turbofan (similar to GE-90).

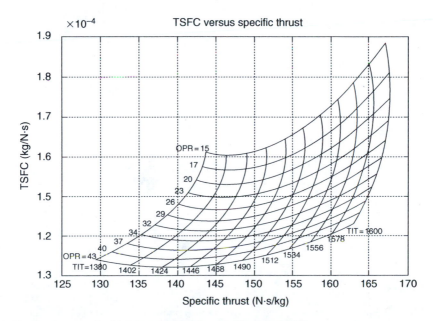

FIGURE 5.40 Engine map for high bypass turbofan (similar the GE-90).

Example 8 A gas generator (Figure 5.41a) is composed of single spool [compressor (C_1) and turbine (T_1)]. It is employed in two versions of turbofan engines, namely, forward fan (Figure 5.41b) and aft fan (Figure 5.41c) to increase the thrust force.

Forward fan

A second turbine (T_2) is added together with the forward-fan (F_1 to form a second low pressure) spool.

Aft fan

A free power turbine (FT) is installed after the turbine T_1, which has outer blade part forming the aft fan (F_2). The operation parameters and engine characteristics are as follows: flight Mach number $M_0 = 0.8$, altitude 33,000 ft with ambient conditions of (1) $P_a = 26.2\,\text{kPa}$ and (2) $T_a = 222\,\text{K}$. Fan pressure ratio (both cases) $\pi_f = 1.8$, compressor pressure ratio $\pi_c = 8$, maximum

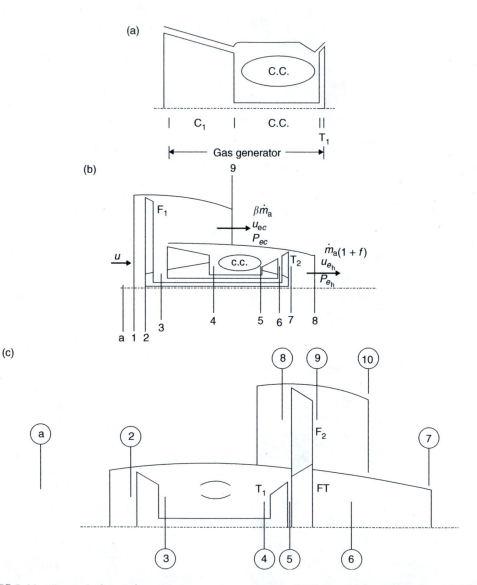

FIGURE 5.41   Two turbofan engines. (a) Common gas generator. (b) Forward fan configuration. (c) Aft fan configuration.

temperature $= 1500$ K, BPR (both cases) $\beta = 5$, and mass flow rate to the compressor $= 50$ kg/s. Efficiencies of the compressor, combustion chamber, fans, turbine, free power turbine, hot and cold nozzles are equal to 0.9. Fuel heating value is 42,000 kJ/kg and combustion chamber pressure loss is at 5%.

$$Cp_c = 1020 \text{ J/kg} \cdot \text{K}, \quad Cp_h = 1160 \text{ J/kg} \cdot \text{K}, \quad R = 287 \text{ J/kg} \cdot \text{K}$$

It is required to

1. Calculate the TSFC in both cases.
2. Compare between both cases clarifying which is to be preferred.

Solution: The given data are

$$P_a = 26.2 \text{ kPa}, \quad T_a = 222 \text{ K}$$

$$Cv_c = Cp_c - R = 1020 - 287 = 733 \text{ J/kg/K}$$

$$\gamma_C = \frac{Cp_c}{Cv_c} = \frac{1020}{733} = 1.391$$

$$\gamma_h = \frac{Cp_h}{Cv_h} = \frac{1160}{873} = 1.328$$

Forward fan

$$T_{02} = T_a \left(1 + \frac{\gamma_c - 1}{2} M^2\right) = 222 \times 1.125$$

$$T_{02} = 249.82 \text{ K}$$

$$P_{02} = P_a \left(1 + \frac{\gamma_c - 1}{2} M^2\right)^{\gamma_c/(\gamma_c - 1)} = 26.2 \times 1.5221 = 39.878 \text{ kPa}$$

$$P_{03} = P_{02} \times \pi_f = 71.78 \text{ kPa}$$

$$T_{03} = T_{02} \left[1 + \frac{\left\{(\pi_f)^{(\gamma - 1)/\gamma} - 1\right\}}{\eta_f}\right] = 249.82 \; [1 + 0.1995] = 299.67 \text{ K}$$

$$P_{04} = 574.24 \text{ kPa}$$

$$T_{04} = T_{03} \left[1 + \frac{\pi_c^{(\lambda - 1)/(\lambda - 1)} - 1}{\eta_c}\right] = 564 \text{ K}$$

$$f = \frac{Cp_h T_{05} - Cp_c T_{04}}{\eta_{cc} Q_R - Cp_h T_{05}} = \frac{1.16 \times 1,500 - 1.02 \times 564}{0.9 \times 42,000 - 1.16 \times 1,500}$$

$$f = 0.0322$$

$$P_{05} = P_{04} (1 - 0.05) = 545.3 \text{ kPa}$$

Energy balance (high-pressure spool)

$$Cp_c \, (T_{04} - T_{03}) = Cp_h (1 + f) \, (T_{06} - T_{05})$$

$$1.02 \, [564 - 299.7] = 1.16 \, (1 + 0.0322) \, [T_{06} - 1500]$$

$$T_{06} = 1274.8 \text{ K}$$

$$\eta_{T1} = \frac{T_{05} - T_{06}}{T_{05} - T_{06s}} = \frac{T_{05} - T_{06}}{T_{05} \left[1 - \left(P_{06} / P_{05} \right)^{(\gamma_h - 1)/\gamma_h} \right]}$$

$$0.9 = \frac{1500 - 1274.8}{1500 \left[1 - \left(P_{06} / P_{05} \right)^{(\gamma_h - 1)/\gamma_h} \right]}$$

$$\left(\frac{P_{06}}{P_{05}} \right)^{(\gamma_h - 1)/\gamma_h} = 1 - 0.1675 = 0.8324$$

$$\frac{P_{06}}{P_{05}} = 0.4757$$

$$P_{06} = 259.4 \text{ kPa}$$

Energy balance (Low-pressure spool)

$$(1 + \beta) Cp_c \, (T_{03} - T_{02}) = Cp_h \, (1 + f) \, (T_{06} - T_{07})$$

$$6 \times 1.02 = 1.16 \, (1.0322) \, (1274.8 - T_{07})$$

$$T_{07} = 1019.7 \text{ K}$$

$$\frac{P_{07}}{P_{06}} = \left[1 - \frac{T_{06} - T_{07}}{\eta_{T2} T_{06}} \right]^{\gamma_h / (\gamma_h - 1)} = \left[1 - \frac{1274.8 - 1062.2}{0.9 \times 1274.8} \right]^{4.05} = 0.3611$$

$$P_{07} = 94 \text{ kPa}$$

$$\frac{P_{07}}{P_a} = 3.59$$

$$\frac{P_{07}}{P_C} = \frac{1}{[1 - ((\gamma_h - 1)/\eta_n (\gamma_h + 1))]^{\gamma_h / (\gamma_h - 1)}} = 1.993$$

Since $P_{07} / P_a > P_{07} / P_C$, then the nozzle is choked

$$P_8 = P_C = \frac{94}{1.993} = 47.16 \text{ kPa}$$

$$T_8 = \left(\frac{2}{\gamma_h + 1} \right) T_{07} = 876 \text{ K}$$

$$V_F = M \sqrt{\gamma R T_a} = 238.25 \text{ m/s}$$

$$V_8 = \sqrt{\gamma R T_8} = 577.8 \text{ K}$$

$$\dot{m}_8 = \rho_8 A_8 V_8$$

$$\dot{m}_a \, (1 + f) = \frac{P_8}{R T_8} A_8 V_8$$

$$A_8 = \frac{\dot{m}_a \, (1 + f) \, R T_8}{P_8 V_8} = \frac{50 \, (1.0332) \, (287) \, (876)}{47.16 \times 10^3 \times 577.8} = 0.477 \text{ m}^2$$

For the hot section

$$\text{Thrust} = T_h = \dot{m}\left[(1+f)\,V_8 - V_F\right] + A_8\,(P_8 - P_a)$$

$$\frac{T_h}{\dot{m}} = \{(1.0322)\,(577.84) - 238.25\} + \frac{0.477}{50}\,(47.16 - 26.2) \times 10^3 = 358.2 + 200$$

$$\frac{T_h}{\dot{m}} = 558.16\frac{\text{N}\cdot\text{s}}{\text{kg}}$$

$$(\text{Thrust})_{\text{hot section}} = 27{,}908\,\text{N}$$

For the cold section

$$\frac{P_{03}}{P_c} = \frac{1}{\left[1 - \left((\gamma_c - 1)/\eta_n\,(\gamma_c + 1)\right)\right]^{\gamma_c/(\gamma_c-1)}} = \frac{1}{0.4899} = 2.04$$

$$\frac{P_{03}}{P_a} = \frac{71.78}{26.2} = 2.739$$

Choked cold nozzle

$$P_9 = P_c = 35.18\,\text{kPa}$$

$$T_9 = \left(\frac{2}{\gamma_c + 1}\right) T_{03} = 299.67 \times 0.836 = 250.66\,\text{K}$$

$$\rho_9 = \frac{P_9}{RT_9} = 0.489\frac{\text{kg}}{\text{m}^3}$$

$$V_9 = \sqrt{\gamma RT_9} = 316.33\ \text{m/s}$$

$$A_9 = \frac{\beta\,\dot{m}}{\rho_9 V_9} = \frac{5 \times 50}{0.489 \times 316.33} = 1.616\,\text{m}^2$$

Thrust of cold section

$$T_c = \beta\dot{m}_a\,(V_9 - V_F) + A_9\,(P_9 - P_a)$$

$$T_c = 5 \times 50\,(316.33 - 283.25) + 1.616\,(35.18 - 26.2) \times 10^3 = 22{,}782\,\text{N}$$

$$\text{Total thrust} = 27{,}908 + 22{,}782 = 50{,}690\,\text{N}$$

$$\text{TSFC} = \frac{\dot{m}_F}{T} = \frac{f \times \dot{m}_a}{T} = \frac{0.0322 \times 50}{50{,}690} = 3.176 \times 10^{-5}\frac{\text{kgFuel}}{\text{N}\cdot\text{s}}$$

Aft fan

Cold section: Since the intake in the fan in its two configurations (Figure 5.41b and 5.41c) have no losses, then the conditions upstream to the fan at both cases are equal. Thus,

$$T_{08} = 249.82\ \text{K} \quad \text{and} \quad P_{08} = 39.878\ \text{kPa} \tag{5.77}$$

Moreover, since the same fan is employed in both cases, then its outlet conditions will be

$$P_{09} = 71.75\ \text{kPa}, \quad T_{09} = 299.67\ \text{K}$$

$$\frac{P_{09}}{P_a} = \frac{71.75}{26.2} = 2.738$$

$$\frac{P_{09}}{P_c} = \frac{1}{\left[1 - \left((\gamma_c - 1)/\eta_n\,(\gamma_c - 1)\right)\right]^{\gamma/(\gamma_c - 1)}} = 2.04$$

Choked nozzle

$$P_{10} = P_c = 35.18 \text{ kPa}$$

$$T_{10} = 250.66 \text{ K}$$

$$V_{10} = 316.33 \text{ m/s}$$

$$A_{10} = 1.616 \text{ m}^2$$

$$T_{fan} = T_{cold} = 22,782 \text{ N}$$

Hot section

$$T_{02} = 299.67 \text{ K}$$

$$P_{02} = 39.878 \text{ kPa}$$

$$P_{03} = \pi_c P_{02} = 8 \times 39.878 = 319.024 \text{ kPa}$$

$$T_{03} = T_{02}\left[1 + \frac{\pi_c^{(\gamma - 1)/\gamma} - 1}{\eta_c}\right] = 470 \text{ K}$$

$$P_{04} = 0.95 \times 319 = 303 \text{ kPa}$$

$$f = \frac{C_{Ph}T_{06} - C_{Pc}T_{03}}{\eta_b Q_R - C_{Ph}T_{06}} = \frac{1.16 \times 1,500 - 1.02 \times 470}{0.9 \times 42,000 - 1.16 \times 1,500} = 0.0322$$

$$P_{05} = P_{04}(1 - 0.05) = 303 \text{ kPa}$$

Energy balance between compressor and turbine

$$\dot{m}_a C_{Pc}\,(T_{03} - T_{02}) = \dot{m}_a\,(1 + f)\,C_{Ph}\,(T_{04} - T_{05})$$

$$T_{04} - T_{05} = (T_{03} - T_{02})\frac{C_{Pc}}{C_{Ph}\,(1 + f)} = \frac{(470 - 300) \times 1.02}{1.03495 \times 1.16} = 144.4 \text{ K}$$

$$T_{05} = 1355.6 \text{ K}$$

$$\frac{P_{05}}{P_{04}} = \left[1 - \frac{T_{04} - T_{05}}{\eta_t T_{04}}\right]^{\gamma/(\gamma - 1)} = \left[1 - \frac{1500 - 1355}{0.9 \times 1500}\right]^{4.05} = (0.893)^{4.05}$$

$$\frac{P_{05}}{P_{04}} = 0.6323$$

$$P_{05} = 191.6 \text{ kPa}$$

Energy balance (Fan and LPT)

$$m_a\,(1 + f)\,C_{Ph}\,(T_{05} - T_{06}) = \beta\,m_a C_{Pc}\,(T_{09} - T_{08})$$

$$(1.0322)\,(1.16)\,(1355 - T_{06}) = 5 \times 1.102\,(299.67 - 222)$$

$$T_{06} = 1355 - \frac{427.96}{1.197} = 997.8 \text{ K}$$

$$\eta_t = \frac{T_{05} - T_{06}}{T_{05} - T_{06S}} = \frac{T_{05} - T_{06}}{T_{05}\left[1 - (P_{06}/P_{07})^{(\gamma_h - 1)/\gamma_h}\right]}$$

$$\frac{P_{06}}{P_{05}} = \left[1 - \frac{T_{05} - T_{06}}{\eta_t T_{05}}\right]^{\gamma_h/(\gamma_h - 1)} = \left[1 - \frac{1355 - 997}{0.9 \times 1355}\right]^{4.05} = (0.7064)^{4.05}$$

$$P_{06} = 46.9 \text{ kPa}$$

$$\frac{P_{06}}{P_a} = 1.79$$

Since $P_{06}/P_c = 1.993$, then the nozzle is unchoked.

$$\eta_n = \frac{T_{06} - T_7}{T_{06} - T_{7S}} = \frac{T_{06} - T_7}{T_{06}\left[1 - (P_7/P_{06})^{(\gamma_h - 1)/\gamma_h}\right]}$$

$$T_7 = T_{06}\left[1 - \eta_n\left\{1 - \left(\frac{P_a}{P_{06}}\right)^{(\gamma_h - 1)/\gamma_h}\right\}\right] = 997.4\left[1 - 0.9\left\{1 - \left(\frac{26.2}{46.9}\right)^{0.247}\right\}\right]$$

$$T_7 = 877.2 \text{ K}$$

$$V_9 = \sqrt{2Cp\,(T_{06} - T_7)} = 530 \text{ m/s}$$

Hot thrust $\equiv T_{hot} = \dot{m}_h\left[(1 + f)\,V_9 - V_F\right] = 15{,}430 \text{ N}$

$(\text{Total thrust})_{\text{aft fan}} = 15{,}430 + 22{,}782 = 37{,}212 \text{ N}$

$$\text{TSFC} = \frac{\dot{m}_F}{T} = \frac{f \times \dot{m}_a}{T} = \frac{0.03495 \times 50}{19{,}417}$$

$(\text{TSFC})_{\text{aft fan}} = 10.434 \times 10^{-5} \dfrac{\text{kg} \cdot \text{Fuel}}{\text{N} \cdot \text{s}}$

$(\text{TSFC})_{\text{forward fan}} < (\text{TSFC})_{\text{aft fan}}$

The previous results can be summarized in the following table:

	Forward Fan	Aft Fan
Maximum pressure (kPa)	474.24	319.02
V_{e_c} (m/s)	316.33	316.33
V_{e_h} (m/s)	577.8	350
P_{e_c} (kPa)	35.18 (Choked)	35.18 (Choked)
P_{e_h} (kPa)	47.16 (Choked)	26.2 (Unchoked)
Cold thrust (N)	22,782	22,782
Hot thrust (N)	27,908	15,430
Total thrust (N)	50,690	37,212
TSFC (kg Fuel/N/s)	3.176×10^{-5}	10.434×10^{-5}

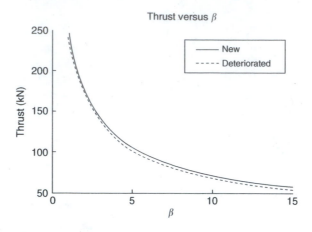

FIGURE 5.42   Variation of thrust with BPR.

From the above results, the forward fan are more advantageous as it generates more thrust and consumes less fuel.

Example 9   The performances of a brand new and a deteriorated engine are dependent on the efficiency of different components. Here the efficiencies of new and deteriorated high BPR turbofan engines are listed. Calculate and plot the variation of the propulsive, thermal, and overall efficiencies as well as the thrust, specific thrust, and TSFC with the BPR if the same operating conditions listed in Example 7 are applied.

Component	New	Deteriorated
η_d (Intake efficiency)	0.98	0.95
η_{fn} (Isentropic fan nozzle efficiency)	0.95	0.95
η_f (Fan polytrophic efficiency)	0.93	0.925
$\eta_{C_{LPC}}$ (LPC efficiency)	0.91	0.88
$\eta_{C_{HPC}}$ (HPC efficiency)	0.91	0.87
η_{CC} (Combustion champers efficiency)	0.99	0.99
ΔP_{cc} Combustion pressure loss (ratio)	0.02	0.05
Q_{HV} (J/kg)	4.80E + 07	4.80E + 07
η_m (Mechanical efficiency)	0.99	0.99
$\eta_{t_{HPT}}$ (HPT efficiency)	0.93	0.91
$\eta_{t_{LPT}}$ (LPT efficiency)	0.93	0.91
η_n (Isentropic nozzle efficiency)	0.95	0.95

Solution: Deterioration of the components of the turbofan engine leads to a decrease in all the efficiencies (propulsive, thermal, and overall). It will also reduce both the thrust and specific thrust. Moreover, higher fuel consumption is expected, which increases the TSFC. These results are depicted in Figures 5.42, 5.43, 5.44, 5.45, 5.46, and 5.47.

Example 10   The most important performance specifications of aircraft engine are the propulsive efficiency of the engine and the TSFC. They are important for the airlines that install this engine on to its aircraft. Both of the propulsive efficiency and TSFC may be used as an indication of the required working cost. Two engine parameters have a direct influence on its performance; namely

- TIT
- The ratio of the bleed air drown from to the air mass flow rate

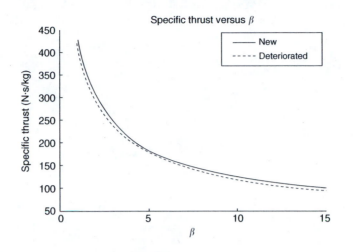

FIGURE 5.43   Variation of specific thrust with BPR.

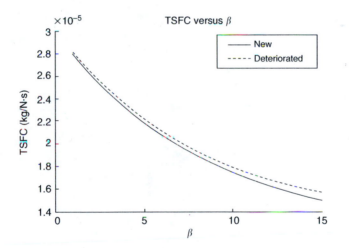

FIGURE 5.44   Variation of specific fuel consumption with BPR.

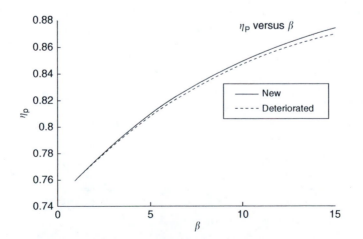

FIGURE 5.45   Variation of propulsive efficiency with BPR.

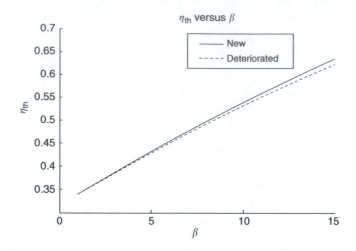

FIGURE 5.46 Variation of thermal efficiency with BPR.

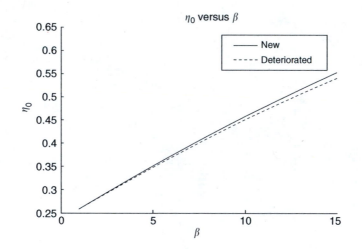

FIGURE 5.47 Variation of overall efficiency with BPR.

Use the data below to calculate and plot the variation of the propulsive efficiency and the TSFC for the engine for successive variation of TIT from 1850 to 2000 K and bleed ratio ranging from 0.05 to 0.15.

η_d (Intake efficiency)	0.98
η_{fn} (Isentropic fan nozzle efficiency)	0.95
η_f (Fan polytrophic efficiency)	0.93
η_C (LPC and HPC efficiencies)	0.91
η_{CC} (Combustion chambers efficiency)	0.99
ΔP_{cc} Combustion pressure loss (ratio)	0.02
Q_{HV} (J/kg)	4.80E + 07
η_m (Mechanical efficiency)	0.99
η_t (HPT and LPT efficiencies)	0.93
η_n (Isentropic nozzle efficiency)	0.95

at (SL), $\pi_{HPC}=4$, $\pi_{LPC}=2$, $\beta=2$, $\pi_f=1.45$, $m^0_{bleed}/m^0_a=0.05$

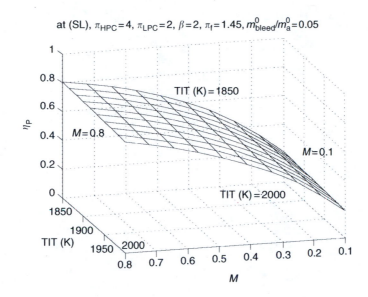

FIGURE 5.48   Variation of propulsive efficiency with TIT.

at (SL), $\pi_{HPC}=4$, $\pi_{LPC}=2$, $\beta=2$, $\pi_f=1.45$, $m^0_{bleed}/m^0_a=0.05$

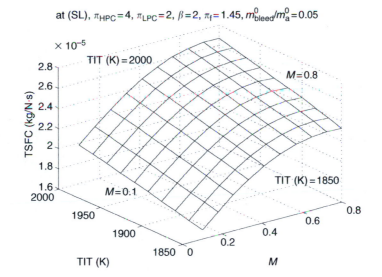

FIGURE 5.49   Variation of specific fuel consumption with TIT.

π_f (Fan pressure ratio)	0.98
π_{LPC} (LPC pressure ratio)	0.95
π_{HPC} (HPC pressure ratio)	0.93
Ambient	Altitude sea level
β (BPR)	2
TIT (for the study of bleed ratio)	1850–2000
$b = \dot{m}_{bleed}/\dot{m}_a$ (for the study of TIT)	0.05–0.15

Solution: The effect of TIT on the propulsive efficiency is plotted in Figure 5.48 and its effect on the TSFC is plotted in Figure 5.49.

at (S.L), 'TIT=2000K, π_f=1.45, π_{LPC}=2.2, π_{HPC}=6, β=2

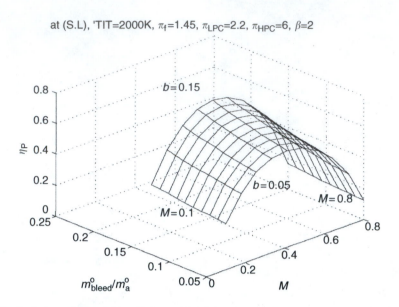

FIGURE 5.50   Variation of propulsive efficiency with b.

at (SL), TIT= 2000 K, π_f=1.45, π_{LPC}=2.2, π_{HPC}=6, β=2

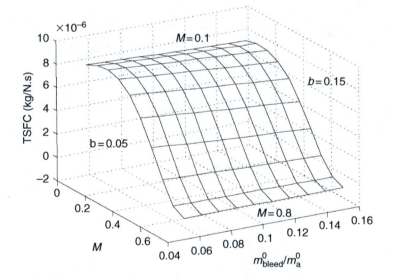

FIGURE 5.51   Variation of specific fuel consumption with b.

The effect of bleed air to air mass flow ratio on the propulsive efficiency is plotted in Figure 5.50 and its effect on the TSFC is plotted in Figure 5.51.

Example 11   For the engine in Example 7 (high bypass engine similar to GE-90) and at the same operating conditions perform a parametric study to examine the changes in engine performance parameters due to the combined effects of

(a) Variation of TIT and OPR while keeping the fan pressure ratio and bypass ratio constants and equal to 1.7 and 8.5, respectively.
(b) Variation of TIT and fan pressure ratio and keeping the OPR and BPR constants and equal to 44 and 8.5, respectively.

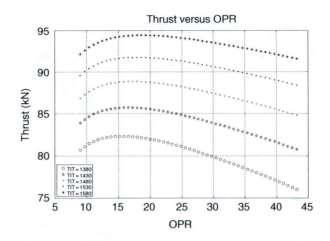

FIGURE 5.52   Variation of thrust with OPR.

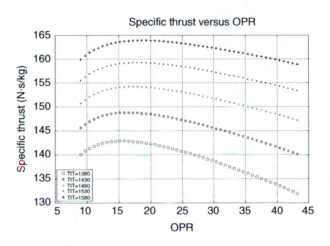

FIGURE 5.53   Variation of specific thrust with OPR.

(c)   Variation of TIT and BPR and keeping the fan pressure ratio and OPR constants and equal to 1.7 and 44, respectively.

Solution: As discussed earlier, there are four thermodynamic parameters that control the performance of turbofan engines. These are

1.   The BPR
2.   The fan pressure ratio
3.   The overall pressure ratio
4.   The TIT

The first parametric study is to examine the effects of TIT and OPR (with $r_f = 1.7$ and $\beta = 8.5$) on the engine performance parameters.

The results are plotted in Figures 5.52 through 5.58. The second study deals with the effects of TIT and fan pressure ratio (with OPR = 44 and $\beta = 8.5$) on engine performance parameters. Some of the results of this analysis are illustrated in Figures 5.59 through 5.62. The last case study is concerned with the effects of TIT and BPR ($r_f = 1.7$ and OPR = 44). Some of the results are plotted in Figures 5.63 through 5.66.

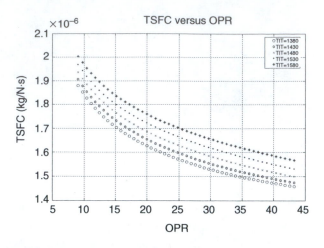

FIGURE 5.54   Variation of specific fuel consumption with OPR.

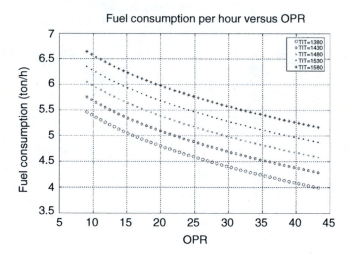

FIGURE 5.55   Variation of rate of fuel consumption with OPR.

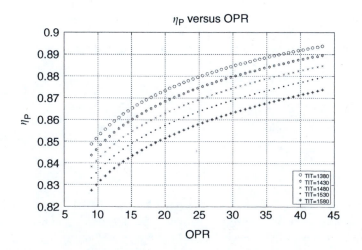

FIGURE 5.56   Variation of propulsive efficiency with OPR.

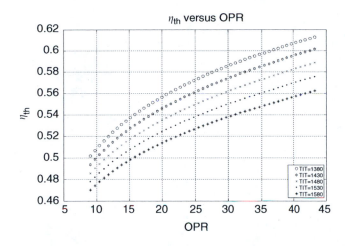

FIGURE 5.57 Variation of thermal efficiency with OPR.

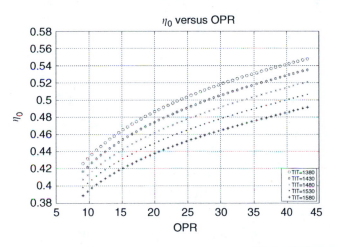

FIGURE 5.58 Variation of overall efficiency with OPR.

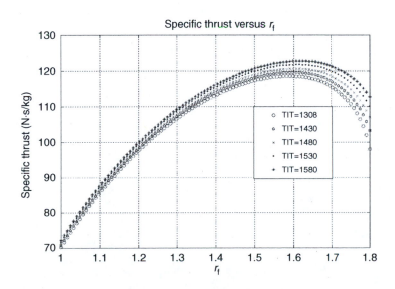

FIGURE 5.59 Variation of specific thrust with fan pressure ratio.

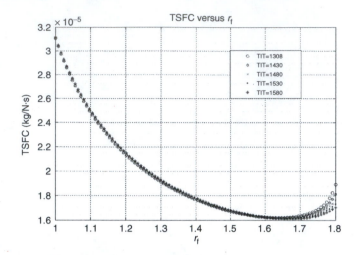

FIGURE 5.60 Variation of TSFC with fan pressure ratio.

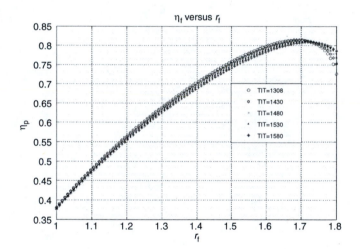

FIGURE 5.61 Variation of propulsive efficiency with fan pressure ratio.

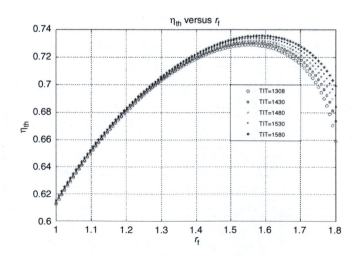

FIGURE 5.62 Variation of thermal efficiency with fan pressure ratio.

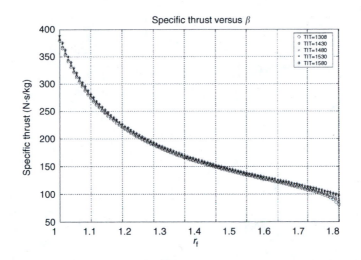

FIGURE 5.63 Variation of specific thrust with bypass ratio.

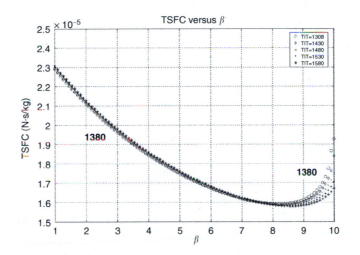

FIGURE 5.64 Variation of TSFC with bypass ratio.

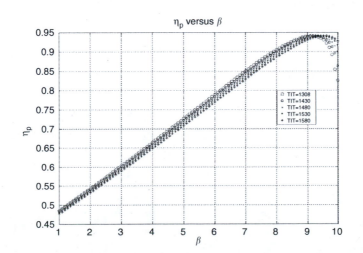

FIGURE 5.65 Variation of propulsive efficiency with bypass ratio.

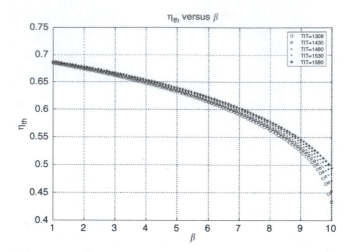

FIGURE 5.66   Variation of thermal efficiency with bypass ratio.

Remarks

1. Increasing the BPR improves TSFC at the expense of a significant reduction in specific thrust
2. The optimum fan pressure ratio increases with increase of TIT
3. The maximum thrust and specific thrust occur at a definite OPR which increases with the increase of TIT.

Aircraft application

1. *Long-range subsonic transport*: The TSFC is of major importance. This is achieved through BPR in the range (4–6), high overall pressure ratio, and low TIT.
2. *Military aircraft*: Maneuverability and supersonic capabilities as well as good subsonic fuel consumption are the major objectives. This is achieved through much smaller BPR ranging from 0.5 to 1.0, which results in a small frontal area as well as adding an afterburner.
3. *Short-haul commercial aircraft*: BPR is similar to the long-range subsonic transport, which also minimizes noise.

SUMMARY

Ideal and actual cycles for a single-spool turbofan engine

Element	Ideal Cycle	Actual Cycle
Diffuser	$\eta_d = 1$	$T_{02} = T_a \left(1 + \frac{\gamma_c - 1}{2} M^2\right)$
		$P_{02} = P_a \left(1 + \eta_d \frac{\gamma_c - 1}{2} M^2\right)^{\gamma_c/(\gamma_c - 1)}$
Fan	$\eta_f = 1$	$T_{03} = T_{02} \left(1 + \frac{\pi_f^{(\gamma_c - 1)/\gamma_c} - 1}{\eta_f}\right)$
		$P_{03} = \pi_f P_{02}$

Continued

Element	Ideal Cycle	Actual Cycle
Compressor	$\eta_c = 1$	$T_{04} = T_{03}\left(1 + \dfrac{\pi_c^{(\gamma_c-1)/\gamma_c}-1}{\eta_c}\right)$ $P_{04} = \pi_c P_{03}$
Combustion chamber	$\eta_b = 1$ $\Delta P\% = 0$	$T_{05} = \text{TIT}$ $f = \dfrac{Cp_h T_{05} - Cp_c T_{04}}{\eta_b Q_R - Cp_h T_{05}}$ $\dot{m}_f = f\dot{m}_a$ $P_{05} = (1 - \Delta P\%)\,P_{04}$
Turbine	$\eta_m = 1$ $\eta_t = 1$	$T_{06} = T_{05} - \dfrac{Cp_c}{\eta_m \lambda(1+f)Cp_h}\left(T_{04} + \beta T_{03} - (1+\beta)T_{02}\right)$ $P_{06} = P_{05}\left(1 - \dfrac{T_{05}-T_{06}}{\eta_t T_{05}}\right)^{\gamma_h/(\gamma_h-1)}$
Nozzle	$\eta_n = 1$	$P_c = P_{06}\left(1 - \dfrac{1}{\eta_n}\dfrac{\gamma_h-1}{\gamma_h+1}\right)^{\gamma_h/(\gamma_h-1)}$ If $P_c \geq P_a$ the nozzle is choked $P_7 = P_c$ $T_7 = T_c = \dfrac{T_{06}}{(\gamma_h+1)/2}$ $V_7 = \sqrt{\gamma_h R T_7}$ If $P_c < P_a$ the nozzle is unchoked $P_7 = P_a$ $T_7 = T_{06}\left(1 - \eta_n\left(1 - \left(\frac{P_7}{P_{06}}\right)^{(\gamma_h-1)/\gamma_h}\right)\right)$ $V_7 = \sqrt{2Cp_h(T_{06}-T_7)}$
Fan nozzle	$\eta_{fn} = 1$	$P_c = P_{03}\left(1 - \dfrac{1}{\eta_{fn}}\dfrac{\gamma_c-1}{\gamma_c+1}\right)^{\gamma_c/(\gamma_c-1)}$ If $P_c \geq P_a$ the nozzle is choked $P_8 = P_c$ $T_8 = T_c = \dfrac{T_{03}}{(\gamma_c+1)/2}$ If $P_c < P_a$ the nozzle is unchoked $P_8 = P_a$ $T_8 = T_{03}\left(1 - \eta_n\left\{1 - \left(\frac{P_8}{P_{03}}\right)^{(\gamma_c-1)/\gamma_c}\right\}\right)$ $V_8 = \sqrt{2Cp_c(T_{03}-T_8)}$

$$T = \dot{m}_a(1+f)V_7 + \beta\dot{m}_a V_8 - (1+\beta)\dot{m}_a V_f + A_7(P_7 - P_a) + A_8(P_8 - P_a)$$

$$\text{Specific thrust} = \frac{T}{\dot{m}_a(1+\beta)}$$

$$\text{TSFC} = \frac{\dot{m}_f}{T}$$

Cp_c Specific heat at constant pressure for cold air
Cp_h Specific heat at constant pressure for hot gases
$\dot{m}_a$ Air mass flow rate through the engine core

$\dot{m}_f$ Fuel mass flow rate
f Fuel-to-air ratio
Q_R Fuel heating value
γ_c Ratio of specific heats for cold air
γ_h Ratio of specific heats for hot gas
β BPR

Layout and governing equations of some types of engines

Engine Type	Governing Equations

Forward fan, single spool, unmixed, no afterburner.

$P_T = P_C + P_F$
$T = \dot{m}_a \left[(1+f)\,u_{e_h} + \beta u_{e_c} - (1+\beta)\,u\right]$
$\quad + A_{e_c}\left(P_{e_c} - P_a\right) + A_{e_h}\left(P_{e_h} - P_a\right)$

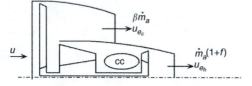

Forward fan, double spool, unmixed, no afterburner

$P_{HPT} = P_{HPC}$
$P_{LPT} = P_F + P_{LPC}$
$T = \dot{m}_a \left[(1+f)\,u_{e_h} + \beta u_{e_c} - (1+\beta)\,u\right]$
$\quad + A_{e_c}\left(P_{e_c} - P_a\right) + A_{e_h}\left(P_{e_h} - P_a\right)$

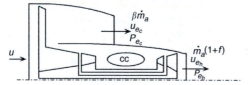

Forward fan, double spool, unmixed, no afterburner

$P_{HPT} = P_{HPC}$
$P_{LPT} = P_F$
$T = \dot{m}_a \left[(1+f)\,u_{e_h} + \beta u_{e_c} - (1+\beta)\,u\right]$
$\quad + A_{e_c}\left(P_{e_c} - P_a\right) + A_{e_h}\left(P_{e_h} - P_a\right)$

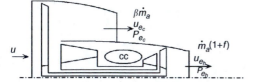

Forward fan, three-spool, unmixed, no afterburner

$P_{LPT} = P_F$
$P_{IPT} = P_{IPC}$
$P_{HPT} = P_{HPC}$
$T = \dot{m}_a \left[(1+f)\,u_{e_h} + \beta u_{e_C} - (1+\beta)\,u\right]$
$\quad + A_{e_c}\left(P_{e_c} - P_a\right) + A_{e_h}\left(P_{e_h} - P_a\right)$

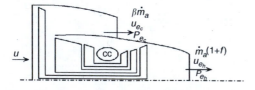

Governing equations

Specific thrust

$$\frac{T}{\dot{m}_a} = \frac{\beta}{1+\beta} u_{e_C} + \frac{1}{1+\beta}(1+f)u_{e_h} - u + \left(\frac{1+f}{1+B}\right)\left(\frac{RT_{e_h}}{P_{e_h} u_{e_h}}\right)(P_{e_h} - P_a) + \frac{\beta}{1+\beta}\frac{RT_{e_c}}{Pu_{e_c}}(P_{e_c} - P_a)$$

Thrust-specific fuel consumption

$$\text{TSFC} = \frac{f}{T/\dot{m}_a}$$

Propulsive efficiency

$$\eta_p = \frac{(T/\dot{m}_a)u}{(T/\dot{m}_a)u + (0.5\times((1+f)/(1+\beta)))(u_{e_h}-u)^2 + 0.5\times(\beta/(1+\beta))(u_{e_c}-u)^2}$$

Thermal efficiency

For choked nozzle

$$\eta_{th} = \frac{(T/\dot{m}_a)u + (0.5\times((1+f)/(1+\beta)))(u_{e_h}-u)^2 + 0.5\times(\beta/(1+\beta))(u_{e_c}-u)^2}{(f/(1+\beta))\times Q_{hv}}$$

For unchoked nozzles, it is reduced to

$$\eta_{th} = \frac{0.5\left[(\beta/(1+\beta))(u_{e_C})^2 + ((1+f)/(1+\beta))(u_{e_h})^2 - (u)^2\right]}{(f/(1+\beta))\times Q_{hv}}$$

Overhaul efficiency

$$\eta_o = \eta_{th} \times \eta_p$$

PROBLEMS

5.1 A single-spool turbofan engine has the following data:

Ambient temperature	288 K
Ambient pressure	101.3 kN/m^2
BPR (β)	0.7
Overall pressure ratio	25
Fan pressure ratio	3.375
Fuel heating value (Q_R)	44,000 kJ/kg
Turbine inlet temperature	2000 K
Diffuser pressure recovery factor (r_d)	0.9
Compressor efficiency (η_c)	0.89
Fan efficiency (η_f)	0.91
Combustion efficiency (η_b)	0.98
Burner pressure recovery factor (r_b)	0.95
Turbine efficiency (η_t)	0.98
Nozzle efficiency (η_n)	1.0
Flight Mach number	0.3

Calculate the thrust per unit mass flow rate and the TSFC.

5.2 A V/STOL aircraft is powered by the tandem fan vectored thrust with a switch-in deflector engine described in Section 5.7.2. The following data apply for the case of takeoff (lift thrust):

$$T_a = 288\text{K}, \quad P_a = 1.01 \text{ bar}, \quad \dot{m}_1 = 80 \text{ kg/s}, \quad Q_R = 45{,}000 \text{ kJ/K}$$

$$\pi_{f1} = 1.6, \quad \pi_{f2} = 1.2, \quad \pi_c = 10, \quad TIT = 1400\text{K}, \quad T_{max} = 2200 \text{ K},$$

$$\frac{P_{04}}{P_{03}} = 97\%, \quad \frac{P_{07}}{P_{06}} = 94\%, \quad \frac{P_{010}}{P_{05}} = 97\%, \quad \Delta P_{ab} = 5\%$$

Bleed from compressor $(b) = 2\%$

Efficiencies

$$\eta_d = 0.97, \quad \eta_{f1} = \eta_{f2} = \eta_c = 0.85, \quad \eta_b = 0.99,$$

$$\eta_{t1} = \eta_{t2} = 0.9, \quad \eta_{fn} = \eta_n = 0.98, \quad \eta_m = 0.99$$

Calculate
   • The mass flow rate $\dot{m}_2$ that gives equal lift forces from both nozzles
   • The total lift force

5.3 For the V/STOL engine described above consider the cruise operation of aircraft. The flight Mach number is 0.8 at an altitude of 32,000 ft. The afterburner is inoperative as described in Section 5.7.2 and the engine is considered as a mixed two-spool configuration. The values of the pressure ratios, fuel heating value, TIT, and pressure losses in the different ducts given as pressure ratios, bleed, and efficiencies are same as in the preceding problem. The total mass flow rate is $\dot{m}_1 = 80$ kg/s and $\Delta P_{ab} = 5\%$. The nozzle is also unchoked.
   Calculate
   (a) The BPR
   (b) The thrust force
   (c) The propulsive, thermal, and overall efficiencies

5.4 (a) Write expressions for the thrust, fan, and core effective jet velocities (V_{Jf}, V_{Jc}).
   (b) Using these two parameters find the expression for the global effective jet velocity $(\bar{V}_e)$, based on the total mass flow rate $(\dot{m}_{fan} + \dot{m}_{core})$.
   (c) Find the expressions for the thermal, propulsive, and overall efficiencies, employing the fan and core effective jet velocities.

5.5 (a) Prove that for a two-stream turbofan engine, with a constant sum of "relative" kinetic power, the thrust and propulsive efficiency are maximum when $V_{e_f} = V_{e_c}$.
   (b) A turbofan engine cruises at $M_o = 0.85$ and altitude of 34,000 ft. The fan and core effective jet velocities are 350 and 470 m/s respectively. The total air mass flow rate is 270 kg/s and the BPR (β) is 4.2. The engine has an overall pressure ratio of 31.3 and a maximum TIT equal to 1405 K. Assuming ideal intake compressors and combustion chamber, determine
   • Combustion chamber "ideal" fuel-to-air ratio
   • Thrust, various efficiencies $(\eta_{th}, \eta_p, \eta_o)$, and TSFC

5.6 Compare the performance of twin-spool turbofan engine with separate fan and compressor to that of a turbojet engine, both with an overall pressure ratio of 30 and a TIT of 1300 K, and designed for an altitude of 12 km and a flight speed of 275 m/s. The BPR is 6 and the fan pressure ratio is 1.6. Assume that the polytropic efficiency of fan, compressor, and turbine efficiencies of 90% and a combustor loss of 3% of the compressor exit total pressure. Compare specific thrusts for engines built from the same core engine. Compare also specific fuel consumption and jet velocities.

5.7 Consider the V/STOL ground attack fighter; Hawker Siddeley Harrier G.R.MK.1 discussed in Example 6. It has the same static takeoff thrust of 98 kN and the other data in the example. However, the real case is considered with the following efficiencies:
- Isentropic efficiencies of the intake, fan, and compressor are respectively, 93%, 88%, and 85%.
- Isentropic efficiencies of both turbines is 91%.
- Isentropic efficiency of the fan nozzle and turbine nozzles is 93%.
- The pressure drop in the combustion chamber is 2%.

Recalculate
1. The air mass flow rate
2. The thermal efficiency
Take $Cp_c = 1.005$ J/kg/K, $Cp_h = 1.148$ J/kg/K, $\gamma_c = 1.4$, $\gamma_h = 4/3$
What are your comments?

5.8 A turbofan engine of the mixing type, where the fan is driven by the LPT and the compressor is driven by the HPT a ground run of the engine, the following readings were recorded $P_a = 100$ kPa, $T_a = 300$ K, fan pressure ratio = 2, compressor pressure ratio = 8.0, TIT = 1600 K, $\eta_c = \eta_f = \eta_t = 0.9, f = 0.065, \gamma_{air} = 1.4, \gamma_{gas} = 1.33$.
Calculate
(i) The BPR on the assumption that the outlet streams of the fan and LPT have the same total pressure just before mixing
(ii) The specific thrust
(iii) TSFC

5.9 A turbofan engine has a fan and low-pressure compressor driven by the LPT. It operates at a flight velocity of 200 m/s at 12,000 m altitude, where the ambient temperature and pressure are 216.65 K and 0.1933 bar, respectively. The overall pressure ratio of the engine is 19, the fan and LPC pressure ratios are 1.65 and 2.5, respectively. The fan, compressors, and turbines have a 90% polytrophic efficiency. Assume an isentropic inlet and separate isentropic convergent nozzles, a fuel heating value of 43,000 kJ/kg, and a combustor total pressure loss of 6.5%. Determine the core engine and bypass duct exit velocities, the engine TSFC, thrust and specific thrust for an engine inlet air flow of 100 kg/s, and BPR of 3.0. Determine whether the nozzles are choked.

5.10 A low-BPR turbofan engine is powering an aircraft flying at a Mach number M. Assuming all the processes are isentropic except for the combustion process in which no pressure drop is encountered, it is required to
(a) Draw the cycle on the T–S diagram.
(b) If the compressor pressure ratio is increased while the maximum cycle temperature is maintained constant, plot the new cycle on the same T–S diagram in (a) above.
(c) If zero flight Mach number is assumed, plot a third graph for the cycle on the same T–S diagram in (a) above.
(d) Calculate the specific thrust and the fuel-to-air ratio in the following case:
- Flight Mach number is 0.8.
- Flying altitude is 34,000 ft where the ambient pressure and temperature are 0.2472 bar and 220 K.
- The fan pressure ratio is 1.5 and the compressor pressure ratio is 6.0.
- The BPR is 1.0.
- Lowest fuel heating value is 45,000 kJ/kg.
- Specific heats for air and gases are respectively 1.005 and 1.148 kJ/(kg.K). Maximum cycle temperature is 1500 K.
Assume full expansion in the nozzles to ambient pressure.

(e)   If the fan and the LPT are removed and the turbofan engine is replaced with a turbojet engine with an afterburner it is required to

- Plot on the same T–S diagram in (a) above, the cycle if the maximum temperature of the afterburner is also 1500 K.
- Calculate the new specific thrust and the fuel-to-air ratio assuming also unchoked nozzle and no pressure drop in the afterburner.
- What are your comments?

5.11   The figure shown illustrates several turbofan engines together with their BPR and flight specific fuel consumptions. Construct a similar figure and plot these engines: CF6, GE90, Trent 700, 900, 1000, F-100, F-101, F-110, RB-I99, RB 211, Nene, RR Pegasus T56, V2500, CFM56, Lyulka AL-7, CF6, TFE 731, TPE331, TV3 VK-1500, AL-31 F.

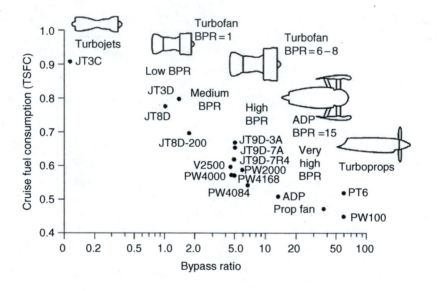

5.12   The figure shown illustrates the mixed high BPR turbofan engine (similar to CFM56-SC). It has the following data:

BPR	6.6
Typical TSFC	0.04 kg/(N · h)
Cruise thrust	29,360 N
Total mass flow rate at cruise	100 kg/s
Flight Mach number	0.8

During takeoff operation

• Takeoff thrust	140,000 N
• Total air mass flow rate at takeoff	474.0 kg
• Overall pressure ratio	37.5
• Total temperature at inlet to the LPT	1251 K
• Exhaust gas temperature	349 K
• Ambient temperature	288.0 K
• Ambient pressure	101 kPa

Assuming that all the processes are ideal and there are no losses in both the combustion chamber and mixing process,

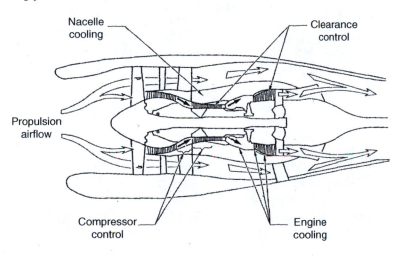

(a) Prove that the thrust force is given by the relation

$$T = \frac{\dot{m}_a\,(1+\beta)\,(u_e - u)}{1 - u_e\,(\text{TSFC})}$$

(b) Calculate
1. The exhaust velocity at both takeoff and cruise (use the above relation)
2. The fuel-to-air ratios at both takeoff and cruise conditions
(c) At takeoff condition, calculate
2. The total temperature at the outlet of the HPC
3. The total temperature at the inlet to the HPT
4. The pressure ratio of the HPC
5. The fan pressure ratio
6. The temperature of the hot gases prior mixing
7. The pressure ratio of the LPC (booster)
    Assume that the nozzle is unchoked

5.13 The figure shown illustrates a supersonic turbofan engine, which has the following characteristic during flight at high altitude ambient conditions 216.7 K and 9.28 kPa. The semi-vertex angle of the fan nose is 16°, the Mach number at the fan is $M_2 = 1.5$, the Mach number at the fan exhaust $M_3 = 2.6$, and maximum cycle temperature 1800 K.

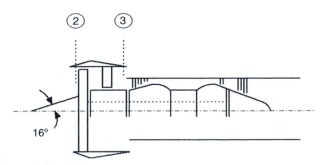

- Fan pressure ratio   2
- Overall pressure ratio   6

- BPR   1.0
- Isentropic efficiency for fan and compressor   0.83
- Isentropic efficiency for cold and hot nozzle   0.95
- Burner efficiency   0.99
- Isentropic efficiency for HPT and LPT   0.9
- Fuel heating value   43,000 kJ/kg
- No pressure drop in the combustion chamber
- Total air mass flow rate   330 kg/s

It is required to

1. Plot the cycle on the T–S diagram.
2. Calculate the flight speed and pressure recovery in the intake.
3. Calculate the fuel-to-air ratio.
4. Calculate the supersonic Mach number of the hot gases.
5. Calculate the thrust force.
6. Calculate the thrust specific fuel consumption.

5.14 The fan shown illustrates the General Electrics CF 700 aft fan Turbofan engine. During a ground test ($M = 0$) assume that

- Ideal compressor and expansion processes in the fan. Compressor, turbines and nozzles
- Negligible fuel-to-air ratio

Required to

1. Prove that the exhaust speeds of the cold air and hot gases are equal. if the BPR (β) is expressed by the relation

$$\beta = \frac{(A - B)(B - 1)}{B(C - 1)} - 1$$

   where $A = T_{04}/T_a$, and $B = \pi_c^{(\gamma-1)/\gamma}$ and $C = \pi_f^{(\gamma-1)/\gamma}$.

2. Now consider the following case study:

$$T_a = 300\,\text{K}, \quad \pi_c = 4.0, \quad \pi_f = 1.5, \quad T_{04} = 1200\,\text{K}$$

   Evaluate the BPR (β) that satisfies equal exhaust speeds for the hot and cold streams.

5.15 Boeing 747 civil transport airplane that is fitted by four high BPR turbofan engine similar to CF6-50C turbofan engines. Each engine has the following data:

- Thrust force   24.0 kN
- Air flow rate $\dot{m}_a$ 125 kg/s
- BPR   5.0
- Fuel mass flaw rate $\dot{m}_r$ 0.75 kg/s
- Operating Mach number   0.8
- Altitude   10 km

- Ambient temperature T_a 223.2 K
- Ambient pressure P_a 26.4 kPa
- Fuel heating value Q_{HV} 42,800 kJ/kg

If the thrust generated from the fan is 75% of the total thrust, determine

1. The specific thrust
2. TSFC
3. The jet velocity of the hot gases
4. Thermal efficiency
5. Propulsive efficiency
6. The overall efficiency

(Assume that the exit pressure is equal to the ambient pressure.)

5.16 One of the techniques used in STOL fighters is the flap blowing turbofan engines. Air is bled from the LPC and ducted to the flap system in the wing trailing edge. Such accelerated air flow creates greater lift than would be possible by the wings themselves. The data for one of these engines is given here. During takeoff at sea level ($T_a = 288$K, $P_a = 101$kPa), the engine has a BPR of 3.0 and the core engine air flow is 22.7 kg/s; 15% of this air is bled at the exit of the LPC. The component efficiencies are given as

$$\eta_c = 0.85, \quad \eta_f = 0.88, \quad \eta_b = 0.96, \quad \eta_t = 0.94, \quad \eta_n = 0.95$$

The total pressure recovery factor of the inlet is $r_d = 0.98$, compressor total pressure ratio is 15, fan total pressure ratio is 2.5 maximum allowable TIT is 1460 K, total pressure loss in the combustor is 3%, and heating value of the fuel is 43,000 kJ/kg. Assume the working medium to be air and treat it as a perfect gas with constant specific heats.

It is required to compute the following:

(a) The properties at each section
(b) The specific thrust
(c) The TSFC
(d) The thermal efficiency

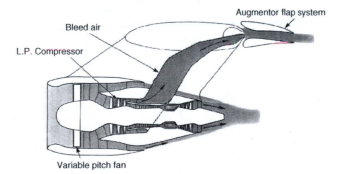

5.17 The figure shown illustrates the Pratt & Whitney Dual-Rotor Turbofan Full Duct Mixed Type JT8D Engine. It has the following characteristics:

Compressor pressure ratio	10
Fan pressure ratio	1.8
Inlet temperature to HPT	1 628 K
Mechanical efficiency	0.85
Fuel-to-air ratio	0.06
Efficiencies of fan/compressors	0.9
Efficiencies of turbines/nozzle	0.95

Ambient temperature   300 K
Ambient pressure   100 kPa
Assume that the outlet streams from the fan and LPT will have the same total pressure before mixing.
(a) Determine

- The BPR
- The specific thrust
- TSFC

(b) Plot the temperature and pressure distributions along the different components of the engine.

$$\text{For the fan and compressor } Cp = 1.0 \text{ kJ/kg/K and } \gamma = 1.4$$

$$\text{For the turbines and nozzle } Cp = 1.2 \text{ kJ/kg/K and } \gamma = 1.33$$

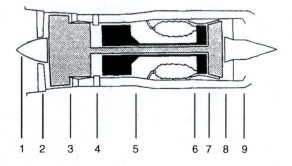

```
1   2   3   4   5        6 7 8 9
```

5.18 For a turbofan engine write down the mathematical expressions for
1. The thrust force
2. The overall efficiency
Prove that the maximum overall efficiency is given by

$$\eta_{0\,\text{max}} = \frac{(1 + \beta)\, u^2}{f Q_R}$$

5.19 The figure shown illustrates a turbofan engine during testing in a test cell. It has the following characteristics:

- $\beta = 3, \quad f = 2\%, \quad V_2 = 3V_1$
- $\dot{m}_a = 200 \text{ kg, thrust} = 60 \text{ kN}$

(a) It is required to
- Calculate $\dot{m}_1, \dot{m}_2,$ and $\dot{m}_f$

(b) Evaluate V_1 and V_2 if (f) is neglected
(c) Calculate the propulsive efficiency
What are your comments?

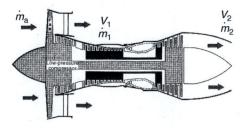

REFERENCES

1. www.boeing.com/commercial/safety/propulsion.html
2. B. Gunston, *The Development of Jet and Turbine Aero Engines*, Patrick Stephens Limited, 1997, p. 67.
3. C. Hess, *Pratt & Whitney Geared Turbofan*, Flug Revue October 1998, p. 54.
4. http://www.pwd.ca, The Pratt & Whitney Canada PW800 Turbofan Family, Status Report, Fact Sheet 2000,
5. http://en.wikipedia.org/wiki/Rolls-Royce_RB211
6. J. Kurzke, *About Simplifications in Gas Turbine Performance Calculations*, ASME Paper GT2007–27620, Montreal, Canada, 2007.
7. http://en.wikipedia.org/wiki/Turbofan
8. *The Jet Engine*, Rolls Royce Plc., 5th edn., 1996, p. 189.
9. H.I.H. Saravanamuttoo, G.F.C. Rogers, and H. Cohen, *Gas Turbine Theory*, Prentice Hall, 5th edn., 2001, p. 146.

6 Turboprop, Turboshaft, and Propfan Engines

This chapter introduces three types of aero engines, namely, turboprop, turboshaft, and propfan engines. Both turboprop and turboshaft have their exhaust gases leaving the engine at a very low or nearly zero speeds. Power, rather than thrust, is the key factor in both. Turboprop engines power many heavy transport aircraft flying at moderate subsonic speeds. The turboshaft engines discussed here are limited to helicopters only. Turboshaft engines employed in other applications are treated in Chapter 8 and identified as gas turbines, which are employed in numerous industrial applications. Propfan engines combine many features of both turbofan and turboprop engines, so it is included here. No separate chapter is allocated to propfan engines until now as these have limited applications in the aviation industry.

6.1 INTRODUCTION TO TURBOPROP ENGINES

Turboprop engine is a hybrid engine that provides jet thrust and also drives a propeller. It is basically similar to a turbojet except that the turbine works through a shaft and speed-reducing gears to turn a propeller at the front of the engine (Figure 6.1).

Many low-speed and small commuter aircraft (King Air A100, Mitsubishi MU-2, Saab 340, and De Havilland Canada Dash 8), military transport (Hercules C-130, Lockheed P-3A, and Antonov AN-38,-140) as well as long-range aircraft (the Russian TU-95 (Figure 6.2) and the CP-140 Aurora) use turboprop propulsion. Turboprop engines are basically propellers driven by a turbine.

Soon after the first turbojets were in the air the turboprop engine was developed. Turboprop is a concept similar to the turbofan though having a higher bypass ratio (BPR). Moreover, instead of the turbine driving a ducted fan, it drives a completely external propeller. The turboprop uses a gas turbine core to turn a propeller. The propeller is driven either by the gas generator turbine (Figure 6.3), or by another turbine denoted as free or power turbine (Figure 6.4). In this case, the free turbine is independent of the compressor-driven turbine, and is free to rotate by itself in the engine exhaust gas stream. Another configuration of the free turbine turboprop has a rather unconventional rear-to-front air and gas direction. This configuration provides great flexibility in the design of nacelle

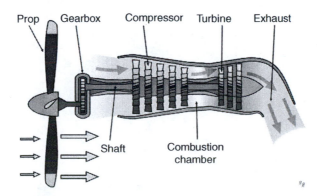

FIGURE 6.1 Turboprop engine.

FIGURE 6.2 Russian TU-95.

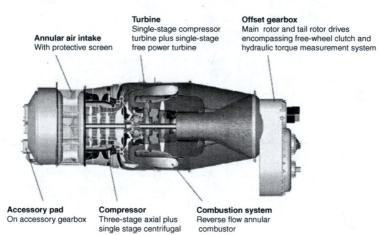

Turbine
Single-stage compressor
turbine plus single-stage
free power turbine

Offset gearbox
Main rotor and tail rotor drives
encompassing free-wheel clutch and
hydraulic torque measurement system

Annular air intake
With protective screen

Accessory pad
On accessory gearbox

Compressor
Three-stage axial plus
single stage centrifugal

Combustion system
Reverse flow annular
combustor

FIGURE 6.3 PT 6B. (Courtesy Pratt & Whitney of Canada.)

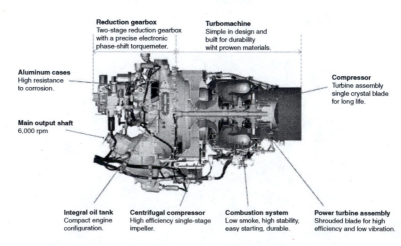

Reduction gearbox
Two-stage reduction gearbox
with a precise electronic
phase-shift torquemeter.

Turbomachine
Simple in design and
built for durability
wiht proven materials.

Aluminum cases
High resistance
to corrosion.

Compressor
Turbine assembly
single crystal blade
for long life.

Main output shaft
6,000 rpm

Integral oil tank
Compact engine
configuration.

Centrifugal compressor
High efficiency single-stage
impeller.

Combustion system
Low smoke, high stability,
easy starting, durable.

Power turbine assembly
Shrouded blade for high
efficiency and low vibration.

FIGURE 6.4 PW200. (Courtesy Pratt & Whitney of Canada.)

installations: the space behind the engine is not used for an exhaust duct end, and can be used for wheel wells or fuel tanks. The compressor is a combination of axial/centrifugal flow design. An example of such an engine is the PT6 manufactured by Pratt & Whitney of Canada.

In either case a heavy gear box is installed between the turbine and the propeller. Such a gear box provides a speed reduction of nearly 1:15. Turboprop engine produces two thrusts, one with the propeller and the other through the exhaust. The thrust developed by the propeller accumulates nearly 85% of the total thrust while the remaining 15% is produced by the jet core exhaust due to the exhaust low speed.

As stated above the propeller rotates at very low speed compared to its driving turbine. The speed reduction may be 1:15. This speed reduction is necessary owing to two reasons:

1. A large centrifugal force arises from the rotation of the large diameter (2–4 m or even more) propeller blades. These blades are fixed to the propeller hub in a cantilever fixed-end configuration. Consequently, such a centrifugal force generates a large tensile stress at blade root. Stress limitations require that the large diameter propeller rotates at a much slow speed. It is a fact that no propeller can withstand the tensile force (and stress) that is generated when it is turned at the same speed of the turbine.

2. Owing to the rotation of the propeller, the relative velocity at the propeller tip will approach the speed of sound before the aircraft approaches the speed of sound. This compressibility effect when approaching the speed of sound limits the design of propellers. At high subsonic flight speeds ($M > 0.7$), the tips of blades may approach supersonic speeds. If this happens, the flow may separate and shock waves may form. As a consequence, the performance of turboprop engine deteriorates due to both the poor propeller efficiency and the decrease in air flow rate into the engine.

The propeller is pitch controlled to be suitable for a wider range of satisfactory applications.

If the shaft of a free turbine is used to drive something other than an aircraft propeller, the engine is called a turboshaft engine. This is one of the turboshaft engines that will be discussed later in this chapter. Turboshaft engines are similar to turboprop engines, except that the hot gases are expanded to a lower pressure in the turbine, thus providing greater shaft power and low exhaust velocity. Examples of turboshaft engines are those used in helicopters.

Now, let us discuss the *advantages* of turboprop engines:

1. Turboprops have high fuel efficiency, even greater than turbofan engines. This is due to the small amount of air flow burned inside the engine. Turboprop engines can then generate a lot of thrust at low fuel consumption.

2. Turboprop engines may find application in vertical takeoff and landing (VTOL). The Osprey V-22 aircraft as shown in Figure 6.5 is one of the famous VTOL aircraft that is powered by a turboprop engine.

3. Turboprop engines have high takeoff thrust that enables aircraft to have a short field takeoff.

4. They have the highest propulsive efficiency for flight speeds of 400 mph compared to turbofan and turbojet engines.

However, turboprop engines have several *disadvantages*:

1. The noise and vibration produced by the propeller is a significant drawback.

2. Turboprop engines are limited to subsonic flights (< 400 mph) and low altitudes (below 30,000 ft).

3. The propeller and its pitch control mechanism as well as the power turbine contribute additional weight, so the turboprop engine may be 1.5 times as heavy as a conventional turbojet of the same gas generator size.

FIGURE 6.5   V-22 Osprey. (Courtesy of the Boeing.)

FIGURE 6.6   Pusher configuration of Jetcruzer 500, AASI airplane.

4. The large-size gearbox connected to the propeller is situated upstream the air path to the engine inlet. Thus the intake shapes are different from those of turbojet and turbofan engines. Moreover, the gearbox (normally a single or double planetary one) has many moving parts that could break and can get in the way of the air stream going into the engine.

6.2 CLASSIFICATION OF TURBOPROP ENGINES

1. Based on engine–aircraft configuration

Turboprop engines may be either of the *tractor* (sometimes identified as *puller*) or pusher types. By the word puller (tractor) it is meant a turboprop engine with a propeller that precedes the intake and compressor. The thrust force (mostly generated by the propeller) is thus a pulling force. Most aircraft are powered by puller (tractor) configuration; Figure 6.2, for example. If the propeller is downstream the inlet and compressor then this turboprop is identified as pusher as shown in Figure 6.6.

Advantages of pusher turboprop engines are as follows:

(a)   A higher quality (clean) airflow prevails over the wing.
(b)   Engine noise in the cabin area is reduced.
(c)   The pilot's front field of view is improved.

Disadvantages are as follows:

(a)   The heavy gearbox is at the back, which shifts the center of gravity rearward and thus reduces the longitudinal stability.

(b)   Propeller is more likely to be damaged by flying debris at landing.

(c)   Engine cooling problems are more severe.

The "pusher" configuration is not very common.

Example   Jetcruzer 500 powered by Pratt & Whitney of Canada (PT6A-66) driving five blades at constant speed with Hartzell propeller.

Advantages of *puller* (*tractor*) turboprop engines are as follows:

(a)   The heavy gearbox is at the front, which helps to move the center of gravity forward and therefore allows a smaller tail for stability considerations.

(b)   The propeller is working in an undisturbed free stream. There is a more effective flow of cooling air for the engine.

Disadvantages are as follows:

(a)   The propeller slipstream disturbs the quality of the airflow over the wing.

(b)   The increased velocity and flow turbulence over the fuselage due to the propeller slipstream increase the local skin friction on the fuselage.

Example   Most (if not all) aircraft powered by turboprops are in the puller configuration. A model for this type is the Pratt & Whitney of Canada (PW 120A) powering the Dash 8 aircraft.

2. Propeller–engine coupling

This classification was already mentioned before, which depends on the way of the coupling between the propeller and the turbine. The propeller may be driven by the gas generator or by a separate turbine (identified as free power turbine).

3. Number of spools

Turboprop engines may be single, double, or triple spools. Propellers driven by the gas generator are found either in a single- or double-spool turboprop engine. On the contrary a propeller driven by a free (power) turbine may be a part of either a double or triple spool. The number of spools is exactly the number of turbines in the engine.

4. Propeller type

Propeller may also be classified, as shown in Figure 6.7, as either fixed pitch or variable pitch. Variable pitch propellers may be further subdivided into adjustable or controllable ones.

5. Type of intake

Intakes may have different shapes as shown in Figure 6.8. These shapes depend on the size and location of the reduction gear box coupled to the propeller. The shapes of the intake may be axial, axisymmetric, axisymmetric through plenum or a scoop, which in turn may have elliptical, rectangular, and annular or a U-shape.

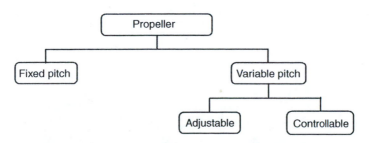

FIGURE 6.7   Classification of propellers.

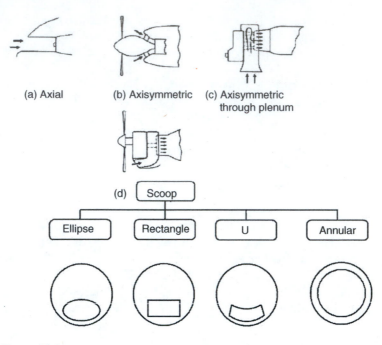

FIGURE 6.8   Types of inlets.

6.3 THERMODYNAMIC ANALYSIS OF TURBOPROP ENGINES

The different modules for a turboprop engines are the intake or inlet, one or two compressors, a combustion chamber and one or more (up to three) turbines, and the exhaust nozzle. For a single- or double-spool engine only one compressor is found. Two compressors are found in a triple-spool engine. One turbine exists for each spool.

6.3.1 SINGLE-SPOOL TURBOPROP

A simplified layout of a single-spool turboprop engines with its different states and the corresponding temperature–entropy diagram is shown in Figures 6.9 and 6.10. The same procedure followed in previous chapters will be followed.

The flight speed is expressed as $U = M_a\sqrt{\gamma R T_a}$.

The thermodynamic properties at different locations within the engine are obtained as follows:

$$\gamma_c = \frac{Cp_c}{(Cp_c - R)}, \quad \gamma_{cc} = \frac{Cp_{cc}}{(Cp_{cc} - R)}, \quad \gamma_t = \frac{Cp_t}{(Cp_t - R)}, \quad \gamma_n = \frac{Cp_n}{Cp_n - R}$$

The different modules of the engine are treated as previously described in Chapters 3, 4, and 5.

1. *Intake:*   The intake has an isentropic efficiency (η_d); the ambient temperature and pressure are T_a and P_a respectively and the flight Mach number is M_a. The temperature and pressure at the intake outlet, T_{02} and P_{02}, are given by the following relations:

$$P_{02} = P_a \left(1 + \eta_d \frac{\gamma_c - 1}{2} M_a^2\right)^{\gamma_c/(\gamma_c - 1)} \tag{6.1}$$

$$T_{02} = T_a \left(1 + \frac{\gamma_c - 1}{2} M_a^2\right) \tag{6.2}$$

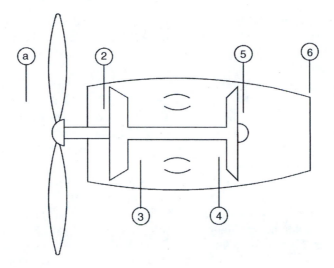

FIGURE 6.9 Layout of a single spool (direct drive turboprop engines).

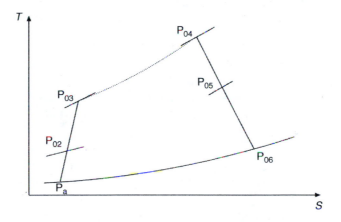

FIGURE 6.10 Temperature–entropy diagram of single spool turboprop.

2. *Compressor:* For a known compressor pressure ratio (π_c) its isentropic efficiency is (η_c); thus the pressure and temperature at the outlet of the compressor as well as the specific power of the compressor are given by the following relations:

$$P_{03} = (P_{02})(\pi_c) \tag{6.3}$$

$$T_{03} = T_{02}\left[1 + \frac{\pi_c^{\frac{\gamma_c-1}{\gamma_c}} - 1}{\eta_c}\right] \tag{6.4}$$

$$\Delta h_c = Cp_c(T_{03} - T_{02})$$

3. *Combustion chamber:* The combustion process takes place in the combustor with an efficiency of (η_b), while the products of combustion experience a pressure drop equal to (ΔP). The pressure at the outlet of the combustion chamber and the fuel-to-air ratio are given by the following:

$$P_{04} = (1 - \Delta P)P_{03}$$

$$f = \frac{Cp_{cc}T_{04} - Cp_c T_{03}}{\eta_b Q_R - Cp_{cc}T_{04}} \tag{6.5}$$

4. *Turbine:*   It is not easy here to determine the outlet pressure and temperature of the turbine. The reason is that the turbine here drives both the compressor and propeller. The portion of each is not known in advance. Let us first examine the power transmission from the turbine to the propeller as illustrated in Figure 6.11.

The output power from the turbine is slightly less than the extracted power owing to friction of the bearings supporting the turbine. This loss is accounted for by the mechanical efficiency of the turbine (η_{mt}). Moreover, the mechanical losses encountered in the bearings supporting the compressor are accounted for by the compressor mechanical efficiency (η_{mc}). The difference between both the turbine and compressor powers is the shaft power delivered to the reduction gear box where additional friction losses are encountered and accounted for by the gearbox mechanical efficiency (η_g). Finally the output power available from the propeller is controlled by the propeller efficiency (η_{pr}). Now, Figure 6.12 illustrates the enthalpy–entropy diagram for the expansion processes through both the turbine and the exhaust nozzle. It has been shown by Lancaster [1] that there is an optimum exhaust velocity that yields the maximum thrust for a given flight speed, turbine inlet temperature, and given efficiencies. Now let us define the following symbols as shown in Figure 6.12. Δh is the enthalpy drop available in an ideal (isentropic) turbine and exhaust nozzle and, $\alpha \Delta h = \Delta h_{ts}$, which is the fraction of Δh that would be available from an isentropic turbine having the actual pressure ratio

$$\Delta h_{ns} = (1 - \alpha)\Delta h$$

which is also the fraction of Δh that may be available from an isentropic nozzle. η_t is the isentropic efficiency of turbine and, η_n is the isentropic efficiency of the exhaust nozzle.

Now to evaluate these values from the following thermodynamic relations:

$$\Delta h = Cp_t T_{04}\left[1 - \left(\frac{P_a}{P_{04}}\right)^{(\gamma_h-1)/\gamma_h}\right] \tag{6.6}$$

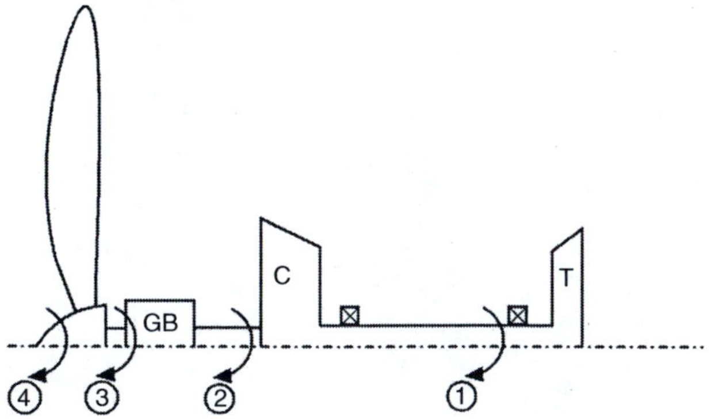

Power transimission

at (1) $\eta_{mt}W_t$

at (2) $\left(\eta_{mt}W_t - \dfrac{W_c}{\eta_{mc}}\right)$

at (3) $\eta_{gt}\left(\eta_{mt}W_t - \dfrac{W_c}{\eta_{mc}}\right)$

at (4) $\eta_{pr}\eta_g\left(\eta_{mt}W_t - \dfrac{W_c}{\eta_{mc}}\right)$

FIGURE 6.11   Power transmission through a single-spool turboprop engine.

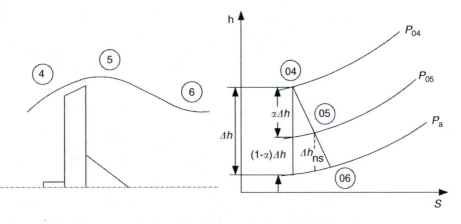

FIGURE 6.12 Expansion in the turbine and nozzle of a single-spool turboprop.

It was assumed in Equation 6.6 that the ratios between specific heats within the turbine and nozzle are constant, or

$$\gamma_t = \gamma_n = \gamma_h$$

The exhaust gas speed (U_e) is given by the following relation:

$$\frac{U_e^2}{2} = \eta_n(1 - \alpha)\Delta h$$

$$U_e = \sqrt{2(1 - \alpha)\Delta h \eta_n} \qquad (6.7)$$

The procedure to be followed here is to deduce a mathematical expression for the thrust force generated by the propeller (T_{pr}) from the power generated by the propeller. Adding this thrust to the thrust generated by the exhaust gases, the total thrust is obtained as a function of (α). Differentiate with respect to (α) to obtain the optimum value of α giving the maximum thrust.

The propeller thrust T_{pr} is correlated to the propeller power by the following relation:

$$T_{pr} = \frac{\dot{m}_a \eta_{pr} \eta_g W_{shaft}}{U}$$

The shaft power is

$$W_{shaft} = \eta_{mt}(1 + f - b)\Delta h_t - \frac{\Delta h_c}{\eta_{mc}}$$

where the turbine specific power, $\Delta h_t = \eta_t \alpha \Delta h$.

$(\dot{m}_a)$ is the air induction rate per second, and the fuel-to-air ratio and the bleed ratio are defined as: $f = (\dot{m}_f/\dot{m}_a)$ and $b = (\dot{m}_b/\dot{m}_a)$

$$T_{pr} = \frac{\dot{m}_a \eta_{pr} \eta_g}{U} \left[(1 + f - b)\eta_{mt}\eta_t \alpha \Delta h - \frac{\Delta h_c}{\eta_{mc}} \right] \qquad (6.8)$$

The thrust force obtained from the exhaust gases leaving the nozzle is abbreviated as (T_n). If the fuel mass flow rate and the air bleed from the compressor are considered then it will be given by the following relation:

$$T_n = \dot{m}_a[(1 + f - b)U_e - U]$$

Total thrust

$$T = T_{pr} + T_n$$

$$\frac{T}{\dot{m}} = \frac{\eta_{pr}\eta_g}{U}\left[(1 + f - b)\eta_{mt}\eta_t\alpha\Delta h - \frac{\Delta h_c}{\eta_{mc}}\right] + \left[(1 + f - b)\sqrt{2(1 - \alpha)\eta_n\Delta h} - U\right] \qquad (6.9)$$

Maximizing the thrust T for fixed component efficiencies, flight speed (U), compressor-specific power Δh_c, and expansion power Δh yields the following optimum value of (α_{opt}):

$$\alpha_{opt} = 1 - \frac{U^2}{2\Delta h}\left(\frac{\eta_n}{\eta_{pr}^2\eta_g^2\eta_{mt}^2\eta_t^2}\right) \qquad (6.10)$$

Substituting this value of (α) in Equation 6.9 gives the maximum value of the thrust force. The corresponding value of the exhaust speed is given by the following equation:

$$U_e = U\frac{\eta_n}{\eta_{pr}\eta_g\eta_{mt}\eta_t} \qquad (6.11)$$

6.3.2 Two-Spool Turboprop

A schematic diagram of a two-spool engine having a free power turbine together with its temperature–entropy diagram is shown in Figures 6.13 and 6.14. The low-pressure spool is composed of the propeller and the free power turbine while the high-pressure spool is composed of the compressor and the high-pressure or gas generator turbine.

The different components are examined here.

1. *Intake:*   The same relations for the outlet pressure and temperature in the single spool; Equations 6.1 and 6.2 are applied here.

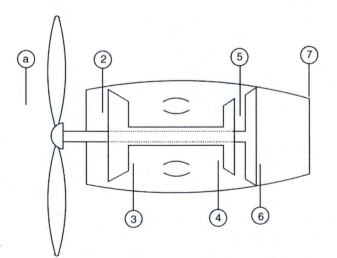

FIGURE 6.13   Layout of a free power turbine turboprop engine.

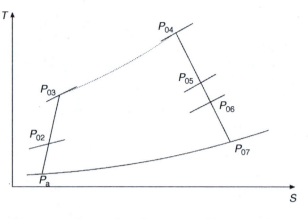

FIGURE 6.14 Temperature–entropy diagram for a free power turbine.

2. *Compressor:* The same relations in Equations 6.3 and 6.4 are applied here. The specific work of compressor (the work per kg of air inducted into the engine) is

$$\Delta h_c = Cp_c(T_{03} - T_{02})$$

3. *Combustion chamber:* The fuel-to-air ratio is obtained from the same relation, namely, Equation 6.5.

4. *Gas generator turbine:* An energy balance between the compressor and this high pressure turbine gives

$$\frac{\Delta h_c}{\eta_{mc}} = \eta_{mt}\Delta h_t \qquad (6.12a)$$

The specific work generated in the turbine of the gas generator is

$$\Delta h_t = Cp_t(T_{04} - T_{05})(1 + f - b) \qquad (6.12b)$$

From Equations 6.12a,b with known turbine inlet temperature, the outlet temperature (T_{05}) is calculated from the following relation:

$$T_{05} = T_{04} - \frac{Cp_c(T_{03} - T_{02})}{Cp_t\eta_{mc}\eta_{mt}(1 + f - b)}$$

Moreover, from the isentropic efficiency of the gas generator turbine, the outlet pressure (P_{05}) is calculated from the relation given below:

$$P_{05} = P_{04}\left[1 - \left(\frac{T_{04} - T_{05}}{\eta_t T_{04}}\right)\right]^{\frac{\gamma_t}{\gamma_t - 1}}$$

5. *Free power turbine:* Figure 6.15 illustrates the power flow from the free turbine to the propeller. The work developed by the free power turbine per unit mass inducted into the engine is

$$\Delta h_{ft} = Cp_{ft}(1 + f - b)(T_{05} - T_{06})$$

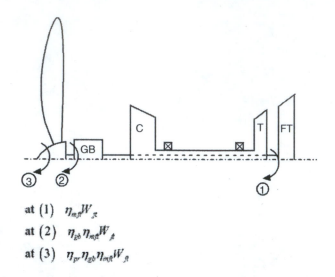

at (1) $\eta_{mp} W_{\mathcal{R}}$

at (2) $\eta_{gb} \eta_{mp} W_{\mathcal{R}}$

at (3) $\eta_{pr} \eta_{gb} \eta_{mp} W_{\mathcal{R}}$

FIGURE 6.15   Power transmission through a double-spool turboprop engine.

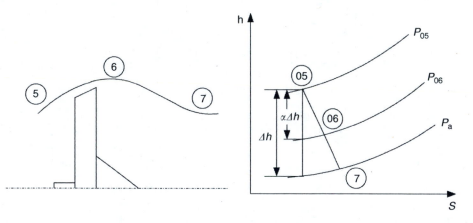

FIGURE 6.16   Expansion in the turbine and nozzle of a double-spool turboprop.

As noticed from the above equation and as previously explained in Section 6.3.1, the temperature (T_{06}) is unknown and cannot be calculated. Then a procedure similar to that described in Section 6.3.1 will be followed. Referring to Figure 6.16, which defines the successive expansion processes in the free power turbine and the nozzle, we have Δh = enthalpy drop available in an ideal (isentropic) turbine and exhaust nozzle; a full expansion to the ambient pressure is assumed in the nozzle $(P_7 = P_a)$. Δh is then calculated as given below:

$$\Delta h = Cp_h T_{05} \left[1 - \left(\frac{P_7}{P_{05}} \right)^{\frac{\gamma_h - 1}{\gamma_h}} \right]$$

where $Cp_h = Cp_t = Cp_n$ and $\gamma_h = \gamma_t = \gamma_n$.

$\alpha \Delta h = \Delta h_{fts}$, which is the fraction of Δh that would be available from an isentropic free power turbine having the actual pressure ratio

$$\Delta h_{ft} = \eta_{ft} \Delta h_{ftS}$$

where η_{ft} is the isentropic efficiency of the free power turbine.

Following the same procedure described above to determine the optimum α, the propeller thrust and the exhaust thrust are determined from the following relations:

$$T_{pr} = \frac{\dot{m}_a \eta_{pr} \eta_g}{U}[(1 + f - b)\eta_{mft}\eta_{ft}\alpha \Delta h] \tag{6.13}$$

$$T_n = \dot{m}_a [(1 + f - b)U_e - U]$$

The total thrust is then given by,

$$T = T_{pr} + T_n$$

$$\frac{T}{\dot{m}} = \frac{\eta_{pr}\eta_g}{U}[(1 + f - b)\eta_{mft}\eta_{ft}\alpha \Delta h] + [(1 + f - b)\sqrt{2(1 - \alpha)\eta_n \Delta h} - U] \tag{6.14}$$

where η_{mft} is the mechanical efficiency of the free power turbine.

Maximizing the thrust T for fixed component efficiencies, flight speed U and Δh yield the following optimum value of (α_{opt})

$$\alpha_{opt} = 1 - \frac{U^2}{2\Delta h}\left(\frac{\eta_n}{\eta_{pr}^2 \eta_g^2 \eta_{mft}^2 \eta_{ft}^2}\right) \tag{6.15}$$

Substituting this value of (α) in Equation 6.14 gives the maximum value of the thrust force. The corresponding value of the exhaust speed is given by the following equation:

$$U_e = U\frac{\eta_n}{\eta_{pr}\eta_g\eta_{mft}\eta_{ft}} \tag{6.16}$$

The outlet conditions at the free turbine outlet are easily calculated from the known value of (Δh) and (α_{opt}).

6.4 ANALOGY WITH TURBOFAN ENGINES

Turboprop engines are analogous to high bypass turbofan engines. The propeller itself is an unducted fan with a bypass ratio equal to or greater than 25. Considering the puller type turboprop engines, the air flow through the propeller is slightly accelerated and thus acquires speed (u_1) slightly higher than the aircraft flight speed (u_0). The momentum difference between the inlet flow and outlet flow through the propeller produces the propeller thrust. Next this accelerated air passes through the engine core and accelerated to higher speeds (u_e). The momentum difference between the outlet and inlet core flow results in the core thrust. The thrust force is given by the relation:

$$T = \dot{m}_0[(u_1 - u_0)] + \dot{m}_a[(1 + f - b)u_e - u_1] \tag{6.17a}$$

Introducing the bypass ratio (β) into Equation (6.17a) to get the following:

$$T = \dot{m}_a[\beta u_1 + (1 + f - b)u_e - (1 + \beta)u_0] \tag{6.17b}$$

The specific thrust related to the engine core mass flow rate is given by

$$\frac{T}{\dot{m}_a} = [\beta u_1 + (1 + f - b)u_e] + ((1 + \beta)u_0) \tag{6.17c}$$

6.5 EQUIVALENT ENGINE POWER

6.5.1 STATIC CONDITION

During testing (on a test bench) or takeoff conditions, the total equivalent horsepower is denoted by tehp and is equal to the shaft horsepower (shp) plus the shp equivalent to the net jet thrust. For estimation purposes it is taken that, under sea level static conditions, one shp is equivalent to approximately 2.6 lb of jet thrust [2]. Thus

$$(\text{tehp})_{\text{takeoff}} = \text{shp} + \frac{\text{jet thrust (lb)}}{2.6} \tag{6.18a}$$

Switching to SI units, experiments have shown also that [3] the total equivalent power (TEP) in kW is related to the shaft power (SP) also in kW by the relation:

$$\{\text{TEP(kW)}\}_{\text{takeoff}} = \text{SP(kW)} + \frac{\text{jet thrust (Newton)}}{8.5} \tag{6.18b}$$

The jet thrust on test bench (ground testing) or during takeoff is given by

$$T = \dot{m}(1 + f - b)U_{\text{e}}$$

6.5.2 FLIGHT OPERATION

For a turboprop engine during flight, the equivalent shaft horsepower (ESHP) is equal to the shp plus the jet thrust power as per the following relation:

$$\dot{\text{ESHP}} = \text{shp} + \frac{T \times U}{\text{constant} \times \eta_{\text{pr}}} \tag{6.19a}$$

where the jet thrust is $T = m[(1 - f - b)U_{\text{e}} - U]$.

The constant in Equation 6.19a depends on the employed units; thus

$$\text{ESHP} = \text{shp} + \frac{T(\text{lb}) \times U(\text{ft/s})}{550 \times \eta_{\text{pr}}} \tag{6.19b}$$

$$\text{ESHP} = \text{shp} + \frac{T(\text{lb}) \times U(\text{mph})}{375 \times \eta_{\text{pr}}} \tag{6.19c}$$

$$\text{ESHP} = \text{shp} + \frac{T(\text{lb}) \times U(\text{knots})}{325 \times \eta_{\text{pr}}} \tag{6.19d}$$

$$\text{ESHP} = \text{shp} + \frac{T(N) \times U(\text{m/s})}{745.7 \times \eta_{\text{pr}}} \tag{6.19e}$$

Normally a value of $\eta_{\text{pr}} \approx 80\%$ is employed as industry standard.

6.6 FUEL CONSUMPTION

As previously explained for turbojet and turbofan engines, the fuel consumption is identified by the thrust-specific fuel consumption (TSFC) defined as TSFC $= \dot{m}f/T$ and expressed in terms of kg fuel/N · h.

For turboprop engines the fuel consumption is identified by the equivalent specific fuel consumption (ESFC) defined as

$$\text{ESFC} = \frac{\dot{m}f}{\text{ESHP}}$$

and expressed in the following units lb fuel/hp.h or kg fuel/kW · h.

Typical values are

$$\text{ESFC} = 0.45 - 0.60 \frac{\text{lb fuel}}{\text{hp} \cdot \text{h}} \quad \text{or} \quad 0.27 - 0.36 \frac{\text{kg fuel}}{\text{kW} \cdot \text{h}}$$

Example 1

1. A two-spool turboprop engine with the propeller driven by a free power turbine. It is required to calculate ESHP and ESFC in the following two cases:

(a) *Static conditions*

$$\dot{m}_f = 1500 \text{ lb/h}, \quad T = 600 \text{ lb}_f \quad \text{and} \quad shp = 2800$$

(b) *Flight conditions*

$$V = 330 \text{ ft/s}, \quad \dot{m}_f = 1000 \text{ lb/h}, \quad T = 300 \text{ lb}_f \quad \text{and} \quad shp = 2000$$

Solution:

(a) For static or ground run conditions, ESHP and ESFC are calculated from the following relations:

$$\text{ESHP} = shp + \frac{T}{2.6} = 2800 + \frac{600}{2.6} = 3030.77 \text{ hp}$$

$$\text{ESFC} = \frac{\dot{m}_f}{\text{ESHP}} = \frac{1500}{3030.77} = 0.49492 \frac{\text{lb}}{\text{ESHP} \cdot \text{h}}$$

(b) During flight

$$\text{ESHP} = shp + \frac{TV}{550} = 2000 + \frac{300 \times 330}{550} = 2180 \text{ hp}$$

$$\text{ESFC} = \frac{\dot{m}_f}{\text{ESHP}} = \frac{1000}{2180} = 0.4587 \quad \frac{\text{lb}}{\text{ESHP} \cdot \text{h}}$$

6.7 TURBOPROP INSTALLATION

Turboprop (or turbo-propeller) engines are installed in one of the following positions:

1. Wing
2. Fuselage, either at nose or empennage of aircraft for a single-engined aircraft
3. Horizontal tail installation

Details of power plant installation will be given in Chapter 9. A brief hint is given here. Most present day turboprops are *wing*-mounted engines to either passenger or cargo transports (Figure 6.17). Either puller or pusher turboprops may be installed to the wing. Typical wing install-ation of the puller type is seen in Fokker F-27 powered by two RR Dart puller engines. In some aircraft a pair of counterrotating propellers is installed to each engine; Antonov AN-20 is an example for these aircraft.

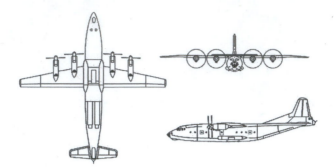

FIGURE 6.17   Wing installation of turboprop engine.

FIGURE 6.18   XF2R-1 aircraft powered by GE XT-31 turboprop engine.

Few aircraft are powered by turboprop engines mounted to the fuselage. Concerning nose install-ation, very rare aircraft are powered by a single turboprop engine fitted to its nose. An example of which is the aircraft XF2R-1 (Dark Shark) which is powered by GE XT-31 turboprop engine (Figure 6.18). Another nose installation was seen in the 1950s in the experimental McDonnell XF-88B aircraft. It was a version of a pure jet XF-88 VOOdoo, but powered by Allison T38 turboshaft engine to conduct propeller research for supersonic planes. Most of the power plants installed to the nose of aircraft are piston engines having single or counter rotating propellers. As an altern-ative a turboprop engine having contra rotating propellers is installed in the tail as seen in the Jetcruzer 500, AASI airplane (Figure 6.6). Figure 6.19 illustrates also the Voronezh Motor Plant M-9F Turboprop engine installed to the top of empennage which powers the aircraft KNAAPO AS-20P single engine light experimental amphibian. The top view represents an installation for the original engine having 450 hp, while the bottom view represents the installation of the upgraded engine having 500 hp.

The Antonov AN-180 medium turboprop airliner (around 175 passengers) is under development. It is powered by two turboprop engines attached to the tips of the horizontal tail and each has contrarotating propellers.

Details of some engines
Some of the data of two turboprop engines is given in Table 6.1.

FIGURE 6.19   KNAAPO AS-20P amphibian powered by Voronezh Motor Plant. M-9F turboprop engine installed to empennage.

Example 2   A single spool turboprop engine when running at maximum rpm at sea level conditions ($P_a = 1$ bar and $T_a = 288\,K$) had the following particulars:

$$m_a = 14.6\,\text{kg/s}, \qquad \pi_c = \frac{P_{03}}{P_{02}} = 8.5, \qquad P_{04} = 0.96\,P_{03}$$

$T_{04} = 1220\,\text{K}, \qquad V_J = 250\,\text{m/s},$

$\eta_c = 0.87, \qquad\qquad \eta_{CC} = 0.98, \qquad\qquad \eta_t = 0.9$

$\eta_n = 0.95, \qquad\qquad \eta_m = 0.98, \qquad\qquad \eta_{pr} \times \eta_g = 0.78$

$C_{P_c} = 1.01\,\text{kJ/kg K}, \qquad C_{P_{CC}} = 1.13\,\text{kJ/kg}\cdot\text{K} \quad C_{P_t} = C_{P_n} = 1.14\,\text{kJ/kg}\cdot\text{K}$

$Q_R = 43{,}000\,\text{kJ/kg}\cdot\text{K}$

It is required to calculate the equivalent brake horsepower (ebhp).

Solution: A sketch for the turboprop engine and cycle on the temperature–entropy diagram is given in Figure 6.20. Successive analysis of the elements will be given here.

Intake: The engine is under ground test (zero flight speed and Mach number); then the total conditions are equal to the static conditions.

TABLE 6.1

Specifications of Two Turboprop Engines (A-53-L13B and LTS101-600A)

Engine	A-53-L13B	LTS101-600A
Power (shp)	1570	640
No. of shafts	1	1
Compressor type	Axial + centrifugal	Axial + centrifugal
Compressor stages	5 Axial + 1 centrifugal	1 Axial + 1 centrifugal
Turbine stages	2	1
Compression ratio	7	14
Mass flow rate (lb/s)	11.5	5
Propeller (rpm)	2200	2000
T/O SFC (lb/ESHP/h)	0.6	0.551
Cruising SFC (lb/ESHP/h)	0.6	0.551
Max PWR. pressure ratio	9.4	8.5
Length (in.)	47.6	37.4
Diameter (in.)	23	21
Weight (lb)	545	290

Moreover, the flow in the diffuser is isentropic, thus

$$T_{02} = T_{0a} = T_a = 288\,\text{K}$$
$$P_{02} = P_{0a} = P_a = 1\,\text{bar}$$

Compressor: The outlet conditions are

$$P_{03} = \pi_c P_{02} = 8.5\,\text{bar}$$

With $C_{P_c} = 1.01\,\text{kJ/kg K}$, then $\gamma_c = 1.397$

$$T_{03} = T_{02}\left[1 + \frac{\pi^{\frac{\gamma_c-1}{\gamma_c}} - 1}{\eta_C}\right] = 567.094\,\text{K}$$

$$W_c = C_{P_c}(T_{03} - T_{02}) = 280\,\text{kJ/kg}$$

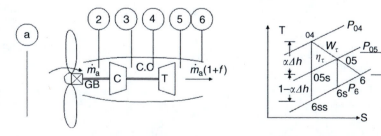

FIGURE 6.20   Sketch for the engine and cycle.

Combustion chamber: The pressure for gases leaving the combustion chamber is

$$P_{04} = 0.96\,P_{03} = 8.16\ \text{bar}$$

The fuel-to-air ratio and fuel mass flow rate are

$$f = \frac{C_{P_{C.C}}T_{04} - C_{P_C}T_{03}}{\eta_{C.C}Q_R - C_{P_{C.C}}T_{04}} = 0.0186$$

$$\dot{m}_f = 0.0186\dot{m}_a = 0.271\ \text{kg/s}$$

Turbine: The enthalpy drop in the turbine and nozzle is

$$\Delta h = C_{P_t}(T_{04} - T_{6s}) = C_{P_t}T_{04}\left(1 - \frac{T_{6s}}{T_{04}}\right) = C_{P_t}T_{04}\left[1 - \left(\frac{P_6}{P_{04}}\right)^{(\gamma_t - 1/\gamma_t)}\right]$$

with $P_6 = P_a$, $\gamma_t = 1.336$, then

$$\Delta h = 571.3\ \text{kJ/kg}$$

From Figure 6.20,

since $\eta_n = \dfrac{V_J^2/2}{(1-\alpha)\Delta h}$ then $\alpha = 1 - \dfrac{V_J^2}{2\eta_n\Delta h} = 0.9424$

The turbine output specific power is

$$W_t = \eta_t\alpha\Delta h = 484.56\ \text{kJ/kg}$$

The specific power delivered to the propeller shaft is

$$W_{shaft} = \eta_m W_t - W_C = 194.87\ \text{kJ/kg}$$

The corresponding shaft power is

$$P_{shaft} = \dot{m}_a W_{shaft} = 2845\ \text{kW}$$

The jet thrust for zero flight speed is

$$T_J = \dot{m}_a[(1+f)\,V_J] = 3717.8\ \text{N}$$

The total power is the sum of propeller and jet thrust contribution

$$P_{total} = \eta_{pr} \times \eta_g \times P_{shaft}(\text{kW}) + \frac{T_J(N)}{8.5} = 2656.5\ \text{kW}$$

The horsepower is then

$$ebhp = \frac{P_{total}(\text{kW})}{0.7457} = 3562\ \text{hp}$$

Example 3 The turboprop considered in Example 2 is reexamined here. The engine is fitted to an aircraft flying at Mach number of 0.6. The maximum temperature is 1120 K and fuel having a heating value of 43,8778 kJ/kg is used. All other data are unchanged. It is required to calculate the specific equivalent shaft horsepower (seshp) when the aircraft is flying at sea level and altitudes of 3, 6, and 9 km.

Solution: The ambient temperature and pressure at the requested altitudes are

At sea level: $T_a = 288$ K $P_a = 1$ bar
3 km $T_a = 268.5$ K $P_a = 0.701$ bar
6 km $T_a = 249$ K $P_a = 0.472$ bar
9 km $T_a = 229.5$ K $P_a = 0.308$ bar

A general formulation for the problem will be followed to account for the variations in the inlet conditions due to altitude changes.

Intake

$$T_{02} = T_{0a} = T_a \left[1 + \frac{\gamma - 1}{2} M^2 \right] = 1.072\, T_a \text{ K}$$

$$P_{02} = P_{0a} = P_a \left[1 + \frac{\gamma - 1}{2} M^2 \right]^{(\gamma_c/\gamma_c - 1)} = 1.072\, P_a \text{ bar}$$

Compressor

$$P_{03} = \pi_C P_{02} = 8.5 \times P_{02} = 8.5 \times 1.276\, P_a = 10.846\, P_a \text{ bar}$$

with $\gamma_c = 1.397$, then the outlet temperature and the compressor specific work are

$$T_{03} = T_{02} \left[1 + \frac{\pi^{(\gamma_c - 1)/\gamma_c} - 1}{\eta_C} \right] = T_{02} \left[1 + \frac{\pi^{0.286} - 1}{0.87} \right] = 1.97\, T_{02} = 2.1118\, T_a$$

$$W_c = \frac{C_{P_c}(T_{03} - T_{02})}{\eta_{mc}} = \frac{1.01 \times 0.97}{0.98} = 1.0716\, T_a \text{ kW/kg}$$

Combustion chamber

$$P_{04} = 0.96\, P_{03}$$

$$f = \frac{C_{Pcc} T_{04} - C_{Pc} T_{03}}{\eta_c Q_R - C_{PCC} T_{04}} = \frac{1265 - 1.01 T_{03}}{43000 - 1265} = \frac{1265 - 1.01 T_{03}}{41735}$$

Turbine: From Figure 6.20,

$$T_{05} - T_{6ss} = T_{05} - T_{6s} = \frac{V_j^2}{2 C_{pn} \eta_n} = \frac{(250)^2}{2 \times 1140 \times 0.95} = 28.8 \text{ K}$$

Since

$$C_{P_t} = C_{P_n} = 1.14, \quad \text{and} \quad \frac{\gamma_n - 1}{\gamma_n} = \frac{R}{C_{p_n}} = \frac{287}{1140} = 0.252, \quad \text{then} \quad \gamma_n = \gamma_t = 1.336.$$

TABLE 6.2
Effect of Flight Altitude on the Available Power of a Turboprop Engine

Altitude (km)	W_C (kJ/kg)	f	W_{shaft} (kJ/kg)	$\dfrac{P_{propeller}}{\dot{m}_a}$	V_f (m/s)	$\dfrac{T}{\dot{m}_a}$	$\dfrac{P_{jet}}{\dot{m}_a}$	$\dfrac{P_{total}}{\dot{m}_a}$ (kJ/kg)	Seshp (hp/(kg/s))
0	308	0.0156	173	135	204	49.9	10.18	145.18	194.7
3	288	0.0166	193	151	197	57	11.23	162.23	217.6
6	267	0.0176	215	168	190	64	12.16	180.16	241.6
9	246	0.0186	236	184	182	73	13.22	197.22	264.5

$$\frac{T_{04}}{T_{6ss}} = \left(\frac{P_{04}}{P_a}\right)^{\frac{\gamma_n-1}{\gamma_n}} = \left(\frac{P_{04}}{P_{03}}\frac{P_{03}}{P_{02}}\frac{P_{02}}{P_a}\right)^{\frac{\gamma_n-1}{\gamma_n}} = [(0.96)(8.5)(1.275)]^{0.252} = 1.805$$

$$T_{6ss} = \frac{1120}{1.805} = 620.5 \text{ K}$$

$$T_{05s} \approx T_{6ss} + (T_{05} - T_{6s}) = 620.5 + 28.8 = 649.3 \text{ K}$$

$$\frac{P_{04}}{P_{05}} = \left(\frac{T_{04}}{T_{05s}}\right)^{\frac{\gamma_t}{\gamma_t-1}} = \left(\frac{1120}{649.3}\right)^{3.97} = 8.709$$

$$W_t = \eta_{mt}\Delta h_t = \eta_{mt}\eta_t Cp_t(T_{04} - T_{05s}) = 0.98 \times 0.9 \times 1.14 \times (1120 - 649.3)$$

$$W_t = 473.3 \text{ kJ/kg}$$

$$W_{shaft} = (1+f)W_t - W_c$$

$$\frac{P_{propeller}}{\dot{m}_a} = \eta_{pr}\eta_g W_{shaft} = 0.78 W_{shaft}$$

$$\frac{T}{\dot{m}_a} = (1+f)V_j - V_f = (1+f)V_j - 0.6\sqrt{\gamma_c RT_a}$$

$$\frac{P_{jet}}{\dot{m}_a} = \frac{T}{\dot{m}_a} \times V_f = \frac{T}{\dot{m}_a} \times \left(0.6\sqrt{\gamma_c RT_a}\right)$$

$$\frac{P_{total}}{\dot{m}_a} = \frac{P_{propeller}}{\dot{m}_a} + \frac{P_{jet}}{\dot{m}_a}$$

$$seshp = \frac{P_{total}}{0.7456\,\dot{m}_a}$$

The results can be arranged in Table 6.2.

Example 4 The Bell/Boeing V-22 Tilt-rotor multimission aircraft is shown in Figure 6.5. It is powered by Allison T406 engine. The T406 engine has the following characteristics:
  Rotor is connected to a free power turbine.

Air mass flow rate	14 kg/s
Compressor pressure ratio	14
Turbine inlet temperature	1400 K
Fuel heating value	43,000 kJ/kg

During landing, it may be assumed that the air entering and gases leaving the engine have nearly zero velocities. The ambient conditions are 288 K and 101 kPa. The propeller efficiency and gear box efficiencies are 0.75 and 0.95. Assuming all the processes are ideal, calculate the propeller power during landing ($\gamma_c = 1.4$ and $\gamma_h = 1.3299$).

Solution: The ambient conditions are

$$T_a = 288\,K, \quad P_a = 101\,kPa$$

The flight and jet speeds are $V_f = V_j = 0$, with $\gamma_c = 1.4$, then $Cp_c = \gamma_c R/(\gamma_c - 1) = 1004.5\,J/kg/K$. Since $\gamma_h = 1.3299$, then $Cp_h = \gamma_h R/(\gamma_h - 1) = 1156.7\,J/kg/K$.

Intake
Since the flight Mach number is zero, then

$$T_{02} = T_{0a} = T_a\left(1 + \frac{\gamma_c - 1}{2}M^2\right) = 288\,K$$

$$P_{02} = P_{0a} = P_a\left(1 + \frac{\gamma_c - 1}{2}M^2\right)^{(\gamma_c/\gamma_c - 1)} = 101\,kPa$$

Compressor: For ideal compressor

$$T_{03} = T_{02}\left(\frac{P_{03}}{P_{02}}\right)^{\frac{\gamma_c-1}{\gamma_c}} = T_{02}(\pi_c)^{\frac{\gamma_c-1}{\gamma_c}} = 612.15\,K$$

$$P_{03} = \pi_c \times P_{02} = 1414\,kPa$$

Combustion chamber

$$T_{04} = 1400\,K \quad and \quad P_{04} = P_{03} = 1414\,kPa$$

Fuel-to-air ratio

$$f = \frac{Cp_h T_{04} - Cp_c T_{03}}{Q_{HV} - Cp_h T_{04}} = 0.02$$

Turbine: Energy balance between the compressor and turbine of the gas generator

$$W_c = W_t$$

$$\dot{m}_a Cp_c(T_{03} - T_{02}) = \dot{m}_a(1+f)Cp_h(T_{04} - T_{05})$$

$$T_{05} = T_{04} - \frac{Cp_c}{Cp_h(1+f)}(T_{03} - T_{02}) = 1125.37\,K$$

$$P_{05} = P_{04}\left(\frac{T_{05}}{T_{04}}\right)^{\frac{\gamma_h}{\gamma_h-1}} = 586.47\,kPa$$

Free power turbine

$$\Delta h = h_{05} - h_{7S} = Cp_h(T_{05} - T_{7S})$$

$$\Delta h = Cp_h T_{05}\left(1 - \left(\frac{P_7}{P_{05}}\right)^{\frac{\gamma_h-1}{\gamma_h}}\right) = 460376.7 \text{ J/kg}$$

$$\alpha_{opt} = 1 - \frac{V_f^2 \eta_n}{2\eta_{pr}^2 \eta_g^2 \eta_m^2 \eta_{fpt}^2 \Delta h} = 1$$

The free power turbine absorbs all the power in the gases leaving the gas generator turbine. Points (6) and (7) in Figure 6.16 are coincident. Thus

$$\Delta h = Cp_h(T_{05} - T_{6S}) = Cp_h(T_{05} - T_{06})$$

from which $T_{06} = 727.36$ K.

Nozzle: There is no thrust generated as

$$Tj = \dot{m}_a((1+f)V_J - V_f) = 0$$
$$P_j = 0$$

The specific power absorbed in the propeller is

$$W_{pr} = \eta_{pr}\eta_g\eta_m\eta_{fpt}\alpha\,\Delta h = 0.75 \times 0.95 \times 1 \times 1 \times 1 \times 460.3767 = 328.02 \text{ kJ/kg}$$

The power consumed in the propeller is then

$$P_{pr} = \dot{m}_a W_{pr} = 4592.26 \text{ kW}$$
$$P_t = P_{pr} + P_j = 4592.26 \text{ kW}$$

The corresponding shaft horsepower is then

$$shp = \frac{P_t}{0.7457} = \frac{4592.26}{0.7457} = 6158.32 \text{ hp}$$

6.8   PERFORMANCE ANALYSIS

A case study for a turboprop fitted with a free power turbine is examined. The flight altitude considered is 6 km. The effect of flight Mach number on the thrust and power of both the propeller and jet is examined. Mach number is increased from 0.1 to 0.8. The fuel is of the hydrocarbon type having a heating value of 43,000 kJ/kg and a compressor pressure ratio of 6 with a turbine inlet temperature of 1200 K. The efficiencies of the diffuser, compressor, and combustion chamber are 0.95, 0.87, and 0.98. The efficiency of the two turbines and nozzle is 0.9. The mechanical efficiency for the compressor and turbine is 0.98. The efficiencies of the gearbox and propeller are 0.95 and 0.8. The air mass flow rate is 25 kg/s. The specific heat ratios for cold section, combustion chamber, and hot section are respectively $\gamma_c = 1.3798$, $\gamma_{cc} = 1.3405$, and $\gamma_h = 1.3365$.

Figure 6.21 illustrates the variation of four parameters; namely, fuel-to-air ratio, enthalpy drop across the free turbine and nozzle in J/kg, optimum α_{opt} and inlet diameter with the flight Mach number.

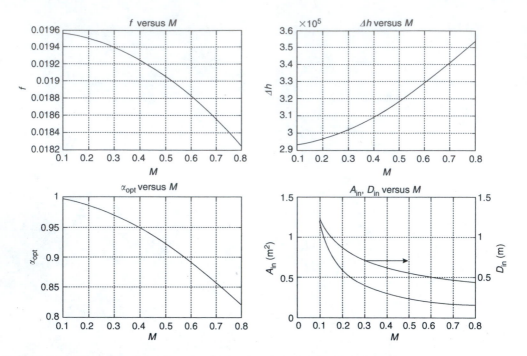

FIGURE 6.21   Variation of f, Δh, α_{opt} and inlet area (A_{in}) with flight Mach number.

Figure 6.22 examines the contribution of both the jet and propeller to the engine thrust and power. It also evaluates the percentage of the jet thrust to the propeller thrust. As seen, this percentage ranges from 1% to 12% as the Mach number increases from 0.1 to 0.8.

The total power expressed in ESHP and the fuel equivalent specific consumption expressed as ESFC are illustrated in Figure 6.23.

To examine the effect of compressor pressure ratio and turbine inlet temperature, the above case is reexamined but for a larger engine having a compressor pressure ratio of 23 and turbine inlet temperature of 1600 K. All other data are unchanged. The variations of fuel-to-air ratio, enthalpy drop across the free turbine and nozzle in J/kg, optimum α_{opt}, and inlet diameter with the flight Mach number are given in Figure 6.24.

Figure 6.25 examines the contribution of both the jet and propeller to the engine thrust and power. It also evaluates the percentage of the jet thrust to the propeller thrust. As seen, this percentage ranges from 1% to 7% as the Mach number increases from 0.1 to 0.8.

The total power expressed in ESHP and the equivalent specific fuel consumption expressed as ESFC is illustrated in Figure 6.26. A peak value at mach number of 0.5 is seen for the ESHP. This point may be selected for the continuous cruise of aircraft.

6.9 COMPARISON OF TURBOJET, TURBOFAN, AND TURBOPROP ENGINES

Some important characteristics of the three engine types are as follows:

1. Turbojet engine
    - Low thrust at low forward speed
    - High thrust-specific fuel consumption
    - Low specific weight (W/T)
    - Small frontal area
    - Good ground clearance

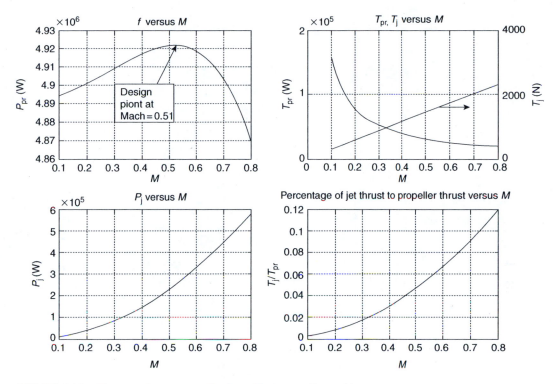

FIGURE 6.22   Thrust and power contribution of both propeller and jet.

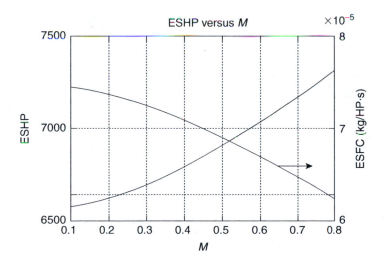

FIGURE 6.23   Variation of ESHP and ESFC with Mach number.

2. Turbofan engine
   - High thrust even at low speed
   - Thrust-specific fuel consumption is better than turbojet engine
   - Less noise compared to turbojet engine
   - Low propulsive efficiency at very high speed, though higher than turbojet. Generally better performance at the range of subsonic and transonic speed
   - Fair ground clearance

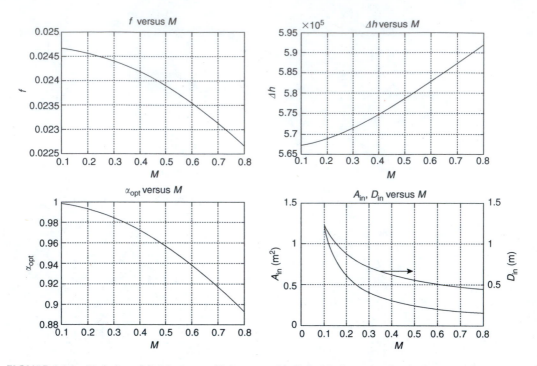

FIGURE 6.24 Variation of f, Δh, α_{opt} and inlet area with flight Mach number for the larger engine.

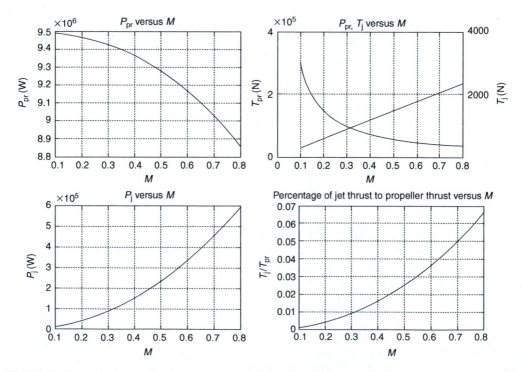

FIGURE 6.25 Thrust and power contribution of both propeller and jet for the larger engine.

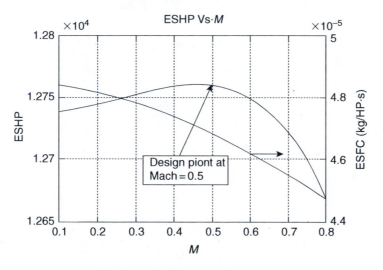

FIGURE 6.26 Variation of ESHP and ESFC with Mach number.

3. Turboprop
   - High propulsive efficiency at low speed
   - More complicated and heavier than turbojet and turbofan engines
   - Low SFC
   - Large ground clearance as high wing design is employed in turboprop powered aircraft
   - Highest noise level

6.10 TURBOSHAFT ENGINES

A turboshaft engine is defined as a gas turbine engine designed to produce only shaft power. Turboshaft engines are similar to turboprop engines, except that the hot gases are expanded to a lower pressure in the turbine, thus providing greater shaft power and little exhaust velocity.

Turboshaft engines are used in helicopters, boats, ships, trains and automobiles, tanks, pumping units for natural gas in cross-country pipelines, and various industrial equipments.

In this chapter, analysis and discussions will be limited to turboshaft engines used in helicopters.

In Chapter 8 all the other applications will be thoroughly analyzed. Many aero engine manufacturers produce two versions from an engine, one as a turboprop and the other as a turboshaft engine. Examples are the PT6B, PW206 produced by Pratt and Whitney of Canada and AlliedSignal LTS101-600A-3 and Allison 250-C30M.

A turboshaft engine also has two configurations; either the load is driven by the same gas generator shaft (Figure 6.27) or driven by the free power turbine (Figure 6.28). The first configuration resembles a single-spool engine while the second is a two-spool one.

The corresponding cycles plotted on the temperature–entropy axes are shown in Figure 6.29 for single-spool and in Figure 6.30 for double-spool engines. It is not that the hot gases expand in the turbine (single spool) or free turbine (double spool) to nearly the ambient pressure. This provides a greater shaft power and little exhaust velocity.

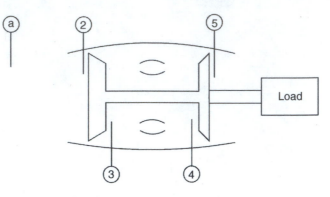

FIGURE 6.27   Single-spool turboshaft engine.

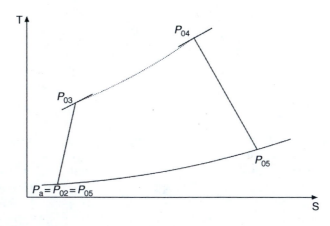

FIGURE 6.28   Two-spool turboshaft engine.

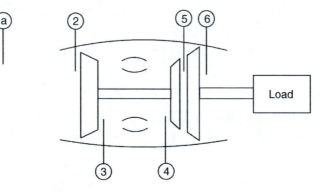

FIGURE 6.29   T–S diagram for a single-spool turboshaft.

6.11   POWER GENERATED BY TURBOSHAFT ENGINES

6.11.1   SINGLE-SPOOL TURBOSHAFT

From Figure 6.29, the following hints may be mentioned.

Diffuser: Due to the nearly zero or small flight speed

$$P_{02} = P_{0a} = P_a \tag{6.20}$$

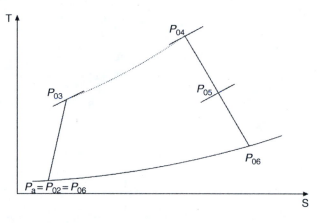

FIGURE 6.30   T–S diagram for a double-spool turboshaft.

and

$$T_{02} = T_{0a} = T_a \qquad (6.21)$$

Compressor: The outlet temperature and pressure are determined from Equations 6.3 and 6.4.

Combustion chamber: The fuel-to-air ratio is evaluated from Equation 6.5.

Turbine: The turbine drives both the compressor and load. The specific power delivered to the output shaft (W_{shaft}) is expressed as

$$W_{shaft} = (1 + f - b)\eta_{mt} W_t - \frac{W_c}{\eta_{mc}} \qquad (6.22)$$

where f is the fuel-to-air ratio and b, the bleed ratio. The mechanical efficiencies for turbine and compressor are respectively η_{mt}, η_{mc}.

This shaft will drive the load (here the rotor of a helicopter) and experience some mechanical losses due to friction in gearbox and bearings, thus

$$W_{load} = \eta_m W_{shaft} \qquad (6.23)$$

The load power is then

$$P_{load} = \dot{m}_a W_{load} \qquad (6.24)$$

6.11.2 DOUBLE-SPOOL TURBOSHAFT

Some hints will also be given here. The diffuser, compressor, and combustion chamber are treated in the same way as in Section 6.11.1.

Gas generator turbine
The turbine provides sufficient energy to drive the compressor; thus the outlet temperature is obtained from the energy balance between the compressor and turbine, or

$$T_{05} = T_{04} - \frac{Cp_c(T_{03} - T_{02})}{Cp_t \eta_{mc} \eta_{mt}(1 + f - b)} \qquad (6.25)$$

and

$$P_{05} = P_{04} \left[1 - \left(\frac{T_{04} - T_{05}}{\eta_t T_{04}} \right) \right]^{\gamma/(\gamma-1)} \tag{6.26}$$

Free power turbine
The gases are assumed to have a complete expansion to the ambient pressure; thus the power delivered to the load is

$$P_{\text{load}} = \dot{m}_a (1 + f - b) \eta_m C p_{ft} T_{05} \left[1 - \left(\frac{P_a}{P_{05}} \right)^{(\gamma-1)/\gamma} \right] \tag{6.27}$$

6.12 EXAMPLES FOR TURBOSHAFT ENGINES

Examples for turboshaft engines used in helicopters are Lycoming T-55-L-7C powering the Boeing Vertol Chinook CH-47 helicopter [6] or Turbomeca Arrius 2K1 powering the Eurocopter EC120 Colibri used in tourism over New York city [7]. Both are two spools where the helicopter rotor is driven by a free power turbine.

Example 5 A single-spool turboshaft engine has an approximately zero exhaust speed and equal compressor and turbine pressure ratios. Prove that if the fuel-to-air ratio f is negligible and constant thermal properties C_p and γ are assumed through the engine, then the power available for driving the load attains a maximum value if the compressor pressure ratio has the value

$$\pi_c = \left(\frac{T_{04}}{T_a} \right)^{\frac{\gamma}{2(\gamma-1)}}$$

where π_c is the compressor pressure ratio, T_{04} is the turbine inlet temperature (TIT), and T_a is the ambient temperature.

The maximum available specific power is then equal to

$$C p T_a \left[\sqrt{\frac{T_{04}}{T_a}} - 1 \right]^2$$

Neglect all mechanical losses and pressure drop and assume that the processes in the compressor and turbine are isentropic.

Solution
From Figure 6.29, the following peculiarities may be listed:

$$P_5 = P_a \quad \text{and} \quad \pi_c = \pi_t$$

The specific power consumed in the compressor is

$$\Delta h_c = Cp(T_{03} - T_{02}) = CpT_{02}(\pi_c^{(\gamma-1)/\gamma} - 1) = CpT_a(\pi_c^{(\gamma-1)/\gamma} - 1)$$

The available specific power from the turbine is

$$\Delta h_t = Cp(T_{04} - T_5) = CpT_{04}\left(1 - \frac{T_5}{T_{04}}\right) = CpT_{04}\left(1 - \frac{1}{\pi_c^{(\gamma-1)/\gamma}}\right)$$

The specific power available to the load with negligible mechanical efficiencies is then

$$P_{load} = \Delta h_t - \Delta h_c$$

$$P_{load} = CpT_{04}\left(1 - \frac{1}{\pi_c^{(\gamma-1)/\gamma}}\right) - CpT_a\left(\pi_c^{(\gamma-1)/\gamma} - 1\right)$$

$$P_{load} = CpT_a\left[\frac{T_{04}}{T_a}\left(1 - \frac{1}{\pi_c^{(\gamma-1)/\gamma}}\right) - \left(\pi_c^{(\gamma-1)/\gamma} - 1\right)\right]$$

Define the following variables:

$$\lambda = CpT_a, \quad x^b = \pi_c^{(\gamma-1)/\gamma}, \quad A = \frac{T_{04}}{T_a}$$

$$P_{load} = \lambda\left[A\left(1 - \frac{1}{x^b}\right) - \left(x^b - 1\right)\right]$$

$$\frac{\partial P_{load}}{\partial x} = \lambda\left[A\left(\frac{b}{x^{1+b}}\right) - bx^{b-1}\right]$$

At maximum power supplied to the load $\partial P_{load}/\partial x = 0$, thus

$$\frac{A}{x^{1+b}} = x^{b-1} \quad \text{or} \quad A = x^{2b}$$

$$x = A^{\frac{1}{2b}} = \left(\frac{T_{04}}{T_a}\right)^{\frac{\gamma}{2(\gamma-1)}}$$

The maximum available power is then

$$(P_{load})_{max} = \lambda\left[A\left(1 - \frac{1}{\sqrt{A}}\right) - \left(\sqrt{A} - 1\right)\right] = \lambda\left[A - \sqrt{A} - \sqrt{A} + 1\right]$$

$$(P_{load})_{max} = \lambda\left[\sqrt{A} - 1\right]^2$$

or finally,

$$(P_{load})_{max} = CpT_a\left[\sqrt{\frac{T_{04}}{T_a}} - 1\right]^2$$

6.13   PROPFAN ENGINES

The propfan engine emerged in the early seventies when the price of fuel began to soar. A propfan is a modified turbofan engine, with the fan placed outside of the engine nacelle on the same axis as

FIGURE 6.31 Contrarotating forward propfan installed to AN-70 aircraft. (Courtesy Antonov.)

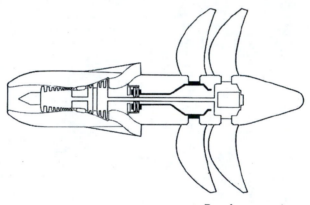

Prop fan concept

FIGURE 6.32 Contrarotating aft propfan. (Reproduced with the permission of Rolls-Royce plc, copyright ©
Rolls-Royce plc 2007.)

the compressor. Propfans are also known as *ultrahigh bypass* (UHB) engines as well as unducted
fan (UDF). The design is intended to offer the speed and performance of a turbofan, with the fuel
economy of a turboprop. There are two main types of propfans:

1. Propfan similar to forward fan with the fan placed outside the engine nacelle. This type is
   next divided into either single or contrarotating; an example is the contrarotating propfan
   installed to the aircraft AN-70 (Figure 6.31).
2. Propfan similar to the aft fan where the fan is coupled to the turbine. It is always of the
   contrarotating type (Figure 6.32).

Propfan engines have the best propulsive efficiency. Single-rotation engines have propulsive
efficiency around 80%, while contrarotating ones have higher efficiency close to 90%.

The main features of the propfan engines versus the turboprop engines are given in Table 6.3.

The only flying aircraft powered by a propfan engine is AN-70. During the 1970s due to the high
escalation in the prices of oil, several projects for developing propfans or unducted fans (UDF) were
introduced. The propfan concept was intended to deliver 35% better fuel efficiency than contemporary
turbofans. In static and air tests on a modified DC-9, propfans reached a 30% improvement. This

TABLE 6.3
Comparison between Turboprop and Propfan Engines

Engine	Turboprop	Propfan
Number of blades	3–6	8–12
Diameter	Large	Smaller
Power/Square of diameter	Small	Larger
Blade shape	Nearly straight	Swept or scimitar-like profile
Maximum thickness	Thick	Thinner
Tip speed	Subsonic	Supersonic
BPR	Nearly 50	25 or more
Propulsive efficiency	High	Higher
ESFC(kg/hp.h)	0.19–0.23	0.17
Mach number	0.5–0.6	0.68
Cruise altitude	6000 m	11,000 m

efficiency comes at a price, as one of the major problems with the propfan is noise, particularly in an era where aircraft are required to comply with increasingly strict Stage III and Stage IV noise limitations.

General Electric (GE) introduced its GE36 UDF; Figure 6.33, which appears similar to a turboprop of the pusher configuration. GE's UDF has a novel direct drive arrangement, where the reduction gearbox is replaced by a low-speed 7-stage turbine. The turbine rotors drive one propeller, while the other prop is connected to the "unearthed" turbine stators and rotates in the opposite direction (Figure 6.34). So, in effect, the power turbine has 14 stages. The fan blades are highly swept which allows high Mach number cruise operation. Moreover, these fan blades are made of advanced composite material that exhibits high strength and stiffness as well as light weight. Moreover, these blades have a replaceable nickel strip bonded to the leading edge and a polyurethane sheet bonded to the surface for erosion protection. GE36 UDF has no thrust reverser. The reverser is eliminated since its function is provided by the reverse-pitch fan. The compressors have high pressure ratios with substantial surge margins and high efficiency. Testing in NASA wind tunnel and GE test cells (Figure 6.35), assured that the fan blades are highly efficient, stable, flutter free in addition to their ruggedness and FOD resistant. Boeing intended to offer GE's pusher UDF engine on the 7J7 platform, and McDonnell Douglas were going to do likewise on their MD-94X airliner. Both airliners were to use rear-fuselage-mounted GE-36 engines.

McDonnell Douglas developed a proof-of-concept aircraft by modifying their MD-80. They removed the JT8D turbofan engine from the left side of the fuselage and replaced it with the GE-36 (Figure 6.33). A number of test flights were conducted that proved the airworthiness, aerodynamic characteristics, and noise signature of the design. The test and marketing flights of the GE-outfitted "Demo Aircraft" concluded in 1988, demonstrating a 30% reduction in fuel burn over turbofan powered MD-80, full Stage III noise compliance, and low-levels of interior noise/vibration. Owing to jet-fuel price drops and shifting marketing priorities, Douglas shelved the program the following year.

In the 1980s, Allison collaborated with Pratt & Whitney on demonstrating the 578-DX propfan. Unlike the competing GE-36 UDF, the 578-DX was fairly conventional, having a reduction gearbox between the LP turbine and the propfan blades. The 578-DX was successfully flight tested on a McDonnell Douglas MD-80.

The Progress D-27 propfan, developed in the erstwhile USSR, is even more conventional in layout, with the propfan blades at the front of the engine in a tractor configuration. Two rear-mounted D-27's propfans propelled the Antonov AN-180, which were scheduled for a 1995 entry into service. Another Russian propfan application was the Yakovlev Yak-46.

FIGURE 6.33   Unducted fan installed in MD-80.

However, none of the above projects came to fruition, mainly because of excessive cabin noise and low fuel prices.

During the 1990s, Antonov also developed the AN-70, powered by four Progress D-27s in a tractor configuration, which still remains available for further investment and production.

SUMMARY OF TURBOPROP RELATIONS

$$T = \dot{m}_a \left((1+f)\sqrt{2\eta_n(1-\alpha)\Delta h} - u \right) + \eta_t \eta_m \eta_g \eta_{pr} \frac{\alpha \Delta h}{u} \dot{m}_a (1+f)$$

$$\frac{T}{\dot{m}_a} = [(1+f)u_e - u] + \eta_t \eta_m \eta_g \eta_{pr} \frac{\alpha \Delta h}{u}(1+f)$$

$$\eta_p = \frac{2u(T/\dot{m}_a)}{(1+f)u_e^2 - u^2}$$

$$\eta_{th} = \frac{(1+f)u_e^2 - u^2}{2fQ_{HV}}$$

$$\eta_o = \frac{u(T/\dot{m}_a)}{fQ_{HV}}$$

PROBLEMS

6.1 The V-22 Osprey aircraft shown in Figure 6.5 takes off and lands as a helicopter. Once airborne, its engine nacelles can be rotated to convert the aircraft to a turboprop aircraft capable of high speed and high altitude flight. It can transport internal and external cargo. The V-22 Osprey is powered by two **turboprop** engines. Each has the following data:

Shaft horsepower	6150
Specific fuel consumption	0.42 lb/shp/h
Compressor pressure ratio	14.1
Compressor isentropic efficiency	0.85
Burner efficiency	0.94
Pressure drop in combustion chamber	2%

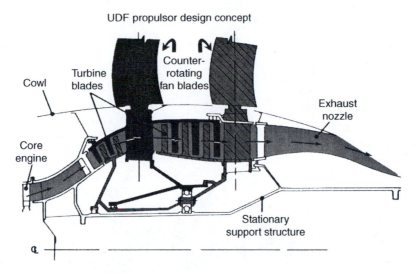

FIGURE 6.34   General Electric UDF Propulsor design concept.

FIGURE 6.35   GE36 UDF during testing in a test cell.

Fuel heating value	45,000 kJ/kg
Turbine inlet temperature	1300 K
Turbines isentropic efficiency	0.92
Gearbox mechanical efficiency	0.94
Propeller efficiency	0.823
Ambient conditions	288 K and 101 kPa

It is required to

(a)   Calculate the air mass flow rate into the engine.

(b)   Check that the power is 6150 hp.

6.2 The total work coefficient (W) in turboprop engines is defined as the ratio of propulsive power to the thermal energy of the airflow inducted into the engine; or

$$W = \frac{\eta_{pr} P_s + U_0 T}{m_a C p T_a}$$

It is required to

(a) Prove that the total work coefficient (W) in the case of an ideal turboprop engine is give by the relation

$$W_{ideal} = (\gamma - 1) M_a^2 \left\{ \sqrt{\left(\frac{\theta_t}{\theta_a \tau_c}\right)\left(\frac{\theta_a \tau_c \tau_t - 1}{\theta_a - 1}\right)} - 1 \right\} + \theta_t(1 - \tau_t) - \theta_a(\tau_c - 1)$$

$$\theta_a = \frac{T_{02}}{T_a}, \quad \tau_c = \frac{T_{03}}{T_{02}}, \quad \tau_t = \frac{T_{05}}{T_{04}}, \quad \theta_t = \frac{T_{04}}{T_a}$$

(b) Prove that the work coefficient is maximum value at takeoff conditions (flight speed is zero).

6.3 As an extension to Example 3, for the same turboprop engine it is required to calculate the specific equivalent specific fuel consumption defined as

$$SESFC = \frac{f}{seshp}$$

at sea level and same three altitudes. Next, to calculate the ESHP and fuel flow rate at different altitudes, the mass flow rate at different altitude must be first determined. Prove that the air mass flow rate at any altitude is related to the air mass flow rate at sea level by the relation

$$\dot{m}_{alt} = \dot{m}_{s.l} \frac{\delta}{\sqrt{\theta}} \equiv \dot{m}_{s.l} \frac{P_{alt}}{P_{s.l}} \sqrt{\frac{T_{s.l}}{T_{alt}}}$$

If the mass flow rate at sea level is 14.6 kg/s, calculate the air mass flow rate, fuel mass flow rate, ESHP, and ESFC at the four levels.

6.4 Turboprop engine with free power turbine has the following data:

$\dot{m}_a = 3 = 15$ kg/s $H = 6$ km
$T_a = 249$ K $P_a = 47.2$ kPa
$M_0 = 0.6$ $\pi_c = \frac{P_{03}}{P_{02}} = 6$
$P_{04} = 0.96 P_{03}$ $T_{04} = 1120$ K
$\eta_d = 0.95$ $\eta_C = 0.87$
$\eta_{C.C} = 0.98$ $\eta_t = \eta_{FPT} = \eta_n = 0.9$
$\eta_g = 0.95$ $\eta_{Pr} = 0.8$
$C_{PC} = 1.01$ kJ/kg K $C_{P_t} = C_{P_{FPT}} = C_{P_n} = 1.14$ kJ/kg K
$Q_R = 43,000$ kJ/kg

The power in the gases leaving the gas generator turbine is divided between the free power turbine and the nozzle based on the optimum condition $\alpha = \alpha_{opt}$. Calculate the ESHP and the ESFC.

6.5 Compare between the thrust force of the following three engines having the same maximum temperature of (1600 K) and compressor pressure ratio of (8), same core mass flow rate $\dot{m}_a$

of 30 kg/s, and same fuel having heating value Q_R of 44,000 kJ/kg. Assume no losses in the fan, compressor, combustion chamber, turbine(s), and nozzle(s). Moreover, all nozzles are unchoked. These engines power aircraft flying at an altitude of 8 km where the ambient conditions are 236.21 K and 35.64 kPa, and with a Mach number of 0.7.

(a) *Turboprop engine:* The propeller is driven by a power turbine. The propeller efficiency is $\eta_{pr} = 0.8$, both the gearbox and power shaft mechanical efficiencies are 0.98.

(b) *Forward fan turbofan engine:* Bypass ratio (BPR) $\beta = 3$ and fan pressure ratio $\pi_f = 1.5$.

(c) *Aft fan turbofan engine: BPR $\beta = 3$ and fan pressure ratio $\pi_f = 1.5$.*

6.6 An aircraft is fitted with four turboprop engines and flying at a Mach number of 0.7 at altitude of 9 km where the ambient conditions are 30.8 kPa and 229.7 K. When engines are running at maximum speed, each engine has an ESHP = 1280. The engine components have the following isentropic efficiencies:

Intake (0.9), Compressor (0.87), Burner (0.98), Turbine (0.89), Nozzle (0.95)

Other data are

Compressor pressure ratio	7.5
Combustion chamber pressure drop	3%
Exhaust velocity	250 m/s
Turbine inlet temperature	1180 K
Fuel Heating Value	43,000 kJ/kg

Calculate the air mass flow rate into the engine.

6.7 A commuter airplane is powered by two turboprop engines. The propeller is driven by compressor-turbine shaft so that expansion process in the turbine and nozzle is governed by the optimum ratio:

$$\alpha_{opt} = 1 - \frac{u^2}{2\Delta h} \frac{\eta_n}{\eta_{pr}^2 \eta_g^2 \eta_{mt}^2 \eta_t^2}$$

Prove that in this case the optimum exhaust speed is given by

$$u_e = \frac{u\eta_t}{\eta_{pr} \eta_g \eta_{mt} \eta_t}$$

Find the corresponding maximum thrust. Explain how to handle such a problem during takeoff; $u = 0$ while (u_e) is greater than zero.

6.8 A short-haul commercial transport (example for which is FOKKER F-27 manufactured in The Netherlands) is powered by two turboprop engines. At sea level condition, each delivers 2250 ESHP and 260 lb thrust. The maximum flight speed is 330 mph at 20,000 ft. Find

(a) The shp at sea level
(b) Corrected thrust at 20,000 ft
(c) Corrected shp at 20,000 ft
(d) ESHP at 20,000 ft

(Ambient conditions at sea level are $T = 288$ K and $P = 14.69$ psia and at altitude 20,000 ft are $T = 248.4$ K, $P = 6.754$ psia)

6.9 A turboprop engine has the following specifications:

Air mass flow rate	15 kg/s
Flight Mach number	$M = 0.6$
Flight altitude	6 km
Ambient temperature and pressure	249 K and 47.2 kPa
Compressor pressure ratio	6
Pressure loss in the combustion chamber	4%
Maximum total temperature	1120 K
Fuel heating value	43,000 kJ/kg

Efficiencies
Diffuser = 0.95, compressor = 0.87, combustion chamber = 0.98, turbines and nozzle = 0.9, gearbox = 0.95, propeller = 0.8.
Specific heats at constant pressure
Compressor = 1.01, combustion chamber = 1.13, turbines = nozzle = 1.14 kJ/kg · K
assuming $\alpha_{opt} = 0.87$.

Calculate

(a) ESHP and (b) ESFC

6.10 Compare between the thrust force of the following three engines having the same maximum temperature of (1600 K) and compressor pressure ratio of (8), same core mass flow rate $\dot{m}_a$ of 30 kg/s, and same fuel having heating value Q_R of 44,000 kJ/kg. Assume no losses in the fan, compressor, combustion chamber, turbine(s), and nozzle(s). Moreover, all nozzles are unchoked. These engines power aircraft flying at an altitude of 8 km where the ambient conditions are 236.21 K and 35.64 kPa, and with a Mach number of 0.7.

(a) *Turboprop Engine:* The propeller is driven by a power turbine. The propeller efficiency is $\eta_{pr} = 0.8$, both the gearbox and power shaft mechanical efficiencies are 0.98.

(b) *Forward Fan Turbofan Engine:* BPR $\beta = 3$ and fan pressure ratio $\pi_f = 1.5$.

(c) *Aft fan Turbofan Engine:* BPR $\beta = 3$ and fan pressure ratio $\pi_f = 1.5$.

REFERENCES

1. O.M. Lancaster (ed.), *Jet Propulsion Engines*, Vol. 12, *High Speed Aerodynamics and Jet Propulsion*, Princeton, NJ, Princeton University Press, pp. 199–267, 1959.

2. *The Jet Engine*, Rolls-Royce plc, 1986, p. 217.

3. H.I.H. Saravanamuttoo, G.F.C. Rogers, and H. Cohen, *Gas Turbine Theory*, 5th edn., 2001, Prentice Hall, p. 137.

4. P. Hill, and C. Peterson, *Mechanics and Thermodynamics of Propulsion*, 2nd edn., Addison Wesley Publication Company, Inc., 1992, p. 155.

5. B. Gunston, *The Development of Jet and Turbine Aero Engines*, Patrick Stephens Limited, an imprint of Haynes Publishing, 2nd edn., 1997, p. 107.

6. http://www.minihelicopter.net/CH47Chinook/index.htm

7. http://www.turbokart.com/about_arrius.htm

7 High-Speed Supersonic and Hypersonic Engines

7.1 INTRODUCTION

In the previous three chapters, four broad categories of aero engines were discussed. The powered vehicles have flight speeds ranging from low subsonic speeds, as in the case of helicopters, and up to moderate supersonic speeds. Supersonic vehicles are mainly military airplanes and the limited number of supersonic transports (SSTs) such as the Concorde and Tu-144 airplanes. Passengers become bored by the long time of trips, say across Atlantic or in general during flights of more than 5–7 hours. Extensive efforts are made these days to satisfy the passenger needs. It is expected in the forthcoming decades that airplanes will satisfy the greatest demand for more comfortable and faster flights. This chapter will concentrate on engines that power such airplanes or airspace planes that will fly at high supersonic and hypersonic flight speeds (Mach 5 and above). The corresponding flight altitudes in such circumstances will be much higher than today's subsonic/transonic civil airliners for some technical reasons, which will be discussed later.

In this chapter, two engines will be analyzed, namely, turboramjet and scramjet. Both were highlighted in Chapter 1. Turboramjet either combines a turbojet/turbofan and a ramjet engine in a single conduit or assembles them in over/under configuration and sometimes identified as turbine-based engine. In Chapters 3 through 5, the ramjet, turbojet, and turbofan engines were analyzed. Here, a brief discussion will be made with a special analysis of their combined layout and combined operation in the dual mode.

Next, scramjet will be analyzed. Similarity with ramjet will be emphasized. Scramjet (sometimes identified as ram-based engine) will be employed in hypersonic vehicles that must be boosted at first by other engines to reach the starting Mach number for its operation.

7.2 SUPERSONIC AIRCRAFT AND PROGRAMS

On October 14, 1947, the Bell X-1 [1] became the first airplane to fly faster than the speed of sound. Piloted by the U.S. Air Force Captain Charles E. Yeager, the X-1 reached a speed of 1127 km (700 mi) per hour, Mach 1.06, at an altitude of 13,000 m (43,000 ft).

It was the first aircraft to have the designation X. The X-1 was air launched at an altitude of approximately 20,000 ft from the Boeing B-29.

With that flight, the supersonic age had started. The second generation of Bell X-1 was modified and redesignated the X-1E (Figure 7.1). It reached a Mach number of 2.44 at an altitude of 90,000 ft. Later on, many aircraft joined the supersonic race. Until now, the fastest manned aircraft is the Blackbird SR-71 (Figure 7.2) long-range, strategic reconnaissance aircraft that can fly at Mach number of 3+ (maximum speed 3.3+ (2200+ mph, 3530+ km/h) at 80,000 ft (24,000 m)). It is powered by 2 Pratt & Whitney J58-1, continuous-bleed afterburning turbojets 32,500 lb force (145 kN) each [2,18]. It was in service from 1964 to 1998.

Continuous R &D programs and testing are performed in the United States, France, Russia, United Kingdom, and Japan. However, the first supersonic programs started by National Advisory Committee for Aeronautics (NACA) first conceptual design of a supersonic commercial transport was designed to carry 10 passengers for a range of about 1500 miles in a cruise speed of 1.5 Mach number.

After that design, there were a lot of other conceptual designs of supersonic commercial aircraft. Very few designs reached completion such as Concorde and Tu-144. Many others were only tested

FIGURE 7.1   The first supersonic aircraft X-1E.

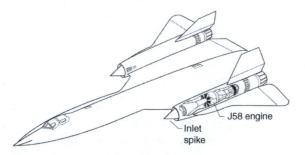

FIGURE 7.2   The fastest manned aircraft SR-71.

in wide tunnels such as Boeing 2707-100, 200, 300, and the Lockheed L-2000. The research on the supersonic commercial air transport started in the United States in 1958 and somewhat earlier in France and Great Britain. The starting date of the Russian project is unknown. A brief description of several programs will be given in the following sections.

7.2.1 ANGLO-FRENCH ACTIVITIES

Concorde

The "supersonic transport aircraft committee (STAC)" was established in 1956. STAC conducted a series of design studies, leading to the Bristol Company's "Bristol 198," which was a slim, delta-winged machine with eight turbojet engines designed to cross the Atlantic at Mach 2. This evolved into the somewhat less ambitious "Bristol 223," which had 4 engines and 110 seats.

In the meantime, Sud-Aviation of France designed the "Super Caravelle," which surprisingly was very similar to Bristol 223. On November 29, 1962, the British and French governments signed a collaborative agreement to develop an Anglo-French SST, which became the "Concorde." It was to be built by the *British Aircraft Corporation* (BAC), into which Bristol had been absorbed in the meantime, and *Rolls-Royce in the United Kingdom*; and *Sud-Aviation* and the *Snecma* engine firm in France. In all, a total 20 Concordes (Figure 7.3) were built between 1966 and 1979 [3]. The first two Concordes were prototype models, one built in France and the other in England. Two more preproduction prototypes were built to further refine the design and test out ground breaking systems before the production runs; thus, only 16 aircraft in total, commenced operations in both countries. Concorde is powered by four turbojet engine (Olympus 593) fixed in the wing. The maximum thrust produced during supersonic cruise per engine is 10,000 lb. Maximum weight without fuel (Zero fuel weight) is 203,000 lb (92,080 kg). Maximum operating altitude is 60,000 ft.

BAe-Aerospatiale AST

The European AST was based on the Concorde design although being much larger and more efficient. It had canard foreplanes for extra stability. The new AST design featured the curved Concorde wing unlike all other delta SST designs to date. It was designed as two aircraft: the first would carry 275 passengers and the second would carry 400 passengers. The first version would weigh the same on takeoff as two Concordes. The second version had a slimmer delta shape and a double-delta leading edge wing shape similar to AST designs coming out of the United States.

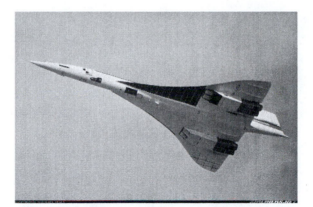

FIGURE 7.3 Concorde supersonic airplane.

FIGURE 7.4 Tupolev TU-144.

7.2.2 RUSSIAN ACTIVITIES

Tupolev TU-144
In the former Soviet Union plans were underway for a trans-Soviet continental SST. The Tupolev Tu-144 "Concord-ski" looked superficially similar to the Anglo-French aircraft (Figure 7.4). It was quite different however. It began life with all four NK144 turbofan engines in a main block underneath the fuselage [4]. This arrangement was changed to a design similar to the Concorde with the four engines in boxes of two each side. The wing design had two delta angles in it. It was withdrawn from service in 1985 after a 10-year line life.

7.2.3 U.S. ACTIVITIES

In the United States, fears about being left behind in the supersonic airliner race by the Soviet and European SSTs started to take form on paper. Boeing and Lockheed competed in the design of an SST paper plane. The Boeing models were the 2707-100, 200, and 300. Lockheed had also proposed an SST of its own, the Lockheed L-2000.

Boeing 2707-100/200
It was a large aircraft 318 ft long from nose to tail and had a complex swing-wing design with a tail plane behind the main swing wing. This tail place formed the back section of the delta wing when the wings were swung back during supersonic flight and held four large General Electric GE-4 turbojet engines. It would fly at a high Mach 2.7 over 3900 mi with a passenger load of at least 300. This was a truly large aircraft—bigger than a Boeing 747 (Figure 7.5).

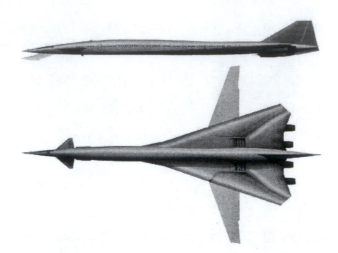

FIGURE 7.5   Boeing 2707-200. (Courtesy Boeing.)

FIGURE 7.6   Lockheed L-2000.

Lockheed L-2000
This jet had a similar range and size to the Boeing jets but was designed as a simpler delta (Figure 7.6).
The Lockheed L-2000 was powered by four General Electric GE4/J5P or Pratt & Whitney JTF17A-21L turbojets with augmentation being fixed in the wing. Empty weight was 23,800 lb (10795.5 kg) and maximum takeoff weight is 590,000 lb (276,620 kg).

Boeing 2707-300
Boeing looked again at the dash 200 and replaced it with the Boeing 2707 dash 300 design. This was a more conventional design although it still had a tail plane but with a conventional SST delta wing and the four GE-3 engines mounted separately underneath the wing.

Convair BJ-58
The Convair BJ-58 was designed to fly at Mach 2.5 and cruise at more than 70,000 ft (Figure 7.7). It was powered by four J-58 engines (each of 23,000 lbs) and has a payload capacity of 52 passengers.

AD-1
The Ames-Dryden-1 (AD-1) was a research aircraft designed to investigate the concept of a wing that could be pivoted obliquely from 0° to 60° during flight. AD-1 was 38.8 ft long and 6.75 ft high

FIGURE 7.7   Convair BJ-58.

FIGURE 7.8   The HSCT.

with a wing span of 32.3 ft, unswept and was constructed of plastic reinforced with fiberglass. It weighed 1,450 pounds, empty and was powered by two small turbojet engines, each producing 220 lb of static thrust at sea level. The AD-1 flew a total of 79 times during the research program. It was limited for reasons of safety to a speed of about 170 mph. If flying at speeds up to Mach 1.4, it would have substantially better aerodynamic performance than aircraft with more conventional wings and might achieve twice the fuel economy of an aircraft with conventional wings.

The high-speed civil transport project

A present project is the high-speed civil transport (HSCT), contracted between NASA subdivision high-speed research (HSR), Boeing, General Electric, and Pratt & Whitney (Figure 7.8). This contract was able to solve a lot of the SST problems such as the emissions problem and the sonic boom problem. However, problem of landing noise, takeoff noise, and materials still remain.

7.3   FUTURE OF COMMERCIAL SUPERSONIC TECHNOLOGY

NASA's HSR Program began in 1985 with the objective "to establish the technology foundation by 2002 to support the U.S. transport industry's decision for a 2006 production of an environmentally acceptable, economically viable, 300 passenger, 5000 nautical mile, and Mach 2.4 aircraft" (NRC, 1997). The first flight of a commercial supersonic aircraft was envisioned around 2010, with the first production aircraft to be operational around 2013. NASA expected that program goals would require a large investment.

The governments of France, Japan, Russia, and the United Kingdom are also sponsoring development of supersonic technology with commercial applications, although none has embarked on a formal program to produce a new commercial supersonic aircraft. The development of a commercial SST that can meet international environment standards and compete successfully with subsonic transports may be a larger effort than the industry of any single nation, including the United States, might wish to undertake. As with many such innovations, the first manufacturer to market will have the potential to dominate the market worldwide. A small supersonic jet could be developed by a single aircraft manufacturer and might lead to important technological innovation.

7.4 TECHNOLOGY CHALLENGES OF THE FUTURE FLIGHT

The key technology challenges that are derived from the customer requirements and vehicle characteristics are related to economics, environment, or certification [5].

1. Environment
   - Benign effect on climate and atmospheric ozone
   - Low landing and takeoff noise
   - Low sonic boom
2. Economics—range, payload, fuel burn, and so on
   - Low weight and low empty weight fraction
   - Improved aerodynamic performance
   - Highly integrated airframe/propulsion systems
   - Low thrust-specific fuel consumption (TSFC)
   - Long life
3. Certification for commercial operations
   - Acceptable handling and ride qualities
   - Passenger and crew safety at high altitudes
   - Reliability of advanced technologies, including synthetic vision
   - Technical justification for revising regulations to allow supersonic operations over land

7.5 HIGH-SPEED SUPERSONIC AND HYPERSONIC PROPULSION

7.5.1 INTRODUCTION

For an aircraft cruising at Mach 6, the aircraft must be fitted with an engine that can give sufficient thrust for takeoff, climb and cruise, and again deceleration, descent and landing [6]. There are two suggestions for such a vehicle and its propulsion.

1. Multistage vehicle
In this method, the aircraft is to be composed of two stages (Figure 7.9). The first stage has turbine engines (either a turbojet or turbofan) to power the aircraft during takeoff, climb, and acceleration to supersonic speed. At such speed the first stage separates and return back to the ground while the engines of the second stage start working, which are either ramjet or scramjet engines.

2. Hybrid cycle engine
The hybrid or combined engine is an engine that works on two or more cycles (Figure 7.10). The first cycle is a cycle of a normal turbojet or turbofan engine. The engine stays working with that cycle till the aircraft reaches a specified speed at which the engine changes its cycle and works on a ramjet, scramjet, or even a rocket cycle depending on the aircraft mission and concept.

For sure, the second concept will be used because in the first concept there will be a lot of problems concerning connecting/disconnecting the stages, how the first stage will be back to airport, and many others.

FIGURE 7.9   Multistage aircraft concept.

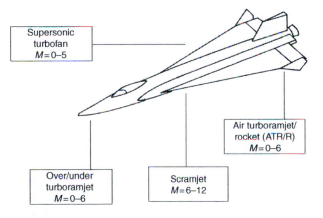

FIGURE 7.10   Hybrid (combined) cycle engine.

7.5.2 Hybrid Cycle Engine

The question arising here is why a hybrid (combined) cycle is needed.

The main engine for aircraft is to be a turbine engine (turbojet or turbofan) but from the performance analysis of those engines (Chapters 4 and 5) it was found that the engine thrust will start decreasing after a certain Mach number (about 2–3) depending on the overall pressure ratio of the engine and the bypass ratio for turbofans. On the contrary, ramjet still have a reasonable thrust for Mach numbers up to 5 or 6 and scramjet engine can provide thrust theoretically up to Mach number of 20. Ramjet and scramjet engines, however, cannot generate thrust at static conditions or low Mach numbers. Ramjet can start generating thrust at Mach numbers close to unity while scramjet develops thrust at Mach numbers greater than 2.5. Therefore, the hypersonic aircraft need both the turbine engine and the ram/scram engine. The above argument is displayed in Figure 7.11 illustrating the specific impulse versus Mach number.

The hybrid engine works on the combined cycle where in the first flight segment the engine works as a turbojet or a turbofan and then switches to a ramjet or a scramjet. It may also have a poster rocket to help the engine in the transition region or drive the vehicle alone if it is to fly in space (the aerospace plane).

In the succeeding sections, both turboramjet and scramjet engines will be treated in detail.

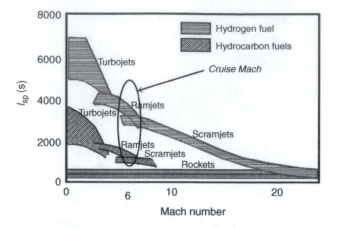

FIGURE 7.11   Specific impulse versus Mach number.

7.6   TURBORAMJET ENGINE

As defined earlier, turboramjet engine is a hybrid engine composed of a ramjet engine in conjunction with either a turbojet or turbofan engines. The turboramjet can be run in turbojet mode at takeoff and during low-speed flight but then switched to ramjet mode to accelerate to high Mach numbers. It is constructed in either of the following forms:

1. Wraparound turboramjet
2. Over/under turboramjet

The differences between them are

1. The position of the ram with respect to turbojet
2. The position of the afterburner of the turbojet with respect to the ramjet.

7.7   WRAPAROUND TURBORAMJET

In that configuration the turbojet is mounted inside a ramjet (Figure 7.12). The turbojet core is mounted inside a duct that contains a combustion chamber downstream of the turbojet nozzle. The operation of the engine is controlled using bypass flaps located just downstream of the diffuser. During low speed flight, these controllable flaps close the bypass duct and force air directly into the compressor section of the turbojet. During high-speed flight, the flaps block the flow into the turbojet, and the engine operates like a ramjet using the aft combustion chamber to produce thrust.

The wraparound turboramjet layout is found in Convair BJ-58 and SR-71 aircraft.

The engine cycle is plotted on the T–S diagram in Figure 7.13. It has two modes of operation: either working as a simple turbojet or as a ramjet. A plot for the cycle on the T–S plane is illustrated in Figure 7.13.

7.7.1   OPERATION AS A TURBOJET ENGINE

In the turbojet mode, a chain description of the different processes through the different modules is described in Table 7.1.

A brief description of the different processes is given subsequently.

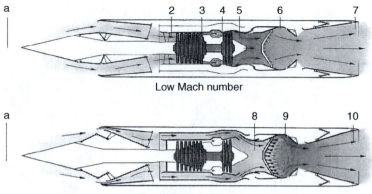

FIGURE 7.12 Wraparound turboramjet.

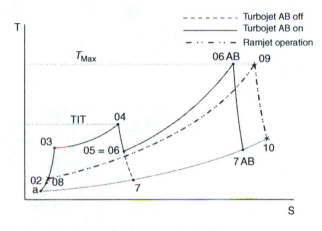

FIGURE 7.13 T–S diagram of the wraparound turboramjet engine.

TABLE 7.1
Description of Different Processes in a Turbojet Engine

Part	States	Processes
Intake	a–2	Compression process with isentropic efficiency η_d
Compressor	2–3	Compression process with isentropic efficiency η_C
Combustion chamber	3–4	Heat addition at constant pressure or with a pressure drop of $\Delta P_{C.C.}$
Turbine	4–5	Expansion process with isentropic efficiency η_t
Afterburner	5–6AB	Heat addition at constant pressure or with a pressure drop ΔP_{AB}
Nozzle	6(or 6AB)–7	Expansion process with isentropic efficiency η_N

Intake

As described previously, the inlet conditions are the ambient ones. For a flight Mach number of (M), the outlet conditions will be

$$T_{02} = T_a \left(1 + \frac{\gamma_c - 1}{2} M^2 \right).$$

(7.1)

$$P_{02} = P_a \left(1 + \eta_d \frac{\gamma_c - 1}{2} M^2 \right)^{\frac{\gamma_c}{\gamma_c - 1}} \tag{7.2}$$

Compressor

With a pressure ratio of π_C the outlet conditions will be

$$P_{03} = \pi_C \times P_{02} \tag{7.3}$$

$$T_{03} = T_{02} \left(1 + \frac{\pi_C^{(\gamma_c - 1)\gamma_c} - 1}{\eta_c} \right) \tag{7.4}$$

Combustion chamber

Owing to friction, pressure losses ΔP_{CC} will be encountered. The output of the combustion chamber is to be controlled by the turbine inlet temperature (TIT) that depends on the turbine material. The outlet pressure and fuel-to-air ratio are then given by the following equations:

$$P_{04} = P_{03} (1 - \Delta P_{CC}) \tag{7.5}$$

$$f = \frac{C_{Ph} T_{04} - C_{Pc} T_{03}}{\eta_{c.c} Q_{HV} - C_{Ph} T_{04}} \tag{7.6}$$

Turbine

As usual it is the main source of power in aircraft so the power of turbine must be equal to the power needed to drive the compressor plus power needed by all systems in aircraft. For a preliminary design, it is assumed that the power of both turbine and compressor are equal. Thus power balance results in the following:

$$T_{05} = T_{04} - \frac{C_{Pc} (T_{03} - T_{02})}{\eta_{mech} C_{Ph} (1 + f)} \tag{7.7}$$

$$P_{05} = P_{04} \left(1 - \frac{T_{04} - T_{05}}{\eta_t T_{04}} \right)^{\frac{\gamma_h}{\gamma_h - 1}} \tag{7.8}$$

Afterburner

Afterburner will be lit if more thrust is needed from the engine. When no extra thrust is needed the afterburner is to be turned off and treated as a pipe. Combustion in the afterburner is associated with pressure loss ΔP_{AB}. The fuel burned in the afterburner and the outlet pressure are calculated:

$$f_{AB} = C_{Ph} (1 + f) \frac{T_{06AB} - T_{05}}{\eta_{AB} Q_{HV} - C_{Ph} T_{06}} \tag{7.9}$$

$$P_{06} = P_{05} (1 - \Delta P_{AB}) \tag{7.10}$$

Nozzle

The gases expand in the nozzle to high velocities to generate the required thrust. The outlet conditions of the gases and nozzle exit area are

$$T_7 = T_{06} - \eta_N T_{06} \left[1 - \left(\frac{P_a}{P_{06}} \right)^{\frac{\gamma_h - 1}{\gamma_h}} \right] \tag{7.11}$$

$$V_7 = \sqrt{2 C_{Ph} (T_{06} - T_7)} \tag{7.12}$$

$$\rho_7 = \frac{P_a}{RT_7} \tag{7.13}$$

$$\frac{A_{ex}}{m_a} = \frac{1 + f + f_{AB}}{\rho_7 V_7}. \tag{7.14}$$

In Equations 7.9, 7.11 and 7.12, the symbol T_{06} is equally used for T_{06} and T_{06AB}.

7.7.2 OPERATION AS A RAMJET ENGINE

In the ramjet mode the flow is passing through the processes and states described in Table 7.2. Analysis of the processes is briefly described as follows.

Intake

The delivery pressure and temperature will be evaluated in terms of the total pressure ratio (r_d), which is to be calculated from Equation 7.15

$$r_d \equiv \frac{P_{08}}{P_{0a}} = \begin{cases} 1 & M \leq 1 \\ 1 - 0.75\,(M-1)^{1.35} & 1 \leq M \leq 5 \\ \dfrac{800}{M^4 + 938} & M \geq 5 \end{cases} \tag{7.15}$$

The outlet temperature and pressure are

$$T_{08} = T_a \left(1 + \frac{\gamma_c - 1}{2} M^2 \right)^{\frac{\gamma_c}{\gamma_c - 1}} \tag{7.16}$$

$$P_{08} = r_d P_{0a} = P_a r_d \left(1 + \frac{\gamma_c - 1}{2} M^2 \right)^{\frac{\gamma_c}{\gamma_c - 1}}. \tag{7.17}$$

The combustion chamber is treated in the same way as the afterburner of turbojet.

$$P_{09} = P_{08}\,(1 - \Delta P_{CCR}) \tag{7.18}$$

$$f_R = \frac{C_{Ph} T_{09} - C_{Pc} T_{08}}{\eta_{C.C.R} Q_{HV} - C_{Ph} T_{09}} \tag{7.19}$$

TABLE 7.2
Description of Different Processes in a Ramjet Engine

Part	States	Processes
Intake	a–8	Compression process with total pressure ratio r_d
Combustion chamber	8–9	Heat addition at constant pressure or with a pressure drop of $\Delta P_{C.C.}$
Nozzle	9–10	Expansion with isentropic efficiency (η_N).

Nozzle

It will be also handled in the same way as in the turbojet engine:

$$T_{10} = T_{09} - \eta_N T_{09} \left[1 - \left(\frac{P_a}{P_{09}} \right)^{\frac{\gamma_h - 1}{\gamma_h}} \right] \tag{7.20}$$

$$V_{10} = \sqrt{2C_{Ph}\left(T_{09} - T_{10} \right)} \tag{7.21}$$

$$\rho_{10} = \frac{P_a}{RT_{10}} \tag{7.22}$$

$$\frac{A_{ex}}{m_a} = \frac{1 + f_R}{\rho_{10}V_{10}} \tag{7.23}$$

7.8 OVER/UNDER TURBORAMJET

As it is clear from the name over/under, the configuration here has separated turbojet and ramjet engines, but may have the same intake (inlet) and nozzle (outlet) as in the wraparound configuration. However, in some cases, each engine has its separate intake and nozzle. In such configuration, the turbojet engine operates at takeoff and low subsonic flight where the movable inlet ramp is deployed to allow for the maximum air flow rate into its intake. At higher Mach number, the engine operates at a dual mode where both turbojet and ramjet engines are operative for a few seconds until the Mach number reaches 2.5 or 3.0. Next, the turbojet engine is shut down and only ramjet becomes operative. An example for such layout is found in the four over/under airbreathing turboramjet engines powering the Mach 5 wave rider [7]. The complexity of both aerodynamic and mechanical design of two variable throat nozzles and necessary flaps led to another design including only one nozzle [8]. The suggested nozzle uses a single-expansion ramp nozzle (SERN) instead of a conventional, two-dimensional, convergent–divergent nozzle.

Figures 7.14 and 7.15 illustrate a layout and T–S diagram of the over/under configuration of the turboramjet engine.

For the over/under configuration of turboramjet engine, the engine height is large compared to the wraparound configuration. The reason is clear as the height in this case is the sum of the heights of turbojet and ramjet engines. Usually, a part of the engine is to be buried inside the fuselage or inside the wing. Since the engine will operate at high Mach number, it needs a long intake. Thus, a part of the aircraft forebody is used as a part of the intake. Concerning installation, the engine is either installed under the fuselage or the wing. For fuselage installation, the turbojet is buried

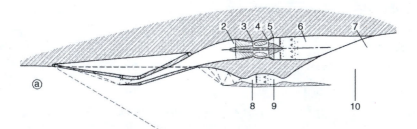

FIGURE 7.14 Over/under layout of turboramjet engine.

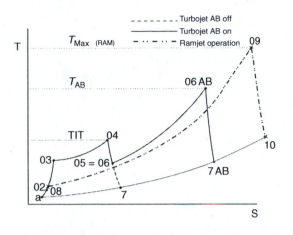

FIGURE 7.15   T–S diagram of the over/under layout of turboramjet engine.

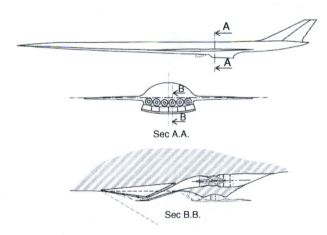

FIGURE 7.16   Hypersonic aircraft (HYCAT-1) [9].

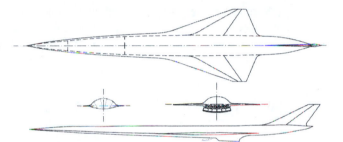

FIGURE 7.17   Turboramjet engine installed under the fuselage.

inside the fuselage and the ramjet is located under the fuselage as in Figures 7.16 and 7.17. For wing installation, the turbojet is situated above the wing and the ramjet is under the wing.

Concerning the thermodynamic cycle of an over/under configuration, there are three modes of operation; namely, a turbojet, a ramjet, or a dual mode. Dual mode represents the combined or transition mode in which both the turbojet and ramjet are operating simultaneously.

7.8.1 TURBOJET MODE

In that region, the engine is working as a simple turbojet engine and develops all the thrust needed by the aircraft. The states and governing equations are the same as the turbojet in the wraparound configuration. Cold air passes through the ramjet in this mode.

7.8.2 DUAL MODE

It is the mode, in which both the turbojet and ramjet are operating simultaneously. Turbojet starts declining and its developed thrust will be intentionally decreased by reducing the inlet air mass flow rate via its variable geometrical inlet. The ramjet starts working by adding fuel and starting ignition in its combustion chamber. The thrust generated in the ramjet is increased by also increasing the air mass flow rate through its variable area inlet. The generated thrust force will then be the sum of both thrusts of turbojet and ramjet engines.

7.8.3 RAMJET MODE

In this mode, turbojet stops working and its intake is completely closed. All the air mass flow is passing through the ramjet intake. With Mach number increase, the forebody acts as a part of the intake with the foremost oblique shock wave located close to the aircraft nose, as shown in Figure 7.18.

7.9 TURBORAMJET PERFORMANCE

Performance parameters: specific thrust, thrust-specific fuel consumption (TSFC), total thrust, and propulsive, thermal, and overall efficiencies are defined here. The following equations are valid for both types of turboramjet engines.

7.9.1 TURBOJET MODE

The specific thrust is

$$\left(\frac{T}{\dot{m}_a}\right)_{TJ} = (1 + f + f_{AB}) V_7 - V \tag{7.24}$$

The TSFC is

$$(TSFC)_{TJ} = \frac{(f + f_{AB})}{T/\dot{m}_a} \tag{7.25}$$

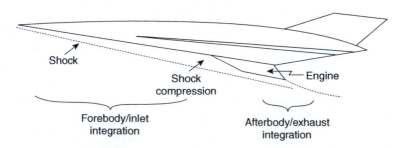

FIGURE 7.18   Intake of the ramjet of the turboramjet engine.

The engine propulsive, thermal, and overall efficiencies are

$$\eta_p = \frac{(T/\dot{m}_a)\,V}{(T/\dot{m}_a)\,V + ((V_7 - V)^2/2)\,(1 + f + f_{AB})} \tag{7.26}$$

$$\eta_{th} = \frac{(T/\dot{m}_a)\,V + ((V_7 - V)^2/2)\,(1 + f + f_{AB})}{Q_{HV}\,(f + f_{AB})} \tag{7.27}$$

$$\eta_o = \eta_p \times \eta_{th} \tag{7.28}$$

7.9.2 RAMJET MODE

For the ramjet engine, the performance parameters will be calculated in the same way. The thrust per unit air mass flow $(T/\dot{m}_a)$ is

$$\left(\frac{T}{\dot{m}_a}\right)_{RJ} = (1 + f_R)\,V_{10} - V \tag{7.29}$$

The TSFC is

$$(\text{TSFC})_{RJ} = \frac{f_R}{((T/\dot{m}_a))} \tag{7.30}$$

The engine propulsive, thermal, and overall efficiencies are

$$\eta_p = \frac{(T/\dot{m}_a)_{RJ}\,V}{(T/\dot{m}_a)_{RJ}\,V + ((V_{10} - V)^2/2)\,(1 + f_R)} \tag{7.31}$$

$$\eta_{th} = \frac{(T/\dot{m}_a)_{RJ}\,V + ((V_{10} - V)^2/2)\,(1 + f_R)}{Q_{HV}\,f_R} \tag{7.32}$$

$$\eta_o = \eta_p \times \eta_{th} \tag{7.33}$$

7.9.3 DUAL MODE

Both turbojet and ramjet are operating simultaneously. The total thrust will be

$$T = (T)_{TJ} + (T)_{RJ} \tag{7.34}$$

$$T = (\dot{m}_a)_{TJ}\,[(1 + f + f_{AB})\,V_7 - V] + (\dot{m}_a)_{RJ}\,[(1 + f_R)\,V_{10} - V] \tag{7.35a}$$

$$T = (\dot{m}_a)_{TJ}\,(1 + f + f_{AB})\,V_7 + (\dot{m}_a)_{RJ}\,(1 + f_R)\,V_{10} - (\dot{m}_a)_{total}\,V \tag{7.35b}$$

The specific thrust is

$$\frac{T}{\dot{m}_a} = \frac{T_{total}}{(\dot{m}_a)_{total}} = \frac{(T)_{TJ} + (T)_{RJ}}{(\dot{m}_a)_{TJ} + (\dot{m}_a)_{RJ}} \tag{7.36}$$

The total fuel mass flow rate is

$$\dot{m}_f = (\dot{m}_a)_{TJ}\,(f + f_{AB}) + (\dot{m}_a)_{RJ}\,f_R \tag{7.37}$$

The specific thrust fuel consumption is

$$\frac{(\dot{m}_f)_{total}}{(T)_{total}} = \frac{(\dot{m}_f)_{TJ} + (\dot{m}_f)_{RJ}}{(T)_{TJ} + (T)_{RJ}} \tag{7.38}$$

7.10 CASE STUDY

A simplified design procedure for a turboramjet engine will be given here. The procedure is the same as followed in any airliner design. It starts by selecting a proper engine for a certain aircraft. It is not an easy task such as selecting a tie for a suit, but it is a very complicated process. The following steps, among many others, are to be followed. The function of the air vehicle is first defined and then its mission is to be specified. During each point on the flight envelop the drag and lift forces are calculated. On the basis of this lift force, the lifting surfaces are designed and a variation of the lift coefficient against Mach number is determined. The corresponding drag force is also calculated and a graph for drag coefficient versus Mach number is also plotted. Now from this last curve, the appropriate power plant is selected. The case of one engine type as described in the previous four chapters is a straightforward process. The case of a hybrid engine is rather a complicated design procedure. The difficulty arises from the proper selection of switching points, or in other words, when one engine is being stopped and another one takes over. If more than two engines are available, the process gets more and more complicated. In addition even for two engines, there is also the combined or dual modes. Such a transition process must be carefully analyzed to keep safe margins of lift and thrust forces above the weight and drag forces. A sophisticated control system controls the air and fuel mass flow rates into different engines is needed. Variable intakes or inlet doors, nozzles, and possibly stator vanes in compressors are activated during such a transition mode.

The following steps summarize the procedure for high supersonic/hypersonic cruise aircraft propulsion integration:

1. Mission or flight envelop has to be selected. The cruise altitude is important from fuel economy point of view, while takeoff, climb, and acceleration are critical from the surplus thrust that must be available over the aircraft drag force.
2. Calculate the drag force based on drag coefficient variations with Mach number.
3. The needed thrust force must be greater than the drag value all over its Mach number operation. A margin of 10% is selected here.
4. Determination of the number of engines based on the maximum thrust is suggested for the hybrid engine at its different modes.
5. The performance of each constituent of the hybrid engine is separately determined. For example, if the hybrid engine includes a ramjet, turbojet, and scramjet, each module is examined separately. Thus, the performance of turbojet engine running alone is considered first. The specific thrust and TSFC are determined for the cases of operative and inoperative afterburner. Next, the performance of the ramjet operating alone is also considered. Since the ramjet operates up to Mach number of 6, liquid hydrogen is considered as its fuel. If a scramjet is available, then its performance is also determined separately.
6. Optimization of the hybrid engine represents the next difficult step. From the specific thrust and TSFC, the switching points are defined. Thus in this step, the designer selects at which Mach number the turbojet operates or stops the afterburner and at which Mach number the turbojet must be completely stopped and replaced by the ramjet. If a scramjet is also available, another decision has to be made concerning the switching Mach number from ramjet to scramjet, the switching process duration, and the accompanying procedures regarding the air and fuel flow rates and inlet doors actuation.

TABLE 7.3
Mission of HYCAT-1-A

Passengers	200
Range	9260 km
Cruise Mach number	6
Field length	3200 m
Fuel	JA-7 For Turbojet LH2 for ramjet
Aircraft life	Commensurate with current aircraft
Fuel reserves	5% of block fuel
Subsonic flight to alternate airport	482 km
Descent Mach at which turbojet turn on	0.8 Mach
Cruise altitude	90,000 ft
Fuselage length	118.26 m
Wing reference area	816.8 m^2
Horizontal Tail Total Area	177.8 m^2
Vertical Tail Area	90.2 m^2
Gross weight	350,953 kg
Payload	19,051 kg
Empty weight	198,729 kg

Since the plan of the twenty-first century is to have hypersonic/high supersonic civil transports, a modified version of the HYCAT-1 aircraft [9] is selected for this analysis as its aerodynamic characteristics up to Mach 6 are available. This selected aircraft may be identified as HYCAT-1-A, which includes a completely buried turbojet engine in the fuselage. Other studies for supersonic aircraft having test results up to Mach 6 are also available [10]. Test results for supersonic aircraft of Mach 2.96, 3.96, and 4.63 are found in Reference 11, while results for a supersonic aircraft flying at Mach number of 4 are discussed in [12]. Some data of the aircraft are given in Table 7.3.

Now after selecting the aircraft, illustrated in Figures 7.16 and 7.17, the drag force is calculated for different Mach numbers. Some important data concerning the turbojet engine are compressor pressure ratio of 10, TIT of 1750 K, maximum temperature in the afterburner of 3000 K, and heating value of hydrocarbon fuel of 45,000 kJ/kg. The efficiencies of the diffuser, compressor, turbine, and nozzle are 0.9, 0.91, 0.92, and 0.98, respectively. The efficiency of the combustion chamber and afterburner are 0.99, and the pressure drops in the combustion chamber and afterburner are 2%. The mechanical efficiency of the compressor turbine shaft is 0.99. Concerning the ramjet, the maximum temperature is 3000 K and heating value of liquid hydrogen is 119,000 kJ/kg. Efficiencies of diffuser, combustion chamber, and nozzle are 0.9, 0.99, and 0.98, respectively. The pressure drop in the combustion chamber is 2%.

The specific thrust and specific fuel consumption for individual turbojet engine and ramjet engine are calculated and plotted in Figures 7.19 and 7.20.

From Figures 7.19 and 7.20, the following points may be selected as the switching points; refer to Figure 7.21:

1. The ramjet starts burning fuel (R1)
2. The turbojet starts its gradual shut down (T1)
3. The ramjet works with its maximum burner temperature (R2)
4. The turbojet completely shuts down (T2)

Details of selections and identification of these points are given in Table 7.4.

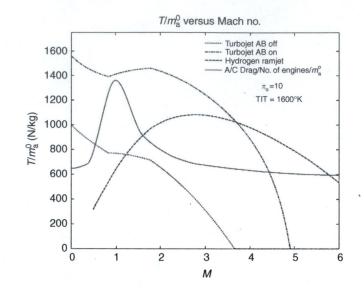

FIGURE 7.19   Specific thrust for individual turbojet and ramjet engines.

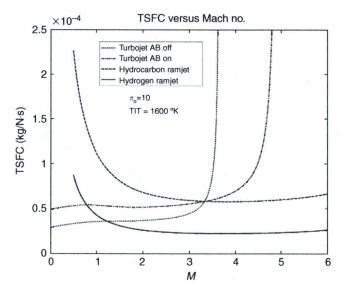

FIGURE 7.20   Thrust-specific fuel consumption for individual turbojet and ramjet engines.

The following two assumptions are identified here concerning both engines:

(a) *Concerning the turbojet engine:* The turbojet has three regions of operation, namely,

1. Constant mass flow rate with afterburner off (from $M = 0$ to $M = 0.5$)
2. Constant mass flow rate with afterburner on (from $M = 0.5$ to $M = 1.5$)
3. Decreasing mass flow rate with afterburner off (from $M = 1.5$ to $M = 2.5$)

The air flow in this last region decreases gradually and linearly from points T1 to T2.

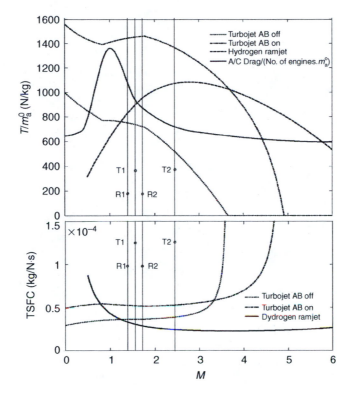

FIGURE 7.21 Regions boundaries.

(b) *Concerning the ramjet engine:* The ramjet has three regions of operation, namely,

1. The cold flow region (burner is off) in which the ramjet does not generate any thrust but generates drag (from $M = 0$ to point R1).
2. The starting region in which the burner temperature increases gradually and linearly. At the end of this region, the temperature reaches its maximum value (from points R1 to R2).
3. The working region in which the burner temperature reaches its maximum value and keeps working (from point R2 to the maximum Mach number).

On the basis of the above explanation, the operation of the hybrid engine is divided into four regions that are identified in Table 7.4 and in Figures 7.21 through 7.25.

Region I is a typical operation of a turbojet engine with an inoperative afterburner.

Region II is also a typical operation of a turbojet engine with an operative afterburner.

Region III (dual region) is important as in this transition region the ramjet engine is starting to build up thrust while the turbojet engine is declining out. A linear variation of the air mass flow rate in the turbojet is assumed during its shut down (region III), and then at any Mach number the thrust generated by the turbojet engine is known. The thrust needed from the ramjet is the difference between the aircraft required thrust and the turbojet thrust. Since the specific thrust of the ramjet is known at this Mach number, the air mass flow rate through the ramjet can be calculated from the following relations:

$$(T)_{RJ} = (T_R)_{A/C} - (T)_{TJ} \tag{7.39a}$$

$$(T)_{TJ} = (T/\dot{m}_a)_{TJ} (\dot{m}_a)_{TJ} \tag{7.39b}$$

TABLE 7.4
Regions of Turboramjet Operation

Region	I	II	III	IV
Mach Number	0–0.5	0.5–1.5	1.5–2.5	2.5–6
Ramjet	Off (cold flow)	Off (cold flow)	Starting (combustion starts and $(\dot{m}_a)_{RJ}$ increases to required value)	On (working alone and stable)
Turbojet	On $(\dot{m}_a)_{TJ} = 160\,kg/s$ (alone and stable)	On $(\dot{m}_a)_{TJ} = 160\,kg/s$ (alone and stable)	On $[(\dot{m}_a)_{TJ}$ starts to decrease to reach zero at $M = 2.5]$	Off $(\dot{m}_a)_{TJ} = 0$
Gas generator				
After burner	Off	On $T_{max} = 3000\,K$	Off	Off

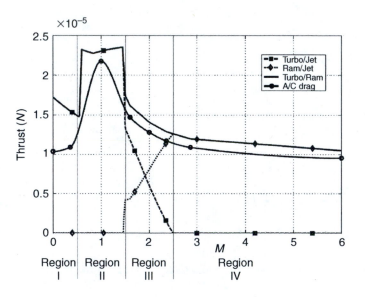

FIGURE 7.22   Thrust versus drag of hybrid engine.

where $(T_R)_{A/C}$ is the thrust required by airplane, say 110% of aircraft drag

$$(\dot{m}_a)_{RJ} = \frac{(T)_{RJ}}{(T/\dot{m}_a)_{RJ}} \qquad (7.40)$$

In region IV, the ramjet is working alone so the desired thrust from ramjet is equal to the aircraft required thrust. In the same way, the ramjet air mass flow rate can be calculated from Equations 7.41 and 7.42:

$$(T)_{RJ} = (T_R)_{A/C} \qquad (7.41)$$

$$(\dot{m}_a)_{RJ} = \frac{(T)_{RJ}}{(T/\dot{m}_a)_{RJ}} \qquad (7.42)$$

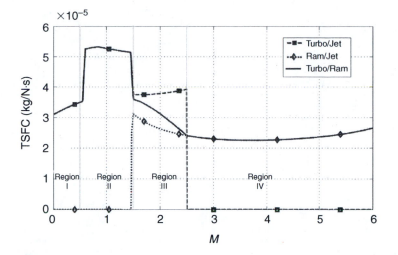

FIGURE 7.23 Thrust-specific fuel consumption (TSFC) of hybrid engine.

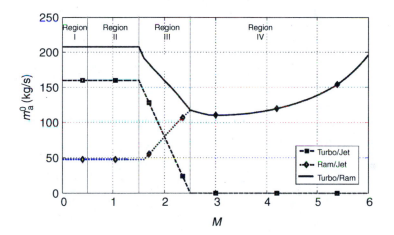

FIGURE 7.24 Air mass flow rate of hybrid engine.

The mass flow rate of the ramjet engine is known all over the engine working interval. Thus, the different performance parameters of the turboramjet engine can be calculated and plotted as illustrated in Figures 7.22 through 7.26.

7.11 EXAMPLES FOR TURBORAMJET ENGINES

Turboramjet engines are not as common as the pure ramjet engine. Two famous known engines may be identified. Most of the others are either secret or only in experimental phases.

1. Pratt & Whitney J58-1, continuous-bleed afterburning turbo (145 kN), two of which are powering the SR-71 strategic reconnaissance aircraft (1964–1998). This enables it to takeoff under turbojet power and switch to ramjets at higher supersonic speeds. Cruise speed is between $M = 3$ and $M = 3.5$.

2. The ATREX turboramjet engine [13]: This engine is developed in Japan. As shown in Figure 7.27 the air flow is pressurized inside the engine by an air fan, while the

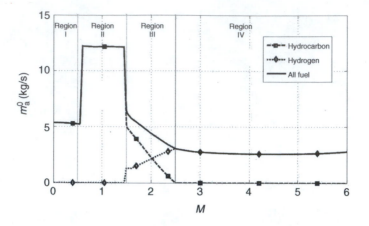

FIGURE 7.25 Fuel mass flow rate of hybrid engine.

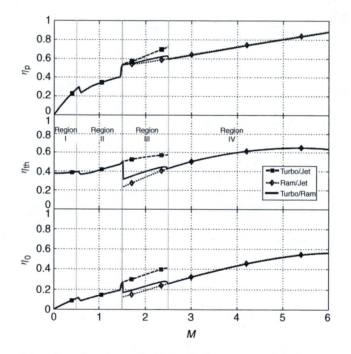

FIGURE 7.26 Propulsive, thermal, and overall efficiencies.

liquid hydrogen (fuel used) is pressurized by a turbopump. Highlights for both flows are given here.

- *Air Flow:* The air intake that reduces velocity of air incoming at high flight Mach number. Next, this air is cooled down by the precooler. The cooled air is pressurized by a fan and injected into the combustion chamber through the mixer together with the hydrogen discharged out of the tip turbine. The plug nozzle provides the effective nozzle expansion over the wide range of flight environment.

- *Hydrogen (H$_2$) Flow:* The hydrogen supplied from storage tank is pressurized by turbopump and heated regeneretively in the precooler.

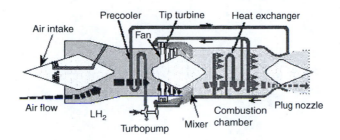

FIGURE 7.27   ATREX turboramjet engine.

7.12 HYPERSONIC FLIGHT

Hypersonic flight is identified as a flight with a Mach number exceeding 5 [14]. The main problem of that speed range is the very large wave drag. To minimize the wave drag, both the aircraft and engine must be completely reconfigured. Until now apart from research work, no civil aircraft can fly at hypersonic speeds. Only rockets can.

7.12.1 History of Hypersonic Vehicles

Some important events will be highlighted here.

WAC Corporal
The first vehicle ever to achieved hypersonic speeds was the WAC Corporal, a U.S. Army second stage rocket mounted atop a captured German V-2 rocket. The origin of the WAC label is a bit unclear, but some sources state it stood for "Without Attitude Control," referring to the fact that the simple rocket had no stabilization and guidance system. On February 24, 1949, the WAC Corporal separated from the V-2 at 3500 mph before its own engine powered it to an altitude of 244 miles and a top speed of 5150 mph upon atmospheric reentry.

Bell aircraft corporation X-2
First flight was on June 27, 1952. The fastest flight recorded a Mach of 3.169. The last flight was on September 27, 1956.

Lockheed Missiles & Space Co (X-7)
First flight was on April 24, 1951. The fastest flight was at a Mach number of 4.31. The last flight was on July 20, 1960.

North American Aviation X-10
First flight on October 14, 1953 and the fastest flight speed was at Mach 2.05. Last flight was on January 29, 1959.

North American Aviation X-15
Robert White, an American flew the X-15 rocket-powered research aircraft at Mach 5.3, thereby becoming the first pilot to reach hypersonic velocities (Figure 7.28). White's mark was later topped by his X-15 co-pilot Pete Knight who flew the craft (denoted X15A-2) to a maximum speed of Mach 6.72 in 1967. The record still stands today as the highest velocity ever reached by an aircraft.

FIGURE 7.28   X-15.

FIGURE 7.29   X-43.

FIGURE 7.30   Booster rocket and X-43.

National Aerospace Plane (NASP); X-30

During the 1980s, NASA began considering a hypersonic single-stage-to-orbit (SSTO) vehicle to replace the Space Shuttle. The proposed national aerospace plane (NASP) would takeoff from a standard runway using some kind of low-speed jet engine. Once the aircraft had reached sufficient speed, airbreathing ramjet or scramjet engines would power the aircraft to hypersonic velocities (Mach 20 or more) and to the edge of the atmosphere. A small rocket system would provide the final push into orbit, but the attractiveness of the concept was using the atmosphere to provide most of the fuel needed to get into space. NASP eventually matured into the X-30 research vehicle, which used an integrated scramjet propulsion system.

X-43 Hyper-X

X-43 is a part of the NASA's Hyper-X project (Figure 7.29). A winged booster rocket with the X-43 itself at the tip, called a "stack," is launched from a carrier plane (B-52). After the booster rocket (a modified first stage of the Pegasus rocket) brings the stack to the target speed and altitude, it is discarded, and the X-43 flies free using its own scramjet engine (Figure 7.30). The first X-43 test flight, conducted in June 2001, ended in failure after the Pegasus booster rocket became unstable and went out of control. The X-43A's successful second flight made it the fastest free flying airbreathing aircraft in the world, though it was preceded by an Australian Hyshot as the first operating scramjet

FIGURE 7.31 The German sänger space transportation systems.

engine flight. The third flight of a Boeing X-43A set a new speed record of 7546 mph (12,144 km/h), or Mach 9.8, on November 16, 2004. It was boosted by a modified Pegasus rocket, which was launched from a Boeing B-52 at 13,157 m (43,166 ft). After a free flight where the scramjet operated for about 10 s, the craft made a planned crash into the Pacific Ocean off the coast of southern California.

LoFlyte
To further the research in this area, the Air Force and Navy sponsored the LoFlyte test vehicle to study the flight characteristics of a wave rider at low-speed conditions, such as takeoff and landing. LoFlyte is another small, unmanned vehicle that can be reconfigured with different types of control surfaces and flight control algorithms to determine the most effective combination. Though designed using a Mach 5.5 conical flowfield, the LoFlyte vehicle is not actually capable of flying at hypersonic speeds. It has instead been tested for basic handling characteristics at low speeds.

The German sänger space transportation systems
The German hypersonic technology program was initiated in 1988, the reference configuration is known as sänger space transportation systems and it is reusable in two stages (Figure 7.31). The piloted first stage is also known as the European hypersonic transport vehicle (EHTV); the EHTV is propelled up to the stage separation Mach number of 6.8 by a hydrogen-fueled turboramjet propulsion system. There are two identical second stages similar to the space shuttle orbiter and powered by hydrogen–oxygen rocket.

7.12.2 HYPERSONIC COMMERCIAL TRANSPORT

Hypersonic vehicles in general and wave riders in particular have long been touted as potential high-speed commercial transports to replace the Concorde. Some aerospace companies, airlines, and government officials have proposed vehicles cruising at Mach 7–12 capable of carrying passengers from New York to Tokyo in less than 2 h. Perhaps one of the most well-known concepts along these lines was the Orient Express, envisioned as a commercial derivative of the NASP project.

The expected flight path for the commercial flight is something unbelievable. Hypersonic vehicles escapes heat buildup on the airframe by skipping along the edge of Earth's atmosphere, much like a rock skipped across water. Hypersonic vehicles would ascend via its power outside the Earth's atmosphere, and then turn off its engines and coast back to the surface of the atmosphere. There, it would again fire its airbreathing engines and skip back into space. The craft would repeat this process until it reached its destination. A flight from the United States to Japan will take 25 such skips (Figure 7.33). The skips will be angled at only 5°. The passengers will feel a force of 1.5 Gs, the same as what one would experience on a child's swing. The plane will power up to 39 km, from where it will coast to double that altitude, before it starts to descend. Each skip will be 450 km long.

FIGURE 7.32   Artist's concept of a hypersonic aircraft (Aurora).

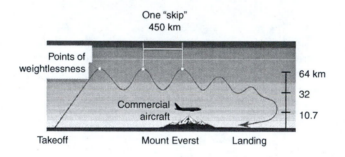

FIGURE 7.33   The expected flight path for hypersonic civil transports.

7.12.3   MILITARY APPLICATIONS

Trends of the 1950s and 1960s indicated that military aircraft had to fly faster and higher to survive, so concepts for high-altitude fighters and bombers cruising at Mach 4 or more were not uncommon. Although the trends soon fizzled and military planners looked to maneuverability and stealth for survival, the military has recently shown renewed interest in hypersonic flight. For example, many have conjectured about the existence of a Mach 5 spy plane, the Aurora (Figures 1.72 and 7.32), may be either under development or perhaps already flying. If so, the Aurora may be scramjet-powered.

7.13   SCRAMJET ENGINES

7.13.1   INTRODUCTION

Scramjet engines find their applications in many recent hypersonic speed vehicles such as rockets, commercial transports in the twenty-first century, as well SSTO launchers. Scramjet is an acronym for supersonic combustion ramjet.

   From a thermodynamic point of view a scramjet engine is similar to a ramjet (Figure 7.34) engine as both consist of an intake, a combustion chamber, and a nozzle. A dominant characteristic of airbreathing engines for hypersonic flight is that the engine is mostly inlet and nozzle (Figure 7.35), and the engine occupies the entire lower surface of the vehicle body.

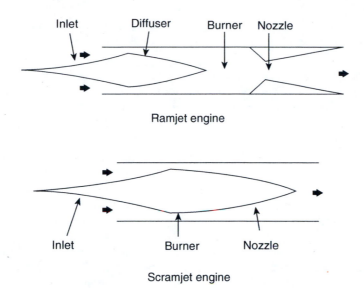

Ramjet engine

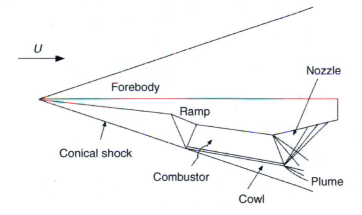

Scramjet engine

FIGURE 7.34 Ramjet and scramjet engines.

FIGURE 7.35 Schematic representation of a hypersonic scramjet engine powering a hypersonic aircraft.

In both engines, the cycle pressure rise is achieved by purely ram compression [15]. The main difference between ramjet and scramjet relies upon the speed of combustion, which is subsonic in a ramjet engine and supersonic in a scramjet engine. A conventional hydrogen-fueled ramjet with a subsonic combustor is capable of operating up to around Mach 5–6 at which the limiting effect of dissociation reduces the effective heat addition to the airflow resulting in a rapid loss in net thrust.

The idea behind the scramjet engine is to avoid dissociation limit by partially slowing the air stream through the intake system (thus reducing the static temperature rise) and hence permitting greater useful heat addition in the now supersonic combustor. Consequently, scramjet engines offer the tantalizing prospect of achieving a high specific impulse up to very high Mach number (Figure 7.11). Scramjet engines have a relatively low specific thrust due to the moderate combustor temperature rise and pressure ratio, and therefore a very large air mass flow rate is required to give adequate vehicle thrust/weight ratio. As outlined in [15], the captured air mass flow reduces for a given intake area as speed rises above Mach 1. Consequently, the entire vehicle frontal area is needed to serve as an intake at scramjet speeds and similarly the exhaust flow has to be re-expanded back into the original stream tube in order to achieve a reasonable exhaust speed. Employing the vehicle forebody and aftbody as part of the propulsion system has the following disadvantages:

1. The forebody boundary layer reaches nearly 40% of the intake flow that upsets the intake flow stability. The conventional solution of bleeding the boundary layer off would be unacceptable due to the prohibitive momentum drag penalty.
2. The vehicle surface must be flat in order to provide the intake with a uniform flowfield. This flattened vehicle cross section is poorly suitable for pressurized tankage and has a higher surface area/volume than a circular cross section.
3. Since the engine and airframe are physically inseparable, little freedom is available to the designer to control the vehicle pitch balance. The single-sided intake and nozzle generate both lift and pitching moments and this adds difficulties to achieve both vehicle pitch balance over the entire Mach number range and center of gravity (CG) movement to trim the vehicle.
4. Normally, scramjet engines are clustered into a compact package underneath the vehicle. This results in interdependent flowfield and thus the failure in one engine with a consequent loss of internal flow is likely to unstart the entire engine installation precipitating a violent change in vehicle pitching moment.
5. To focus the intake shock system and generate the correct duct flow areas over the whole Mach number, variable geometry intake/combustor and nozzle surfaces are required. The large variation in flow passage shape forces the adoption of a rectangular engine cross section with flat moving ramps thereby incurring a severe penalty in the pressure vessel mass.
6. To maximize the installed engine performance requires a high dynamic pressure trajectory with the high Mach number imposing severe heating rates on the airframe. Active cooling of significant portions of the airframe will be necessary with further penalties in mass and complexity.
7. The net thrust of a scramjet is very sensitive to the intake, combustion, and nozzle efficiencies due to the exceptionally poor work ratio of the cycle. Since the exhaust velocity is only slightly greater than the incoming free stream velocity, a small reduction in pressure recovery is likely to convert a small net thrust into a small net drag.

7.13.2 THERMODYNAMICS

A typical layout of a scramjet engine is shown in Figure 7.36. Three processes only are found, namely, compression, combustion, and expansion. Both compression and expansion take place internally and externally. The first part of compression takes place external to the engine while the last part of expansion occurs outside the engine. The same governing equations described in Section 7.7.2 are applied here. Thus, Equations 7.15 through 7.23 are also applicable here. The five performance parameters, namely, specific thrust, TSFC, and the propulsive, thermal, and overall efficiencies, are also calculated from equations similar to Equations 7.29 through 7.33.

The previously described pressure recovery factors in Chapter 3 for ramjet engine may be also employed here. Thus, Equations 3.23 through 3.29 may also be applied.

The different modules will be briefly described in the following sections.

7.14 INTAKE OF A SCRAMJET ENGINE

The design of an air intake (inlet) of any airbreathing engine is crucial in achieving optimal engine performance. The flow pattern for supersonic flow is more complicated due to shock waves/turbulent boundary layer interaction in an adverse pressure gradient. It is necessary to design such intake for a maximum pressure recovery $r_d = P_{02}/P_{0a}$.

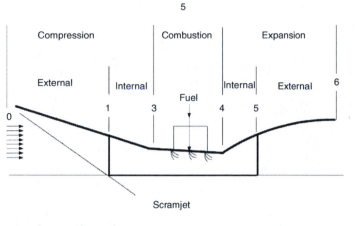

FIGURE 7.36 Layout of a scramjet engine.

Supersonic intakes are classified into three basic types, characterized by the location of the supersonic compression wave system, namely, internal compression, external compression, and mixed compression. The first type (*internal compression*) achieves compression through a series of internal oblique shock waves. The second (*external compression*) achieves compression through either one or a series of external oblique shock waves. The last (*mixed compression*) type makes oblique shock waves in various external and internal blends in order to achieve compression and flow turn. Apart from scramjet engine, in all other types of airbreathing engines the oblique shock waves described previously are followed by a normal shock to convert the flow into a subsonic one. The intake of a scramjet engine is unique by itself. It is of the supersonic–supersonic type. Both inlet and outlet flow are supersonic. The flow is decelerated through several shock waves in the intake, but all of them are of the oblique type. The flow is to be decelerated to the supersonic speed suitable for the succeeding supersonic combustion chamber. The outlet Mach number could be three or more [16].

The pressure recovery of the intake may be calculated from Equation 7.15 based on the flight Mach number or calculated from the product of the pressure ratio across different shock waves.

The details of intake design are given in Chapter 9 together with the governing equations. The pressure recovery for scramjet engines attains very small values. It depends on both the flight Mach number and the number of shocks in the intake. It reaches small values for large Mach numbers. It is in the range of 0.1–0.2 for Mach numbers close to 10 and 0.01–0.02 for Mach numbers close to 20 [16].

Case study
The intake of a scramjet engine is designed in two cases, 4 and 8 external shock waves. The free stream Mach number is 6. The intake shape together with the shock train is shown in Figure 7.37. The intake aspect ratio (height/length) is 1:5.0 and 1:10.65, respectively. The efficiency is 76.82% and 86.45% for the 4 and 8 shock intakes, respectively. The pressure recovery factor (r_d) is 0.451 and 0.642 for the four and eight shocks respectively. The intake length increases continuously with the increase of the number of shocks.

7.15 COMBUSTION CHAMBER

Combustion process here represents a heat addition process at high-speed flow (supersonic flow). Generally, the Mach number at the inlet to combustion chamber is nearly one-third of the flight Mach

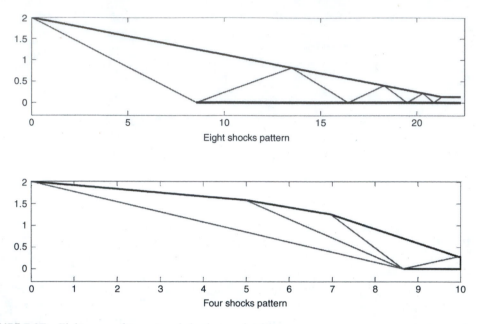

FIGURE 7.37   Eight versus four external shock wave intakes.

number. The cross-sectional area of the combustion chamber is assumed constant for simplicity. The flow may be treated as a Rayleigh flow as a first approximation. In such a simplified model, both the specific heat at constant pressure and the specific heat ratio are assumed constant. Supersonic combustion will be treated in detail in Chapter 11. Here some highlights will be given. First, the practical problems of employing supersonic combustion may be stated as follows:

1. Capture a stream tube of supersonic air.
2. Inject fuel into the air stream.
3. Achieve a fairly uniform mixture of fuel and air.
4. Carry out the combustion process within a reasonable length without causing a normal shock within the engine.

For a hydrocarbon fuel C_xH_y, the maximum combustion temperature occurs when hydrocarbon fuel molecules are mixed with just enough air so that all of the hydrogen atoms form water vapor and all of the carbon atoms form carbon dioxide. The stoichiometric fuel-to-air ratio is

$$f_{st} = \frac{36x + 3y}{103\,(4x + y)} \text{ kg fuel/kg air}$$

Fuel mixtures could be rich or lean depending on the equivalence ratio, the ratio of the actual fuel-to-air ratio to the stoichiometric fuel-to-air ratio, or

$$\phi = \frac{f}{f_{st}}$$

Combustion includes successive processes including fuel injection, mixing, and ignition.

The fuel injector for supersonic combustion should be capable of generating vertical motions, which cause the fuel–air interface area to increase rapidly and enhance the micromixing of molecular

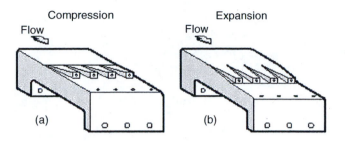

FIGURE 7.38 (a) Compression ramps. (b) Expansion ramps. (From W.H. Heiser and D.T. Pratt, *Hypersonic Airbreathing Propulsion*, AIAA Education Series, 1994, p. 278. With permission.)

level, hence leading to efficient combustion. Injection and mixing are correlated. There are two types of injection that exist either in parallel or normal streams.

(a) *Fuel mixing in parallel stream*

Ramp Injectors

Ramp injectors are passive mixing devices that produce counterrotating streamwise vortices to enhance the mixing of fuel and air (Figure 7.38).

Compression ramps, configuration (a), is associated with an oblique shock standing at the base of the ramp while the compressed air above the ramp spills over the sides into the lower pressure, forming a counterrotating pair of axial vortices. Concerning expansion ramps, configuration (b), the wall is turned away from the flow while the top surface of the ramp remains in the plane of the upstream wall, thus leading to a Prandtl–Meyer expansion so the pressure difference between the flow on the upper ramp surface and the expanded flow is produced. The location of the shock was found to affect the mixing performance, and the expansion design produces higher combustion efficiencies despite having lower overall mixing performance. This was attributed to the improved local, small-scale mixing near the fuel injection port.

(b) *Fuel mixing in normal stream*

Performance of high-speed combustor systems requires fuel and air mixing at the molecular level in the near field of the fuel injection. One of the simplest approaches is the transverse (normal) injection of fuel from a wall orifice. As the fuel jet, sonic at the exit, interacts with the supersonic cross flow, an interesting but rather complicated flowfield is generated.

Fuel–air mixing is so important and numerous studies for such process are available. Two types of mixing are examined, namely, laminar and turbulent mixing. Upon ignition and flame generation, it is necessary to have a flame holder to reduce the ignition delay time and to provide a continuous source of radicals for the chemical reaction to be established in the shortest distance possible.

Analysis of the combustion chamber may be simplified and treated as one-dimensional flow. If the flow is supersonic, adding heat lowers the Mach number and vice versa. A case study examined in [16] outlined that if the inlet Mach number to the combustion chamber is 3.0 and the heat added due to fuel burning increases the total temperature by 30%, then the outlet Mach number will be 1.7 and the total pressure ratio between the outlet and inlet states will be 0.37 [16]. Finally, two points may be added here:

1. Hydrogen fuel is mostly used in scramjet engines, which generally has heating value per unit mass some 2.3 times the conventional hydrocarbon or jet fuel.
2. Computational fluid dynamics (CFD) are extensively used in the analysis of combustion chambers to verify and overcome any difficulties in performing a complete experimental analysis.

7.16 NOZZLE

The nozzle is very similar to the intake. It has an internal part and an external part formed by the vehicle body. The internal part is a typical two-dimensional nozzle while the aftbody is the external part, which is a critical part of the nozzle component.

A two-dimensional (or planar rather than axisymmetric or circular) nozzle is used in scramjet engines planned for powering hypersonic aircraft; Mach number is greater than 5, because of the following reasons:

First, circular nozzles are comparatively heavy and do not lend themselves easily to variable geometry, and the need for tight integration.

Second, the flow at the exhaust nozzle entry is supersonic, rather than the sonic or choked throat condition of convergent–divergent nozzles.

Third, the design produces uniform and parallel flow at the desired exit Mach number because that maximizes the resulting thrust.

Fourth, the design is a minimum length exhaust nozzle. This is achieved by placing sharp corners that generate centered simple or Prandtl–Meyer expansion fans at the nozzle entry.

Case Study
An axisymmetric nozzle of a scramjet has the following data:
  Inlet Mach number = 1.56
  Outlet Mach number = 3.65
  Inlet radius 0.366 m

An analytical procedure is followed for drawing the nozzle contour and the Mach number within the nozzle. The procedure is described in full details in Section 11.3.2.1 of the nozzle chapter. Here, only the final results are given in Table 7.5. The nozzle contours are plotted in Figure 7.39 and the Mach number within the nozzle is also given in Figure 7.40.

7.17 PERFORMANCE PARAMETERS

The generated thrust force is expressed by the following relation: as the nozzle is unchoked

$$\frac{T}{\dot{m}_a} = [(1+f)\,u_e - u_a]$$

TABLE 7.5
Nozzle Characteristics

	1	2	3	4	5	6	7	8	9	10
M	1.56	1.79	2.02	2.25	2.49	2.72	2.95	3.18	3.42	3.65
μ	39.94	33.96	29.64	26.33	23.71	21.58	19.81	18.34	17.02	15.91
θ_0	47.21	47.21	47.21	47.21	47.21	47.21	47.21	47.21	47.21	47.21
θ	47.21	40.38	33.83	27.69	22.01	16.79	12.01	7.65	3.65	0
r_c	0.366									
r	0.366	0.429	0.517	0.632	0.782	0.974	1.215	1.515	1.886	2.340
x	0	0.146	0.394	0.789	1.391	2.272	3.527	5.270	7.638	10.798
y	0.404	0.542	0.728	0.961	1.237	1.544	1.863	2.16	2.389	2.484

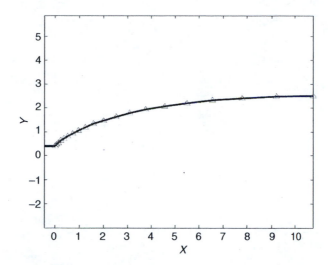

FIGURE 7.39 Nozzle geometry for a scramjet engine.

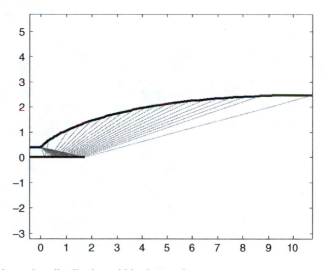

FIGURE 7.40 Mach number distribution within the nozzle.

The fuel-to-air ratio can be given by

$$f = \frac{\left(C_{p4} T_{04} / C_{p2} T_{0a}\right) - 1}{\left(\eta_b Q_R / C_{p2} T_{0a}\right) - \left(C_{p4} T_{04} / C_{p2} T_{0a}\right)}$$

The propulsive, thermal, and overall efficiencies are calculated using the appropriate relations.

Example 1 Figure 7.41 illustrates a scramjet engine powering an airplane flying at Mach number equal to 5.0 at an altitude of 55,000 ft where $T_a = 216.67$ K and $P_a = 9.122$ kPa. Two oblique shock waves are formed in the intake before entering the combustion chamber at supersonic speed and having a deflection angle $\delta = 10°$. Hydrogen fuel is burned that gives rise a maximum temperature of 2000 K. The fuel-to-air ratio is 0.025. The nozzle has an expansion ratio $A_5/A_4 = 5$. The inlet

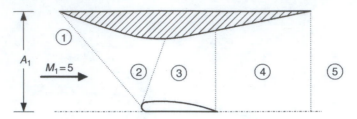

FIGURE 7.41 Scramjet layout.

and exit areas of the engine are equal, $A_1 = A_5 = 0.2\,\mathrm{m}^2$, and the hydrogen fuel heating value is 120,900 kJ/kg. It is required to

1. Calculate the inlet Mach number to the combustion chamber.
2. Calculate exhaust jet velocity.
3. Calculate the overall efficiency.

$(C_p)_{cc} = 1.51$ kJ/kg K, $\gamma_n = 1.238$, and burner efficiency is 0.8.

Solution:
Diffuser section
Flight speed $V_f = M_1 \sqrt{\gamma R T_a} = 5\sqrt{1.4 \times 287 \times 216.7} = 1475.4$ m/s
$$T_{01} = T_a \left(1 + \tfrac{\gamma-1}{2}M^2\right) = 1300\ \mathrm{K}$$

An oblique shock wave separates region (1) and (2) with $M_1 = 5$ and $\delta = 10°$, then either from the governing equations listed in Chapter 3 or from oblique chock wave table/figure, the shock wave inclination angle is $\sigma_1 = 19.38°$. The normal component of Mach number ahead of the shock is

$$M_{1n} = M_1 \sin(\sigma - \delta) = 1.659$$

The normal component of Mach number behind (downstream) the shock wave is $M_{2n} = 0.65119$. The corresponding Mach number is then

$$M_2 = \frac{M_{2n}}{\sin(\sigma - \delta)} = 4.0$$

The static temperature ratio across the shock is $T_2/T_1 = 1.429$.
A second oblique shock wave is developed between regions (2) and (3). With inlet Mach number and wedge angle having the values $M_2 = 4.0$ and $\delta = 10°$, then the shock angle is $\sigma_2 = 22.23°$ and the normal component of the inlet Mach number is $M_{2n} = M_2 \sin \sigma = 1.513$.

From oblique shock tables, the normal component of the outlet Mach number is obtained from tables or normal shock relations as

$$M_{3n} = 0.698$$

$$M_3 = \frac{M_{3n}}{\sin(\sigma - \delta)} = \frac{0.698}{\sin 12.23} = 3.295$$

$$\frac{T_3}{T_2} = 1.328$$

The inlet Mach number to combustion chamber is $M_3 = 3.295$.

Combustion chamber (Rayleigh flow)
Within the combustion chamber (from 3 to 4), heat is added. The flow is a Rayleigh flow with $M_3 = 3.316$, from Rayleigh flow tables

$$\frac{T_{03}}{T_0^*} = 0.6282, \quad \frac{P_{03}}{P_0^*} = 4.53$$

Since $T_{01} = T_{02} = T_{03} = 1300\,\text{K}$ and $T_{04} = 2000\,\text{K}$, then

$$\frac{T_{04}}{T_0^*} = \frac{T_{04}}{T_{03}} \frac{T_{03}}{T_0^*} = \frac{2000}{1300} \times 0.6282 = 0.966$$

From Rayleigh flow tables

$$M_4 = 1.26 \quad \text{and} \quad \frac{P_{04}}{P_0^*} = 1.0328$$

The outlet Mach number from combustion chamber is $M_4 = 1.26$. This confirms the fact that adding heat to a supersonic flow reduces its speed.
The pressure ratio in the combustion chamber (P_{04}/P_{03}) is calculated as

$$\frac{P_{04}}{P_{03}} = \frac{P_{04}}{P_0^*} \frac{P_0^*}{P_{03}} = \frac{1.0328}{4.53} = 0.2279$$

Note that this value is rather a small value, as normally $(P_{04}/P_{03}) \approx 0.3 - 0.4$.
The fuel- to-air ratio

$$f = \frac{C_p T_{04} - C_p T_{03}}{\eta_b Q_R - C_p T_{04}} = \frac{1.51\,(2000 - 1300)}{0.8 \times 120{,}900 - 1.51 \times 2000} = 0.01128$$

sometimes the following simple formula is used for (f):

$$f = \frac{C_p\,(T_{04} - T_{03})}{\eta_b Q_R} = \frac{1.51\,(2000 - 1300)}{0.8 \times 120{,}900} = 0.011$$

Nozzle
The flow in the nozzle is isentropic. From isentropic flow tables with inlet Mach number $M_4 = 1.26$, then $A_4/A^* = 1.05$ and $T_{04}/T_4 = 1.317$.

$$\frac{A_5}{A^*} = \frac{A_5}{A_4} \frac{A_4}{A^*} = 5 \times 1.05 = 5.25$$

The outlet Mach number and temperature ratio are

$$M_5 = 3.23 \quad \text{and} \quad \frac{T_{05}}{T_5} = 3.11$$

$$T_5 = \frac{T_5}{T_{05}} \frac{T_{05}}{T_{04}} \frac{T_{04}}{T_4} \frac{T_4}{T_3} \frac{T_3}{T_2} \frac{T_2}{T_1} T_1$$

with $T_{05} = T_{04}$, then

$$T_5 = \frac{1}{3.11} \times 1.317 \times 3.744 \times 1.333 \times 1.429 \times 216.7 = 654.5\,\text{K}$$

The exhaust speed is $V_5 = M_5\sqrt{\gamma_h R T_5} = 1558$ m/s.

Note also that the difference between flight and exhaust speeds are very small as identified previously, which is one of the drawbacks of scramjet engines.

The air mass flow rate is

$$\dot{m} = \rho_1 V_1 A_1 = \frac{P_1}{RT_1} V_1 A_1 = \frac{9122}{287 \times 216.67} \times 1475.4 \times 0.2 = 43.3\,\text{kg/s}$$

Thrust Force

For unchoked nozzle as usual in scramjet engines, the thrust force is

$$T = \dot{m}\left[(1 + f)\,V_5 - V\right]$$

$$T = (43.3)\left[1.012 \times 1558 - 1475.4\right] = 4386\,\text{N}$$

This value is small and it is appropriate for the high-altitude flights as the drag force on air vehicles is also small.

Overall efficiency

$$\eta_0 = \frac{TV_F}{\dot{m}_f Q_{HV}} = \frac{TV_F}{\dot{m}_a Q_{AV}} = \frac{4386 \times 1475}{0.012 \times 43.3 \times 120.9 \times 10^6} = 10.29\%$$

A summary for Mach numbers at different stations in the engine is given as

$$M_1 = 5, \quad M_2 = 4, \quad M_3 = 3.295, \quad M_4 = 1.26, \quad M_5 = 3.23$$

PROBLEMS

7.1 The drag of a hypersonic future businesses jet is to be simplified by the following equations:

Flight Mach Number	Drag (N)
$M = 0$–0.75	$D = 10^5 \cdot \left(4.1M^2 - 1.1M + 2.1\right)$
$M = 0.75$–1.25	$D = 10^5 \cdot \left(-9.7M^2 + 20M - 5.9\right)$
$M = 1.25$–6	$D = 10^5 \cdot \left(\dfrac{21}{M^4} - \dfrac{37}{M^3} + \dfrac{26}{M^2} - \dfrac{5.9}{M} + 2.3\right)$

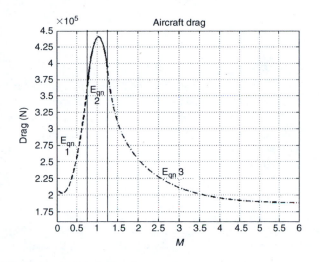

PROBLEM 1 Drag versus Mach number.

If that aircraft is to be driven by two turbojet engines with afterburner, each having the following parameters:

Turbine inlet temperature (TIT) = 1750 K

Maximum temperature T_{max} = 2500 K

Pressure ratio π_o = 10

Efficiencies of different components:

$\eta_d = 0.9, \quad \eta_N = 0.98, \quad \eta_C = 0.91$

$\eta_T = 0.91, \quad \eta_m = 0.99, \quad \eta_b = 0.99$

Pressure losses

$\Delta P_{CC} = 0.02, \qquad \Delta P_{AB} = 0.02$

The fuel used is a hydrocarbon fuel having a heating value of 45,000 kJ/kg. It is required to consider the following two cases:
  i. Afterburner is off
  ii. Afterburner is on
In each case calculate and plot for a Mach number varying from 0 to 4
- The air mass flow rate ($\dot{m}_a$)
- The fuel flow rate ($\dot{m}_f$)
- The thrust-specific fuel consumption (TSFC)

7.2 The aircraft of Problem 7.1 is to be driven by two ramjet engine having the following parameters:

$$T_{max} = 2500\,K$$

$$\eta_d = 0.9, \quad \eta_N = 0.98$$

$$\eta_b = 0.99, \quad \Delta P_{CC} = 0.02$$

Calculate and plot
- The air mass flow rate ($\dot{m}_a$)
- The fuel flow rate ($\dot{m}_f$)
- The thrust-specific fuel consumption (TSFC)

If the engine is using the same fuel of Problem 7.1 and if it is using a hydrogen fuel having a heating value of 119,000 kJ/kg for a Mach range: $M = 0.75$ to $M = 6$.

7.3 Compare between the four cases of Problems 7.1 and 7.2 at the following values of Mach number $M = 0.9, M = 1.5$, and $M = 3.5$ and write your comment.

7.4 If the mass flow through the turbojet engine is limited to $\dot{m}_a = 200$ kg/s find the point at which the afterburner will start burning fuel. Then plot

- The air mass flow rate ($\dot{m}_a$)
- The fuel flow rate ($\dot{m}_f$)
- The TSFC

for the case having a Mach number range from $M = 0$ to $M = 4$.

7.5 The aircraft of Problem 7.1 is to be fitted with an over/under turboramjet, which is constructed from both of the turbojet described in Problem 7.1 and the ramjet described in Problem 7.2. The air mass flow in the turbojet engine is to be limited to $\dot{m}_a = 200$kg/s as in Problem 7.4. During the flight instants where the turbojet needs to develop more thrust and needs extra air flow rate ($\dot{m}_a > 200$kg/s), the following switching mode is selected: the afterburner is put off and the ramjet starts to operate with an appropriate mass flow that increases the total thrust to overcome the drag of the aircraft (110% drag). For that transition case calculate and plot

- The air mass flow rate ($\dot{m}_a$)
- The fuel flow rate ($\dot{m}_f$)
- The thrust-specific fuel consumption (TSFC)

7.6 Compare between the TSFC in Problems 7.1, 7.2, and 7.5 write your comment.

7.7 If the aircraft is to be driven with the over/under turboramjet of Problem 7.5 and works on the following regimes

Regime	Engine Condition
$M = 0$–0.5	Turbojet with AB off
$M = 0.5$–1.5	Turbojet with AB on
$M = 1.5$–3	Turbojet with AB off
	Mass flow in turbojet is decreasing from 100% $\dot{m}_a$ at
	$M = 1.5$ to 0% $\dot{m}_a$ at $M = 2.5$
	Ramjet is operating as a support for the turbojet engine
$M = 3$–6	The turbojet is off
	The ramjet is driving the A/C alone

For that case calculate and plot
- The air mass flow rate ($\dot{m}_a$)
- The fuel flow rate ($\dot{m}_f$)
- The thrust-specific fuel consumption (TSFC)

Write your comment.

7.8 For the wraparound turboramjet engine, the following data were recorded when the turbojet engine was operating with operative afterburner at an altitude of 60,000 ft

Flight Mach number 2.5
TIT 1300 K

Maximum temperature 2000 K

Air mass flow rate 60 kg/s

If the pilot is to planning to switch operation to ramjet and stop the turbojet, what is the procedure he or she has to follow for this transition time?

If the maximum temperature of the ramjet is 2500 K, what will be the needed air mass flow rate to keep the same thrust as the turbojet mode?

Do you expect a dual mode is visible? Explain why or why not.

7.9 A military airplane is powered by a scramjet engine. The airplane is flying at an altitude of 60,000 ft where the ambient temperature and pressure are 216.5 K and 7.17 kPa. The flight Mach number is 7.0. The fuel used is hydrogen where its gas constant $R = 4.124$ kJ/kg · K, its heating value is 130,000 kJ/kg and an average specific heat of 1.4 may be assumed all over the engine components. The maximum temperature in the engine is 3000 K. The exhaust gases leave the combustion chamber at a Mach number equal to unity. The efficiency of combustion is 0.98. The exit pressure is equal to the ambient pressure. Assume that in the nozzle the outlet total pressure to the inlet total pressure is 0.1.

It is required to calculate:

(a) The flight speed
(b) The fuel-to-air ratio (f)
(c) The Mach number at the inlet to the combustion chamber
(d) The total pressure ratio at the diffuser ($P_{ooutlet}/P_{oinlet}$)
(e) The total pressure ratio at the combustion chamber
(f) The exhaust jet speed
(g) The specific thrust
(h) TSFC

7.10 Compare between turboramjet and scramjet engines with respect to the following points:
- Flight altitude
- Maximum Mach number
- Fuel consumption

7.11 When do you expect to fly via a hypersonic vehicle?

REFERENCES

1. http://en.wikipedia.org/wiki/Bell_X-1
2. http://en.wikipedia.org/wiki/SR-71_Blackbird
3. http://www.concordesst.com/
4. http://home.comcast.net/~yoshac/TU144/index.html
5. S.J. Morris, Jr., K.A. Geiselhart, and P.G. Coen, *Performance Potential of an Advanced Technology Mach 3 Turbojet Engine Installed on a Conceptual High-Speed Civil Transport*, NASA-TM-4144, November 1989.
6. J.C. Ellison, *Investigation of The Aerodynamic Characteristics of A Hypersonic Transport Model at Mach Numbers to 6*, NASA TN D-6191, April 1971.
7. R.J. Pegg et al., *Design of Hypersonic Waverider-Derived Airplane*, AIAA Paper 93-0401, January 1993.
8. D.W. Lam, *Use of the PARC Code to Estimate the Off-Design Transonic Performance of an Over/Under Turboramjet Nozzle*, NASA TM-106924, AIAA-95-2616.
9. R.E. Morris and G.D. Brewer, *Hypersonic Cruise Aircraft Propulsion Integration Study*, Volume I and Volume II, NASA-CR-158926-1 and NASA-CR-158926-2, September, 1979.
10. G.D. Riebe, *Aerodynamic Characteristics Including Effect of Body Shape of A Mach 6 Aircraft Concept*, NASA-TP-2235, December, 1983.

11. J.L. Pittman and G.D. Riebe, *Experimental and Theoretical Aerodynamic Characteristics of Two Hypersonic Cruise Aircraft Concepts at Mach Numbers of 2.96, 3.96, and 4.63*, NASA-TP-1767, December 1980.
12. C.S. Domack, S.M. Dollyhigh, F.L. Beissner, Jr., K.A. Geiselhart, M.E. McGraw, Jr., E.W. Shields, and E.E. Swanson, *Concept Development of a Mach 4 High-Speed Civil Transport*, NASA-TM-4223, December, 1990.
13. http://en.wikipedia.org/wiki/ATREX
14. J.D. Anderson, Jr., *Introduction to Flight*, 3rd edn., McGraw-Hill, 1989, p. 108.
15. R. Varvill and A. Bond, A comparison of propulsion concepts for SSTO reusable launchers, *JBIS*, 56, 108–117, 2003.
16. J.L. Kerrebrock, *Aircraft Engines and Gas Turbines*, MIT Press, 1992, p. 405.
17. W.H. Heiser and D.T. Pratt, *Hypersonic Airbreathing Propulsion*, AIAA Education Series, 1994, p. 278.
18. T.R. Conners, *Predicted Performance of a Thrust—Enhanced SR-71 Aircraft with an External Payload*, NASA TM 104330, 1997.

8 Industrial Gas Turbines

8.1 INTRODUCTION

Industrial gas turbines are defined as all gas turbines other than aircraft gas turbines or aero engines [1]. Industrial gas turbine is basically a turboshaft engine. Historically, industrial gas turbine had its roots planted much further back in time than its aero engine counterpart and again these roots sprouted in European soil [2].

Industrial gas turbine proves itself as an extremely versatile and cost-effective prime mover as it has been used for a variety of applications in land and sea. Examples for land applications are electric power generation for land and offshore platforms. It may be used as load topping because of its rapid start-up ability. It is used in conjunction with a steam turbine in a "combined cycle plant," which is at present the most efficient form of large-scale power generation. Moreover, it has many applications in chemical industry. In oil industry, gas turbines drive compressor/pump for natural gas or liquid oil transmission pipelines. Land transport also employs gas turbines in freight trains as well as starters for large trains. Some military tanks are powered by gas turbines such as the tank M1A1. In marine propulsion, gas turbines are also used in numerous naval vessels, cargo ships, hydrofoils, patrol boats, frigates, and others.

Industrial gas turbines differ from aero engines as they have no limitations on their size and weight. They have nearly zero exhaust speeds. Moreover, they have a long time between overhauls, which is of the order of 100,000 operating hours.

Currently, there are gas turbines, which run on natural gas, diesel fuel, naphtha, crude, low-Btu gas, vaporized fuel oils, and biomass.

The principal manufacturers for large industrial gas turbines are Honeywell, Alstom, Aviadvigatel, Bharat Heavy Electricals, Centrax Gas Turbine, Dresser Rand, Ebara, Fiat Avio, General Electric, Hitachi, Japan Gas Turbine, Kawasaki Heavy Industries, Mitsubishi Heavy Industries, MTU, Motoren, Nuovo Pignone-Turbotecnicaall, Pratt &Whitney, Pratt & Whitney Canada, Rolls-Royce, Ruston Turbines, Siemens, Siemens-Westinghouse, Solar Turbines, Sulzer Turbines, Toshiba, Turbomeca, and Volvo Aero.

Now in the market, there are many single-shaft engines delivering more than 250 MW per unit. The most important advantages of industrial gas turbines may be described as follows:

1. Compactness (small mass and volume) for a given power output, that is, high W/m^3 or W/kg. This makes it convenient for its usage in naval application.
2. Low vibration as parts are rotating and not reciprocating as in internal combustion engines.
3. Reliable: low dynamic stresses from near constant rotational speed operation.
4. Low cost per kg at larger sizes.
5. Gas turbines have low emissions, low installed cost, and low-cost power generation [3].
6. Changeover from primary to secondary fuel with constant load is automatically achieved.
7. Transportation is easily done by air, land, or sea.
8. An on-site maintenance is performed with inspection intervals from 25,000 to 30,000 h.

Now we consider the advantages in special applications:

1. Units used in generating electricity have the ability to produce full power from cold in less than 2 min. Moreover, they can be erected in deserts (or in general away from water sources) since they do not need cooling water.

2. Gas turbines may be installed on mobile trailers as a mobile source for generating electricity.
3. Engines used in naval applications are of compact size and have high power density and low noise in comparison with diesel engines.
4. Electrical power units have very compact size with respect to other power sources. For example, the ABB ALSTOM power unit GT 26, which generates 265 MW, has the dimensions of 12.3 × 5.0 × 5.5 m [4].

Disadvantages:

1. Lower thermal efficiency compared with diesel engines
2. Relatively poor dynamic performance—less suitable for cars and other "stop–start" applications

8.2 CATEGORIES OF GAS TURBINES

The simple cycle gas turbine is classified into five broad groups [5] (Figure 8.1):

1. *Heavy-duty gas turbines:* These are large power generation units ranging from 3 to 480 MW in a simple cycle configuration, with efficiencies ranging from 30% to 46%.

2. *Aircraft-derivative gas turbines (aeroderivatives):* As the name indicates, these power generation units are originally aircraft engines, which were later on adapted in electrical generation industry by adding a power turbine to absorb the kinetic energy of the exhaust. An engine of this class is the General Electric LM2500. For example, the aero RR RB211 turbofan engine is modified by removing the fan and replacing the bypass fans and adding a power turbine at their exhaust. Other examples for the aeroderivatives are Avco Lycoming TF25, rated at 2000 hp and weighing approximately 1020 lb. The power ranges from 3 MW to about 50 MW. The efficiencies of these units range from 35% to 45%. The outstanding characteristic of the aircraft-derived engines is their lightness. The LM2500 WEIGHS 34,000 lb, including typical mounts, base, and enclosure, for approximately 1.5 pounds per horsepower [lb/hp] is an example. For marine applications, the material of these jet engines is either changed or coated to resist sea salt corrosion.

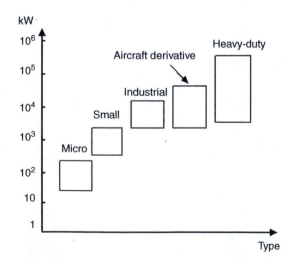

FIGURE 8.1   Categories of gas turbine engines.

3. *Industrial type gas turbines:* These vary in range from 2.5 to 15 MW. These are used extensively in petrochemical plants for compressor drive trains. The efficiencies of these units are less than 30%.

4. *Small gas turbines:* These gas turbines are in the range 0.5–2.5 MW. They usually have centrifugal compressors and radial inflow turbines. Efficiencies in the simple cycle applications are in the range from 15% to 25%.

5. *Microturbines:* These turbines are in the range 20–350 kW. These microturbines have had a dramatic growth since the late 1990s.

8.3 TYPES OF INDUSTRIAL GAS TURBINES

Open-cycle engines take a number of different forms. They may be either of the following:

1. Single-shaft engine
2. Multiple-shaft engine

These different layouts are illustrated in Figures 8.2 through 8.4. The single shaft (or spool); Figure 8.2 has also two forms as is shown in Figures 8.2a and 8.2b. In both figures, the gas generator (compressor, combustion chamber, and turbine) is coupled to an industrial load. This industrial load may be coupled to the cold end (compressor side) (Figure 8.1a) or to the hot end (turbine side) (Figure 8.1b). The second type is the multiple shaft, which is either a two-shaft or a three-shaft engine. The two-shaft configuration may be found in one of these two layouts. The two shafts may be concentric (sometimes identified as split-shaft or compound form) (Figure 8.3a) and the load may be coupled to either shaft. The other two-shaft configuration (Figure 8.3b) includes a power turbine driving the load through a separate shaft downstream the main shaft comprising the gas generator. The three-shaft engines are similar either to the three-spool turbofan, where the load is; driven by one of the three shafts Figure 8.4a, mostly the low-pressure spool, or a two-spool (low- and high-pressure shafts) where a free power turbine drives the load; Figure 8.4b.

The single shaft promotes simplicity in design, compactness, and has high inertia to handle large load step up and down (which is favorable in electrical generation units). However, the two-shaft configuration is normally quicker to start and needs less starting power.

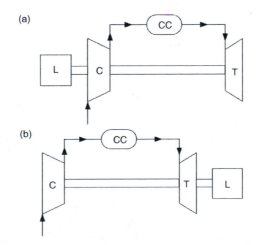

FIGURE 8.2 (a) Single spool with load coupled to the cold side. (b) Single spool with load coupled to the hot side.

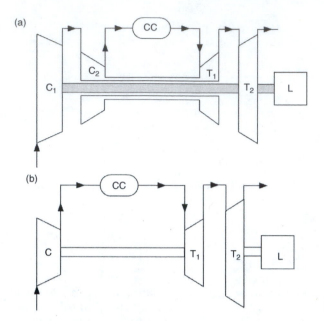

FIGURE 8.3   (a) Two spool with load coupled to low-pressure spool. (b) Two spool with free power turbine.

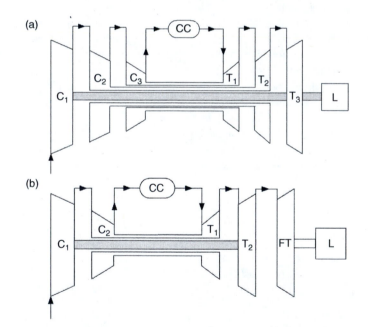

FIGURE 8.4   (a) Three spool with load coupled to low-pressure spool. (b) Three spool with free power turbine.

8.4 SINGLE-SHAFT ENGINE

The basic gas turbine cycle is named after the Boston engineer, George Brayton, who first proposed the Brayton cycle around 1870. The gas turbine first successfully ran in 1939 at the Swiss National Exhibition at Zurich. The early gas turbines built in the 1940s and even 1950s had simple cycle efficiencies of about 17% because of the low compressor and turbine efficiencies and low turbine inlet temperatures due to metallurgical limitations of those times. The gas turbine has experienced

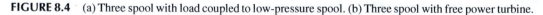

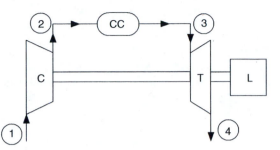

FIGURE 8.5 A typical single-spool gas turbine.

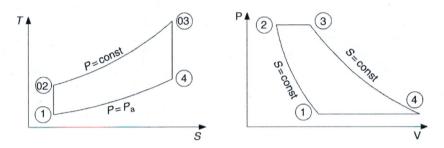

FIGURE 8.6 Temperature–entropy and pressure–specific volume diagrams.

phenomenal progress and growth since its first successful development in the 1930s. The efforts to improve the cycle efficiency concentrated in three areas:

1. Increasing the turbine inlet (or firing) temperatures
2. Increasing the efficiencies of turbomachinery components
3. Adding modifications to the basic cycles

These added modules include regeneration (or reproduction), intercooling, and reheating. It will be discussed in the succeeding sections.

An example for this single spool is Siemens (SGT-800) Gas Turbine (45 MW).

8.4.1 Single Compressor and Turbine

8.4.1.1 Ideal Cycle

Gas turbines usually operate on an open Brayton cycle, as shown in Figure 8.5. The T–S and P–V diagrams of an ideal Brayton cycle are shown in Figure 8.6.

Frequently gas turbines are designated symbolically by several letters that identify their constituents. The letter "C" refers to compressor, letter "B" refers to burner or combustion chamber, and letter "T" refers to turbine; thus the simple Brayton cycle is denoted as "**CBT.**"

The following assumptions are considered:

1. All processes are steady.
2. The compression and expansion processes are isentropic.
3. No pressure drop in the combustion chamber.
4. Constant properties (specific heat and specific heat ratio Cp, γ).
5. The working fluid is assumed to be air.
6. The amount of fuel added is also neglected in the energy balance relations.

The different modules are described as follows:

Air inlet assembly

The engine inlet sucks air from the nearby area and delivers it to the compressor. It is assumed that there are no changes in either the total temperature or pressure from its ambient values, $T_{01} = T_a$ and $P_{01} = P_a$.

Compressor

Fresh air at ambient conditions is drawn into the compressor. The compressor is either axial or centrifugal. The axial compressor has higher efficiency, higher air flow rate, and more stages than centrifugal compressor. The centrifugal compressor is used in small and midsize industrial engines. It has fewer stages. Moreover, it is simple and rugged. The most recent axial compressor are Electron Beam (EB)-welded rotor which gives low vibration, straightforward torque transfer, good control of blade tip clearance, and still the possibility to replace individual blades *in situ*, provided that the casings in the compressor section have a longitudinal split.

Next, the air temperature and pressure are raised in an isentropic compression process. Since the compression process is isentropic, then

$$\frac{T_{02}}{T_{01}} = \left(\frac{P_{02}}{P_{01}}\right)^{(\gamma-1)/\gamma} = \pi_c^{(\gamma-1)/\gamma}$$

where $\pi_c = P_{02}/P_{01}$. The specific work in the compressor is

$$W_c = Cp(T_{02} - T_{01}) = Cp(T_{02} - T_a) = CpT_a \left(\pi_c^{(\gamma-1)/\gamma} - 1\right) \qquad (8.1)$$

Combustion chamber

The combustion chamber, also known as the "Burner," must be compact and provide even temperature distribution of the hot gases as delivered next to the turbines. The three basic configurations of the combustion chamber are annular, can, and can-annular. The annular is a donut-shaped, single, continuous chamber that encircles the turbine. It requires less cooling air due to less hot surface area and gives a better flow inlet into the turbine. It also has simple cross-ignition during start-up. However, there is a great difficulty for acoustic control due to flame instability at ultra-low emission levels. The can-annular comprises multiple, single burner "cans" evenly spaced around the rotor shaft. Single cans can be removed without removal of turbine module. Cans are either one or more combustion chambers mounted external to the gas turbine body. Generally, the high-pressure air proceeds into the combustion chamber, where the fuel is burned at constant pressure. The heat added per unit air mass flow rate is

$$Q_{in} = h_{03} - h_{02} = Cp\,(T_{03} - T_{02}) \qquad (8.2)$$

Turbine

The resulting high-temperature gases then enter the turbine, where they expand in an isentropic process to atmospheric pressure, thus producing power. The exhaust gases leaving the turbine are thrown out (not recirculated), causing the cycle to be classified as an open cycle. The expansion process in the turbine is also isentropic. Moreover, since $P_{02} = P_{03}$ and $P_{04} = P_{01}$, then

$$\frac{T_{03}}{T_{04}} = \left(\frac{P_{03}}{P_{04}}\right)^{(\gamma-1)/\gamma} = \left(\frac{P_{02}}{P_{01}}\right)^{(\gamma-1)/\gamma} = \pi_c^{(\gamma-1)/\gamma}$$

The specific work generated in the turbine is

$$W_t = Cp(T_{03} - T_{04}) = CpT_{03}\left[1 - \frac{1}{\pi_c^{(\gamma-1)/\gamma}}\right] \qquad (8.3)$$

The net specific work of the cycle that is delivered to the load is

$$W_{net} = W_t - W_c = Cp\,(T_{03} - T_{04}) - Cp\,(T_{02} - T_{01}) \qquad (8.4)$$

The thermal efficiency of the ideal Brayton cycle under the cold-air-standard assumptions becomes

$$\eta_{th,\,Brayton} = \frac{W_{net}}{Q_{in}} = 1 - \frac{Cp\,(T_{04} - T_{01})}{Cp\,(T_{03} - T_{02})} = 1 - \frac{T_{01}\,(T_{04}/T_{01} - 1)}{T_{02}\,(T_{03}/T_{02} - 1)}$$

Since processes 1–2 and 3–4 are isentropic, and $P_{02} = P_{03}$ and $P_{04} = P_{01}$, thus,

$$\frac{T_{02}}{T_{01}} = \left(\frac{P_{02}}{P_{01}}\right)^{(\gamma-1)/\gamma} = \left(\frac{P_{03}}{P_{04}}\right)^{(\gamma-1)/\gamma} = \frac{T_{03}}{T_{04}} \quad \text{or} \quad \frac{T_{04}}{T_{01}} = \frac{T_{03}}{T_{02}}$$

Substituting these equations into the thermal efficiency relation and simplifying gives

$$\eta_{th,\,Brayton} = 1 - \frac{1}{\pi_c^{(\gamma-1)/\gamma}} \qquad (8.5)$$

Thus, the thermal efficiency depends only on the compressor pressure ratio and the specific heat ratio. As will be described below, the thermal efficiency of real cycle depends on other parameters as well.

The specific power ($\mathscr{P}$) is the net power per unit mass flow rate (W · s/kg), which can be rearranged from Equation 8.4 as follows:

$$\mathscr{P} = W_t - W_c = Cp(T_{03} - T_{04}) - Cp(T_{02} - T_{01})$$

$$\mathscr{P} = W_t - W_c = CpT_{03}\left(1 - \frac{T_{04}}{T_{03}}\right) - CpT_{01}\left(\frac{T_{02}}{T_{01}} - 1\right)$$

$$\mathscr{P} = CpT_{03}(1 - x^{-1}) - CpT_{01}(x - 1)$$

where $x = \pi_c^{(\gamma-1)/\gamma}$.

From the above equation, the specific power depends on the ambient temperature, compressor pressure ratio, and maximum temperature of the turbine. The specific work attains a maximum value when its derivative with respect to the compressor pressure ratio is equal to zero.

$$\frac{d\mathscr{P}}{dx} = 0$$

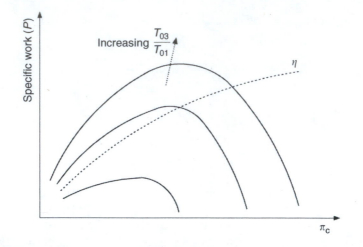

FIGURE 8.7   Variation of specific power and thermal efficiency with the compressor pressure ratio.

$$\frac{CpT_{03}}{x^2} - CpT_{01} = 0$$

$$\frac{T_{03}}{T_{01}} = x^2 = \pi_c^{2(\gamma-1)/\gamma} \tag{8.6a}$$

or

$$\pi_c = \left(\frac{T_{03}}{T_{01}}\right)^{\gamma/2(\gamma-1)} \tag{8.6b}$$

A plot for the above relation is given in Figure 8.7.

8.4.1.2 Real Cycle

The real case is associated with losses due to friction. The changes in specific heat at constant pressure are considered. A schematic diagram for the engine and its processes on the temperature–entropy diagram is shown in Figure 8.8. The different modules will be again be analyzed as follows.

Compressor

The inlet conditions are also the ambient conditions, thus the outlet pressure and temperature are given by the relations:

$$P_{02} = (P_{01})(\pi_c) = P_a \times \pi_c \tag{8.7}$$

$$T_{02} = T_a \left[1 + \frac{\pi_c^{(\gamma_c-1)/\gamma_c} - 1}{\eta_c} \right] \tag{8.8}$$

where η_c is the isentropic efficiency of the compressor and γ_c is the specific heat ratio of ambient air.

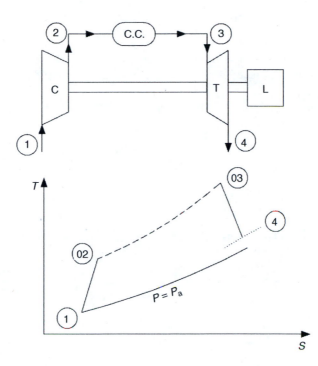

FIGURE 8.8   Real Brayton cycle.

The specific work of the compressor is

$$W_c = Cp_c(T_{02} - T_a)$$

where Cp_c is the air specific heat.

Combustion chamber
The combustion process is associated with a pressure drop; thus the outlet pressure is given by
the relation:

$$P_{03} = P_{02} - \Delta P_{cc} \qquad (8.9a)$$

Some gas turbine manufacturers employ a fraction pressure loss (fpl), (about 0.05% or 5%) to
account for combustor or burner losses; thus fpl $= (P_{02} - P_{03})/P_{02}$.
    Thus

$$P_{03} = P_{02}(1 - \text{fpl}) \qquad (8.9b)$$

The maximum temperature is known (T_{03}); thus the mass of burned fuel is calculated from the
energy balance in the combustion chamber. The resulting fuel-to-air ratio is given by the relation:

$$f = \frac{Cp_h T_{03} - Cp_c T_{02}}{\eta_b Q_R - Cp_h T_{03}} \qquad (8.10)$$

where η_b is the burner efficiency, Q_R is the fuel heating value, and Cp_h is the hot gases specific heat.

Turbine

The hot gases expand in the turbine to a pressure very near to the ambient pressure. The exhaust gases must leave with some velocity so that it can get clear of the engine before its kinetic energy is dissipated in the atmosphere. Therefore, a typical pressure ratio [6] would be

$$\frac{P_{04}}{P_a} = 1.1 - 1.2 \tag{8.11a}$$

It was suggested in [1] that the outlet pressure will be very close to the ambient pressure or

$$P_{04} - P_a = 0.01 - 0.04 \text{ bar} \tag{8.11b}$$

Thus, the turbine outlet pressure is known. The temperature drop in the turbine is

$$T_{03} - T_{04} = \eta_t T_{03} \left[1 - \frac{1}{(P_{03}/P_{04})^{(\gamma_h - 1)/\gamma_h}} \right] \tag{8.12}$$

where η_t is the isentropic efficiency of the turbine.

The specific turbine work is given by the relation:

$$W_t = Cp_h(T_{03} - T_{04}) \tag{8.13}$$

The specific power to the load is given by the relation:

$$P = \left[(1 + f)\eta_{mt}W_t - \frac{1}{\eta_{mc}}W_c \right] \tag{8.14}$$

where η_{mc} and η_{mt} are the mechanical efficiencies of the compressor and turbine, respectively. The units of the specific power in Equation 7.13 are kJ/kg or kW · s/kg.

The specific fuel consumption is defined as

$$\text{SFC} = \frac{f}{P} \tag{8.15}$$

The unit of the specific fuel consumption in the above equation is as usual kg/(kW · s).

The thermal efficiency is expressed as

$$\eta_{th} = \frac{P}{fQ_R} = \frac{1}{\text{SFC} \times Q_R} \tag{8.16}$$

If the following assumptions are considered:

1. All processes are steady.
2. The mechanical efficiencies of the compressor and turbine are assumed unity.
3. No pressure drop in the combustion chamber.
4. Constant properties (specific heat and specific heat ratio Cp,γ).
5. The working fluid is assumed to be air.

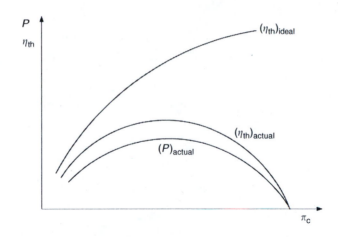

FIGURE 8.9   Real versus actual specific power and thermal efficiency.

6. The amount of fuel added is also neglected in the energy balance relations; then the thermal efficiency will be expressed by the relation:

$$\eta_{th} = \left(\frac{\alpha \eta_c \eta_t - \beta}{\alpha \eta_c - \beta + (1 - \eta_c)} \right) \left(1 - \frac{1}{\beta} \right) \tag{8.17}$$

where $\alpha = T_{03}/T_{01}$ and $\beta = \pi_c^{(\gamma-1)/\gamma}$.

The specific power is expressed as

$$P = C_p T_{01} (\beta - 1) \left(\frac{\eta_t \alpha}{\beta} - \frac{1}{\eta_c} \right) \tag{8.18}$$

Equation 8.17 will be reduced to the form in Equation 8.5 if the compressor and turbine efficiency are set equal to unity. A plot of Equations 8.17 and 8.18 versus the compressor pressure ratio is shown in Figure 8.9. A comparison between Figures 8.7 and 8.9 outlines that

1. The thermal efficiency is dependent on the turbine inlet temperature and thus the turbine material limits the maximum thermal efficiency.
2. The loss due to compressor efficiency is partially compensated for by the reduction in heat added due to fuel burning in the combustion chamber. Thus the compressor efficiency appears both in the numerator and denominator.
3. The loss in turbine is a loss from the system; thus the turbine efficiency appears only in the numerator.

8.4.2   REGENERATION

In a gas turbine engine, the temperature of the exhaust gas leaving the last turbine (if there is more than one turbine) is often considerably higher than the temperature of the air leaving the last compressor (also if there is more than one compressor). Therefore, the high-pressure air leaving the compressor can be heated by transferring heat to it from the hot exhaust gases in counterflow heat exchanger, which is also known as a regenerator or a recuperator. A sketch of the single-shaft gas turbine engine utilizing a regenerator together with its T–S diagram of the new cycle is shown in Figure 8.10. The regenerator is normally referred to by the letter "X"; thus the engine is referred to as "CXBT."

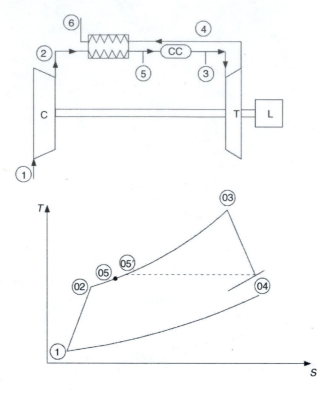

FIGURE 8.10 Single spool with regeneration.

The highest temperature occurring within the regenerator is (T_{04}), the temperature at which the exhaust gases leave the turbine and enter the regenerator. Under no conditions can the air be preheated in the regenerator to a temperature above this value [7]. Air normally leaves the regenerator at a lower temperature (T_{05}). In the limiting (ideal) case, the air will exit the generator at the inlet temperature of the exhaust gases (T_{04}). The actual heat transfers from the exhaust gases to the air can be expressed as

$$q_{regen,\,act} = h_{05} - h_{02}$$

Assuming the regenerator to be well insulated and any changes in kinetic and potential energies to be negligible, the maximum heat transfers from the exhaust gases to the air can be expressed as

$$q_{regen,\,max} = h_{05'} - h_{02} = h_{04} - h_{02}$$

The extent to which a regenerator approaches an ideal regenerator is called the effectiveness ε and is defined as

$$\varepsilon = \frac{q_{regen,\,act}}{q_{regen,\,max}} = \frac{h_{05} - h_{02}}{h_{04} - h_{02}}$$

When the cold-air-standard assumptions are utilized, it reduces to

$$\varepsilon = \frac{T_{05} - T_{02}}{T_{04} - T_{02}} \tag{8.19}$$

A regenerator with a higher effectiveness will obviously save a greater amount of fuel since it will preheat the air to a higher temperature before combustion. However, achieving a higher effectiveness requires the use of a larger regenerator of higher price and causes a larger pressure drop. The effectiveness of most regenerators used in practice is below 0.85.

Thus, the new value for the fuel-to-air ratio is now:

$$f = \frac{Cp_h T_{03} - Cp_c T_{05}}{\eta_b Q_R - Cp_h T_{03}} \tag{8.20}$$

Concerning the thermal efficiency of an ideal Brayton cycle with regeneration under the cold-air-standard assumptions, its value is

$$\eta_{th,\,regen} = 1 - \left(\frac{T_{01}}{T_{03}}\right)(\pi_c)^{(\gamma-1)/\gamma} \tag{8.21}$$

Therefore, the thermal efficiency of an ideal Brayton cycle with regeneration depends on the ratio of the minimum to maximum temperatures as well as the compressor pressure ratio.

Concerning the actual cycle, the specific work in the compressor and the turbine are the same, as in Section 8.3.1.2. The actual cycle efficiency will then be obtained from Equation 8.16 by substituting the new value of the fuel-to-air ratio from Equation 8.15.

The main advantage of regeneration is its reduction in fuel consumption and consequently, the reduction in the fuel-to-air ratio. This leads to an increase in the thermal efficiency as depicted from Equation 8.16.

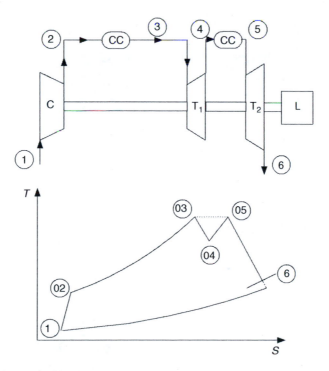

FIGURE 8.11 Single spool with reheat.

8.4.3 REHEAT

In this case reheat is proposed at some point in the expansion process for a single-shaft configuration. Thus, a second combustion chamber is installed between the two turbines. The engine layout together with its temperature–entropy diagram is shown in Figure 8.11. Reheat is referred to by the letter "R"; thus the engine is identified by "CBTRT." Again an analysis of the cycle is considered here.

The compressor and the first combustion chamber (downstream of the compressor) are treated in the same way as above. But, since there are two combustion chambers, then the fuel burned in this combustion chamber is denoted by $\dot{m}_{f1}$ and the corresponding fuel-to-air ratio is also denoted by f_1 where

$$f_1 = \frac{Cp_h T_{03} - Cp_c T_{02}}{\eta_{b1} Q_R - Cp_h T_{03}} \tag{8.22}$$

The subsequent elements are analyzed hereafter.

High-pressure turbine (HPT) or turbine (1)
The inlet total temperature of the turbine is defined from metallurgical capabilities of the turbine material. Moreover, its total pressure is known from the pressure drop in the first combustion chamber. However, a problem arises; namely, how to determine the outlet conditions of this first turbine. Theoretically, the optimum pressure ratio for each turbine is equal and its value would be equal to the square root of the overall pressure ratio of both turbines. Now, since the outlet pressure of the second turbine is slightly higher than the ambient pressure (say 0.01–0.04 bar greater than the ambient pressure), then the outlet pressure of the second turbine (P_{06}) is known. It can be shown also that to maximize the total turbine work output, the pressure ratio of both turbines must be equal. Thus, for optimum operation we have

$$\frac{P_{03}}{P_{04}} = \frac{P_{05}}{P_{06}} = \sqrt{\frac{P_{03}}{P_{06}}} \tag{8.23}$$

The temperature drop in the first (high pressure) turbine is

$$T_{03} - T_{04} = \eta_{t1} T_{03} \left[1 - \frac{1}{(P_{03}/P_{04})^{(\gamma_h - 1)/\gamma_h}} \right] \tag{8.24}$$

The specific work of the first turbine is

$$W_{t1} = Cp_h (T_{03} - T_{04}) \tag{8.25}$$

Second combustion chamber (reheat)
Here for simplicity, the pressure drop in this combustion chamber is neglected. The corresponding fuel-to-air ratio is

$$f_2 = \frac{Cp_h (T_{05} - T_{04})}{\eta_{b2} Q_R - Cp_h T_{05}} \tag{8.26}$$

The above equation is written in a general form where the second burner has a new efficiency (η_{b2}). The maximum temperature in both combustion chambers is assumed equal;

$$T_{05} = T_{03} \tag{8.27}$$

Low-pressure turbine or turbine (2)
The temperature drop in the turbine is

$$T_{05} - T_{06} = \eta_{t2} T_{05} \left[1 - \frac{1}{(P_{05}/P_{06})^{(\gamma_h - 1)/\gamma_h}} \right] \tag{8.28}$$

where η_{t2} is the isentropic efficiency of the second turbine. The specific work of the second turbine is

$$W_{t2} = Cp_h (T_{05} - T_{06}) \tag{8.29}$$

The specific power to the load is given by the relation:

$$P = \left[(1 + f_1) \eta_{mt1} W_{t1} + (1 + f_1 + f_2) \eta_{mt2} W_{t2} - \frac{1}{\eta_{mc}} W_c \right] \tag{8.30}$$

The specific fuel consumption is defined as

$$\text{SFC} = \frac{f_1 + f_2}{P} \text{ kg/ (kW} \cdot \text{s)} \tag{8.31}$$

The thermal efficiency is expressed as

$$\eta_{th} = \frac{P}{(f_1 + f_2) Q_R} = \frac{1}{\text{SFC} \times Q_R} \tag{8.32}$$

The main advantage here is that reheat increases in the power output. This is very similar to reheat in aircraft engines, which also increases the thrust force. However, there is an increase in turn in the fuel consumption and the fuel-to-air ratio. The outcome of both points is a slight reduction in the thermal efficiency of the engine.

8.4.4 INTERCOOLING

The third module which may be added is the intercooler. The single-shaft gas turbine is now composed of two compressors (low pressure and high pressure) with an intercooler in between, the second compressor is followed by a combustion chamber, a turbine driving the two compressors, and the load (Figure 8.12). The cycle is also illustrated on the T–S diagram in the same figure. Intercooler is referred to by the letter "I" and this engine is then referred to as "CICBT." Air is cooled in station (2) before proceeding to the high-pressure compressor. The minimum possible temperature is again the ambient temperature. The effectiveness of the intercooler can thus be defined as

$$\varepsilon_{IC} = \frac{T_{02} - T_{03}}{T_{02} - T_{01}}$$

This lower temperature means that the high-pressure compressor will be compact or in other words will be smaller than its normal size in nonintercooled engine. Another advantage for the intercooled gas turbine is that the high-pressure compressor power requirement is reduced. Thus, higher power will be available for driving the load. However, the reduced high-pressure compressor delivery temperature leads to substantial more fuel to attain the same maximum temperature of a normal gas turbine engine. The analysis of the cycle will be briefly discussed.

The low-pressure compressor is typical of the basic case described in Section 8.3.1.2 with appropriate pressure ratio and isentropic efficiency (π_{c1}, η_{c1}).

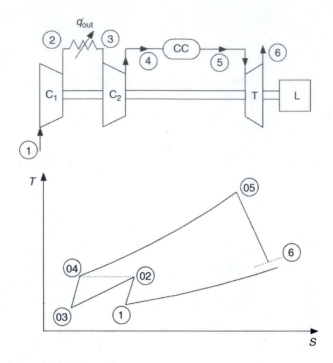

FIGURE 8.12 Single spool with intercooler.

High-pressure compressor
The high-pressure compressor has a pressure ratio and isentropic efficiency (π_{c2}, η_{c2}). The output conditions are

$$P_{04} = (P_{03})(\pi_{c2}) \tag{8.33}$$

$$T_{04} = T_{03}\left[1 + \frac{\pi_{c2}^{(\gamma_c-1)/\gamma_c} - 1}{\eta_{c2}}\right] \tag{8.34}$$

Combustion chamber
More fuel will be needed to reach the maximum temperature of the basic cycle engine. The fuel-to-air ratio is then

$$f = \frac{Cp_h T_{05} - Cp_c T_{04}}{\eta_b Q_R - Cp_h T_{05}}$$

Turbine
The outlet pressure (P_{06}) is known through either a ratio to the ambient pressure or a specified difference between its value and the ambient pressure. Then the outlet temperature is obtained from the relation:

$$T_{06} = T_{05}\left[1 - \eta_t\left\{1 - \left(\frac{P_{06}}{P_{05}}\right)^{(\gamma_h-1)/\gamma_h}\right\}\right]$$

Load

The specific power available for driving the load is

$$P = \eta_{mt}(1+f)Cp_h(T_{05} - T_{06}) - Cp_c \left(\frac{T_{02} - T_a}{\eta_{mc1}} + \frac{T_{04} - T_{03}}{\eta_{mc2}} \right) \qquad (8.35)$$

where η_{mc1}, η_{mc2}, and η_{mt} are the mechanical efficiencies of the low-pressure and high-pressure compressors and the turbine, respectively.

8.4.5 COMBINED INTERCOOLING, REGENERATION, AND REHEAT

Figure 8.13 illustrates this combined case together with its temperature–entropy diagram. The engine is then symbolically defined by "**CICXBTRT**."

Example 1 A preliminary design analysis is to be made for a stationary gas turbine power plant. The proposed power plant is to incorporate one stage of intercooling and a regenerator "**CICXBT**" having an effectiveness of $\varepsilon_R = 0.75$ as shown in the figure. The inlet temperature to the turbine at the maximum load condition is limited to 650°C. Derive a general relation for the intermediate pressure (P_{02}) as a function of the inlet and outlet pressures of the compression process $P_{02} = f(P_{01}, P_{04})$ that gives the maximum thermal efficiency. Prove that the specific work of the two compressors of the intercooler is minimal if the intermediate pressure of the intercooler is related to the inlet and outlet pressure of the intercooler set by the relation

$$P_{02}^2 = \left(\frac{\eta_{c1} + \varepsilon - 1}{\eta_{c2} + \varepsilon - 1} \right)^{\gamma/(\gamma-1)} P_{01} P_{04}$$

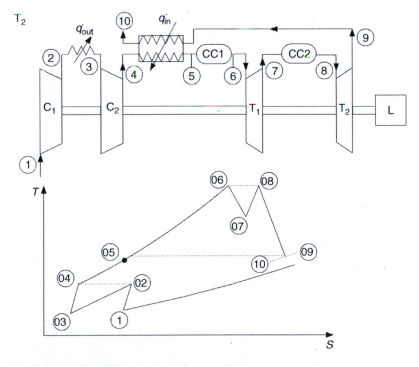

FIGURE 8.13 Single spool with intercooler, reheat, and regenerator.

Next, if the isentropic efficiency of both compressors is equal, then the minimum specific work attains a maximum value when the pressure ratio of both compressors is equal; thus

$$\frac{P_{02}}{P_{01}} = \frac{P_{04}}{P_{02}} = \sqrt{\frac{P_{04}}{P_{01}}}$$

Determine this pressure ratio and also the corresponding thermal efficiency if the component efficiencies are as follows:

$$\eta_{c1} = 0.83, \quad \eta_{c2} = 0.83, \quad \eta_t = 0.85, \quad \eta_b = 0.98, \quad \eta_{mc1} = \eta_{mc2} = \eta_{mt} = 0.98, \quad \varepsilon_{IC} = 0.9$$

Air inlet temperature to the first compressor is 20°C and maximum temperature in the cycle is 650°C. For overall pressure ratios of 2, 3, 4, 6, and 8, calculate and plot the specific power and thermal efficiency for optimum pressure ratios in the compression system ($P_{02}/P_{01} = P_{04}/P_{02} = \sqrt{P_{04}/P_{01}}$) and determine the value of the overall pressure ratio corresponding to the maximum specific power and maximum thermal efficiency.

Solution: Figure 8.14 illustrates a layout of the gas turbine together with its temperature–entropy diagram.

The efficiencies of different modules are

$$\eta_{c_1} = \eta_{c_2} = 0.83, \quad \eta_t = 0.85, \quad \eta_{mc_1} = \eta_{mc_2} = \eta_t = 0.98, \quad \eta_B = 0.98$$

The effectiveness of the intercooler and regenerators are $\varepsilon_{IC} = 0.9, \varepsilon_R = 0.75$.

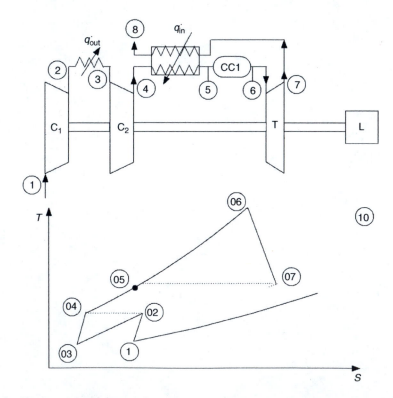

FIGURE 8.14   Single spool with intercooler and regenerator.

The operating conditions are

$$T_{06} = 650 + 273 = 823 \text{ K}, \quad T_{01} = 20 + 273 = 293 \text{ K}, \quad P_{01} = 1.0 \text{ bar}$$

For general procedure we first assume that η_{c_1} and η_{c_2} have any different values

$$\varepsilon_{IC} = \frac{T_{02} - T_{03}}{T_{02} - T_{01}}$$

$$\therefore T_{03} = T_{02} - \varepsilon_{IC} (T_{02} - T_{01}) = T_{02} (1 - \varepsilon) + \varepsilon T_{01}$$

But the outlet temperature of the first compressor (C_1) is

$$T_{02} = T_{01} \left[1 + \frac{1}{\eta_{c_1}} \left\{ \left(\frac{P_{02}}{P_{01}} \right)^{(\gamma-1)/\gamma} - 1 \right\} \right]$$

$$T_{03} = T_{01} \left[\varepsilon + (1 - \varepsilon) \left\{ 1 + \frac{1}{\eta_{c_1}} \left(\frac{P_{02}}{P_{01}} \right)^{(\gamma-1)/\gamma} - \frac{1}{\eta_{c_1}} \right\} \right]$$

$$T_{03} = T_{01} \left[1 - \left(\frac{1-\varepsilon}{\eta_{c_1}} \right) + \left(\frac{1-\varepsilon}{\eta_{c_1}} \right) \left(\frac{P_{02}}{P_{01}} \right)^{(\gamma-1)/\gamma} \right]$$

Define $B = (1 - \varepsilon)/\eta_{c_1}$ and $\lambda = 1 - B$

$$\therefore T_{03} = T_{01} \left[\lambda + B \left(P_{02}/P_{01} \right)^{(\gamma-1)/\gamma} \right]$$

The specific work of compressors (C_1) and (C_2) are

$$W_{c1} = \frac{C_P}{\eta_{m_1} \eta_{c_1}} T_{01} \left\{ \left(\frac{P_{02}}{P_{01}} \right)^{(\gamma-1)/\gamma} - 1 \right\}$$

$$W_{c2} = \frac{C_P}{\eta_{m2} \eta_{c_2}} T_{03} \left\{ \left(\frac{P_{04}}{P_{02}} \right)^{(\gamma-1)/\gamma} - 1 \right\}$$

For $\eta_{m2} = \eta_{m1} = \eta_m$, and the total specific work of the compressor W_{CT} defined as $W_{CT} = W_{C1} + W_{C2}$, then

$$W_{CT} = \frac{C_P T_{01}}{\eta_m \eta_{c1}} \left[\left(\frac{P_{02}}{P_{01}} \right)^{(\gamma-1)/\gamma} - 1 \right] + \frac{C_P T_{01}}{\eta_m \eta_{c2}} \left[\left\{ \lambda + B \left(\frac{P_{02}}{P_{01}} \right)^{(\gamma-1)/\gamma} \right\} \left\{ \left(\frac{P_{04}}{P_{02}} \right)^{(\gamma-1)/\gamma} - 1 \right\} \right]$$

$$W_{CT} = \frac{C_P T_{01}}{\eta_m} \left[\frac{1}{\eta_{c1}} \left(\frac{P_{02}}{P_{01}} \right)^{(\gamma-1)/\gamma} - \frac{1}{\eta_{c1}} + \frac{\lambda}{\eta_{c2}} \left(\frac{P_{04}}{P_{02}} \right)^{(\gamma-1)/\gamma} - \frac{\lambda}{\eta_{c2}} + \frac{B}{\eta_{c2}} \left(\frac{P_{04}}{P_{01}} \right)^{(\gamma-1)/\gamma} \right.$$

$$\left. - \frac{B}{\eta_{c2}} \left(\frac{P_{02}}{P_{01}} \right)^{(\gamma-1)/\gamma} \right]$$

$$W_{CT} = A \left[\left(\frac{1}{\eta_{c1}} - \frac{B}{\eta_{c2}} \right) \left(\frac{P_{02}}{P_{01}} \right)^{(\gamma-1)/\gamma} + \frac{\lambda}{\eta_{c2}} \left(\frac{P_{04}}{P_{02}} \right)^{(\gamma-1)/\gamma} + \frac{B}{\eta_{c2}} \left(\frac{P_{04}}{P_{01}} \right)^{(\gamma-1)/\gamma} - \frac{\lambda}{\eta_{c2}} - \frac{1}{\eta_{c1}} \right]$$

For minimum work, differentiate with respect to the intermediate pressure (P_{02}) and equate the differential to zero or $\partial W_{CT}/\partial P_{02} = 0$

$$\frac{\partial W_{CT}}{\partial P_{02}} = A\left[\left(\frac{1}{\eta_{c1}} - \frac{1-\varepsilon}{\eta_{c1}\eta_{c2}}\right)\left(\frac{\gamma-1}{\gamma}\right)\frac{(P_{02})^{-1/\gamma}}{(P_{01})^{(\gamma-1)/\gamma}} - \frac{\lambda}{\eta_{c2}}\left(\frac{\gamma-1}{\gamma}\right)(P_{04})^{(\gamma-1)/\gamma}(P_{02})^{-2+1/\gamma}\right] = 0$$

$$\therefore \left(\frac{1}{\eta_{c1}} - \frac{1-\varepsilon}{\eta_{c1}\eta_{c2}}\right)\frac{P_{02}^{-1/\gamma}}{(P_{01})^{(\gamma-1)/\gamma}} = \frac{1}{\eta_{c2}}\left(1 - \frac{1-\varepsilon}{\eta_{c1}}\right)(P_{04})^{(\gamma-1)/\gamma}(P_{02})^{-2+1/\gamma} = 0$$

$$\left(\frac{\eta_{c2}-1+\varepsilon}{\eta_{c1}\eta_{c2}}\right)(P_{02})^{2(\gamma-1)/\gamma} = \left(\frac{\eta_{c1}-1+\varepsilon}{\eta_{c1}\eta_{c2}}\right)(P_{01}P_{04})^{(\gamma-1)/\gamma}$$

Then the general relation requested is

$$P_{02}^2 = \left(\frac{\eta_{c1}+\varepsilon-1}{\eta_{c2}+\varepsilon-1}\right)^{\gamma/(\gamma-1)} P_{01}P_{04}$$

Then if the efficiency of both compressors are equal, $\eta_{c1} = \eta_{c2} = \eta_c$. The above equation is reduced to the well-known formula

$$P_{02}^2 = P_{01}P_{04}$$

Thus, the requested relation is proved as

$$\frac{P_{02}}{P_{01}} = \frac{P_{04}}{P_{02}} = \sqrt{\frac{P_{04}}{P_{01}}}$$

Now, returning to the numerical values in this example:

$$T_{02} = T_{01}\left[1 - \frac{1}{\eta_c} + \frac{1}{\eta_c}\left(\frac{P_{02}}{P_{01}}\right)^{(\gamma-1)/\gamma}\right]$$

Then for $\gamma = 1.4$, $T_{01} = 293°K$, and $\eta_c = 0.83$

$$T_{02} = 293\left[1.202\left(P_{02}/P_{01}\right)^{0.286} - 0.202\right] \tag{1}$$

$$T_{03} = T_{01}\left[1 - \left(\frac{1-\varepsilon}{\eta_c}\right) + \left(\frac{1-\varepsilon}{\eta_c}\right)\left(\frac{P_{02}}{P_{01}}\right)^{(\gamma-1)/\gamma}\right]$$

$$T_{03} = 293\left[0.88 + 0.12\left(P_{02}/P_{01}\right)^{0.286}\right] \tag{2}$$

$$T_{04} = T_{03}\left[1 - \frac{1}{\eta_c} + \frac{1}{\eta_c}\left(\frac{P_{04}}{P_{02}}\right)^{(\gamma-1)/\gamma}\right]$$

$$T_{04} = T_{03}\left[1.202\left(\frac{P_{04}}{P_{02}}\right)^{0.286} - 0.202\right] \tag{3}$$

The total compression work of the two compressors (W_{CT}) is reformulated as

$$W_{CT} = \frac{C_p T_{01}}{\eta_m \eta_c} \left[(1-B) \left\{ \left(\frac{P_{02}}{P_{01}}\right)^{(\gamma-1)/\gamma} + \left(\frac{P_{04}}{P_{02}}\right)^{(\gamma-1)/\gamma} \right\} + B \left(\frac{P_{04}}{P_{01}}\right)^{(\gamma-1)/\gamma} - (1+\lambda) \right]$$

Since $P_{02}^2 = P_{01} P_{04}$, then $\dfrac{P_{02}}{P_{01}} = \dfrac{P_{04}}{P_{02}} = \sqrt{\dfrac{P_{04}}{P_{01}}}$.

Then $W_{CT} = \dfrac{293}{(0.83)(0.98)} \left[(0.88)(2) \left(\dfrac{P_{02}}{P_{01}}\right)^{0.286} + 0.12 \left(\dfrac{P_{04}}{P_{01}}\right)^{0.286} - 1.88 \right]$ kJ/kg

$$W_{CT} = 360 \left[1.76 \left(\frac{P_{02}}{P_{01}}\right)^{0.286} + 0.12 \left(\frac{P_{04}}{P_{01}}\right)^{0.286} - 1.88 \right] \tag{4}$$

The specific power generated in the turbine is (W_t), which is expressed as

$$W_T = \eta_m \eta_t T_{06} \left[1 - \left(\frac{P_{01}}{P_{04}}\right)^{0.286} \right] = 770 \left[1 - \left(\frac{P_{04}}{P_{01}}\right)^{-0.286} \right] \text{ kJ/kg} \tag{5}$$

The heat added in the combustion chamber per unit air mass flow rate is

$$q = \frac{1}{\eta_B} (h_{06} - h_{05}) = \frac{C_p}{\eta_B} (T_{06} - T_{05})$$

$$q = 1.02 \left[(T_{06} - T_{04}) - \varepsilon_R (T_7 - T_{04}) \right] \text{ kJ/kg} \tag{6}$$

$$T_7 = 923 \left[0.15 + 0.85 \left(\frac{P_{04}}{P_{01}}\right)^{-0.286} \right]$$

P_{04}/P_{01}	$P_{02}/P_{01} = P_{04}/P_{02}$ $= \sqrt{P_{04}/P_{01}}$	T_{03} (K)	T_{04} (K)	W_{CT} (kJ/kg)	T_7 (K)
2	1.414	296	330	75.2	785
3	1.732	299	360	124.1	714
4	2	300	376	158	668
6	2.45	304	410	212	608
8	2.729	306	444	254	572

P_{04}/P_{01}	$T_7 - T_{04}$ (K)	$T_6 - T_{04}$ (K)	q (kJ/kg)	W_T (kJ/kg)	$W_{net} = W_T - W_{CT}$ (kJ/kg)	η_{th} (%)
2	455	593	257	138.2	63	24.8
3	356	565	300	207	83	27.7
4	292	547	326	251.8	93	28.6
6	199	514	371	308	96	25.9
8	127	478	390	344	90	23.3

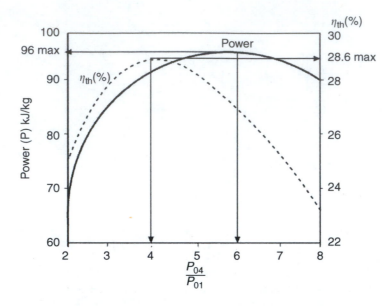

FIGURE 8.15   Plot of specific work and thermal efficiency versus pressure ratio.

The thermal efficiency of the power plant is

$$\eta_{th} = \frac{W_T - W_{CT}}{q}$$

The results in the above table are displayed in Figure 8.15 which shows a plot of specific work and thermal efficiency versus pressure ratio.

The maximum thermal efficiency is obtained when $P_{04}/P_{01} = 4.0$ and the maximum thermal efficiency is $\eta_{th} = 0.286$. The maximum power is 96 kJ/kg at a compression ratio of 6.0.

8.5   DOUBLE-SHAFT ENGINE

The double-shaft engines—as described in Section 8.1—have two main configurations; the split shafts and the free power turbine.

8.5.1   FREE POWER TURBINE

Figure 8.16 illustrates the case of a two spool having a free turbine. The compressor is directly connected and driven by the turbine (high-pressure turbine, HPT), while the load is coupled and driven by the free power turbine, thus denoted by "**CBTT**." Only a single combustion chamber is found, which is located as part of the gas generator.

The different processes are described as follows.

Compressor
The same treatment described in Section 8.3.1 is followed here for real case. Thus, the outlet pressure and temperature are given by Equations 8.7 and 8.8.

Combustion chamber
Also, the same methodology is followed to obtain the outlet pressure and the fuel-to-air ratio. The maximum temperature is known in advance. The fuel-to-air ratio is also calculated as above.

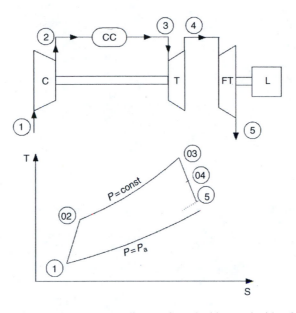

FIGURE 8.16 Layout and temperature–entropy diagram for a double spool with a free power turbine.

High-pressure turbine (the compressor turbine)

The power generated in the turbine is just sufficient to drive the compressor. Thus, the energy balance for the compressor turbine coupling is

$$(1 + f)\eta_{mt}W_t = \frac{1}{\eta_{mc}}W_c$$

$$T_{04} = T_{03} - \frac{1}{\eta_{mc}\eta_{mt}(1 + f)}\frac{Cp_c}{Cp_h}(T_{02} - T_a)$$

The turbine pressure ratio is calculated from the relation:

$$\frac{P_{04}}{P_{03}} = \left(1 - \frac{T_{03} - T_{04}}{\eta_t T_{03}}\right)^{\gamma_h/(\gamma_h - 1)}$$

Free power turbine

Assuming that there is no pressure loss in the ducts between the compressor turbine and the free power turbine, then the inlet pressure to the free power turbine is also (P_{04}). The outlet pressure (P_{05}) is also very close to the ambient pressure. It is defined as either a ratio to the ambient pressure $(P_{05}/P_a \approx 1.1)$ or $P_{05} = P_a + 0.04$ bar.

The exhaust gas temperature (T_{05}) at the outlet to the free power turbine is given by the relation:

$$T_{05} = T_{04}\left[1 - \eta_{pt}\left\{1 - \left(\frac{P_{05}}{P_{04}}\right)^{(\gamma_h - 1)/\gamma_h}\right\}\right] \tag{8.36}$$

The specific power, in kJ/kg (or kW · s/kg), delivered to the load is

$$P = (1 + f)Cp_h(T_{04} - T_{05}) \tag{8.37}$$

The specific fuel consumption is defined as

$$SFC = \frac{f}{P} \text{ kg/(kW} \cdot \text{s)}$$

The thermal efficiency is expressed as

$$\eta_{th} = \frac{P}{fQ_R} = \frac{1}{(SFC)Q_R}$$

8.5.2 Two Discrete Shafts (Spools)

The two-shaft arrangement here is composed of two compressors, two turbines, and a single combustion chamber, thus denoted by "**CCBTT**." It is similar to the two-spool turbofan engine. The low-pressure spool, as demonstrated in Figure 8.17, is composed of compressor ($C1$) and turbine ($T1$). The high-pressure spool is composed of compressor ($C2$) and turbine ($T2$). The load is driven by the low-pressure spool.

Low-pressure spool
The inlet conditions are the ambient pressure and temperature. The outlet conditions are obtained from the relations 8.7 and 8.8.

The specific work of the low-pressure compressor (C_1) is

$$W_{c1} = Cp_c(T_{02} - T_a)$$

High-pressure spool
The outlet conditions are obtained from the relations:

$$P_{03} = (P_{02})(\pi_{c2}) \tag{8.38}$$

$$T_{03} = T_{02}\left[1 + \frac{\pi_{c2}^{(\gamma_c - 1)/\gamma_c} - 1}{\eta_{c2}}\right] \tag{8.39}$$

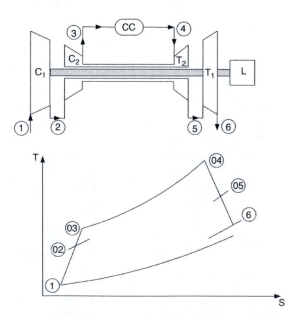

FIGURE 8.17 Layout of a two-spool engine (load coupled to the low-pressure spool) with temperature–entropy diagram.

The specific work of the high-pressure compressor (C_2) is

$$W_{c2} = Cp_c(T_{03} - T_{02}) \tag{8.40}$$

Combustion chamber
The combustion process is associated with a pressure drop, thus the outlet pressure is given by the relation:

$$P_{04} = P_{03} - \Delta P_{cc}$$

The maximum temperature is known (T_{04}), thus the mass of burned fuel is calculated from the energy balance in the combustion chamber. The resulting fuel-to-air ratio is given by the relation:

$$f = \frac{Cp_h T_{04} - Cp_c T_{03}}{\eta_b Q_R - Cp_h T_{04}} \tag{8.41}$$

High-pressure turbine
The HPT (T_2) drives the high-pressure compressor (C_2). The energy balance gives the outlet temperature of the turbine (T_2) as:

$$T_{05} = T_{04} - \frac{1}{\eta_{mc2}\eta_{mt2}(1+f)} \frac{Cp_c}{Cp_h}(T_{03} - T_{02}) \tag{8.42}$$

The turbine pressure ratio is calculated from the relation:

$$\frac{P_{05}}{P_{04}} = \left(1 - \frac{T_{04} - T_{05}}{\eta_{t2} T_{04}}\right)^{\gamma_h/(\gamma_h - 1)} \tag{8.43}$$

Low-pressure turbine (T_1)
The outlet pressure (P_{06}) is known to be slightly higher than the ambient pressure. The outlet temperature (T_{06}) is determined from the relation:

$$T_{06} = T_{05}\left[1 - \eta_{t2}\left\{1 - \left(\frac{P_{06}}{P_{05}}\right)^{(\gamma_h - 1)/\gamma_h}\right\}\right] \tag{8.44}$$

The specific power consumed in the load is

$$P = \eta_{mt1}(1+f)Cp_h(T_{05} - T_{06}) - \frac{1}{\eta_{mc1}}Cp_c(T_{02} - T_a) \tag{8.45}$$

The specific fuel consumption is also given by the

$$\text{SFC} = \frac{f}{P} \text{ kg/(kW} \cdot \text{s)}$$

The thermal efficiency is expressed as

$$\eta_{th} = \frac{P}{fQ_R} = \frac{1}{(\text{SFC})Q_R}$$

Example 2 A two-spool industrial gas turbine in which the load is coupled to a free power turbine, "**CBTT**," operates at the following conditions:

Ambient temperature	288 K
Ambient pressure	101.3 kPa
Compressor pressure ratio	9
Turbine inlet temperature	1380 K

The efficiencies of the different elements are

$$\eta_c = 0.87, \quad \eta_b = 0.99, \quad \eta_t = \eta_{ft} = 0.89$$

The fuel heating value = 43,000 kJ/kg
The delivery pressure of the free turbine = 120 kPa
Calculate:

1. The fuel-to-air ratio
2. The net specific power
3. The thermal efficiency

Solution: The engine examined is illustrated in Figure 8.16. The successive processes are described as follows:

Compressor
The outlet conditions from the compressor are determined from Equations 8.7 and 8.8 as

$$P_{02} = \pi \times P_a = 9 \times 101.3 = 911.7 \text{ kPa}$$

$$T_{02} = 288 \left[1 + \frac{9^{0.286} - 1}{0.87} \right] = 577.53 \text{ K}$$

Combustion chamber
The fuel-to-air ratio

$$f = \frac{C_{Ph} T_{03} - C_{Pc} T_{02}}{\eta_b Q_R - C_{Ph} T_{03}} = \frac{1.148 \times 1380 - 1.005 \times 577.53}{0.99 \times 43,000 - 1.148 \times 1380} = 0.0245$$

Since no pressure drop is assumed in the combustion chamber, then

$$P_{03} = P_{02} = 911.7 \text{ kPa}$$

Gas generator turbine
From the energy balance between the compressor and turbine ($W_t = W_c$), the turbine outlet temperature and pressure are

$$T_{04} = T_{03} - \frac{C_{Pc}(T_{02} - T_{01})}{(1+f)C_{Ph}} = 1380 - \frac{1.005}{1.0245 \times 1.148}(577.53 - 288) = 1132.6 \text{ K}$$

$$P_{04} = P_{03} \left(1 - \frac{T_{03} - T_{04}}{\eta_t T_{03}} \right)^{\gamma/(\gamma-1)} = 911.7 \times \left(1 - \frac{247.4}{0.89 \times 1380} \right)^4 = 370.8 \text{ kPa}$$

Free power turbine
The outlet pressure is 120 kPa, thus

$$T_{05} = T_{04} \left[1 - \eta_{ft} \left\{ 1 - \left(\frac{P_{05}}{P_{04}} \right)^{(\gamma-1)/\gamma} \right\} \right] = 1128.77 \left[1 - 0.89 \left\{ 1 - \left(\frac{120}{370.8} \right)^{0.25} \right\} \right]$$

$$T_{05} = 884.9 \, \text{K}$$

The specific power is equal to the specific power available in the free power turbine as no mechanical looses are considered; thus

$$W_{ft} = (1 + f) C_{P_h} (T_{04} - T_{05}) = 291.35 \, \text{kJ/kg}$$

The corresponding thermal efficiency is calculated as follows:

$$\eta_{th} = \frac{W_{ft}}{f Q_R} = \frac{291.35}{0.0245 \times 43,000} = 0.2765 = 27.65\%$$

Example 3 It is required to improve the thermal efficiency of the gas described in Example 2 by splitting the compressor into two compressors and adding an intercooler in between, that is, a "**CICBTT**" configuration. The delivery pressure of the second compressor is kept at 911.7 kPa or the same as in Example 2. The intercooler pressure is assumed to be the optimum one. Both compressors have the same efficiency (87%). The inlet total temperature of the second compressor is the same as that of the first compressor (288 K).
    Calculate:

1. The new value of the fuel-to-air ratio
2. The net specific power
3. The thermal efficiency

Solution: A layout of the engine is shown in Figure 8.18. Since the intercooler operates at the optimum pressure, then

$$P_{02} = \sqrt{P_{01} P_{04}} = 303.9 \, \text{kPa}$$

$$\text{and} \quad \frac{P_{02}}{P_{01}} = \frac{P_{04}}{P_{03}} = \sqrt{\frac{P_{04}}{P_{01}}} = 3$$

Compressor (1)

$$T_{02} = T_a \left[1 + \frac{\eta_{c_1}^{(\gamma_c-1)/\gamma_c} - 1}{\eta_{c_1}} \right] = 288 \left[1 + \frac{3^{0.286} - 1}{0.87} \right] = 410.2 \, \text{K}$$

Compressor (2)

$$P_{03} = P_{02} = 303.9 \, \text{kPa}$$

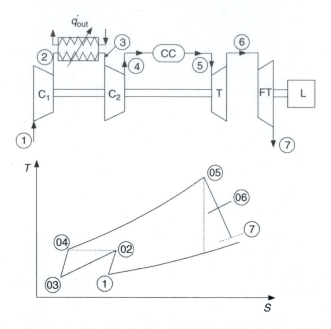

FIGURE 8.18   Layout of a two-spool engine having a free power turbine and an intercooler.

Since $T_{03} = T_a$, and $\eta_{c_1} = \eta_{c_2}$, then

$$T_{04} = T_{02} = 410.2 \text{ K}$$

Combustion chamber
The fuel-to-air ratio is

$$f = \frac{C_{P_h}T_{05} - C_{P_c}T_{04}}{\eta_b Q_R - C_{P_h}T_{05}} = \frac{1.148 \times 1380 - 1.005 \times 410.2}{0.99 \times 43{,}000 - 1.148 \times 1380} = \frac{1172}{40985.8} = 0.0286$$

Gas generator turbine
The turbine here drives both compressors, thus

$$W_t = W_{c_1} + W_{c_2}$$

$$T_{06} = T_{05} - \frac{C_{P_c}[T_{04} - T_{03} + T_{02} - T_{01}]}{(1+f)C_{P_h}} = T_{05} - \frac{2C_{P_c}(T_{02} - T_a)}{(1+f)C_{P_h}}$$

$$T_{06} = 1380 - \frac{2 \times 1.005 \times (410.2 - 288)}{1.0286 \times 1.148} = 1380 - 208 = 1172 \text{ K}$$

$$\frac{P_{06}}{P_{05}} = \left(1 - \frac{T_{05} - T_{06}}{\eta_t T_{05}}\right)^{\gamma/(\gamma-1)} = \left(1 - \frac{1380 - 1172}{0.98 \times 1380}\right)^4 = 0.476$$

$$P_{06} = 911.7 \times 0.476 = 434 \text{ kPa}$$

Free power turbine

$$T_{07} = T_{06}\left[1 - \eta_{\text{ft}}\left\{1 - \left(\frac{P_{07}}{P_{06}}\right)^{(\gamma-1)/\gamma}\right\}\right] = 1276\left[1 - 0.89\left\{1 - \left(\frac{120}{434}\right)^{0.25}\right\}\right] = 963.9 \text{ K}$$

$$W_{\text{ft}} = (1+f)C_{P_h}(T_{06} - T_{07}) = 1.0286 \times 1.148 \times (1276 - 963.9) = 368.6 \text{ kJ/kg}$$

$$\eta_{\text{th}} = \frac{W_{\text{ft}}}{fQ_R} = \frac{368.6}{0.0286 \times 43,000} = 0.2996 = 29.96\% \approx 30\%$$

The increase in fuel consumption is $\dfrac{0.0286 - 0.0245}{0.0245} = 14.3\%$.

While the increase in thermal efficiency is $\dfrac{30 - 26.76}{26.76} = 12.1\%$.

Example 4   A two-shaft industrial gas turbine operates on the regenerative cycle. The efficiency of the regenerator is 0.85. The data in Example 2 is applied here. Determine

1. The thermal efficiency
2. Fuel-to-air ratio
3. The improvement in thermal efficiency and fuel-to-air ratio

Solution: A layout for the regenerative cycle and its temperature–entropy diagram are shown in Figure 8.19; "**CXBTT**" configuration. From Example 2, the following information is known:

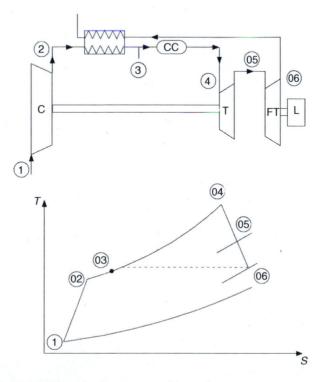

FIGURE 8.19   Two-spool engine with a free power turbine and regenerator.

State	1	2	3	4	5	6
P_0(KPa)	101.3	911.7	911.7	911.7	364.7	120
T_0(K)	288	577.5	?	1380	1128.8	885

The regenerator efficiency is

$$\varepsilon = \frac{T_{03} - T_{02}}{T_{06} - T_{02}}$$

$$\therefore T_{03} = T_{02} + \varepsilon(T_{06} - T_{02}) = 795.5 \text{ K}$$

$$W_{ft} = 291.35 \text{ kJ/kg}$$

$$f = \frac{C_{P_h}T_{04} - C_{P_c}T_{03}}{\eta_b Q_R - C_{P_h}T_{04}} = \frac{1.148 \times 1380 - 1.005 \times 795.5}{0.99 \times 43,000 - 1.148 \times 1380} = 0.091$$

$$\eta_{th} = \frac{W_{ft}}{fQ_R} = \frac{291.35}{0.0191 \times 43,000} = 35.47\%$$

The improvement in fuel consumption is $\dfrac{0.0191 - 0.0245}{0.0245} = -22\%$.

This represents a substational improvement in fuel consumption.

The improvement in thermal efficiency is $\dfrac{0.3547 - 0.2676}{0.2676} = 32.6\%$.

Example 5 A two-shaft gas turbine engine operates with reheat between the gas turbine and the free power turbine, that is, the "**CBTRT**" configuration. Assuming that the data of the engine is same as in Example 2 calculate

1. The net specific power.
2. The fuel-to-air ratio.
3. The thermal efficiency.
4. Compare these values with the basic gas turbine results in Example 2.
5. If the power of the engine is 20 MW, what will be the air mass flow rate?

Solution: A layout and temperature–entropy diagram is shown in Figure 8.20.

Since reheat means adding a new combustor, this means an additional heat is burnt. The temperature is raised from the outlet temperature of the gas generator, 1129 K, to turbine to the same maximum temperature of 1380 K. Thus, the additional fuel-air ratio is

$$f_1 = \frac{C_{P_h}(T_{05} - T_{04})}{\eta_b Q_R - C_{P_h}T_{05}} = \frac{1.148(1380 - 1129)}{0.99 \times 43,000 - 1.148 \times 1380} = 0.007$$

$$\therefore \text{Total fuel-to-air ratio} = f + f_1 = 0.0245 + 0.007 = 0.0315.$$

Free power turbine
Since $P_{05} = P_{04} = 364.68$ kPa

$$T_{06} = T_{05}\left[1 - \eta_{ft}\left\{1 - \left(\frac{P_{06}}{P_{05}}\right)^{(\gamma-1)/\gamma}\right\}\right]$$

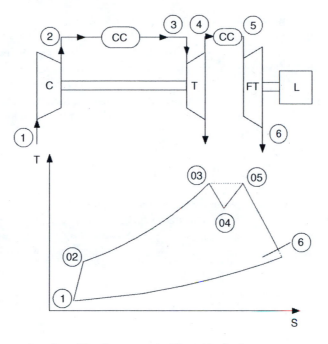

FIGURE 8.20   Two-spool engine with a free power turbine and reheat.

$$T_{06} = 1380 \left[1 - 0.89 \left\{ 1 - \left(\frac{120}{364.68} \right)^{0.25} \right\} \right] = 947.6 \, \text{K}$$

$$W_{ft} = (1 + f + f_1) \, C_{P_h} (T_{05} - T_{06}) = 1.0315 \times 1.148 \times (1380 - 947.6) = 512 \, \text{kJ/kg}$$

$$\eta_{th} = \frac{W_{ft}}{f_{total} \times Q_R} = \frac{512}{0.0315 \times 43,000} = 37.8\%$$

Thus, an improvement in thermal efficiency equal to $\dfrac{37.8 - 26.76}{26.76} = 41.3\%$.

The increase in fuel consumption is $\dfrac{0.0315 - 0.0245}{0.0245} = 28.6\%$.

For a power of 20 MW, then the air mass flow rate $= \dfrac{\text{Power}}{W_{ft}} = \dfrac{20 \times 10^3}{512} = 39 \, \text{kg/s}$.

Summary of the results of the basic and new added modules:

Description	Basic Engine	Regeneration	Intercooling	Reheat
Thermal efficiency%	27.65	35.47	30	37.8
Fuel-to-air ratio	0.0245	0.0191	0.0286	0.0315
Net specific power kJ/kg	291.35	291.35	368.6	512

8.6   THREE SPOOL

The basic layout of three-spool gas turbines is illustrated earlier in Figures 8.4a and 8.4b. In Figure 8.4a three compressors and three turbines are seen. The load is derived by the low-pressure spool. This engine is denoted by "**CCCBTTT.**" In Figure 8.4b two gas turbines drive the corresponding two compressors. The load is driven by the free power turbine. The engine is denoted by "**CCBTTT**." As described in the single- and double-spool configurations, intercooler may be added between the compressors. Thus, one intercooler may be added in the layout of Figure 8.4b or two in the configuration of Figure 8.4a. Moreover, one or two reheat stages may be added between turbines.

If more than one intercooler is added between the compressors, then, for equal inlet temperature of all compressors, the optimum pressure ratio for each compressor [8] that yields a minimum overall compression work will be given by the relation

$$\pi_{c,stage} = \sqrt[Nc]{\pi_{c,overall}} \tag{8.46}$$

where (Nc) is the number of compressors. Thus, for an overall pressure ratio of 27 and when three compressors are employed, the pressure ratio for each compressor is $\sqrt[3]{27} = 3$ (not $27/3 = 9$).

Moreover, if more than one reheat stages are used in between turbines, then, for equal inlet temperature for all turbines, the optimum turbine pressure ratio that yields maximum power is given by the relation

$$\pi_{t,stage} = \sqrt[Nt]{\pi_{t,overall}} \tag{8.47}$$

where (Nt) is the number of turbines.

The corresponding maximum specific power for layout of Figure 8.4a, if the compressors have equal efficiencies and the turbines have also equal efficiencies, will be

$$\frac{P}{\dot{m}Cp} = T_{0\,max}\eta_t(n_{RH}+1)\left(1 - \frac{1}{\pi_t^{(\gamma-1)/\gamma}}\right) - T_a\frac{(n_{IC}+1)}{\eta_C}\left(\pi_c^{(\gamma-1)/\gamma} - 1\right) \tag{8.48}$$

where n_{RH} is the number of reheat stages and n_{IC} is the number of intercoolers. In Equations 8.47 and 8.48 the working fluid is assumed to be air with constant properties throughout the thermodynamic cycle.

The rate of heat added is given by the relation

$$\frac{Q}{\dot{m}Cp} = T_{0\,max}\left\{(n_{RH}+1) - (n_{RH}+\varepsilon_R)\left[1 - \eta_t\left(1 - \frac{1}{\pi_t^{(\gamma-1)/\gamma}}\right)\right]\right\}$$
$$- T_a(1-\varepsilon_R)\left[1 + \frac{\left(\pi_c^{(\gamma-1)/\gamma} - 1\right)}{\eta_C}\right] \tag{8.49}$$

where ε_R is the effectiveness of the regenerator installed between the high-pressure compressor and HPT.

Example 6 A three-shaft industrial gas turbine with a free power turbine is modified by incorporating an intercooler between the two compressors, a regenerator, and a reheater. The intercooler pressure is assumed the optimum one. The inlet temperature to the two compressors is 288 K. The inlet pressure to the first compressor is 101.3 kPa. The overall pressure ratio of both compressors is 25. The inlet temperature to both gas turbines is 1500 K. The exhaust pressure is 110 kPa. The pressure drop in the regenerator and first combustor is 2% and 3% of the delivery pressure of the second compressor. The pressure drop in the second combustor is 4% of the delivery pressure of the second gas turbine. Assume equal pressure ratios of both gas turbines. The efficiencies of the

different modules are

$$\eta_{c_1} = \eta_{c_2} = 88\%, \quad \varepsilon_R = 80\%, \quad \eta_{b_1} = \eta_{b_2} = 99\%, \quad \eta_{t_1} = \eta_{t_2} = \eta_{ft} = 90\%,$$

$$Q_R = 43{,}000 \, \text{kJ/kg}$$

Calculate

1. The net power
2. The thermal efficiency

If the power required is 300 MW, what will be the air and fuel mass flow rate?

Consider variable specific heat based on the relation: $C_P = 950 + 0.21\,T_{01}$, where T_{01} is the total temperature at the start of the process.

Solution: The three-spool layout with the three elements (intercooler, regenerator, and reheater) together with temperature–entropy diagram is shown in Figure 8.21, which is denoted by **"CICXBTRTT."**

Since the intercooler operates at the optimum pressure, then

$$\frac{P_{02}}{P_{01}} = \frac{P_{04}}{P_{03}} = \sqrt{\frac{P_{04}}{P_{01}}} = \sqrt{25} = 5$$

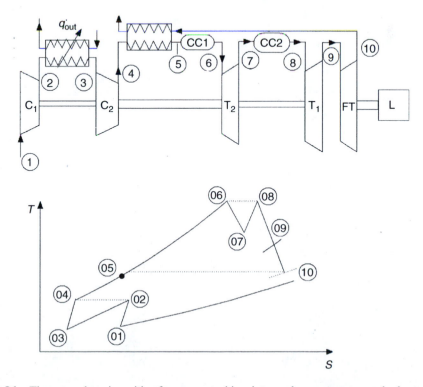

FIGURE 8.21 Three-spool engine with a free power turbine, intercooler, regenerator, and reheater.

Compressor (1)

$$C_{P_1} = 950 + 0.21\,T_{01} = 950 + 0.21 \times 288 = 1010.5\ \text{J/kg/K}$$

$$\therefore \frac{\gamma}{\gamma - 1} = \frac{C_{P_1}}{R} = 3.521 \quad \text{and} \quad \frac{\gamma - 1}{\gamma} = 0.284$$

$$\therefore T_{02} = T_a \left[1 + \frac{5^{0.286} - 1}{0.88} \right] = 477.6\ \text{K}$$

$$P_{02} = 5P_{01} = 506.5\ \text{kPa}$$

Compressor (2)

Since $T_{03} = T_{01}$, then same values for C_P and γ as Compressor (1) are obtained.
    Also $T_{04} = T_{02} = 477.6$ K

$$P_{03} = P_{02} = 506.5 \ \text{kPa} \quad \text{and} \quad P_{04} = 5P_{03} = 2532.5\ \text{kPa}$$

Gas turbine (2)

Turbine (2) is driving Compressor 2 and together these form the high-pressure spool.

$$P_{06} = P_{04}(1 - \Delta P_R - \Delta P_{CC_1}) = 2532.5(1 - 0.02 - 0.03) = 2405.9\ \text{kPa}$$

Moreover,

$$Cp_6 = Cp_8 = 950 + 0.21 \times 1200 = 1265\ \text{J/kg/K}$$

$$\left(\frac{\gamma}{\gamma - 1} \right)_6 = \left(\frac{\gamma}{\gamma - 1} \right)_8 = 4.4$$

Energy balance for both compressors and turbines forming the low- and high-pressure spools gives

$$W_{C_1} + W_{C_2} = W_{t_1} + W_{t_2}$$

$$2Cp_1(T_{02} - T_{01}) = Cp_6(T_{06} - T_{07}) + Cp_8(T_{08} - T_{09})$$

$$= \eta_t Cp_6 T_{06} \left[\left(1 - \frac{1}{\pi_{t_1}^{[(\gamma-1)/\gamma]_6}} \right) + \left(1 - \frac{1}{\pi_{t_2}^{[(\gamma-1)/\gamma]_8}} \right) \right]$$

For a maximum power available from the turbines, both turbines must have equal pressure ratios.

$$\pi_{t_1} = \pi_{t_2} = \pi_t$$

Solving, we get $\pi_t = 1.688$.
    The outlet turbine from the HPT is then

$$P_{07} = \frac{P_{06}}{\pi_t} = 1425.3\ \text{kPa}$$

$$T_{07} = T_{06} \left[1 - \eta_{t_1} \left\{ 1 - \left(\frac{1}{\pi_t} \right)^{[(\gamma-1)/\gamma]_6} \right\} \right] = 1500 \left[1 - 0.9 \left\{ 1 - \left(\frac{1}{1.688} \right)^{0.227} \right\} \right] = 1348\ \text{K}$$

Gas turbine (1)
The inlet pressure is

$$P_{08} = P_{07}(1 - \Delta P_{cc_2}) = 1423.6 \times 0.96 = 1294.5 \text{ kPa}$$

The outlet pressure is $P_{09} = \dfrac{P_{08}}{\pi_t} = 766 \text{ kPa}$.

The outlet temperature is then $T_{09} = T_{08} \left[1 - \eta_{t_2} \left\{ 1 - \left(\dfrac{1}{\pi_t} \right)^{[(\gamma-1)/\gamma]_8} \right\} \right] = 1348 \text{ K}$.

Free power turbine

$$Cp_9 = 950 + 0.21 \times 1348 = 1233 \text{ J/kg/K} \quad \left(\frac{\gamma}{\gamma - 1} \right)_9 = 4.296$$

$$T_{010} = T_{09} \left[1 - \eta_{ft} \left\{ 1 - \left(\frac{T_{010}}{T_{09}} \right)^{[(\gamma-1)/\gamma]_9} \right\} \right] = 1348 \left[1 - 0.9 \left\{ 1 - \left(\frac{110}{766} \right)^{0.233} \right\} \right] = 906.7 \text{ K}$$

$$W_{net} = Cp_9(T_{09} - T_{010}) = 1233(1348 - 906.7) = 544 \text{ kJ/kg}$$

Regenerator

$$\varepsilon_R = \frac{T_{05} - T_{04}}{T_{010} - T_{04}}$$

$$T_{05} = T_{04} + \varepsilon_R(T_{010} - T_{04}) = 477.6 + 0.8(906.7 - 477.6) = 820.9 \text{ K}$$

$$Cp_5 = 950 + 0.21 \times 820.9 = 1122.4 \text{ J/kg/K}$$

First combustor
The fuel-to-air ratio in the first combustor

$$f_1 = \frac{Cp_6 T_{06} - Cp_5 T_{05}}{\eta_b Q_R - Cp_6 T_{06}} = \frac{1.265 \times 1500 - 1.1224 \times 820.9}{0.99 \times 43{,}000 - 1.265 \times 1500} = 0.02399$$

$$Cp_7 = 950 + 0.21 \times 1348 = 1233 \text{ J/kg/K}$$

Second combustor
The fuel-to-air ratio in the second combustor

$$f_2 = \frac{Cp_8 T_{08} - Cp_7 T_{07}}{\eta_b Q_R - Cp_8 T_{08}} = \frac{1.265 \times 1500 - 1.233 \times 1348}{0.99 \times 43{,}000 - 1.265 \times 1500} = 0.0058$$

The total fuel-to-air ratio $f_{total} = f_1 + f_2 = 0.02977$.
For a power of 300 MW, the needed air mass flow rate is

$$\dot{m} = \frac{\text{Power}}{(1 + f_{total})W_{net}} = \frac{300 \times 10^3}{544 \times 1.0298}$$

$$= 535.5 \text{ kg/s}$$

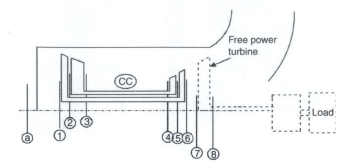

FIGURE 8.22   Aeroderivative engine.

Example 7   Consider an aeroderivative gas turbine that is developed from the turbofan engine described in Example 3 of Chapter 5. The fan is removed and the bypass duct of the engine is cancelled. The gases leaving the IPT are used to drive a power turbine, which in turn drives the load as shown in Figure 8.22. The efficiency of the free power turbine is $\eta_{Ft} = 0.9$. Calculate the power delivered to the load. All other data are kept same as in the previous example.

Solution: Inlet mass flow to the whole engine

$$\dot{m}_a = 100\,\text{kg/s}, \quad P_{01} = P_a = 1.0\,\text{bar}, \quad T_{01} = T_a = 288\,\text{K}$$

1. Air inlet

$$T_{01} = T_a = 288\,\text{K}, \quad P_{01} = P_a = 1.0\,\text{bar}$$

2. Intermediate-pressure compressor

$$P_{02} = \pi_{\text{IPC}} \times P_{01} = 5 \times 1.0 = 5.0\,\text{bar}$$

$$T_{02} = T_{01}\left(1 + \frac{\pi_{\text{IPC}}^{(\gamma-1)/\gamma} - 1}{\eta_{\text{IPC}}}\right) = 288 \times 1.656 = 477.2\,\text{K}$$

3. High-pressure compressor

$$P_{03} = \pi_{\text{HPC}} \times P_{04} = 5 \times 5 = 25\,\text{bar}$$

$$T_{03} = T_{02}\left\{1 + \left[\frac{\pi_{\text{HPC}}^{(\gamma-1)/\gamma} - 1}{\eta_{\text{HPC}}}\right]\right\} = 477.2\left\{1 + \left[\frac{(5)^{0.286} - 1}{0.90}\right]\right\} = 477.2 \times 1.6489 = 787\,\text{K}$$

4. Combustion chamber
    For all the hot elements including combustion chamber, turbines, and nozzles $\gamma = 1.33, Cp = 1.147\,\text{kJ/kg}\cdot\text{K}$.

Energy balance of combustion chamber:

A bleed from the outlet of the high-pressure compressor is considered here with

$$b = \dot{m}_b/\dot{m}_a \quad \text{and} \quad f = \dot{m}_f/\dot{m}_a$$

$$\dot{m}_a[(1-b)Cp_aT_{03}] + \eta_{cc}\dot{m}_fQ = [(1-b)\dot{m}_a + \dot{m}_f]Cp_gT_{04}$$

$$(1-b)Cp_aT_{05} + \eta_{cc}fQ = (1+f-b)Cp_gT_{06}$$

$$T_{04} = [(1-b)Cp_aT_{03} + \eta_{cc}fQ]/[(1+f-b)Cp_g] = 1359.4\,\text{K}$$

$$P_{04} = 0.98\,P_{03} = 0.98 \times 25 = 24.5\,\text{bar}$$

5. High-pressure turbine (HPT)

$$\eta_m W_{HPT} = W_{HPC}$$

$$\eta_m\dot{m}_{hot}Cp_g(T_{04} - T_{05}) = \dot{m}_aCp_a(T_{03} - T_{02})$$

$$\dot{m}_{hot} = (1+f-b)\dot{m}_a = 0.9977\dot{m}_a$$

$$\Delta T_{045} = T_{04} - T_{05} = \frac{Cp_a(T_{03} - T_{02})}{0.9977 \times \eta_m \times Cp_g} = \frac{1.005(787 - 477.2)}{0.9977 \times 0.99 \times 1.147} = 274.82\,\text{K}$$

$$T_{05} = T_{04} - \Delta T_{045} = 1084.6\,\text{K}$$

$$P_{05} = P_{04}\left[1 - \frac{[1 - (T_{05}/T_{04})]}{\eta_{HPT}}\right]^{\gamma_g/(\gamma_g-1)} = 24.5\left[1 - \frac{[1 - (1084.6/1359.4)]}{0.9}\right]^{\frac{1.33}{0.33}} = 8.855\,\text{bar}$$

6. Intermediate-pressure turbine (IPT)

From energy balance:

$$\eta_m W_{IPT} = W_{IPC}$$

$$\eta_m\dot{m}_{hot}Cp_g(T_{05} - T_{06}) = \dot{m}_aCp_a(T_{02} - T_{01})$$

$$\Delta T_{056} = \frac{Cp_a(T_{02} - T_{01})}{0.9977 \times \eta_m \times Cp_g} = \frac{1.005(477.2 - 288)}{0.9977 \times 0.99 \times 1.147} = 167.84\,\text{K}$$

$$T_{06} = T_{05} - \Delta T_{056} = 1084.6 - 167.84 = 916.76\,\text{K}$$

$$P_{06} = P_{05}\left[1 - \frac{[1 - (T_{06}/T_{05})]}{\eta_{IPT}}\right]^{\gamma_g/(\gamma_g-1)} = 0.47P_{05} = 4.1632\,\text{bar}$$

Expansion in the free power turbine to ambient pressure

$$T_7 = T_{06}\left\{1 - \eta_{ft}\left[\left(\frac{P_{07}}{P_8}\right)^{(\gamma_h-1)/\gamma_h} - 1\right]\right\} = 916.7\left\{1 - 0.9\left[(4.1632)^{0.25} - 1\right]\right\} = 563.2\,\text{K}$$

The power delivered to the load $= \mathscr{P}$

$$\mathscr{P} = \dot{m}_a(1+f-b)Cp_g(T_{07} - T_8) = 100 \times 0.997 \times 1.147(916.76 - 563.2)$$

$$= 40340\,\text{kW} = 40.34\,\text{MW}$$

8.7 COMBINED GAS TURBINE

Though gas turbine cycles have a greater potential for higher thermal efficiencies because of the higher average temperature at which gas is supplied, they have an inherent disadvantage, namely, the gas leaves the gas turbine at very high temperatures (usually above 500° C), which wipes out any potential gains of thermal efficiency. The continued quest for higher thermal efficiencies has resulted in rather innovative modifications to conventional power plants. A popular modification is to use the high temperature exhaust gases of the gas turbine as the energy source for bottoming such steam power cycle. Thus the suggested power plant here is a gas power cycle topping a vapor power cycle, which is called the combined gas-vapor cycle, or just the combined cycle. The combined cycle of greatest interest is the gas turbine (Brayton) cycle topping a steam turbine (Rankine) cycle, which has a higher thermal efficiency than either of the cycles executed individually.

Gas turbine cycles typically operate at considerably higher temperatures than steam cycles. The maximum fluid temperature at the turbine inlet is about 620°C (1150°F) for modern steam power plants, but over 1150°C (2100°F) for gas turbine power plants. The use of higher temperatures in gas turbines is made possible by recent developments in cooling the turbine blades and coating the blades with high-temperature-resistant materials such as ceramics. A combined gas turbine-Rankine cycle may be implemented in several ways. The open cycle gas turbine may be linked to a steam cycle through what may be considered a gas turbine heat exchanger containing an economizer and a boiler. This combined gas-steam plant is called a heat-recovery steam generator (HRSG), which is shown in Figure 8.23. The T–S diagram for the steam bottoming cycle is shown in Figure 8.24. In this cycle, the energy is recovered from the exhaust gases by transferring it to the steam in a heat exchanger that serves as the boiler.

Assuming constant specific heat for the gases leaving the turbine, the energy balance gives

$$\dot{m}_s(h_{12} - h_{11}) = \dot{m}_g C p_g (T_{06} - T_{07}) = \dot{m}_s A_1 \tag{8.50}$$

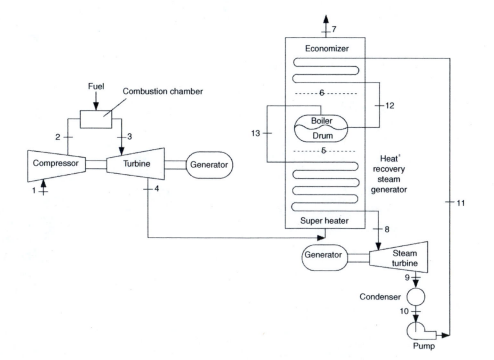

FIGURE 8.23   Combined cycle gas and steam turbines.

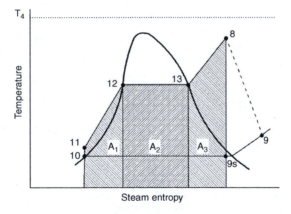

FIGURE 8.24 Temperature–entropy diagram for steam bottoming cycle.

$$\dot{m}_s(h_{13} - h_{12}) = \dot{m}_g Cp_g(T_{05} - T_{06}) = \dot{m}_s A_2 \tag{8.51}$$

$$\dot{m}_s(h_8 - h_{13}) = \dot{m}_g Cp_g(T_{04} - T_{05}) = \dot{m}_s A_3 \tag{8.52}$$

where $\dot{m}_s$ and $\dot{m}_g$ are the steam and gas mass flow rates. Combining Equations 8.50 through 8.52 gives

$$\dot{m}_s(h_8 - h_{11}) = \dot{m}_g Cp_g(T_{04} - T_{07}) = \dot{m}_s(A_1 + A_2 + A_3) \tag{8.53}$$

The total power from the combined cycle is $W_{gt} + W_{st}$, while the heat input is only from the gas turbine (Q) which is unchanged. The overall efficiency is then given by

$$\eta = \frac{W_{gt} + W_{st}}{Q}$$

The combined cycle increases the efficiency without appreciably increasing the initial cost. Consequently, many new power plants operate on combined cycles and many more existing steam or gas turbine plants are being converted to combined cycle power plants. Thermal efficiencies well over 40% are reported as a result of conversion.

8.8 MARINE APPLICATIONS

The marine gas turbine like most of the engines used for marine propulsion is usually an adaptation of a machine that was originally developed for some other purpose. Aeroderivative gas turbine is a prominent example, where aircraft derivative gas turbine engines are used for both main propulsion and the ship's service electrical power. A high degree of plant automation is achieved with an integrated system of control and monitoring consoles.

A wide range of vessels is powered by marine gas turbine engines ranging from modest-sized pleasure craft requiring several hundred horsepower to major naval vessels and the largest merchant ships which may require 20,000–50,000 hp by shaft [9]. These vessels belong to the following categories—merchant ships, coast guard vessels, naval vessels, and small crafts.

The advantages of a gas turbine plant as compared to a steam plant of comparable horsepower include

1. Weight reduction of 70%
2. Simplicity (fewer propulsion auxiliaries)

3. Reduced manning due to automated propulsion plant control
4. Quicker response time
5. Faster acceleration/deceleration

The load to be driven by a marine propulsion engine is usually a screw propeller, a jet pump, an electrical generator, or a hydraulic pump. The load changes rapidly in maneuvering situations and especially when an emergency stop is required. The gas turbine is the optimum choice that satisfies these requirements due to its small inertia of the moving part compared to diesel engines. Moreover, for a rearward motion of the vessel, the propulsion load must be reversed. Gas turbines use some auxiliary methods this requirement such a reversing the propeller pitch or using a reverse train in the reduction gear to satisfy.

Marine gas turbines operate in a wet and salty environment. Vessel motions have widely ranging magnitudes and frequencies. Air sucked into the compressor is wet with salty sprays that may deposit on the engine with consequent corrosive effects and obstructions of air passages. Moreover, the products of combustion have a polluting effect on the surrounding environment.

There are two types of gas turbines used in marine applications, namely, the single-shaft engine (as an example Allison 501-K17 engine) and double- or two-spool engine. The latter type is sometimes identified as split-shaft engine. An example of the latter type is General Electric LM2500. The power turbine is aerodynamically coupled to the gas generator but the two shafts are not mechanically connected. The power turbine converts the thermal energy from the gas generator to mechanical energy to drive the load. The output speed is varied by controlling the speed of the gas generator that determines the amount of exhaust gases sent to the power turbine. Split-shaft gas turbine engines are suitable for main propulsion applications. The advantages in this application are

1. The gas generator is more responsive to load demands because the compressor is not restricted in operation by the load on the power turbine.
2. The gas generator section and power turbine section operate near their most efficient speeds throughout a range of load demands.

To cope with environmental protection, extensive research is also performed to design super marine gas turbine (SMGT) in a variety of ship classes that develop a low NOx (below 1 g/kWh) and high efficiency (at least 38%) and application of marine diesel oil [10] gas turbines. The engine is a two-split-shaft gas turbine consisting of gas generator and the power system and incorporating a regenerative (recuperator) unit.

8.8.1 Additional Components for Marine Applications

For naval applications, gas turbines have the same components as described previously. However, three important components (modules) are added; namely, the intake and exhaust systems as well as the propeller. Gas turbines use high rate of air, nearly 5 ft^3/hp-min, which is roughly twice that of diesel ones. This large mass flow rate means high velocity, high pressure drop, and increase of salt-laden droplets of sea spray. Other requirements for inlet and exhaust systems are silencing, avoidance of reingestion or injection into the intake, and finally, keeping the hot exhaust gases away from the wooden structure of the vessel if applicable.

As shown in Figure 8.25 the intake system is composed of the following elements: high hat assembly, intake duct, anti-icing manifold, silencers, and expansion unit. The *high hat assembly* supports the moisture separators in removing water droplets and dirt from the intake air to prevent erosion of compressor components. Electric strip heaters prevent ice formation on the louvers. Moreover, it houses the blow-in door that prevents engine air starvation when the moisture separators become dirty [11].

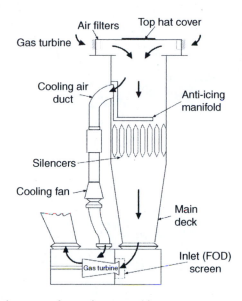

FIGURE 8.25 A typical intake system for marine gas turbine.

The *intake duct* provides combustion air for the engine and cooling air for the module. The module cooling system routes a portion of the intake air to the engine enclosure for module ventilation and external engine cooling. The module cooling air swirls about the engine removing heat and ventilating the module before exiting via a small air gap around the aft end of the power turbine. The *anti-icing manifold* is to inject hot bleed air into the intake trunk, below the module cooling air duct, to prevent ice formation, which may occur when outside air temperature drops to 38°F. Icing at the compressor inlet can restrict airflow causing a stall and also presents serious foreign object damage (FOD) hazard to the engine. *Intake silencers* are located halfway down the intake duct to reduce airborne noise. The silencers consist of vertical vanes of sound-deadening material encased in perforated stainless steel sheets. The module cooling air duct contains a single bullet-shaped silencer to silence the noise created by the cooling air. The *expansion joint* is a rubber boot connecting the intake duct to the module inlet plenum. This prevents the noise of the module from being transmitted to the hull of the ship.

The exhaust duct system routes engine exhaust gases to the atmosphere while reducing both the heat and noise of the exhaust (Figure 8.26). It is composed of an exhaust collector, uptake ducting, silencing, exhaust eductors, and boundary layer infrared suppression system (BLISS). The *exhaust elbow* directs exhaust gases into the exhaust uptake duct. A gap between the exhaust elbow and ship's uptake causes an *eductor* effect drawing module cooling air into the exhaust uptake. Exhaust uptake ducting is insulated to control heat and noise as the exhaust is passed to the atmosphere. A vane type *silencer* is located in the center of the duct. These silencers are the same as those in the intake ducting but are permanently mounted. Exhaust eductors are located at the uppermost end of the exhaust ducting. It cools the exhaust gases by mixing with cool ambient air to reduce the infrared signature of the ship. *Bliss* caps are installed on the top of each mixing tube to further cool the exhaust air by mixing it with layers of ambient air. This is accomplished by use of several louvers that are angled to create an eductor effect. This allows cool ambient air to mix with the hot exhaust gases.

The propeller in marine gas turbine is similar to the propeller in turboprop engines. It is the propulsor element that converts the mechanical energy into thrust power. First, the gas turbine converts the thermal energy into mechanical energy. Next, the transmission system transmits the mechanical energy to its point of use. It often includes a speed-reducing gear set and may include clutches, brakes, and couplings of various types. The propulsor is most often the familiar screw propeller. The propeller is either of the fixed- or variable-pitch type. The rotational speed of the

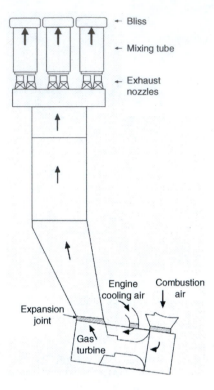

FIGURE 8.26 A typical exhaust system for marine engine.

propeller is proportional to the vessel speed. The propeller torque and power are proportional to the square and cube of its rotational speed, respectively.

8.8.2 EXAMPLES FOR MARINE GAS TURBINES

1. General Electric LM2500

It has various series of somewhat different power. The following operating conditions are considered for different series:

- No inlet or exhaust losses, ambient temperature 59°F and 60% relative humidity at sea level.
- Fuel is natural gas.
- Direct load drive (no gear box) at a speed of 3000 rpm (50 Hz).
- Generator losses are assumed.

The corresponding available power is here stated for different series: LM2500PE (22,346 kW), LM2500PH (26,463 kW), LM2500RC (32,916 kW), LM2500RC (35,842 kW), and LM2500RD (32,689 kW).

Figure 8.27 illustrates the details of LM2500. A brief description of the LM2500 is given here.
(a) Gas generator components

Compressor section

The LM2500 has a 16-stage axial flow compressor (17 for the advanced LM2500+). Compressor casing contains one stage of inlet guide vanes (IGV), six stages of variable stator vanes (VSV), to prevent compressor stall, and 10 stages of stationary stator vanes. Bleed air is extracted from the compressor for use in the ship's bleed air system and for internal use in the engine.

FIGURE 8.27 LM2500. (Courtesy GE Energy.)

Combustor
The combustor is an annular type with 30 fuel nozzles and two spark ignitors. Of the air from the compressor, approximately 30% is mixed with fuel to support combustion. The other 70% is used to cool and center the flame within the combustion liner.

High-pressure turbine section
The HPT is a two-stage axial flow type. It uses approximately 65% of the thermal energy from the combustor to drive the compressor and engine-mounted accessories.

Power turbine
The power turbine is a six-stage axial flow type turbine. The power turbine extracts the remaining 35% of useable energy and uses this to drive the main reduction gear. The power turbine drives the reduction gear through a high-speed flexible coupling shaft and clutch assembly. For different series, the power-to-weight ratio is 5.86–5.95 kW/kg, the power-to-foot print ranges from 2019 to 2310 kW/m^2, and the specific fuel consumption ranges from 200 to 215 (g/(kW · h)).

2. Rolls-Royce MT30 [12]

It delivers a broad band of powers for commercial marine and naval applications in an exceptionally lightweight package. It is derived from Trent 800, which efficiently powers Boeing 777 aircraft. Its power is 36 MW at 26°C. The MT30 is also designed to burn commercially available distillate fuels, giving a high degree of flexibility and associated through-life cost benefits. It is a twin-spool high-pressure ratio gas generator with an eight-stage variable geometry intermediate-pressure compressor (IPC) and a six-stage high-pressure compressor (HPC). Both IPT and HPT have only one stage. The four-stage free power turbine is derived from Trent 800 incorporating the latest cooling technique. The key parts are protectively coated for service in the marine environment to reduce maintenance and ensure long service life. The annular combustor similar to its parent aero engine is employed to meet all current and anticipated legislation on emission and smoke. The gas turbine change unit (GTCU) includes the power turbine and weighs 6200 kg (dry).

Table 8.1 outlines a comparison between gas turbine and diesel engine. The values given represent average values. Comparison is in favor of gas turbines if the power per weight and power per footprint are examined. Diesel engines have better specific fuel consumption. The difference in fuel consumption decreases as the power rating increases.

8.9 OFFSHORE GAS TURBINES

Gas turbines are widely used offshore for a variety of purposes including power generation, gas and water injection, pumping, gas lift, waterflood, and export compression. Gas turbines have been adapted in place of diesel generation systems on significant numbers of platforms because they can be locally powered (self-fueled) and since 70% of the operational costs are in fuel, they can be very cost-effective on a platform where the gas supply comes at well-head prices. Moreover, since they are lightweight (thus have high power to weight ratio) and compact in size, gas turbines are well suited to such tasks. In addition, availability, reliability, and ruggedness are additional key factors for

TABLE 8.1

Comparison Between Gas Turbine and Diesel Engines Used in Marine Applications

Power Rating (MW)	5–10	10–20	20–30	30–40
Power/weight (kW/kg)				
GT	4.0	3.5	3.0	2.9
Diesel	0.3	0.2	0.15	0.15
Power/footprint (kW/m^2)				
GT	1000	1100	1500	1750
Diesel	250	250	250	250
SFC (gr/(kW · hr))				
GT	240	210	200	195
Diesel	190	175	170	168

Source: N. Gee, *Fast Ferry Powering and Propulsors—The Options.*
http://www.ngal.co.uk/downloads/techpapers/paper7.pdf

selecting gas turbines as offshore power plants. The only operational constraints are on efficiency, as there is no combined cycle efficiency gain [14].

Aeroderivative gas turbines are increasingly favored offshore because of the requirements for low weight, simple change out, and ease of maintenance. For example, the GE LM series are now favored over General Electric's older industrial frame series. Industrial gas turbines of compact modular system utilizing aerospace rotor technology are very different from the earlier bulkier technology commonly used in onshore power generation.

Factors that need to be considered in designing turbines offshore include weight and dimensions, minimizing vibration, and resistance to saltwater, and resistance to pitch and roll, particularly in floating installations. They are prone to material degradation such as corrosion, environmental attack, erosion, oxidation, and FOD. Thus, turbines are made of nimonic alloys, while Co–Ni-based alloys are used in compressors. Ceramics have been used in some Rolls-Royce designs [15]. Offshore engines are either single-shaft or two-shaft gas turbines.

Offshore main suppliers are Rolls-Royce (RB211, Avon, Coberra), Solar Turbines (Saturn, Mars, Titan), General Electric (GE Industrial Gas Turbine, Nuovo Pigneone, Partners like John Brown), Siemens-Westinghouse, Pratt & Whitney, ABB, Allison, Dresser, and MAN.

8.10 MICRO GAS TURBINES (μ-GAS TURBINES)

Though micro gas turbines are defined as those generating less than 350 kW power [16], the same name is extensively used recently as gas turbines whose dimensions are of the order of centimeters. The current trend toward miniaturization, portability, and more in general ubiquitous intelligence, has led to the development of a wide range of new tiny products. The power units appropriate for such products may be fuel cells, thermoelectric devices, combustion engines, and gas turbines. While fuel cells are expected to offer the highest efficiency, micro gas turbines are expected to offer the highest power density. Gas turbine was invented in the twentieth century and micro gas turbines are considered the key to powering twenty-first century technology. Though no bigger than a regular shirt button, micro gas turbine engine uses the same process for power generation as its big brother power plant. Continuous work is running now for a shortcoming production. Extensive work for design, development, and production of micro gas turbines is carried out in many institutions worldwide. Epstein [17] and his coworkers in MIT (the United States) have completed making micro gas turbine engines in which each of the individual parts functions inside a tiny combustion chamber, fuel and air quickly mix and burn at the melting point of steel. Turbine blades (spanning an area less than a dime) made of low-defect, high-strength microfabricated materials spin at 20,000 revolutions per second (rps)—100 times faster than those in jet engines. A minigenerator produces 10 W of power. A little compressor raises the pressure of air in preparation for combustion. And cooling (always a challenge in hot microdevices) appears manageable by sending the compression air around the outside of the combustor. The resulting device is sealed all around, with holes on the top and bottom for air intake, fuel intake, and exhaust.

8.10.1 Microturbines versus Typical gas Turbines

The main difference between a small and large gas turbine is the amount of gas submitted to an almost unchanged thermodynamic cycle. Velocity and pressure levels remain the same when scaling down a gas turbine. Only the dimensions are smaller. The work exchange between compressor or turbine and fluid is proportional to the peripheral speed, such that the rotational speed should scale inversely proportional to the diameter, resulting in speeds of more than 500,000 rpm for rotor diameters below 20 mm.

A major problem with miniaturization of microturbines is a large decrease in Reynolds number, resulting in higher viscous losses and a lower overall cycle efficiency.

Also the required temperature is a problem. As a pressure ratio of 3 is envisaged, the turbine inlet temperature should be at least 1200 K to obtain positive cycle efficiency, and higher temperatures would considerably boost the overall efficiency. In large turbines, the blades are cooled by internal cooling channels and protected by thermal barrier coatings. In case of microturbines, internal cooling of blades a few millimeters in size is unrealistic. Therefore, temperatures of 1200 K and higher can only be reached with ceramic materials.

Another major consequence of the small dimensions is the extreme temperature gradient between the hot turbine and colder compressor. The resulting massive heat flux provokes a nonnegligible decrease of both compressor and turbine efficiency.

A gas turbine net power output is the small difference between the large turbine power output and the large compressor energy requirements. The deterioration of aeroperformance of the components with decreasing dimensions and increased heat transfer results in larger decrease of power output and cycle efficiency. Therefore, careful optimization is needed to guarantee positive output also at the smallest dimensions. However, little knowledge is available on aerothermodynamics at these small scales. There is also no guarantee that the existing flow solvers and turbulence models are still accurate for these extremely low Reynolds numbers [18].

8.10.2 Design Challenges

1. Manufacturing

Gas turbines are among the most advanced systems as they combine extreme conditions in terms of rotational speed with elevated gas temperatures (up to 2100 K for military engines). Miniaturization of such a system poses tremendous technical problems as it leads to extremely high rotational speeds (e.g., 10^6 rpm). Moreover, scaling down the system unfavorably influences the flow and combustion process. Fabricating such devices requires new materials to be explored (such as Si_3N_4 and SiC) and also requires three-dimensional micromanufacturing processes. Combining the "best-of-both-worlds," that is, photolithographic micromanufacturing techniques with more traditional micromechanical manufacturing will be essential for reaching the envisaged objectives.

2. Selection and design of bearings

The bearings must operate throughout the whole domain of possible temperature conditions during start-up and in steady-state operation. Maximum temperatures between 100°C and 1000°C can be expected depending on the exact location of the bearings. Rotor unbalance can result in dangerously high dynamic radial loading, and therefore, the eccentricity of the mass center should be balanced within a few micrometers.

It is clear that conventional ball bearings are not feasible regarding speed and temperature. Magnetic bearings could offer a solution regarding speed, but the high temperature dissuades the use of permanent magnets as these could demagnetize. Consequently, such bearings should be constructed with electromagnets, which consume a considerable amount of electrical energy.

Air bearings seem most suited for this application. Aerostatic as well as aerodynamic ones can be used. Aerostatic bearings can be fed by tapping a small amount of compressed air from the compressor. This results in problems at start-up and moreover decreases the overall efficiency.

Aerodynamic bearings are self-pressurizing and therefore need no external supply. However, the phase of dry friction during start-up is a major drawback. Additionally, self-excited instabilities (half-speed whirl) limit the maximum attainable speed. Nevertheless, aerodynamic bearings are the most promising choice on condition that the issue of instability is tackled.

Several stabilizing techniques exist, but most promising for this application are bearings with conformable surfaces, more specifically aerodynamic foil bearings. These bearings are virtually immune to half-speed whirl and suffer less from centrifugal and thermal rotor growth. Current work focuses on designing new foil bearing concepts, suitable for small dimensions [18].

3. Compressor and turbine

As stated before, the efficiency of all components is critical. This is especially true for the compressor and turbine, requiring efficiencies of at least 60–70. Thus, it is clearly a challenge to obtain the required efficiency despite the low Reynolds numbers, increased heat transfer, and lower relative geometric accuracy of the components.

8.10.3 Applications

1. Electric generation

The micro gas turbine developed [18] will be in the centimeter range (2 cm diameter and 5 cm long) and will produce a power output of about 100 W. The system basically consists of a compressor, regenerator, combustion chamber, turbine, and electrical generator, as illustrated in Figure 8.28.

Generally micro gas turbines can produce enough electricity to power handheld electronics. In the foreseeable future, these tiny turbines will serve as a battery replacement. A micro gas turbine could run for ten or more hours on a container of diesel fuel slightly larger than a D battery; when the

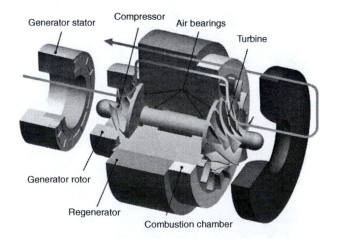

FIGURE 8.28   Micro gas turbine layout. (From http://www.powermems.be/gasturbine.html)

FIGURE 8.29   Illustrates the microcentrifugal compressor for the turbopump.

fuel cartridge ran out, a new one could be easily swapped in. Each disposable cartridge would pack as much energy as a few heavy handfuls of lithium-ion batteries. Consequently, a small pack of the cheap and light cartridges could power a personal digital assistant (PDA) or cell phone through several days of heavy usage, with no wall-outlet recharging required; this is a highly attractive feature for soldiers in remote locations or for travelers. Moreover, a cluster of tiny engines, each capable of producing up to a hundred watts, could supply a home with an efficient and reliable source of electricity.

2. Turbopumps for liquid rocket propellant

A micro turbopump (Figure 8.29) has the following components and systems:

- Micro axial/centrifugal compressor
- Microburners
- Micro radial inflow/axial turbines
- Gearboxes and reduction drives and associated shaft dynamics analysis [19]

3. Unmanned aerial vehicle (UAV) and scaled aircraft/helicopters

A miniature aircraft engine may be used for as small as 15-centimeter-long planes that could carry small cameras for surveillance. Moreover, micro turboshaft engines may be used for scaled helicopters [19].

PROBLEMS

In the following problems unless otherwise stated, use the following data:
    Ambient conditions are 288 K and 101 kPa and fuel heating value is 42,000 kJ/kg.

8.1 A single-spool gas turbine is composed of a compressor, combustion chamber, and turbine. The isentropic efficiencies of the compressor and turbine are η_c and η_t, respectively. Assume that there is no pressure drop in the combustion chamber, the working fluid has constant properties in all the components, fuel flow rate is negligible, and gasses expand in the turbine to the ambient pressure. Prove that the power attains a maximum value if the compressor pressure ratio satisfies the following relation:

$$\pi_c^* = (\eta_c \eta_t T_{03}/T_{01})^{\gamma/2(\gamma-1)}$$

Plot the relation of the specific power and thermal efficiency versus the compressor pressure ratio for compression ratios in the range 4–10. The compressor and turbine efficiencies are 86% and 89%, respectively, while the turbine inlet temperature is 1100 K.

What will be the value of (π_c^*) if real conditions are assumed instead of the above assumption?

8.2 A regenerator is added to the gas turbine in Problem 8.1. The compressor pressure ratio is 6, while the pressure drops in the regenerator and in the combustion chamber are 2% and 4%, respectively. Other data remain the same as above. It is required to:

(a) Plot the processes in the four modules on the temperature–entropy diagram

(b) Calculate and plot the specific power versus the regenerator effectiveness ranging from 20% to 85%

(c) What is the effect of regenerator effectiveness on the fuel consumption?

8.3 The compressor in a single-spool gas turbine has a pressure ratio of 15 and isentropic efficiency of 85%. The pressure drop is 2% of the compressor delivery pressure. The turbine inlet temperature and isentropic efficiency are 1100 K and 90%. Gases leave the turbine at a pressure of 110 kPa.

(a) Calculate:

    (i) Fuel-to-air ratio

    (ii) Specific power

    (iii) SFC

    (iv) Thermal efficiency

(b) If all the gas turbine components are assumed ideal, there are no pressure drops in the combustion chamber, and the gases leave the turbine at the ambient conditions, what will be the ideal efficiency?

8.4 For the same gas turbine in Problem 8.3 calculate and plot the thermal efficiency and specific power, if the compressor pressure ratio attains the values of 6, 8, 10, 12, and 14.

8.5 For the same data of Problem 8.3 repeat the above procedure for turbine inlet temperatures of 1000, 1200, and 1400 K. Plot the thermal efficiency and specific power versus the compressor pressure ratio.

8.6 A single-spool gas turbine incorporating an intercooler, regenerator, and a reheat stage. The compression and expansion processes in the compressors and turbines are isentropic. The pressure ratio in both compressors is equal and the expansion ratio at both turbines is equal. Full expansion is assumed for the gases leaving the second turbine to ambient pressure.

The ambient temperature and pressure are 288 k and 101.3 kPa. The compression pressure ratio is 9. The turbine inlet temperature is 1300 K. The fuel heating value is 43,000 kJ/kg.

(a) Sketch a layout for the industrial gas turbine and draw the T–S diagram.

(b) Calculate:

    (i) Fuel-to-air ratio

    (ii) Specific power

    (iii) SFC

    (iv) Thermal efficiency

8.7 During testing of the gas turbine illustrated in Figure 8.13 the following values for temperature are recorded:

$T_{03} = 288$, $T_{04} = 750$, $T_{06} = 1400$, $T_{07} = 1270$ K. The regenerator effectiveness is 75%. The isentropic efficiency for the compressor and turbine are 0.87 and 0.91.

Calculate:

- The compression ratio in both compressors
- The expansion ratio in both turbines
- The temperature of air leaving the regenerator
- The specific power
- The thermal efficiency

8.8 The Ruston TA2500 gas turbine is a two-shaft engine rated at 2500 bhp at ISO conditions. The compressor is a 14-stage axial type. The combustion system employs one large diameter chamber. It is possible for fuels other than natural gas and distillate oil to be burned. The gas turbine (compressor turbine) is a two-stage axial turbine. The free power turbine is also a two-stage one. The conditions to the elements are as follows:

Module	Compressor Inlet	Compressor Outlet	Turbine Inlet	Free Power Inlet	Exhaust Duct Inlet
Temperature	59(°F), 15°C	415(°F), 213°C	1564(°F) 851°C	1203(°F) 651°C	944(°F) 507°C
Pressure	14.7(psia) 1.03 ata	72.0(psia) 5.06 ata	70.1(psia) 4.93 ata	29.2(psia) 2.05 ata	14.7(psia) 1.03 ata

It is required to calculate:

(a) The isentropic efficiency of the compressor, compressor turbine, and free power turbine

(b) The pressure drop in the combustion chamber

(c) The fuel-to-air ratio

(d) The thermal efficiency of the engine

(e) The air mass flow rate

REFERENCES

1. H.I.H. Saravanamuttoo, G.F.C. Rogers, and H. Cohen, *Gas Turbine Theory*, Prentice Hall, 5th edn., 2001.

2. Ruston, *A Range Of Industrial Gas Turbines Below 10,000 B.H.P*, Publication 028, Revised July 1981.

3. Secareanu, D. Stankovic, L. Fuchs, V. Mllosavljevic, and J. Holmborn, *Experimental Investigation of Airflow and Spray Stability in an Air Blast Injector of an Industrial Gas Turbine*, GT2004-53961, ASME TURBO EXPO, June 14–17, 2004 Vienna, Austria.

4. Gas Turbine World 1999–2000 Performance Specs, *Performance Rating of Gas Turbine Power Plants for Project Planning, Engineering, Design & Procurement*, Vol. 19, 1999, pp. 10–48.

5. M.P. Boyce, *Gas Turbine Engineering Handbook*, 2nd edn., Gulf-Professional Publishing, 2002, p. 16.

6. R.T.C. Harman, *Gas Turbine Engineering-Applications, Cycles and Characteristics*, Halsted Press Book, John Wiley & Sons, 1981, p. 31.

7. Y.A. Cengel and M.A. Boles, *Thermodynamics: An Engineering Approach*, McGraw-Hill, Inc., 3rd edn., 1998, p. 517.

8. M.M. El-Wakil, *Powerplant Technology*, McGraw-Hill International Edition, 1984, p. 325.

9. J.B. Woodward, *Marine Gas Turbines*, John Wiley & Sons, 1975, p. 7.

10. M. Arai, T. Sugimoto, K. Imai, H. Miyaji, K. Nakanishi, and Y. Hamachi, *Research and Development of Gas Turbine for Next-Generation Marine Propulsion System (Super Marine Gas Turbine)*, Proceeding of the International Gas Turbine Congress, 2003 Tokyo, November 2–7, 2003, IGTC 2003 Tokyo OS-202.

11. P.P. Walsh and P. Fletcher, *Gas Turbine Theory*, 2nd edn. ASME Press, pp. 25–31.

12. www.rolls-royce.com (MP/46/Issue May 8, 03).

13. N. Gee, *Fast Ferry Powering and Propulsors—The Options*, http://www.ngal.co.uk/downloads/techpapers/paper7.pdf

14. http://www.iso.ch/iso/en/commcentre/pdf/Oilgas0002.pdf

15. M. Wall, R. Lee, and S. Frost, *Offshore Gas Turbines (and Major Driven Equipment) Integrity and Inspection Guidance Notes*, ESR Technology Ltd, 2006.

16. A.S. Rangwala, *Turbomachinery Dynamics*, McGraw-Hill, 2005, p. 62.

17. http://web.mit.edu/newsoffice/2006/microengines.html

18. http://www.powermems.be/gasturbine.html

19. http://www.m-dot.com

Part II

Component Design

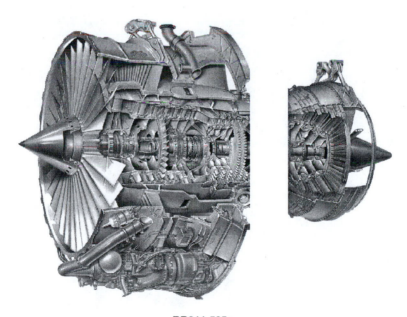

RB211-535c

9 Power Plant Installation and Intakes

9.1 INTRODUCTION

All air breathing engines installed in an aircraft must be provided with an air intake and a ducting system (which is also identified as inlet or diffuser) to diffuse the air from freestream velocity to a lower velocity acceptable for further processing by other engine components [1]. The inlet component is designed to capture the exact amount of air required to accomplish the diffusion with maximum static pressure rise, minimum total pressure loss, deliver the air with tolerable flow distortion (as uniform as possible), and contribute the least possible external drag to the system. For gas turbine engine (turbojet, turbofan, and turboprop), the airflow entering subsonic compressors or fans must be of low Mach number, of the order 0.4–0.5 or less even if the aircraft speed is supersonic. For a ramjet also, the inlet reduces the speed to a subsonic value to have a subsonic combustion. Thus, the entrance duct usually acts as a diffuser.

Inlets may be classified as

1. Subsonic or supersonic
2. Fixed or variable geometry

The intake geometry and performance are closely related to the method of power plant installation. The position of the power plant must not affect the efficiency of the air intake. In the following sections, the different methods of power plant installation will be first discussed for both civil and military aircraft; next subsonic and supersonic intakes will be analyzed.

9.2 POWER PLANT INSTALLATION

Power plant installation is sometimes referred to as propulsion airframe integration. It is the process of locating the power plants and designating their installation to meet many operating requirements while minimizing drag and weight penalties [2]. The installation of the engines influences aircraft safety, structural weight, flutter, drag, control, maximum lift, performance and reliability of engine, maintainability, and aircraft growth potential. Power plant installation influences the design of air-inlet system, exhaust system (which will be discussed in Chapter 11), cooling arrangement and mounting structure. These elements are largely the responsibility of the airframe manufacturer. However, the engine manufacturer is also involved, as the engine data must be available to the airframe manufacturer to design a suitable installation.

Power plant installation for both subsonic and supersonic aircraft will be discussed in this chapter.

9.3 SUBSONIC AIRCRAFT

Power plant installations for both civil and military aircraft are similar as long as their flight speeds are subsonic or transonic. However, for supersonic aircraft there are great differences in the installation methods of civil and military aircraft as will be described later. Turbojet and turbofan engines have the same methods for installation while turboprops employ other methods.

9.3.1 TURBOJET AND TURBOFAN ENGINES

Though most of the present day transport (passenger or cargo) aircraft are powered by turbofan engines, turbojet engines still have small avenues. The following methods are frequently seen for turbojet and turbofan engines:

1. Wing installation (Even number of engines: 2, 4, 6, or 8)
2. Fuselage (Even number of engines 2 or 4)
3. Wing and tail combination (Three engines).
4. Fuselage and tail combination (Three engines)

9.3.1.1 Wing Installation

There are three forms for power plants to be installed to wings, namely

1. Buried in the wing (2 or 4 engines)
2. Pod installation (2, 4, 6, or 8 engines)
3. Above the wing (2 or 4 engines)

The first type of the wing installation is *engines buried in the wing*, which are found in early aircraft like the DeHaviland Comet 4 and the Hawker Siddeley Nimrod. In both aircraft, four engines are buried in the wing root (Figures 9.1a,b). The B-2's four General Electric F118-GE-110 nonafterburning turbofans are buried in the wings, with two engines clustered together inboard on each wing, Figure 9.1c. The intakes of the B-2 aircraft have a zigzag lip to scatter radar reflections, and there is a zigzag slot just before each intake to act as a "boundary layer splitter," breaking up the stagnant turbulent airflow that tends to collect on the surface of an aircraft.

This method of installation had the following advantages:

1. Minimum parasite drag and probably the minimum weight.
2. Minimum yawing moment in case of engine failure that counteracts the asymmetric thrust; thus the pilot can easily maintain a straight level flight.

However, it has the following disadvantages:

1. It poses a threat to the wing structure in the case of failure of a turbine blade or disk.
2. It is very difficult to maximize the inlet efficiency.
3. If a larger diameter engine is desired in a later version of the airplane, the entire wing may have to be redesigned.
4. It poses difficult accessibility for maintenance and repair.
5. It eliminates the flap in the region of the engine exhaust, therefore reducing the maximum lift coefficient $C_{L_{MAX}}$ [2].

The second type of wing installation is the *pod installation* where engines are attached to the wings via pylons (Figure 9.2). Most engines nowadays have pod installation; either two or four engines are found in the following aircraft Boeing 707, 737, 747, 757, 767, 777, MD-11 and the future 787, Airbus A 300, 320, 330, 340, 350, and 380 and Antonov AN-148. Six engines power the Antonov AN-225, while eight P&W TF33 turbofan engines power the B-52.

This type of engine installation has the following advantages:

1. It minimizes the risk of wing structural damage in case of blade/disk failure.
2. It is simple to obtain high ram recovery in the inlet since the angle of attack at the inlet is minimized and no wakes are ingested.
3. Easy engine maintenance and replacement are possible.

FIGURE 9.1 (a) Buried wing installation. (b) Nimrod aircraft as a typical buried wing installation, and (c) B-2 aircraft. (Courtesy Nortrop Grumman.)

However, it has the following disadvantages:

1. For low-wing aircraft like Boeing 737, the engines are mounted close to ground and thus tend to suck dirt, pebbles, rocks, snow, and so on into the inlet. This is known as foreign object damage (FOD), which may cause serious damage to the engine blades.
2. The high temperature and high dynamic pressure of the exhaust impinging on the flap increases flap loads and weight, and might require titanium structure, which is more expensive.
3. Pylon-wing interference affects the local velocities near the wing leading edge, thus increasing the drag and reducing the maximum lift coefficient. However, this drawback can be remedied by choosing the nacelle locations sufficiently forward and low with respect to the wing as shown in Figure 9.3.

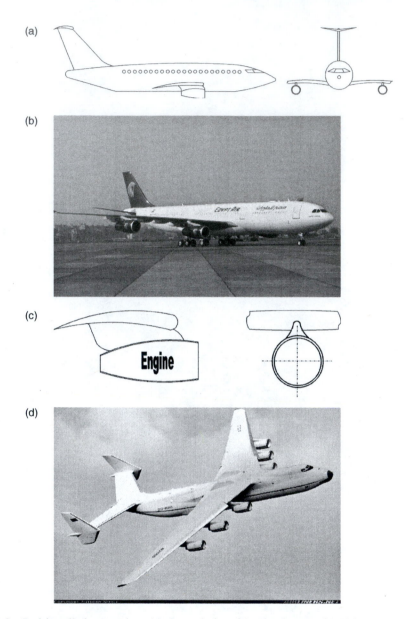

FIGURE 9.2   Pod installation to wing. (a) General aircraft/engine layout. (b) Airbus A340-600 EgyptAir airliner. (c) Close view of pod. (d) Antonov An-225. (Courtesy Airbus industries and EgyptAir company.)

Engines installed *above the wing* represent the third type of wing installation. Some of these are sea planes like A-40 Albatros powered by two Soloviev D-30KPV engines of 117.7 kN thrust each plus two Klimov RD-60K booster engines of 24.5 kN thrust (Figures 1.61 and 9.4) and others are conventional aircraft like Antonov AN-74TK-200 aircraft powered by two turbofan engines D-36, series 3-A mounted over the wing. This type of installation has the following advantages:

1. It prevents (or greatly reduces) foreign object ingestion into the engines from the surface of a runway during takeoff and landing.
2. It prevents water ingestion into the engine for seaplanes during takeoff and landing.

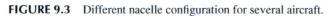
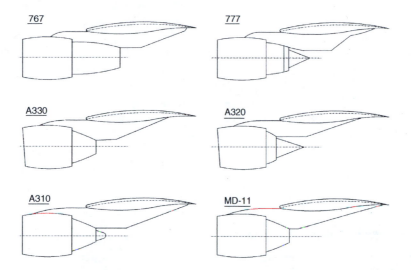

FIGURE 9.3 Different nacelle configuration for several aircraft.

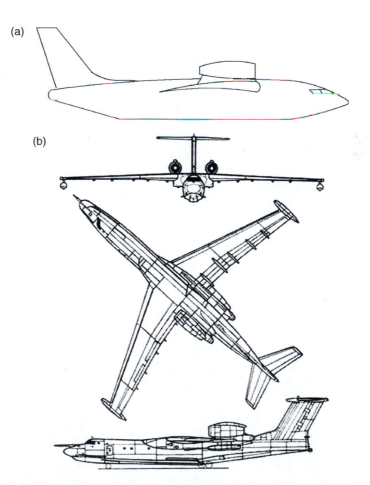

FIGURE 9.4 Above the wing installation. (a) Typical layout. (b) Three views for A-40 Albatros aircraft.

3. It improves wing lift due to the blowing over of wing upper surface and inboard flaps by engine streams.
4. It reduces noise level at the terrain due to screening engine streams by the wing, which copes with the increasingly stringent civilian aviation noise regulations.

As at present and in the future, increasingly stringent civilian aviation noise regulations will require the design and manufacture of extremely quiet commercial aircraft. In addition, the large fan diameters of modern engines with increasingly higher bypass ratios pose significant packaging and aircraft installation challenges. The design approach that addresses both these challenges is to mount the engines above the wing. In addition to allowing the performance trend toward large diameters and high bypass ratio turbofan engines to continue, this approach allows the wing to shield much of the engine noise from people on the ground.

9.3.1.2 Fuselage Installation

As described in Reference 2, fuselage installation (which is sometimes identified as *aft-engine* arrangements) is most suitable for small aircraft where it is difficult to install engines under the wing and maintain adequate wing-nacelle and nacelle ground clearances. Same as wing, pod engine mounting is used. Engine pods are mounted on struts to allow an air gap between the pod and the fuselage (and wing in wing installation). Thus, there is a minimum of aerodynamic interference with the smooth flow of air over the structure adjacent to the engine pod. The engines are attached to the sides of the rear fuselage with short stub wings. aircraft are powered by either two engines (Figure 9.5a) or four engines. Examples for two-engine aircraft are the DC-9, MD-80, Gates Learjet Model 25, Fokker F-28, Cessna 550 Citation II, TU-324 and 334, and Sud aviation Caravelle 12. Examples for the four engines are the Jetstar and BAC Super VC-10 (Figure 9.5b).

Fuselage installation has the following advantages [2]:

1. Greater maximum lift coefficient and less drag due to elimination of wing–pylon and exhaust–flap interference.
2. Less asymmetric yaw after engine failure as the engines are close to the fuselage.
3. Lower fuselage height above the ground, which permits shorter landing gear and air stair length.

FIGURE 9.5   (a) Two engine fuselage installation and (b) four engine fuselage installation; Super VC-10.

However, this installation has the following disadvantages [3]:

1. The center of gravity of the empty airplane is moved aft, well behind the center of gravity of the payload. Thus a greater center of gravity range is required, which leads to difficult balance problems and larger tail.
2. On wet runway, the wheels kick water and the special deflectors on the landing gear may be needed to avoid water ingestion into the engine.
3. At very high angles of attack, the nacelle wake blankets the T-tail and may cause stall, which needs a larger tail span.
4. Vibration and noise isolation is a difficult problem.
5. It appears that for DC-9 size aircraft, the fuselage installation is to be slightly preferred. In general, smaller aircraft employ fuselage arrangement.

9.3.1.3 Combined Wing and Tail Installation (Three Engines)

Tail installation represents a *center engine* installation. Only one turbofan engine is installed in all available tail combinations with either wing or fuselage arrangement. Examples are found in Lockheed Tristar and DC-10 aircraft. There are four arrangements of this type of installation as seen in Figure 9.6 for center (or tail) engine installation, where layout (a) represents bifurcated inlet, (b) represents a long inlet, (c) represents long tail pipe, and (d) represents "S" bend inlet. Each possibility entails compromises of weight, inlet total pressure loss, inlet flow distortion, drag, thrust reverser effectiveness, and maintenance accessibilities.

The two usually used are the S-bend and the long inlet configurations. Both installation methods have the following advantages:

1. Mounted very far aft, so a ruptured turbine blade or disc will not impact on the basic tail structure.
2. High-thrust reverser without interfering with control surface effectiveness. This is achieved by shaping and tilting the cascades used to reverse the flow.

Moreover, both arrangements have the following disadvantages:

1. Large inlet losses due to the long length of intake
2. Difficult accessibility for maintenance and repair

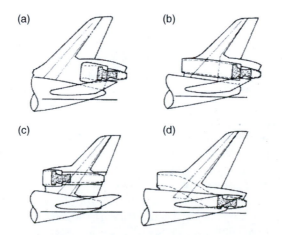

FIGURE 9.6   Tail engine installation (DC-10) transport aircraft. (Copyright Boeing, All rights reserved.)

FIGURE 9.7   Wing and tail installation.

FIGURE 9.8   Fuselage and tail installation.

The S-bend has lower engine location and uses the engine exhaust to replace part of the fuselage boat-tail (saves drag) but has several disadvantages; namely, a distortion risk, a drag from fairing out the inlet, and the serious drawback of cutting a huge hole in the upper fuselage structure.

The long (or straight through) inlet has the engine mounted on the fin, which has an ideal aerodynamic inlet free of distortion, but does have an increase in fin structural weight to support the engine. Figure 9.7 illustrates a typical wing and tail installation.

9.3.1.4   Combined Fuselage and Tail Installation

Examples for this type of installation are the Boeing 727 aircraft, TU-154, and Yak-42D. A typical layout for this configuration is shown in Figure 9.8.

9.3.2   Turboprop Installation

Turboprop (or turbopropeller) engines, however, have limited options. The propeller requirements always place the engine in one of the following positions:

1.   Installation in the wing
2.   Installation in the fuselage, either at nose or empennage of aircraft for a single-engined aircraft
3.   Horizontal tail installation

Most present day turboprops are wing-mounted engines to either passenger or cargo transports. Either puller or pusher turboprops may be installed to the wing. Typical wing installation of the puller type is shown in Figure 9.9. Examples for wing installation are DeHavill and DHC-8 Commuter airplane powered by two PW120 puller engines, Fokker F-27 powered by two RR Dart puller engines and Beech Starship powered by two PT6A-67A pusher engines (Figure 9.10). In some aircraft, a pair of counterrotating propellers is installed to each engine. Antonov AN-70 is an example for these aircraft where the aircraft is powered by four engines.

Small aircraft may be powered by a single turboprop engine fitted to its nose (Figure 9.11).

Nose-mounted turboprop engines may have two contrarotating propellers. Example for this aircraft is the Westland Wyvern TF MK2 long-distance ship airplane developed for the Royal navy in 1944. This was the last fixed-wing aircraft at the Westland Company, and it was the first ship airplane in the world equipped with the turboprop engines having contrarotating propeller (Figure 9.12).

FIGURE 9.9 Puller turboprop engines installed to wing.

FIGURE 9.10 Starship aircraft powered by two PT6A-67A pusher engines.

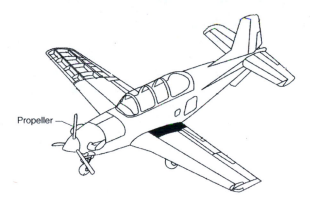

Propeller

FIGURE 9.11 Nose-mounted turboprop engine.

Westland Wyvern TF Mk 2

FIGURE 9.12 Contrarotating propeller of a turboprop (nose installation).

FIGURE 9.13   Contrarotating propeller of a turboprop (empennage installation).

FIGURE 9.14   Turboprop engines mounted in the horizontal tail.

Another engine fitted with a turboprop engine having contrarotating propellers in the tail was the XB-42, which allowed the wing to have a clean unfettered design (Figure 9.13).

The Antonov AN-180 medium turboprop airliner (around 175 passengers) is under development. It is powered by two turboprop engines attached to the tips of the horizontal tail and each has contrarotating propellers (Figure 9.14).

9.4   SUPERSONIC AIRCRAFT

Supersonic aircraft are mostly military ones. Few civil aircraft are supersonic ones including the Anglo-French Concorde (powered by turbojet engines) and the Russian TU-144 (powered by turbofan engines). All the remaining supersonic aircraft are military ones.

9.4.1   CIVIL TRANSPORTS

The engines of the two civil supersonic transports (SSTs) are installed in the wing. Concorde is powered by four Rolls-Royce/SNECMA Olympus 593 turbojets (Figure 9.15), while TU-144 is powered by four turbofan engines. All engines are installed to the lower surface and aft part of the wing. Engine nacelle is flush to wing surface and has rectangular inlet (Figure 9.16).

FIGURE 9.15   Concorde aircraft.

FIGURE 9.16   Inlet of Olympus 593 engine of the Concorde.

FIGURE 9.17   Nose intake for a single engine fuselage-mounted engine.

9.4.2 MILITARY AIRCRAFTS

Most, if not all, military aircraft are powered by fuselage installation engines. Turbojet or turbofan engines are installed at the aft end of the fuselage. Fighters powered by a single engine may have a single intake (Figure 9.17), or divided intakes (Figure 9.18). Nose intake involves the use of either a short duct as in the case of F-86 or a long duct as in F-8 and F-16. Nose intake enjoys good characteristics through a wide range of angle of attack and sideslip. It is free from aerodynamic interference effects, such as flow separation from other parts of the aircraft. Perhaps the largest drawback of the nose inlet, however, is that neither guns nor radar can be mounted in the front of the fuselage. Moreover, the long internal duct leading from the inlet to the engine generates excessive friction and thus has relatively high pressure losses. In addition, interference between the duct and the pilot's cockpit may be encountered. For this reason, the divided intake is often used where the inlets are located at the roots of aircraft wings.

Another type of the nose inlet is the chin inlet employed on the F-8 airplane, which has many of the advantages of the simple nose inlet but leaves space in the front of the fuselage for radar or guns

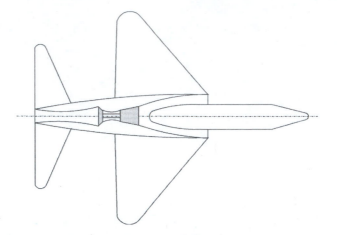

FIGURE 9.18   Divided intake for a single engine fuselage-mounted engine.

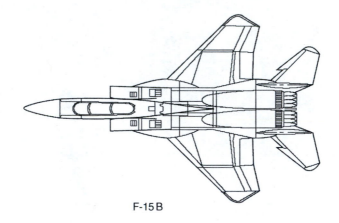

F-15 B

FIGURE 9.19   F-15.

and has a somewhat shorter internal duct. Care should be taken in such a design to ensure that at no important flight condition does separate or unsteady flow enter the inlet from the nose of the aircraft. The proximity of the inlet to the ground introduces a possible risk of foreign object ingestion, and, obviously, the nose wheel must be located behind the inlet. The chin inlet, however, is employed on the General Dynamics F-16.

Twin-engine fighters have mostly wing-root inlet installation, as in the case of F-15 and F-101 fighters. Inlets located in this manner offer several advantages. Among these are short, light, internal flow ducts, avoidance of fuselage boundary layer air ingestion, and freedom to mount guns and radar in the nose of the aircraft. Further, no interference between the cockpit and internal ducting is encountered in this arrangement.

Twin-engine fighters like McDonnell Douglas F-15 have a particular problem fairing the aft end of the fuselage around twin exhausts (Figure 9.19). Different types of inlet will be discussed later. A unique wing installation is found in the SR-71 aircraft (Figure 9.20).

9.5   AIR INTAKES OR INLETS

The intake is the first part of all air breathing propulsion systems. Both the words intake and inlet are used alternatively. Intake is normally used in Britain while inlet is used in the United States. Air intake

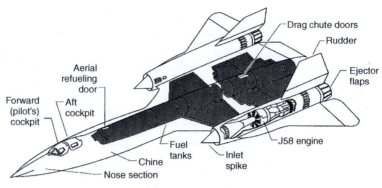

General arrangement of the SR-71 aircraft (NASA TM 104330)

FIGURE 9.20　SR-71.

duct is designed and manufactured by airframe manufacturer and not by the engine manufacturer. Both manufacturers cooperate in testing air intakes. An aircraft will require one or more intakes to capture (collect) the atmospheric air (working fluid) at free stream Mach number, sometimes change its direction of flow, and supply this flow to the engine with as little distortion as is possible, to ensure smooth running and efficient propulsion. Moreover, the intake has to achieve all this with minimum disturbance to the external flow around the aircraft and thus minimum external drag. Thus for a successful operation of the engine within the desired flight envelope, the engine-intake compatibility is essential. Subsonic inlets in the present aircraft contain noise-absorbing materials to cope with the international acoustics limitations [4].

9.6　SUBSONIC INTAKES

Subsonic intakes are found in the turbojet or turbofan engines powering most of the present civil transports (commercial and cargo aircraft). Examples of these engines are the JT8, JT9, PW 4000 series, RB211, Trent series, and V2500 powering many of the Boeing and Airbus aircraft transports. The surface of the inlet is a continuous smooth curve where the very front (most upstream portion) is called the inlet lip. A subsonic aircraft has an inlet with a relatively thick lip. Concerning turboprop engines, the intakes are much complicated by the propeller and gearbox at the inlet to the engine.

Subsonic inlets have fixed geometry, although inlets for some high bypass ratio turbofan engines are designed with blow-in-doors. These doors are spring-loaded parts installed in the perimeter of the inlet duct designed to deliver additional air to the aero engine during takeoff and climb conditions when the highest thrust is needed and the aircraft speed is low [5]. The most common type of subsonic intake is the pitot intake. This type of intakes makes the fullest use of ram due to forward speed, and suffers the minimum loss of ram pressure with changes of aircraft altitude [4]. However, as sonic speed is approached, the efficiency of this type of air intake begins to fall because of the formation of a shock wave at the intake lip. It consists of a simple forward entry hole with a cowl lip. The three major types of pitot intakes as shown in Figure 9.21 are as follows:

1.　Podded intakes
2.　Integrated intake
3.　Flush intakes

Podded intake is common in transport aircraft (civil or military). Examples of the commercial aircraft are Boeing 707, 767, 777, 787 and Airbus A330, 340, 350, and 380. B-52 is an example for the military aircraft. The integrated intake is used in combat (military) aircraft, an example for which is the British Aerospace Harrier. For integrated intakes, the internal flow problems are of dominant

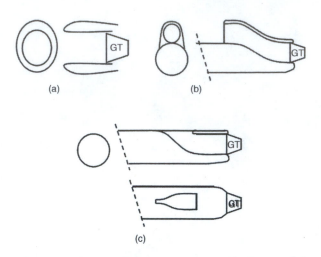

FIGURE 9.21   Types of pitot intakes: (a) podded pitot, (b) integrated pitot, and (c) flush pitot.

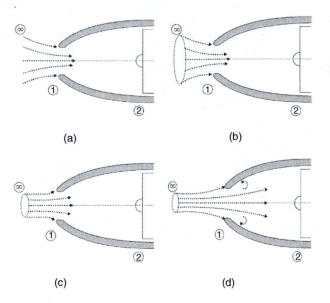

FIGURE 9.22   Flow characteristics of podded intakes: (a) ground run, (b) climb, (c) high-speed cruise, and (d) top speed.

concern, due to (a) the duct being longer, usually containing bends and shape changes and (b) the presence of aircraft surface ahead of the intake, wetted by the internal flow [6]. The flush intake is usually used in missiles since they can be more readily accommodated into missile airframes.

For typical podded intakes, the friction losses are insignificant while the flow separation is of prime importance. The internal flow has the shortest and most direct route possible to the engine and its pressure recovery is almost 100% [6]. From aerodynamics point of view, the flow in intake resembles the flow in a duct. The duct "captures" a certain stream tube of air, thus dividing the air stream into an internal flow and an external flow. The external flow preserves the good aerodynamics of the airframe, while the internal flow feeds the engine. The flow characteristics in podded intakes are illustrated in Figure 9.22, for four flow conditions. In ground running (Figure 9.22a), there will be no effective free stream velocity that results in a large induced flow capture area causing the

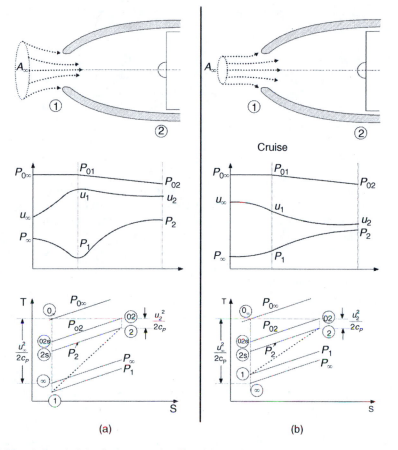

FIGURE 9.23   Subsonic inlet during (a) takeoff and (b) cruise.

streamlines to converge into the intake area. The ratio between the upstream capture area to the inlet area approaches infinity. The stream tube has a bell-shaped pattern. During climb (Figure 9.22b), the free stream velocity will be lower than the intake velocity due to the requirement of high mass flow rates. This will also result in a larger entry stream tube area than the intake area (a convergent stream tube pattern). At high speed cruise (where $M = 0.85$, Figure 9.22c), the entry stream tube will be smaller than the intake area and diffusion partially takes place outside the intake and partially inside and hence the air velocity attains lower values in the intake with a small resultant rise in pressure (15%). At top speed (higher than cruise where $M = 0.95$, Figure 9.22d) the high pressure gradient on the intake lip can cause separation and an unstable flow into the intake.

9.6.1   Inlet Performance

Depending on the flight speed and the mass flow demanded by the engine, the inlet might have to operate with a wide range of incident stream conditions. Figures 9.23a and b show the performances of subsonic intake during two typical subsonic conditions, takeoff and cruise, respectively.

   For each operating condition, three plots are given in Figure 9.23. The first illustrates the stream tube, while the second depicts the pressure and speed variation and the third is a temperature–entropy diagram. The flow in intake is identified by three states, namely far upstream that is denoted as (∞), at the duct entry denoted by (1) and at the engine face denoted by (2). The flow outside the engine (from state ∞ to 1) is an isentropic one, where no losses are associated with the total temperature and pressure. For high speed, or cruise condition (Figure 9.23b), the stream tube will have a divergent shape and following conditions can be stated:

$$u_1 < u_\infty, \quad P_1 > P_\infty, \quad P_{01} = P_{0\infty}, \quad T_{01} = T_{0\infty}$$

During low speed high-thrust operation (e.g., during takeoff and climb), as shown in Figure 9.23a, the same engine will demand more mass flow and the air stream upstream the intake will be accelerated. The stream tube will have a converging shape and the following conditions are satisfied:

$$u_1 > u_\infty, \quad P_1 < P_\infty, \quad P_{01} = P_{0\infty}, \quad T_{01} = T_{0\infty}$$

For both cases of takeoff and cruise, there will be internal diffusion within the intake up to the engine face. The static pressure will rise and the air speed will be reduced. The total pressure will also decrease owing to skin friction while the total temperature remains unchanged, as the flow through diffuser is adiabatic. Thus, for both takeoff and cruise conditions

$$P_2 > P_1, \quad P_{02} < P_{0\infty}, \quad u_2 < u_1$$

Since the inlet speed to the engine (compressor/fan) should be nearly constant for different operating conditions, then

$$\left(\frac{P_2}{P_1}\right)_{\text{takeoff}} > \left(\frac{P_2}{P_1}\right)_{\text{cruise}}$$

If this pressure increase is too large, the diffuser may stall due to boundary layer separation. Stalling usually reduces the stagnation pressure of the stream as a whole [7]. Conversely, for cruise conditions (Figure 9.23b) to avoid separation, or to have a less severe loading on the boundary layer, it is recommended to have a low velocity ratio (u_1/u_∞) and consequently less internal pressure rise [8]. Therefore, the inlet area is often chosen so as to minimize external acceleration during takeoff with the result that external deceleration occurs during level-cruise operation. Under these conditions the "upstream capture area" A_∞ is less than the inlet area A_1, and some flow is spilled over the inlet.

The serious problem is the change in M_∞ from zero at takeoff to about 0.8 at cruise.

If optimized for the $M_\infty = 0.8$ cruise, the inlet would have a thin lip to minimize the increase in Mach number as the flow is divided.

However, this inlet would separate badly on the inside at takeoff and low subsonic conditions because the turn around the sharp lip would impose severe pressure gradients. To compromise, the lip is rounded making it less sensitive to flow angle, but incurring some loss due to separation in the exterior flow.

When fully developed a good inlet will produce a pressure recovery $P_{02}/P_{0a} = 0.95$–0.97 at its optimum condition.

Example 1   Prove that the capture area (A_∞) for a subsonic diffuser is related to the free stream Mach number (M_∞) by the relation:

$$A_\infty = \lambda/M_\infty$$

where

$$\lambda = \frac{\dot{m}_\infty}{P_\infty}\sqrt{\frac{RT_\infty}{\gamma}}$$

A turbofan engine during ground ingests airflow at the rate of $\dot{m}_\infty = 500$ kg/s through an inlet area (A_1) of 3.0 m^2. If the ambient conditions (T_∞, P_∞) are 288 K and 100 kPa, respectively, calculate the area ratio (A_∞/A_1) for different free stream Mach numbers. What is the value of the Mach number where the capture area is equal to the inlet area? Draw the air stream tube for different Mach numbers.

Solution: The mass flow rate is

$$\dot{m}_\infty = \rho_\infty V_\infty A_\infty = \frac{P_\infty}{RT_\infty}\left(M_\infty\sqrt{\gamma RT_\infty}\right) A_\infty$$

Then

$$A_\infty = \frac{\dot{m}_\infty}{P_\infty M_\infty}\sqrt{\frac{RT_\infty}{\gamma}}$$

or

$$A_\infty = \frac{\lambda}{M_\infty} \tag{a}$$

where

$$\lambda = \frac{\dot{m}_\infty}{P_\infty}\sqrt{\frac{RT_\infty}{\gamma}}$$

From the given data, then

$$\lambda = \frac{\dot{m}_\infty}{P_\infty}\sqrt{\frac{RT_\infty}{\gamma}} = \frac{500}{100\times10^3}\sqrt{\frac{287\times288}{1.4}} = 1.215$$

With $A_1 = 3.0$ m^2, then from relation (a):

$$\frac{A_\infty}{A_1} = \frac{0.405}{M_\infty} \tag{b}$$

From (b), the capture area is equal to the engine inlet area ($A_\infty/A_1 = 1$) when $M_\infty = 0.405$
From relation (b), the following table is constructed:

M_∞	0.05	0.1	0.2	0.3	0.4	0.5
A_∞/A_1	8.1	4.05	2.025	1.35	1.0125	0.81
D_∞/D_1	2.84	2.0125	1.42	1.162	1.006	0.9

Figure 9.24 illustrates the stream tube for different free stream Mach numbers.

9.6.2 Performance Parameters

Two parameters will be discussed here:

1. Isentropic efficiency (η_d)
2. Stagnation pressure ratio (r_d)

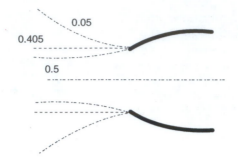

FIGURE 9.24   Stream tubes for different free stream Mach numbers.

1. Isentropic efficiency (η_d)

The isentropic efficiency of the intake η_d is a static-to-total efficiency and as previously defined in Chapters 3 through 7, it is a measure for the losses from the far upstream conditions to the engine face. The efficiency is then expressed by the following relation (refer to Figure 9.23a):

$$\eta_d = \frac{h_{02s} - h_\infty}{h_{02} - h_\infty} = \frac{T_{02s} - T_\infty}{T_{02} - T_\infty} = \frac{(T_{02s}/T_\infty) - 1}{(T_{02}/T_\infty) - 1}$$

Since

$$\frac{T_{02}}{T_\infty} = \frac{T_{0\infty}}{T_\infty} = 1 + \frac{\gamma - 1}{2} M_\infty^2$$

$$\therefore \eta_d = \frac{(P_{02}/P_\infty)^{(\gamma-1)/\gamma} - 1}{[(\gamma - 1)/2] M_\infty^2} \tag{9.1}$$

$$\frac{P_{02}}{P_\infty} = \left(1 + \eta_d \frac{\gamma - 1}{2} M_\infty^2\right)^{\gamma/(\gamma-1)} \tag{9.2}$$

2. Stagnation pressure ratio (r_d)

Modern jet transports may cruise with values of the pressure recovery of 97%–98%. Supersonic aircraft with well-designed, practical inlet and internal flow systems may have pressure recoveries of 85% or more for Mach numbers in the 2.0–2.5 range.

Stagnation pressure ratio is defined as the ratio between the average total pressure of the air entering the engine to that of the free stream air, or

$$r_d = P_{02}/P_{0\infty}$$

$$\frac{P_{02}}{P_\infty} = \frac{P_{02}}{P_{0\infty}} \frac{P_{0\infty}}{P_\infty} = r_d \left[1 + \frac{\gamma - 1}{2} M_\infty^2\right]^{\gamma/(\gamma-1)} \tag{9.3}$$

From Equations 9.2 and 9.3

$$\therefore \eta_d = \frac{(r_d)^{(\gamma-1)/\gamma} \left[1 + \frac{\gamma-1}{2} M_\infty^2\right] - 1}{\left(\frac{\gamma-1}{2}\right) M_\infty^2} \tag{9.4}$$

Moreover, the pressure recovery can be expressed as

$$r_d = P_{02}/P_{0\infty} = \left[\frac{1 + \eta_d \frac{\gamma-1}{2} M_\infty^2}{1 + \frac{\gamma-1}{2} M_\infty^2} \right]^{\gamma/(\gamma-1)} \tag{9.5}$$

Thus if the isentropic efficiency is known, the pressure recovery can be obtained from Equation 9.5, while if the pressure recovery is known, the diffuser efficiency can be determined from Equation 9.4.

However, since the flow upstream of the intake is isentropic, then all the losses are encountered inside the intake, or from state (1) to state (2). In some cases, the efficiency is defined for the *internal part* of diffuser (η_d'). In this case, the diffuser efficiency is defined as

$$\eta_d' = \frac{h_{02s} - h_1}{h_{02} - h_1} = \frac{T_{02s} - T_1}{T_{02} - T_1} = \frac{(T_{02s}/T_1) - 1}{(T_{01}/T_1) - 1}$$

$$\eta_d' = \frac{(P_{02}/P_1)^{(\gamma-1)/\gamma} - 1}{(P_{01}/P_1)^{(\gamma-1)/\gamma} - 1} = \frac{(P_{02}/P_1)^{(\gamma-1)/\gamma} - 1}{\frac{\gamma-1}{2} M_1^2}$$

$$(P_{02}/P_1) = \left(1 + \eta_d' \frac{\gamma-1}{2} M_1^2 \right)^{\gamma/(\gamma-1)} \tag{9.6}$$

Example 2 Consider the turbofan engine described in Example 1 during flight at a Mach number of 0.9 and altitude of 11 km where the ambient temperature and pressure are respectively $-56.5°C$ and 22.632 kPa. The mass ingested into the engine is now 235 kg/s. If the diffuser efficiency is 0.9 and the Mach number at the fan face is 0.45, calculate the following:

1. The capture area
2. The static pressures at the inlet and fan face
3. The air speed at the same stations as above
4. The diffuser pressure recovery factor

Solution:

1. The ambient static temperature is $T_\infty = -56.5 + 273 = 216.5$ K
   The free stream speed is $u_\infty = M_\infty \sqrt{\gamma R T_\infty} = 271.6$ m/s
   The free stream density $\rho_\infty = P_\infty/RT_\infty = 0.3479$ kg/m^3
   The capture area is $A_\infty = \dot{m}/\rho_\infty u_\infty = 2.486$ m^2
   Which is smaller than the inlet area?
2. To calculate the static pressures at states (1) and (2) two methods may be used, namely,
   (a) Gas dynamics tables
   (b) Isentropic relations

The first method

3. From gas dynamics tables, at Mach number equal to 0.9, the area ratio $A/A^* = 1.00886$
   Then $A^* = 2.486/1.0086 = 2.465$ m^2

$$A_1/A^* = 3/2.465 = 1.217 \text{ m}^2$$

From tables, $M_1 = 0.577$.
The corresponding temperature and pressure ratios are $P_1/P_{01} = 0.798$, $T_1/T_{01} = 0.93757$.
Since $T_{01} = T_{0\infty}$, $P_{01} = P_{0\infty}$, then $P_1 = 30.547$ kPa, $T_1 = 246.9$ K.
Moreover, $u_1 = M_1 \sqrt{\gamma R T_1} = 181.7$ m/s.

Since $M_2 = 0.45$, then $P_2/P_{02} = 0.87027$.

Now from the diffuser efficiency, from Equation 9.2, the pressure ratio

$$\frac{P_{02}}{P_\infty} = (1 + 0.9 \times 0.2 \times 0.9^2)^{3.5} = 1.6102$$

$$P_{02} = 36.442 \text{ kPa}$$

But

$$P_2 = \frac{P_{02}}{\left(1 + \frac{\gamma - 1}{2} M_2^2\right)^{\gamma/(\gamma - 1)}} = 31.714 \text{ kPa}$$

and

$$T_2 = \frac{T_{02}}{\left(1 + \frac{\gamma-1}{2} M_2^2\right)} = 253.1 \text{ K}$$

Then $u_2 = M_2 \sqrt{\gamma R T_2} = 143.5$ m/s.

The pressure recovery factor $r_d = P_{02}/P_{0\infty} = 36.442/38.278 = 0.952$.

The above results can be summarized in the following table:

State	∞	1	2
T (K)	226.65	246.9	253.1
T_0 (K)		263.36	
P (kPa)	22.632	30.547	31.71
P_0 (kPa)	38.278	38.278	36.442
u (m/s)	271.6	181.7	143.5
M	0.9	0.577	0.45

The second method

The continuity equation

$$\frac{\dot{m}}{A} = \rho u = \frac{P}{RT} M \sqrt{\gamma R T} = PM \sqrt{\frac{\gamma}{RT}}$$

$$\frac{\dot{m}}{A} = \frac{MP_0}{\left(1 + \frac{\gamma-1}{2} M^2\right)^{\gamma/(\gamma-1)}} \sqrt{\frac{\gamma}{R}} \sqrt{\frac{\left(1 + \frac{\gamma-1}{2} M^2\right)}{T_0}}$$

$$\frac{\dot{m}}{A} = \frac{MP_0}{\left(1 + \frac{\gamma-1}{2} M^2\right)^{(\gamma+1)/2(\gamma-1)}} \sqrt{\frac{\gamma}{RT_0}}$$

Applying the above equation at station (1), then

$$\frac{\dot{m}}{A_1 P_{01}} \sqrt{\frac{RT_{01}}{\gamma}} = \frac{M_1}{\left(1 + \frac{\gamma-1}{2} M_1^2\right)^{(\gamma+1)/2(\gamma-1)}}$$

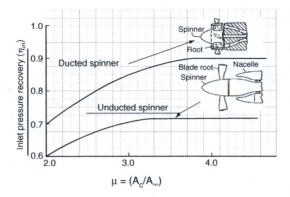

FIGURE 9.25　Efficiency of inlets of turboprop engine. (Courtesy of John Seddon [6].)

The left-hand side is known, and the Mach number M_1 is unknown, which can be determined iteratively. From M_1 determine T_1

$$T_1 = \frac{T_{0\infty}}{\left(1 + \frac{\gamma-1}{2}M_1^2\right)}$$

Determine $P_1 = P_{0\infty}/\left(1 + \frac{\gamma-1}{2}M_1^2\right)^{\gamma/(\gamma-1)}$ and u_1 as above.

Determine P_{02} from Equation 9.2, thus determine the recovery pressure ratio.

Determine $P_2 = P_{02}/\left(1 + \frac{\gamma-1}{2}M_2^2\right)^{\gamma/(\gamma-1)}$ and $T_2 = T_{0\infty}/\left(1 + \frac{\gamma-1}{2}M_2^2\right)$.

Next, determine the air speed at the fan face from the relation $u_2 = M_2\sqrt{\gamma R T_2}$.

9.6.3 TURBOPROP INLETS

The intake for turboprop engines is much complicated due to the bulk gearbox. However, an aerodynamically efficient turboprop intake can be achieved by use of a ducted spinner [9] as shown in Figure 9.25. Improvement of intake performance is achieved by choosing large entry area. In addition, minimum cylindrical roots are encased in fairings of low thickness/chord, which form the structural members supporting the spinner cowl [6]. The efficiency of the intake, defined as the ratio between the difference between the static pressure at the engine face and the free stream value to the dynamic pressure is plotted versus the ratio between the capture area and the duct entry.

9.7　SUPERSONIC INTAKES

The design of inlet systems for supersonic aircraft is a highly complex matter involving engineering trade-offs between efficiency, complexity, weight, and cost. A typical supersonic intake is made up of a supersonic diffuser, in which the flow is decelerated by a combination of shocks and diffuse compression, and a subsonic diffuser, which reduces the Mach number from high subsonic value after the last shock to the value acceptable to the engine [10]. Subsonic intakes that have thick lip are quite unsuitable for supersonic speeds. The reason is that a normal shock wave ahead of the intake is generated, which will yield a very sharp static pressure rise without change of flow direction and correspondingly big velocity reduction. The adiabatic efficiency of compression through a normal shock wave is very low as compared with oblique shocks. At Mach 2.0 in the stratosphere adiabatic efficiency would be about 80% or less for normal shock waves, whereas its value will be about 95% or even more for an intake designed for oblique shocks.

Flight at supersonic speeds complicates the diffuser design for the following reasons [10]:

1. The existence of shock waves that lead to large decrease in stagnation pressure even in the absence of viscous effects.
2. The large variation in capture stream tube area between subsonic and supersonic flight for a given engine, as much as a factor of four between $M_\infty = 1$ and $M_\infty = 3$.
3. As M_∞ increases, the inlet compression ratio becomes a larger fraction of the overall cycle compression ratios and as a result, the specific thrust becomes more sensitive to diffuser pressure ratio.
4. It must operate efficiently both during the subsonic flight phases (takeoff, climb, and subsonic cruise) and at supersonic design speed.

Generally, supersonic intake may be classified as follows.

1. *Axisymmetric or two-dimensional intakes (Figure 9.26)*

The axisymmetric intakes use axisymmetric central cone to shock the flow down to subsonic speeds. The two-dimensional inlets have rectangular cross sections as found in the F-14 and F-15 fighter aircraft.

2. *Variable or fixed geometry (Figure 9.27)*

For variable geometry axisymmetric intakes, the central cone may move fore-and-aft to adjust the intake area. Alternatively, the inlet area is adjusted in the case of rectangular section through hinged flaps (or ramps) that may change its angles. For flight at Mach numbers much beyond 1.6, variable

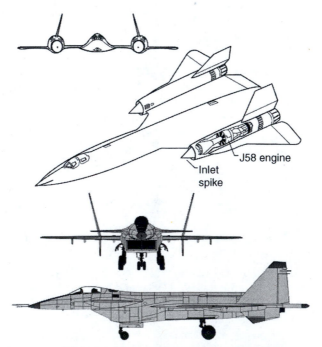

SR-71-A & MIG-MFI Axisymmetric & 2-D Intakes

FIGURE 9.26   Axisymmetric and two-dimensional supersonic intakes.

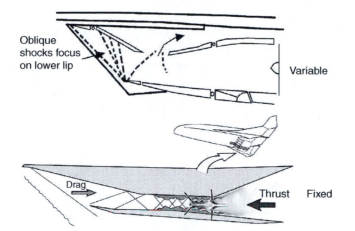

FIGURE 9.27   Variable and fixed geometry supersonic intakes.

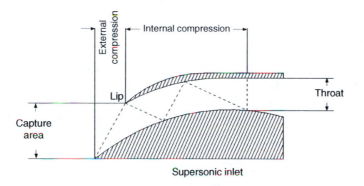

FIGURE 9.28   External and internal compression supersonic intake.

geometry features must be incorporated in the inlet to achieve high inlet pressure recoveries together with low external drag. The General Dynamics F-111 airplane has a quarter-round inlet equipped with a translating center body or spike. The inlet is bounded on the top by the wing and on one side by the fuselage. An installation of this type is often referred to as an "armpit" inlet. The spike automatically translates fore and aft as the Mach number changes. The throat area of the inlet also varies with Mach number. This is accomplished by expansion and contraction of the rear part of the spike.

3. *Internal, external or mixed compression (Figure 9.28)*

As shown in Figure 9.28 the set of shocks situated between the forebody and intake lip are identified as external shocks, while the shocks found between the nose lip and the intakes throat are called internal shocks. Some intakes have one type of shocks either external or internal and given the same name as the shocks, while others have both types and denoted as mixed compression intakes.

A constant area (or fixed geometry) intake may be either axisymmetric or two-dimensional. The supersonic pitot intake is a constant cross-sectional area intake similar to the subsonic pitot intake but it will require a sharp lip for shock wave attachment. This will substantially reduce intake drag compared with that of the subsonic round-edged lip operating at the same Mach number. Sharp edges unfortunately result in poor pressure recovery at low Mach numbers. However, this design presents a very simple solution for supersonic intakes. Variable geometry intakes may involve the use of translating center body, variable geometry center body, and the use of a cowl with a variable

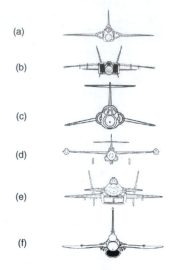

(a)

(b)

(c)

(d)

(e)

(f)

FIGURE 9.29 Intakes for military aircrafts: (a) F5D-1 Skylancer, (b) MIG-25, (c) F-104, (d) F-89, (e) MIG-FMI, and (f) F-16-A.

lip angle, variable ramp angles, and/or variable throat area. Using variable geometry inlets requires the use of sensors, which add complexity and weight to the inlet.

The use of axisymmetric or two-dimensional inlets is dependent on the method of engine installation on the aircraft, the cruise Mach number, and the type of the aircraft (i.e., military or civil) as clearly illustrated in Figures 9.15 and 9.29.

9.7.1 REVIEW OF GAS DYNAMIC RELATIONS FOR NORMAL AND OBLIQUE SHOCKS

9.7.1.1 Normal Shock Waves

For a normal shock wave, denoting the conditions upstream and downstream the shock by subscripts 1 and 2, respectively, the following relations give the downstream Mach number, static temperature and pressure ratios, density ratio, and total pressure ratio across the shock:

$$M_2^2 = \frac{(\gamma - 1)M_1^2 + 2}{2\gamma M_1^2 - (\gamma - 1)} \tag{9.7}$$

$$\frac{T_2}{T_1} = \left[\frac{2}{(\gamma + 1)M_1^2} + \left(\frac{\gamma - 1}{\gamma + 1} \right) \right] \left[\left(\frac{2\gamma}{\gamma + 1} \right) M_1^2 - \left(\frac{\gamma - 1}{\gamma + 1} \right) \right] \tag{9.8}$$

$$\frac{P_2}{P_1} = \left[1 + \left(\frac{2\gamma}{\gamma + 1} \right) \left(M_1^2 - 1 \right) \right] \tag{9.9}$$

$$\frac{\rho_2}{\rho_1} = \left[\frac{(\gamma + 1)M_1^2}{2 + (\gamma - 1)M_1^2} \right] \tag{9.10}$$

$$\frac{P_{02}}{P_{01}} = \left[\frac{(\gamma + 1)M_1^2}{2 + (\gamma - 1)M_1^2} \right]^{\gamma/(\gamma-1)} \left[\left(\frac{2\gamma}{\gamma + 1} \right) M_1^2 - \left(\frac{\gamma - 1}{\gamma + 1} \right) \right]^{-(1/(\gamma-1))} \tag{9.11}$$

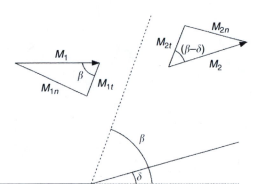

FIGURE 9.30 Nomenclature of oblique shock wave.

9.7.1.2 Oblique Shock Waves

The shown oblique shock (Figure 9.30) has a shock angle β and a deflection angle δ. Oblique shock can be treated as a normal shock having an upstream Mach number $M_{1n} = M_1 \sin \beta$ and a tangential component $M_{1t} = M_1 \cos \beta$. The tangential velocity components upstream and downstream the shocks are equal. From References 11 and 12, the following relations are given:

$$\tan \delta = \frac{2 \cot \beta (M_1^2 \sin^2 \beta - 1)}{(\gamma + 1)M_1^2 - 2(M_1^2 \sin^2 \beta - 1)} \tag{9.12a}$$

For $\gamma = 7/4$, then

$$\tan \delta = 5 \frac{M_1^2 \sin 2\beta - 2 \cot \beta}{10 + M_1^2(7 + 5 \cos 2\beta)} \tag{9.12b}$$

$$M_2^2 \sin^2(\beta - \delta) = \frac{(\gamma - 1)M_1^2 \sin^2 \beta + 2}{2\gamma M_1^2 \sin^2 \beta - (\gamma - 1)} \tag{9.13a}$$

For $\gamma = 7/4$, then

$$M_2^2 = \frac{36M_1^4 \sin^2 \beta - 5(M_1^2 \sin^2 \beta - 1)(7M_1^2 \sin^2 \beta + 5)}{(7M_1^2 \sin^2 \beta - 1)(M_1^2 \sin^2 \beta + 5)} \tag{9.13b}$$

$$\frac{P_2}{P_1} = \left[1 + \left(\frac{2\gamma}{\gamma + 1}\right)\left(M_1^2 \sin^2 \beta - 1\right)\right] \tag{9.14a}$$

For $\gamma = 7/4$, then

$$\frac{P_2}{P_1} = \left(\frac{7M_1^2 \sin^2 \beta - 1}{6}\right) \tag{9.14b}$$

$$\frac{T_2}{T_1} = \left[\frac{2}{(\gamma + 1)M_1^2 \sin^2 \beta} + \left(\frac{\gamma - 1}{\gamma + 1}\right)\right]\left[\left(\frac{2\gamma}{\gamma + 1}\right)M_1^2 \sin^2 \beta - \left(\frac{\gamma - 1}{\gamma + 1}\right)\right] \tag{9.15a}$$

For $\gamma = 7/4$, then

$$\frac{T_2}{T_1} = \frac{(7M_1^2 \sin^2 \beta - 1)(M_1^2 \sin^2 \beta + 5)}{36M_1^2 \sin^2 \beta} \tag{9.15b}$$

$$\frac{\rho_2}{\rho_1} = \left[\frac{(\gamma + 1)M_1^2 \sin^2 \beta}{2 + (\gamma - 1)M_1^2 \sin^2 \beta} \right] \tag{9.16a}$$

For $\gamma = 7/4$, then

$$\frac{\rho_2}{\rho_1} = \left[\frac{6M_1^2 \sin^2 \beta}{M_1^2 \sin^2 \beta + 5} \right] \tag{9.16b}$$

$$\frac{P_{02}}{P_{01}} = \left[\frac{(\gamma + 1)M_1^2 \sin^2 \beta}{2 + (\gamma - 1)M_1^2 \sin^2 \beta} \right]^{\gamma/(\gamma-1)} \left[\left(\frac{2\gamma}{\gamma + 1} \right) M_1^2 \sin^2 \beta - \left(\frac{\gamma - 1}{\gamma + 1} \right) \right]^{-(1/(\gamma-1))} \tag{9.17a}$$

For $\gamma = 7/4$, then the relation for total pressure ratio will be

$$\frac{P_{02}}{P_{01}} = \left[\frac{6M_1^2 \sin^2 \beta}{M_1^2 \sin^2 \beta + 5} \right]^{7/2} \left[\frac{6}{7M_1^2 \sin^2 \beta - 1} \right]^{5/2} \tag{9.17b}$$

9.7.2 External Compression Intake (Inlet)

External compression intakes complete the supersonic diffusion process outside the covered portion of the inlet where the flow is decelerated through a combination of oblique shocks (may be a single, double, triple, or multiple). These oblique shocks are followed by a normal shock wave that changes the flow from supersonic to subsonic flow. Both the normal shock wave and the throat are ideally located at the cowl lip. The supersonic diffuser is followed by a subsonic diffuser, which reduces the Mach number from high subsonic value after the last shock to the value acceptable to the engine. The simplest form of staged compression is the single oblique shock, produced by a single-angled wedge or cone that projects forward of the duct, followed by a normal shock as illustrated in Figure 9.31. The intake in this case is referred to as a *two-shock* intake. With a wedge, the flow after the oblique shock wave is at constant Mach number and parallel to the wedge surface. With a cone the flow behind the conical shock is itself conical, and the Mach number is constant along rays from the apex and varies along streamline. Forebody intake is frequently used for "external compression intake of wedge or cone form" [6].

The capture area (A_c) for supersonic intakes is defined as the area enclosed by the leading edge, or "highlight," of the intake cowl, including the cross-sectional area of the forebody in that plane. The maximum flow ratio is achieved when the boundary of the free stream tube (A_∞) arrives undisturbed at the lip. This means

$$\frac{A_\infty}{A_c} = 1.0$$

This condition is identified as the *full flow* [6] or the critical flow [5]. This condition depends on the Mach number, angle of the forebody and the position of the tip. In this case, the shock angle θ is equal to the angle subtended by lip at the apex of the body and corresponds to the maximum possible

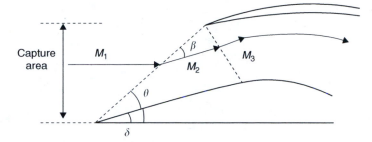

FIGURE 9.31   Single oblique shock external compression intake.

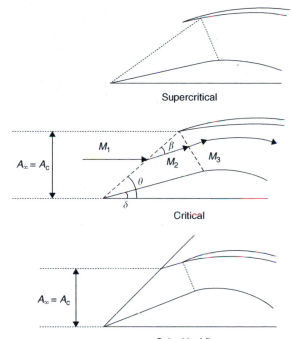

FIGURE 9.32   Types of flow in an external compression intake.

flow through the intake (Figure 9.32). This is the design point for constant area intake and the goal of the variable area intake as to stay at that condition over the flight operating range. At Mach numbers (or speeds) below the value of the critical (design) value described above, the mass flow is less than that at the critical condition and the normal shock wave occurs in front of the cowl lip and this case is identified as *subcritical*. It is to be noted here that

$$\frac{A_\infty}{A_c} < 1.0$$

Moreover, the outer drag of the intake becomes very large and smaller pressure recovery is obtained. If the at air speeds is greater than the design value, then the oblique shock will impinge inside the cowl lip and the normal shock will move to the diverging section. This type of operation is referred to as the *supercritical* operation.

The two-shock intake is only moderately good at Mach 2.0 and unlikely to be adequate at higher Mach numbers [4]. The principle of breaking down an external shock system can be extended to

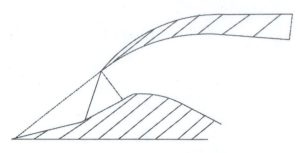

FIGURE 9.33   Three shocks for double-cone (or double-wedge) geometry.

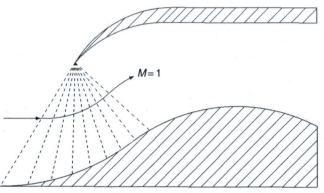

Isotropic compression supersonic inlet

FIGURE 9.34   Isotropic compression supersonic inlet.

any desired number of stages. The next step is the three-shock intake where two oblique shocks are followed by a normal shock, where the double-wedge and double-cone are the archetypal forms (Figure 9.33).

Continuing the process of breaking down the external shock system, three or more oblique shocks may be used ahead of the normal shock. For a system of $(n-1)$ oblique shocks, the pressure recovery factor will be

$$r_d \frac{P_{02}}{P_{0\infty}} = \frac{P_{02}}{P_{01}} \frac{P_{03}}{P_{02}} \frac{P_{04}}{P_{03}} \cdots \frac{P_{0n}}{P_{0n-1}} \left(\frac{P_{0n}}{P_{0n-1}} \right)_{\text{normal shock}} \tag{9.18}$$

Two remarks are to be mentioned here:

1. As the number of oblique shocks increases, the pressure recovery factor increases.
2. Up to Mach 2, equal deflections of the successive wedge angles give the best results, while for higher Mach numbers the first deflection angle needs to be the smallest and the last the largest [4].

Extending the principle of multishock compression to its limit leads to the concept of *isentropic compression,* in which a smoothly contoured forebody produces an infinitely large number of infinitely weak oblique shocks, Figure 9.34. In this case, the supersonic stream is compressed with no losses in the total pressure.

Example 3   A fighter aircraft is flying at an altitude of 50,000 ft where the ambient pressure and temperature are respectively 1.682 psia and 390°R. The aircraft has the shown two-dimensional inlet where air enters the intake inlet with a Mach number of 2.5. Within the intake diffuser, a

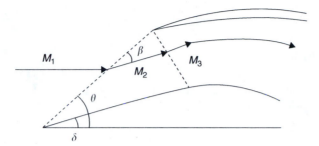

FIGURE 9.35   Two-shock intake.

single oblique shock wave is developed followed by a normal shock. The two-dimensional oblique shock has an oblique shock angle of 30°. Assuming constant specific heats ($C_P = 0.24$ Btu/Ib$_m$ °R), determine the following:

1. The total temperature and pressure of the air entering the oblique shock
2. The static and total conditions after the oblique shock
3. The Mach number after the oblique shock
4. The flow deflection angle (δ)
5. The Mach number after the normal shock
6. The static and total conditions after the normal shock (Figure 9.35)

Solution:

1. With $\gamma = 1.4$

$$V_1 = 2.5\sqrt{\gamma g_C R T_1}$$

$$V_1 = 2.5\sqrt{(1.4)(32.174)(53.35)(390)}$$

$$V_1 = 2420 \text{ ft/s}$$

$$T_{01} = T_1 + \frac{V^2}{2g_C C_P J} = 390 + \frac{(2420)^2}{(2)(32.174)(0.24)(778)} = 390 + 487$$

$$T_{01} = 877°R$$

$$P_{01} = P_1 \left(\frac{T_{01}}{T_1}\right)^{\gamma/(\gamma-1)} = 1.682 \left(\frac{877}{390}\right)^{1.4/0.4}$$

$P_{01} = 28.68$ psia

2. Since oblique shock wave can be treated as a normal shock wave with the normal component of the Mach number as the Mach number upstream the normal shock wave, then for a shock angle of 30°, the normal and tangential components of the Mach number are

$$M_{1n} = 2.5(\sin 30) = 1.25$$

$$M_{1t} = 2.5(\cos 30) = 2.165$$

The conditions downstream the oblique shock waves are determined either from tables of normal shock waves or from the mathematical expressions of normal shock waves, noting that in both methods the normal component of Mach number is used.

From normal shock wave tables with $M_{1n} = 1.25$, then

$$M_{2n} = 0.814 \quad \frac{P_2}{P_1} = 1.659 \quad \frac{P_{02}}{P_{01}} = 0.986 \quad \frac{T_2}{T_1} = 1.159$$

As an alternative, relations 9.7 through 9.11 are used.
From Equation 9.7 with $M_{1n} = 0.814$, then

$$M_{2n} = 0.813$$

and

$$\frac{P_2}{P_1} = 1.656, \quad \frac{P_{02}}{P_{01}} = 0.987, \quad \frac{T_2}{T_1} = 1.159$$

Both procedures give nearly the same results.
The total pressure after the oblique shock wave is

$$P_{02} = \frac{P_{02}}{P_{01}} P_{01} = (0.986)(26.68)$$
$$= 26.31 \text{ psia}$$

The static pressure after the oblique shock wave

$$P_2 = (P_2/P_1)P_1 = (1.695)(1.682)$$
$$P_{02} = 2.7904 \text{ psia}$$

The static temperature after the oblique shock wave

$$T_2 = (T_2/T_1)T_1 = (1.159)(390)$$
$$= 452.2°R$$

3. The tangential velocity across an oblique shock does not change but since the static temperature changes, the tangential Mach numbers before and after the shock are unequal. Now since

$$V_{1t} = V_{2t}$$

$$\therefore \frac{V_{1t}}{\sqrt{\gamma RT_1}}\sqrt{T_1} = \frac{V_{2t}}{\sqrt{\gamma RT_2}}\sqrt{T_2}$$

$$M_{1t}\sqrt{T_1} = M_{2t}\sqrt{T_2}$$

$$M_{2t} = M_{1t}\sqrt{\frac{T_1}{T_2}}$$

$$\therefore M_{2t} = 2.165\sqrt{\frac{1}{1.159}} = 2.0106$$

$$\therefore M_2 = \sqrt{M_{2n}^2 + M_{2t}^2} = \sqrt{(0.814)^2 + (2.0106)^2}$$

$$= 2.169$$

4. The angle of the streamline leaving the oblique shock is β where

$$\tan \beta = \frac{M_{2n}}{M_{2t}} = \frac{0.814}{2.0106} = 0.4048$$

$$\therefore \beta = 22.04°$$

$\therefore$ The deflection angle $\delta = \theta - \beta = 30 - 22.04 = 7.96°$.

5. The Mach number after the normal shock wave is to be calculated from Equation 9.6.

$$M_3^2 = \frac{(0.4)(2.0106)^2 + 2}{(2)(1.4)(2.0106)^2 - 0.4}$$

$$M_3 = 0.5755$$

6. The total pressure ratio across the normal shock wave may be obtained from Equation 9.11.

$$\frac{P_{03}}{P_{02}} = \left[\frac{(2.4)(2.0106)^2}{2 + (0.4)(2.0106)^2} \right]^{3.5} \left[\left(\frac{2.8}{2.4} \right)(2.0106)^2 - \frac{0.4}{2.4} \right]^{-2.5}$$

$$\frac{P_{03}}{P_{02}} = 0.7159$$

$$P_{03} = 0.7159 \times 26.31 = 18.835 \text{ psia}$$

7. The static temperature is calculated from Equation 9.8.

$$\frac{T_3}{T_2} = \left[\frac{2}{(2.4)(2.0106)^2} + \frac{0.4}{2.4} \right] \left[\left(\frac{2.8}{2.4} \right)(2.0106)^2 - \frac{0.4}{2.4} \right]$$

$$\frac{T_3}{T_2} = 1.696$$

$$T_3 = (1.696)(452.2) = 767°\text{R}$$

The static pressure is obtained from Equation 9.9.

$$\frac{P_3}{P_2} = 1 + \frac{2.8}{2.4}[(2.0106)^2 - 1] = 4.55$$

$$P_3 = (4.55)(2.7904) = 12.695 \text{ psia}$$

9.7.3 INTERNAL COMPRESSION INLET (INTAKE)

The internal compression inlet locates all the shocks within the covered passageway (Figure 9.36). The terminal shock wave is also a normal one, which is located near or at the throat.

A principal difference between internal and external compression intakes is that with internal compression, since the system is enclosed, oblique shocks are reflected from an opposite wall, which have to be considered. The simplest form is a three-shock system. The single-wedge turns the flow toward the opposite wall. The oblique shock is reflected from the opposite wall and the flow passing the reflected shock is restored to an axial direction. A normal shock terminates the supersonic as usual. For a symmetrical intake, as shown in Figure 9.34, there are four oblique shocks. The reflected shocks are part of the symmetrical four-shock intersection system.

Again, there are three flow patterns resembling critical (design), subcritical, and supercritical cases. The inlet is said to be operating in the critical mode, when the normal shock wave is located

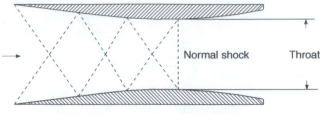

All internal compression supersonic inlet

FIGURE 9.36   Internal compression supersonic intake.

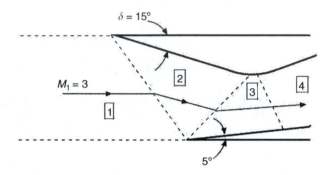

FIGURE 9.37   Three-shock system.

at the throat, resulting in the highest pressure recovery factor. In subcritical operation, the terminal (normal) shock wave is located upstream of the throat, or in the converging part of the diffuser. Supercritical mode occurs when the terminal shock is located downstream of the throat, or in the diverging part of the diffuser.

Subcritical flow operation is an unstable mode [5]. The terminal shock may move upstream or ahead of the cowl lip producing a condition called unstart. During unstart condition, the pressure recovery factor will drop and flow will spill over the cowl, which will produce high drag.

For the above reason, internal supersonic compression intakes operate in the supercritical mode rather than in the critical one. This represents a margin of stability if the inlet flow is suddenly changed.

9.7.4 MIXED COMPRESSION INTAKES

These inlets use a combined external and internal compression system (Figure 9.28). In the present days, all the supersonic intakes are of the mixed compression type.

Example 4   For the mixed compression two-dimensional supersonic inlet shown, the free stream Mach number is 3.0. Three-shock system (two oblique and normal shocks) reduces the speed from supersonic to subsonic speeds. Calculate the following:

1. The overall total pressure ratio
2. The overall static pressure ratio (Figure 9.37)

Solution:

1. The first oblique shock wave has an upstream Mach number $M_1 = 3$ and $\delta_1 = 15°$; then from tables $\beta_1 = 32.25°$.

Alternatively, from Equation 9.12, the angle (β_1) is determined iteratively. The same value of $\beta_1 = 32.25°$ is obtained.

The Mach number after the shock wave is determined from the relation

$$M_2^2 = \frac{(\gamma - 1)M_1^2 \sin^2 \beta + 2}{2\gamma M_1^2 \sin^2 \beta - (\gamma - 1)} \frac{1}{\sin^2(\beta - \delta)}$$

With $\beta_1 = 32.25°$ and $\delta_1 = 15°$, then

$$M_2 = 2.25$$

From Equation 9.17, the total pressure ratio is

$$\frac{P_{02}}{P_{01}} = 0.8935$$

The flow in the region 2 is deflected at angle

$$\delta_2 = 15 + 5 = 20°$$

Then with $M_2 = 2.25$, from either tables or equations, $\beta_2 = 46.95°$

$$M_2^2 = \frac{36M_1^4 \sin^2 \beta - 5(M_1^2 \sin^2 \beta - 1)(7M_1^2 \sin^2 \beta + 5)}{(7M_1^2 \sin^2 \beta - 1)(M_1^2 \sin^2 \beta + 5)}$$

$$M_3 = 1.444$$

The total pressure ratio is then

$$\frac{P_{03}}{P_{02}} = 0.878$$

The flow in region 4 is downstream a normal shock wave

$$M_4 = 0.7219$$

$$\frac{P_{04}}{P_{03}} = 0.9465$$

$$\therefore \frac{P_{04}}{P_{01}} = \frac{P_{04}}{P_{03}} \times \frac{P_{03}}{P_{02}} \times \frac{P_{02}}{P_{01}}$$

$$= 0.9465 \times 0.878 \times 0.8935$$

The intake pressure recovery factor is

$$\frac{P_{04}}{P_{01}} = 0.7865$$

2. To evaluate the static pressure ratio, tables of the normal shock waves will be used (or Equations 9.7 through 9.11).

For the first oblique shock, with $M_1 = 3$ and $\beta_1 = 32.25$

$$\therefore\ M_{1n} = M_1 \sin \beta_1$$
$$= 3 \sin 32.25 = 3 \times 0.5446$$
$$= 1.6008$$

The static pressure ratio $P_2/P_1 = 2.82$
For $M_2 = 2.25$ and $\beta_2 = 47$

$$\therefore\ M_{2n} = M_2 \sin \beta_2 = 2.25 \times 0.682$$
$$= 1.6455$$

The static pressure ratio $P_3/P_2 = 2.992$
For the last normal shock

$$M_3 = 1.444, \text{ then } P_4/P_3 = 2.333$$

The overall static pressure ratio is

$$\frac{P_4}{P_1} = \frac{P_4}{P_3} \times \frac{P_3}{P_2} \times \frac{P_2}{P_1} = 2.333 \times 2.992 \times 2.82$$

The overall static pressure ratio is

$$\frac{P_4}{P_1} = 19.691$$

9.8 MATCHING BETWEEN INTAKE AND ENGINE

The intake, as in other engine modules, is designed for certain operating conditions. However, the aircraft operates through a flight envelop containing enormous operating conditions. Therefore, the intake has to cope with these different situations. This is accomplished by different capabilities for ingesting extra air quantity in some cases (takeoff) and dump unneeded air through bleeding in other cases. Some engines, particularly supersonic ones, are fitted with variable geometry intakes. The engine control unit (ECU) controls both air and fuel flow into the engine that fulfills the requirements of different operating conditions.

Example 5 A ramjet engine is powering an aircraft flying at an altitude 15 km where the ambient temperature and pressure are 216.5 K and 12.11 kPa. The maximum temperature is 2500 K and the fuel heating value is 43,000 kJ/kg. To examine the effect of intake design on the performance of the engine, it is assumed that only one normal shock converts the supersonic flow into subsonic flow. Moreover, all the other components are ideal. Plot the specific thrust and thrust specific fuel consumption versus the Mach number.

Solution: Since only one normal shock exists in the intake, then the pressure recovery is the ratio between the total pressures upstream and downstream the shock wave, or

$$r_{\mathrm{d}} = \frac{P_{02}}{P_{01}}$$

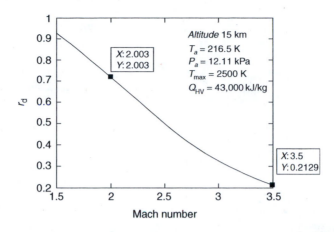

FIGURE 9.38 Pressure recovery due to a normal shock versus the Mach number for ramjet engine.

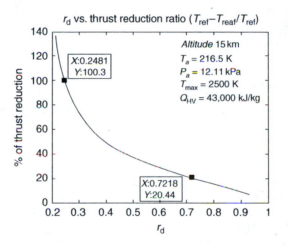

FIGURE 9.39 Percentage of thrust loss versus pressure recovery for a ramjet engine.

This ratio is obtained either from the normal shock tables or from the Equation 9.11. The procedure for obtaining the specific thrust is described in Chapter 3 and here only the obtained results will be plotted. Figure 9.38 illustrates the relation between the free stream Mach number and the pressure recovery factor. At Mach number of nearly 2.0, the pressure recovery is nearly 0.72, while for a Mach number of 3.5 the pressure recovery is 0.212. This assures that the efficient intake cannot use only one normal shock wave to reduce the air speed from supersonic speed to low subsonic speed suitable for subsonic combustors.

Figure 9.39 illustrates the drastic drop in thrust with the increase in Mach number if the intake employs a one normal shock wave. For the same case of Mach number of 2.0, and pressure recovery of 0.72, the thrust drops by nearly 20%. Moreover, at Mach number of 3.5, the shock losses would become so severe that all the engine thrust would be lost.

Comment

A similar case was discussed for a turbojet engine in Reference 13; the examined intake also had only one normal shock wave. The thrust was completely lost at Mach number of 3.5.

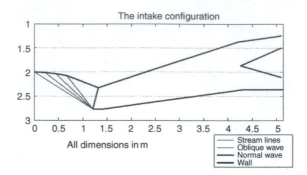

FIGURE 9.40   Intake geometry.

9.9 CASE STUDY

The turbo ramjet engine previously described in Chapter 7 is analyzed here. The geometry of the intake for the turbojet engine operating at low supersonic speeds (up to Mach number 2.0) is shown in Figure 9.40. Two designs are examined; namely, four oblique shocks and eight oblique shocks. In both cases, a normal shock wave exists downstream the oblique shocks. The operating conditions are listed in the following table:

The design point of the turbojet intake

	Intake Design Point
Mach number (M)	2
Altitude	12.507 km
Ambient temperature (T_a)	216.66 K
Ambient pressure (P_a)	17.857 kPa
Ambient density (ρ_a)	0.2872 kg/m^3
Flight speed (V_∞)	590.0983 m/s
Mass flow rate	130 kg/s

The following table presents the shock characteristics as well as the Mach number and speed downstream successive shocks:

Four oblique shocks intake

Position	$\delta(°)$	$\beta(°)$	M_e	V_e (m/sec)	P_{0e}/P_{0i}
After 1st wave	3	32.5055	1.8924	571.8126	0.9995
After 2nd wave	3.1026	34.6033	1.7839	552.0081	0.9995
After 3rd wave	4.0353	37.8574	1.6454	524.6129	0.9991
After 4th wave	4.4098	41.9685	1.4953	492.0425	0.9989
After normal wave	—	—	0.7027	265.3856	0.93127
After diffuser	—	—	0.3386	132.5240	0.98620

The pressure recovery factor for the four oblique shocks is 91.57% and the intake efficiency is 94.41%. The static pressure, temperature, and density variation along the diffuser is plotted in Figure 9.41. The Mach number, speed, and overall total pressure ratio are plotted along the diffuser in Figure 9.42.

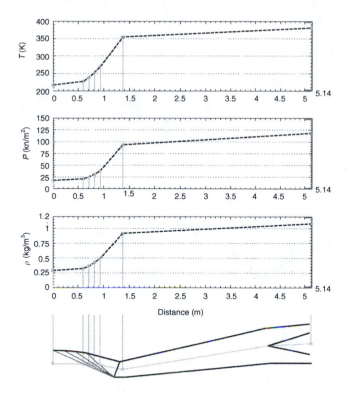

FIGURE 9.41   Static pressure and temperature as well as density distribution for a four oblique shock case.

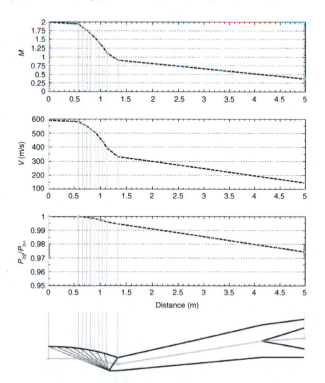

FIGURE 9.42   Mach number, speed, and total pressure ratio distribution for a four oblique shock case.

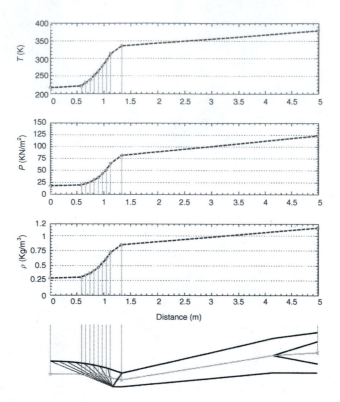

FIGURE 9.43   Static temperature and pressure as well as density distribution for an eight oblique shock case.

The table below presents the shock wave characteristics for the ten oblique shock case:

Ten oblique shocks intake

Position	$\delta(°)$	$\beta(°)$	M_e	V_e (m/sec)	P_{0e}/P_{0i}
After 1st wave	1.5	31.2259	1.9459	581.0759	0.9999
After 2nd wave	2.0263	33.1751	1.8518	564.5681	0.9998
After 3rd wave	2.6578	34.4534	1.7814	551.5333	0.9997
After 4th wave	2.74903	37.1048	1.6713	529.9305	0.9996
After 5th wave	3.2082	39.4409	1.5781	510.3819	0.9995
After 6th wave	3.5027	43.5466	1.4449	480.4076	0.9994
After 7th wave	3.9039	48.2073	1.3219	450.4606	0.9993
After 8th wave	4.0155	55.8417	1.1659	409.2317	0.9991
After normal wave	—	—	0.8642	319.0882	0.9956
Subsonic diffuser	—	—	0.3704	144.6517	0.9813

Pressure recovery for the eight oblique shocks case is 97.37% while the intake efficiency is 98.29%. The static temperature, pressure, and density variation along the diffuser is plotted in Figure 9.43. The Mach number, speed, and overall total pressure ratio are plotted along the diffuser in Figure 9.44.

From both cases, it is clarified that increasing the number of shocks increases the effectiveness of the intake and consequently the pressure recovery and adiabatic efficiency of the intake. Ultimately, by a continuous increase of the number of oblique shocks, there will be no need for the normal shock and the case of isentropic diffuser will be reached.

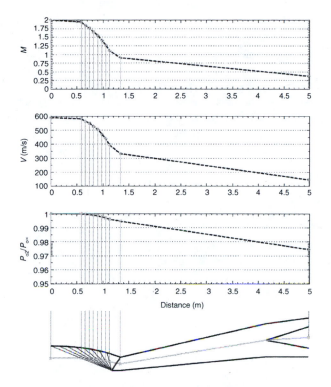

FIGURE 9.44 Mach number, speed, and total pressure ratio distribution for an eight oblique shock case.

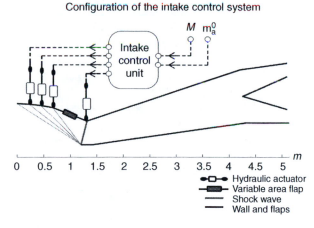

FIGURE 9.45 Actuation system for the variable geometry four oblique shock intake.

The intake is of the variable geometry type. A simplified layout for the actuation system is shown in Figure 9.45.

PROBLEMS

9.1 A fighter airplane is flying at an altitude where the ambient pressure and temperature are 11.6 kPa and 216.7 K. The intake is composed of a supersonic part followed by a subsonic part. Within the supersonic part two shock waves are generated; namely an oblique shock followed by a normal shock wave. After the normal shock wave, the Mach number is 0.504. Within the subsonic part pressure recovery is 98.7% and the total temperature remains constant. The static pressure and temperature after the normal shock wave is 111.5 kPa and 600 K. It is required to

(a) Calculate the turning angle of the supersonic diffuser δ.
(b) Calculate the Mach number at the inlet of the supersonic diffuser.

9.2 The figure shown illustrates a four-shock type intake where three oblique shocks are followed by a normal shock. The Mach number after the normal shock is 0.6. Find

(a) The free stream Mach number
(b) The static pressure ratio after the four shocks
(c) If the total pressure ratio within the subsonic diffuser is 0.99, what will be the pressure recovery factor for the intake

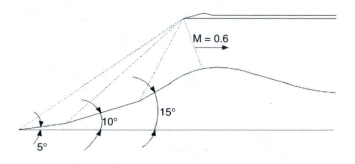

9.3 A supersonic fighter has the shown intake, which is fitted with a movable spike. The circular intake has an inlet radius $R = 1$ m. During ground run, the following data was recorded:

Air mass flow rate	350 kg/s
Ambient temperature	288 K
Ambient pressure	101 kPa
Free stream velocity	100 m/s

What will be the distance x of the outer portion of the spike?

If the needed mass flow rate becomes 380 m/s, two options are suggested; either retracting the spike inward to increase the inlet area or to open the 8 blow up doors. Calculate the new position of the spike (x'). If blow up doors are opened, calculate the length of doors, if the subtended angle to each door is 30° and the length (L) is equal to 1.6 m.

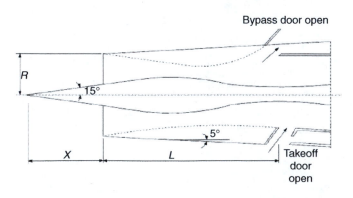

9.4 The two-dimensional or wedge-type diffuser consists of two external oblique shocks, a normal shock and a subsonic diffuser as shown in the figure below. Assuming that the external losses are due to the shock losses and the pressure drop in the subsonic part is 4.5%,
  (a) Determine the total pressure ratio for the shown Mach number of 3.5.
  (b) If the height of intake is 0.5 m, what will be the external length of the wedge?

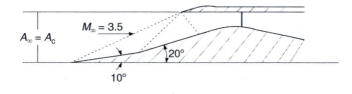

9.5 A two-dimensional wedge-type diffuser of a military aircraft flying at a Mach number of 2.5 and at an altitude, where the ambient conditions are $-56.5°C$ and 12 kPa is shown in figure P01. The Mach number of the flow leaving the throat is 0.7; what is the value of the wedge angle δ_1? Design the shown two-dimensional supersonic inlet by calculating the static and total temperatures and pressures at different states of the inlet.

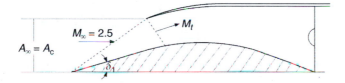

9.6 It is required to examine the case of a subsonic inlet if installed to a fighter airplane flying at a Mach number of 1.8 as shown in the figure. Calculate the Mach number downstream of the normal shock wave. If the Mach number at the lips of the subsonic intake is 0.5, what will be the ratio between the mass flow rate entering the engine and the approaching mass upstream the normal shock and having the same area ($A_\infty = A_c$).

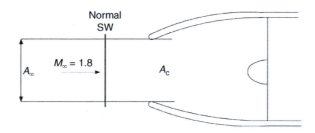

9.7 The supersonic axysymmetric intake shown is fitted to an aircraft flying at Mach number of 2.1 at an altitude of 12 km. If the supersonic turbofan engine powering this fighter consumes 35 kg/s and the spike length outside the intake is 0.8 m and intake radius is 0.6 m, determine whether the intake side doors are inject or eject airflow from the intake. What will be the spike length that allows exact airflow rate into the engine if the side doors are closed?

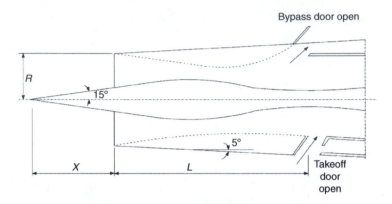

9.8 The figure shown illustrates a supersonic intake of an engine installed to the lower surface of the wing of a fighter airplane. The airplane has a maximum Mach number of 2.3. Two shocks are generated (oblique and normal) and the long duct upstream of the engine is a subsonic part having a pressure drop of 10%. The Mach number at the engine inlet is 0.5, which is 15% less than the Mach number downstream of the normal shock wave. Calculate the wedge angle of the oblique shock wave. What will be the overall pressure ratio of this intake? Have you any suggestions?

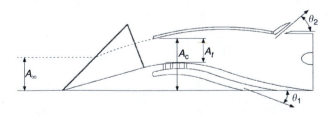

9.9 Is the intake shown in its design or off-design conditions? The flight Mach number is 2.5 and the flight altitude is 18 km. The wedge angle is 10°. Calculate the ratio A_∞/A_c if the spike length outside the engine is 1.0 m and the intake width is 1.0 m. What is the total pressure ratio of the intake if the subsonic part develops 15% losses.

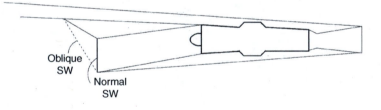

9.10 The supersonic two-dimensional intake shown has three oblique shocks and a final normal shock. The first wedge angle is $\delta = 5°$. Each subsequent wedge has an additional 1°. The flight

Mach number is 2.2; what will be the efficiency of the intake. For a flight altitude of 9 km, calculate the static and total pressures downstream each shock.

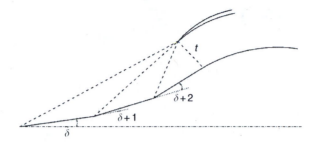

REFERENCES

1. J.J. Mahoney, *Inlet for Supersonic Missiles*, AIAA Education Series, 1990, p. 1.
2. R.S. Shevel, *Fundamentals of Flight*, Prentice-Hall Inc., 1983, p. 348.
3. I. Kroo and R. Shevel, *Aircraft Design: Synthesis and Analysis, Digital Textbook version 0.99*, Stanford University, January 2001, p. 374.
4. *The Jet Engine*, Rolls-Royce plc, 5th edn., Reprinted 1996 with revisions, p. 245.
5. W.W. Bathie, *Fundamentals of Gas Turbines*, 2nd edn., John Wiley & Sons, Inc., 1996, p. 330.
6. J. Seddon and E.L. Goldsmith, *Intake Aerodynamics*, AIAA Education Series, 1989, p. 19.
7. P. Hill and C. Peterson, *Mechanics and Thermodynamics of Propulsion*, 2nd edn., Addison Wesley Publication Company, Inc., 1992, p. 219.
8. *Power plant*, Volume II, Boeing publication D6-1420, pp. 2–36.
9. J. Roskam, *Airplane Design, Volume VI: Preliminary Calculation of Aerodynamic, Thrust and Power Characteristics*, Roskam Aviation and Engineering Corporation, 1990, p. 152.
10. J.L. Kerrebrock, *Aircraft Engines and Gas Turbines*, 2nd edn., MIT Press, 1992, p. 121.
11. *Compressible Flow Data Book for Data Book for Engineering Tripos*, 2004 edn., Cambridge University Engineering Department.
12. Equations, Tables and Charts for Compressible Flow, NACA Report 1153, 1953, Ames Research Staff.
13. D.D. Wyatt, Air intake problems in supersonic propulsion. In *A Review of Supersonic Air Intake Problems*, J. Fabri (ed.), Pergamon Press, 1958, p. 21.

10 Combustion Systems

10.1 INTRODUCTION

The combustion process in aircraft engines and gas turbines is a heat addition process to the compressed air in the combustor or burner. Thus, the combustion is a direct-fired air heater in which fuel is burned. The combustor is situated between the compressor and turbine, where it accepts air from the compressor and delivers it at elevated temperature to the turbine. Some engines have a second combustion system which either reheats the flow for the later turbine stages (as described in Chapter 8) or burns more fuel in an afterburner behind the turbines to provide the high exhaust velocity required for propulsion of supersonic aircraft (as described in Chapters 4 and 5). A part of the supplied energy provides useful work absorbed by the turbine(s) and the remainder goes to waste as heat in the exhaust gas [1]. Combustion in gas turbines is a continuous process that takes place at high pressure in a smaller space and usually at a very high temperature. Fuel is burned in the air supplied by the compressor; an electric spark is required for initiating the combustion process, and thereafter the flame must be self-sustaining.

The design of combustors is a very complicated process where aerodynamics, chemical reactions, and mechanical design are linked together [2]. The interweaving of the various processes and the hardware geometry is reflected in the empirical nature of much of the design process. Though the individual processes are well known, their combination into a working combustion system owes much to experimentation and experience rather than to mathematical modeling. Recent developments in computational fluid mechanics (CFD) have helped reduce this experimentation. To understand how complex the design of a combustion chamber is, it is necessary to identify the main requirements from gas turbine combustors. These requirements may be summarized as follows [3]:

1. Its length and frontal area to remain within the limits set by other engine components, that is, size and shape compatible with engine envelop.
2. Its diffuser minimizes the pressure loss.
3. The presence of a liner to provide stable operation [i.e., the flame should stay alight over a wide range of air–fuel ratios (AFRs)].
4. Fulfills the pollutant emissions regulations (low emissions of smoke, unburned fuel, and gaseous pollutant species).
5. Ability to utilize much broader range of fuels.
6. Durability and relighting capability.
7. High combustion efficiency at different operating conditions: (1) altitude ranging from sea level to 11 km for civil transport and higher for some military aircraft and (2) Mach numbers ranging from zero during ground run to supersonic for military aircraft.
8. Design for minimum cost and ease of maintenance.
9. An outlet temperature distribution (pattern form) that is tailored to maximize the life of the turbine blades and nozzle guide vanes.
10. Freedom from pressure pulsations and other manifestations of combustion-induced instabilities.

11. Reliable and smooth ignition both on the ground (especially at very low ambient temperature) and in the case of aircraft engine flameout at high altitude.
12. The formation of carbon deposits (coking) must be avoided, particularly the hard brittle variety. Small particles carried into the turbine in the high velocity gas stream can erode the blades. Furthermore, aerodynamically excited vibration in the combustion chamber might cause sizeable pieces of carbon to break free, resulting in even worse damage to the turbine.

According to the present aircraft, combustors may be classified as either *subsonic* or *supersonic*, depending on the velocity of combustion. Moreover, subsonic combustors may be subdivided into *axial* flow, *reverse* flow, and *cyclone* types. The last type is little used with present gas turbines. Axial flow combustors may be subdivided into *tubular*, *tubo-annular*, and *annular* types. Subsonic combustion chambers have three zones: (1) a recirculation zone, (2) a burning zone, and (3) a dilution zone [4].

10.2 SUBSONIC COMBUSTION CHAMBERS

The combustion process occurs at subsonic speeds; air normally leaves the compressor at 150 m/s or at Mach numbers less than unity (0.3–0.5). Apart from scramjet engines, all the available aero engines and gas turbines have subsonic combustors. There are three main types of subsonic combustion chambers in use in gas turbine engines, namely, multiple chamber (tubular or can type), tubo-annular chamber, and the annular chamber.

10.2.1 TUBULAR (OR MULTIPLE) COMBUSTION CHAMBERS

The first turbojet engines invented by Sir Whittle and von Ohain had subsonic tubular combustor. Tubular type is sometimes identified as multiple- or can-type combustion chamber.

As shown in Figure 10.1 this type of combustor is composed of cylindrical chambers disposed around the shaft connecting the compressor and turbine. Compressor delivery air is split into a number of separate streams, each supplying a separate chamber [5]. These chambers are interconnected to allow stabilization of any pressure fluctuations. Ignition starts sequentially with the use of two igniters. The Rolls-Royce Dart uses tubular type combustor.

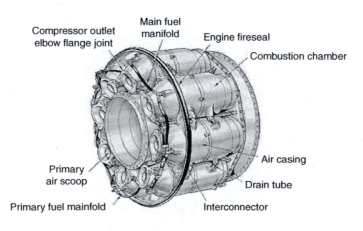

Mutiple can

FIGURE 10.1 Multiple combustion chambers. (Reproduced from Rolls-Royce plc, copyright © Rolls-Royce plc 2007. With permission.)

The number of combustion chambers varies from 7 to 16 per engine. The present trend is to have fewer numbers (8 or 10) and even a single tubular chamber may be preferred. On American-built engines these chambers are numbered in a clockwise direction facing the rear of the engine with the number (1) chamber at the top. The can-type combustion chamber is typical of the type used on both centrifugal and axial-flow engines. It is particularly well suited for the centrifugal compressor engine since the air leaving the compressor is already divided into equal portions as it leaves the diffuser vanes. It is then a simple matter to duct the air from the diffuser into the respective combustion chambers arranged radially around the axis of the engine. The advantages of tubular type are as follows:

- Mechanically robust
- Fuel flow and airflow patterns are easily matched
- Rig testing necessitates only a small fraction of total engine air mass flow
- Easy replacement for maintenance

The disadvantages are as follows:

- Bulky and heavy
- High pressure loss
- Requires interconnectors
- Incur problem of light-round
- Large frontal area and high drag

For these reasons, tubular type is no longer used in current designs. Small gas turbines used in auxiliary power units (APUs) and automotives are designed with a single can.

10.2.2 TUBO-ANNULAR COMBUSTION CHAMBERS

This type may also be identified as can-annular or cannular. It consists of a series of cylindrical burners arranged within common single annulus as is shown in Figure 10.2. Thus, it bridges the evolutionary gap between the tubular (multiple) and annular types. It combines the compactness of the annular chamber with the best features of the tubular type.

The combustion chambers are enclosed in a removable shroud that covers the entire burner section. This feature makes the burners readily available for any required maintenance. Cannular combustion chambers must have fuel drain valves in two or more of the bottom chambers. This ensures drainage of residual fuel to prevent its being burned at the next start. The flow of air through the holes and louvers of the can-annular system is almost identical to the flow through other types of burners. Reverse-flow combustors are mostly of the can-annular type. Reverse-flow combustors make the engine more compact. Pratt & Whitney use can-annular-type combustion chamber in their JT3 axial-flow turbojet engine. Moreover, General Electric and Westinghouse use this type in their industrial gas turbines. Advantages of can-annular types are as follows:

- Mechanically robust.
- Fuel flow and airflow patterns are easily matched.
- Rig testing necessitates only a small fraction of total engine air mass flow.
- Shorter and lighter than tubular chambers.
- Low pressure loss.

Their disadvantages are as follows:

- Less compact than annular
- Requires connectors
- Incur a problem of light around

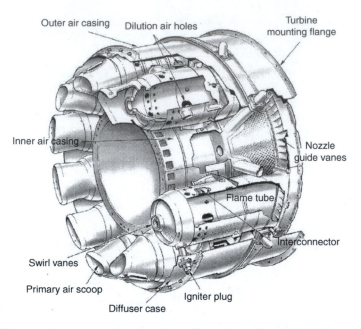

FIGURE 10.2 Tubo-annular combustion chamber. (Reproduced from Rolls-Royce plc, copyright © Rolls-Royce plc 2007. With permission.)

10.2.3 ANNULAR COMBUSTION CHAMBERS

In this type an annular liner is mounted concentrically inside an annular casing. This combustor represents the ideal configuration for combustors since its "clean" aerodynamic layout results in compact dimensions (and consequently an engine of small diameter) (see Figure 10.3) and lower pressure loss than other designs. Usually, enough space is left between the outer liner wall and the combustion chamber housing to permit the flow of cooling air from the compressor. Normally, this type is used in many engines using axial-flow compressor and also others incorporating dual-type compressors (combinations of axial flow and centrifugal flow). Currently, most aero engines use annular type combustors, examples of which are the V 2500 engine, Rolls-Royce Trent series, General Electric's GE-90, and Pratt & Whitney's PW4000 series of aero engines. Moreover, several industrial gas turbines such as ABB and Siemens plants of over 150 MW fall into this category.

The advantages of annular type may be summarized as follows:

- Minimum length and weight (its length is nearly 0.75 of cannular combustor length).
- Minimum pressure loss.
- Minimum engine frontal area.
- Less wall area than cannular and thus cooling air required is less; thus the combustion efficiency raises as the unburnt fuel is reduced.
- Easy light-round.
- Design simplicity.
- Combustion zone uniformity.
- Permits better mixing of the fuel and air.
- Simple structure compared to can burners.
- Increased durability.

However, it has the following disadvantages:

- Serious buckling problem on outer liner.
- Rig testing necessitates full engine air mass flow.

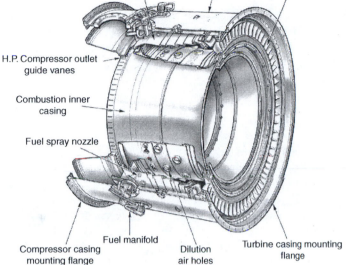

FIGURE 10.3   Annular type combustor. (Reproduced from Rolls-Royce plc, copyright © Rolls-Royce plc 2007. With permission.)

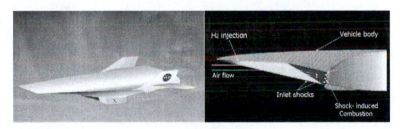

FIGURE 10.4   Supersonic combustion in X-43 aircraft.

- Difficult to match fuelflow and airflow patterns.
- Difficult to maintain stable outlet temperature transverse.
- Must remove the engine from aircraft to disassemble for maintenance and overhaul.

10.3 SUPERSONIC COMBUSTION CHAMBER

Supersonic combustion exists only in scramjet engines. Supersonic combustor is mechanically simple but aerodynamically vastly more complex than a subsonic one. Hydrogen is normally the fuel used. Scramjet engine has a 2D combustor. A typical layout is given in Figure 10.4 for the combustor of X-43 aircraft powered by a scramjet engine. Details of supersonic combustion will be discussed later in this chapter.

10.4 COMBUSTION PROCESS

The objectives of the combustion process are to introduce and burn a fuel in the compressed air flowing through the combustor with the minimum pressure loss and with as complete a utilization of fuel as possible. From a thermodynamic viewpoint the process occurring in the combustor may take place in two major steps:

1. The introduction of the liquid fuel in the form of a fine spray of droplets, mixing it with the primary air and vaporizing it.
2. The combustion (chemical reaction) of vaporized fuel and the thorough mixing of the resulting combustion products with the secondary air.

Step 1 above results in a decrease in the temperature of the working fluid before combustion since both the enthalpy used in raising the temperature of the liquid fuel to its boiling point and the latent heat of evaporation of the fuel are absorbed from the enthalpy of the warm compressed air.

In Step 2 the combustion process raises the temperature of the mixture of combustion products and secondary air to the desired temperature for the gases entering the turbine nozzle ring. That temperature is limited by the permissible operating temperature for the turbine blades. For a complete combustion process, the rate of reaction must be defined as the process efficiency depends on it.

The above steps will be explained in detail here considering the simple combustion chamber illustrated in Figure 10.5.

Air from the engine compressor enters the combustion chamber at a velocity of the range of 150 m/s. This high air speed is far too high for combustion. The first step is to decelerate (diffuse) this airflow to say 20–30 m/s and raise its static pressure. Since the speed of burning kerosene at normal mixture ratios is only a few meters per second, any fuel lit even in the diffused air steam would be blown away [5]. A region of low axial velocity has therefore to be created in the chamber. In normal operation, the overall AFR of a combustion chamber can vary between 45:1 and 130:1. However, kerosene will only burn efficiently at, or close to, a ratio of 15:1, so the fuel must be burned with only part of the air entering the chamber in what is called *primary combustion zone*. Approximately 20% of the air mass flow is taken in by the snout or entry section Figure 10.5. Immediately downstream of the snout are swirl vanes and a perforated flare, through which air

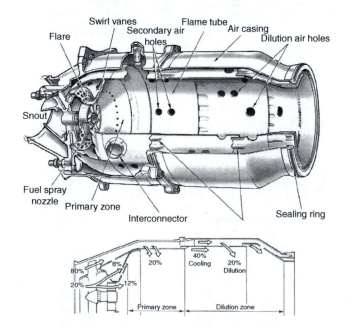

FIGURE 10.5 Combustion chamber and the corresponding air flow rates. (Reproduced from Rolls-Royce plc, copyright © Rolls-Royce plc 2007. With permission.)

passes into the primary combustion zone. The swirling air induces a flow upstream of the center of the flame tube and promotes the desired recirculation. The air not picked up by the snout flows into the annular space between the flame tube and the air casing. Through the wall of the flame tube body, adjacent to the combustion zone, are a selected number of secondary holes through which a further 20% of the main flow of air passes into the primary zone. The air from the swirl vanes and that from the secondary air holes interacts and creates a region of low velocity recirculation. This takes the form of a toroidal vortex, similar to a smoke ring, which has the effect of stabilizing and anchoring the flame. The recirculating gases hasten the burning of freshly injected fuel droplets by rapidly bringing them to ignition temperature. It is arranged such that the conical fuel spray from the nozzle intersects the recirculation vortex at its center, which assists in breaking up the fuel and mixing it with the incoming air. The temperature of the gases released by combustion is about 1800–2000°C, which is far too hot for entry into the nozzle guide vanes of the turbine. The air not used for combustion, which amounts to about 60% of the total airflow, is therefore introduced progressively into the flame tube. Approximately a third of this is used to lower the gas temperature in the dilution zone before it enters the turbine, and the remainder is used in film cooling of the walls of the flame tube. An electric spark from an igniter plug initiates combustion and the flame is then self-sustained.

10.5 CHEMISTRY OF COMBUSTION

Under most operating conditions, the rate of combustion is limited by the rate at which the fuel is vaporized and mixed with air. In most combustors, the fuel is injected as an atomized liquid droplet spray into the hot reaction zone where it mixes with air and hot combustion gases.

The rate of reaction depends on both static pressures "P" and temperature "T" in a very complex way. For many situations, the reaction rate can be approximated by a form of the Arrhenius equation written for the mass rate of reaction as,

$$\text{Reaction rate} \propto P^n f(T) e^{-E/RT} \tag{10.1}$$

where "n" is an exponent that depends on the number of molecules involved in a reactive collision (e.g., $n = 2$ for two molecules, for hydrocarbon–air combustion, $n = 1.8$), $f(T)$ is a function that relates the reaction rate to the forms of energy (translation, rotation, and vibration). The term $e^{-E/RT}$ accounts for the number of molecular collisions in which the energy of one molecule relative to another exceeds the active energy (E) and R is the universal gas constant [6].

At low pressure, the reaction rate becomes slow and can become limiting for aircraft engines at very high altitudes. However, under most operating conditions, the rate of combustion is limited by the rate at which the fuel is vaporized and mixed with air.

If the temperature and pressure in the reaction zone are sufficiently high, the reaction rate will be fast and the fuel vapor will react as it comes in contact with sufficient oxygen. Thus, for fast or more acceptable reaction, the combustion should be occurring under possible sufficient oxygen and this named by **stoichiometric ratio**.

The stoichiometric ratio is the required ratio for complete combustion of a hydrocarbon fuel to convert the fuel completely to carbon dioxide (CO_2) and the hydrogen to water vapor (H_2O).

Since 23% by mass of oxygen in the air participates in combustion, the stoichiometric AFR can be calculated from the reaction equation as follows. Consider the complete combustion of a general hydrocarbon fuel of average molecular composition $C_a H_b$ with air. The overall complete combustion equation is

$$C_a H_b + \left(a + \frac{b}{4}\right)(O_2 + 3.773 N_2) = a CO_2 + \frac{b}{2} H_2 O + 3.773\left(a + \frac{b}{4}\right)N_2 \tag{10.2}$$

where the molecular weights of oxygen, atmospheric nitrogen, atomic carbon, and atomic hydrogen are, respectively, 32, 28.16, 12.011, and 1.008. The fuel composition could have been written CH_y where $y = b/a$.

The combustion equation defines the stoichiometric (or chemically correct or theoretical) proportions of fuel and air (just enough oxygen for conversion of all the fuel into completely oxidized products). The stoichiometric AFR or fuel–air ratio (FAR) depends on fuel composition and can be defined as,

$$\left(\frac{A}{F}\right)_S = \left(\frac{F}{A}\right)_S^{-1} = \frac{(1+y/4)\,(32+3.773\times 28.16)}{12.011+1.008y} = \frac{34.56\,(4+y)}{12.011+1.008y} \tag{10.3}$$

Note that if the fuel is burned at a numerically large AFR, the mixture is referred to as lean or weak, and if the combustion at an AFR lower than the stoichiometric value implies a deficiency of oxygen and hence combustion is incomplete, then the fuel is partially burned, resulting in carbon monoxide (CO) and unburned hydrocarbons.

Example 1   A hydrocarbon fuel of composition 84.1% by mass **C** and 15.9 percent by mass **H** has a molecular weight of 114.15. Determine the number of moles of air required for stoichiometric combustion and the number of moles of products produced per mole of fuel. Calculate air-to-fuel ratio (A/F).

Solution: Assume a general form for fuel composition, namely, C_aH_b. The molecular weight relation gives

$$114.15 = 12.011a + 1.008b$$

The gravimetric analysis of the fuel gives

$$\frac{b}{a} = \frac{15.9/1.008}{84.1/12.011} = 2.25$$

Thus, $a = 8$ and $b = 18$; therefore, the fuel is octane C_8H_{18}. The burning equation then becomes

Fuel	Air	Products

$$C_8H_{18} + \overline{12.5(O_2 + 3.773N_2)} = \overline{8CO_2 + 9H_2O + 47.16N_2}$$

Relative mass:
$$114.15 + 59.66 \times 28.96 = 8 \times 44.01 + 9 \times 18.02 + 47.16 \times 28.16$$
$$114.15 + 1727.8 = 1842.3$$

Per unit mass fuel:
$$1 + 15.14 = 16.14$$

The stoichiometric air-to-fuel ratio is then (A/F) = 15.14 and fuel-to-air ratio (F/A) = 0.0661.

Fuel–air mixtures with more than or less than the stoichiometric air requirement can be burned. With excess air or fuel-lean combustion, the extra air appears in the products in uncharged form. With less than the stoichiometric air requirement, that is, with fuel-rich combustion, there is insufficient oxygen to oxidize fully the fuel constituents (C) and (H) to CO_2 and H_2O. The products are a mixture of CO_2 and H_2O with carbon monoxide CO and hydrogen H_2 (as well as N_2). The product composition cannot be determined from an element balance alone and an additional assumption about the chemical composition of the product species must be made.

Because the composition of the combustion products is significantly different for fuel-lean and fuel-rich mixtures and because the stoichiometric FAR depends on fuel composition, the ratio of the actual FAR to the stoichiometric ratio (or its inverse) is a more informative parameter for defining mixture composition. The fuel-air equivalence ratio Φ,

$$\Phi = \frac{(F/A)_{actual}}{(F/A)_{Stoich}} \tag{10.4a}$$

For fuel-lean mixtures: $\Phi < 1$
For stoichiometric mixtures: $\Phi = 1$
For fuel-rich mixtures: $\Phi > 1$
    Equation 10.4a can be rewritten in terms of the F/A (f) as

$$\Phi = \frac{f}{f_{stoich}} \tag{10.4b}$$

To prevent excessive temperatures at the exit of the main burner or the afterburner and protect its walls the overall F/A ratio must be much less than stoichiometric ratio where $\Phi < 1$.

Example 2   An aero engine is using octane $C_{12}H_{24}$ as a fuel. The temperature of the working fluid at the inlet and outlet of combustion chamber are $T_{01} = 650$ K and $T_{02} = 1600$ K. The fuel heating value is $Q_{HV} = 42{,}800$ kJ/kg. The specific heats of cold and hot streams are $Cp_c = 1.005$ kJ (kg $\cdot$ K) and $Cp_h = 1.23$ kJ/ (kg $\cdot$ K). Calculate the equivalence ratio.

Solution: The chemical reaction of fuel is

$$2C_{12}H_{24} + 18\left(O_2 + \frac{79}{21}N_2\right) \rightarrow 12CO_2 + 12H_2O + 18\left(\frac{79}{21}\right)N_2$$

Thus, the stoichiometric FAR is found to be

$$f_{stoich} = \frac{12 \times 12.011 + 24 \times 1.008}{18 \times 32 + 18 \times (79/21) \times 28.16} = 0.067795$$

The F/A is given by the relation

$$f = \frac{Cp_h T_{02} - Cp_c T_{01}}{Q_{Hv} - Cp_h T_{02}} = \frac{1.23 \times 1600 - 1.005 \times 650}{42800 - 1.23 \times 1600} = 0.032199$$

The equivalence ratio is

$$\Phi = \frac{f}{f_{stoich}} = 0.4749$$

Figure 10.6 illustrates the flammability limits for air mixture. Fuel–air mixtures will react over only a rather narrow range of F/A ratio (corresponding also to a narrow range of equivalence ratio). The equivalence ratio for hydrocarbon is in the range from 0.5 to 3, while hydrogen fuel has a wider range from 0.25 to 6 nearly [6].

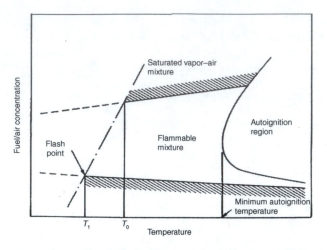

FIGURE 10.6 Effect of temperature on flammability limits. (Adapted from A.H. Lefebvre, *Gas Turbine Combustion*, 2nd ed., Philadelphia, Taylor & Francis, 1999. With permission.)

10.6 COMBUSTION CHAMBER PERFORMANCE

A combustion chamber must be capable of allowing fuel to burn efficiently over a wide range of operating conditions without incurring a large pressure loss. So, the combustion chamber performance can be evaluated by some conditions or performance as follows:

1. Pressure loss
2. Combustion efficiency
3. Combustion stability
4. Combustion intensity

10.6.1 PRESSURE LOSSES

The sources of pressure drop or loss are either cold or hot losses. Cold losses arise from sudden expansion, wall friction, turbulent dissipation, and mixing. Cold losses can be measured by flowing air without fuel through all the slots, holes, orifices, and so on. The hot losses (fundamental losses) are due to temperature increase. Generally, the fundamental loss due to heat addition in an aero engine will be low compared with the losses due to friction and mixing. Experiments have shown that the overall pressure loss can be expressed by the relation:

$$\text{Pressure loss factor (PLF)} = \frac{\Delta P_0}{m^2/\left(2\rho_1 A_m^2\right)} = \overset{\text{Cold}}{\overline{K_1}} + \overset{\text{Hot}}{\overline{K_2\left(\frac{T_{02}}{T_{01}} - 1\right)}} \qquad (10.5)$$

where m is the air mass flow rate, A_m is the maximum cross-sectional area of chamber, ρ_1 is the inlet density, and K_1 and K_2 are determined by measurement of the performance of the system. Equation 10.5 is plotted in Figure 10.7.

Typical values of pressure loss factor (PLF) at design operating conditions for can, can-annular, and annular combustion chambers are 35, 25, and 18, respectively.

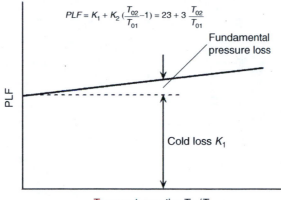

$$PLF = K_1 + K_2 \left(\frac{T_{02}}{T_{01}} - 1\right) = 23 + 3\,\frac{T_{02}}{T_{01}}$$

Fundamental pressure loss

Cold loss K_1

PLF

Temperature ration T_{02}/T_{01}

FIGURE 10.7   Variation of pressure loss factor.

10.6.2 COMBUSTION EFFICIENCY

The main objective of the combustor is to transfer all the energy of the fuel to the gas stream. In practice this will not occur for many reasons; for example, some of the fuel may not find oxygen for combustion in the very short time available. Therefore, it is necessary to define the efficiency of the combustion.

To obtain high efficiency from the combustor, the following conditions must be achieved:

- Fuel and air have adequate time and adequate space to mix and react
- Completely vaporized fuel and mixed with air before burning

The combustion efficiency may be expressed as [3]

$$\eta_b = f(\text{airflow rate})^{-1} \left(\frac{1}{\text{evaporation rate}} + \frac{1}{\text{mixing rate}} + \frac{1}{\text{reaction rate}}\right)^{-1} \quad (10.6)$$

As stated in Reference 5 and illustrated in Figure 10.8 the combustion efficiency of most gas turbine engines at sea level takeoff conditions is almost 100%. It is reduced to 98% at altitude cruise.

Another definition for efficiency is discussed in Reference 1 and introduces it in terms of the heat release during combustion. Thus

$$\eta_b = \frac{(h_{02} - h_{01})_{\text{actual}}}{(h_{02} - h_{01})_{\text{ideal}}} = \frac{(T_{02} - T_{01})_{\text{actual}}}{(T_{02} - T_{01})_{\text{ideal}}} = \frac{\dot{m}_{f\ \text{ideal}}}{\dot{m}_{f\ \text{actual}}} \quad (10.7)$$

Since heat release from combustion arises from the formation of (CO_2, H_2O) the combustion efficiency is determined by the quantity of these two constituents in the exhaust gas relative to the maximum possible quantity. Before combustion is completed, the gas contains various molecules and radicals of hydrocarbon as well as carbon monoxide. These represent unused chemical energy and, if allowed to remain in the exhaust gas, constitute two of the major pollutants.

10.6.3 COMBUSTION STABILITY

Combustion stability means smooth burning and the ability of the flame to remain alight over a wide operating range. For any particular type of combustion chamber there is both a rich and a weak limit to the AFR beyond which the flame is extinguished.

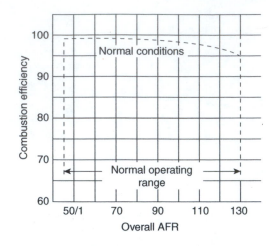

FIGURE 10.8 Combustion efficiency. (Reproduced from Rolls-Royce plc, copyright © Rolls-Royce plc 2007. With permission.)

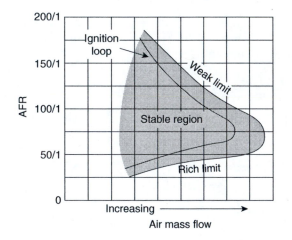

FIGURE 10.9 Combustion stability limits. (Reproduced from Rolls-Royce plc, copyright © Rolls-Royce plc 2007. With permission.)

Actually when a new system is tested the stability limits, which represent the highest and lowest FARs at which combustion will take place, will be investigated. Normally the limits will be well away from the normal operating point. The range of stable operation will reduce with the increase of air velocity, and if the air mass flow rate is increased beyond a certain value, flame will be extinguished. A typical stability loop is illustrated in Figure 10.9.

There are two reasons that cause instability of combustion. The first is, if the fuel–air mixture becomes too lean or too rich, the temperatures and reaction rates drop below the level necessary to effectively heat and vaporize the incoming fuel and air. The second reason is, if the velocity of gas stream "U" becomes higher than the flame speed "S," it causes a blow out of the flame. Stability loops provide two basic kinds of information.

10.6.4 COMBUSTION INTENSITY

The heat released by a combustion chamber is dependent on the volume of the combustion area. So, the combustion area must be increased. Therefore, the average velocities increasing in the combustor will make efficient burning more and more difficult. However, it is found that if the intensity of combustion is kept within certain limiting values and the fuel is atomized adequately, it is possible to obtain combustion efficiencies of 98%–100% at design conditions with conventionally designed combustion chamber.

The larger volume will achieve a low-pressure drop, high efficiency, good outlet temperature distribution, and satisfactory stability characteristics.

The design problem is also eased by an increase in the pressure and temperature of the air entering the chamber.

The combustion intensity (CI) can be defined as

$$CI = \frac{\text{heat release rate}}{\text{combustion volume} \times \text{pressure}} \quad kW/(m^3 atm) \qquad (10.8)$$

Certainly the lower the value of the CI the easier it is to design a combustion system that will meet all the design requirements.

For aircraft system the CI is in the region of $(2 - 5) \times 10^4$ kW/m^3 atm.

10.7 COOLING

Cooling air must be used to protect the burner liner and dome from the high radiative and connective heat loads produced within the burner. The heat release from burning the fuel in the combustor introduces products with a high temperature that can reach 2000°C. Thus, as shown in Figure 10.5 about 60% of the total airflow is introduced progressively into the flame tube for cooling. Approximately one-third of this is used to lower the gas temperature in the dilution zone before it enters the turbine, while the remainder is used for cooling the flame tube.

More than one cooling technique is used in cooling as illustrated subsequently.

Louver cooling

Many early gas turbine combustors used the louver cooling technique whereby the liner was fabricated in the form of cylindrical shells that, when assembled, formed annular passages at the shell intersection points as shown in Figure 10.10.

These passages permitted a film of cooling air to be injected along the hot side of the liner wall to provide a protective thermal barrier.

Splash-cooling

Through splash cooling, air enters the liner through a row of holes of small diameter as is shown in Figure 10.11. The air jets impinge on a cooling skirt, which then directs the flow so as to form a film along the inside of the liner wall.

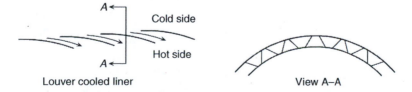

FIGURE 10.10 Louver cooling. (Reprinted from Jack D. Mattingly, *Elements of Gas Turbine Propulsion*, American Institute of Aeronautics and Astronautics, Inc. With permission.)

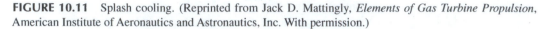

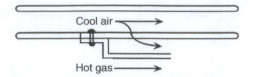

FIGURE 10.11 Splash cooling. (Reprinted from Jack D. Mattingly, *Elements of Gas Turbine Propulsion*, American Institute of Aeronautics and Astronautics, Inc. With permission.)

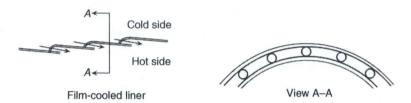

FIGURE 10.12 Film cooling. (Reprinted from Jack D. Mattingly, *Elements of Gas Turbine Propulsion*, American Institute of Aeronautics and Astronautics, Inc. With permission.)

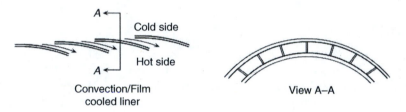

FIGURE 10.13 Convection/film cooling. (Reprinted from Jack D. Mattingly, *Elements of Gas Turbine Propulsion*, American Institute of Aeronautics and Astronautics, Inc. With permission.)

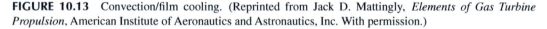

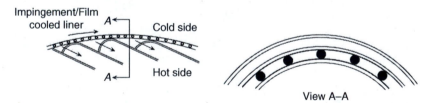

FIGURE 10.14 Impingement/film cooling. (Reprinted from Jack D. Mattingly, *Elements of Gas Turbine Propulsion*, American Institute of Aeronautics and Astronautics, Inc. With permission.)

Film cooling

The air is normally introduced through liner in a film of air that is formed between the combustion gases and liner hardware as is shown in Figure 10.12.

Convection-film cooling

Convection-film cooling utilizes simple but controlled convection cooling enhanced by roughened walls while providing a protective layer of cool air along the hot side of the wall at each cooling panel (Figure 10.13).

Impingement-film cooling

It is well suited to high temperature combustion as shown in Figure 10.14. However, its complex construction poses difficulties in manufacture and repair.

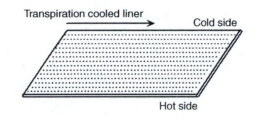

FIGURE 10.15 Transpiration cooling. (Reprinted from Jack D. Mattingly, *Elements of Gas Turbine Propulsion*, American Institute of Aeronautics and Astronautics, Inc. With permission.)

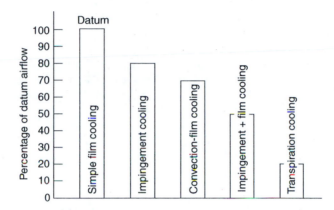

FIGURE 10.16 Comparison between different cooling techniques.

Transpiration cooling

Transpiration is the most advanced form of wall cooling developed now (Figure 10.15). It has the potential of reducing the required amount of cooling air by as much as 50%. With this scheme, the cooling air flows through a porous liner wall, first removing heat from the wall itself and then providing a barrier between the wall and the hot combustion gases.

Effective cooling

The following parameter is introduced to evaluate the effectiveness of different cooling techniques. It is the cooling effectiveness and is defined by

$$\Phi = \frac{T_g - T_m}{T_g - T_c} \tag{10.9}$$

where T_g is the main stream gas temperature, T_m the average metal temperature, and T_c is the cooling air temperature.

Figure 10.16 illustrates a comparison between the quantities of air needed for cooling with respect to that employed in a simple film cooling. The inlet total pressure and temperature to the combustion chamber are 2.425 MPa and 850 K, respectively, while the outlet temperature of the hot products is 2000 K.

10.8 MATERIAL

The walls and internal parts of the combustion chamber must be capable of resisting the very high gas temperature in the primary zone. In practice, this is achieved by using the best heat-resisting materials available. Normally the selected material must withstand severe operating conditions. For example, the combustor is designed for a lifetime of some 9000 hot hours at an average temperature around 2800°F. Moreover, the combustor liners are subjected to high thermal stresses due to temperature gradient and transient stresses during takeoff. Thus, they must withstand these critical conditions together with the possible corrosion. The most commonly used materials at present are nickel-based alloy with ceramic composite planned for the future.

10.9   AIRCRAFT FUELS

The fuel specification for aircraft engines is more severe than for all other types of gas turbine.
   Several requirements by different sources have to be fulfilled.

Operation requirements set by aircraft and engine
   1. Low cost and high availability
   2. Low fire risk (which implies low vapor pressure, low volatility, high flash point, and high conductivity to minimize the buildup of static electricity during fueling)
   3. High heat content for maximum range and/or payload
   4. High thermal stability to avoid filter plugging, sticking of control valves, and so on
   5. Low vapor pressure to minimize vaporization losses at high altitude
   6. High specific heat to provide effective heat absorption on high-speed aircraft

Fuel system requirements
   1. Pumping ability (fuel must remain liquid phase and flow freely to the atomizer)
   2. Freedom from filter clogging by ice or wax crystals
   3. Freedom from vapor locking
   4. High lubricity for minimum pump wear

Combustion chamber requirements
   1. Freedom from contaminants that cause blockage of small passage in fuel nozzle
   2. Good atomization
   3. Rapid vaporization
   4. Minimum carbon formation for low flame radiation and freedom from coke deposition and exhaust smoke
   5. Less carbon content to avoid excessive carbon deposits in the combustion chamber
   6. Relatively small amount of ash
   7. Freedom from alkali metals (Such as sodium, which combine with sulfur in the fuel to form corrosive sulfates)
   8. Freedom from vanadium

Nowadays, provided an aircraft combustion chamber has been properly designed, its performance and operating characteristics are fairly insensitive to the type of fuel employed.
   Probably the only flight condition during which fuel properties become significant is the relight and pull away phase after a flameout at high altitude.

Safety fuels
One of the greatest hazards in aviation is the possibility of fire after a crash landing. Fuel released from ruptured tanks may be ignited by friction sparks, hot surfaces, or flames. Once a fire has started, it will spread rapidly to all regions where the vapor–air ratio is within the limits of flammability. One method of reducing this fire hazard is by the use of thickened fuels, which have a reduced tendency to flow and a lower rate of evaporation than liquid fuels; if a tank containing thickened fuel is ruptured,

then both the spread of fuel and the rate of formation of flammable mists are drastically reduced, and more time is available for passenger evacuation (45–50 min). Thickened fuels may be produced by gelling or emulsification.

10.10 EMISSIONS AND POLLUTANTS

Recently, control of emissions has become one of the most important factors in the design of aero engines and industrial gas turbines. Solutions for emissions from aero engines are quite different from those for land-based gas turbines.

The pollutants appearing in the exhaust will include oxides of nitrogen (NOx), CO and unburned hydrocarbons (UHC); any sulfur in the fuel will result in oxides of sulfur (SO_x), the most common of which is SO_2. Although all these represent a very small proportion of the exhaust, the large flow of exhaust gases produces significant quantities of pollutants that can become concentrated in the area close to the plant. The oxides of nitrogen, in particular, can react in the presence of sunlight to produce "smog," which can be seen as a brownish cloud. This led to major efforts to clean up the emissions from vehicle exhausts and stringent restrictions on emissions from all types of power plant. NOx also cause acid rain, in combination with moisture in the atmosphere, and ozone depletion at high altitudes, which may result in a reduction in the protection from ultraviolet rays provided by the ozone layer, leading to increases in the incidence of skin cancer. UHC may also contain carcinogens, and CO is fatal if inhaled in significant amounts. With an ever-increasing worldwide demand for power and transportation, control of emissions is becoming essential.

10.10.1 POLLUTANT FORMATION

For many years the attention of combustion engineers was focused on the design and development of high-efficiency combustors that were rugged and durable, followed by a relatively simple solution to the problem of smoke. When the requirements for emission control emerged, much basic research was necessary to establish the fundamentals of pollutant formation. The single most important factor affecting the formation of NOx is the flame temperature; this is theoretically a maximum at stoichiometric conditions and will fall off at both rich and lean mixtures. Unfortunately, while NOx could be reduced by operating well away from stoichiometric conditions, these result in increasing formation of both CO and UHC, as shown in Figure 10.17.

NOx Emissions
NOx emissions are a considerable environmental problem. They also react with hydrocarbons to form ground level ozone. They are the main contributors to acid rain together with sulfur dioxide

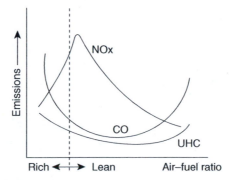

FIGURE 10.17 Effect of fuel–air ratio on emission. (Courtesy from Saravanamuttoo et al. *Gas Turbine Theory*, 5th ed., 2001, p. 109. With permission.)

(SO_2). NOx comes from many natural resources; the majority of the NOx emissions come from the combustion of biomass and fossil fuels. The combustion of the fossil fuels accounts for approximately 42% of the world's total NOx emissions. These NOx emissions have steadily increased since the 1970s, but controlling them can be simply accomplished by changing the combustion process.

There are several methods used to reduce atmospheric NOx pollution. There are two primary solutions: the first method involves combustion modification techniques, while the second utilizes new technologies. Selective catalytic reduction (SCR) is currently the most prominently technology. SCR uses a process that mixes the exhaust air with gaseous reagents such as ammonia or urea and carries that mixture over a catalyst bed at air stream temperature. Another technology currently being used is called selective noncatalytic reduction (SNCR). This technology uses the same basic plan as SCR, only it utilizes gas-phase free-radical reactions instead of a catalyst. In order to reduce NOx in the SNCR method, the gas must be kept within the range of 1600–2100°F.

New technologies have combined the SCR and SNCR processes in order to create even greater reductions. This technology reduces NOx emissions by using the low temperature injections as well as the ammonia reagent from the SCR system.

General Electric approaches the NOx issue by trying to avoid NOx-producing temperature spikes. Its first attack on NOx was its low emissions combustor (LEC) on the CF6-80C, which used the existing case and fuel nozzles, optimized mixing for reduced NOx and smoke, and controlled fuel spray and cooling to cut unburned hydrocarbons. This LEC combustor is used in the GE90, CF34, and GP7000 engine families.

For a larger NOx cut, General Electric introduced the dual annular combustor (DAC) as an option for CFM56-5B and CFM56-7B engines. The DAC has two combustion zones or stages, a rich pilot stage optimized for emissions during ignition and low power conditions, and then a lean main stage optimized for low NOx production for high power conditions.

Sulfur dioxide (SiO_2) emissions

The amount of SO_2 emissions are rising considerably every year, causing damage to the environment and the human population as sulfur acid, formed when sulfur dioxide combines with water, is the primary component in acid rain. Many respiratory problems can be a result of SO_2 inhalation. Bronchial constriction can be caused by any amount of SO_2 larger than 1 ppm, and 10 ppm can cause eye, throat, and nose irritation, which can lead to chronic bronchitis. Much larger amounts of sulfur dioxide can be lethal. SO_2 emissions are a result of the combustion of sulfur and oxygen, and one sulfur and two oxygen combine.

In conclusion, gas turbine emissions are regulated to increasingly stringent levels in both aircraft and industrial applications. As regards limitations for industrial gas turbines, NOx will be limited to 15/9 ppm by 2010 and to 9/5 ppm by 2020. The approaches suggested are as follows:

1. Advanced fuel mixers
2. Novel configurations (i.e., variable geometry, fuel staging)
3. Advanced cooling
4. Instability and noise control methods

10.11 THE AFTERBURNER

The afterburner, as described previously, is a second combustion chamber located downstream of the turbine and is fitted to supersonic aircraft (mostly military) to develop a greater thrust force. Afterburner consists of only two fundamental parts:

1. The spray bars or the fuel nozzle
2. The flame holder

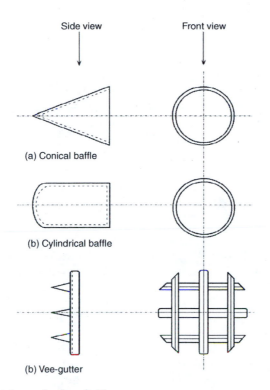

FIGURE 10.18   Different shapes for flameholder.

Afterburner is quite similar to the combustion chamber in ramjet engines as both have a maximum temperature much greater than conventional combustion chambers. For the case of afterburners (in turbojet and turbofan engines), the gases leave the turbine at very high speeds (250–400 m/s), which is far too high for a stable flame to be maintained. Though the flow is diffused before entering the afterburner combustion zone, the gas speed is still high and the diffused air stream would be blown away. For this reason, the AB is provided with flame holders downstream of the spray bars. The flame holders have different forms shown in Figure 10.18. The vee-gutter flame holder type has been widely used as it has the advantages of low flow blockage, low total pressure loss (dry loss), being simple and lightweight, and having a good development history.

The specific design requirements for afterburner are larger temperature rise and easy ignition.

First, the afterburner is not subjected to the physical and temperature limits of combustion chamber as the outlet of AB passes through a fixed element (nozzle) rather than a rotating element (turbine) as in the case of combustion chamber. The temperature rise is limited by the amount of oxygen that is available for combustion and limited by the liner and nozzle cooling air. Moreover, combustion takes place easily because 75% of the air (oxygen) entering the AB, with no longer need for cooling, is still available for burning with fuel. Ignition in an afterburner is either by a special spark igniter or by a flash of flame through the turbines of the basic engine or arc igniter or a pilot burner.

10.12   SUPERSONIC COMBUSTION SYSTEM

As described in Chapter 7, the scramjet engine has a supersonic combustion to avoid the possible large total pressure loss and high static temperature rise that would be caused by the deceleration of hypersonic flow to subsonic. Incomplete combustion occurs in scramjet engine because of short flowtime available for combustion. Similar to subsonic combustion, there are two scales of mixing:

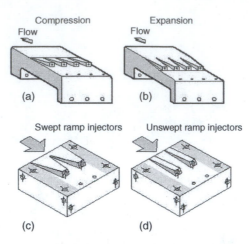

FIGURE 10.19   Supersonic injectors. (Reprinted from W.H. Heiser, *Hypersonic Airbreathing Propulsion*, American Institute of Aeronautics and Astronautics, Inc. 1994. With permission.)

macro and micro. In macromixing, fluids are mixed to a very fine scale but no molecular mixing, while in micromixing, molecular mixing occurs.

As a typical value the ratio of burner length L to entry duct height H, which we might call the burner aspect ratio, is $L/H \sim 10$. The fuel injector for supersonic combustion should be capable of generating vertical motions that cause the fuel–air interface area to increase rapidly and enhance the micromixing at molecular level, hence leading to efficient combustion. Different injection strategies have been proposed with particular concern for rapid, near-field mixing; these injection strategies typically rely on the generation of strong streamwise counterrotating vortices. Fuel is both injected and mixed in parallel or normal streams [7]. In parallel mixing, ramp injectors are used. These injectors are passive mixing devices that produce counterrotating streamwise vortices to enhance the mixing of fuel and air (Figure 10.19). Figures 10.19a and b represent the compression and expansion ramps, respectively. Figures 10.19c and d represent the swept and unswept ramp types, respectively.

Finally, swept expansion ramp has the highest performance in parallel of all the designs investigated.

Now, the other type of mixing, that is, normal mixing, is discussed. One of the simplest approaches is the transverse (normal) injection of fuel from a wall orifice. As the fuel jet interacts with the supersonic cross flow, an interesting but rather complicated flow field is generated. Figure 10.20 illustrates the general flow features of an underexpanded transverse jet injected into a cross flow. As the supersonic cross flow is displaced by the fuel jet a 3-D bow shock is produced due to the blockage produced by the flow. The bow shock causes the upstream wall boundary layer to separate, providing a region where the boundary layer and jet fluids mix subsonically upstream of the jet exit. This region confined by the separation shock wave formed in front of it, is important in transverse injection flow fields owing to its flameholding capability in combusting situations.

The stabilization of flames in supersonic flow is a difficult task for two primary reasons: the first is that the flow time is merely a short second. The second is that the strain rates tend to be high in compressible flows, which can suppress combustion.

The primary objective of a flameholder in supersonic combustion, therefore, is to reduce the ignition delay time and to provide a continuous source of radicals for the chemical reaction to be established in the shortest distance possible. In general, flameholding is achieved by three techniques:

1. Organization of a recirculation area where the fuel and air can be mixed partially at low velocities

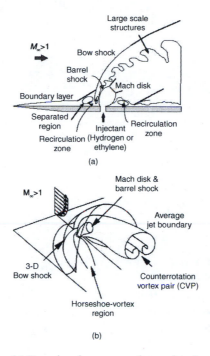

FIGURE 10.20 Normal mixing. (a) Boundary layer separation and recirculation zone. (b) 3-D Bow shock. (Reprinted from W.H. Heiser, *Hypersonic Airbreathing Propulsion*, American Institute of Aeronautics and Astronautics, Inc. With permission.)

2. Interaction of a shock wave with partially or fully mixed fuel and oxidizer
3. Formation of coherent structures containing unmixed fuel and air, wherein a diffusion flame occurs as the gases are convected downstream

Cavity-based fuel injection is one of the simplest approaches in flameholding. It uses the backward facing step. While generating a recirculation zone behind the step, that contains the hot gases in it, it serves as a continuous ignition source. This approach can provide sustained combustion, but it has a disadvantage of relatively high stagnation pressure loss. In recent years, a cavity flameholder, which is an integrated fuel injection flameholding approach, has been proposed as a new concept for flameholding and stabilization in supersonic combustor.

Burners are of two types: either a constant area type or a constant pressure type.

PROBLEMS

10.1 Describe the three basic designs for a combustion chamber.

10.2 Explain afterburning (or reheat) operation.

10.3 Describe the gas flow through a typical combustion chamber.

10.4 How it is possible to burn additional fuel in the exhaust system?

10.5 With the aid of tree illustration, identify the different types of combustion chamber.

10.6 Describe briefly the possible configuration for short annular combustors.

10.7 Describe the combustion chamber of the following engines:

    a. Turbojet engine P&W JT3

    b. Turboprop P&W C PT6

    c. Turbofan engines GE CF6, RR Trent 700, PW 4000

10.8 Describe afterburners of the following engines:
    a. Turbojet GE J47
    b. Turbofan PW F135
10.9 Compare between the different types of subsonic combustion chambers giving the advantages and disadvantages of each type.
10.10 Compare between subsonic and supersonic combustion chambers.
10.11 What are the design requirements for a good combustion chamber?
10.12 The temperature rise in combustion chamber (ΔT) depends on the fuel to air ratio (f) and air temperature upstream of the combustion chamber. For an upstream temperature of 650 K, this relation may be expressed as

$$\Delta T(K) = af + b = 3,393 \times 10^4 f + 21.08$$

For C_8H_{18} fuel described in example (1)
    a. Calculate the fuel to air ratio if the temperature rise in combustion chamber is 950 K.
    b. Calculate the equivalence ratio ϕ.
    c. Write down the chemical reaction equation.
10.13 As an extension to example (2), consider a lean mixture ($\phi < 1.0$). The chemical reaction is expressed by the relation:

$$\phi C_{12}H_{24} + 18\left(O_2 + \frac{79}{21}N_2\right) \rightarrow 12\phi(CO_2 + H_2O) + 18(1 - \phi)O_2 + 18\left(\frac{79}{21}\right)N_2$$

Calculate the percentage of unburnt oxygen.
10.14 If the combustion efficiency (η_c) is also considered in the above case of lean mixture, then the chemical reaction equation becomes

$$\phi C_{12}H_{24} + 18\left(O_2 + \frac{79}{21}N_2\right) \rightarrow 12\eta_c\phi(CO_2 + H_2O) + (1 - \eta_c)\phi C_{12}H_{24} + 18(1 - \eta_c\phi)O_2$$
$$+ 18\left(\frac{79}{21}\right)N_2$$

For a combustion efficiency of 0.98 and the same equivalence ratio, calculate the percentage of
    a. Unburnt fuel
    b. Unused oxygen
10.15 Consider the case of a rich mixture ($\phi > 1.0$) for the same fuel discussed in Example (2). The chemical reaction for a combustion efficiency of (η_c) is given by

$$\phi C_{12}H_{24} + 18\left(O_2 + \frac{79}{21}N_2\right) \rightarrow 12\eta_c\phi(CO_2 + H_2) + (1 - \eta_c)\phi C_{12}H_{24} + 18(1 - \eta_c\phi)O_2$$
$$+ 18\left(\frac{79}{21}\right)N_2$$

For combustion efficiency $\eta_c = 98\%$ and equivalence ratio $\phi = 1.02$, calculate
    a. The percentage increase of fuel injected into the combustion chamber
    b. The percentage of unburnt fuel
    c. The percentage of unused oxygen

10.16 In a two-year study for low emission combustor technology (LOWNOX I) by numerous aero engine manufacturers including RR, MTU, and Volvo as well as other institutions and agencies [8], the following data were deduced for a turbofan engine:

Condition	Idle	Takeoff	Cruise
Inlet temperature to combustion chamber	550 (K)	917 (K)	846 (K)
A/F ratio	114.8	36.9	39.9

The fuel used is C_8H_{18}; it is required for the three operating conditions to

a. Write down the chemical reaction equation.
b. Calculate the equivalence ratio.
c. Calculate the temperature rise in combustion chamber assuming that

$$Q_{HV} = 42,800 \text{ kJ/kg}, \ Cp_c = 1.005 \text{ kJ/(kg} \cdot \text{K)}, \ Cp_h = 1.23 \text{ kJ/(kg} \cdot \text{K)}$$

d. Calculate the combustion efficiency.

10.17 What are the operation requirements in aircraft fuels set by aircraft and engines?

10.18 If the combustion gases have a temperature of 1400 K and the coolant temperature is 500 K, calculate the average metal temperature if the effective cooling parameter (ϕ) has the values of 0.2, 0.4, and 0.6.

10.19 A turbofan engine powering a military aircraft is fitted with an afterburner. The inlet condition to the afterburner are $T_{01} = 600$ K, $M_1 = 0.2$ and the outlet temperature is $T_{02} = 3000$ K. Using Rayleigh flow equations (or tables)

a. Calculate the Mach number at the outlet of the afterburner M_2.
b. If the afterburner is designed to have an outlet Mach number $M_2 = 1.0$, what will be the value of the outlet temperature T_{02}?
c. If the Mach number at the outlet of afterburner is $M_2 = 1.0$ and the inlet and outlet total temperatures are $T_{01} = 600$ K and $T_{02} = 3000$ K respectively, what will be the value of the inlet Mach number M_1?

REFERENCES

1. R.T.C. Harman, *Gas Turbine Engineering- Applications, Cycles and Characteristics*, Halsted Press Book, John Wiley & Sons, New York, 1981, p. 88.
2. H.I.H. Saravanamuttoo, G.F.C. Rogers, and H. Cohen, *Gas Turbine Theory*, 5th edn., Pearson Education, Essex, England, 2001.
3. A.H. Lefebvre, *Gas Turbine Combustion*, 2nd edn., Philadelphia, Taylor & Francis, 1999, p. 39.
4. M.P. Boyce, *Gas Turbine Engineering Handbook*, 2nd edn., Butterworth-Heinemann, TX, USA, 2002, p. 370.
5. *The Jet Engine*, Rolls-Royce plc, 5th edn., Derby, England, 1996, p. 35.
6. J.D. Mattingly, *Elements of Gas Turbine Propulsion*, AIAA Education Series, 2005, p. 817.
7. W.H. Heiser and D.T. Pratt, *Hypersonic Airbreathing Propulsion*, AIAA Education Series, 1994.
8. R. Dunker (Ed.), *Advances in Engine Technology, European Commission Directorate—General XII, Science, Research and Development*, John Wiley & Sons, pp. 105–152.

11 Exhaust System

11.1 INTRODUCTION

Aero gas turbines have an exhaust system that passes the turbine discharge gases to atmosphere at a velocity and direction appropriate to the flight condition. Thus, it provides the thrust force required for all flight conditions except in landing where it may provide through the thrust-reverse system the dragging force needed for aircraft stopping at an appropriate distance. For turbopropeller engines, the major part of thrust is developed by the propeller, and the exhaust gases have little to contribute in that activity.

The main components of the exhaust system for nonafterburning engines are the tail cone and exhaust duct, which are often referred to as the tail pipe and the exhaust nozzle (Figure 11.1).

Gas from the engine turbine enters the exhaust system at velocities from 750 to 1200 ft per s [1]. This high velocity produces high friction losses; thus these high speeds have to be reduced by diffusion. Diffusion is accomplished by increasing passage area between the exhaust cone and the outer wall as shown in Figure 11.1. The tail cone is fastened to the casing by turbine rear support struts. These struts straighten out the flow before the gases pass into the jet pipe. The exhaust gases pass to atmosphere through the propelling nozzle. In a nonafterburning turbojet engine, the exit velocity of the exhaust gases is subsonic at low thrust conditions and reaches the speed of sound in most operating conditions. As shown in Figure 11.1a, the nozzle is of the convergent type. For supersonic flow, the nozzle is always of the convergent–divergent (C-D) type (Figure 11.1b). Both figures apply for simple ramjet and turbojet engines. In the case of afterburning engines, the afterburner fuel nozzle and flame holder are located in the tail pipe, which is also identified as the jet pipe (Figure 11.1c). Figure 11.1d illustrates the case of low-bypass ratio turbofan engine of the mixed type with an afterburner. Both hot and cold streams are mixed before entering the propelling nozzle or the afterburner.

Many of these engines use separate nozzles for exhausting core and fan streams. Unmixed turbofan engines have two separate nozzles for exhausting core and fan streams as described in Chapter 5. A hot nozzle is located downstream the turbine or low-pressure turbine for multispool engines, which is similar to that in Figure 11.1. The cold nozzle is downstream the fan section and is identified as fan nozzle. The vast majority of jet-powered aircraft use what are known as axisymmetric nozzles. These nozzles, like those illustrated in Figure 11.1, direct thrust purely along the axis of the engine; hence the name axisymmetric. However, in some aircraft, two-dimensional nozzles are used (Figure 11.2). These nozzles are capable of not only directing the thrust along the axis of the engine but also deflecting to vector the thrust and producing a force that points the nose of the plane in a different direction. This type of nozzle is at the heart of what is known as *thrust vectoring*. The nozzles depicted in Figure 11.2 are capable of vectoring the thrust up or down to produce an up force or down force. Down force will force the aircraft nose to pitch upward, while up force has the opposite effect. The two nozzles can also be deflected differentially (one producing up force, the other down force) to provide roll control.

Two important topics are incorporated with the exhaust system, namely thrust reversers and noise suppressor. Thrust reversing is an engine system used for bringing to a stop all the civil and some military airplanes in service nowadays. These devices reverse the engine thrust by closing in when deployed by the pilot, pushing the air out the front of the engine rather than the back. This motion decreases the speed of the aircraft and is the loud noise always heard when landing.

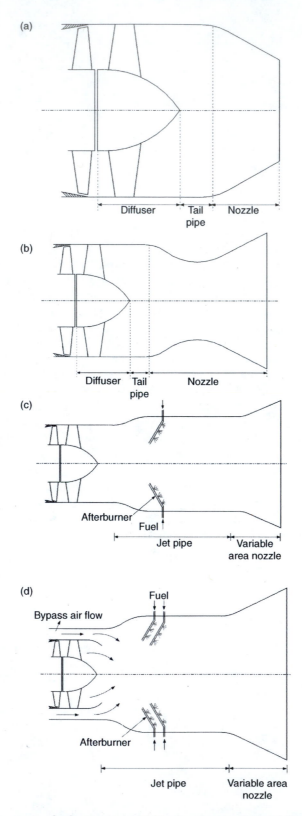

FIGURE 11.1 Arrangements of exhaust system. (a) Simple exhaust system. (b) Supersonic aircraft exhaust. (c) Exhaust system for turbojet afterburning engines. (d) Exhaust system for low bypass ratio mixed afterburning turbofans.

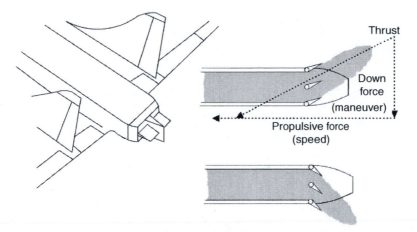

FIGURE 11.2   Two-dimensional nozzles for thrust vectoring.

Presently regulations specify certain noise levels that must not be exceeded by new aircraft. The noise sources are the fan or compressor, the turbine, and the exhaust jet or jets. The noise of these sources increases to varying degrees with greater relative airflow velocity. Exhaust jet noise varies by a larger factor than the compressor or turbine noise, therefore a reduction in exhaust jet velocity has a stronger influence than an equivalent reduction in compressor or turbine blade speeds. The amount of energy in the exhaust that is transformed into noise varies as approximately tile eighth power of the exhaust velocity. Details of noise suppression are discussed later in this chapter.

11.2 NOZZLE

The exhaust nozzles may be classified as

1. Convergent or C-D types
2. Axisymmetric or two-dimensional types
3. Fixed geometry or variable geometry types

The simplest form is the fixed geometry convergent type, as no moving parts and control mechanisms are needed. It is found in subsonic commercial aircraft. Sophisticated nozzle mechanisms are normally found in thrust vectoring types such as that in F-22 Raptor and F-15 S, X-36, and Sukhoi Su-35.

Nozzles should fulfill the following requirements [2]:

1. Be matched to other engine components for all engine-operating conditions.
2. Provide the optimum expansion ratio.
3. Have minimum losses at design and off-design conditions.
4. Permit afterburner operation (if available) without affecting main engine operation.
5. Allow for cooling of walls if necessary.
6. Provide reversed thrust when necessary.
7. Suppress jet noise and infrared radiation (IR) if desired.
8. Provide appropriate force for takeoff or landing for vertical takeoff and landing (VTOL) or vertical/short takeoff and landing (V/STOL) aircraft.

9. Provide necessary maneuvering for military aircraft fitted with thrust vectoring systems.
10. Do all the above with minimal cost, weight, and boat tail drag while meeting life and reliability goals.

Convergent nozzles are used almost in all the present subsonic transports. Moreover, in most cases also these convergent nozzles are choked and incomplete expansion of the flowing gases to the ambient pressure is encountered. At nozzle outlet, the gases exit at sonic speed, while the pressure is greater than the ambient pressure. Thus, a pressure thrust force is developed. On the contrary, C-D nozzle satisfies a full expansion to the ambient pressure.

Thus, the exit and ambient pressures are equal and the exit velocity is higher than the sonic speed. C-D nozzle develops higher momentum thrust that is greater than the pressure thrust of a convergent nozzle operating at the same inlet conditions. Thus, if both types are examined for the case of subsonic civil transports, C-D nozzle provides higher thrust. However, it has the penalties of increased weight, length, and diameter, leading to an increase of aircraft weight, possibly drag [3].

11.2.1 Governing Equations

The flow of a calorically perfect gas in a nozzle is considered here. The mass flow rate $\dot{m}$ may be determined in terms of the local area from the relation

$$\dot{m} = \rho u A = \left(\frac{P}{RT}\right)\left(M\sqrt{\gamma R T}\right)A = (MA)\left(\frac{P}{P_0}\right)P_0\frac{\sqrt{\gamma}}{\sqrt{RT}}\sqrt{\frac{T_0}{T_0}}$$

$$\dot{m} = \frac{\sqrt{\gamma}P_0}{\sqrt{T_0 R}}MA\frac{\{1 + ((\gamma - 1)/2)M^2\}^{1/2}}{\{1 + ((\gamma - 1)/2)M^2\}^{\gamma/(\gamma - 1)}}$$

Finally, the mass flow is

$$\dot{m} = \frac{AP_0}{\sqrt{T_0}}\sqrt{\frac{\gamma}{R}}\frac{M}{\{1 + ((\gamma - 1)/2)M^2\}^{(\gamma + 1)/2(\gamma - 1)}} \tag{11.1a}$$

11.2.1.1 Convergent–Divergent Nozzle

The flow in nozzles can be assumed adiabatic as the heat transfer per unit mass of fluid is much smaller than the difference in enthalpy between inlet and exit. Figure 11.3 illustrates a C-D nozzle together with its T-S diagram. Three states are identified, namely, inlet (i), throat (t), and exit (e). Flow is assumed to be isentropic from the inlet section up to the throat, thus $P_{0i} = P_{0t}, T_{0i} = T_{0t}$. Flow from the throat and up to the exit is assumed to be adiabatic but irreversible due to possible boundary layer separation, thus $P_{0t} > P_{0e}$ but $T_{0t} = T_{0e}$. Gases expand to the static pressure P_e with an adiabatic efficiency η_n in the range from 0.95 to 0.98.

This efficiency is used to evaluate the pressure ratio in the nozzle (P_{0i}/P_{0e}) as follows. The efficiency is defined as

$$\eta_n = \frac{h_{0i} - h_e}{h_{0i} - h_{es}} = \frac{T_{0i} - T_e}{T_{0i} - T_{es}} = \frac{1 - T_e/T_{0e}}{1 - T_{es}/T_{0i}}$$

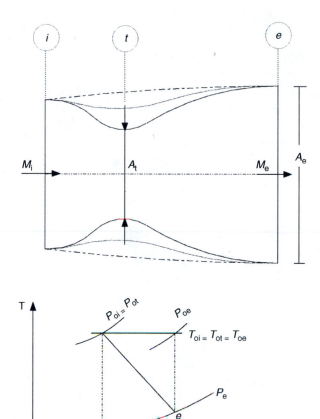

FIGURE 11.3   Convergent–divergent nozzle.

Since $T_{0i} = T_{0e}$, then

$$\eta_n = \frac{1 - (P_e/P_{0e})^{(\gamma-1)/\gamma}}{1 - (P_e/P_{0i})^{(\gamma-1)/\gamma}}$$

Thus

$$\frac{P_e}{P_{0e}} = \left[1 - \eta_n \left\{ 1 - \left(\frac{P_e}{P_{0i}} \right)^{(\gamma-1/\gamma)} \right\} \right]^{\gamma/(\gamma-1)}$$

Rewriting $\dfrac{P_{0i}}{P_{0e}} = \dfrac{P_e}{P_{0e}} \dfrac{P_{0i}}{P_e}$, then the nozzle pressure ratio is

$$\frac{P_{0i}}{P_{0e}} = \frac{P_{0i}}{P_e} \left[1 - \eta_n \left\{ 1 - \left(\frac{P_e}{P_{0i}} \right)^{(\gamma-1)/\gamma} \right\} \right]^{\gamma/(\gamma-1)} \tag{11.2}$$

The exit velocity is evaluated from the relation $u_e = \sqrt{2(h_{0i} - h_e)} = \sqrt{2\eta_n(h_{0i} - h_{es})}$

$$u_e = \sqrt{2C_p\eta_n(T_{0i} - T_{es})} = \sqrt{2C_p\eta_nT_{0i}\left\{1 - \left(\frac{P_e}{P_{0i}}\right)^{(\gamma-1)/\gamma}\right\}}$$

$$u_e = \sqrt{\frac{2\gamma R}{(\gamma-1)}\eta_nT_{0i}\left\{1 - (P_e/P_{0i})^{(\gamma-1)/\gamma}\right\}}$$

The exit Mach number is then $M_e^2 = \dfrac{u_e^2}{a_e^2} = \dfrac{u_e^2}{\gamma RT_e}$

Since

$$\frac{T_{0e}}{T_e} = 1 + \frac{\gamma-1}{2}M_e^2 = \frac{T_{0i}}{T_e}$$

$$M_e^2 = \frac{2\eta_n}{\gamma-1}\left\{1 + \frac{\gamma-1}{2}M_e^2\right\}\left\{1 - \left(\frac{P_i}{P_{0i}}\right)^{(\gamma-1)/\gamma}\right\}$$

$$M_e^2 = \frac{2}{\gamma-1}\left[\frac{\eta_n\left\{1 - (P_e/P_{0i})^{(\gamma-1)/\gamma}\right\}}{1 - \eta_n\left\{1 - (P_e/P_{0i})^{(\gamma-1)/\gamma}\right\}}\right] \tag{11.3}$$

From Equation 11.1a, and noting that isentropic flow is assumed from nozzle inlet until the throat ($P_{0i} = P_{0t}$), the ratio between the throat area and exit area variation with Mach number can be expressed as

$$\frac{A_t}{A_e} = \frac{P_{0e}}{P_{0t}}\frac{M_e}{M_t}\left[\frac{1 + ((\gamma-1)/2)M_t^2}{1 + ((\gamma-1)/2)M_e^2}\right]^{(\gamma+1/2(\gamma-1))} = \frac{P_{0e}}{P_{0i}}\frac{M_e}{M_t}\left[\frac{1 + ((\gamma-1)/2)M_t^2}{1 + ((\gamma-1)/2)M_e^2}\right]^{(\gamma+1)/2(\gamma-1)}$$

$$\tag{11.4a}$$

Since in most cases the throat will be choked, or $M_t = 1$, then the above equation will be

$$\frac{A^*}{A_e} = \frac{P_{0e}M_e}{P_{0i}^*}\left[\frac{(\gamma+1)/2}{1 + ((\gamma-1)/2)M_e^2}\right]^{(\gamma+1)/2(\gamma-1)} \tag{11.4b}$$

The mass flow rate in this case will be expressed as

$$\dot{m} = \frac{A^*P_{0i}^*}{\sqrt{T_{0i}^*}}\sqrt{\frac{\gamma}{R}}\frac{1}{((\gamma+1)/2)^{(\gamma+1)/2(\gamma-1)}} \tag{11.1b}$$

The velocity and static pressure along the longitudinal axis of nozzle are plotted in Figure 11.4. C-D nozzle is designed to have the exit pressure equal to the ambient pressure. If the nozzle operates at higher or lower altitudes, then the nozzle will exhibit different behavior as shown in Figure 11.5. Figure 11.5a illustrates the design conditions. This may be the case of a nozzle fitted to a turbojet and having an area ratio of 4 and flight Mach number $M \approx 2$.

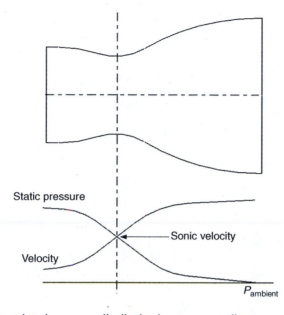

FIGURE 11.4 Velocity and static pressure distribution in convergent–divergent nozzle.

When the exit pressure exceeds ambient pressure (underexpansion case, $M > 2$), expansion waves exist on nozzle lip (Figure 11.5b). In the opposite case, when the ambient pressure exceeds the exit pressure (overexpansion, $M < 2, P_e > P_a/2$), oblique shock waves are generated at the exit plane (Figure 11.5c). If the ambient pressure is much greater than the exit pressure ($M < 2, P_e \gg P_a$), the oblique shock waves exist within the nozzle (Figure 11.5d).

11.2.1.2 Convergent Nozzle

The convergent nozzle is very similar to the convergent part of the previously discussed C-D nozzle. However, the main difference is that the flow in the C-D nozzle is assumed isentropic, while the flow in the convergent nozzle is assumed only adiabatic.

If the nozzle is choked, then the exit Mach number is unity and all conditions are denoted by subscript (c). The temperature ratio is then

$$\frac{T_{0e}}{T_e} = \frac{T_{0e}}{T_c} = \frac{\gamma + 1}{2} \tag{11.5}$$

The efficiency of the nozzle is also defined as

$$\eta_n = \frac{h_{0i} - h_e}{h_{0i} - h_{es}} = \frac{T_{0i} - T_e}{T_{0i} - T_{es}} = \frac{1 - (T_e/T_{0e})}{1 - (T_{es}/T_{0i})} = \frac{1 - (2/(\gamma + 1))}{1 - (P_c/P_{0i})^{(\gamma - 1)/\gamma}}$$

Thus, the pressure ratio is expressed as

$$\frac{P_{0i}}{P_c} = \frac{1}{(1 - (1/\eta_n)((\gamma - 1)/(\gamma + 1)))^{\gamma/(\gamma - 1)}} \tag{11.6}$$

In general, an estimate for the critical value is obtained by assuming that the nozzle is isentropic; thus Equation 11.6 is reduced to $P_{0i}/P_c = ((\gamma + 1)/2)^{\gamma/(\gamma - 1)}$, and for $\gamma = 4/3$ this ratio will be $P_{0i}/P_c = 1.853$. If $(P_{0i}/P_a) > (P_{0i}/P_c)$, then the nozzle is choked.

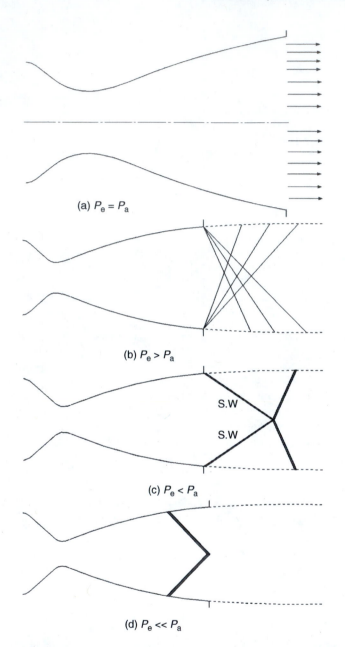

FIGURE 11.5 Behavior of convergent–divergent nozzle. (a) Design condition. (b) Exit pressure exceeds ambient pressure. (c) Ambient pressure exceeds exit pressure. (d) Ambient pressure greatly exceeds exit pressure.

11.2.2 Variable Geometry Nozzles

Variable area nozzle, which is sometimes identified as adjustable nozzle, is necessary for engines fitted with afterburners. Generally, as the nozzle is reduced in area, the turbine inlet temperature increases and the exhaust velocity and thrust increase. Three methods are available, namely:

1. Central plug at nozzle outlet
2. Ejector type nozzle
3. IRIS nozzle

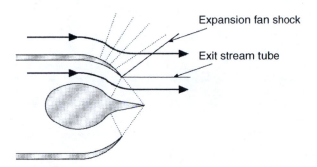

FIGURE 11.6 Plug nozzle at design point. (Courtesy J.L. Kerreebrock, *Aircraft Engines and Gas Turbines*, 2nd ed., The MIT Press, MA, USA, 1992, p. 142.)

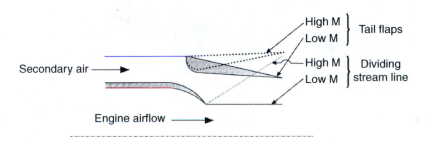

FIGURE 11.7 Variable geometry ejector nozzle. (Courtesy J.L. Kerreebrock, *Aircraft Engines and Gas Turbines*, 2nd ed., The MIT Press, MA, USA, 1992, p. 142.)

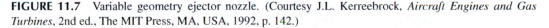
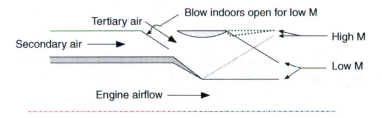

FIGURE 11.8 Ejector nozzle with blow-in doors for tertiary air. (Courtesy J.L. Kerreebrock, *Aircraft Engines and Gas Turbines*, 2nd ed., The MIT Press, MA, USA, 1992, p. 142.)

Central plug (Figure 11.6) is the exact analogue of spike at the inlet as described in Chapter 9. The plug employs external expansion fan as the spike employs external oblique shocks [4].

The second type is the "ejector nozzle," which creates an effective nozzle through a secondary airflow and spring-loaded petals. At subsonic speeds, the airflow constricts the exhaust to a convergent shape. As the aircraft speeds up, the two nozzles dilate, which allows the exhaust to form a C-D shape, speeding the exhaust gases past Mach 1 (Figure 11.7). More complex engines can actually use a tertiary airflow to reduce exit area at very low speeds. Advantages of the ejector nozzle are relative simplicity and reliability. Disadvantages are average performance (compared to the other nozzle type) and relatively high drag due to the secondary airflow (Figure 11.8). Some of the aircraft that have utilized this type of nozzle are the SR-71, the Concorde, the F-111, and the Saab Viggen.

The third type is an iris nozzle, which is used for higher performance nozzles (Figure 11.9). This type uses overlapping, hydraulically adjustable "petals." Although more complex than the ejector

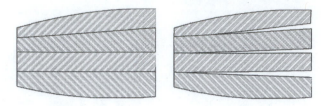

FIGURE 11.9   Iris variable nozzle.

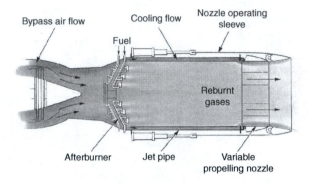

FIGURE 11.10   Variable geometry nozzle for afterburning engine. (Reproduced from Rolls-Royce plc, copyright © Rolls-Royce plc 2007. With permission.)

nozzle, it has significantly higher performance and smoother airflow. It is employed primarily on high-performance fighters such as the F-14, F-15, F-16, and high-speed bomber B-1B. Some modern iris nozzles have the ability to change the angle of the thrust.

11.2.3 Afterburning Nozzles

Afterburning is a method for augmenting the basic thrust of an engine to improve the takeoff, rate of climb, and combat performance of military aircraft. Supersonic civil transports are also fitted with afterburners. The thrust of an afterburning engine, with inoperative afterburner is slightly less than that of a similar engine not fitted with afterburning equipment due to the added restrictions in the jet pipe (Figure 11.1). The temperature of the afterburner flame may exceed 1700°C. The area of the afterburning jet pipe is larger than a normal jet pipe to obtain a reduced velocity gas stream. The rear fan duct and turbine exhaust cone form a diffuser section to reduce the velocity of fan discharge air and engine exhaust gases [5]. Fuel is introduced through a series of perforated spray bars located inside the forward section. A flameholder is mounted behind the manifolds and reduces the velocity of gases to ensure flame stability.

The afterburning jet pipe is made of heat resistant nickel alloy and requires more insulation than the normal jet pipe. The jet pipe may be of a double-skin construction with a cooling fan bypass air often induced between both inner and outer skins [1]. A circular heat shield of similar material to the jet pipe is often fitted to the inner wall of the jet pipe to improve cooling at the rear of the burner section. The propelling nozzle is of similar material and construction as the jet pipe. Variable geometry nozzle is achieved by either a two-position nozzle or a nozzle whose area varies automatically in proportion to the total thrust produced by the basic engine and the afterburner together. A two-position propelling nozzle has two movable eyelids that are operated by actuators (Figure 11.10), or pneumatic rams, to give an open or closed position.

The most common type of variable area propelling nozzle for axial-flow compressor engines has a ring of interlocking flaps spaced around the perimeter of the nozzle and actuated either by powered rams [1] or pneumatic actuators [5]. The iris-type camera shutter (Figure 11.9) is not often used today. Balanced-beam nozzle is the recent most sophisticated exhaust nozzle, which automatically varies the size of nozzle opening. The F100 engine is fitted with this type of nozzle.

Example 1 A turbojet engine powering an aircraft flying at an altitude of 11,000 m where $T_a = 216.7$ K and $P_a = 24.444$ kPa. The flight Mach number is 0.9. The inlet conditions to the nozzle are 1000 K and 60 kPa. The specific heat ratio of air and gases at nozzle are 1.4 and 4/3. The nozzle efficiency is 0.98.

1. Determine the thrust per inlet frontal area in the following two cases:
   a. C-D nozzle
   b. Convergent nozzle
2. Calculate the ratio between the thrust forces in both cases.
3. Calculate the exit total pressure in both cases.

Neglect the mass of added fuel.

Solution:
Case 1: C–D nozzle
  From Equation 11.3, with

$$\frac{P_a}{P_{0i}} = \frac{24.475}{60} = 0.4074$$

Then

$$M_e^2 = \frac{2}{\gamma - 1}\left[\frac{\eta_n\left\{1 - (P_a/P_{0i})^{(\gamma-1)/\gamma}\right\}}{1 - \eta_n\left\{1 - (P_a/P_{0i})^{(\gamma-1)/\gamma}\right\}}\right] = \frac{2}{(1/3)}\left[\frac{0.98\left\{1 - (0.4074)^{0.25}\right\}}{1 - 0.98\left\{1 - (0.4074)^{0.25}\right\}}\right]$$

$$M_e = 1.27276$$

Since

$$\dot{m} = \rho u A = \frac{P}{RT}uA$$

Then

$$\frac{\dot{m}}{A} = \rho u = \frac{P}{RT}u = \frac{P\sqrt{\gamma}}{\sqrt{RT}}\frac{u}{\sqrt{\gamma RT}} = M\sqrt{\frac{\gamma}{R}}\frac{P}{\sqrt{T}}$$

The mass flow per unit area of the inlet of intake is

$$\left(\frac{\dot{m}}{A}\right)_a = M_a\sqrt{\frac{\gamma_a}{R}}\frac{P_a}{\sqrt{T_a}} \quad \text{(a)}$$

Similarly, the mass flow rate per unit area of the exit of nozzle is

$$\left(\frac{\dot{m}}{A}\right)_e = M_e\sqrt{\frac{\gamma_e}{R}}\frac{P_e}{\sqrt{T_e}} \quad \text{(b)}$$

From (a) and (b) the ratio between the areas of the nozzle exit and inlet of intake, assuming negligible fuel mass

$$\frac{A_e}{A} = \frac{M_a}{M_e} \sqrt{\frac{\gamma_a}{\gamma_e}} \sqrt{\frac{T_e}{T_a}} \frac{P_a}{P_e} \quad (c)$$

With $T_e = \dfrac{T_{0e}}{1 + ((\gamma - 1)/2) M_e^2} = \dfrac{1000}{1 + (0.333/2)(1.27276)^2} = 787.4 \text{ K}$

From (a), (b), (c)

$$\frac{A_e}{A} = \frac{0.9}{1.273} \sqrt{\frac{1.4}{1.333}} \sqrt{\frac{787.4}{216.7}} \times 1 = 1.381$$

The thrust force is

$$T = \dot{m}_a [(1+f) u_e - u] + A_e (P_e - P_a) \quad (d)$$

With $f = 0$ and $P_e = P_a$, then the thrust equation becomes

$$T = \dot{m}_e u_e - \dot{m}_a u = \rho_e u_e^2 A_e^2 - \rho u^2 A \quad (11.7)$$

where A is the area of the inlet of intake section

With $\rho u^2 = (P/RT) u^2 = (\gamma P / \gamma RT) u^2 = \gamma P M^2$, then from Equation (d)

$$T = \gamma_e P_e M_e^2 A_e - \gamma P M^2 A$$

$$\frac{T}{A} = \gamma_e P_e M_e^2 \frac{A_e}{A} - \gamma P M^2 = \frac{4}{3} \times 24.475 \times (1.273)^2 \times 1.381 - 1.4 \times 24.475 \times (0.9)^2$$

$$\left(\frac{T}{A}\right)_1 = 45.27 \quad \text{KN/m}^2 \quad (e)$$

From Equation (11.2),

$$\frac{P_{0i}}{P_{0e}} = \frac{P_{0i}}{P_e} \left[1 - \eta_n \left\{ 1 - \left(\frac{P_e}{P_{0i}}\right)^{(\gamma-1)/\gamma} \right\} \right]^{\gamma/(\gamma-1)} = \frac{60}{24.475} \left[1 - 0.98 \left\{ 1 - \left(\frac{24.475}{60}\right)^{0.25} \right\} \right]^4$$

$$\frac{P_{0i}}{P_{0e}} = 1.02$$

The total pressure at the nozzle exit is $P_{0e} = 58.8$ kPa

Case 2: Convergent Nozzle

The nozzle is first checked for choking. From Equation 11.6

$$\frac{P_{0i}}{P_c} = \frac{1}{[1 - (1/\eta_n)((\gamma-1)/(\gamma+1))]^{\gamma/(\gamma-1)}} = \frac{1}{[1 - (1/0.98)(1/7)]^4} = 1.878$$

Since

$$\frac{P_{0i}}{P_a} = \frac{60}{24.444} = 2.45 P_a$$

Then

$$\frac{P_{0max}}{P_a} > \frac{P_{0max}}{P_c}$$

Consequently, $P_c > P_a \Rightarrow$ Choked nozzle

$$P_e = P_c = \frac{60}{1.878} = 31.95 \text{ kPa}$$

The pressure thrust term is no longer zero; thus

$$T = \dot{m}_e u_e - \dot{m}_a u_e + (P_e - P_a) A_e$$

From Equation (c) and noting that $M_e = 1$, then the nozzle to inlet area ratio is

$$\frac{A_e}{A} = \frac{M_a}{M_e} \sqrt{\frac{\gamma_a}{\gamma_e}} \sqrt{\frac{T_e}{T_a}} \frac{P_a}{P_e} = \frac{0.9}{1.0} \sqrt{\frac{1.4}{1.333}} \sqrt{\frac{787.4}{216.7}} \frac{24.444}{31.95}$$

$$\frac{A_e}{A} = 1.3451$$

(Note that this ratio is less than the case of C-D nozzle.)
The thrust per unit area of the inlet of intake is

$$\left(\frac{T}{A}\right)_2 = \gamma_e P_e M_e^2 \frac{A_e}{A} - \gamma_a P_a M^2 + (P_e - P_a) \frac{A_e}{A}$$

$$\left(\frac{T}{A}\right)_2 = \frac{4}{3} \times 31.95 \times 1.3451 - 1.4 \times 24.444 \times (0.9)^2 + 10.11 = 39.692 \text{ kN/m}^2$$

$$\frac{(T/A)_1}{(T/A)_2} = \frac{45.27}{39.692} = 1.1405$$

Total pressure ratio in nozzle is determined also from Equation 11.2 but for choked nozzle, it becomes

$$\frac{P_{0i}}{P_{0e}} = \frac{P_{0i}}{P_c} \left[1 - \eta_n \left\{ 1 - \left(\frac{P_c}{P_{0i}}\right)^{(\gamma-1)/\gamma} \right\} \right]^{\gamma/(\gamma-1)} = 1.0137$$

$$P_{0e} = 59.19 \text{ kPa}$$

11.3 CALCULATION OF THE TWO-DIMENSIONAL SUPERSONIC NOZZLE

The profile of a subsonic nozzle is totally convergent while supersonic nozzle has a convergent part followed by a divergent part. In this section, a preliminary design for the subsonic convergent nozzle and the supersonic divergent nozzle will be given.

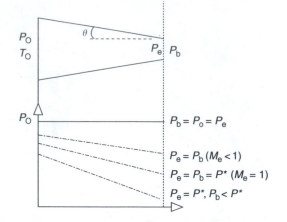

FIGURE 11.11   Convergent nozzle.

11.3.1 CONVERGENT NOZZLE

Convergent nozzle is used in all the present subsonic/transonic civil transports. The convergent part is assumed to have a linear shape in the present preliminary design. Curved nozzle profiles need rather a computational fluid dynamics programming. The linear nozzle shape and the relation between the exit pressure and back pressure are plotted in Figure 11.11.

Design steps:

1. Calculate the nozzle radius at inlet (r_i) or exit (r_e) from the relation:

$$r = \sqrt{\frac{\dot{m}}{\pi \rho V}}$$

2. Calculate the length of nozzle (L) from the relation

$$L = \frac{r_i - r_e}{\tan \theta}$$

where $\theta = 11 - 15°$

3. The area at any section along the nozzle length is obtained as follows:
    Assume that the nozzle radius decreases linearly along its length; thus

$$r_x = ax + b$$

where $a = (r_e - r_i)/L$ and $b = r_i$. Thus the corresponding area

$$A_x = \pi r_x^2$$

4. The area ratio at any station (A_x/A^*) is related to the corresponding Mach number at this location by the isentropic relation

$$\frac{A_x}{A^*} = \frac{1}{M_x} \left[\frac{2}{\gamma + 1} \left(1 + \frac{\gamma - 1}{2} M_x^2 \right) \right]^{(\gamma+1)/[2(\gamma-1)]}$$

The local Mach number M_x is obtained iteratively from the relation:

$$M_x = \frac{A^*}{A_x} \left[\frac{2}{\gamma + 1} \left(1 + \frac{\gamma - 1}{2} M_x^2 \right) \right]^{(\gamma+1)/[2(\gamma-1)]} \tag{11.8}$$

5. Next, the corresponding static to total pressure, density, and temperature ratios are obtained as follows:

$$\frac{P}{P_0} = \left[1 + \frac{\gamma - 1}{2} M_x^2 \right]^{-\gamma/(\gamma-1)} \tag{11.9a}$$

$$\frac{\rho}{\rho_0} = \left[1 + \frac{\gamma - 1}{2} M_x^2 \right]^{-1/(\gamma-1)} \tag{11.9b}$$

$$\frac{T}{T_0} = \left[1 + \frac{\gamma - 1}{2} M_x^2 \right]^{-1} \tag{11.9c}$$

6. Since the total temperature (T_0) is constant through the nozzle, the static temperature (T) is obtained at any location (x)
7. The velocity is calculated from the Mach number and static temperature as

$$V = M\sqrt{\gamma R T}$$

8. The static density is then calculated from the continuity equation

$$\rho = \frac{\dot{m}}{VA}$$

9. The static pressure is calculated from equation of state

$$P = \rho R T$$

10. The total pressure and density are calculated from Equations 11.9a and 11.9b.

Example 2
A convergent nozzle is to be designed based on the procedure mentioned above. A linear shape is assumed. The following data is given at the inlet and exit sections:

$\dot{m} = 64.36$ kg/s, $\theta = 14°$ $\gamma = 4/3$
$P_i = 63.23$ kPa, $T_i = 747.5$ K, $V_i = 131.56$ m/s
$P_e = 36.13$ kPa $T_e = 647.1$ K $V_e = 497.617$ m/s

Solution:
The density at inlet and outlet sections are calculated from the given static pressures and temperatures as:

$$\rho_i = 0.2994 \text{ kg/m}^3 \text{ and } \rho_e = 0.1945 \text{ kg/m}^3$$

The inlet and outlet areas are calculated as:

$$A_i = \frac{\dot{m}}{\rho V_i} = 1.66 \text{ m}^2 \text{ and } A_e = \frac{\dot{m}}{\rho V_e} = 0.665 \text{ m}^2$$

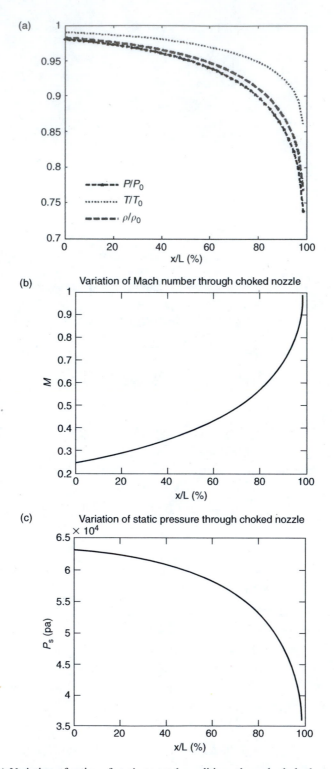

FIGURE 11.12 (a) Variation of ratios of static to total conditions through choked nozzle. (b) Variation of Mach number through choked nozzle. (c) Variation of static pressure through choked nozzle. (d) Variation of static temperature through choked nozzle. (e) Variation of static density through choked nozzle. (f) Variation of total pressure through choked nozzle. (g) Variation of total density through choked nozzle.

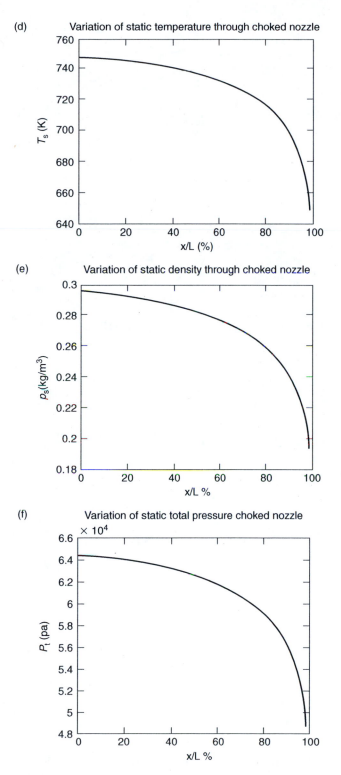

(d) Variation of static temperature through choked nozzle

(e) Variation of static density through choked nozzle

(f) Variation of static total pressure choked nozzle

FIGURE 11.12   (Continued)

(g)

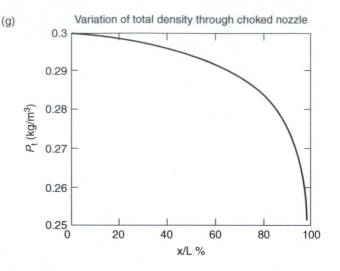

FIGURE 11.12   (Continued)

Then, the inlet and exit radii are $r_i = 0.7269$ m $r_e = 0.4601$ m.
The nozzle length is then $L = 1.0703$ m.
The inlet and exit Mach numbers are calculated as

$$M_i = \frac{V_i}{\sqrt{\gamma R T_i}} = 0.2463 \qquad M_e = \frac{V_e}{\sqrt{\gamma R T_e}} = 1.00$$

The nozzle is choked as the exit Mach number equals unity.
    Assuming a linear relation for the radius, or $r_x = ax + b$
Then

$$a = \frac{r_e - r_i}{L} = -0.2493 \text{ m}^{-1} \text{ and } b = r_i = 0.7269 \text{ m}$$

The corresponding area along any axial position is $A_x = \pi r_x^2$.
The results of calculations are listed in Table 11.1.
The following figures illustrate the variations of the different parameters inside the nozzle (Figure 11.12).

11.3.2 Divergent Nozzle

A supersonic nozzle that produces uniform, parallel flow is shown in Figure 11.13. Since it is required that the flow be parallel and supersonic, the wall must first curve outward (AD) and then curve in again (DE), so that the wall is again parallel to the initial flow [6]. Point D is the point where the wall has its maximum slope. It is usually assumed that the flow at throat is uniform, parallel, and sonic $M = 1$. Because of the nozzle symmetry, the nozzle axis is a streamline and may for design processes be replaced by a solid boundary. Therefore, only one half of the longitudinal section of the nozzle is designed.
    The flow region ABCDA is called the expansion zone. The curved wall AD generates expansion waves, which reflect off the centerline. The region DCED is called the straightening section, and the wall in this section is curved so that the incoming expansion waves are canceled. The flow past CE is uniform, parallel and supersonic. Point D is the inflection point and is the point where the wall has the maximum slope.

TABLE 11.1

Properties of Flowing Gases within the Convergent Nozzle Passage

Station	x/L%	M	T(K)	V(m/s)	$\rho \left(\text{kg/m}^3\right)$	P(kPa)	P_0(kPa)	T_0(K)	$\rho_0 \left(\text{kg/m}^3\right)$
1	0	0.246	747.46	131.55	0.294	63.22	64.50	755	0.299
2	10	0.27	746.17	142.35	0.293	62.78	64.28	755	0.299
3	20	0.29	744.58	154.66	0.291	62.25	64.00	755	0.298
4	30	0.32	742.59	168.82	0.289	61.58	63.66	755	0.297
5	40	0.35	740.05	185.29	0.286	60.74	63.22	755	0.296
6	50	0.39	736.74	204.75	0.282	59.67	62.66	755	0.294
7	60	0.43	732.32	228.19	0.277	58.25	61.91	755	0.292
8	70	0.49	726.18	257.25	0.270	56.32	60.88	755	0.289
9	80	0.56	717.08	295.06	0.260	53.55	59.36	755	0.285
10	90	0.67	701.84	349.37	0.244	49.14	56.86	755	0.277
11	100	1	647.13	497.63	0.194	36.13	48.60	755	0.251

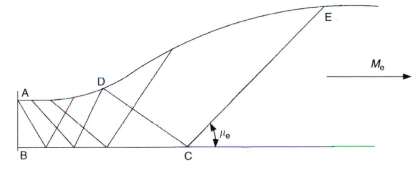

FIGURE 11.13   Divergent nozzle producing parallel flow.

The most important parameter is the Mach number in the exit section (M_e). It depends on how much expansion occurs between A and D. For large exit Mach number, a nozzle of this type may be too long for jet engine applications. In these cases, a limiting form of this nozzle is used. A nozzle with a sharp-edged throat, as shown in Figure 11.14, produces uniform, parallel flow in the shortest possible length. It is a limiting form of the nozzle shown in Figure 11.13, where points A and D coincide. The flow expands around a sharp edge, producing waves that are reflected by the centerline. Cancellation of the reflected waves is again used to obtain uniform parallel flow at the exit. Thus the straightening contour extends from the corner to the exit.

Now, from symmetry if the flow deflection corresponding to the expansion is accomplished by the means of the two walls is (θ), then expansion corresponding to a single wall is ($\theta/2$).

The waves are reflected on the axis, producing a total expansion of the same intensity as that of the incident wave; thus the expansion referring to one wall is ($\theta/2$), with the remaining expansion to (θ) being completed by the waves reflected by the axis of symmetry.

It must be further noted that all the reflected waves complete the restoring of the stream to its initial direction, since they are of opposite sense.

Consider any expansion corresponding to a deflection (θ); see Figure 11.15. The incoming Mach number is M and the leaving Mach number is M_e. The wave angles corresponding to M and M_e are μ and $\mu^\backslash$, forming what is called a Mach fan at the corner.

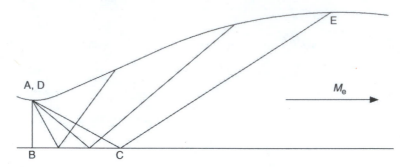

FIGURE 11.14   Sharp corner nozzle.

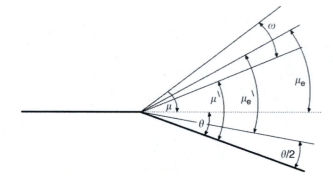

FIGURE 11.15   Mach angles distribution.

The angle ω is given by the relation:

$$\omega = \mu - \mu^{\backslash} + \theta = -\Delta\mu + \theta \tag{11.10}$$

Since

$$\tan\mu = \frac{1}{\sqrt{M^2 - 1}} \tag{11.11}$$

then

$$\frac{d\mu}{\cos^2\mu} = -\frac{M^2}{M^2 - 1}\frac{dM}{M} = -\frac{1}{\cos^2\mu}\frac{dM}{M}$$

or

$$d\mu = -\frac{dM}{M} \tag{11.12}$$

If the deflection angle θ is divided into several incremental $d\theta$, then the incremental increase in Mach number due to this small expansion is expressed by the relation

$$d\theta = \frac{\sqrt{M^2 - 1}}{1 + ((n-1)/2)M^2}\frac{dM}{M} \tag{11.13}$$

It follows from Equations 11.11 through 11.13 that

$$-d\mu = \left(1 + \frac{n-1}{2}M^2\right)\tan\mu.d\theta \qquad (11.14)$$

From Equations 11.10 and 11.14 by considering small finite increments, replacing $d\mu$ by $\Delta\mu$ and $d\theta$ by θ, then:

$$\omega = \theta + \theta\left(1 + \frac{n-1}{2}M^2\right)\tan\mu$$

$$\omega = \theta\left[1 + \left(1 + \frac{n-1}{2}M^2\right)\tan\mu\right] \qquad (11.15)$$

For small angle (θ) the Mach fan in Figure 11.15 can be replaced by a single line; a mean line is taken. In that case, the average Mach angle will be

$$\mu_e = \mu - \frac{\omega}{2} \qquad (11.16)$$

This angle corresponds approximately to a stream deflected by $\theta/2$, since the limits of approximation:

$$\omega_e = \frac{\theta}{2}\left[1 + \left(1 + \frac{n-1}{2}M^2\right)\tan\mu\right] = \frac{\omega}{2} \qquad (11.17)$$

The mean angle μ_e will correspond to a stream deflected by μ_e' and

$$\mu_e' = \mu_e + \frac{\theta}{2} \qquad (11.18)$$

If the deflection is so great that the approximation entails a great error, the deflection is divided into several parts, applying the above procedure to each separate part.

The determination of the inclination of the waves after intersection is of primary importance for the graphical calculation of the nozzle, since small errors in the orientation of the waves after intersection could lead to major errors on the contour of the walls.

Two methods are used in nozzle design, namely, the graphical and analytical solutions. Graphical solution is best illustrated in [7]. Although the graphical design of a nozzle is simple, it entails laborious calculations and yields approximate results. For this reason, only the analytical solution is considered here [8].

11.3.2.1 Analytical Determination of the Contour of a Nozzle

Although the graphical design of a nozzle is simple, it entails laborious calculations and yields approximate results. Atkin [8] had indicated a method through which the contour can be determined as an analytical curve with parameter representation. The basic idea in this method is the transformation of radial supersonic system due to a source into a uniform parallel stream. If the single expansion slope is prolonged, the point of intersection with the axis is the center of the source and also the origin of the polar system of coordinates (r, θ). Such a flow can only exist from a radius r_c onward, for which the Mach number is unity; hence it is close to the sonic throat (Figure 11.16).

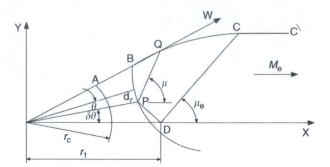

FIGURE 11.16   Notations for analytical treatment.

The Mach number varies along a radius vector (supersonic source flow) according to the formula:

$$\frac{r}{r_c} = \frac{1}{M}\left(\frac{\gamma-1}{\gamma+1}M^2 + \frac{2}{\gamma+1}\right)^{(\gamma+1)/2(\gamma-1)}$$

(11.19)

For a Mach number M_e in the nozzle outlet (the corresponding radius r_1 where $r_1 = OD$) can be derived from Equation 11.9 by setting $M = M_e$. The wave DC is inclined at an angle μ_e where $\mu_e = \sin^{-1}(1/M_e)$. Beyond DC the streamlines are uniform and the wall $CC^{\backslash}$ is parallel to the axis of the nozzle. The slope begins to decrease at point B and the curve BQC must ensure the transition from the radial flow AB to the parallel flow $CC^{\backslash}$. At each point Q on this curve, a wave Q arises along which the velocity is parallel to the tangent at the wall and to the radial velocity at point P. Two characteristics pass through P; the first is the wave PQ at an angle μ with direction of the velocity; the second is PD at an angle $(-\mu)$. The locus of P is a characteristics curve, defined by the equation:

$$\frac{dy}{dx} = \tan(\theta \pm \mu)$$

In polar coordinates this equation has the form:

$$\frac{rd\theta}{dr} = \pm\tan\mu = \pm\frac{1}{\sqrt{M^2-1}}$$

which can be integrated. From Equation 11.13 setting (r) and (dr) in terms of M, then:

$$\frac{dr}{r} = \frac{M^2-1}{1+((\gamma-1)/2)M^2}\frac{dM}{M}$$

$$\text{and } \pm d\theta = \frac{M^2-1}{1+((\gamma-1)/2)M^2}\cdot\frac{dM}{M\sqrt{M^2-1}}$$

Setting $\sqrt{M^2-1} = y$, then

$$\frac{MdM}{\sqrt{M^2-1}} = dy$$

The equation is readily integrated resulting in

$$\pm d\theta = \frac{dy}{1 + [(\gamma - 1)/(\gamma + 1)]\, y^2} - \frac{dy}{1 + y^2}$$

$$\text{and } \pm\theta = \sqrt{\frac{\gamma + 1}{\gamma - 1}} \tan^{-1}\left[\sqrt{\frac{\gamma - 1}{\gamma + 1}}\sqrt{M^2 - 1}\right] - \tan^{-1}\sqrt{M^2 - 1} + C \qquad (11.20a)$$

as obtained in continuous expansion around a corner. The constant C is determined by setting $\theta = 0$ for $M = M_e$

$$C = \tan^{-1}\sqrt{M_e^2 - 1} - \sqrt{\frac{\gamma + 1}{\gamma - 1}} \tan^{-1}\left[\sqrt{\frac{\gamma - 1}{\gamma + 1}}\sqrt{M_e^2 - 1}\right] \qquad (11.20b)$$

The curve required is BQC whose coordinates are

$$X = r\cos\theta + PQ\cos(\mu + \theta)$$
$$Y = r\sin\theta + PQ\sin(\mu + \theta)$$

For the length PQ we can put

$$PQ = \frac{r(\theta_0 - \theta)}{\sin\mu}$$

So that

$$X = r\cos\theta + r(\theta_0 - \theta)\frac{\cos(\mu + \theta)}{\sin\mu} \qquad (11.21a)$$

$$Y = r\sin\theta + r(\theta_0 - \theta)\frac{\sin(\mu + \theta)}{\sin\mu} \qquad (11.21b)$$

For the practical construction of this curve, the angle μ will be first taken from the equation or from tables of characteristics as function of M, decreasing from M_e. The parameter of variation is thus M and the plot of the curve will be more accurate, the more numerous the values of M that are taken. This method being easy to apply and yielding fairly exact result is recommended compared with the graphical methods, especially in view of the errors generally associated with geometrical constructions.

11.3.2.2 Design Procedure for a Minimum Length Divergent Nozzle

1. Decide the number of characteristics used for calculations.
2. Divide Mach number from M^* (nozzle entry) to M_e (nozzle exit) into the number of characteristics.
3. Calculate the Mach wave angle μ from the relation.

$$\mu = \sin^{-1}\left(\frac{1}{M}\right)$$

4. Calculate the maximum turning angle θ_0 and deflection angle θ for each characteristic.
   where $2\theta_0 = \sqrt{\frac{\gamma+1}{\gamma-1}} \tan^{-1}\left[\sqrt{\frac{\gamma-1}{\gamma+1}}\sqrt{M_e^2-1}\right] - \tan^{-1}\sqrt{M_e^2-1}$
   From Equations 11.20a and 11.20b then

$$\theta = \sqrt{\frac{\gamma+1}{\gamma-1}} \tan^{-1}\left[\sqrt{\frac{\gamma-1}{\gamma+1}}\sqrt{M^2-1}\right] - \tan^{-1}\sqrt{M^2-1}+$$

$$\tan^{-1}\sqrt{M_e^2-1} - \sqrt{\frac{\gamma+1}{\gamma-1}} \tan^{-1}\left[\sqrt{\frac{\gamma-1}{\gamma+1}}\sqrt{M_e^2-1}\right]$$

5. Calculate r_c where,
$$r_c = \frac{h_0}{\tan\theta_0}, \text{ and}$$
$$h_0 = \frac{D_{\text{throat}}}{2} = \text{the height of the throat section.}$$

6. Calculate the radius (r)

$$r = \frac{r_c}{M}\left(\frac{\gamma-1}{\gamma+1}M^2 + \frac{2}{\gamma+1}\right)^{(\gamma+1)/2(\gamma-1)}$$

7. Calculate X and Y from Equations 11.21a and 11.21b.

11.3.2.3 Procedure of Drawing the Expansion Waves Inside the Nozzle

1. Decide number of characteristics used for calculations.
2. Assign (θ) for each characteristic, normally uniformly distributed starting with a small value up to θ_{max} with constant incremental step $\Delta\theta$.
3. Calculate the Mach number and Mach angle μ at every (θ) from Prandtl-Meyer tables.

Note: we assume that there is a small expansion fan and the angle of this fan starts with μ_1 and ends with μ_2. Then, get the equivalent μ of the expansion fan (Figure 11.17) where

$$\mu = \frac{1}{2}[\mu_1 + \mu_2 - \Delta\theta]$$

4. Draw the characteristics lines.

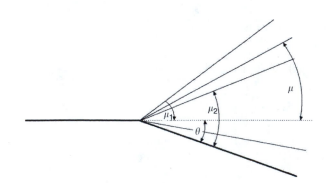

FIGURE 11.17   Expansion fan notation.

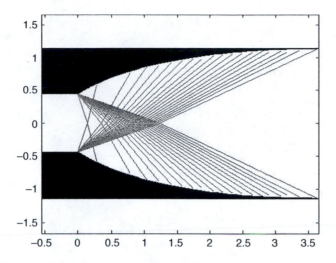

FIGURE 11.18   Illustrates the nozzle contour together with the expansion Mach number.

TABLE 11.2
Geometry of the Divergent Supersonic Nozzle

	1	2	3	4	5	6	7	8	9	10
M	1	1.156	1.311	1.467	1.622	1.778	1.933	2.089	2.244	2.4
μ	90	59.93	49.70	42.90	38.06	34.23	31.15	28.60	26.46	24.62
θ_0	36.75	36.75	36.75	36.75	36.75	36.75	36.75	36.75	36.75	36.75
θ	36.75	34.24	30.27	25.82	21.23	16.67	12.22	7.947	3.87	0
r_c	0.738	0.738	0.738	0.738	0.738	0.738	0.738	0.738	0.738	0.738
r	0.738	0.752	0.790	0.852	0.936	1.044	1.179	1.343	1.539	1.773
x	0	0.027	0.112	0.262	0.491	0.820	1.27	1.871	2.656	3.663
y	0.441	0.461	0.514	0.593	0.692	0.804	0.919	1.025	1.105	1.137

Example 3
An axisymmetric nozzle has a throat of 1.0 m diameter and an outlet Mach number of 2.4. Calculate and draw the profile of the divergent part of the nozzle. Draw also the expansion Mach numbers (Figure 11.18).

Solution
The above mentioned procedure is followed here. Ten characteristics are selected between the inlet and exit Mach numbers and the corresponding nozzle profile is considered. All the results are tabulated in Table 11.2 and all dimensions are in meter.

11.4   THRUST REVERSAL

Stopping an aircraft after landing is not an easy problem due to the increases in its gross weight, wing loadings and landing speeds [5]. The amount of force required for stopping an aircraft at a given distance after touchdown increases with the gross weight of the aircraft and the square of the landing speed. The size of modern transport aircraft, which results in higher wing loadings and increased landing speeds, makes the use of wheel brakes alone unsatisfactory for routine operations [5]. Moreover, in the cases of wet, icy, or snow-covered runways, the efficiency of aircraft brakes may be reduced by the loss of adhesion between aircraft tire and the runway [1]. Thus, for present large

FIGURE 11.19   Boeing XB-70A Valkyrie landing with triple drag chutes. (Copyright Boeing, All rights reserved.)

aircraft and under such runway conditions, there is a need for additional methods for augmenting the stopping power provided by the brakes to bring the aircraft to rest within the required distance. The reversible pitch propeller has solved the problem for reciprocating-engine and turboprop-powered aircraft. Turbojet and turbofan aircraft, however, must rely upon either some device such as parabrake or runway arrester gear or some means for reversing the thrust produced by their engines. For military aircraft, parabrake or drag parachutes are in use. Arrester gears are primarily for aircraft-carrier-deck, although sometimes used at military basis. One or more parachutes may be deployed in some military aircraft as well as space shuttle after touchdown. Figure 11.19 illustrates the three chutes used in braking the XB-70 aircraft. Following each landing, the parachutes must be detached from the aircraft and repacked. The use of these devices for deceleration is not an attractive alternative for any type of routine operations and is completely unacceptable for commercial airline operations.

However, a number of Soviet transport aircraft, including early versions of the Tupelov Tu-134 twinjet transport, were equipped with braking parachutes. For civilian transport, reversing the engine thrust is the most simple and effective way to reduce the aircraft landing run on both dry and slippery runways. Thus, turbojet- and turbofan-powered transport aircraft are equipped with some form of diverter that, when activated, reverses the thrust and thus provides a powerful stopping force. The direction of the exhaust (gas or air) is reversed; thus the engine power is used as a deceleration force. The landing runs could be reduced by some 500 ft by applying thrust reversal [2]. Engine thrust reverser provides an effective braking force on the ground. Some reversers are suitable for use in flight to reduce airspeed during descent to slow the aircraft's rate of descent, allowing it to land at steeper angles. However, they are not commonly used in modern aircraft. Ideally, the gas should be directed in a completely forward direction, which is not possible due to aerodynamic reasons. Actually, a discharge angle of approximately 45° is appropriate.

On turbojet engines, low-bypass turbofan engines, whether fitted with afterburner or not, and mixed turbofan engines, the thrust reverser is achieved by reversing the exhaust gas flow (hot stream). On high-bypass ratio turbofan engines, reverse thrust is achieved by reversing the fan (cold stream) airflow. Mostly, in this case it is not necessary to reverse the hot stream as the majority of the engine thrust is derived from the fan, although some engines use both systems.

A good thrust reverser must fulfill the following conditions [5]:

1. Must not affect the engine operation whether the thrust reverser is applied or stowed
2. Withstand high temperature if it is used in the turbine exhaust
3. Mechanically strong

4. Relatively light in weight
5. When stowed should be streamlined into the engine nacelle and should not add appreciably to the frontal area of the engine
6. Reliable and fail safe
7. Cause few increased maintenance problems
8. Provide at least 50% of the full forward thrust

11.4.1 CLASSIFICATION OF THRUST REVERSER SYSTEMS

The most commonly used reversers are clamshell-type, external-bucket type doors and blocker doors, as shown in Figure 11.20.

Clamshell door system—sometimes identified as preexist thrust reverser [9]—is a pneumatically operated system. When reverse thrust is applied, the doors rotate to uncover the ducts and close the normal gas stream. Sometimes clamshell doors are employed together with cascade vanes; type (A) in Figure 11.20. Clamshell type is normally used for non-afterburning engines.

The bucket target system; type (B) in Figure 11.20, is hydraulically actuated and uses bucket-type doors to reverse the hot gas stream. Sometimes it is identified as postexit or target thrust reverser [9]. In the forward (stowed) thrust mode, the thrust reverser doors form the convergent–divergent final nozzle for the engine. When the thrust reverser is applied, the reverser automatically opens to form a "clamshell" approximately three-fourth to one nozzle diameter to the rear of the engine exhaust nozzle. When the thrust reverser is applied, the reverser automatically opens to form a "clamshell" approximately three-fourth to one nozzle diameter to the rear of the engine exhaust nozzle. The thrust reverser in Boeing 737-200 aircraft is an example for this type of thrust reverser.

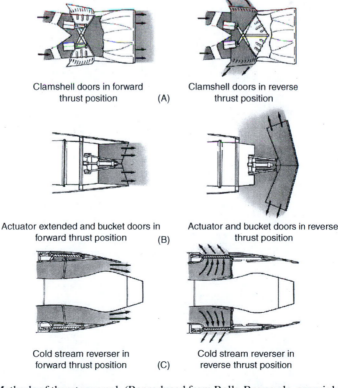

Clamshell doors in forward thrust position   (A)   Clamshell doors in reverse thrust position

Actuator extended and bucket doors in forward thrust position   (B)   Actuator and bucket doors in reverse thrust position

Cold stream reverser in forward thrust position   (C)   Cold stream reverser in reverse thrust position

FIGURE 11.20 Methods of thrust reversal. (Reproduced from Rolls-Royce plc, copyright © Rolls-Royce plc 2007. With permission.)

High by-pass turbofan engines normally use blocker doors to reverse the cold stream airflow; type (C) in Figure 11.20. Cascade-type reverser uses numerous turning vanes in gas path to direct the gas flow outward and forward during operation. Some types utilize a sleeve to cover the fan cascade during forward thrust. Aft movement of the reverse sleeves causes blocker doors to blank off the cold stream final nozzle and deflect fan discharge air forward through fixed cascade vanes, producing reverse thrust. In some installations, cascade turning vanes are used in conjunction with clamshell to reverse the turbine exhaust gases. Both the cascade and the clamshell are located forward of the turbine exhaust nozzle. For reverse thrust, the clamshell blocks the flow of exhaust gases and exposes the cascade vanes, which act as an exhaust nozzle. Some installations in low-bypass turbofan unmixed engines use two sets of cascade: forward and rearward. For the forward cascade, the impinging exhaust is turned by the blades in the cascade into the forward direction. Concerning the rearward cascade, the exhaust from the hot gas generator strikes the closed clamshell doors and is diverted forward and outward through a cascade installed in these circumferential openings in the engine nacelle.

Thrust reversers of all types are not used if the forward speed is approximately 60 knots; otherwise recirculating exhaust air or gases together with any foreign objects will be ingested into the engine.

Example 4

Prove that the force developed by thrust reverser during landing is given by the relation

$$F = \dot{m}_a \left[(1 + f) V_j \cos \beta + V_f \right]$$

where V_f is the flight speed and V_j is the exhaust speed, f is the fuel-to-air ratio, and β is the inclination of the exhaust gases leaving the buckets measured from the engine longitudinal axis. Next, calculate the dragging force developed by thrust reversers of the *two* engine aircraft in the following case $\dot{m}_a = 50$ kg/s, $f = 0.02$, $\beta = 60°$, $V_j = 600$ m/s and $V_f = 80$ m/s.

Solution:

For the control volume shown in Figure 11.21a, the gases leave the engine at (j) with a jet velocity (V_j) and rate $(\dot{m}_j)$. These gases impact the buckets and leave the control volume at (1) and (2).

From the continuity equation assuming constant density, then

$$\rho V_j A_j = \rho V_1 A_1 + \rho V_2 A_2$$

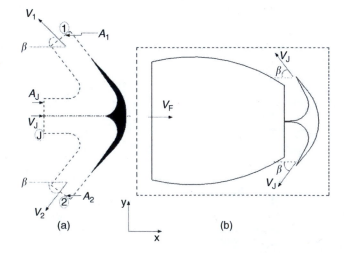

(a)   (b)

FIGURE 11.21   (a) Control volume for the exhaust jet. (b) Control volume for aircraft engine.

or

$$V_j A_j = V_1 A_1 + V_2 A_2 \quad \text{(a)}$$

Applying Bernoulli's equation twice between states (j) and (1) and between states (j) and (2) gives

$$\frac{P_j}{\rho} + \frac{V_j^2}{2} = \frac{P_1}{\rho} + \frac{V_1^2}{2}$$

and

$$\frac{P_j}{\rho} + \frac{V_j^2}{2} = \frac{P_2}{\rho} + \frac{V_2^2}{2}$$

Assuming $P_j = P_1 = P_2$, yields $V_j = V_1 = V_2$, then from Equation (a)

$$A_j = A_1 + A_2 \quad \text{(b)}$$

Writing the y-momentum equation

$$F_y = 0 = \rho A_1 V_1^2 \sin \beta - \rho A_2 V_2^2 \sin \beta$$

Then $A_1 = A_2$

From Equation (b) $A_1 = A_2 = A_j/2$

Next, for the control volume shown in Figure 11.21b and assuming $P_j = P_a$

$$F_x = -\dot{m}_e V_j \cos \beta - \dot{m}_a V_f$$

$$F_x = -\dot{m}_a \left[(1 + f) V_j \cos \beta + V_f \right]$$

This force is the force on the control volume. Then the force on the engine is in its opposite direction and so has a positive value. This means that it is in the positive x-direction or a dragging force.

$$F_x = \dot{m}_a \left[(1 + f) V_j \cos \beta + V_f \right]$$

For (n) engines, the force applied on the aircraft due to thrust reverses is

$$F_x = n \dot{m}_a \left[(1 + f) V_j \cos \beta + V_f \right]$$

Substituting the given data

$$F_x = (2)(50) \left[(1 + 0.02) \times 600 \times \cos 60 + 80 \right]$$

$$F_x = 38600 \, \text{N} = 38.6 \, \text{kN directed rearward}$$

Example 5

During landing, the pilot applies thrust reverse to shorten the landing distance (Figure 11.22). The thrust reverse exhaust is inclined at an angle (β) to the normal flight speed. If the aircraft is powered

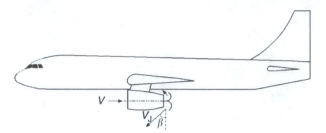

FIGURE 11.22   Thrust reversal.

by (n) engines, and assuming the average drag force on the airplane during landing is D_{av}, it is required to prove that the time necessary to decelerate the airplane from the touchdown speed (V_0) to any value (V) during landing is given by

$$t = -\frac{m}{n\dot{m}_a} \ln \frac{V + V_j \sin \beta + (D_{av}/n\dot{m}_a)}{V_0 + V_j \sin \beta + (D_{av}/n\dot{m}_a)}$$

where m: airplane total mass
$\dot{m}_a$: mass flow rate of air entering or leaving one engine (i.e., negligible fuel mass flow rate)
 Next, for a particular aircraft having the following data:
 $n = 2, m = 50,000$ kg, $D_{av} = 5000$ N, $\dot{m}_a = 50$ kg/s, $V_0 = 40$ m/s, $V_j = 600$ m/s, $\beta = 30°$
calculate the time needed for complete stop and distance needed.
 From the control surface shown in Figure 11.22, the force generated by each engine when the thrust reverser is applied is

$$F = m_e V_j \sin \beta + m_a V \qquad\qquad\qquad\text{(a)}$$

Neglecting the fuel mass flow rate, then

$$\dot{m}_a \approx \dot{m}_e$$

Then the force is

$$F = \dot{m}_a \left(V + V_j \sin \beta\right)$$

The thrust force in this case is a resistance force that adds to the drag force.
For an aircraft powered by (n) engines, the RHS of equation (a) is multiplied by (n).
Since acceleration during landing is negative (i.e., deceleration)

$$\therefore -m\frac{dV}{dt} = n\dot{m}_a \left(V + V_j \sin \beta\right) + D_{av}$$

$$\frac{dV}{dt} + \frac{n\dot{m}_a}{m} V = -\left(\frac{n\dot{m}_a}{m} V_j \sin \beta + \frac{D_{av}}{m}\right)$$

The above equation may be written in the form

$$\dot{x} + Ax = B$$

where

$$A = \frac{n\dot{m}_a}{m} \text{ and } B = -\left(\frac{n\dot{m}_a}{m}V_j \sin\beta + \frac{D_{av}}{m}\right)$$

The general solution for the above differential equation is

$$x = \lambda e^{-At} + \frac{B}{A}$$

where (λ) is a constant and (t) is the time

$$\text{Thus } V = \lambda e^{-\frac{n\dot{m}_a}{m}t} - \left(V_j \sin\beta + \frac{D_{av}}{n\dot{m}_a}\right)$$

To determine the value of the constant (λ), assuming that at touchdown ($t = 0$) the aircraft speed is V_0, then,

$$V_0 = \lambda - \left(V_j \sin\beta + \frac{D_{av}}{n\dot{m}_a}\right)$$

$$\lambda = V_0 + V_j \sin\beta + \frac{D_{av}}{n\dot{m}_a}$$

$$\text{Thus } V = \left(V_0 + V_j \sin\beta + \frac{D_{av}}{n\dot{m}_a}\right) e^{-\frac{n\dot{m}_a}{m}t} - \left(V_j \sin\beta + \frac{D_{av}}{n\dot{m}_a}\right)$$

or

$$e^{\frac{-n\dot{m}_a}{m}t} = \frac{V + V_j \sin\beta + (D_{av}/n\dot{m}_a)}{V_0 + V_j \sin\beta + (D_{av}/n\dot{m}_a)}$$

The time is then expressed as

$$t = -\frac{m}{n\dot{m}_a}\ln\frac{V + V_j \sin\beta + (D_{av}/n\dot{m}_a)}{V_0 + V_j \sin\beta + (D_{av}/n\dot{m}_a)}$$

To determine the time and distance between touchdown and aircraft stop, the given data are substituted in the above equation. For two engines, then

$$t = -\frac{50,000}{2 \times 50}\ln\frac{0 + 600 \times \sin 30 + (5000/2 \times 50)}{40 + 600\sin 30 + (5000/2 \times 50)} = 54.1\,\text{s} = 0.9\,\text{min}$$

The deceleration is determined from the relation

$$a = -\frac{V_0}{t} = -\frac{40}{54.1} = -0.739\,\text{m/s}^2$$

The distance is then

$$s = -\frac{V_0^2}{2a} = 1082\,\text{m}$$

11.4.2 Calculation of Ground Roll Distance

As the aircraft approaches the ground, the aircraft goes in what is known as ground roll just before its touch down with the ground. The normal landing procedure after touchdown may be described as follows. Immediately after touchdown at the speed of V_{TD} the aircraft goes into what is identified as free roll for a few seconds. It is next followed by applying thrust reversal system for some 600–800 m. Finally, the thrust lever is returned back to idle position, giving a small amount of forward thrust, and the pilot applies the brake intermittently until a final stop. Thus, the three landing distances will be analyzed here at first and next it will be applied to a large civil transport aircraft such as Boeing 777 or Airbus A330.

The landing distance may then be written as

$$S = S_1 + S_2 + S_3 \tag{11.22}$$

Where the free roll distance is

$$S_1 = S_{fr} = tV_{TD} \tag{11.23}$$

And the time (t) is one second for small airplanes and three seconds for large airplane [10]. The touchdown speed is related to the stall speed of aircraft given by the relation:

$$V_{TD} = iV_{stall} = i\sqrt{\frac{2W}{\rho SC_{L_{max}}}}$$

where (i) is equal to (1.15) for civil airplanes and (1.1) for military airplanes [10]. The wing area is (S).

The second landing distance (S_2) is governed by the aircraft equation of motion

$$F = T_{rev} + D + \mu_r(W - L) = -m\frac{dV}{dt}$$

The forces here are the thrust reverse force (T_{rev}), aerodynamic drag (D), and frictional force with the runway coefficient of friction ($\mu_r = 0.01 - 0.04$) depending on the runway condition.

The resultant force is correlated to the aircraft speed and landing distance as follows:

$$F = -m\frac{dV}{dt} = -m\frac{dV}{dS}\frac{dS}{dt} = -mV\frac{dV}{dS} = -\frac{m}{2}\frac{dV^2}{dS}$$

The above equation is integrated, keeping in mind the landing procedure described above. Thus, the integration is from the instant of application of the thrust reversal until stowing the thrust reversal barrel in turbofan engines, for example, or in general, ending thrust reversal application.

$$S_2 = -\frac{m}{2F}\left[\int_{V_{TD}}^{V_{brake}} dV^2\right]$$

$$S_2 = \frac{m}{2}\frac{V_{TD}^2 - V_{brake}^2}{[T_{rev} + D + \mu_r(W - L)]_{0.7V_{TD}}} \tag{11.24}$$

The total reverse thrust force for (n) engines is given by the relation:

$$T_{rev} = n\dot{m}_{ac}[(V_e \cos\beta) + V_{TD}]$$

Here the lift and drag forces are calculated at an average speed equal to seven-tenth of the touch down speed ($V_{av} = 0.7V_{TD}$).

The third segment is when the brakes are applied and the thrust is again in the normal direction but at idle condition.

$$S_3 = \frac{m}{2} \frac{V_{brake}^2}{[-T + D + \mu_r(W - L)]_{0.35V_{TD}}} \tag{11.25}$$

where

$$V_{brake} \approx 0.3 - 0.5V_{TD}$$

11.5 THRUST VECTORING

Thrust vectoring is the ability of an aircraft or other vehicle to direct the thrust from its main engine(s) in a direction other than parallel to the vehicle's longitudinal axis. Thrust vectoring is a key technology for current and future air vehicles. The primary challenge is to develop a multiaxis thrust-vectored exhaust nozzle that can operate efficiently at all flight conditions while satisfying the design constraint of low cost, low weight, and minimum impact on radar cross-section signature. The technique was originally envisaged to provide upward vertical thrust as a means to give the aircraft VTOL or short takeoff and landing (STOL) capability. Subsequently, it was realized that the use of vectored thrust in combat situations enabled an aircraft to perform various maneuvers and have better rates of climb not available to conventional-engined planes. In addition, and most important, thrust vectoring can control the aircraft by engine forces, even beyond its stall limit, that is, during "impossible" post-stall (PS) maneuvers at extremely high nose turn rates [11]. An interesting definition for thrust vectoring is introduced in Reference 12 as a *maneuver effector which can be used to augment aerodynamic control moments throughout and beyond the conventional flight envelop*. Rockets or rocket-powered aircraft can also use thrust vectoring. Examples of rockets and missiles which use thrust vectoring are the space shuttle SRB, S-300P, UGM-27 Polaris nuclear ballistic missile, and Swingfire small battlefield.

Aircraft that do not use thrust vectoring have to rely on ailerons or flaps to facilitate turns, while aircraft that do still have to use ailerons, but to a lesser extent. For pure-vectored propulsion, the flight-control forces generated by conventional aerodynamic control surfaces of the aircraft have been replaced by the stronger internal thrust forces of the jet engines [11]. These forces may be simultaneously or separately oriented in all directions, that is, in the yaw, pitch, roll, thrust reversal, and forward thrust directions of the aircraft. Thrust vectoring is also used as a control mechanism for airships, particularly modern nonrigid airships. In this application, the majority of the load is typically supported by buoyancy, and vectored thrust is used to control the motion of the aircraft.

There are two methods for achieving thrust vectoring either mechanically or by fluidic control. Mechanical thrust vectoring includes deflecting the engine nozzle and thus physically changing the direction of the primary jet. Fluidic thrust vectoring involves injecting fluid into or removing it from the boundary layer of a primary jet to enable vectoring. Mechanical thrust vectoring system is heavy and complex, while fluidic thrust vectoring is simple, lightweight, inexpensive, and free from moving parts (fixed geometry).

Mechanical thrust vectoring can be further subdivided into engine/nozzle internal thrust vectoring (ITV) and engine/nozzle external thrust vectoring (ETV). ITV system is illustrated in Figure 11.23, where only pitch control is available. Figure 11.23a illustrates unvectored engine operation, while Figure 11.23b illustrates a down-pitch thrust vectoring operation. ETV is based on postnozzle exit, that is, (three or four) jet deflecting vanes that deflect exhaust jet(s) in the yaw and pitch coordinates, and in a few designs, also in roll coordinates [13]. Figure 11.24 illustrates three and four vane (pedals) ETV [14].

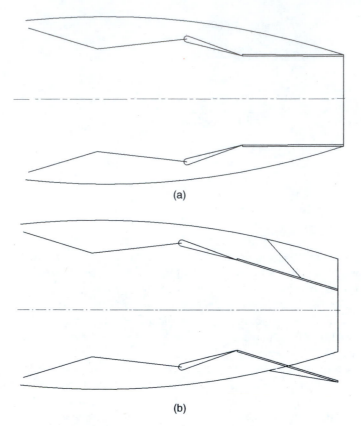

FIGURE 11.23   Example for pitch control internal thrust vectoring nozzle (ITV). (a) Normal (Unvectored) operation. (b) Down-pitch operation.

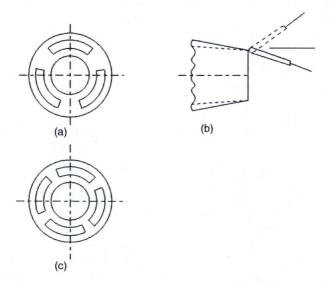

FIGURE 11.24   External thrust vectoring. (a) 3 Pedals. (b) Side view. (c) 4 Pedals.

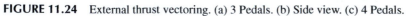

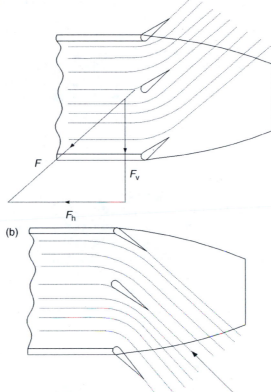

(a)

F

F_v

F_h

(b)

FIGURE 11.25   Thrust vectoring nozzle. (a) Up pitch. (b) Down-pitch.

The two-dimensional nozzle depicted in Figure 11.25 is capable of vectoring the thrust up or down to produce an up force or down force.

If the two nozzles of a twin-engined plane can also be deflected differentially (one producing up force, the other down force), then it will provide roll control. This type of nozzle may be identified as three-dimensional thrust vectoring. Examples for aircraft using two-dimensional nozzle are Harrier, Boeing X-32, F-35, Yak-38, and Yak-141.

Examples for aircraft employing three-dimensional thrust vectoring (pitch and yaw moments) are F-15, F-16, F-18, MiG-29, MiG-35, Sukhoi Su-30, Su-37, and Su-47.

For example, the F-22 carries two Pratt & Whitney F119- PW-100, reheated engines. The exhaust nozzles move 20° in all directions, providing it with three-dimensional thrust vectoring capabilities. Heat-resistant components give the nozzles the durability needed to vector thrust, even in afterburner conditions. With precision digital controls, the nozzles work like another aircraft flight control surface. Thrust vectoring is an integrated part of the F-22's flight control system, which allows for seamless integration of all components working in response to pilot commands.

Concerning MiG-29M OVT, nozzle vectoring mechanism employs three hydraulic actuators mounted 120° apart around the engine; the nozzle can deflect the engine's full thrust by up to 18° in any direction [15].

Most currently operational vectored thrust aircraft use turbofans with rotating nozzles or vanes to deflect the exhaust stream. This method can successfully deflect thrust through as much as 90°, relative to the aircraft centerline.

Next, let us take a look at fluidic nozzles, which divert thrust via fluid effects. Tests have shown that air forced into the exhaust stream can affect deflected thrust of up to 15°. Extensive experimental studies for utilizing a fluidic injection system instead of the heavy and mechanically complex door system are now performed by scientists in academia, research centers, and by military aircraft manufacturers. In addition to the lightweight and simplicity, it has minimal aircraft observability penalty (or lower radar cross-section). For this reason, some studies were performed to utilize it in low observable unmanned air vehicle operating in the subsonic flight regime [12]. However, the main challenges with fluidic vectoring are in obtaining an effective, efficient system with reasonably linear control response.

Fluidic thrust vectoring must be implemented at the outset of the design process unlike mechanical systems that can be retrofitted to existing aircraft currently in use today [13]. The basic idea of fluidic thrust vectoring systems is to use a secondary air jet to control the direction of the primary jet. It has several types, namely, shock vector control, coflow, counterflow, synthetic jet actuators, and sonic throat skewing. Figure 11.26 illustrates the first three cases.

Shock thrust vector control involves injecting a secondary jet into the primary jet from one of the divergent flaps. An oblique shock wave is formed that deflects the primary jet in the pitch plane. Both coflow and counterflow concepts involve the use of a secondary jet together with the Coanda effect. Curved reaction surfaces are positioned downstream of the nozzle exit to which the jet may attach according to the Coanda effect. The side onto which the jet attaches is controlled by tangential injection of a secondary jet of air upstream of the surfaces. The resulting thrust vector force generated on the reaction surface can then be used to provide useful moments for aircraft control.

11.5.1 Governing Equations

Here only two-dimensional thrust vectoring will be discussed. As shown in Figure 11.27, the four forces that influence the aircraft motion are the lift (L), weight (W), drag (D), and thrust (T). For a case of thrust vectoring, the thrust force is inclined at an angle (θ) to the horizontal. Thus, the

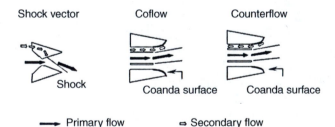

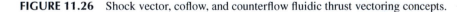

FIGURE 11.26 Shock vector, coflow, and counterflow fluidic thrust vectoring concepts.

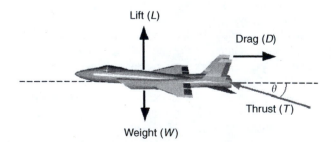

FIGURE 11.27 Forces on aircraft for thrust vectoring.

governing equations in the horizontal and vertical directions, respectively, are

$$T \cos\theta - D = ma_x \tag{11.26a}$$

$$L - W + T \sin\theta = ma_y \tag{11.26b}$$

where a_x and a_y are the acceleration in the horizontal and vertical directions, respectively. The first acceleration controls the forward motion, while the second acceleration controls the rate of climb or descent. The angle (θ) is the inclination of the exhaust jet stream from the axis of aircraft. For a vertical flight this angle is 90°. This angle also develops a pitching moment necessary for nose up or down depending on deflecting the jet nozzle upward or downward.

11.6 NOISE

11.6.1 INTRODUCTION

Since airline travel has become much more common and widely available, resulting in an increased amount of air traffic, an increased number of people are affected by aircraft noise, both as passengers and as members of communities located near airports. Aero engine manufacturers have to stick to the international legislation norms concerning noise by measuring noise around an engine on a bench in an open environment. Noise may be defined as *unwanted and usually irritating sound* [16]. The unit used to express noise annoyance is the Effective Perceived Noise deciBel (EPNdB), where Bel is the basic unit and Db is one-tenth of Bel.

Aircraft noise derives primarily from engines (or the propulsion systems) and from airframe structure. Figure 11.28 illustrates the contribution of several noise sources for an aircraft, where the inlet fan stands for fan in turbofan engines and compressor in turbojet engines. The biggest source of aircraft noise is the engines (although the air rushing over the airframe also creates noise). Concerning the engine modules, the significant noise originates in the fan or compressor, the turbine, and the exhaust jet or jets. Exhaust jet noise varies by a larger factor than the compressor or turbine noise, therefore a reduction in the exhaust jet velocity has a stronger influence than an equivalent reduction in compressor and turbine blade speed [1]. The intensity of the sound produced by a jet engine at takeoff may attain a value of 155 dB (15.5B), while the threshold of feeling for a human being is 130. The different noise levels of jet engines with and without noise suppressors are plotted in Figure 11.29.

Jet exhaust noise is caused by turbulent mixing of the exhaust gases "efflux" with the lower velocity surrounding air. The small eddies near the exhaust duct cause high frequency noise, while the larger eddies downstream the exhaust jet create low frequency noise. Accelerating the mixing rate or reducing the velocity of the exhaust jet reduces the noise level. In addition, when the exhaust jet velocity exceeds the local speed of sound, a regular shock pattern is formed within the exhaust core. This produces a discrete (single frequency) tone and selective amplification of the mixing noise as depicted in Figure 11.30.

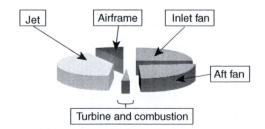

FIGURE 11.28   Different sources of noise in aircraft.

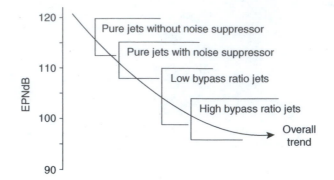

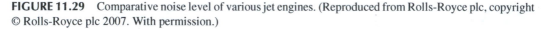

FIGURE 11.29   Comparative noise level of various jet engines. (Reproduced from Rolls-Royce plc, copyright © Rolls-Royce plc 2007. With permission.)

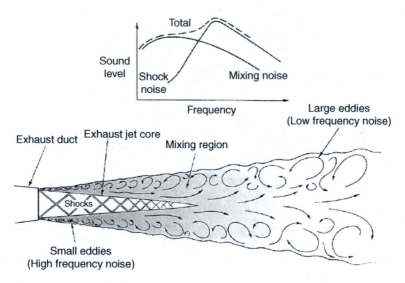

FIGURE 11.30   Exhaust mixing and shock structure. (Reproduced from Rolls-Royce plc, copyright © Rolls-Royce plc 2007. With permission.)

In turbojet engine, there are three types of noise: noise radiated from the air intake, noise radiated from vibrations of the engine shell, and exhaust noise. For turbofan engine fan exhaust has lower velocity, resulting in quieter operation for a given total thrust [17]. The second type of noise, namely, the airframe noise, includes the noise that is produced by the passing of air around the aircraft. Under normal cruising conditions, this type of noise predominates, especially among higher frequencies. However, propulsion noise tends to predominate at the lower takeoff and landing speeds [17].

In subsonic flight, which is typical for the vast majority of passenger flights, aircraft noise peaks when the aircraft is approximately overhead and then gradually diminishes to the ambient noise level. In addition, noise that is heard as the aircraft approaches tends to be dominated by higher frequency sound, whereas noise that is heard after the aircraft has passed tends to include more lower-frequency sound [17].

As illustrated in Figure 11.31, for turbojet engines (which dominated in 1996s), lower-frequency exhaust noise predominates over the rearward-propagating engine, while higher-frequency noise is produced by the compressor and turbine. In turbofan engines (which have been dominating since the 1990s), on the other hand, higher-frequency sound radiating from the fan, compressor, and turbine predominate over exhaust noise and propagate both forward and rearward from the aircraft.

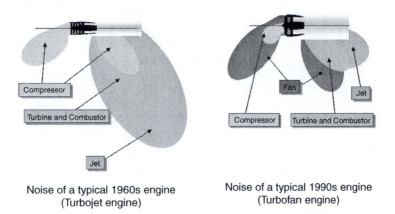

Noise of a typical 1960s engine
(Turbojet engine)

Noise of a typical 1990s engine
(Turbofan engine)

FIGURE 11.31   Noise of the 1960s and 1990s aero engines. (Courtesy Airbus Industries.)

This may be a moot point, however, since the vast majority of commercial aircraft now use turbofan engines. Thus, contemporary aircraft noise pollution generally includes more higher-frequency noise than aircraft noise pollution of yesteryear.

Supersonic airplanes produce a unique noise phenomenon: the sonic boom. In most cases, a sonic boom is experienced as a shock wave sweeping across the ground below the aircraft. However, sonic booms can also occur during acceleration periods such as diving, and are more transient in such cases. Though sonic booms are obviously much more damaging than noise produced during subsonic flight, they are generally restricted to military aircraft and are banned above most urban areas, and thus contribute less to the noise pollution experienced by most people.

Historically, the first complaints about "noise pollution" arose from Boeing 707, first delivered in 1958, which was powered by turbojet engines. The establishment of the Federal Aviation Administration (FAA) noise regulations in the early 1970s called for a 25%–50% reduction in maximum noise generated by the existing long-range Boeing 707, Douglas DC-8, and Vickers VC-10 aircraft. Larger aircraft then entering service such as the Lockheed L-1011, Douglas DC-10, and Airbus A300 were designed with more stringent noise requirements in mind, while Boeing 747 needed some modification to cope with the new regulations. In 1971, the U.S. FAA implemented its own rules in Federal Aviation Regulations (FAR) Part 36, in 1971, FAR Part 36, which limited maximum noise that could be produced at an airport at three points—two on either end of the runway beneath takeoff and landing paths and one at the middle and sides of the runway. The FAA regulations on aircraft noise were tightened several times after the initial rules were set. Naturally, this has prompted aircraft/engine manufacturers to try to develop quieter aircraft/engines. The development of high-bypass turbofans resulted in a beneficial reduction in noise. Between 1966 and 2000 an improvement of more than 20 dB in lateral noise level corrected to aircraft thrust has been achieved (Figure 11.32). The main challenge for aircraft industry is to reduce the noise level by a factor of two (10 EPNdB) from 1997 by 2007 for subsonic aircraft and by a factor of four (20 EPNdB) by 2022.

The maximum perceived noise level of different turbofan modules at approach and takeoff are illustrated in Figure 11.33. The jet has a larger noise level during takeoff than approach, while the fan exhaust is nearly equal.

11.6.2 ACOUSTICS MODEL THEORY

The main model used in the study of the jet noise is the Ffowcs Williams and Hawkings (FW-H) Model. The FW-H equation is essentially an inhomogeneous wave equation that can be derived by manipulating the continuity equation and the Navier-Stokes equations [18]. This equation is capable of addressing three-dimensional, nonlinear, and unsteady flow problems of compressible and viscous

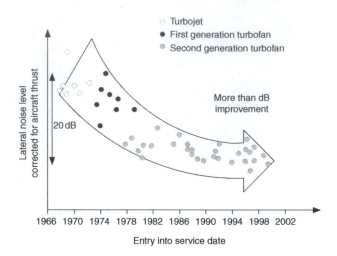

FIGURE 11.32   Improvement of engine noise level.

fluids. An alternative equation is the Kirchhoff (K) equation [19]. Using generalized function theory, both the FW-H and the K equations can be written with inhomogeneous source terms involving the Dirac delta function.

11.6.3 Methods Used to Decrease the Jet Noise

Noise suppression of internal sources is approached in two ways: by basic design to minimize noise originating within or from engine, and by the use of acoustically absorbent lining [20] through [22]. Noise suppressor is usually combined with the engine thrust reverser and is called suppressor-reverser [5].

As described above, exhaust jet is the major source of noise in turbojet and low bypass ratio turbofan engines, and this can be reduced by inducing a shorter mixing region. This is achieved by increasing the contact area of atmosphere with the exhaust gas stream by using a propelling nozzle incorporating either *corrugated* or *lobe-type* (multiple-tube) noise suppressors. Corrugated perimeter type is warped or angled, often looking like a flower (Figure 11.34). It breaks up the single main jet exhaust into a number of smaller jet streams. The shape of the suppressor increases the total perimeter of the nozzle area and reduces the size of eddies created as the gases are discharged to the open air, which promote rapid mixing. The size of eddies scale down linearly with the size of the exhaust stream. This has two effects. First, the change in frequency may put some noise above the audibility range of human ear. Second, high frequencies within the audible range are more highly attenuated by atmospheric absorption than are low frequencies.

In the lobe-tube nozzle, the exhaust gases are divided to flow through the lobes and a small central nozzle. These form a number of separate exhaust jets that rapidly mix with the air entrained by the suppressor lobes (Figure 11.35). This principle can be extended by the use of a series of tubes to give the same overall area as the basic circular nozzle. This type of noise suppressors was used by military aircraft and early commercial aircraft. But both these methods increased drag and reduced engine performance, and the multitube approach also increased weight, sometimes substantially.

Another solution that emerged in the 1960s was the "ejector-suppressor." This was essentially a large tube fitted aft of the engine around the exhaust nozzle that allowed air from outside the engine to mix with the exhaust, reducing the final efflux velocity.

Turbofan engines are quieter than turbojet engines as the exhaust velocities of both streams of turbofans are less than those for turbojets. Mixing of the flows in turbofan engines for the purpose of noise reduction was examined as early as 1970s [23]. Extensive research was performed by air

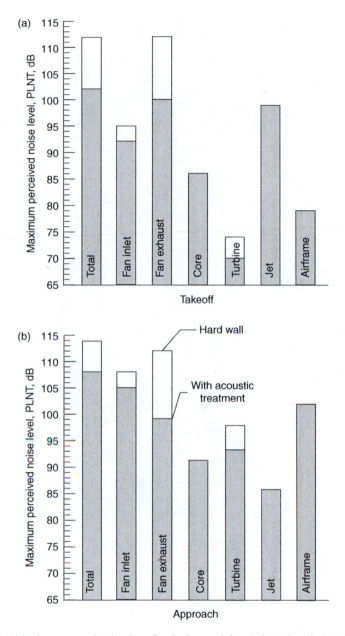

FIGURE 11.33   (a) Maximum perceived noise of turbofan modules during takeoff. (b) Maximum perceived noise of turbofan modules during approach. (Courtesy Airbus Industries.)

forces in the 1980s as they started looking for ways to reduce fighters' infrared signature. It was noticed that those same nozzles had an effect on noise emissions as well. Several nozzle designs were tested to reduce the fully expanded jet velocity by mixing (1) core flow with fan flow only, (2) fan flow with ambient flow only, or (3) both flows simultaneously. Depending on the type of mixing attempted, these designs fell into two broad categories: tabs and chevrons. Tabs are severe protrusions into the flow at the nozzle exit plane. Chevrons are also protrusions but of much less severity than tabs. The aggressive mixing produced by the tabs greatly reduced low-frequency noise but with the penalty of tab-induced high-frequency noise. Chevrons, which provided a more balanced approach

Corrugated internal mixer

FIGURE 11.34 Corrugated perimeter noise suppressor (rear view). (Reproduced from Rolls-Royce plc, copyright © Rolls-Royce plc 2007. With permission.)

Lobe-type nozzle

FIGURE 11.35 Lobe-type nozzle. (Reproduced from Rolls-Royce plc, copyright © Rolls-Royce plc 2007. With permission.)

to mixing, reduced low-frequency noise without significant chevron-induced high-frequency noise. Other nozzle designs attempted to shield the core flow by using a scarf fan nozzle and an offset fan nozzle for mixed flow turbofan engines; mixing of both hot and cold streams greatly reduces the noise.

Generally, the most successful method employed by numerous aero engine manufacturers is to mix the hot and cold exhaust streams within the confines of the engine and expel the lower velocity exhaust gas flow through a single nozzle. This single nozzle may also have the corrugated or lobe-type shape [1]. Numerous numerical research works are performed to examine the mixing process for turbofan engines. For example, refer to Reference 24.

Another technique for noise suppression that may be applied to both turbojet and turbofan engines, which has been recently investigated, is the eccentric coannular nozzle [25]. Eccentric nozzle has nearly 10% improvement for low-frequency noise and nearly 5% for high-frequency noise.

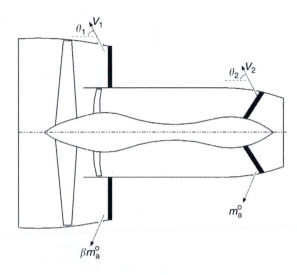

FIGURE 11.36 Layout of a turbofan engine with two thrust reverse flow streams.

PROBLEMS

11.1 A turbojet engine is powering an aircraft flying at an altitude of 8000 m where the temperature and pressure are 236.2 K and 35.6 kPa. At a flight Mach number of 0.8, the inlet conditions to the nozzle were 600 K and 60 kPa. For a convergent nozzle, check whether the nozzle is choked or not. Calculate the thrust per unit frontal area. Next the pilot lit the afterburner to increase the Mach number to 1.6. The nozzle's geometry was changed to a C-D shape. The inlet temperature and pressure to the nozzle were 1200 K and 60 kPa. Recalculate the thrust per unit frontal area. Compare the nozzle outlet areas in both cases.
Calculate the throat area of the C-D nozzle.

11.2 Figure 11.36 illustrates a turbofan engine during application of thrust reverse. The exhaust from the fan is leaving through reverse cascade at a speed of $V_1 = 300$ m/s and an inclination angle with respect to the longitudinal axis of engine of $\theta_1 = 50°$. The hot gases are also leaving the engine through reverse cascades at a speed of $V_2 = 600$ m/s and an inclination angle of $\theta_2 = 60°$ to the engine axis. The touchdown speed is $V_f = 80$ m/s. Prove that the force developed by thrust reversers is

$$F = \beta \dot{m}_a V_1 \cos\theta_1 + \dot{m}_a (1+f) V_2 \cos\theta_2 + (1+\beta) \dot{m}_a V_f$$

Calculate the resulting drag force due to the application of thrust reverser in the following case: fuel-to-air ratio $f = 0.02$ and $\dot{m}_a = 50$ kg/s, bypass ratio $\beta = 5$.

11.3 The time needed for landing when the thrust reverse is applied up to a complete stop and assuming a mean value for drag force was derived in Example 5 given by:

$$t = -\frac{m}{n\dot{m}_a} \ln \frac{V + V_j \sin\beta + (D_{av}/n\dot{m}_a)}{V_0 + V_j \sin\beta + (D_{av}/n\dot{m}_a)}$$

where V_0 = touchdown velocity
V = velocity at any time during landing and is equal to zero at aircraft stop
V_j = exhaust jet velocity
m = mass of airplane during landing
$\dot{m}_a$ = mass flow rate of air sucked and exhausted by engine (assume $\dot{m}_f \approx 0$)

n = number of engines

For a twin engines airplane, with $m = 50,000$ kg, $m_a = 50$ kg/s, $V_j = 600$ m/s, $V_0 = 80$ m/s, examine

  a. The effect of the angle β on the time and distance for landing. Assume β changes from $0°$ to $60°$ in intervals of $10°$.
  b. For the above airplane with the angle $\beta = 60$, examine the effect of exhaust speed V_j; assume it varies from 200 to 600 in one hundred intervals.

11.4 A wide-body civil transport aircraft powered with two turbofan engines is examined during landing. The pilot is applying the thrust reverse to shorten the ground landing distance and the thrust reverse is inclined at an angle $\beta = 60°$ to the axial direction. The runway is dry concrete/asphalt where the coefficient of friction with brakes-on and -off is 0.4 and 0.04, respectively. The aircraft has the following data:

Maximum weight = 187 tons, wing span = 60.3 m, wing area = 361.6 m^2, fan air mass flow rate $\dot{m}_{ac} = 500$ kg/s, engine maximum thrust = 320 kN, idle thrust = 50 kN, lift and drag coefficient factors are $C_{DO} = 0.02$ and $C_{L\,max} = 3.2$, span efficiency factor (e) = 0.9

Calculate

  a. The ground landing distance with applied thrust reverse
  b. The ground landing distance without thrust reversal

11.5 The convergent nozzle discussed in Example 2 is to be reexamined to determine the effect of the angle θ on the length of the nozzle. What will be the length of the nozzle if θ changes in $1°$ increment from 11 to 14? Follow the same procedure and recalculate the variation of the gas properties through the nozzle for an angle $\theta = 11°$.

11.6 Follow the same procedure in Example 3 and calculate and plot the nozzle contour for an outlet Mach number of 4.0 and the same throat diameter. How does the length of the nozzle change with the outlet Mach number?

11.7 Discuss why fluidic thrust reverse is not used in large aircraft.

11.8 Discuss how the runway length will be drastically longer if aircraft land without application of thrust reverse. Consider the wide-body civil transport in Problem 11.4 and recalculate the landing distance without thrust reversal force.

11.9 Why is thrust reversal not applied at low speeds? Consider again the civil transport in Problem 11.4 and calculate the needed distance for aircraft to reach zero speed if the pilot depends totally on the thrust reverse and does not use the brakes at all.

REFERENCES

1. *The Jet Engine*, Rolls-Royce plc. 3, 5th ed., Reprinted 1996 with revisions, p. 159.
2. W.W. Bathie, *Fundamentals of Gas Turbines*, 2nd ed., John Wiley & Sons, Inc., NY, USA, 1996, p. 343.
3. H.I.H. Saravanamuttoo, G.F.C. Rogers, and H. Cohen, *Gas Turbine Theory*, 5th ed., Prentice Hall, Harlow, Essex, England, 2001, p. 109.
4. J.L. Kerreebrock, *Aircraft Engines and Gas Turbines*, 2nd ed., The MIT Press, MA, USA, 1992, p. 142.
5. United Technologies Pratt Whitney, *The Aircraft Gas Turbine Engine and Its Operation*, P&W Operation Instruction 200, 1988, pp. 3–33.
6. A. Glassman, *Turbine Design and Application*, NASA SP-290, Vol. III, 1975, p. 14.
7. A.H. Shapiro, *The Dynamics and Thermodynamics of Compressible Flow Field*, Vol. I, 1953, p. 507.
8. A. McCabe, *Design of a Supersonic Nozzle*, Ministry of Aviation, R&M No. 3440, London, March 1964.
9. I.E. Treager, *Aircraft Gas Turbine Engine Technology*, 3rd ed., McGraw-Hill, Inc., NY, USA, 1999, p. 220.
10. J.D. Anderson, Jr., *Aircraft Performance and Design*, McGraw-Hill Book Co., 1999, NY, USA, p. 370.
11. B. Gal-Or, Fundamental concepts of vectored propulsion, *Journal of Propulsion and Power*, AIAA, 6, 1990, 747–757.

12. M.S. Mason and W.J. Crowther, *Fluidic Thrust Vectoring of Low Observable Aircraft*, CEAS, Aerospace Aerodynamic Research Conference, June 10–12, 2002, Cambridge, UK.

13. B.L. Berrier and M.L. Mason, *Static Performance of an Axisymmetric Nozzle with Post-exit Vanes for Multiaxis Thrust-Vectoring*, NASA TP-2800, May 1988.

14. E.A. Bare and D.E. Reubush, *Static Internal Performance of a Two-Dimensional Convergent–Divergent Nozzle with Thrust Vectoring*, NASA TP-2721, July 1987.

15. Farnborough: MiG-29 Thrust Vectoring Rekindles Spirit; *Flight International Magazine*, Paris Issue, 25/07/2006.

16. M.J. Kroes and T.W. Wild, *Aircraft Powerplants*, Glencoe/McGraw-Hill, 7th ed., *International Magazine*, Paris Issue, NY, USA, 2002, p. 303.

17. J.P. Raney and J.M. Cawthorn, Aircraft noise, In *Handbook of Noise Control* (2nd ed.), C.M. Harris (Ed.), McGraw-Hill Book Company, NY, USA, 1979, Chapter 34.

18. J.E.W. Ffowcs and D.L. Hawings, Sound generation by turbulence and surfaces in arbitrary motion, *Philosophical Transactions of the Royal Society*, A264, 321–542.

19. F. Farassat and K.S. Brentner, *A Study of Supersonic Surface Sources—The Ffowcs Williams-Hawkings Equation and the Kirchhoff Formula*, NASA-AIAA-98-2375, 1998.

20. M.J.T. Smith, *Aircraft Noise.* Cambridge: Cambridge University Press, 1989.

21. D. Papamoschou and M. Debiasi, Directional suppression of noise from a high-speed jet, *AIAA Journal*, 39, 2001, 380–387.

22. M. Harper-Bourne, Physics of jet noise suppression. *Proceedings of Jet Noise Workshop*, NASA/CP–2001-211152, 2001, pp. 701–719.

23. G.C. Paynter and S.C. Birch, *An Experimental and Numerical Study of the 3-D Mixing Flows of a Turbofan Engine Exhaust System*, 15th Aerospace Science Meeting, AIAA-77-204, Los Angeles, CA, 1977.

24. N.J. Cooper and P. Merati, *Numerical Simulation of the Vortical Structures in a Lobed Jet Mixing Flow*, AIAA-2005-0635, 2005.

25. K. Zaman, *Noise- and Flow-Field of Jets from an Eccentric Coannular Nozzle*, AIAA-2004-0005, 2004.

12 Centrifugal Compressors

12.1 INTRODUCTION

This chapter and the succeeding three chapters treat the rotating modules of both aero engines and gas turbines. These four chapters discuss the compressors, fans, and turbines. Since all of these are identified as *turbomachines*, it is necessary to introduce a definition for turbomachines here. The word (*turbo*) or (*turbines*) is of Latin origin, meaning "that which spins or whirls around." Turbomachines are often referred to as rotor dynamic devices that transfer energy to or from a working fluid through the forces generated by a rotor. For compressors or fans, energy is transferred from the rotor to the fluid. The reverse occurs in turbines that deliver shaft power in exchange of thermal energy taken from the working fluid.

In this chapter as well as in Chapter 13, details about dynamic compressors will be discussed. Brown [1] classified compressors as either intermittent flow or continuous flow ones (Figure 12.1). Intermittent flow compressors are positive displacement ones. They are classified as either reciprocating or rotary types. Such positive displacement compressors achieve increase in pressure by trapping fluid in a confined space and transporting it to the region of higher pressure. Continuous flow compressors are classified as either ejectors or dynamic compressors. Dynamic compressors are further classified as radial flow, mixed flow, or axial flow compressors.

Dynamic compressors have close similarity to propellers and propfans. In both, energy is transferred from the blades to the working fluid. However, they differ in influenced fluid quantity. Dynamic compressors accommodate a finite quantity of working fluid that flows steadily in an annular duct that comprises a hub and casing. Consequently, it is defined as *enclosed* machines. On the contrary, propellers and propfans influence unbounded quantity of air; in other words, they are turbomachines without shrouds or annulus walls (or casing) near the tip. Thus, these machines are termed as extended turbomachines.

Dynamic compressors work by converting velocity to pressure in a continuous flow. They are efficient, compact, and handle vast quantities of working fluid. Considering air as the working fluid at 1.013 bar and 288 K, the inlet flow is typically 150 kg/s (122 m^3/s or 259,000 ft^3/min) per square

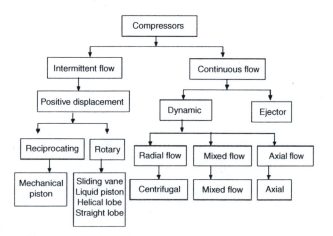

FIGURE 12.1 Classification of compressors.

meter of inlet area at Mach 0.4 [2]. The intake diameters of various types range from 0.1 m (4 in.) to 2.44 m (8 ft). The range of mass flow varies from about 1.2 to 700 kg/s. The isentropic efficiency ranges typically from 0.7 to 0.9.

In this chapter, the aerodynamic and mechanical designs of centrifugal compressors are discussed. Axial compressors will be treated in the succeeding chapter. Mixed-flow compressors are rarely used in aero engines or gas turbines.

Centrifugal compressors and radial inflow turbines are identified as radial turbomachines. Surprisingly, radial turbomachines have a long history and predate axial flow turbomachines, both compressors and turbines [3]. In 1705, Papin [33] published descriptions of the centrifugal pumps and blowers that he had developed. In the closing years of the nineteenth century, relatively crude centrifugal pumps and blowers were used for ventilation. Gas turbine engine came to prominence in the twentieth century, which pushed turbomachine design to the limits of technology and materials. The highest efficiency turbomachines available to the early gas turbine engineers were the centrifugal compressors and axial turbines [3]. For this reason, centrifugal compressors were employed in the first jet engines developed by Frank Whittle and von Ohain due to the experience gained in the design of superchargers. With the development of aero engines, axial compressors have nearly replaced centrifugal compressors. Thus, centrifugal compressors are found in small turbofan engines such as AlliedSignal 731 turbofan engines and small turboprop/turboshaft engines such as the Rolls-Royce DART, Allied Signal 331, and the famous engine manufactured by Pratt & Whitney of Canada, PT6 (Figure 12.2). They are also found in auxiliary power units (APU) of many civil transport airplanes such as Garrett GTCP-85 for DC8 and 9, MD-80, Boeing 707, 737, as well as Sunstrand Turbomach APS 200 for B737-500 and A321.

Other applications for the centrifugal compressors are the aircraft cabin air conditioning such as the Hamilton Standard 777 (ACTCS) and the multipurpose small power units (MPSPU) represented by SunStrand Turbomach T-100. In a few gas turbine engines, centrifugal compressor is used for the final stage of compression downstream of multistage of axial compressor. This arrangement is called *axial–centrifugal* compressor similar to those in the General Electric T700 engine, very popular PT6 Pratt & Whitney Canada engine, and T53 Honeywell engine.

Centrifugal compressors in aero engines provide small to moderate air flows (up to nearly 50 kg/s). They have either single or double stages (sometimes identified as impellers in tandem). The pressure ratio per stage ranges from 4:1 to 8:1, while for double (two) stages in series a pressure ratio of 15:1 is found in Pratt & Whitney PW100. Since Frank Whittle engines, Kenny [4] reviewed the evolution of centrifugal compressors since Frank Whittle engines up to the mid-1980s. It is expected, based on the efforts paid in Pratt & Whitney Canada as well as other manufacturers, that

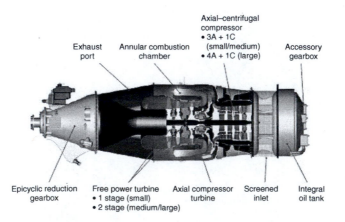

FIGURE 12.2   Centrifugal compressor in the turboprop engine PT6. (Courtesy Pratt & Whitney of Canada.)

higher value of pressure ratios (16:1 or more) will be achieved by the beginning of the twenty-first century.

Centrifugal compressors are used in many industrial applications such as refineries, chemical and petrochemical plants, natural gas processing and transmission plants, very large-scale refrigeration, and iron and steel mills. These centrifugal compressors may have single or multiple stages (up to five stages) and can operate at very high inlet pressures. Rolls-Royce RFA36 and RFA24 giant compressors may have pressures up to 155 bar and flow rate of 200 m^3/min at design speed from 3,600 to 12,800 rpm. The General Electric Power Systems, Oil and Gas, Nuovo Pignone centrifugal compressors for reinjection services, have discharge pressure up to 500 bar. Hitachi had manufactured 700 sets of centrifugal compressors over the past 85 years employed in different applications. Earlier types of compressors manufactured by Hitachi in 1978 had a pressure ratio of ~260 for CO_2 applications. On the contrary, centrifugal compressors are employed in the microgas turbine generator by PCA Engineers.

Centrifugal compressors have the following advantages: light weight, low cost, rigidity, high pressure ratio per stage, easily made in relatively small sizes thus suitable for handling small volume flows, simplicity, better resistance to foreign object damage (FOD), less susceptibility to loss of performance by build up of deposits on the blade surfaces, and the ability to operate over a wider range of mass flow at a particular speed. The disadvantages of the centrifugal compressor are that it is generally less efficient than axial compressor (perhaps 4–5%) and it has a larger cross section compared with the cross section of the inlet flow. A detailed comparison between centrifugal and axial compressors will be given in the next chapter.

12.2 LAYOUT OF COMPRESSOR

Figure 12.3 illustrates a typical centrifugal compressor [2]. Nomenclature for a single-sided compressor with channel-type diffuser is given here.

F:	Impeller	G:	Inducer (rotating guide vane)
H:	Diffuser	J:	Impeller vane
K:	Half vane (or splitter)	L:	Impeller tip
M:	Impeller hub	N:	Shroud (casing)
O:	Impeller eye	P:	Vaneless gap

Generally, centrifugal compressor is called so because the flow through the compressor is turned perpendicular to the axis of rotation. This type of compressor is composed of three main elements, namely, the rotating part or impeller, the stationary part or stator, and a manifold or a collector (denoted by scroll or volute). The impeller has several blades referred to as vanes. The diffuser may have vanes or may be vaneless. Other arrangements for the diffuser will be discussed later on in Section 12.5.

An important part upstream of the impeller is the *inducer duct*. The nature of flow through the impeller is strongly dependent on the duct arrangement upstream. This could be a simple straight duct, a curved inlet duct, a curved return passage of a multistage compressor, or the final stage of an axial compressor. These alternative configurations can lead to distorted flow conditions at the impeller inlet, which in turn will lead to an overall performance deterioration of the compressor.

12.2.1 IMPELLER

Impeller scoops in the working fluid (air/gas). Air is drawn at the center or eye of the impeller, then accelerated through the fast spinning speed of the impeller and finally thrown out at the tip. The forces exerted on the air are centripetal. At the eye (inlet), the vanes are curved to induce the flow: this axial portion is called the inducer or rotating guide vane and may be integral with or

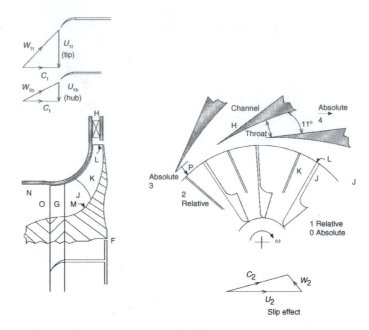

FIGURE 12.3 Layout of a typical centrifugal compressor.

separated from the main impeller. The divergence (increasing cross-sectional area) of these passages diffuses (slows) the flow to a lower relative velocity and higher static pressure. The impeller is a complicated diffuser compared with the conventional straight conical diffuser as the passage is doubly curved first in the axial plane and then in the radial plane (Figure 12.4). The impeller vanes are aligned radially, but may be curved near the tip, having forward or backward configurations. The impeller may have double sides (sometimes denoted double entry), as shown in Figure 12.5, to pass the maximum flow rate relative to its diameter and to balance the stresses of single-sided impeller. The vanes are cast, forged, or machined integrally with the impeller hub. The hub is the curved surface of revolution extending from the eye root to the outlet. The number of vanes is usually a prime number, typically from 19 to 37, to avoid vibration problems [2]. Half vanes (splitter blades) are sometimes used toward the tip to improve the flow pattern where the full vanes are widely spaced (denoted K in Figure 12.3). The outer curve of the vanes is sealed by the shroud, which may be part of the stationary structure or may rotate with the rotor. Typical impeller proportions are that the eye root diameter is about half the eye tip diameter, and the tip (outlet) diameter is nearly twice the eye tip diameter.

The impeller material is often aluminum, with titanium or steel for smaller, high-duty machines.

To conclude this section, it is worth mentioning that the part of the compression process achieved in the impeller is caused by moving the fluid outward in a centrifugal force field produced by the rotation of the impeller. This part of pressure rise differs from the pressure rise in axial-flow compressor rotors and stators; instead of arising from the exchange of kinetic energy for thermal energy in a diffusion process, it arises from the change in potential energy of the fluid in the centrifugal force field of the rotor. It is therefore less limited by the problems of boundary layer growth and separation in adverse pressure gradients. Probably for this reason, the centrifugal compressor first attained a range of pressure ratio and efficiency useful for turbojet engines, and it was used in the von Ohain engine (1939) and Whittle engine (1941).

12.2.2 DIFFUSER

The impeller blades sling the air radially outward where it is once again collected (at higher pressure) before it enters the diffuser. The diffuser represents a part of the fixed structure of the compressor.

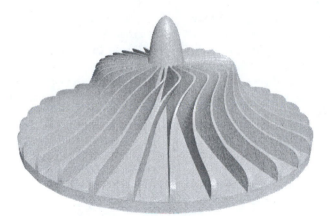

FIGURE 12.4   Impeller shape.

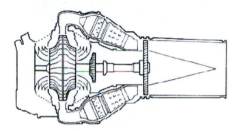

Double-entry single-stage
centrifugal turbojet

FIGURE 12.5   Double-sided centrifugal compressor. (Reproduced from Rolls-Royce plc, copyright © Rolls-Royce plc 2007. With permission.)

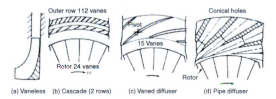

FIGURE 12.6   Different types of diffuser.

It discharges air from the compressor impeller with a high absolute velocity and the role of diffusion is to reduce the kinetic energy, thereby increasing the static pressure. The diffuser is either a vaneless passage or a vaneless passage followed by a vaned section. The vaned diffuser represents a large group including the vanes together with the cascade, channel, and pipe types (Figure 12.6) [2].

All diffusers have an initial vaneless gap outboard of the impeller in which the flow settles. The vaneless diffuser is bulky and can be inefficient in comparison to the other types. The cascade may involve one, two, or three rows of vanes, with performance similar to that of axial cascades. Vaned diffusers may use curved or straight vanes, having a longer path length than the cascade diffuser. Vaned diffuser may be made variable, the vanes pivoting about an axis as shown in Figure 12.6 to accommodate different incident flow angles through the speed range.

Vaneless, cascade, and vaned diffusers collect the flow in a one volute type collector, which is suitable for industrial applications. Channel and pipe diffusers are suitable for aero engines and gas turbines.

The channel and pipe diffusers collect the flow in separate passages, sometimes with corner vanes at their outlet. The passage diverges slowly, the pipe consisting of a conical hole and the channel type being rectangular in section with one pair of opposed walls diverging at 10–11°, the others being parallel.

The pipe and channel diffusers may feed separate combustors or may merge downstream. Recently, it had been shown that the pipe-type diffuser (diffuser comprising channel of circular cross section) is more efficient than the conventional type with rectangular channels. The passages are formed by drilling holes tangentially through a ring surrounding the impeller, and after the through the passages are conical. Overall isentropic efficiencies of over 85% are claimed for compressors using this new pipe diffuser.

12.2.3 SCROLL OR MANIFOLD

The final element of centrifugal compressors is either a manifold or a scroll. Centrifugal compressors with manifolds are used when the compressor is a part of a gas generator—in either a gas turbine or an aero engine—and thus the compressor is followed by a combustion chamber. In this case, the diffuser is bolted to the manifold, and often the entire assembly is referred to as the diffuser. In case of industrial applications such as oil and gas industry (up-, mid-, and downstream), chemical industries and petrochemicals, iron and steel mining, and paper industry, the working fluid is collected in a scroll or volute. The working fluid leaving the stators is collected in a spiral casing surrounding the diffuser called a *volute* or *scroll*. The area of the cross section of the volute increases along the flow path in such a way that the velocity remains constant. Finally, the total fluid leaving all the stator passages is ejected from a pipe/duct normal to the inlet duct. The possible shapes of volute are shown in Figure 12.7.

12.3   CLASSIFICATION OF CENTRIFUGAL COMPRESSORS

Centrifugal compressors may be classified as

1. *Single or multiple stages:* For aero engines, centrifugal compressors have either single stage (Figure 12.8) or double (two or tandem) stages (Figure 12.9). In some cases two (or double) stages are mounted on the same shaft, handling the fluid in series to boost

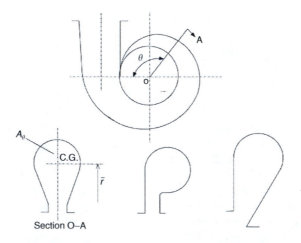

FIGURE 12.7   Different shapes of volute. (From H.I.H. Saravanamuttoo, *GFC Rogers and H Cohen, Gas Turbine Theory*, 5th edn., Prentice Hall, 2001, p. 167.)

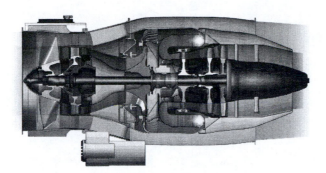

FIGURE 12.8   Single-stage centrifugal compressor. (Courtesy Pratt & Whitney of Canada.)

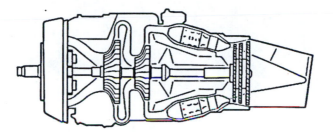

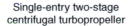

Single-entry two-stage
centrifugal turbopropeller

FIGURE 12.9   Two-stage centrifugal compressor. (Reproduced from Rolls-Royce plc, copyright © Rolls-Royce plc 2007. With permission.)

its final pressure. In industrial gas turbines such as pipeline compressors, there are up to five stages.

2. *Single entry (or single-face) or double (dual) entry (or double-face):* The principal differences between the single entry and dual entry are the size of the impeller and the ducting arrangement. The single-entry impeller permits ducting directly to the inducer vanes, as opposed to the more complicated ducting needed to reach the rear side of the dual-entry type. Although slightly more efficient in receiving air, single-entry impellers must be of greater diameter to provide sufficient air. Dual-entry impellers are smaller in diameter and rotate at higher speeds to ensure sufficient airflow. Most gas turbines of modern design use the dual-entry compressor to reduce engine diameter.

3. *Shrouded or unshrouded impeller:* Unshrouded impeller means that there is a clearance between the ends of the impeller vanes and a stationary shroud, while shrouded impeller means that there is a rotating shroud fixed to the impeller vanes. Shrouding reduces the losses due to leakage of air from the pressure side to the suction side of the blade. However, it increases the friction and weight of the impeller. For gas turbine and turbocharger applications, the impeller is usually unshrouded and the stationary cover is referred to as the shroud. For large process compressors, the impeller is often constructed with an integral shroud.

4. The impeller may have a non-, semi-, and full-inducer section as shown in Figure 12.10. For full-inducer impeller, the impeller vanes are continued around into the axial direction and the compressor resembles an axial compressor at inlet.

5. The impeller vanes may be radial at outlet or they may be inclined backward or forward, thus identified as backward-leaning (commonly now described as backswept) or forward-leaning compressor (Figure 12.11).

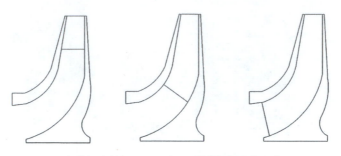

Left to right: non-, semi-, and full-inducer impeller

FIGURE 12.10 Types of inducers.

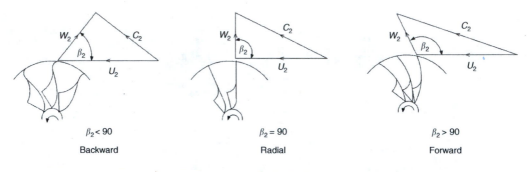

Velocity triangle for different impeller types

FIGURE 12.11 Different types of impeller.

6. The impeller layout may be open, semiclosed, or closed. Open type is built with blades in a radial direction without enclosing covers on either front or backsides, normally found in turbochargers. Semiclosed types are built with blades in a radial direction with an enclosed cover on the backside that extends to the periphery of the blades, normally found in air compressors. The closed type is built with backward- or forward-leaning blades and has enclosing covers on both the front and backside, normally found in multistage compressors.

7. Intake or inlet passage to the compressor may or may not be fitted with inlet guide vanes (IGV). IGVs are designed to give either positive or negative prewhirl (Figure 12.12). IGVs giving a positive prewhirl are frequently found in high-speed compressors, say in aero engines, to minimize the possibility of air reaching or exceeding the sonic speed. In this case, possible shock waves formed will lead to great losses in the intake. Negative prewhirl is employed to increase the inlet relative speed at some applications. These IGVs may also be set at fixed angle or they can be rotated to vary the inlet flow angle.

8. Outlet through a collector (denoted as scroll or volute) or a manifold.
    In the aviation field, the air leaving the compressor must be directed toward the combustion chamber with the minimum ducting losses. The diffuser is either constructed or ended by a 90° bend to satisfy this requirement. If the compressor is used in other industrial applications, then the air leaving the successive passages of the stators is collected in a scroll or volute and ducted from a single pipe.

9. Exit may or may not be fitted with exit guide vanes (EGVs).

10. The casings of the compressor may be horizontally split casings, vertically split casings, or bell casings.

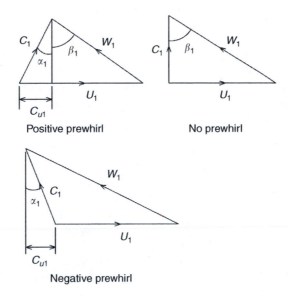

Positive prewhirl

No prewhirl

Negative prewhirl

FIGURE 12.12 Positive and negative prewhirl.

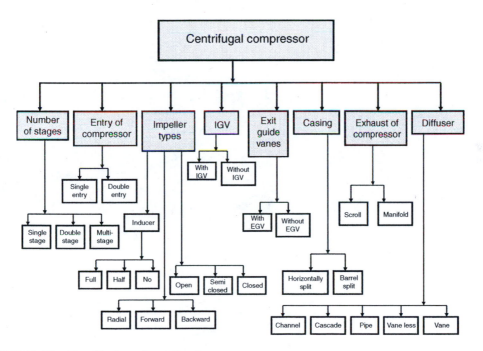

FIGURE 12.13 Classification of centrifugal compressors.

To conclude this section of classification, a breakdown of different classifications is illustrated in Figure 12.13.

12.4 GOVERNING EQUATIONS

Hereafter, the details of the governing equations will be discussed. The thermodynamic state of an ideal fluid is described by two properties, which may be the pressure and temperature. For fluid in motion, its velocity (magnitude and direction) must also be specified. The centrifugal compressor illustrated in Figure 12.14 defines three states, namely, the impeller inlet (state 1), impeller outlet (state 2), and diffuser outlet (state 3). These states are also plotted on the T–S diagram in Figure 12.15. Each state is defined by two points, namely, the static and total (stagnation) conditions. Thus, two pressure lines (static and total) are seen for each state. The compression processes are assumed adiabatic irreversible; thus entropy increase is found in each process.

The velocity triangles at the inlet and outlet of the impeller are shown in Figure 12.16. The velocity triangles in any turbomachine are governed by the relation

$$\vec{C} = \vec{U} + \vec{W} \tag{12.1}$$

where $\vec{C}$, $\vec{U}$, and $\vec{W}$ are the absolute, rotational, and relative velocities, respectively. As shown in figure, the flow at approach to the eye may be axial (C_1) with a typical velocity of 150 m/s for air or may have a swirl angle (in case of prewhirl as will be discussed later). The flow velocity when added to the impeller tangential (rotational) velocity (U_1) will give the relative velocity (W_1), which varies in magnitude and direction from eye hub to tip since (U_1) varies with radii. This requires the inducer vane to be twisted to align with the flow at all radii. It is normal for the relative Mach number at the eye tip to be below 0.9, which falls to below 0.5 for most of the subsequent passage length to minimize frictional losses. This inlet velocity is drawn in the axial–tangential plane. At the impeller outlet, the air leaves the passage with a relative velocity (W_2), which when added vectorially to the tangential speed (U_2), will give the absolute velocity (C_2), which is greater than (C_1) and represents

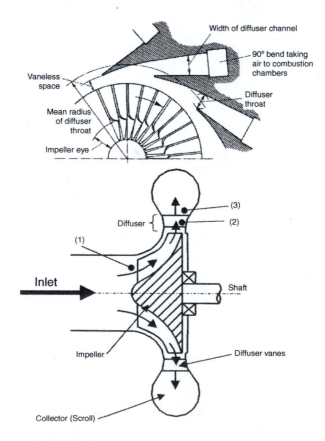

FIGURE 12.14 Layout of centrifugal compressor.

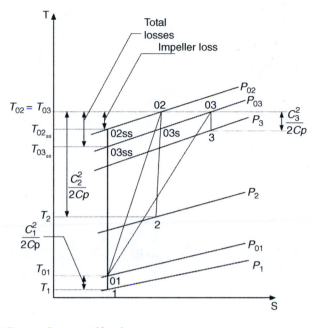

FIGURE 12.15 T–S diagram for a centrifugal compressor.

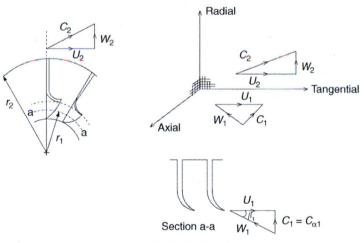

Layout and velocity triangles

FIGURE 12.16 Inlet and outlet velocity triangles at space.

the effect of work input to the rotor shaft. The tip tangential velocity (U_2) at outlet is slightly greater than the whirl velocity (C_{u2}); the difference is called the *slip*. It is interesting to note here that the outlet velocity triangle is at the radial–tangential plane. Thus, the inlet and outlet velocity triangles at the inlet and outlet of impeller are in two different planes as shown in Figure 12.16.

Returning back to the thermodynamics and velocities at the three states of the compressor, State (1) impeller inlet:

The static and total temperatures are T_1 and T_{01}. The static and total pressures are P_1 and P_{01}. The absolute, relative, and rotational velocities are C_1, W_1, and U_1

State (2); Tip (outlet) of impeller:

The static and total temperatures are T_2 and T_{02}. The static and total pressures are P_2 and P_{02}.

The absolute, relative, and rotational velocities are C_2, W_2, and U_2, respectively. The relations between inlet and outlet conditions are

$$P_2 > P_1, \quad T_2 > T_1, \quad C_2 \gg C_1, \quad P_{02} \gg P_{01}, \quad T_{02} \gg T_{01}.$$

Both the total temperature and total pressure increase as all the energy is added to the air (or in general to the fluid) in the impeller. Moreover, since the impeller is a rotating diffuser then the static pressure rises while the relative speed decreases ($W_2 < W_1$).

State (3) stator outlet:

The outlet pressures, temperatures, and absolute velocity are P_3, P_{03}, T_3, T_{03}, and C_3.

The relations between the inlet and outlet conditions in the stator are

$$T_3 > T_2, \quad T_{02} = T_{02}, \quad P_3 > P_2, \quad P_{03}P_{02}, \quad C_3 < C_2, \quad (C_3 \approx C_1).$$

In the diffuser, there is merely a conversion of the kinetic energy into pressure energy. Thus, the total temperature at the inlet and outlet of the stator are equal from the first law. The outlet total pressure is less than the inlet total pressure due to friction. The static pressure rises in the diffuser while the absolute velocity decreases as found in any diffusing passage.

The governing equations for a moving fluid are

1. Conservation of mass or continuity equation
2. Momentum equation or Newton's second law
3. Conservation of energy or the first law of thermodynamics

12.4.1 THE CONTINUITY EQUATION

The equation is simply a statement that in steady flow the mass rates of fluid entering and leaving a control volume are identical. Thus, considering a control volume between the impeller inlet (State 1) and outlet (State 2), the continuity equation is written as

$$\dot{m}_1 = \dot{m}_2$$

Air mass flow rate ($\dot{m}$) is at inlet:

$$\dot{m} = \rho_1 C_{a1} \times \text{Inlet area} = \rho_1 C_{a1} \left[\left(\frac{\pi}{4} \right) \left(D_{1t}^2 - D_{1r}^2 \right) \right] \qquad (12.2)$$

where ρ_1 and C_{a1} are the density and axial velocity at the inlet section, and D_{1r} and D_{1t} are the eye root and tip diameters at inlet.

At exit:

$$\dot{m} = \rho_2 \, C_{r2} A_2 = \rho_2 \, C_{r2} [b(\pi \, D_2 - nt)] \qquad (12.3)$$

where ρ_2 and C_{r2} are the density and radial velocity at the impeller tip, n is the number of blades, and t is the thickness of blade at impeller outlet and (b) is the axial width of the impeller.

12.4.2 THE MOMENTUM EQUATION OR EULER'S EQUATION FOR TURBOMACHINERY

Newton's second law of motion applied to the fluid passing through a control volume states that the rate of change of momentum of the fluid is equal to the net applied force on the fluid in the control volume, or

$$\vec{F} = \frac{\mathrm{d}}{\mathrm{d}t}\left(m\vec{C}\right)$$

For a steady flow between States 1 and 2, this can be written as

$$\vec{F} = \dot{m}\left(\vec{C}_2 - \vec{C}_1\right)$$

The above equation is used to obtain the torque arising from the net force, which is related to the angular momentum of the fluid within the control volume by multiplying the above equation by the appropriate radii; thus

$$\vec{T} = \vec{r} \otimes \vec{F} = \dot{m}\left(\vec{r_2} \otimes \vec{C}_2 - \vec{r_1} \otimes \vec{C}_1\right)$$

The above vectorial equation is reduced to the following scalar relation for the torque about the axis of rotation:

$$T = \dot{m}[r_2 C_{u2} - r_1 C_{u1}]$$

Or the final form known as Euler's equation is given by the relation

$$T = \dot{m}[(rC_u)_2 - (rC_u)_1] \tag{12.4}$$

where C_{u1} and C_{u2} are the swirl velocities at the inlet and outlet to the impeller. The mean inlet radius and the impeller tip radius are r_1 and r_2, respectively. For the case when the flow enters the impeller axially, there is no swirl velocity at the inlet ($C_{u1} = 0$), then Euler equation is reduced to

$$T = \dot{m}r_2 C_{u2}. \tag{12.4a}$$

Since the power is equal to torque times the rotational speed, the power P is expressed as

$$P = T \times \omega = \dot{m}(\omega r_2)C_{u2} = \dot{m}U_2 C_{u2} \tag{12.5}$$

The power needed to drive the compressor is greater than the power derived above. A part of this power is consumed in overcoming the friction between the casing and the air carried round by the vanes, the disk friction or the windage, and the bearing friction. Thus, the power input to the compressor (P_i) is equal to the power derived above but multiplied by a power input factor, ψ

$$P_i = \psi P = \psi \dot{m}C_{u2}U_2 \tag{12.6}$$

where

$$\psi = 1.035 - 1.04$$

12.4.3 THE ENERGY EQUATION OR THE FIRST LAW OF THERMODYNAMICS

The first law of thermodynamics states that the net change of energy of a fluid undergoing any process is equal to the net transfer of work and heat between the fluid and its surroundings.

Thus

$$\frac{\dot{Q} - \dot{W}}{\dot{m}} = (h_2 - h_1) + \frac{1}{2}\left(c_2^2 - c_1^2\right) + g(z_2 - z_1)$$

In most turbomachines, the potential energy is negligible. Moreover, turbomachines are adiabatic. Thus the heat term is negligible, by adjusting the sign of the rate of work, then

$$\frac{\dot{W}}{\dot{m}} = \left(h_2 + \frac{1}{2}c_2^2\right) - \left(h_1 + \frac{1}{2}c_1^2\right) = H_{02} - H_{01} = H_{03} - H_{01}$$

Since the rate of work is the power ($P_i = \dot{W}$), the left-hand side, which is the mechanical input power per unit mass flow rate, is equal to the specific enthalpy rise through the compressor. Thus from Equation 12.6

$$Cp(T_{03} - T_{01}) = \psi C_{u2} U_2$$

The temperature rise in the compressor is then

$$\Delta T_0 = T_{03} - T_{01} = \frac{\psi \, C_{u2} U_2}{Cp} \tag{12.7}$$

The pressure ratio of compressor (π_c) is defined as

$$\pi_c \equiv \frac{P_{03}}{P_{01}} = \left(\frac{T_{03s}}{T_{01}}\right)^{\gamma/(\gamma-1)}$$

The isentropic efficiency of the compressor is defined as

$$\eta_c = \frac{T_{03s} - T_{01}}{T_{03} - T_{01}} = \frac{T_{01}[(T_{03s}/T_{01}) - 1]}{T_{03} - T_{01}}$$

The pressure ratio of the compressor is

$$\pi_c \equiv \frac{P_{03}}{P_{01}} = \left[1 + \eta_c \frac{T_{03} - T_{01}}{T_{01}}\right]^{\gamma/(\gamma-1)}$$

Then from Equation 12.7

$$\therefore \pi_c \equiv \frac{P_{03}}{P_{01}} = \left[1 + \eta_c \frac{\psi C_{u2} U_2}{Cp T_{01}}\right]^{\gamma/(\gamma-1)} \tag{12.8}$$

Since in the ideal case

$$U_2 = C_{u2}$$

then, the torque equation 12.4a will be expressed as

$$T = \dot{m} \, r_2 U_2$$

The power from Equation 12.5 is

$$P = \dot{m}U_2^2.$$ (12.5a)

The input power from Equation 12.6 is

$$P_i = \psi P = \psi \dot{m}U_2^2.$$ (12.6a)

The temperature rise from Equation 12.7 is

$$\Delta T_0 = \frac{\psi U_2^2}{Cp}$$ (12.7a)

and the pressure ratio from Equation 12.8a is

$$\pi_c = \left[1 + \eta_c \frac{\psi U_2^2}{CpT_{01}} \right]^{\gamma/(\gamma-1)}.$$ (12.8a)

From the above relation, it is justified to say that centrifugal compressor has the following performance characteristics, often called "the three fan laws":

1. Flow is proportional to impeller speed.
2. Differential pressure across an impeller is proportional to the square of the impeller speed.
3. Power absorbed by the impeller varies with the cube of the impeller speed.

Example 1   A two-stage centrifugal compressor, as shown in Figure 12.17, has an overall pressure ratio of π_c. The isentropic efficiencies of both stages are η_1 and η_2, respectively. If no losses are assumed in the passage between the two stages, derive a mathematical expression for T_{03} in terms of π_c, T_{01}, T_{06}, η_1, and η_2.

Next, if $\pi_0 = 11$, $T_{01} = 290$ K, $T_{06} = 660$ K, $\eta_1 = \eta_2 = 0.8$, then calculate T_{03}.

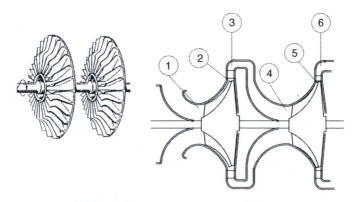

FIGURE 12.17   Two-spool centrifugal compressor.

Solution

The two-stage centrifugal compressor has an overall pressure ratio, π_c, which is defined as

$$\pi_c = \frac{P_{06}}{P_{01}} \quad \text{and} \quad \frac{P_{04}}{P_{03}} = 1$$

$$\frac{P_{06}}{P_{01}} = \frac{P_{03}}{p_{01}} \times \frac{P_{06}}{P_{04}}$$

$$\frac{P_{03}}{P_{01}} = \left(1 + \eta_1 \frac{\Delta T_{0s1}}{T_{01}}\right)^{\gamma/(\gamma-1)}$$

$$\Delta T_{0s1} = T_{03} - T_{01}$$

$$\frac{P_{03}}{P_{01}} = \left(1 + \left[\eta_1 \frac{T_{03}}{T_{01}} - 1\right]\right)^{\gamma/(\gamma-1)}$$

$$\frac{P_{06}}{p_{04}} = \left(1 + \eta_2 \frac{\Delta T_{0s2}}{T_{04}}\right)^{\gamma/(\gamma-1)}$$

$$T_{04} \equiv T_{03}$$

$$\Delta T_{0s2} = T_{06} - T_{04} \equiv T_{06} - T_{03}$$

$$\frac{P_{06}}{P_{04}} = \left(1 + \eta_2 \left[\frac{T_{06}}{T_{03}} - 1\right]\right)^{\gamma/(\gamma-1)}$$

$$\pi_c = \frac{P_{06}}{P_{01}} = \left(1 + \eta_1 \left[\frac{T_{03}}{T_{01}} - 1\right]\right)^{\gamma/(\gamma-1)} \times \left(1 + \eta_2 \left[\frac{T_{06}}{T_{03}} - 1\right]\right)^{\gamma/(\gamma-1)}$$

$$\pi_c^{(\gamma-1)/\gamma} = \left(1 + \frac{T_{03}}{T_{01}}\eta_1 - \eta_1\right)\left(1 + \frac{T_{06}}{T_{03}}\eta_2 - \eta_2\right)$$

$$\pi_c^{(\gamma-1)/\gamma} = 1 + \frac{T_{06}}{T_{03}}\eta_2 - \eta_2 + \frac{T_{03}}{T_{01}}\eta_1 + \frac{T_{06}}{T_{01}}\eta_1\eta_2 - \frac{T_{03}}{T_{01}}\eta_1\eta_2 - \eta_1 - \frac{T_{06}}{T_{03}}\eta_1\eta_2 + \eta_1\eta_2$$

$$\pi_c^{(\gamma-1)/\gamma} = T_{03}\left(\frac{\eta_1}{T_{01}} - \frac{\eta_1\eta_2}{T_{01}}\right) + \frac{1}{T_{03}}(T_{06}\eta_2 - T_{06}\eta_1\eta_2) + 1 - \eta_1 - \eta_2 + \eta_1\eta_2\left(\frac{T_{06}}{T_{01}} + 1\right)$$

$$\pi_c^{(\gamma-1)/\gamma} = A_1 T_{03} + \frac{A_2}{T_{03}} + A_3,$$

where

$$A_1 = \frac{\eta_1}{T_{01}} - \frac{\eta_1\eta_2}{T_{01}}$$

$$A_2 = T_{06}\eta_2 - T_{06}\eta_1\eta_2$$

$$A_3 = 1 - \eta_1 - \eta_2 + \eta_1\eta_2\left(\frac{T_{06}}{T_{01}} + 1\right)$$

$$\pi_c^{(\gamma-1/\gamma)} = \frac{A_1 T_{03}^2 + A_2 + A_3 T_{03}}{T_{03}}$$

$$A_1 T_{03}^2 + A_2 + A_3 T_{03} = \pi_c^{(\gamma-1)/\gamma} T_{03}$$

$$T_{03}^2 + \left(\frac{A_3 - \pi_c^{(\gamma-1)/\gamma}}{A_1}\right) T_{03} + \frac{A_2}{A_1} = 0.$$

Define

$$\lambda_1 = \frac{A_3 - \pi_c^{(\gamma-1)/\gamma}}{A_1} \quad \text{and} \quad \lambda_2 = \frac{A_2}{A_1}$$

$$T_{03} = \frac{-\lambda_1 \pm \sqrt{\lambda_1^2 - 4\lambda_2}}{2} \tag{a}$$

Now substitute the given data in Equation (a) to get the value of T_{03}

$$\pi_c = 11, \quad T_{01} = 290 \text{ K}, \quad T_{06} = 660 \text{ K}, \quad \eta_1 = \eta_2 = 0.8$$

The values of the constants are then

$$A_1 = 5.517 \times 10^{-4}, \quad A_2 = 105.6, \quad A_3 = 1.497, \quad \text{and} \quad \pi_c^{(\gamma-1)/\gamma} = 1.984$$

$$\frac{\left(A_3 - \pi_c^{(\gamma-1/\gamma)}\right)}{A_1} = -882.726 \quad \text{and} \quad \frac{A_2}{A_1} = 19.141 \times 10^4$$

From Equation (a)

$$T_{03} = \frac{+882.726 \pm \sqrt{882.726^2 - 4 \times 19.141 \times 10^4}}{2}$$

$$T_{03} = 441.363 \pm 58.235$$

$$T_{03} = 499.598 \text{ K} \quad \text{or} \quad T_{03} = 383.128 \text{ K}$$

12.4.4 SLIP FACTOR σ

If the flow at impeller discharge is perfectly guided by the impeller blades then the tangential component of the absolute velocity (swirl velocity) is equal to the rotational velocity ($C_{u2} = U_2$) in the case of radial type impeller. In practice, the flow cannot be perfectly guided by a finite number of blades and it is said to slip. Thus, at the impeller outlet, the swirl velocity is less than the impeller rotational speed ($C_{u2} < U_2$). The classical explanation for the slip phenomenon [3] uses the concept of the relative eddy. As the flow into the impeller is normally irrotational, that is, has no initial rotation, then at the impeller discharge the flow relative to the impeller must rotate with an angular velocity equal and opposite to the impeller (Figure 12.18).

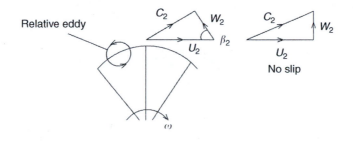

Slip due to relative eddies

FIGURE 12.18   Velocity triangles with slip.

By combining this relative eddy and the radial through flow, the resultant relative velocity vector has a component in the direction opposite to the impeller rotation.

The slip factor is denoted by σ and is defined as

$$\sigma = \frac{C_{u2}}{U_2} \tag{12.9}$$

Though Stodola and Busemann derived formulae for this slip factor, the correlation proposed by Staintz [7] has a wide range of application. Stanitz formula is expressed as

$$\sigma = 1 - \frac{0.63\pi}{n} \tag{12.10}$$

where n is the number of blades. It was considered satisfactory for both radial blades, forward- and backward-leaning blades having blade angles ranging from $-45°$ to $+45°$. As the number of vanes of impeller is increased, then the slip factor is also increased; slip lag at the tip of the impeller reduces as shown in Figure 12.19.

The effect of slip is to reduce the magnitude of the swirl velocity from the ideal case. This in turn will reduce the pressure ratio. In this way, slip will also reduce the compressor power consumption. The detrimental effect of slip is that the impeller must be larger or the rotational speed must be higher to deliver the same pressure ratio. This leads to increased stress level and it will also lead to increased relative velocity that will lead to increased friction loss and a reduced efficiency.

Then from Equation 12.7, the temperature rise will be

$$\Delta T_0 = \frac{\sigma \psi U_2^2}{Cp} \tag{12.7b}$$

The pressure ratio will be reduced and has the value

$$\pi_c = \left(1 + \frac{\eta_c \psi \sigma U_2^2}{CpT_{01}}\right)^{(\gamma/\gamma-1)} \tag{12.8b}$$

From Equation 12.8b, $Cp = \gamma R/\gamma - 1$ and $a_{01}^2 = \gamma RT_{01}$, the pressure ratio can be expressed as

$$\pi_c = \left(1 + \frac{(\gamma - 1)\eta_c \psi \sigma U_2^2}{a_{01}^2}\right)^{\gamma/(\gamma-1)} \tag{12.8c}$$

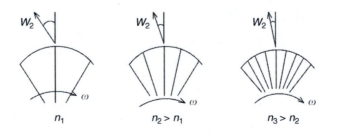

Effect of the number of blades

FIGURE 12.19   Effect of the number of blades on the outlet relative velocity.

where a_{01} is the total sonic speed or the sonic speed referred to the total temperature at the compressor inlet.

Note that when there is no slip, the slip factor σ is equal to unity and this yields the maximum temperature rise and pressure ratio that may be achieved by this compressor. These values are expressed by the relations

$$(\Delta T_0)_{max} = \frac{\psi U_2^2}{Cp}$$

$$(\pi_c)_{max} = \left(1 + \frac{\eta_c \psi U_2^2}{CpT_{01}}\right)^{\gamma/(\gamma-1)}.$$

Example 2 At inlet of a centrifugal compressor eye, the relative Mach number is to be less than 0.97 to eliminate the formation of shock waves and its attendant losses. The hub to tip radius ratio of the inducer is 0.4. There is no IGV and the inlet velocity is axial and uniform over the impeller eye. The eye tip diameter is 20 cm.

Determine

a. The maximum mass flow rate throughout for a given rotational speed of 486 rps
b. The blade angle at the inducer tip corresponding to this mass flow rate (Take air inlet conditions to be 101.3 kPa, 288 K.)

Solution

The given data are

$$M_{1_{rel_t}} = 0.97, \quad \zeta = \frac{r_h}{r_t} = 0.4, \quad D_{e_t} = 20 \text{ cm}$$

$$N = 486 \text{ rps}, \quad P_{01} = 101.3 \text{ kPa}, \quad T_{01} = 288 \text{ K}$$

The rotational speed at the inducer tip is

$$U_{1_t} = \pi D_{e_t} N = \pi \times 0.2 \times 486 = 305.363 \text{ m/s}$$

The maximum relative Mach number is at the eye tip (refer to Figure 12.20) will be:

$$M_{1_{rel}} = \frac{W_{1_t}}{\sqrt{\gamma RT_1}} = \frac{\sqrt{C_1^2 + U_{1_t}^2}}{\sqrt{\gamma R\left(T_{01} - (C_1^2/2C_P)\right)}} = \frac{\sqrt{C_1^2 + U_{1_t}^2}}{\sqrt{\gamma RT_{01} - ((\gamma-1)C_1^2/2)}}$$

$$0.97^2 = \frac{C_1^2 + 305.363^2}{115,718.4 - 0.2C_1^2}$$

$$C_1^2 + 93, \quad 246.443 = 108, \quad 879.44 - 0.19C_1^2$$

$$1.19C_1^2 = 15,633$$

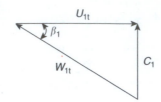

FIGURE 12.20   Inlet velocity triangle at tip.

The absolute velocity at inlet is

$$C_1 = 114.62 \text{ m/s}$$

$$T_1 = T_{01} - \frac{C_1^2}{2C_P} = 281.464 \text{ K}$$

$$\frac{P_{01}}{P_1} = \left(\frac{T_{01}}{T_1}\right)^{\gamma/(\gamma-1)}$$

$$P_1 = 93.48 \text{ kPa}$$

$$\rho_1 = \frac{P_1}{RT_1} = 1.157 \text{ kg/m}^3$$

$$A_1 = \frac{\pi}{4}\left(D_{e_t}^2 - D_{e_r}^2\right) = \frac{\pi}{4}D_{e_t}^2(1 - \zeta^2)$$

$$A_1 = 0.0264 \text{m}^2$$

Since the flow is axial, $C_{a1} = C_1$. The mass flow rate is calculated from the inlet conditions as

$$\dot{m} = \rho_1 C_{a1} A_1 = 1.157 \times 114.62 \times 0.0264 = 3.5 \text{ kg/s}$$

The blade inlet angle at the tip β_{1_t} is given from the relation

$$\tan \beta_{1_t} = \frac{C_1}{U_{1_t}}$$

$$\beta_{1_t} = 20.57°$$

12.4.5 PREWHIRL

The tangential component of velocity at the inlet to the impeller is usually zero as the flow enters the impeller axially. If prewhirl (or inlet guide) vanes are installed to the inlet duct before the impeller, then the incoming air has a tangential component of velocity. This velocity component depends on the vane outlet angle. Prewhirl may be positive or negative as shown in Figure 12.12. Positive prewhirl reduces the inlet relative velocity, while negative prewhirl increases the inlet relative velocity. In aero engines, positive prewhirl is frequently used to reduce the inlet relative speed. The objective here is to avoid the formation of shock waves on the blade suction side. The designer of compressor seeks for a small inlet area of the engine to reduce the drag force. At the same time, the air mass flow rate is chosen as maximum as possible to maximize the thrust force. Both factors led to an increase in the axial absolute velocity at inlet, and this in turn increases the relative velocity. Since the relative velocity is maximum at the tip radius of the inlet than when accelerated, there is always a tendency for the air to break away from the convex face of the curved part of the impeller vane. Here a shock wave might occur, which upon interaction with the boundary layer over the convex surface of blades, causes a large increase in boundary layer thickness. Thus, excessive pressure loss occurs.

The value of the inlet relative Mach number must be in the range from 0.8 to 0.85 to avoid the shock wave losses described previously, or

$$M_1 = \frac{W_1}{\sqrt{\gamma R T_1}} \approx 0.8 - 0.85$$

where T_1 is the static inlet temperature. Though this Mach number may be satisfactory on ground operation, it may be too high at altitude as the ambient temperature decreases with altitude. For this reason, IGVs are added to decrease the Mach number. These IGVs are attached to the compressor casing to provide a positive prewhirl that decreases the magnitude of the maximum relative velocity (W_1) at the eye tip.

The IGVs, sometimes denoted as swirl vanes, may swivel about radial axes to vary the swirl angle with speed. Keeping W_1 aligned with the inducer leading edge results in a minimum incidence or shock losses. Thus, they widen the operating speed range, increase efficiency, and reduce the relative Mach number at the eye tip, but they also reduce the mass flow by reducing the inlet area.

Assuming that the axial velocity is uniform from the root to the tip of the eye, then the maximum relative Mach number at the eye tip (Figure 12.20) is given by the relation:

$$M_{1\,max} = \frac{W_{1\,max}}{a_1} = \sqrt{\frac{C_{a1}^2 + (U_{1t} - C_{a1} \sin \zeta)^2}{\gamma R T_1}} \tag{12.11}$$

where U_{1t} is the rotational speed at the eye tip and ζ is the prewhirl angle, which is the angle between the absolute velocity leaving the IGV and the axial direction.

Another advantage of the prewhirl is that the curvature of the impeller vanes at the inlet will be reduced and so the radius of curvature will increase. Thus, the bending stress is decreased. The reason is that the inducer of the impeller resembles a curved beam (Figure 12.21). The bending stress for curved beams is inversely proportional to the radius of curvature [5]. The bending stress

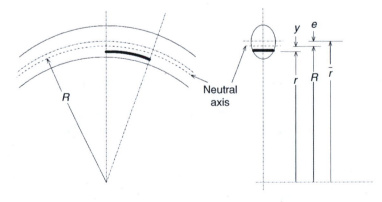

Nomenclature for curved beams

FIGURE 12.21   Curved-beam nomenclature.

is obtained from the relation

$$\sigma = \frac{My}{Ae\,(R - y)} \tag{12.12}$$

where

$$e = \bar{r} - R \quad \text{and} \quad R = \frac{A}{\displaystyle\int dA/r}$$

with M as the the bending moment, A the cross-sectional area, e the distance from centroidal axis to the neutral axis, R the radius of the neutral axis, and $\bar{r}$ the radius of centroidal axis.

r is radius of a general point at a distance (y) from the neutral axis at which the stress is to be calculated. The critical stresses occur at the inner and outer surfaces of the curved beam. Since the pressure side here is the concave side of the inducer, the bending stresses on this side are greater than that on the convex (suction) side. Moreover, prewhirl increases the radius of curvature of the inducer at its inlet ($\bar{r}$); thus, the stress decreases (Equation 12.12).

Prewhirl also increases the relative angle (β_1) measured between the relative velocity (W_1) and the tangent. Thus, $\beta_{1\text{prewhirl}} > \beta_{1\,\text{prewhirl}}$.

However, prewhirl leads to a decrease in the work capacity of the compressor as the angular momentum is reduced, leading to a decrease in the pressure ratio of the compressor. Since prewhirl creates a swirl component of the absolute inlet velocity, it reduces the temperature rise in the compressor and the compressor pressure ratio

$$\Delta T_0 = \frac{\psi\,(C_{u2}U_2 - C_{u1}U_{1m})}{Cp} \tag{12.13}$$

where $C_{u1} = C_1 \sin \zeta$. Also, U_{1m} is the rotational speed at the mean radius of the eye. The pressure ratio is then

$$\pi_c = \left\{ 1 + \eta_c \frac{\psi\left[\sigma U_2^2 - U_{1m}C_1 \sin \zeta\right]}{CpT_{01}} \right\}^{\gamma/(\gamma-1)} \tag{12.14}$$

Example 3 An aircraft engine is fitted with a single-sided centrifugal compressor. The aircraft flies at a Mach number 0.85 and at an altitude of 35,000 ft. The inlet duct of the impeller is fitted with IGV. The hub and shroud diameters of the eye are 14 and 28 cm, respectively, and the rotational speed is 270 rps. Estimate the prewhirl angle if the mass flow is 3.6 kg/s and the maximum relative inlet Much number is 0.8.

Solution: The given data are

$$d_{\text{eh}} = 14 \text{ cm}, \quad d_{\text{et}} = 28 \text{ cm}, \quad N = 270 \text{ rps}, \quad \dot{m} = 3.6 \text{ kg/s}$$

At altitude 35,000 ft, the ambient conditions are

$$P_a = 23.844 \text{ kPa} \quad T_a = 218.67 \text{ K}$$

$$T_{01} = T_a \left[1 + \frac{\gamma - 1}{2} M_{\text{flight}}^2 \right]$$

$$T_{01} = 218.67 \left[1 + \frac{0.4}{2}(0.85)^2 \right] = 250.67 \text{ K}$$

$$P_{01} = P_a \left(\frac{T_{01}}{T_a} \right)^{\gamma/(\gamma-1)} = 23{,}844.24 \left(\frac{250.27}{218.67} \right)^{1.4/0.4} = 38{,}243 \text{ Pa}$$

$$U_{1t} = \pi D_t \frac{N}{60}$$

$$U_{1t} = \pi \times 0.28 \times 270 = 237.5 \text{ m/s}$$

$$M_{1t} = \frac{W_{1T}}{\sqrt{\gamma RT_1}}$$

$$W_{1t}^2 = M_{1t}^2 \gamma RT_1$$

From Figure 12.22, the velocity triangle at inlet with slip

$$C_{a1}^2 + (U_{1_t} - C_{a1} \tan \zeta)^2 = M_{1t}^2 \gamma RT_1$$

$$(U_{1t} - C_{a1} \tan \zeta)^2 = M_{1t}^2 \gamma RT_1 - C_{a1}^2$$

$$U_{1t} - C_{a1} \tan\zeta = \sqrt{M_{1t}^2 \gamma RT_1 - C_{a1}^2}$$

$$C_{a1} \tan \zeta = U_{1t} - \sqrt{M_{1t}^2 \gamma RT_1 - C_{a1}^2}$$

$$\tan \zeta = \frac{U_{1_t}}{C_{a1}} - \sqrt{\frac{M_{1t}^2 \gamma RT_1}{C_{a1}^2} - 1}$$

$$\zeta = \tan^{-1} \left(\frac{U_{1_t}}{C_{a1}} - \sqrt{\frac{M_{1t}^2 \gamma RT_1}{C_{a1}^2} - 1} \right)$$

$$\zeta = \tan^{-1} \left(\frac{237.5}{C_{a1}} - \sqrt{\frac{(0.8)^2 \times 1.4 \times 287 \, T_1}{C_{a1}^2} - 1} \right)$$

$$\zeta = \tan^{-1} \left(\frac{237.5}{C_{a1}} - \sqrt{\frac{257.15 T_1}{C_{a1}^2} - 1} \right)$$

$$A_1 = \frac{\pi}{4} \left(D_t^2 - D_h^2 \right) = \frac{\pi}{4} \left[(0.28)^2 - (0.14)^2 \right] = 0.0461814 \text{ m}^2$$

As a first assumption, calculate the inlet density from the total temperature and pressure, thus

$$\rho_1 = \frac{P_{01}}{RT_{01}} = \frac{38{,}243}{287 \times 250.27} = 0.532428461 \text{ kg/m}^3 \qquad \text{(a)}$$

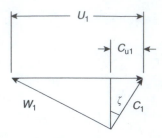

FIGURE 12.22 Velocity triangle at inlet with slip.

Now, calculate the axial inlet velocity

$$C_{a1} = \frac{\dot{m}}{\rho_1 A_1} = \frac{3.5}{0.53242861 \times 0.0461814} = 142.344 \text{m/s}$$

The prewhirl angle is now calculated from the relation

$$\zeta = \tan^{-1}\left(\frac{237.5}{C_{a1}} - \sqrt{\frac{257.15 T_{01}}{C_{a1}^2} - 1}\right)$$

$$\zeta = \tan^{-1}\left(\frac{237.5}{142.344} - \sqrt{\frac{257.15 \times 250.27}{(142.344)^2} - 1}\right) = 10.94°$$

The inlet absolute velocity is now calculated as

$$C_1 = \frac{C_{a1}}{\cos \zeta} = \frac{142.244}{\cos 10.94} = 144.98 \text{m/s}$$

The static inlet temperature is

$$T_1 = T_{01} - \frac{C_1^2}{2C_p} = 250.27 - \frac{(144.98)^2}{2 \times 1005} = 239.81 \text{K}$$

The static inlet pressure is calculated as

$$P_1 = P_{01}\left(\frac{T_1}{T_{01}}\right)^{\gamma/(\gamma-1)} = 32,840.86 \text{ Pa}$$

The density is recalculated from the static temperature and pressure just obtained as

$$\rho_1 = \frac{P_1}{RT_1} = 0.47755 \text{ kg/m}^3 \qquad\qquad\qquad (b)$$

A large difference is noticed between this value and the value obtained in (a); thus iteration is performed until equal values for the density are obtained in two successive iterations as shown.

Iteration Number	1	2	10	11
$C_{a1} = \dfrac{\dot{m}}{\rho_1 A_1}$	142.344	158.702	167.04	167.04
$\zeta = \tan^{-1}\left(\dfrac{237.5}{C_{a1}} - \sqrt{\dfrac{257.15 T_1}{C_{a1}^2} - 1}\right)$	13.44	16.375	18.9395	18.94
$C_1 = \dfrac{C_{a1}}{\cos\zeta}$	146.35	165.411	176.6	176.6
$T_1 = T_{01} - \dfrac{C_1^2}{2 C_P}$	239.614	236.66	234.754	234.75
$P_1 = P_{01}\left(\dfrac{T}{T_{01}}\right)^{\gamma/(\gamma-1)}$	32,840.86	31,444.4	30,567.98	30,567.8
$\rho_1 = \dfrac{P_1}{RT_1}$	0.47755	0.46295	0.453703	0.4537

Eleven iterations were needed for convergence. Only some were displayed in the above table. The final results are

$$C_{a1} = 167.04 \text{m/s}, \quad \zeta = 18.94°, \quad C_1 = 176.6 \text{ m/s}$$

$$P_1 = 30,567.8 \text{ Pa}, \quad T_1 = 234.75 \text{ K}, \quad \rho_1 = 0.4537 \text{ kg/m}^3$$

Example 4   To examine how prewhirl reduces the bending stress for the blades of centrifugal compressors, consider the following case where the blade shape at tip of the eye is shown in Figure 12.23 for the two cases with prewhirl and without prewhirl.

$$(r_i)_{\text{prewhirl}} = 15 \text{ cm}, \quad (r_i)_{\text{no prewhirl}} = 5\text{cm}, \quad t = 0.4\text{cm}, \quad A = 0.4 \text{ cm}^2$$

If the bending moment $M = 10.0$ Nm, calculate the maximum stresses in both cases and the percentage of stress reduction when prewhirl is employed.

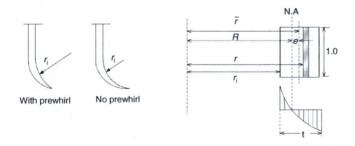

FIGURE 12.23   Stresses over the section of a curved beam of the impeller with prewhirl.

Solution

The stress in such curved beams is given by

$$\sigma = \frac{My}{Ae(R - y)}$$

where $R = A/(\int dA/r)$ is the radius of the neutral axis, which, for a rectangular cross section is expressed by

$$R = \frac{t}{\ln\left(1 + \frac{t}{r_i}\right)} \quad \text{and} \quad e = \bar{r} - R$$

$\bar{r}$ is the radius of the center of gravity of the section,

$$\bar{r} = r_i + \frac{t}{2}$$

$$r = R - y$$

r is any radius defining any point within the rectangular cross section.

Since the maximum stress occurs at the point having minimum radius, the critical point is at the concave side of the blades ($r = r_i$).

Calculation for the two cases will be tabulated here as follows:

	No prewhirl	With prewhirl
$R = \dfrac{t}{\ln\left(1 + \dfrac{t}{r_i}\right)}$	5.1974 cm	15.199 cm
$e = \bar{r} - R$	2.565×10^{-3} cm	0.001 cm
$y_{\text{inner}} = R - r_i$	0.1974 cm	0.199 cm
$r_{\text{inner}} = R - y$	5.0 cm	15.0 cm
$\sigma = \dfrac{My}{Ae\,(R - y)}$	384.8 MPa	331.6 MPa

The above results shows that prewhirl reduced the maximum stresses by a percentage of 13%.

Example 5   A centrifugal compressor has the following data:

$$D_{eh} = 0.12 \text{ m}, \quad D_{et} = 0.24 \text{ m}, \quad D_t = 0.5 \text{ m}$$
$$C_1 = 100 \text{ m/s}, \quad N = 300 \text{ rev/s}, \quad T_{01} = 300 \text{ K}$$
$$\psi = 1.04, \quad \sigma = 0.92, \quad \eta_c = 0.85$$

Now IGVs are added upstream to the compressor. Two designs are suggested:

(A) IGVs having constant prewhirl angle ($\zeta = 25°$)
(B) IGVs having constant inlet relative angle (β_1) as its value at the hub if no prewhirl is employed

Assuming that the axial velocity is kept constant and equal to 100 m/s, in the following three cases:

(i) No prewhirl
(ii) Design (A)
(iii) Design (B)

1. Calculate the inlet relative Mach number ($M_{1\text{rel}}$) at the hub, mean, and tip sections of the eye.
2. Calculate the compressor pressure ratio.
3. Write down your comments and suggestions.

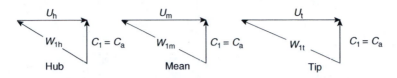

FIGURE 12.24 Velocity triangles with no prewhirl.

Solution

$$U_{1h} = \pi D_{eh}\frac{N}{60} = \pi \times 0.12 \times 300 = 113.1 \text{ m/s}$$

$$U_{1m} = \pi D_{em}\frac{N}{60} = \pi \times 0.18 \times 300 = 169.65 \text{ m/s}$$

$$U_{1t} = \pi \times D_{et}\frac{N}{60} = \pi \times 0.24 \times 300 = 226.2 \text{ m/s}$$

$$U_2 = \pi \times D_{et}\frac{N}{60} = \pi \times 0.5 \times 300 = 471.24 \text{ m/s}$$

Case (A) No prewhirl:
Figure 12.24 illustrates the velocity triangles with no prewhirl.

$$T_1 = T_{01} - \frac{C_1^2}{2C_P} = 300 - \frac{(100)^2}{2 \times 1005} = 295 \text{K}$$

(i) At the hub

$$W_{1h} = \sqrt{U_{1h}^2 + C_a^2}$$

$$W_{1h} = \sqrt{(113.1)^2 + (100)^2} = 150.97 \text{ m/s}$$

$$(M_{1rel})_h = \frac{W_{1h}}{\sqrt{\gamma R T_1}} = \frac{150.97}{\sqrt{1.4 \times 287 \times 295}} = 0.4385$$

$$\tan \beta_{1h} = \frac{C_a}{U_{1h}} = \frac{100}{113.1}$$

$$\beta_{1h} = 41.48$$

(ii) At the mean

$$W_{1m} = \sqrt{U_{1m}^2 + C_a^2}$$

$$W_{1m} = \sqrt{(169.65)^2 + (100)^2} = 196.93 \text{ m/s}$$

$$(M_{1rel})_m = \frac{W_{1m}}{\sqrt{\gamma R T_1}} = \frac{196.93}{\sqrt{1.4 \times 287 \times 295}} = 0.572$$

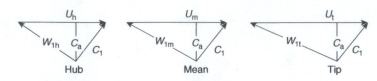

FIGURE 12.25 Velocity triangles with constant prewhirl angle.

(iii) At the tip

$$W_{1t} = \sqrt{U_{1t}^2 + C_a^2}$$

$$W_{1t} = \sqrt{(226.2)^2 + (100)^2} = 247.32 \text{ m/s}$$

$$(M_{1rel})_t = \frac{W_{1t}}{\sqrt{\gamma R T_1}} = \frac{247.32}{\sqrt{1.4 \times 287 \times 295}} = 0.7184$$

$$\pi_c = \frac{P_{03}}{P_{03}} = \left(1 + \eta_c \frac{\sigma U_2^2 - U_1 C_{u1}}{C_P T_{01}}\right)^{\gamma/(\gamma-1)}$$

$$\pi_c = \left(1 + 0.85 \frac{0.92 \times (471.24)^2 - 0}{1005 \times 300}\right)^{1.4/0.4} = 4.914$$

Case (B) Constant prewhirl angle ($\zeta = 25$):
Figure 12.25 illustrates the velocity triangles with constant prewhirl angle. Since the same axial velocity is kept constant in all cases,

$$C_1 = \frac{C_a}{\cos \zeta} = \frac{100}{\cos 25} = 110.34 \text{ m/s}$$

$$T_1 = T_{01} - \frac{C_1^2}{2C_P} = 300 - \frac{(110.34)^2}{2 \times 1005} = 293.94 \text{ K}$$

(i) At the hub

$$C_{u1h} = C_a \tan \zeta = 100 \ \tan 25 = 46.63 \text{ m/s} = C_{u1m} = C_{u1h}$$

Note that

$$C_{u1h} = C_{u1m} = C_{u1t}$$

$$W_{1h} = \sqrt{(U_{1h} - C_{u1h})^2 + C_a^2}$$

$$W_{1h} = \sqrt{(113.1 - 46.63)^2 + (100)^2} = 120.07 \text{ m/s}$$

$$(M_{1rel})_h = \frac{W_{1h}}{\sqrt{\gamma R T_1}} = \frac{120.07}{\sqrt{1.4 \times 287 \times 293.94}} = 0.3494$$

(ii) At the mean

$$W_{1m} = \sqrt{(U_{1m} - C_{u1m})^2 + C_a^2}$$

$$W_{1h} = \sqrt{(169.65 - 46.63)^2 + (100)^2} = 158.54 \text{ m/s}$$

$$\left(M_{1_{rel}}\right)_m = \frac{W_{1m}}{\sqrt{\gamma R T_1}} = \frac{158.54}{\sqrt{1.4 \times 287 \times 293.94}} = 0.4613$$

(iii) At the tip

$$W_{1t} = \sqrt{(U_{1t} - C_{u1t})^2 + C_a^2}$$

$$W_{1t} = \sqrt{(226.2 - 46.63)^2 + (100)^2} = 205.54 \text{ m/s}$$

$$\left(M_{1_{rel}}\right)_t = \frac{W_{1t}}{\sqrt{\gamma R T_1}} = \frac{205.54}{\sqrt{1.4 \times 287 \times 293.94}} = 0.598$$

$$\pi_c = \frac{P_{03}}{P_{03}} = \left(1 + \eta_c \frac{\sigma U_2^2 - U_1 C_{u1}}{C_P T_{01}}\right)^{(\gamma/\gamma - 1)}$$

$$\pi_c = \left(1 + 0.85 \frac{0.92 \times (471.24)^2 - 169.65 \times 46.63}{1005 \times 300}\right)^{(1.4/0.4)} = 4.6747.$$

Case (C) Constant inlet relative angle ($\beta_1 = 41.48$):
Figure 12.26 illustrates the velocity triangles with constant inlet relative angle.

(i) At the hub

$$T_1 = T_{01} - \frac{C_1^2}{2C_P} = 300 - \frac{(100)^2}{2 * 1005} = 295 \text{ K}$$

$$W_{1h} = \sqrt{U_{1h}^2 + C_a^2}$$

$$W_{1h} = \sqrt{(113.1)^2 + (100)^2} = 150.97 \text{ m/s} = W_{1m} = W_{1t}$$

$$\left(M_{1_{rel}}\right)_h = \frac{W_{1h}}{\sqrt{\gamma R T_1}} = \frac{150.97}{\sqrt{1.4 \times 287 \times 295}} = 0.4385$$

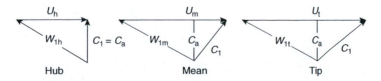

FIGURE 12.26   Velocity triangles with constant inlet relative angle.

(ii) At the mean

$$(C_{u1})_m = U_m - U_h = 169.65 - 113.1 = 56.55$$

$$C_{1m} = \sqrt{C_{u1m}^2 + C_a^2} = \sqrt{(56.55)^2 + (100)^2} = 114.88 \text{ m/s}$$

$$T_1 = T_{01} - \frac{C_1^2}{2C_P} = 300 - \frac{(114.88)^2}{2 \times 1005} = 293.43 \text{ K}$$

$$(M_{1_{rel}})_m = \frac{W_{1m}}{\sqrt{\gamma R T_1}} = \frac{150.97}{\sqrt{1.4 \times 287 \times 293.43}} = 0.4397$$

(iii) At the tip

$$(C_{u1})_t = U_t - U_h = 226.2 - 113.1 = 113.1 \text{ m/s}$$

$$C_{1t} = \sqrt{C_{u1t}^2 + C_a^2} = \sqrt{(113.1)^2 + (100)^2} = 150.96 \text{ m/s}$$

$$T_1 = T_{01} - \frac{C_1^2}{2C_P} = 300 - \frac{(150.96)^2}{2 \times 1005} = 288.66 \text{ K}$$

$$(M_{1_{rel}})_t = \frac{W_{1t}}{\sqrt{\gamma R T_1}} = \frac{150.97}{\sqrt{1.4 \times 287 \times 288.66}} = 0.4433$$

$$\pi_c = \frac{P_{03}}{P_{01}} = \left(1 + \eta_c \frac{\sigma U_2^2 - \overline{U_1 C_{u1}}}{C_P T_{01}}\right)^{\gamma/\gamma-1}$$

$$\overline{U_1 C_{u1}} = \frac{\int_{r_h}^{r_t} U C_u \, dr}{r_t - r_h} = \frac{\int_{r_h}^{r_t} \omega r C_u \, dr}{r_t - r_h}$$

but

$$C_u = U - C_a \tan \beta \quad \text{or} \quad C_u = U - A$$

$$\overline{U_1 C_{u1}} = \frac{\omega \int_{r_h}^{r_t} r (U - A) dr}{r_t - r_h} = \frac{\omega \int_{r_h}^{r_t} r (\omega r - A) dr}{r_t - r_h}$$

$$\overline{U_1 C_{u1}} = \frac{\omega \int_{r_h}^{r_t} (\omega r^2 - A r) dr}{r_t - r_h}$$

$$\overline{U_1 C_{u1}} = \frac{\omega}{r_t - r_h} \left[\omega \frac{r^3}{3} - A \frac{r^2}{2}\right]_{r_h}^{r_t}$$

$$\overline{U_1 C_{u1}} = \frac{\omega}{r_t - r_h} \left[\frac{\omega}{3} \left(r_t^3 - r_h^3 \right) - \frac{A}{2} \left(r_t^2 - r_h^2 \right) \right]$$

$$\overline{U_1 C_{u1}} = \omega \left[\frac{\omega}{3} \left(r_t^2 + r_t r_h + r_h^2 \right) - \frac{A}{2} (r_t + r_h) \right]$$

$$\overline{U_1 C_{u1}} = \frac{1}{6} \left[2U_t^2 + 2U_t U_h + 2U_h^2 - 3AU_t - 3AU_h \right]$$

$$\overline{U_1 C_{u1}} = \frac{1}{6} \left[2U_t (U_t - A) + 2U_h (U_h - A) + U_t (U_h - A) + U_h (U_t - A) \right]$$

$$\overline{U_1 C_{u1}} = \frac{1}{6} \left[2U_t C_{ut} + 2U_h C_{uh} + U_t C_{uh} + U_h C_{ut} \right]$$

But $C_{uh} = 0$

$$\therefore \ \overline{U_1 C_{u1}} = \frac{1}{6} \left[2U_t C_{ut} + U_h C_{ut} \right] = \frac{1}{6} \left[2 \times 226.2 \times 113.1 + 113.1 \times 113.1 \right] = 10659.675 \ (\text{m/s})^2$$

$$\pi_c = \left(1 + \eta_c \frac{\sigma U_2^2 - \overline{U_1 C_{u1}}}{C_P T_{01}} \right)^{\gamma/(\gamma-1)}$$

$$\pi_c = \left(1 + 0.85 \frac{0.92 (471.24)^2 - 10,659.675}{1005 \times 300} \right)^{1.4/0.4} = 4.5936$$

The above results can be summarized in Table 12.1.

Table 12.1 shows that the selection will be in favor of the constant relative angle (β) as it achieves the lowest relative Mach number at tip. Moreover, the pressure ratio is close to the constant prewhirl angle (ζ).

12.4.6 Types of Impeller

As described previously, the impeller may be shrouded or unshrouded. In general, shrouded impellers are common with multistage configuration. A shroud fixed to the impeller reduces the axial load on the thrust bearing and may be very important in high-pressure applications. However, the shroud complicates the design of impeller and increases its mass, which means that lower peripheral speeds are required for stress reasons. For steel impellers in industrial applications, the maximum tip speed range for shrouded impellers is 300–360 m/s while for unshrouded it is 380–430 m/s [8].

The impeller may have full-, half-, or no-inducer at inlet as classified previously. The inducer improves the aerodynamic performance [8]. For the impeller of Eckardt, the inducer is capable of increasing the maximum mass flow rate. For a compressor without inducer running at 16,000, the

TABLE 12.1

Variations of Mach Number and Pressure Ratio for Different Design Methods

	$(M_{1rel})_h$	$(M_{1rel})_m$	$(M_{1rel})_t$	π_c
No prewhirl	0.4385	0.572	0.7184	4.914
Constant prewhirl angle ($\zeta = 25$)	0.3494	0.4613	0.598	4.6747
Constant inlet relative angle ($\beta_1 = 41.48$)	0.4385	0.4397	0.4433	4.5936

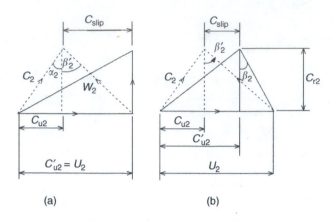

FIGURE 12.27   Slip for (a) radial and (b) backward impellers.

flow appears choked at 6 kg/s, whereas the same sized machine with an inducer is able to pass 7 kg/s with no sign of choking [8].

The impeller may also be classified based on blading into three types, namely, straight radial, forward leaning, or backward leaning. Figure 12.11 shows three types schematically, together with typical velocity triangles in the radial plane for the outlet of each type. The three velocity triangles have the same rotational speed U_2 and same radial velocity component $C_r = W_r$. As a first approximation, the relative fluid velocity leaving the impeller W_2 is assumed parallel to the blade. The angle of the blade at exit with respect to the radial direction is denoted by β'_2, and is positive for a backward-leaning blade (Figure 12.27). In the same figure, the velocity triangles for radial impeller with and without slip are shown.

The swirl (or tangential) component of the absolute velocity is

$$C_{u2} = U_2 - W_{u2} \tan \beta'_2 \tag{12.15}$$

From Equation 12.13,

$$\frac{T_{03} - T_{01}}{T_{01}} = \frac{\Delta T_0}{T_{01}} = \frac{(C_{u2}U_2 - C_{u1}U_{1m})}{C_p T_{01}}$$

$$\frac{T_{03} - T_{01}}{T_{01}} = \frac{(\gamma - 1)\,(C_{u2}U_2 - C_{u1}U_{1m})}{\gamma R T_{01}} = \frac{(\gamma - 1)U_2^2}{a_{01}^2}\left[\frac{C_{u2}}{U_2} - \left(\frac{U_1}{U_2}\right)^2 \frac{C_{u1}}{U_1}\right]$$

$$\frac{T_{03} - T_{01}}{T_{01}} == (\gamma - 1)\left(\frac{U_2}{a_{01}}\right)^2\left[\frac{C_{u2}}{U_2} - \left(\frac{r_1}{r_2}\right)^2 \frac{C_{u1}}{U_1}\right]$$

If the flow enters the impeller axially, then $C_{u1} = 0$ and the temperature relation above is expressed as

$$\frac{T_{03} - T_{01}}{T_{01}} = (\gamma - 1)\left(\frac{U_2}{a_{01}}\right)^2 \frac{C_{u2}}{U_2} = (\gamma - 1)\left(\frac{U_2}{a_{01}}\right)^2\left(1 - \frac{W_{r2}}{U_2}\tan \beta'_2\right) \tag{12.16}$$

$$\frac{T_{03} - T_{01}}{(\gamma - 1)\,(U_2/a_{01})^2\,T_{01}} = \left(1 - \frac{W_{r2}}{U_2}\tan \beta'_2\right) \tag{12.16a}$$

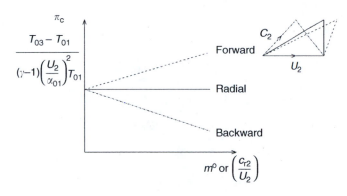

Pressure ratio for different impeller blading

FIGURE 12.28 Temperature rise versus the mass flow rate for different types of impellers.

From Equation 12.3, neglecting the blade thickness (t), thus

$$W_{r2} = \frac{\dot{m}}{\pi \rho_2 b D_2}$$
(12.17)

The variable,

$$\frac{T_{03} - T_{01}}{(\gamma - 1)(U_2/a_{01})^2 T_{01}}$$

represents the relative stage temperature rise and consequently an index for pressure ratio, while the variable, W_{r2}/U_2 (which is equivalent to C_{r2}/U_2), represents the mass flow parameter for a certain rotational speed. Plotting relation 12.15a yields a straight line provided that the blade angle β'_2 is constant (Figure 12.28). At the same mass flow rate and tip speed, U_2, the rotor with *forward-leaning* blades will do more work on the fluid. It might at first sight be expected to produce the highest pressure ratio compared to the radial or backward-leaning blades. However, as will be explained later in the next chapter for axial compressor, positive slope is not as advantageous as the higher pressure ratio and temperature rise will lead to dynamic instability [9]. Moreover, the resulting absolute velocity at exit C_2 is maximum (for the same W_{r2} and U_2), which increases the necessary pressure rise in the diffuser. For these reasons, forward-leaning blades are not suitable for high-speed compressors and never used for aircraft engines.

For radial blades ($\beta'_2 = 0$) a horizontal line is obtained, which means that the pressure ratio is independent from the mass flow rate.

For backward-leaning (backswept) blades, the above relation yields a straight line having a negative slope. Since the absolute velocity (C_2) for backward is less than that for straight radial, backward-leaning blades often have higher efficiency than straight radial blades. This means less stringent diffusion requirements in both the impeller and the diffuser, tending to increase the efficiency of both components [6]. The backswept angle may be in the range 30–40°. The use of backswept vanes gives the compressor a wider operating range of the airflow at a given rotational speed, which is important for matching the compressor to its driving turbine.

Since the pressure ratio is given by the relation

$$\frac{P_{03}}{P_{01}} = \left[1 + \eta_c \frac{T_{03} - T_{01}}{T_{01}}\right]^{\gamma/(\gamma - 1)}$$

with

$$\frac{T_{03} - T_{01}}{T_{01}} = (\gamma - 1)\left(\frac{U_2}{a_{01}}\right)^2 \frac{C_{u2}}{U_2}$$

the slip factor as defined by Dixon [10] is

$$\sigma_s = \frac{C_{u2}}{C'_{u2}} = 1 - \frac{0.63\pi/N_b}{1 - \phi_2 \tan \beta'_2} \qquad (12.18)$$

where N_b is the number of blades, thus $C_{u2}/U_2 = \sigma C'_{u2}/U_2$ and

$$C'_{u2} = U_2 - C_{r_2} \tan \beta'_2 = U_2 \left(1 - \frac{C_{r_2}}{U_2} \tan \beta'_2\right)$$

The swirl velocity ratio is then $C_{u2}/U_2 = \left(1 - \phi_2 \tan \beta'_2\right)$ where $\phi = C_{r_2}/U_2$ or

$$\frac{C_{u2}}{U_2} = \sigma_s \left(1 - \phi_2 \tan \beta'_2\right)$$

Since the pressure ratio is expressed as

$$\frac{P_{03}}{P_{01}} = \left[1 + \eta_c(\gamma - 1)\frac{U_2 C_{u2}}{a_{01}^2}\right]^{\gamma/(\gamma-1)}$$

Thus

$$\frac{P_{03}}{P_{01}} = \left[1 + (\gamma - 1)\eta_c \sigma_s \left(\frac{U_2}{a_{01}}\right)^2 \left(1 - \phi_2 \tan \beta'_2\right)\right]^{\gamma/(\gamma-1)} \qquad (12.19)$$

Other correlations for the slip factor are found in References 3 and 8.

Only the pressure ratio of the *impeller* is defined in a way similar to the whole compressor and expressed as

$$\frac{P_{02}}{P_{01}} = \left[1 + \eta_{imp} \frac{\psi C_{u2} U_2}{Cp T_{01}}\right]^{\gamma/(\gamma-1)} \qquad (12.20)$$

This value is greater than the pressure ratio of the compressor due to the losses in the diffuser.

The impeller isentropic efficiency is related to the compressor efficiency. Now, if the losses in the impeller are equal to a certain fraction of the total losses, say λ, then from the T–S diagram (Figure 12.15), the impeller loss and the compressor loss are, respectively, $(T_{02} - T_{02SS})$ and $(T_{02} - T_{03SS})$.

The impeller efficiency is defined as $\eta_{imp} = (T_{02SS} - T_{01})/(T_{03} - T_{01})$. From this definition and compressor efficiencies

$$\frac{T_{02} - T_{02SS}}{T_{02} - T_{03SS}} = \frac{1 - \eta_{imp}}{1 - \eta_c} = \lambda$$

$$\eta_{imp} = 1 - \lambda + \lambda \eta_c \qquad (12.21)$$

For the case of equal losses in impeller and diffuser

$$\eta_{imp} = 0.5 \left(1 + \eta_c\right).$$
(12.21a)

Example 6 A single-sided centrifugal compressor is to deliver 14 kg/s of air when operating at a pressure ratio of 4:1 and a speed of 200 rps. The inlet stagnation conditions may be taken as 288 K and 0.1 MPa. The number of blades is 20, power input factor is 1.04, the compressor isentropic efficiency is 0.8. If the Mach number is not to exceed unity at the impeller tip, then 60% of the losses are assumed to occur in the impeller.

Find

1. The overall diameter of the impeller if its blades are radial.
2. The minimum possible axial depth of the diffuser for the impeller is described in (a).
3. If the radial impeller in (a) is to be replaced by a backward-leaning impeller having the same diameter, same flow coefficient at impeller outlet $\phi_2 = C_{r2}/U_2$ and the backswept angle $\beta_2' = 30°$, find the corresponding pressure ratio and temperature rise.
4. Same as in (c) but a forward-leaning impeller is employed with $\beta_2' = -30°$.
5. If the three impellers (radial-, forward-, and backward-leaning) are to have the same pressure ratio, same flow coefficient ϕ_2, same rotational speed ($N = 200$ rps), and same isentropic efficiency, then find the corresponding impeller diameter d_2 and tip speed U_2.

Solution: 1. For a radial impeller having 20 blades, the slip factor is

$$\therefore \; \sigma_S = 1 - \left(\frac{0.63\pi}{N_b}\right) = 0.901$$

The compressor pressure ratio is related to the impeller tip speed U_2 by the relation

$$\pi_c^{(\gamma-1)/\gamma} = \left(1 + \frac{\eta_c \sigma \psi U_2^2}{C_P T_{01}}\right)$$

$$\therefore \; 4^{(1.4-1)/1.4} = 1 + \frac{0.8 \times 1.04 \times 0.901 \times U_2^2}{1005 \times 288}$$

$$\therefore \; U_2 = 435.16 \, \text{m/s}$$

Then, the impeller tip diameter is

$$d_2 = \frac{U_2}{\pi \times N} = \frac{435.16}{\pi \times 200} = 0.69 \text{m}$$

The temperature rise across the impeller is

$$\Delta T_0 = T_{02} - T_{01} = \frac{\sigma \psi U_2^2}{C_P} = \frac{0.901 \times 1.04 \times 18.94 \times 10^4}{1005} = 175.6 \text{K}.$$

The total temperature at the impeller outlet is thus $T_{02} = T_{03} = 463.6$ K. Figure 12.29 illustrates a typical T-S diagram that illustrates both of the impeller and diffuser losses.

From Equation 12.21, the impeller efficiency

$$\eta_{imp} = 1 - \lambda + \lambda \eta_c$$

$$\eta_{imp} = 1 - 0.6 + 0.6 \times 0.8 = 0.88$$

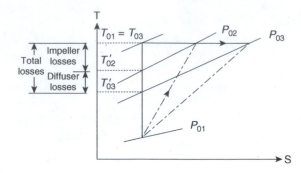

FIGURE 12.29 Impeller and diffuser losses.

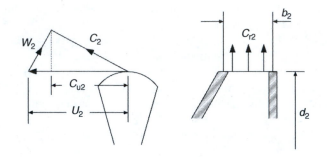

FIGURE 12.30 Outlet flow from impeller.

The pressure ratio across the impeller is

$$\frac{P_{02}}{P_{01}} = \left[1 + \eta_{imp}\left(\frac{T_{02} - T_{01}}{T_{01}}\right)\right]^{\gamma/(\gamma-1)}$$

$$\frac{P_{02}}{P_{01}} = \left[1 + 0.88\left(\frac{175.6}{288}\right)\right]^{1.4/0.4} = (1.5366)^{3.5} = 4.498$$

The total pressure at the impeller outlet $P_{02} = 4.498$ bar.

2. Since the Mach number at impeller outlet must not exceed unity, the absolute velocity at outlet is taken equal to the sonic speed, then C_2 = sonic speed = $\sqrt{\gamma R T_2}$. The outlet flow from impeller is illustrated by Figure 12.30. The total to static temperature in this case is expressed as $T_{02}/T_2 = \gamma + 1/2 = (1.4 + 1)/2 = 1.2$.

Then, the static temperature is $T_2 = 386.8$ K, and the absolute velocity is

$$C_2 = \sqrt{1.4 \times 287 \times 386.8} = 394 \text{ m/s}$$

The total and static pressure ratio is also

$$\frac{P_{02}}{P_2} = \left(\frac{\gamma + 1}{2}\right)^{\gamma/(\gamma-1)} = (1.2)^{3.5} = 1.8929$$

Thus, $P_2 = 4.498/1.8929 = 2.376$ bar.

The air density at the impeller outlet is

$$\rho_2 = \frac{P_2}{RT_2} = \frac{2.376 \times 10^5}{287 \times 386.8} = 2.14 \text{ kg/m}^3$$

The mass flow rate is $\dot{m} = \rho_2 C_{r_2} A_2$

$$\dot{m} = \rho_2 C_{r_2} (\pi \times d_2 \times b_2) \tag{a}$$

$$\dot{m} = \rho_2 \sqrt{C_2^2 - (\sigma U_2)^2} \, (\pi d_2 b_2) = \rho_2 \sqrt{C_2^2 - (\sigma U_2)^2} \, (\pi d_2 b_2)$$

With a constant $\dot{m}$ and rotational speed at maximum (C_2), the impeller width b_2 will be a minimum.

$$C_{r_2}^2 = C_2^2 - (\sigma U_2)^2 = (394)^2 - (0.9 \times 435)^2$$

$$C_{r_2} = 45.5 \text{ m/s}$$

From Equation (a)

$$\therefore b_2 = \frac{\dot{m}}{\rho_2 C_{r_2} \pi \, d_2} = \frac{14}{(2.14)(45.5)(\pi \times 0.69)} = \frac{14}{211.0} = 0.066 \text{ m}$$

3. Backward-leaning impeller $(\beta_2') = 30°$

$$\phi_2 = \frac{C_{r_2}}{U_2} = \frac{W_{r_2}}{U_2} = \frac{45.5}{435} = 0.1046$$

Slip factor is given by the relation

$$\sigma_S = 1 - \frac{0.63(\pi/N_b)}{1 - \phi_2 \tan \beta_2'} = 1 - \frac{0.63\pi/20}{1 - (0.1046) \times \tan 30°} = 0.8951$$

The total sonic speed is $a_{01} = \sqrt{\gamma R T_{01}} = \sqrt{1.4 \times 287 \times 288} = 340 \text{ m/s}$.

The pressure ratio is determined from Equation 12.19 as follows:

$$\frac{P_{03}}{P_{01}} = \left\{ 1 + 0.4 \times 0.8 \times 0.8951 \times \left(\frac{435}{340}\right)^2 (1 - 0.1046 \times 0.5774) \right\}^{3.5} = 3.5876$$

The temperature ratio is

$$\frac{T_{02}}{T_{01}} = \frac{T_{03}}{T_{01}} = 1 + \frac{\pi_c^{(\gamma-1)/\gamma} - 1}{\eta_c} = 1 + \frac{1.4405 - 1}{0.8}$$

Then, the impeller and compressor outlet temperature is $T_{02} = 446.6$ K.

4. Forward-leaning impeller $(\beta_2') = -30°$

The slip factor is

$$\sigma_S = 1 - \frac{0.63\pi/N_b}{1 - \phi_2 \tan \beta_2'} = 1 - \frac{0.63\pi/20}{1 - 0.1046 \times (-0.5774)} = 0.9067.$$

The pressure ratio calculated from Equation 12.19 is

$$\frac{P_{03}}{P_{01}} = \left\{ 1 + 0.4 \times 0.8 \times 0.9067 \times \left(\frac{435}{340}\right)^2 (1 + 0.1046 \times 0.5774) \right\}^{\gamma/(\gamma-1)} = 4.1687$$

The corresponding temperature ratio is

$$\frac{T_{03}}{T_{01}} = 1 + \frac{\pi_c^{\gamma/(\gamma-1)} - 1}{\eta_c} = 1 + \frac{1.50365 - 1}{0.8} = 1.6296.$$

Then, the outlet temperature is $T_{03} = 469.3$ K. Figure 12.31 illustrates the temperature rise for different impellers having a constant diameter.

5. If the pressure rise in backward-leaning impeller = pressure rise in radial impeller

$$\left(\frac{P_{03}}{P_{01}}\right)^{(\gamma-1)/\gamma} = \left[1 + (\gamma-1)\eta_c\sigma_{S_b}\left(\frac{U_2}{a_{01}}\right)_b^2 (1 - \phi_2 \tan \beta_2')_b\right] = \left[1 + (\gamma-1)\eta_c\sigma_{S_r}\left(\frac{U_2}{a_{01}}\right)_r\right].$$

Subscripts (b) and (r) stand for backward-leaning and radial impellers.

$$\therefore \ \sigma_{S_b}U_{2_b}^2 \left(1 - \phi_2 \tan \beta_2'\right)_b = \sigma_{S_r} U_{2_r}^2$$

$$\left(\frac{U_{2_b}}{U_{2_r}}\right)^2 = \frac{\sigma_{S_r}}{\sigma_{S_b}\left(1 - \phi_2 \tan \beta_2'\right)_b} = \frac{0.901}{0.8951\,(1 - 0.1046 \times 0.5774)} = 1.0713$$

$$U_{2_b} = 1.035, \quad U_{2_r} = 450 \, \text{m/s}$$

The diameter ratio between the backward and radial impellers is $d_{2_b}/d_{2_r} = U_{2_b}/U_{2_r} = 1.035$. Then, the diameter of the backward impeller is $d_{2_b} = 0.714$ m.

Moreover, $\phi_2 = 0.1046 = C_{r_2}/U_2$.

The radial velocity at the impeller outlet is $C_{r_2} = 47.07$ m/s.

For a forward-leaning impeller, by equating the pressure rise in forward-leaning and radial impellers, $\sigma_{S_f}(U_2/a_{01})_f^2(1 - \phi_2 \tan \beta_2')_f = \sigma_{S_r}(U_2/a_{01})_r^2$.

Where subscript (f) stands for forward-leaning impeller, the rotational speed is

$$U_{2_f} = U_{2_r}\left\{\frac{\sigma_{S_r}}{\sigma_{S_f}\left(1 - \phi_2 \tan \beta_2'\right)_f}\right\}^{1/2} = 435\left\{\frac{0.901}{0.9067\,(1 + 0.1046 \times 0.5774)}\right\}^{1/2} = 421.09 \, \text{m/s}$$

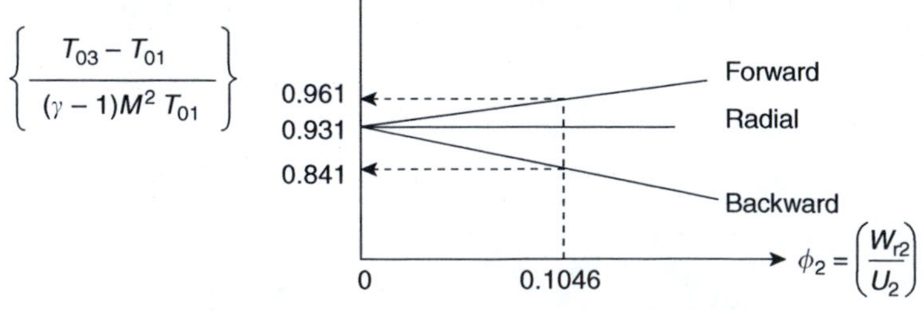

FIGURE 12.31   Temperature rise for different impellers having a constant diameter.

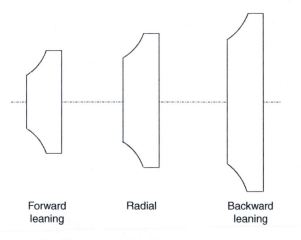

Forward   Radial   Backward
leaning   leaning

FIGURE 12.32 Impeller sizes for different impellers having a constant pressure ratios.

Then, the tip diameter is $d_{2_f} = d_{2_r}(U_{2_f}/U_{2_r}) = 0.69 \times 0.968 = 0.6679$ m.

$$C_{r_2} = \phi_2 \times U_2 = 0.1046 \times 421.09 = 44.05 \text{ m/s}$$

A comparison for the sizes of three compressors are shown in Figure 12.32.

12.5 DIFFUSER

The function of the diffuser is to decelerate the flow leaving the impeller, thus reducing the absolute velocity of the gas at impeller exit from (C_2) to a lower velocity (C_3) as shown in the T–S diagram. The static pressure of the gas the impeller exit is further raised by passing through the diffuser. The amount of deceleration and the static pressure rise $(P_3–P_2)$ in the diffuser depend on the degree of reaction and the efficiency of the diffusion process. An efficient diffuser must have minimum losses $(P_{02}–P_{03})$ but maximum efficiency.

There are broadly two types of diffuser, namely, vaneless and vaned. The vaned diffuser has many types, including not only the diffuser having fixed or pivoting vanes but also the cascade, channel, and pipe types. All diffusers have an initial vaneless gap outboard of the impeller in which the flow settles. The vaneless diffuser is bulky and can be inefficient compared with the other types. The cascade may involve one, two, or three rows of vanes, with performance similar to that of axial cascades. Vaned diffusers may use curved or straight vanes, having a longer path length than the cascade diffuser. Vaneless, cascade, and vaned diffusers collect the flow in a one volute type collector. The channel and pipe diffusers collect the flow in separate passages, sometimes with corner vanes at their outlet. The passage diverges slowly, the pipe consisting of a conical hole and the channel type being rectangular in section with one pair of opposed walls diverging at 10°–11°, the others being parallel.

The pipe and channel diffusers may feed separate combustors or may merge downstream. The channel type is sometimes known as the vane-island or wedge-type diffuser. The blunt end causes as much loss as 5% of the overall efficiency. The pipe-type diffuser was originally proposed and patented by Pratt & Whitney of Canada [8]. It had been shown that the pipe-type diffuser is more efficient than the conventional type with rectangular channels. The passages are formed by drilling holes tangentially through a ring surrounding the impeller, and after the throat the passages are conical. Overall isentropic efficiencies of over 85% are obtained for compressors using this pipe diffuser. Of all the types of diffuser, the vaned diffuser may be made variable, the vanes pivoting about an axial axis.

In general, the performance of diffuser is examined by considering the pressure coefficient

$$C_p = \frac{P_3 - P_2}{P_{02} - P_2}$$

where stations 2 and 3 are the inlet and outlet of the diffuser, respectively. The pressure coefficient is sometimes identified as the pressure rise coefficient and given by the symbol C_{pr}.

12.5.1 VANELESS DIFFUSER

1. *Incompressible flow:*

As the name implies, the gas in a vaneless diffuser is diffused in the vaneless space around the impeller before it leaves the stage of compressor through a volute casing. The gas in the vaneless diffuser gains static pressure rise simply due to the diffusion process from a smaller diameter (d_2) to a larger diameter (d_3). The corresponding areas of cross sections in the radial direction are

$$A_2 = \pi d_2 b_2 = 2\pi r_2 b_2$$
$$A_3 = \pi d_3 b_3 = 2\pi r_3 b_3 \tag{12.22}$$

Such a flow in the vaneless space is axisymmetric inviscid. Since no external torque is applied, the angular momentum is constant, or

$$r_2 C_{u2} = r_3 C_{u3} \tag{12.23}$$

From the continuity equation, at the entry and exit section of the vaneless diffuser,

$$\rho_2 C_{r_2} A_2 = \rho_3 C_{r_3} A_3$$
$$\rho_2 C_{r_2} (2\pi\, r_2 b_2) = \rho_3 C_{r_3} (2\pi\, r_3 b_3)$$
$$\rho_2 C_{r_2} r_2 b_2 = \rho_3 C_{r_3} r_3 b_3 \tag{12.24}$$

For a small pressure rise across the diffuser, $\rho_2 \approx \rho_3$, therefore

$$C_{r_2} r_2 b_2 = C_{r_3}\, r_3 b_3$$

For a constant width (parallel wall) diffuser $b_2 = b_3$

$$C_{r_2} r_2 = C_{r_3} r_3 \tag{12.24a}$$

From Equations 12.23 and 12.24a,

$$\therefore \frac{C_{u3}}{C_{u2}} = \frac{C_{r_3}}{C_{r_2}} = \frac{C_3}{C_2} = \frac{r_2}{r_3} \tag{12.25}$$

Equation 12.25 gives

$$\alpha_2 = \alpha_3 = \tan^{-1} \frac{C_{u2}}{C_{r2}} = \tan^{-1} \frac{C_{u3}}{C_{r3}} = \alpha \tag{12.26}$$

This equation is valid for incompressible flow through a constant width diffuser. Equation 12.20 clearly shows that the diffusion is directly proportional to the diameter ratio (d_3/d_2). This leads to a relatively large-sized diffuser, which is a serious disadvantage of the vaneless type. This leads to an impractically large compressor that prohibits the use of vaneless diffuser in aeronautical applications.

The streamlines for incompressible flow in a vaneless diffuser of constant axial width make a constant angle with the radial direction α. Thus, its shape is sometimes called logarithmic spiral.

$$\theta_3 - \theta_2 = \tan \alpha_2 \ln \frac{r_3}{r_2} \tag{12.27}$$

2.   *Compressible flow:*
Assuming that the flow is reversible, the fluid properties at any point are related to their values at a point where the Mach number is unity (denoted here after by *).

From Equation 12.19c, the continuity equation gives

$$\rho_2 C_{r_2} r_2 = \rho^* C_r^* r^*$$

Conservation of angular momentum requires

$$r_2 C_{u_2} = r^* C_u^*$$

Combining the above two equations, with $\alpha = \tan^{-1}(C_u/C_r)$

$$\frac{\tan \alpha^*}{\tan \alpha} = \frac{\rho^*}{\rho}$$

$$\frac{\rho^*}{\rho} = \frac{\rho_0}{\rho} \frac{\rho^*}{\rho_0} = \frac{(1 + ((\gamma - 1)/2)M^2)^{(1/\gamma - 1)}}{((\gamma + 1)/2)^{1/(\gamma - 1)}}$$

$$\frac{\tan \alpha^*}{\tan \alpha} = \frac{\rho^*}{\rho} = \left[\frac{2}{\gamma + 1} \left(1 + \frac{\gamma - 1}{2} M^2 \right) \right]^{1/(\gamma - 1)} \tag{12.28}$$

From the conservation of angular momentum

$$\frac{r^* \sin \alpha^*}{r \sin \alpha} = \frac{C}{C^*} = M \sqrt{T/T^*}$$

With

$$\frac{T}{T^*} = \frac{T_0}{T^*} \frac{T}{T_0} = \left[\frac{(\gamma + 1)/2}{1 + ((\gamma - 1)/2)M^2} \right]$$

Then

$$\frac{r^* \sin \alpha^*}{r \sin \alpha} = M \left\{ \frac{(\gamma + 1)/2}{1 + ((\gamma - 1)/2)M^2} \right\}^{1/2}. \tag{12.29}$$

From Equations 12.28 and 12.29 since α^* is constant, for a desired Mach number at the diffuser outlet, the corresponding angle and radius can be obtained from the relations

$$\tan \alpha_3 = \tan \alpha_2 \left[\frac{1 + ((\gamma - 1)/2)M_2^2}{1 + ((\gamma - 1)/2)M_3^2} \right]^{1/\gamma - 1} \tag{12.30}$$

and

$$r_3 = r_2 \frac{\sin \alpha_2}{\sin \alpha_3} \frac{M_2}{M_3} \left[\frac{1 + ((\gamma - 1)/2)M_3^2}{1 + ((\gamma - 1)/2)M_2^2} \right]^{1/2}. \tag{12.31}$$

The vaneless diffuser is a very efficient device with an essentially limitless operating range, since the angle of attack sensitivity is absent. Unfortunately, however, the desired diffusion usually requires an excessively large radius ratio across the vaneless diffuser. This is especially critical in aircraft applications where large engine diameters are to be avoided.

12.5.2 Vaned Diffuser

To save space, vanes are employed to reduce the fluid angular momentum and thereby reduce C_u more rapidly. The layout of a channel diffuser is shown in Figure 12.33. It is composed of three domains: vaneless, semi-vaneless, and channel. The channel section represents a low aspect ratio diffuser. It is extensively studied and summarized by Runstadler [11] and Japikse [12]. The crucial parameters are the inlet (throat) Mach number, inlet Reynolds number, stagnation pressure level, and inlet boundary layer blockage in addition to some geometrical parameters. The following dimensional parameters for a typical channel diffuser (Figure 12.23) may be defined as [3]

1. Aspect ratio $AS = b_{th}/W_{th}$
2. Area ratio $AR = A_{exit}/A_{throat}$
3. Nondimensional length $LWR = L/W_{th}$
4. Divergence angle 2θ, where $\tan \theta = (AR - 1)/2LWR$
5. Inlet boundary layer blockage BK where

$$BK = 1 - \frac{\text{Effective flow area}}{\text{Geometric flow area}}$$

6. Inlet Reynolds number, $R_e = C_{in}d_h\rho/\mu$, where d_h is the hydraulic diameter and is defined as

$$d_h = 4 \times \frac{\text{Cross-sectional area}}{\text{Passage perimeter}}$$

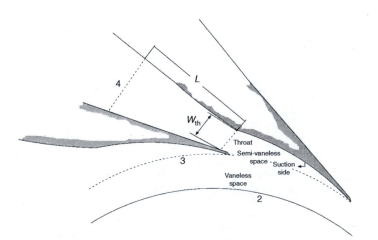

FIGURE 12.33   Channel diffuser.

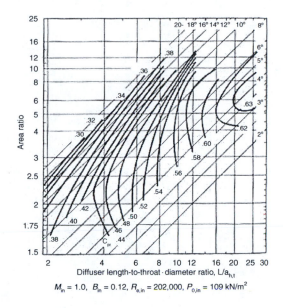

Diffuser length-to-throat · diameter ratio, $L/a_{h,t}$

$M_{in} = 1.0$, $B_{in} = 0.12$, $R_{e,in} = 202,000$, $P_{0,in} = 109$ kN/m²

FIGURE 12.34   A typical diffuser performance map.

A series of performance maps was presented in Reference 11 to show the effect of inlet blockage, Mach number, and Reynolds number on the diffusers with different aspect ratios. A typical map is shown in Figure 12.34. As outlined in Reference 13, the static pressure rise can be approximated by the values of total pressures and absolute velocities at inlet and outlet.

At any state, the isentropic relation between the static and total pressures is given as

$$\frac{P_S}{P_0} = \left(\frac{T_S}{T_0}\right)^{\gamma/(\gamma-1)} = \left(\frac{T_S}{T_0}\right)^{C_p/R} = \left(\frac{T_0 - (C^2/2C_p)}{T_0}\right)^{C_p/R} = \left(1 - \frac{C^2}{2C_pT_0}\right)^{C_p/R}.$$

Using only two terms for the binomial expansion gives

$$\frac{P_S}{P_0} \approx 1 - \left(\frac{C_p}{R}\right)\left(\frac{C^2}{2C_pT_0}\right) \approx 1 - \frac{C^2}{2RT_0}$$

With $P_0/\rho_0 = RT_0$, then $P_S/P_0 \approx 1 - (C^2/2)(\rho_0/P_0)$ or

$$P_S \approx P_0 - (\rho_0 C^2/2)$$

Then, the static pressure rise coefficient for a conical diffuser ($C_{P_r} \equiv (P_3 - P_2/P_{02} - P_2)$) may be approximated by

$$C_{P_r} \approx \frac{(P_{03} - (\rho_{03}C_3^2/2)) - (P_{02} - (\rho_{02}C_2^2/2))}{P_{02} - (P_{02} - (\rho_{02}C_2^2/2))} = \frac{P_{03} - P_{02}}{\rho_{02}C_2^2/2} + \left[1 - \frac{\rho_{03}}{\rho_{02}}\left(\frac{C_3}{C_2}\right)^2\right]$$

or

$$C_{P_r} \approx \frac{P_{03} - P_{02}}{\rho_{02}C_2^2/2} + \left[1 - \frac{\rho_{03}}{\rho_{02}}\left(\frac{C_3}{C_2}\right)^2\right]$$

Since for an adiabatic diffuser $T_{02} = T_{03}$, $P_{02}/\rho_{02} = P_{03}/\rho_{03}$

$$C_{P_r} \approx 2\frac{(P_{03} - P_{02})}{\rho_{02}C_2^2} + \left[1 - \frac{P_{03}}{P_{02}}\left(\frac{C_3}{C_2}\right)^2\right] \qquad (12.32)$$

Example 7   The performance of a conical diffuser is illustrated in Figure 12.34. The diffuser to be examined has a length-to-throat diameter ratio of 8 and a conical angle (2θ) equal to $8°$. The absolute velocity at impeller outlet, $C_2 = 356$ m/s, and the impeller outlet diameter is 0.5 m. The rotational speed of impeller $N = 290$ rps. The number of diffuser conical passages is 12 and the inlet diffuser area per passage is 7.8×10^{-4} m^2. The total temperature and pressure at diffuser inlet are 488 K and 5.48 bar, respectively. Using the approximate static pressure rise, coefficient relation is given by

$$C_{P_r} \approx \frac{2(P_{03} - P_{02})}{\rho_{02}C_2^2} + \left[1 - \frac{P_{03}}{P_{02}}\left(\frac{C_3}{C_2}\right)^2\right]$$

It is required to calculate (a) mass flow rate and (b) outlet velocity C_3.

Solution: From Figure 12.34, the pressure coefficient and area ratio are

$$C_{P_r} \approx 0.52 \quad \text{and} \quad \text{Area ratio} = 4.5$$

The static temperature and pressure are

$$T_2 = T_{02} - \frac{C_2^2}{2C_P} = 488 - \frac{(356)^2}{2 \times 1005} = 424.95 \text{ K}$$

$$P_2 = P_{02}\left(\frac{T_2}{T_{02}}\right)^{\gamma/(\gamma-1)} = 3.376 \text{ bar}$$

The density is

$$\rho_2 = \frac{P_2}{RT_2} = 2.768 \text{ kg/m}^3.$$

The mass flow rate $\dot{m}$ is given by $\dot{m} = \rho_2 C_2 (A_d \times 12)$, with $A_d = 7.8 \times 10^{-4}$ m^2,

$$\dot{m} = 2.768 \times 356 \times 7.8 \times 10^{-4} \times 12 = 9.223 \text{ kg/s}$$

An approximate value for C_3 can be evaluated based on the following arguments:
The continuity equation is $\dot{m} = \rho_2 C_2 A_2 = \rho_3 C_3 A_3$.
For incompressible flow $A_2 C_2 = A_3 C_3$ or $C_3 = C_2 (A_2/A_3) = C_2/\text{Area ratio}$.
However, since the flow is compressible, the static pressure and consequently the density increase in the diffuser, that is, $\rho_3 > \rho_2$, thus $C_3 < C_2/\text{Area ratio}$.
Now for our case, the area ratio is 4.5, then $C_3 < 356/4.5$, or $C_3 < 79.1$ m/s.

$$\text{Try } C_3 = 70 \text{ m/s}$$

$$C_{P_r} \approx \frac{2(P_{03} - P_{02})}{\rho_{02}C_2^2} + \left[1 - \frac{P_{03}}{P_{02}}\left(\frac{C_3}{C_2}\right)^2\right]$$

$$C_{P_r} = \frac{2P_{02}}{\rho_{02}C_2^2}\left[\frac{P_{03}}{P_{02}} - 1\right] + \left[1 - \frac{P_{03}}{P_{02}}\left(\frac{C_3}{C_2}\right)^2\right]$$

$$= \frac{2RT_{02}}{C_2^2}\left[\frac{P_{03}}{P_{02}} - 1\right] + \left[1 - \frac{P_{03}}{P_{02}}\left(\frac{C_3}{C_2}\right)^2\right]$$

Define $\lambda = P_{03}/P_{02}$, then

$$C_{P_r} = \frac{2 \times 287 \times 488}{(356)^2}[\lambda - 1] + \left[1 - \lambda\left(\frac{70}{356}\right)^2\right]$$

or

$$0.52 = 2.21(\lambda - 1) + (1 - 0.039\lambda) \doteq 2.171\lambda - 1.21$$

Thus, $\lambda = 0.796$ and $P_{03} = 4.366$ bar.

Then, the static properties at the diffuser outlet are

$$T_3 = T_{03} - \frac{C_3^2}{2C_P} = 488 - \frac{(70)^2}{2 \times 1005} = 485.6 \text{ K}$$

$$P_3 = P_{03}\left(\frac{T_3}{T_{03}}\right)^{\gamma/\gamma-1} = 4.366 \times \left(\frac{485.6}{488}\right)^{3.5} = 4.291 \text{ bar}$$

$$\rho_3 = \frac{P_3}{RT_3} = \frac{4.291 \times 10^5}{287 \times 485.6} = 3.079 \text{ kg/m}^3$$

The outlet speed is

$$C_3 = \frac{\dot{m}}{\rho_3 A_3} = \frac{9.223}{(4.079 \times 4.5 \times 7.8 \times 10^{-4} \times 12)} = 71.1 \text{ m/s}.$$

Example 8   In the early phase of design, two possibilities for the diffuser of a centrifugal compressor, namely, vaned and vaneless diffusers, are considered. The diffuser has an inlet radius of 0.42 and the flow has inlet Mach number of 0.9 and at an angle of 68° with respect to the radial direction.

1. First, a vaned diffuser is selected, it is required to calculate the Mach number at a diffuser; outlet if the outlet radius of the vanes is 0.525 m, their axial width is constant. Considering an isentropic flow with a pressure coefficient,

$$c_{pd} = \frac{p_3 - p_2}{p_{02} - p_2} = 0.65$$

what would be the outlet blade angle?
2. If a vaneless diffuser is replaced the vaned diffuser, then what would be the outlet radius and flow outlet angle if the outlet Mach number is maintained the same as for a vaned diffuser?
3. If the flow in the vaned diffuser is no longer isentropic but a drop in stagnation pressure of 10% is encountered, and the same value of $c_{pd} = 0.65$ is maintained, then determine the required exit blade angle.

(Assume as a first approximation that the vane and flow angles are the same.)

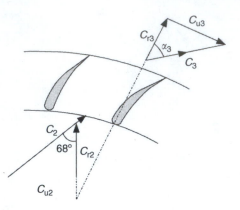

FIGURE 12.35   Vaned diffuser.

Solution:

Figure 12.35 illustrates a vaned diffuser.

1. *Isentropic vaned diffuser:* Since

$$c_{pd} = \frac{P_3 - P_2}{P_{02} - P_2} = \frac{(P_3/P_2) - 1}{(P_{02}/P_2) - 1} = 0.65$$

$$\frac{P_3}{P_2} = 1 + 0.65 \left[\left(\frac{P_{02}}{P_2} \right) - 1 \right] = 1 + 0.65 \left[\left(1 + \frac{\gamma - 1}{2} M^2 \right)^{\gamma/\gamma - 1} - 1 \right] = 1.449.$$

Since

$$P_{03} = P_{02}, \quad \frac{P_3}{P_2} = \frac{P_3}{P_{03}} \frac{P_{02}}{P_2} = \frac{(1 + ((\gamma - 1)/2)M_2^2)^{\gamma/(\gamma - 1)}}{(1 + ((\gamma - 1)/2)M_3^2)^{\gamma/(\gamma - 1)}} = \left(\frac{1.162}{1 + 0.2 M_3^2} \right)^{3.5}$$

Or $(1.449)^{1/3.5} = 1.162/1 + 0.2M_3^2$ or $1 + 0.2M_3^2 = 1.162/1.112 = 1.044964$.
The outlet Mach number is $M_3 = 0.474$.
Since the flow is isentropic in the vaned diffuser, that is, $T_{03} = T_{02}$ and $P_{03} = P_{02}$

$$\frac{P_3}{P_2} = \frac{\rho_3}{\rho_2} \frac{T_3}{T_2} = \frac{\rho_3}{\rho_2} \frac{(T_3/T_{03})}{(T_2/T_{02})} = \frac{\rho_3}{\rho_2} \frac{1 + ((\gamma - 1)/2)M_2^2}{1 + ((\gamma - 1)/2)M_3^2}$$

$$\therefore \ 1.449 = \frac{\rho_3}{\rho_2} \frac{1.162}{1.04496}$$

$$\frac{\rho_3}{\rho_2} = 1.303$$

From the continuity equation

$$\rho_2 c_{r2} (\pi \ d_2 \ b_2) = \rho_3 c_{r3} (\pi \ d_3 \ b_3)$$

With $b_2 = b_3$,

$$\frac{C_3}{C_2} = \frac{\rho_2 \ d_2}{\rho_3 \ d_3} \tag{1}$$

From the conservation of angular momentum in the vaned diffuser

$$r_2 \, C_{u2} = r_3 \, C_{u3}, \quad \text{thus} \frac{C_{u3}}{C_{u2}} = \frac{r_2}{r_3} = \frac{d_2}{d_3} \tag{2}$$

From Equations 1 and 2

$$\frac{C_{u3}/C_{r3}}{C_{u2}/C_{r2}} = \frac{\rho_3}{\rho_2}$$

or

$$\frac{\tan \alpha_3}{\tan \alpha_2} = \frac{\rho_3}{\rho_2} = 1.303$$

$$\tan \alpha_3 = 1.303 \, \tan \alpha_2 = 1.303 \times 2.4751$$

$$\alpha_3 = 72.766°$$

2. *Vaneless diffuser:* The following data are also assumed for the vaneless diffuser

$$M_2 = 0.9, \quad r_2 = 0.42 \text{ m}, \quad M_3 = 0.474, \quad \text{and} \quad \alpha_2 = 68°.$$

The flow is compressible and isentropic, and from Equation 12.30

$$\alpha_3 = 72.7833$$

Also from Equation 12.31

$$r_3 = 0.744 \text{ m}$$

It is clear from cases (1) and (2) that the outlet diameter of the vaneless diffuser (0.744 m) is much greater than the outlet diameter of the vaned diffuser (0.525 m); thus, vaneless diffuser enlarges the outer diameter of the compressor, which in turn increases the frontal area of aero engines and increases the drag force.

If $P_{03} = 0.9 P_{02}$ & $T_{03} = T_{02}$ and $P_3/P_2 = 1.449$

$$\therefore \frac{P_{02}/P_2}{P_{03}/P_3} = \left(\frac{1 + ((\gamma - 1)/2)M_2^2}{1 + ((\gamma - 1)/2)M_3^2} \right)^{\gamma/(\gamma-1)} = \left(\frac{1.162}{1 + 0.2M_3^2} \right)^{3.5}$$

But

$$\frac{P_{02}/P_2}{P_{03}/P_3} = \frac{P_{02}}{P_{03}} \frac{P_3}{P_2} = \frac{1.449}{0.9} = 1.61.$$

Then $M_3 = 0.2805$

$$\frac{P_3}{P_2} = \frac{\rho_3}{\rho_2} \frac{T_3}{T_2} = \frac{\rho_3}{\rho_2} \frac{(T_3/T_{03})}{(T_2/T_{02})} = \frac{\rho_3}{\rho_2} \frac{1 + ((\gamma - 1)/2)M_2^2}{1 + ((\gamma - 1)/2)M_3^2} = \frac{\rho_3}{\rho_2} \frac{1.0157}{1.162} = \frac{\rho_3}{\rho_2}(0.87413)$$

$$\frac{\rho_3}{\rho_2} = \frac{1.449}{0.87413} = 1.6576$$

Since $\dfrac{\tan \alpha_3}{\tan \alpha_2} = \dfrac{\rho_3}{\rho_2}$, then $\tan \alpha_3 = 1.6576 \times 2.4751 = 4.1028$ or $\alpha_3 = 76.3°$.

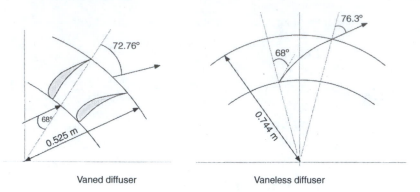

Vaned diffuser   Vaneless diffuser

FIGURE 12.36   Vaned versus vaneless diffuser.

Figure 12.36 illustrates the vaned versus vaneless diffuser. Note that the outlet diameter of the vaneless diffuser is much greater than the vaned diffuser. In addition, the outlet angle from the radial direction is greater in the case of vaneless diffuser than vaned diffuser.

12.6 DISCHARGE SYSTEMS

The discharge system as described earlier is either a manifold, a crossover duct (an inverted U-shaped passage), or a collector (scroll or volute). Aero engines and many gas turbines require the discharged air to be turned into axial direction for transmission to the combustion chamber. Multistage compressors have inverted U-shaped passage to turn the flow from radially outward to radially inward, and then to axial direction for transmission to the inlet of the next stage. In many industrial gas turbines and turbochargers, the fluid is collected in one pipe for delivery to the subsequent modules of the system.

Very little work treated the volute. Figure 12.37 illustrates the volute shape. In general, the cross-sectional area of the volute must increase with the azimuth angle to accommodate the increasing flow. The cross-sectional area of the volute may be very close to the circular shape. However, other shapes for the volute are shown in the figure. The simplest method of volute design assumes that the angular momentum of the flow remains constant ($rC_u = K$). The area at any azimuth angle A_θ at any angle θ is expressed as

$$\frac{A_\theta}{\bar{r}} = Q\frac{\theta}{2\pi K} \tag{12.33}$$

where $\bar{r}$ and Q are the radius of the centroid of the cross section of the volute and volume flow, respectively.

12.7 CHARACTERISTIC PERFORMANCE OF A CENTRIFUGAL COMPRESSOR

In this section, the characteristic behavior of a centrifugal compressor operating between inlet station 1 and outlet station 2 will be discussed. Changing the nomenclature a little, the overall stagnation pressure ratio will be P_{02}/P_{01}. For any compressor, the stagnation pressure at outlet and the overall compressor efficiency are dependent on other physical properties as follows:

$$P_{02}, \eta_c = f\ (\dot{m},\ P_{01},\ T_{01},\ \gamma,\ R,\ D,\ N,\ \text{and}\ \upsilon)$$

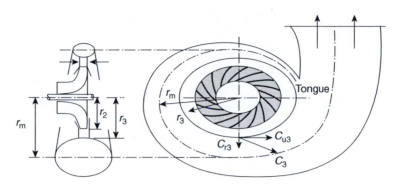

FIGURE 12.37   Volute geometry.

where subscripts 01 and 02 denote the compressor inlet and outlet total conditions, the mass flow rate is denoted by $\dot{m}$, specific heat ratio and gas constant are γ and R, and the kinematic viscosity υ. On the basis of Buckingham's theory, there are nine physical quantities and four basic dimensions; hence, the number of dimensionless quantities is $9 - 4 = 5$. The dimensionless analysis will give the following relationship:

$$\frac{P_{02}}{P_{01}}, \eta_C = f\left(\frac{\dot{m}\sqrt{\gamma R T_{01}}}{P_{01}D^2}, \frac{ND}{\sqrt{RT_{01}}}, \frac{ND^2}{\upsilon}, \gamma\right). \tag{12.34}$$

The specific heat ratio γ does not vary much with compressing the air. The viscous effects are insensitive to the variation of Reynolds number, ND^2/υ, if the Reynolds number itself is high. Since the Reynolds number range is sufficiently high, the compressor performance can be adequately described by

$$\frac{P_{02}}{P_{01}}, \eta_C = f\left(\frac{\dot{m}\sqrt{\gamma R T_{01}}}{P_{01}D^2}, \frac{ND}{\sqrt{RT_{01}}}\right) \tag{12.35}$$

Moreover, for a given air compressor if both the diameter and gas constant are fixed, then the performance can be described by the following pseudo-dimensionless variables:

$$\frac{P_{02}}{P_{01}}, \eta_C = f\left(\frac{\dot{m}\sqrt{T_{01}}}{P_{01}}, \frac{N}{\sqrt{T_{01}}}\right) \tag{12.36}$$

To account for the air properties at different altitudes (when this compressor is part of an aero engine), the right-hand side variables refer to the standard day conditions. Thus, the performance relation above is described by

$$\frac{P_{02}}{P_{01}}, \eta_C = f\left(\frac{\dot{m}\sqrt{\theta}}{\delta}, \frac{N}{\sqrt{\theta}}\right) \tag{12.37}$$

$$\theta = \frac{T_{01}}{(T_{01})_{\text{std day}}} \quad \text{and} \quad \delta = \frac{P_{01}}{(P_{01})_{\text{std day}}} \tag{12.38}$$

where $(T_{01})_{\text{std day}} = 288.15$ K and $(P_{01})_{\text{std day}} = 101.325$ kPa.

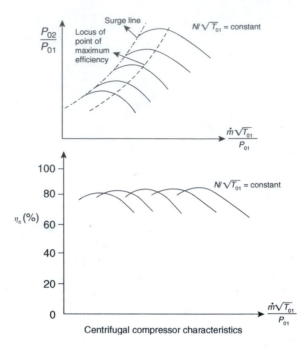

FIGURE 12.38   Compressor map.

The compressor performance is defined (either experimentally or numerically) and plotted in what is called a *compressor map*. This map has the vertical axis as the pressure ratio, or the compressor efficiency, and the mass flow parameter as the vertical axis for varying rotational speed parameter. If the performance of a compressor is to be defined experimentally, then the test rig provides capabilities for both speed and air mass flow rate. For each rotational speed, the air mass flow is varied and the corresponding pressure ratio and efficiency are determined. Next, the rotational speed is changed and another set of experiments is followed by changing the mass flow rate and so on. A schematic representation of the compressor map is shown in Figure 12.38.

In general, three forms for this map are found:

1. The actual values for the mass and speed parameters are used $\left(\dot{m}\sqrt{T_{01}}/P_{01},\ N/\sqrt{T_{01}} \right)$,
2. The values of the mass and speed parameters but referred to the design point values, or,

$$\left(\frac{\left(\dot{m}\sqrt{T_{01}}/P_{01} \right)}{\left(\dot{m}\sqrt{T_{01}}/P_{01} \right)_{d}} \right) \quad \text{and} \quad \left(\frac{\left(N/\sqrt{T_{01}} \right)}{\left(N/\sqrt{T_{01}} \right)_{d}} \right)$$

in such a case at the design point the values of both parameters are unity and other range is from 0.4 to 1.1 (Figure 12.39).

3. The values of these parameters are referred to the standard day temperature and pressure, or $\left(\left(\dot{m}\sqrt{\theta}/\delta \right),\ N/\sqrt{\theta} \right)$. This form of map is usually used by manufacturers of aero engines (Figure 12.40) [14].

Each speed line (or corrected speed) has two limiting points, namely, surge and choke points. The choking point is the extreme right point on the curve. It represents the maximum delivery obtainable at a certain mass flow rate, beyond which no further mass flow rate can be obtained and choking is said to occur. The left point on the constant speed curve represents *surging*. Surging is associated with a drop in delivery pressure with violent aerodynamic pulsation, which is propagated into the whole machine [6]. Surging may be

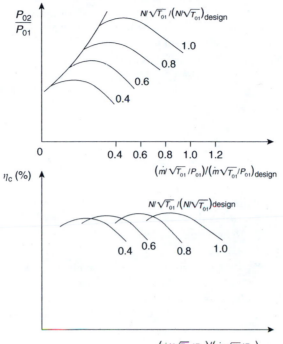

Compressor characteristics referred to design value

FIGURE 12.39   Compressor map with reference to design point.

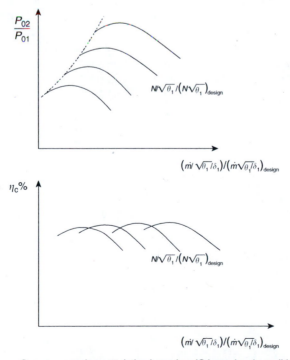

Compressor characteristics based on ISA sea level conditions

FIGURE 12.40   Compressor map based on ISA standard conditions.

described as follows. If the operating point is on the part of the speed line having a positive slope, then a decrease in the mass flow will be accompanied by a drop in the delivery pressure. If the pressure downstream the compressor does not fall quickly enough, the air will tend to reverse and flow back in the direction of the resulting pressure gradient. This leads to a rapid drop in the pressure ratio. Meanwhile, the pressure downstream the compressor has also fallen, so that the compressor will be able to pick up again and repeat the cycle of events that occurs at high frequency.

12.8 EROSION

12.8.1 INTRODUCTION

Erosion studies have been of great importance due to the increased usage of gas turbines in polluted or dusty environment. This is primarily a result of the stepped-up employment of the helicopters, which has the inherent ability to land on dusty terrains.

Extensive studies for particulate flows in the aviation field and the resulting erosion/fouling of turbomachines were performed at least in the past four decades. These studies examined gas turbines as well as aircraft propulsion components in helicopters, hovercrafts, V/STOL, and some conventional airplanes like 737–200, which have the ability to operate from 5000 ft unpaved runways of small cities and remote communications [15]. In such circumstances, billions of high velocity abrasive dust particles are ingested into the core of turbomachines, eroding away the thin metal areas of engine hardware like blade tips. For this reason, theoretical studies were initiated for analyzing particulate flows in axial turbomachines [16], radial turbines [17], and centrifugal compressors [18], respectively. However, very rare experimental results are available until now. The missing experimental results may be compensated by the good agreement between the in-field and theoretical erosion results.

Operation of small gas turbine-powered helicopters over dusty unimproved land areas routinely ingest up to 25 lb of sand and dust for every hour of low altitude operation. Consequently, many billions of high velocity, abrasive dust particles impact the surfaces of compressor blades eroding away the thin metal areas. The inevitable result is progressive and rapid degradation of engine performance often to the extent of premature engine failure within 10%–20% of the normal overhaul time [19]. This is in addition to the numerous in-flight stall/power losses and unscheduled maintenance.

Centrifugal compressors are found in many turboshaft engines powering helicopters and turboprop engines. Particle ingestion into these compressors results in erosion damage of many critical parts of the blades of both rotating and stationary elements. This structural wear results in aerodynamic performance losses and possible boundary layer separation. Consequently, both the compressor pressure ratio and efficiency are reduced.

As the compressor efficiency is reduced, an excessive heating occurs to produce constant power level. Moreover, the compressor operating line becomes closer to the surge line resulting in a smaller surge margin and stable range of operation.

Sand/dust particles are ingested into the core of aero engines in four different circumstances. During takeoff, climb, cruise, and descent of aircraft, particles of variable sizes, materials, and concentrations are sucked by their intakes. Next, during landing ground roll, the thrust reverser operation leads to re-ingestion of hot gases and debris ingestion into the intakes of the engines as described in [20] for military aircraft. The video tape [21] prepared by Pratt & Whitney illustrated the same problem for civil transports. The third circumstance—though rarely encountered—sometimes happens when volcanic ash is ingested into the engines of commercial airplanes. The first recorded volcanic eruption that had its impact on aviation occurred on March 22, 1944, when Mount Vesuvius erupted. Thus, 88 North American B-25 Mitchells had enough damage to be scrapped [22]. Later on, several incidents were encountered by a Lockheed C-130 Hercules on May 25, 1980, and two Boeing 727 airplanes in the same year. Several incidents for Boeing 747 airplanes were next recorded where they encountered volcanic debris. Some of these airplanes were powered by Rolls-Royce RB-211-524 engines while others were powered by P & W JT 9D-7As and GE CF6-80C2. In all of the above cases, volcanic dust caused rapid erosion and damage to the internal components of engines.

The science and technology of multiphase flows has been advanced in the past few years because of the enhanced computational and experimental capabilities. Three steps followed in theoretical treatment of erosion problems are

1. Gas flow analysis
2. Particle trajectories
3. Erosion calculations

Generally, particulate flows are two-phase flow problems. They are treated through either as a continuum or as a discrete flow. In the continuum approach, each phase is considered homogeneous. The particle phase is treated as a continuum phase that interacts with the fluid phase. The mechanisms of the two phases are solved simultaneously through an Eulerian solution technique. The main disadvantage of this approach is its limited ability to provide information for individual particle trajectory. Moreover, it does not account for particle to particle or particle to surface interactions. These disadvantages made this approach unsuitable for the prediction of erosion due to particulate flow. In the discrete approach, the particles are treated in a discrete manner and are solved based on Lagrangian approach. The particle collisions and particle–wall interactions can be considered as well. Coupling between the two phases is considered through mass, momentum, and energy exchange. This coupling may be considered as one- or two-way coupling. In one-way coupling (dilute flow), the particles have no influence on the gas flow and the fluid properties can be evaluated as if no particles are carried. In the two-way (dense flow) coupling, each phase is mutually affected by the other.

The particle trajectories can be obtained by solving the equations of motion of solid particles in the predefined flow field. These trajectories depend on many factors such as the air flow field, domain geometry, wall and particle material combination, particle size and shape, particle density, and particle initial injection velocity and location [18]. Particles are assumed to have a spherical shape in most of the published papers. The trajectories of nonspherical particles (ellipsoid, cylinder, disc, etc.) were also investigated [23]. After defining the particle trajectories, the internal surfaces of turbomachines subjected to excessive impacts are determined and the erosion rates are defined.

The erosion rate is defined as the mass removed from a surface per unit mass of particles impinging on the surface. According to most of the experimental results that have been conducted, the erosion depends on

1. Impact velocity and impact angle
2. Gas temperature
3. Material properties of erodent and target
4. Particle concentration
5. Mass of the particle
6. Particle composition
7. Erodent size
8. Other factors such as the time duration, the shape, and orientation of the particles during impacts

The erosion phenomenon can be classified into two categories, namely, ductile and brittle erosion depending on the target materials [24]. For ductile mode, maximum erosion occurs at

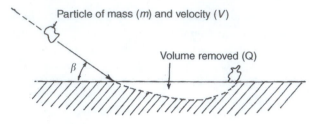

FIGURE 12.41 Ductile erosion simulation.

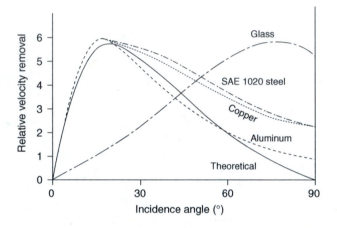

FIGURE 12.42 Effect of impingement angle on the erosion for different target material.

someintermediate angle between 20° and 30°. Brittle mode is characterized by the increase in erosion rate with the ascending impingement angle up to a maximum at normal impingement. In this case, the erosion leads to surface and subsurface cracking and spalling of the target.

In one of the earliest analysis, ductile erosion [25], the particles were likened to the cutting edge of a tool that moves into the surface causing plastic deformation of the material and removal of the debris so formed (Figure 12.41). The erosion parameter was interrelated as a function of impact velocity and impact angle.

Another model that accounts for the erosion at all attack angles considered erosion to be consisting of two simultaneous processes, indentation deformation wear for high angle of attack and cutting wear for low angles [26]. A two-stage mechanism of erosion model [27] proposed that particles impacting at near-normal incidence may fragment and the fragments may subsequently erode exposed surface features. Moreover, a decrease in erosion with decreasing particle size was noticed and the concept of a minimum particle size for effective erosion was introduced.

The model proposed in [28] assumes that erosion occurs as a result of Hertzian contact stresses during impact. These stresses cause cracks to grow from preexisting flaws in the target surface. The volume removed per impact is set proportional to both the penetration depth and the area of the crack.

Both the impact angle and the impact velocity have a dominant effect on the erosion pattern. Figure 12.42 shows the effect of the impact angle on the erosion for different target materials

$$\varepsilon = k_1 f(\beta)\, V^2 \cos^2 \beta \left(1 - R_T^2\right) + k_3 (V \sin \beta)^4 \qquad (12.39)$$

TABLE 12.2
Constants Used in Erosion Prediction by Equation 12.39 [18]

Material Particle	Surface	k_1 gm/gm/(m/s)^2	k_{12}	k_3 gm/gm/(m/s)^4
SiO_2	2024 Al	3.95×10^{-8}	0.585	6.95×10^{-13}
Al_2O_3	2024 Al	5.73×10^{-8}	0.585	10.43×10^{-13}
SiO_2	410 SS	6.53×10^{-8}	0.293	8.94×10^{-13}
Al_2O_3	410 SS	12.46×10^{-8}	0.293	25.3×10^{-13}
SiO_2	Ti-6Al-4v	5.32×10^{-8}	0.192	8.94×10^{-13}
Al_2O_3	Ti-6Al-4v	11.13×10^{-8}	0.192	24.9×10^{-13}

where ε is the erosion parameter (gm/gm), V is the impact velocity (m/s), β is the impact angle (degree), and

$$R_T = 1.0 - 0.00525 \, V \sin \beta$$

$$f(\beta) = 1.0 + CK \left(k_{12} \sin \frac{\pi}{2} \frac{\beta}{\beta_0} \right)^2 \tag{12.40}$$

where

$$CK = \begin{cases} 1 & \beta \leq \beta_0 \\ 0 & \beta > \beta_0 \end{cases} \quad \beta_0 \approx 20°$$

The values for the constants k_1, k_2, k_{12} are given in Table 12.2 for different particles and target material combinations.

In Equations 12.40, β_0 is the impact angle at which the maximum erosion occurs. It has a value near 20° in case of the ductile erosion.

12.8.2 THEORETICAL ESTIMATION OF EROSION

A theoretical estimation of the erosion rates of the internal surfaces of a centrifugal compressor is presented [29]. This estimation is based on an analytical study of the trajectories of particles through the impeller [30]. The analytical treatment of this problem may be described as follows. The particles are assumed to have no influence on the gas flow field. Thus, the gas flow properties are obtained as if no particles were suspended. Next, the particle motion is obtained through the integration of the equations of motion of a particle superimposed on the fluid flow field. Both silicon dioxide and aluminum oxide particles were considered.

Figures 12.43 illustrates the trajectories of very small (3 μm) and small (15 μm) silicon dioxide particles. Figure 12.44 illustrates the trajectories of large diameter (60 μm) particles made of silicon dioxide and aluminum oxide material.

Reviewing the different particle trajectories allowed the definition of the areas that are subjected to severe erosion rates. These will be the areas that are subjected to higher number of collisions. Figure 12.45 or Figure 12.46 summarizes the data on the number and location of particle impacts with the impeller surfaces. Four critical areas were identified, namely, casing surface close to the inlet, two areas on the blade pressure surface, and hub surface near the outlet of impeller. The blade suction surface was occasionally impacted by only massive particles. Even these few impacts occurred at relatively low to moderate incidence velocities and relatively small incidence angles. On the contrary, the blade pressure surface was subjected to a greater number of collisions. It is

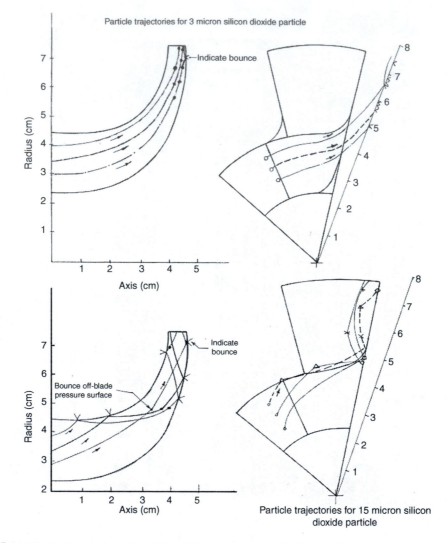

FIGURE 12.43 Particle trajectories of 3 and 15 micron silicon dioxide particles.

worth noting that the leading edge of the inducer and the curved section connecting to the straight blade of the impeller was subjected to a low number of impacts. The straight blade of the impeller was repeatedly impacted by small- and large-sized particles. The shroud surface was subjected to a moderate number of collisions. Two materials for the impeller were considered, namely, 410 stainless steel and 2024 aluminum alloy.

The rates of erosion penetration of these different surfaces were calculated from a specified concentration of particulate flow. Particle diameters ranged from 3 μm to 60 μm. From the rates of erosion over the four selected areas, the depth of penetration after certain operating hours were defined. For the shroud surface, the depth of erosion of an aluminum alloy after 1000 operating hours was 0.688 mm, while this depth will be 0.4 mm for the 410 stainless steel (SS). Concerning the blade pressure surface (region 1), the depth of penetration of a 2024 Al material was 1.3025 mm, while this depth was 1.166 mm for a 410 SS material after 100 operating hours as shown in Figure 12.34. Concerning region 2 of the blade pressure surface, after 100 operating hours, the depth of erosion of 2024 Al and 410 SS materials were, respectively, 1.81 mm and 1.228 mm. The hub surface was subjected to the most severe erosion rates where after only one operating hour the depths of

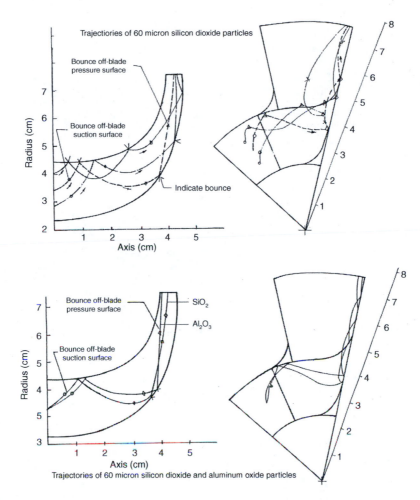

FIGURE 12.44   Particle trajectories of 60 micron silicon dioxide and aluminum oxide particles.

erosion would be 5.6 mm for 2024 Al material and 3.87 mm for 410 SS material. These results are summarized in Figure 12.47 for both 410 SS and 2024 Al alloy. The experimental work of Montgomery [19] outlined that after 25 operating hours, there was visual evidence of severe erosion of the impeller. Tip wear at the top of the blades in addition to the undercutting of the blade near its root was reported. The hub surface was subjected to severe erosion damage as it was reported that particles with 0–5 μm diameter polished the leading side of the compressor blades and the area between the blades to an excellent micro finish. Further, the area near the blade root suffers from the worst erosion damage. This fact complies with the particle trajectory patterns, as most particles tended to strike the hub surface near the blade pressure surface. Similar results were reported in Reference 31 for a centrifugal compressor with splitter blades.

The severe erosion rates described above outlined that the material of impeller must be changed to a more erosion-resistant one. Moreover, coatings may be added to increase the life time as well as time between overhaul of centrifugal compressor.

The titanium impeller of the engine powering the V-22 tilt rotor aircraft (Ospery) was coated by three different coatings. High-velocity sand erosion tests were conducted with sand particles impacting these coatings [32]. Erosion rates were reduced to 10%–30% of its original uncoated base material.

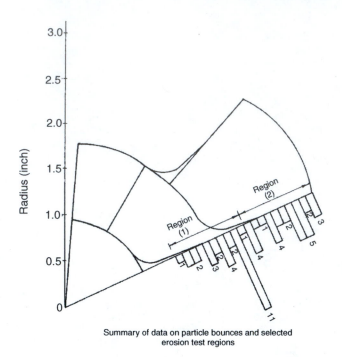

Summary of data on particle bounces and selected
erosion test regions

FIGURE 12.45   Summary of particle bounces and selected erosion test area (front view).

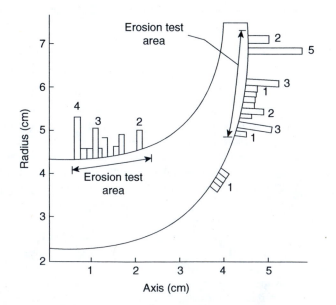

FIGURE 12.46   Summary of particle bounces and selected erosion test area (side view).

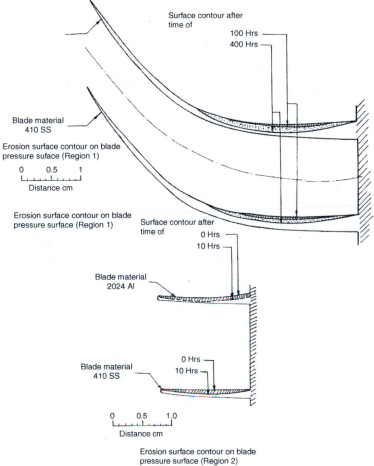

Surface contour after
time of
100 Hrs
400 Hrs

Blade material
410 SS

Erosion surface contour on blade
pressure suface (Region 1)

0   0.5   1

Distance cm

Erosion surface contour on blade
pressure surface (Region 1)

Surface contour after
time of
0 Hrs
10 Hrs

Blade material
2024 Al

0 Hrs
10 Hrs

Blade material
410 SS

0   0.5   1.0

Distance cm

Erosion surface contour on blade
pressure surface (Region 2)

FIGURE 12.47   Erosion results for aluminum Al 2024 and 410 stainless steel impellers.

PROBLEMS

12.1 Choose the appropriate answer or answers:
   (a) Types of diffuser in centrifugal compressors (reaction–cascade–impulse–
        vaneless)
   (b) Types of prewhirl (positive–negative–zero–pipe)
   (c) The number of stages in centrifugal compressors in aircraft engines are
        (1–2–5–9)
12.2 (a) Prove that the torque applied to the shown shaft of a centrifugal compressor shown
        in Figure 12.48 is given by

$$\text{Torque } \tau = \dot{m}[(rC_u)_2 - (rC_u)_1]$$

where C_u is the tangential component of the absolute velocity C.
      (b) For the centrifugal compressor diffuser if the free vortex flow is assumed
(rC_u = constant) and the static pressure and temperature (P and T) are constant,
then prove that the absolute Mach number is inversely proportional to the radius r

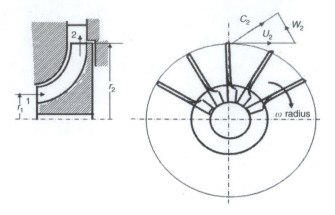

FIGURE 12.48   Problem_3.

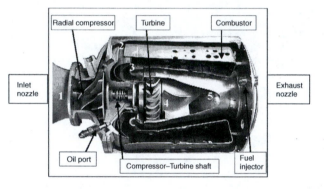

FIGURE 12.49   Problem_4.

$(M \propto 1/r)$. Also, prove that $\alpha = \tan^{-1} C_r/C_u = $ constant implies that the streamlines within the diffuser have a logarithmic spiral shape.

12.3   A turbojet engine incorporates a centrifugal compressor having the following data:

$$T_1 = 283 \text{ K}, \quad P_{01} = 1 \text{ bar}, \quad M_1 = 0.3$$

$$\Psi = 1.04, \quad \sigma = 0.9, \quad \frac{P_{03}}{P_{01}} = 4.0 \text{ bar}$$

$$D_{e_r} = 0.15 \text{ m}, \quad D_{e_t} = 0.3 \text{ m}, \quad D_2 = 0.5 \text{ m}, \quad C_p = 1.005 \text{ kJ/kg} \cdot \text{K}$$

(a.) It is required to calculate
  1. Temperature rise ΔT_0
  2. The velocities U_2, W_2, C_1 (assume flow enters axially and $C_{a_1} = C_{r_2}$)
  3. Relative Mach number at impeller tip
  4. Mass flow rate $\dot{m}$
(b.) Draw the velocity triangles at impeller inlet and outlet
(c.) Plot this operating point on the compressor map

12.4   Figure 12.49 illustrates a cut-away view of the SR-30 gas turbine engine, which represents a small turbojet engine. It contains an inlet nozzle (intake), centrifugal compressor, combustion chamber, and an axial turbine and exhaust nozzle.
    The centrifugal compressor has the following data:

$$T_{01} = 300 \text{ K}, \quad P_{01} = 100 \text{ kPa}, \quad P_{03}/P_{01} = 3.0$$
$$P_2/P_1 = 1.94, \quad P_3/P_2 = 1.48, \quad C_1 = 106 \text{ m/s}$$

Maximum inlet relative Mach number $W_{1\,max}/a_1 = 0.84$
Maximum absolute Mach number at outlet $C_2/a_2 = 0.9$

$$D_h = 20.6 \text{ mm}, \quad D_s = 60.3 \text{ mm}, \qquad D_2 = 102 \text{ mm}$$
$$\psi = 1.04, \qquad \text{Number of blades} = 18$$

Assuming no prewhirl, it is required to
(a) Calculate the air mass flow rate.
(b) Calculate the rotational speed.
(c) Calculate the temperature rise in the compressor.
(d) Prove that

$$C_2 = \sqrt{\frac{2\gamma R M_2^2 T_{02}}{2 + (\gamma - 1)M_2^2}}.$$

(e) Calculate the compressor efficiency.
(f) Calculate the percentage of losses in the impeller to the total losses in the compressor.
(g) Calculate the impeller efficiency.

12.5 A centrifugal compressor has a noninducer impeller (Figure 12.10). It has 30 straight radial blades extending from an inlet radius of 0.12 m to outer radius of 0.2 m that have a thickness of 0.5 cm. The velocity triangles have $C_{u1} = U_1$ and $C_{u2} = U_2$. The air mass flow rate is 30 kg/s. The impeller rotates at 300 rps. The impeller is made up of an aluminum alloy with a yield shear strength—based on the distortion energy theorem—of 60 MPa. The blade roots are considered as 30 rectangles of dimensions 8×0.5 cm and its center has a radius of 16 cm.

Calculate the maximum shear stress at the roots of the blades and the corresponding factor of safety using the relation

$$\tau_{max} = \frac{T \cdot r_{max}}{J_0}$$

12.6 The centrifugal compressor shown in Figure 12.50 is used in a small turbojet engine, which is fitted with a can-type combustion chamber.

Measurement of the pressure, temperature, and velocity components at different locations in the compressor is listed in Table 12.3:
Other data are
Power input factor $\psi = 1.04$
Rotational speed $N = 200$ rps
Eye tip diameter $= 0.3$ m
eye hub diameter $= 0.14$ m
If the flow in the stator is assumed to be a free vortex flow, it is required to
(a) Draw the velocity triangles at impeller inlet and outlet.
(b) Calculate the percentage of the impeller losses out of the total losses in compressor.
(c) Calculate the air mass flow rate into the compressor.
(d) Calculate the diffuser outlet diameter and width.
(e) Calculate the power needed to drive the compressor.

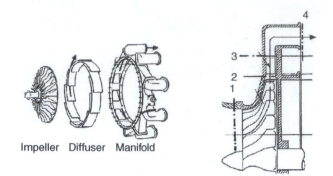

FIGURE 12.50 Problem_6.

TABLE 12.3
Thermodynamic Properties and Velocities at Different Locations

	Location			
	1	**2**	**3**	**4**
T_0 (K)	290	374.7	374.7	374.7
P_0 (kPa)	85	188.5	172	170
T (K)	288.4	339.8	351.1	374.5
P (kPa)	83.36	133.9	150.1	169.8
C_u (m/s)	—	263.4	216.8	—
C_r (m/s)	—	27.1	21.86	—
C_a (m/s)	57.0	—	—	16.8

(f) If air leaving the compressor is assembled in eight circular manifolds to direct the air to the eight cans of the combustion chamber, calculate the diameter of each manifold.

12.7 Three cases for a centrifugal compressor, namely, radial, forward, and backward impellers, Figure 12.51, are considered. The design conditions are as follows:

Inlet total temperature	$T_{01} = 288$ K	Impeller outlet diameter	$D_2 = 0.50$ m
Inlet total pressure	$P_{01} = 101$ kPa	Eye hub diameter	$d_{1h} = 0.15$ m
Mass flow rate	$\dot{m} = 9.0$ kg/s	Eye tip diameter	$d_{1t} = 0.30$ m
Rotational speed	$N = 270$ rev/s	Impeller isentropic efficiency	$\eta = 0.90$
Power input factor	$\psi = 1.04$		

Outlet angle for radial, backward, and forward impellers are 90°, 70°, and 110°, respectively. The impeller losses are equal to 50% of the total compressor losses. Assume that the inlet axial velocity is equal to the radial velocity ($C_a = C_r$).

(a) Calculate the pressure ratio for the three types of compressors.

(b) If the mass flow rate is decreased such that the new axial velocity at the inlet is reduced by 30% ($C'_a = 0.7C_a$) and due to losses at the inducer, the impeller adiabatic efficiency is decreased to 0.87, calculate the new pressure ratio and power for the three cases.

Comment on the results.

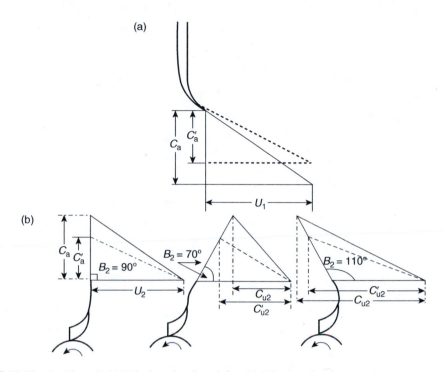

FIGURE 12.51 Problem_7. (a) Velocity triangle at inlet. (b) Velocity triangles at outlet.

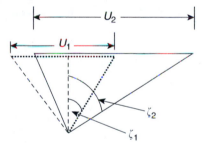

FIGURE 12.52 Problem_8.

12.8 The velocity triangles at inlet and outlet of the impeller of a centrifugal compressor is illustrated in Figure 12.52 drawn to a common apex.
   (a) Identify the velocities C and W at the inlet and outlet.
   (b) What is the physical meaning of ζ_1 and ζ_2.
   (c) Prove that the power necessary for driving the compressor if the power input factor ψ is unity is given by

$$P = \frac{\omega \dot{m}^2}{\rho_1 \pi \left(r_{et}^2 - r_{eh}^2\right)} \left(r_2 \tan \zeta_2 - r_{em} \tan \zeta_1\right)$$

   where subscripts (et, eh, and em) refer to the eye tip, eye hub, and eye mean, respectively, and subscripts 1 and 2 refer to the compressor inlet and outlet.
   (d) What are the conditions of maximum power? Draw the velocity triangles in this case.

(e) Prove that the compressor pressure ratio π_c is expressed as

$$\pi_C = \left[1 + \frac{\eta_C \omega \dot{m}}{\rho_1 \pi \left(r_{et}^2 - r_{eh}^2\right) C_P T_{01}} \left(r_2 \tan \zeta_2 - r_{em} \tan \zeta_1\right)\right]^{(\gamma/\gamma-1)}.$$

(f) Write three disadvantages for each of the centrifugal and axial compressors.

12.9 (a) For a centrifugal compressor, PLOT the process from the impeller inlet (state) to the volute outlet (state 4) on the T–S diagram.

(b) Compare between the geometry of centrifugal compressors used in aircraft engines and those used in wind tunnels.

(c) Explain why total-to-total isentropic efficiency cannot be employed for the diffuser of a centrifugal compressor.

(d) Compare between the compressor maps for axial and centrifugal compressors.

(e) If the isentropic efficiency of the diffuser in a centrifugal compressor is defined as

$$\eta_d = \frac{T_{3s} - T_2}{T_3 - T_2}$$

where

Prove that $\eta_d = \dfrac{AC - BD}{BE}$,

$$A = 1 + \eta_c \frac{\Delta T_0}{T_{01}}, \quad B = 1 + \eta_{imp} \frac{\Delta T_0}{T_{01}}$$

$$C = T_{03} - \frac{C_3^2}{2C_P}, \quad D = T_{03} - \frac{C_2^2}{2C_P}$$

$$E = \frac{C_2^2 - C_3^2}{2C_P}$$

12.10 A two-stage centrifugal compressor has the following data:

	First stage	Second stage
Power input factor	1.04	1.04
Slip factor	0.90	0.90
Overall diameter of the impeller (m)	0.60	0.60
Eye root diameter (m)	0.16	0.16
Eye tip diameter (m)	0.32	0.28
Isentropic efficiency	0.78	0.80

The rotational speed is 240 rps, the air mass flow rate is 10 kg/s, and inlet stagnation conditions are $T_{01} = 290$ K and $P_{01} = 1.1$ bar. The total pressure drop in the inverted U-shape duct between the two stages is 5%. The first stage is identified by states from 1 to 3, while the second stage is identified by states from 4 to 6.

Calculate

(a) The overall pressure ratio of the compressor.

(b) If $\rho_1 = 1.234$ kg/m^3, $C_{a1} = C_{r2}$, $C_{a4} = C_{r5}$ calculate the diffuser depth for each stage.

(c) The power needed to drive the compressor.

12.11 The following results were obtained from a test on a small single-sided centrifugal compressor:

Compressor delivery stagnation pressure	2.97 bar
Compressor delivery stagnation temperature	429 K
Static pressure at impeller tip	1.92 bar
Mass flow rate	0.6 kg/s
Rotational speed	766 rps
Ambient pressure	0.99 bar
Ambient temperature	288 K

Calculate the overall isentropic efficiency of the compressor. The diameter of the impeller is 16.5 cm, the axial depth of the vaneless diffuser is 1.0 cm and the number of the impeller vanes is 17. Making use of Stanitz equation for the slip factor, calculate the stagnation pressure at the impeller tip and hence find the fraction of the overall losses that occur in the impeller.

12.12 A single-stage centrifugal compressor receives air at a stagnation pressure of 1.013 bar and a stagnation temperature of 288 K. The rotor tip speed is 450 m/s and the velocity at the impeller eye is 100 m/s. The inlet (eye) hub and shroud diameters are 6 cm and 12 cm, respectively. The isentropic efficiency of the compressor is 0.85. Determine the vane angle, β_2^1 if the slip factor is 0.95. If the compressor is fitted with inlet guide vanes, calculate the prewhirl angle required to obtain a stagnation pressure ratio across the compressor of 4.0. The rotational speed is 36,000 rpm. Assume that the slip factor is 0.95.

12.13 A double-sided centrifugal compressor has the following characteristics:
- Mass flow rate   50 kg/s, which is divided equally between the front and back faces of impeller
- Power input factor ψ 1.04
- Number of impeller blades n 20
- Rotational speed N 300
- Isentropic efficiency η_c 0.78
- Inlet stagnation temperature T_{01} 295 K
- Inlet stagnation pressure P_{01} 1.05 bar
- Power required to drive the compressor   9.8 MW
- Absolute Mach number at impeller outlet $M_2 = 1.1$
- Eye tip diameter is double the eye root diameter
- Assuming that $C_{r2} = C_{a1}$

Calculate:
1. Temperature rise in the compressor $(T_{03} - T_{01})$
2. Overall diameter of impeller D_2
3. The absolute velocity at impeller outlet C_2
4. Axial velocity at impeller inlet C_{a1}
5. Eye root and tip diameters
6. Pressure drop in the stator $(P_{02} - P_{03})$

12.14 In what case, the specific work of a centrifugal compressor is expressed by the relation:

$$w = U_2^2 - U_1^2$$

Next, it is required to
(a) Draw the impeller shape
(b) Draw the corresponding velocity triangles

(c)   Prove that if $C_{a1} = C_{r2}$, then $W_1 = W_2$
(d)   if $W_1/U_1 = 0.6$ and $U_1/U_2 = 0.4$, prove that:

$$\frac{C_2}{C_1} = 2.2$$

12.15   A light business executive transport is powered by two turboprop engines. Each engine is fitted with two centrifugal compressors. The engines are running in a polluted weather, hence they are subjected to sand particles ingestion that erodes the blades of the low-pressure compressor. Such erosion of the impeller blades at inlet causes separation of flow from the suction side of impeller blades at inlet and increases eddies and vortices, in the flow leaving the impeller. Thus, the circumferential component of the absolute velocity leaving the impeller is reduced by 10% ($C'_{u2} = 0.9\,C_{u2}$). It is required to
(a)   Calculate the ratio between the temperature rise before and after erosion:

$$\frac{\Delta T'_0}{\Delta T_0}$$

(b)   Prove that the drop in the pressure ratio due to erosion is given by

$$\pi_c - \pi'_c \approx 0.1\frac{\gamma}{\gamma-1}\eta_c\frac{\Delta T_0}{T_{01}}$$

Now to restore the same pressure ratio of the eroded compressor, the rotational speed of the compressor has to be increased.
(c)   What will be the ratio between this rotational speed and the original one?
(d)   Next, if the angle of the relative velocity (W_1) at impeller inlet is maintained constant with axial direction before and after increasing the rotational speed, what will be the increase in the mass flow rate in this case?
(e)   Plot the original and final state on the compressor map.

REFERENCES

1.   R.N. Brown, *Compressors-Selection and Sizing*, Gulf Publication Co., 1981, Chapter 1.
2.   R.T.C. Harman, *Gas Turbine Engineering, Applications, Cycles and Characteristics*, John Wiley & Sons, 1981, p. 7.
3.   A. Whitfield and N.C. Baines, *Design of Turbomachines*, Longman Scientific & Technical, 1990, p. 3.
4.   D.P. Kenny, *The History and Future of the Centrifugal Compressor in Aviation Gas Turbines*, 1st Cliff Garrett Turbomachinery Award Lecture, SAE paper SAE/SP-802/602, Dec. 1984.
5.   J.E. Shigley and C.R. Mischke, *Mechanical Engineering Design*, 5th edn., McGraw-Hill Book Company, 1989, p. 65.
6.   H.I.H. Saravanamuttoo, *GFC Rogers and H Cohen, Gas Turbine Theory*, 5th edn., Prentice-Hall, 2001, p. 167.
7.   J.D. Staintz, Some theoretical aerodynamic investigations of impellers in radial and mixed flow centrifugal compressors, *Trans ASME* 74, 1952, 473.
8.   N.A. Cumpsty, *Compressor Aerodynamics*, Longman Scientific & Technical, 1989, p. 71.
9.   P. Hill and C. Peterson, *Mechanics and Thermodynamics of Propulsion*, 2nd edn., Addison Wesley Publication Company, Inc., 1992, p. 431.
10. S.L. Dixon, *Fluid Mechanics and Thermodynamics of Turbomachinery*, 4th edn., Butterworth-Heinemann, 1998, p. 213.

11. P.W. Runstadler, F.X. Dolan, and R.C. Dean, *Diffuser Data Book*, Creare TN, 1975, 186.

12. D. Japikse, *Turbomachinery Diffuser Design Technology*, DTS-1, Concepts ETI, 1984.

13. D.G. Wilson and T. Korakianiti, *The Design of High-Efficiency Turbomachinery and Gas Turbines*, 2nd edn., Prentice-Hall, 1998, p. 180.

14. *The Aircraft Gas Turbine Engine and Its Operation*, P&W Oper. Instr. 200, Part No. P&W 182408, 1988, pp. 3–81.

15. Boeing staff, "Unpaved Runways Operation-737" *Boeing Airliner Magazine*, pp. 2–6, April 1975.

16. M.F. Hussein, *Dynamic Characteristics of Solid Particulate Flow in Rotating Turbomachinery*, PhD Dissertation, University of Cincinnati, OH, 1972.

17. W.B. Clevenger and W. Tabakoff, *Erosion in Radial Inflow Turbines, Volume III: Trajectories of Erosive particles in Radial Inflow Turbines*, NASA-CR 134700, 1974B.

18. A.F. El-Sayed, *Aerodynamics of Air Carrying Sand-Particulate Flow in Centrifugal Compressors*, PhD Dissertation, Cairo University, Cairo, Egypt, 1980.

19. J.E. Newhart, Evaluating and Controlling Erosion in Aircraft Turbine Engines, Paper No. 21, 14 pp, *Proc. of the National Conference on Environmental Effects on Aircraft and Propulsion Systems*, Trenton, NJ, United States, 21–23 May, 1974.

20. W. Kurz, *Comparison of Model and Flight Testing with Respect to Hot Gas Reingestion and Debris Ingestion During Thrust Reversal*, Paper No. 13, 5th Int. Symposium on Airbreathing engines (5th ISABE), February 1981, Bangalore, India.

21. United Technologies, Pratt & Whitney, Reingestion Caused by Thrust Reverser Operation, Videotape P/N 807833, December 1988, © 1989 United Technology Corporation.

22. A.T. Lloyd, *Vulcan's Blast, Boeing Airliner*, April–June, 1990, pp. 15–21.

23. A.F. Elsayed and W.T. Rouleau, *The Effect of Different Parameters on the Trajectories of Particulate in a Stationary Turbine Cascade*, 3rd Multi-phase Flow and Heat Transfer Symposium Workshop, Miami Beach, FL, 1983.

24. C.E. Smeltzer, M.G. Gulden, and W.A. Compton, Mechanisms of metal removal by impacting dust particles, *J. Basic Engineering Transactions (ASME)*, 92, 639–654, 1970.

25. I. Finnie, The mechanism of erosion of ductile materials, *Proc. of 3rd National Congress of Applied Mechanics, Trans. ASME*, 1958, pp. 527–532.

26. J.G.A. Bitter, A study of erosion phenomena, Part I and II, *Wear*, 6, 1963, 5–21 and 169–190.

27. G.P. Tilly, A two stage mechanism of ductile erosion, *Wear*, 23, 1973, 87–96.

28. G.L. Sheldon and I. Finnie, The mechanism of material removal in the erosive cutting of brittle materials, *Trans. ASME J. Eng. Ind.*, 88, 1966, 393–400.

29. M.I.I. Rashed and A.F. Abd-el-Azim El-Sayed, Erosion in centrifugal compressor, *Proc. 5th Int. conf. on Erosion by Solid and Liquid Impact*, ELSI V, Cambridge, U.K., September 1979.

30. A.F. Abdel Azim Elsayed and M.I.I. Rashed, Particulate trajectories in centrifugal compressors, *Proc. of Gas Borne Particles Conference, Inst. of Mechanical Engineers*, Paper C61/81, I. Mech. E., Oxford, U.K., 1981.

31. S. Elfekki and W. Tabakoff, Erosion study of radial flow compressor with splitter, *ASME J. Turbo-Machinery*, 109, 62–69, 1987.

32. G.Y. Richardson, C.S. Lei, and W. Tabakoff, Erosion testing of coatings for V-22 aircraft applications, *Int. J. Rotating Machinery*, 9, 35–40, 2003.

33. Rhys. Jenkins, The heat engine idea in the 17th century, *Trans. Newcomen Soc.*, 17, 1–11, 1936–37.

13 Axial-Flow Compressors and Fans

13.1 INTRODUCTION

Since day one of the gas turbine engine, the compressor was the key component for the success and always required high development efforts and costs. The importance of the compressors for modern high-bypass ratio engines is demonstrated by the fact that 50%–60% of the engine length and up to 40%–50% of the weight, 35%–40% of the manufacturing, and 30% of the maintenance costs are covered by the compression system. The advances in compressors allow the aero engines to operate with core engine thermal efficiency in the range of 50% while the propulsive efficiency approaches 80% [1].

In axial compressors, the air flows mainly parallel to the rotational axis. Axial-flow compressors have large mass flow capacity, high reliability, and high efficiency, but have a smaller pressure rise per stage (1.1:1 to 1.4:1) than centrifugal compressors (4:1 to 5:1). However, it is easy to link together several stages and produce a multistage axial compressor having pressure ratios up to 40:1 in recent compressors. Integrally bladed rotors permit blade speeds significantly above conventional rotors and hence stage pressure ratios greater than 1.8 [1]. Axial compressors are widely used in gas turbines, notably jet engines, wind tunnels, air blowers, and blast furnaces. Engines using an axial compressor are known as axial flow engines; for example, axial flow turbofan. Almost all present-day jet engines use axial-flow compressors, the notable exception being those used in helicopters, where the smaller size of the centrifugal compressor is useful. The fan in turbofan engines is also an axial-compression module, which is treated as an axial compressor having a fewer number of blades of very large height, wide chord, and large twist. These fans may be single or up to three stages in low-bypass turbofan engines. Nowadays, jet engines use two or three axial compressors for higher pressure ratios. Nearly all the turbojet engines are of the two-spool type, which have two compressors identified as low- and high-pressure compressors. Two-spool turbofan engines have the fan as the low-pressure spool. Sometimes, this fan represents the first stage of the low-pressure compressor. The fan in this case is followed by few stages of an axial compressor that represent the first module of the engine core. In other turbofans, the fan by its own represents the low-pressure compressor, which is followed by the high-pressure compressor. Three-spool turbofan engines have the fan, intermediate pressure, and high-pressure compressors. Low-pressure compressor turns at the lowest rpm while high-pressure compressor turns at the highest speed.

A typical axial compressor depicted in Figure 13.1 has a series of rotating "*rotor*" blades followed by a stationary "stator" set of blades that are concentric with the axis of rotation. The compressor blades/vanes are relatively flat in section. Each pair of rotors and stators is referred to as a *stage*, and most axial compressors have a number of such stages placed in a row along a common power shaft in the center. The stator blades are required to ensure reasonable efficiency; without them, the gas would rotate with the rotor blades, resulting in a large drop in efficiency. The axial compressor compresses the working fluid (here, only air will be treated) by first accelerating the air and then diffusing it to obtain an increase in pressure. The air is accelerated in the rotor and then diffused in the stator. This is illustrated in Figure 13.1, where the absolute velocity (C) increases in the rotor but decreases in the diffuser. For successive stages, a saw-teeth pattern for the velocity is obtained, while the static pressure continuously increases in both the rotor and stator rows of all stages.

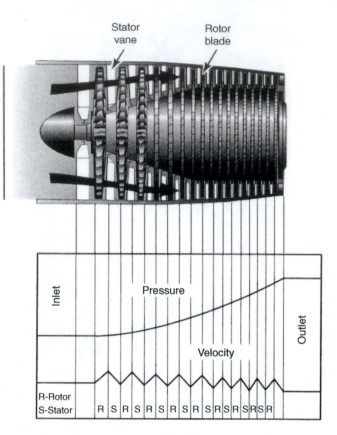

FIGURE 13.1 Layout of an axial compressor.

The axial compressor is built up of a series of stagers, each consisting of a disc of rotor blades followed by a ring of stator vanes. The axial compressor is generally composed of four main elements: front frame, casing with inlet (stator) vanes, rotor with rotor blades, and rear frame [2].

The *front frame* in turbojet engines is a ring-shaped single piece lightweight structure made up of aluminum alloy or steel, usually cast and then machined. It is composed of an outer ring, an inner hub, and 6–8 streamlined supporting struts. If the compressor is a part of the turbofan engine, then this front frame is replaced by a row of inlet guide vanes (IGVs). The compressor *casing* is a tube-like construction split lengthwise to facilitate engine assembly and maintenance. To retain the stator blades and variable IGVs in modern engines, the inner surfaces of the casing are machined with circumferential T-section grooves. The final ring of the stator blades (vanes) may be called outlet guide vanes (OGVs), as they guide the flow to the axial direction to suit the compressor outlet. After installing the rotor, both casing halves are bolted together through longitudinal flanges. The compressor casing is made up of lightweight titanium. In recent engines such as Pratt & Whitney 4084 engine powering the Boeing 777, for example, the casing is made up of "Thermax" alloy of Inconel to allow for the expansion of the casing and keeping acceptable tip clearance margin. The third constituent of the compressor is the *rotor assembly*. The rotor blades (similar to the stator ones) have aerofoil section similar to the aircraft wing, but they are highly twisted from root to tip to obtain the optimum angle-of-attack to the flow everywhere along the blade length. The reason for that twist is that the root section travels much slower than the tip section. Thus for a constant axial velocity, the relative velocity and angle change from root to tip.

The length of the blades decreases progressively downstream in the same proportion as the pressure increases. The rotor blades are attached by a dovetail root, pin fixing, fir tree, and straddle

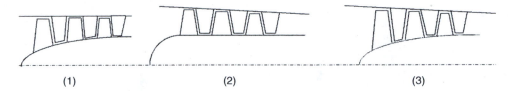

(1)   (2)   (3)

FIGURE 13.2   Variations of the annulus section.

T root [3]. These sets of rotor are mounted on the rotor shaft through drums, individual discs, bolted discs, or clamps. The discs are assembled with a number of tie-bolts, with the bolt-circle diameter selected to produce a dynamically stiff rotor and good torque transmission. The last constituent is the compressor *rear frame* that guides and delivers the pressurized air stream to the combustion chamber. The center of the rear frame accommodates the rearward bearing of the rotor that absorbs the longitudinal thrust of the rotor. Moreover, the struts of the rear frame provide structural strength, lubrication, and venting of the bearing as well as supply of bleed air. As the flow usually has a constant axial velocity through the compressor, the cross-sectional area of the annular passage reduces toward the outlet (or from the front to the rear). This is due to the decrease in the volume of air (thus increasing the density) as compression progresses from stage to stage. Three different layouts are possible (Figure 13.2). The first layout maintains a constant casing diameter of successive stages and increases the hub diameter of successive stages. This design provides the maximum Mach number and pressure ratio. In the second pattern, the hub radius is kept constant and the casing diameter is increased, which provides the best geometry of a compressor. The final pattern has a constant mean diameter, which is associated with a decrease in casing diameter and an increase in the hub diameter. This typical layout is generally employed in deriving the basic theory of axial compressors.

13.2   COMPARISON OF AXIAL AND CENTRIFUGAL COMPRESSORS

The centrifugal compressor achieves part of the compression process by causing the fluid to move outward in the centrifugal force field produced by the rotation of the impeller. Thus, the pressure rises due to the change in the potential energy of the fluid in the centrifugal force field of the rotor. On the contrary, the pressure rise in axial-flow compressor rotors and stators is achieved by the exchange of kinetic energy with thermal energy in a diffusion process. The pressure rise in a centrifugal compressor is therefore less limited by the problems of boundary layer growth and separation in adverse pressure gradients. Probably for this reason, the centrifugal compressor, which first attained a range of pressure ratio and efficiency useful for turbojet engines, was used in the von Ohain engine (1939) and Whittle engine (1941).

The essential feature of airflow in the impeller of a centrifugal compressor is that all the fluid leaves the rotor at the tip radius rather than over a range of radii as in the axial compressor.

13.2.1   ADVANTAGES OF THE AXIAL-FLOW COMPRESSOR OVER THE CENTRIFUGAL COMPRESSOR

1. Smaller frontal area for a given mass rate of flow (may be 1/2 or 1/3); thus, the aerodynamic drag or nacelle housing the engine is smaller as shown in Figure 13.3.
2. Much greater mass flow rates (e.g., present day axial compressors); have mass flow rates up to 200 kg/s (up to 900 kg/s for high-bypass ratio turbofan engines [2]), while centrifugal compressors have mass flow rates less than 100 kg/s.
3. Flow direction of discharge is more suitable for mitigating; thus, suitable for large engines.

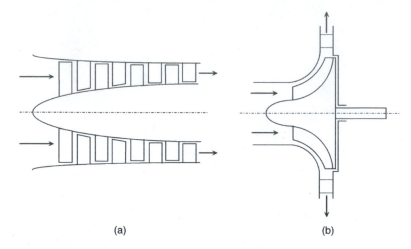

FIGURE 13.3 Layouts of (a) axial- and (b) centrifugal-flow compressors.

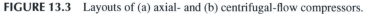

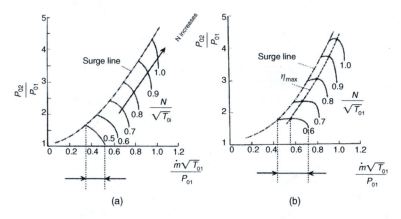

FIGURE 13.4 Compressor maps for axial (a) and centrifugal (b) compressors.

4. May use cascade experiment research in developing compressor.
5. Somewhat higher efficiency at high pressure ratio (perhaps 4%–5% higher than centrifugal compressor).
6. Higher maximum pressure ratio, which was about 17 in the 1960s and achieved up to 45 for the present transonic compressors [1].

13.2.2 ADVANTAGES OF CENTRIFUGAL-FLOW COMPRESSOR OVER THE AXIAL-FLOW COMPRESSOR

1. Higher stage pressure ratio (5:1 or even 10:1).
2. Simplicity and ruggedness of construction.
3. Shorter length for the same overall pressure ratio.
4. Generally less severe stall characteristics.
5. Less drop in performance with the adherence of dust to blades.
6. Cheaper to manufacture for equal pressure ratio.
7. Flow direction of discharge air is convenient for the installation of an intercooler and/or heat exchanger in gas turbines.
8. Wider range (margin) of stable operation between surging and choking limits at a given rotational speed as shown in Figure 13.4.

13.2.3 MAIN POINTS FOR COMPARISON OF CENTRIFUGAL AND AXIAL COMPRESSORS

Table 13.1 lists the main points for comparison between axial and centrifugal compressors.

13.3 MEAN FLOW (TWO-DIMENSIONAL APPROACH)

The flow is considered to occur in the tangential plane at the mean blade height where the peripheral rotation velocity is U (Figure 13.5). This two-dimensional approach means that the flow velocity

TABLE 13.1

Comparison Between Axial and Centrifugal Compressors

	Axial Compressor	Centrifugal Compressor
Dimensions (frontal area)	Smaller	Larger
Dimensions (length)	Longer	Shorter
Number of blades per stage	Great	Less
Weight	Heavier	Lighter
rpm	Low	High
Efficiency	High (85% or greater)	Moderate (75–80%)
Pressure ratio per stage	Small	High
Overall pressure ratio	Higher (30:1)	Much smaller (<10)
Mass flow rate	Higher	Smaller
Flow direction at outlet	Axial	Radial
Manufacturing and its cost	Difficult and expensive	Simple and cheaper
Balance	Difficult	Easier
Construction	Complex	Simple
Rapid change of inlet mass flow	Cannot accept	Can accept
Maintenance and its cost	Difficult and expensive	Simple and cheap
Balancing	More difficult due to blade scatter effect	Easier owing to support bearings
FOD	Less resistant	Good resistant
Fouling	Greatly influenced	Less influenced
Reliability	Lower	Higher
Operating range (between choking and surge points)	Narrow	Wide
Ruggedness	Fragile	Strong
Material	Impeller (aluminum alloy), Inlet section (steel)	Rotor: Front stages (steel alloy), Mid stages (aluminum alloy), Rear stages (titanium alloys); Drum or discs: steel; Casing: Magnesium alloy at the front, Steel at the rear
Engine operation	1. Trouble usually experienced when starting due to lack of initial axial flow (blade stall) 2. Reluctant to respond to rapid acceleration	1. Works satisfactorily at all speeds provided turbine is properly matched 2. Can accept rapid changes of mass air flow (acceleration)
Starting power requirement	Higher	Lower
Engine thrust	Higher	Lower
Engine-specific fuel consumption (SFC)	Lower	Higher
Aircraft altitude	Higher	Lower
Application	All aero engines	Small aero engines and APU

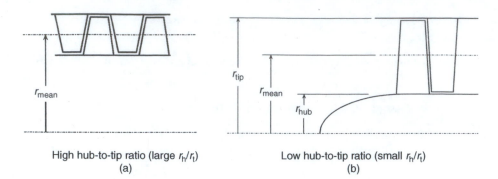

High hub-to-tip ratio (large r_h/r_t)   Low hub-to-tip ratio (small r_h/r_t)
(a)   (b)

FIGURE 13.5   Mean tangential plane for both axial turbine (a) and axial compressor (b).

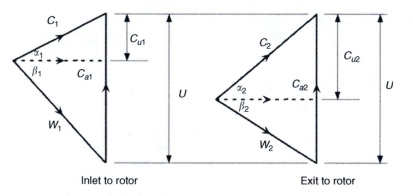

FIGURE 13.6   Variation of mean radius for a constant casing multistage axial compressor.

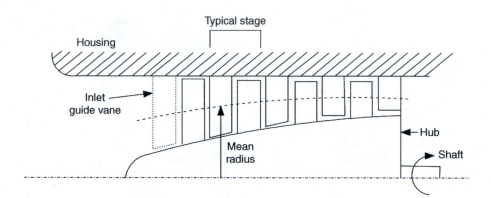

Inlet to rotor   Exit to rotor

FIGURE 13.7   Velocity triangles at inlet and outlet of a constant mean radius.

will have two velocity components: (1) axial denoted by subscript (a) and (2) peripheral denoted by subscript (u). Here, the radial velocity component is neglected. The mean section for both the axial compressor and axial turbine is shown side by side in Figure 13.5, which illustrates how the hub-to-tip ratio of axial compressor is much smaller than its value in axial turbine. For an axial compressor having a constant casing diameter, the radius of the mean section is increasing in the rearward direction as shown in Figure 13.6. The velocity triangles at the inlet and outlet of a single stage are shown in Figure 13.7.

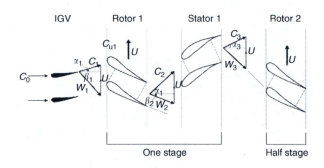

FIGURE 13.8 Airflow in IGV as well as one and half-stage of axial compressor.

The air approaches the rotor with an absolute velocity (C_1), at an angle (α_1) to the axial direction; when combined with the rotational speed (U), the relative velocity will be (W_1), at an angle (β_1) to the axial (see Figure 13.8). After passing through the diverging passages formed between the rotor blades that do work on the air, its absolute velocity (C_2) will increase ($C_2 > C_1$) while its relative velocity W_2 will decrease ($W_2 < W_1$) and the flow exit angle β_2 will be less than the flow inlet angle $\beta_1 (\beta_2 < \beta_1)$. This turning of flow toward the axial direction is necessary to increase the effective flow area.

Since the relative velocity is decreased due to diffusion ($W_2 < W_1$), the static pressure increases ($P_2 > P_1$), that is, a pressure rise will be developed in the rotor.

The absolute velocity of the flow leaving the rotor is C_2. The flow then passes through the stator passages, which are also diverging; thus, the absolute velocity is decreased due to diffusion and the static pressure is increased, that is, $C_3 < C_2$ and $P_3 > P_2$. The flow angle at the outlet of the stator will be equal to the inlet angle to the rotor in most design (i.e., $\alpha_3 = \alpha_1$).

From Figure 13.8, the following kinematical relations are derived:

$$\frac{U}{C_{a1}} = \tan \alpha_1 + \tan \beta_1 \qquad (13.1a)$$

$$\frac{U}{C_{a2}} = \tan \alpha_2 + \tan \beta_2 \qquad (13.1b)$$

If the axial velocity is constant at inlet and outlet of the stage, then $C_{a1} = C_{a2} = C_a$.

The kinematical relations in Equations 13.1a and b will be

$$\frac{U}{C_a} = \tan \alpha_1 + \tan \beta_1 = \tan \alpha_2 + \tan \beta_2 \qquad (13.1c)$$

13.3.1 Types of Velocity Triangles

These velocity diagrams assume equal axial velocities at inlet and outlet, implying that the density is varying inversely with area.

The separate velocity triangles are superimposed either on the common base, namely, the blade rotational speed (U) (Figure 13.9a) or superimposed with a common apex (Figure 13.9b).

$$(C_{u2} - C_{u1}) \equiv \Delta C_u$$

which is the essential element in the energy transfer. Another useful concept is that of the mean fluid angle and mean velocities demonstrated by the mean velocity triangle (Figure 13.9c). The velocities of this mean triangle are C_m, W_m, as well as U. The angles are α_m and β_m. With constant axial

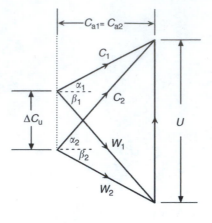

(a)

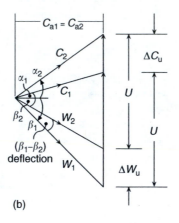

(b)

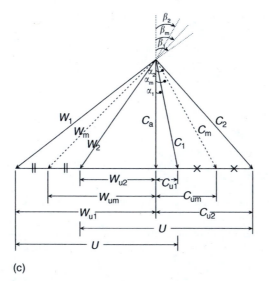

(c)

FIGURE 13.9   Different layouts for mean velocity triangle. (a) Common Apex velocity triangles. (b) Common base velocity triangles. (c) Mean velocity triangle.

and rotational velocities, the absolute and relative swirl velocities are calculated from the following relations:

$$\left.\begin{array}{l} C_{\text{um}} = \dfrac{C_{\text{u}1} + C_{\text{u}2}}{2} \\[2mm] W_{\text{um}} = \dfrac{W_{\text{u}1} + W_{\text{u}2}}{2} \end{array}\right\} \tag{13.2}$$

$$\left.\begin{array}{l} \tan \alpha_{\text{m}} = \dfrac{C_{\text{um}}}{C_{\text{a}}} \\[2mm] \tan \beta_{\text{m}} = \dfrac{W_{\text{um}}}{C_{\text{a}}} \end{array}\right\} \tag{13.3a}$$

Thus, the mean absolute and relative angles are obtained from the relations:

$$\left.\begin{array}{l} \tan \alpha_{\text{m}} = \dfrac{\tan \alpha_1 + \tan \alpha_2}{2} \\[2mm] \tan \beta_{\text{m}} = \dfrac{\tan \beta_1 + \tan \beta_2}{2} \end{array}\right\} \tag{13.3b}$$

13.3.2 VARIATION OF ENTHALPY VELOCITY AND PRESSURE OF AN AXIAL COMPRESSOR

The gain in energy of the gas is shown by the enthalpy achieved only in the rotor blade row. Thus, the total enthalpy increases in the rotor and remains constant in the stator as shown in Figure 13.10. On the same figure, the variations of the absolute velocity and static pressure are plotted for any two successive stages. While energy can only be added at the rotor blades as shown by the horizontal line at the stator, the gain in *static* pressure can be divided between the rotor and stator, as indicated by the continuous increase in pressure as the gas moves from the rotor through the stator.

The energy equation (or the first law of thermodynamics) is expressed in terms of the inlet and outlet conditions of a stage, states (1) and (3), respectively, as

$$Q_{13} + h_{01} = W_{13} + h_{03}.$$

Since an adiabatic process is assumed, the change in total enthalpy will be equal to the work done on the rotor, or

$$\Delta h_0 = U_2 C_{\text{u}2} - U_1 C_{\text{u}1}. \tag{13.4}$$

For an axial compressor with constant mean radius at inlet and outlet of the rotor (or $U_1 = U_2 = U$), Euler's equation for turbomachinery is again expressed for axial compressors as

$$\Delta h_0 = U\left(C_{\text{u}2} - C_{\text{u}1}\right) = W_{13}. \tag{13.5}$$

All the energy transfer between the fluid and is rotor accounted for by the difference between the two UC_{u} terms. Now from the velocity triangles shown in Figure 13.11, Euler equation can be mathematically expressed in a different form:

$$C_{\text{a}}^2 = C^2 - C_{\text{u}}^2$$
$$W_{\text{a}}^2 = W^2 - W_{\text{u}}^2 = W^2 - (U - C_{\text{u}})^2$$

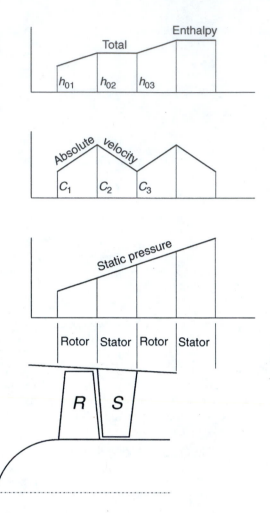

FIGURE 13.10   Variations of the enthalpy, absolute velocity, and static pressure across two successive stages.

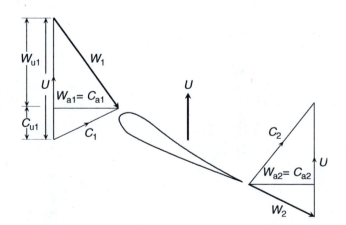

FIGURE 13.11   Detailed velocity triangles.

Since $C_a = W_a$

$$\therefore C^2 - C_u^2 = W^2 - U^2 + 2UC_u - C_u^2$$

$$\therefore UC_u = \frac{1}{2}\left(C^2 + U^2 - W^2\right)$$

$$\therefore U_1 C_{u1} = \frac{1}{2}\left(C_1^2 + U_1^2 - W_1^2\right)$$

$$\therefore U_2 C_{u2} = \frac{1}{2}\left(C_2^2 + U_2^2 - W_2^2\right)$$

$$\therefore \Delta h_0 = (UC_u)_2 - (UC_u)_1$$

$$\therefore \Delta h_0 = \frac{1}{2}\left(C_2^2 - C_1^2 + U_2^2 - U_1^2 + W_1^2 - W_2^2\right) \tag{13.6}$$

This equation is an alternative form of the basic energy by definition.

$$\Delta h_0 = h_{02} - h_{01} = h_2 + \frac{C_2^2}{2} - h_1 - \frac{C_1^2}{2}$$

$$\therefore \Delta h = h_2 - h_1 = \frac{1}{2}\left(U_2^2 - U_1^2 + W_1^2 - W_2^2\right) \tag{13.7}$$

Thus, the terms (U^2 and W^2) represent the change in static enthalpy across the rotor while the C^2 term represents the absolute kinetic energy across the rotor.

From Equation 13.5, the specific input work is equal to the change in the total enthalpy across the stage, or

$$W_s \equiv h_{03} - h_{01} = h_{02} - h_{01} = U(C_{u2} - C_{u1})$$

where h_0 is the total enthalpy, or

$$W_s \equiv Cp\,(T_{02} - T_{01}) = U\Delta C_u \equiv Cp\Delta T_0$$

$$\therefore \Delta T_0 = \frac{U\Delta C_u}{Cp}$$

or

$$\frac{\Delta T_0}{T_{01}} = \frac{U\Delta C_u}{CpT_{01}} \tag{13.8}$$

From Figure 13.12, Equation 13.8 can be expressed in terms of the axial and rotational velocities and the absolute and relative angles as follows:

$$\Delta C_u = C_{a1} \tan \beta_1 - C_{a2} \tan \beta_2$$

$$\text{or} \quad \Delta C_u = C_{a2} \tan \alpha_2 - C_{a1} \tan \alpha_1$$

$$\text{or} \quad \Delta C_u = U - C_{a1} \tan \alpha_1 - C_{a2} \tan \beta_2$$

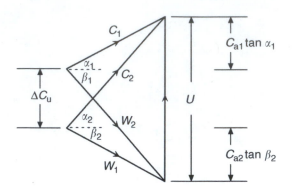

FIGURE 13.12   Velocity triangles over a stage.

If $C_{a1} = C_{a2} = C_a$

$$\therefore \Delta C_u = U \left[1 - \frac{C_a}{U} (\tan \alpha_1 + \tan \beta_2) \right] \text{ or}$$

$$= C_a (\tan \beta_1 - \tan \beta_2) \equiv C_a (\tan \alpha_2 - \tan \alpha_1)$$

The stage temperature rise is now expressed as

$$\Delta T_0 = \frac{UC_a}{Cp} (\tan \alpha_2 - \tan \alpha_1) = \frac{UC_a}{Cp} (\tan \beta_1 - \tan \beta_2) \qquad (13.9a)$$

and

$$\frac{\Delta T_0}{T_{01}} = \frac{U^2}{CpT_{01}} \left[1 - \frac{C_a}{U} (\tan \alpha_1 + \tan \beta_2) \right] \qquad (13.10a)$$

or

$$\frac{\Delta T_0}{T_{01}} = \frac{UC_a}{CpT_{01}} (\tan \beta_1 - \tan \beta_2) = \frac{UC_a}{CpT_{01}} (\tan \alpha_2 - \tan \alpha_1) \qquad (13.10b)$$

The specific work is also given by the relation

$$W_s = UC_a (\tan \alpha_2 - \tan \alpha_1) = UC_a (\tan \beta_1 - \tan \beta_2) \qquad (13.11)$$

Figure 13.13 illustrates the temperature–entropy variation for the compression process through the stage. If the whole processes were isentropic, the final stagnation pressure for the same work input would be $P_{0 \max}$. However, due to losses in the rotor and stator the pressure at the outlet of the rotor is less than the maximum value and the pressure for air leaving the stator is also less than the pressure at the outlet of the rotor, or $P_{03} < P_{02} < P_{0 \max}$.

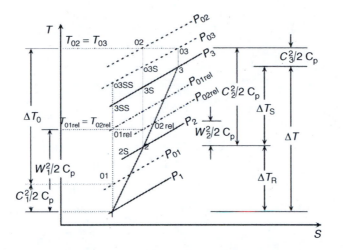

FIGURE 13.13 Temperature–entropy diagram for a compressor stage.

If the stage total-to-total efficiency (η_{st}) is defined as

$$\eta_{st} = \frac{h_{03ss} - h_{01}}{h_{03} - h_{01}}$$

$$\therefore \eta_{st} = \frac{T_{03ss} - T_{01}}{T_{03} - T_{01}} = \frac{T_{03ss} - T_{01}}{\Delta T_0} = \frac{(T_{03ss}/T_{01}) - 1}{(\Delta T_0/T_{01})}$$

$$\therefore \frac{T_{03ss}}{T_{01}} - 1 = \eta_{st}\frac{\Delta T_0}{T_{01}}$$

As

$$\frac{P_{03}}{P_{01}} = \left(\frac{T_{03ss}}{T_{01}}\right)^{\gamma/(\gamma-1)},$$

$$\pi_{stage} \equiv \frac{P_{03}}{P_{01}} = \left(1 + \eta_{st}\frac{\Delta T_0}{T_{01}}\right)^{\gamma/(\gamma-1)}$$

But

$$\left(\frac{\Delta T_0}{T_{01}}\right)_{stage} = \frac{UC_a}{CpT_{01}}(\tan\alpha_2 - \tan\alpha_1)$$

so

$$\pi_{stage} = \left[1 + \eta_{st}\frac{UC_a(\tan\alpha_2 - \tan\alpha_1)}{CpT_{01}}\right]^{\gamma/(\gamma-1)} \tag{13.12a}$$

In actual stage, temperature rise will be less than the above value due to the three-dimensional (3D) effect in compressor annulus. Analysis of experimental results showed that it is necessary to multiply the results by a factor less than unity identified as the work-done factor (λ), due to the radial variation of the axial velocity. The axial velocity is not constant across the annulus but is peaky as

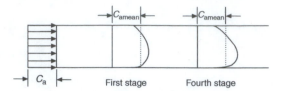

FIGURE 13.14   Variations of the axial velocity through successive stages.

the flow proceeds, settling down to a fixed profile at about the fourth stage, as shown in Figure 13.14. The stage temperature rise pressure ratio, Equation 13.12, should be modified to be

$$\therefore \Delta T_0 = \frac{\lambda}{Cp} U C_a \left(\tan \alpha_2 - \tan \alpha_1 \right) = \frac{\lambda}{Cp} U C_a \left(\tan \beta_1 - \tan \beta_2 \right) \qquad (13.9b)$$

$$\pi_S = \left[1 + \frac{\lambda \eta_{st}}{CpT_{01}} U C_a \left(\tan \alpha_2 - \tan \alpha_1 \right) \right]^{\gamma/(\gamma-1)} \qquad (13.12b)$$

Thus, the pressure ratio across the stage could be determined. It is entirely possible to achieve a very high work input per stage, but it is not possible to achieve a high $\Delta T_0/T_{01}$ while maintaining high efficiency. For a single-stage axial compressor, the temperature rise for efficient operation is always limited such that $\left(\Delta T_0/T_{01} \right) \ll 1$. Then Equation 13.12b may be approximated to

$$\frac{P_{03}}{P_{01}} = \pi_{stage} \approx 1 + \frac{\gamma}{\gamma - 1} \eta_{st} \frac{\Delta T_0}{T_{01}} = 1 + \frac{\gamma}{\gamma - 1} \lambda \eta_{st} \frac{UC_a}{CpT_{01}} (\tan\alpha_2 - \tan\alpha_1) \qquad (13.12c)$$

It is highly desirable to achieve high work input per compressor stage to minimize the total number of stages necessary to provide the desired overall compression ratio.

Example 1   Air at 1 bar and 15°C enters a three stage axial compressor with a velocity of 120 m/s. There are no IGVs and constant axial velocity is assumed throughout. In each stage, the rotor turning angle is 25°. The annular flow passages are shaped in such a way that the mean blade radius is 20 cm everywhere. The rotor speed is 9000 rpm. The polytropic efficiency is constant at 0.9. The blade height at the inlet is 5 cm.

Draw the velocities diagram and calculate

1. Specific work for each stage
2. The mass flow rate
3. The power necessary to run the compressor
4. Stage pressure ratios
5. The overall compressor pressure ratio
6. The blade height at the exit from the third stage
7. The degree of reaction of the first stage

Solution: The given data are $P_{01} = 1$ bar, $r_m = 0.2\,\mathrm{m}$, $N = 9000$ rpm, $\eta_S = 0.9$, $\beta_1 - \beta_2 = 25°$, $h = 0.05\,\mathrm{m}$, Number of stages $= 3$, and $C_1 \equiv C_{a1} = 120\,\mathrm{m/s}$

1. Since the flow is axial, $\alpha_1 = 0$, $C_{u1} = 0$, and $\Delta C_u \equiv C_{u2}$. A typical velocity triangle for this case is shown in Figure 13.15.

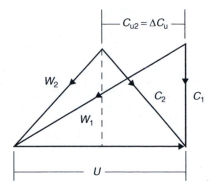

FIGURE 13.15   Velocity triangles of the first stage.

If the inlet total temperature is $T_{01} = 15 + 273 = 288$ K with $U_m = 2\pi r_m N = 188.496$ m/s, then

$$\tan \beta_1 = \frac{U}{C_{a1}}$$

$\beta_1 = 57.52° = $ constant for all stages

$\beta_2 = \beta_1 - 25 = 32.52° = $ constant also for all stages.

Since

$$\tan \beta_2 = \frac{U - C_{u2}}{C_{a2}}$$

then $C_{u2} = 111.989$ m/s $\equiv \Delta C_u$.

The axial compressor is composed of three identical stages as for the three stages:

1. The flow enters axially.
2. The rotor turning angle is the same, $\beta_1 - \beta_2 = $ constant, thus $\Delta C_u = $ constant.
3. The polytropic efficiency is constant.
4. The compressor has a constant mean radius layout $U_m = $ constant.

Thus, the temperature rise per stage and the specific work are constant.

The specific work per stage is $w_S = U \Delta C_u = 21.11$ kJ/kg $\equiv$ Constant for all stages [or $w_S = U \Delta C_u = U C_a (\tan \beta_1 - \tan \beta_2) = 21.11$ kJ/kg].

The specific work for the three-stage compressor is

$$W_s = 3w_s = 3 \times 21.11 = 63.33 \text{ kJ/kg}$$

2. The mass flow rate may be calculated from the relation $\dot{m} = \rho_1 C_{a1} A_1$.

At the compressor inlet, the static temperature

$$T_1 = T_{01} - \frac{C_1^2}{2C_p} = 280.836 \text{ K}$$

and the static pressure

$$P_1 = \frac{P_{01}}{(T_{01}/T_1)^{\gamma/(\gamma-1)}} = 0.916 \text{ bar}$$

The density

$$\rho_1 = \frac{P_1}{RT_1} = 1.136 \text{ kg/m}^3$$

The annulus area $A = 2\pi r_m h = 0.0628 \text{ m}^2$.

$$\therefore \dot{m} = \rho_1 C_{a1} A = 8.565 \text{ kg/s}$$

3. The power necessary to drive the compressor is then

$$\text{Power} = \dot{m} w_S = 180.81 \text{ kW}$$

4. The temperature rise per stage is $\Delta T_{0S} = w_S/Cp = 21 \text{ K} \equiv$ constant for all stages.

Pressure ratio of stage # 1

$$\pi_1 = \left(1 + \eta_S \frac{\Delta T_{0S}}{T_{01}}\right)^{\gamma/(\gamma-1)} = 1.249$$

Pressure ratio of stage # 2

$$\pi_2 = \left(1 + \eta_S \frac{\Delta T_{0S}}{T_{01} + \Delta T_{0S}}\right)^{\gamma/(\gamma-1)} = 1.231.$$

Pressure ratio of stage # 3

$$\pi_3 = \left(1 + \eta_S \frac{\Delta T_{0S}}{T_{01} + 2\Delta T_{0S}}\right)^{\gamma/(\gamma-1)} = 1.215$$

5. Overall pressure ratio $\pi_{comp} = \pi_1 \times \pi_2 \times \pi_3 = 1.87$.
6. To calculate the blade height at the exit, the air density must be first calculated. The pressure at the outlet of the compressor is $(P_{03})_{S3} = \pi_C \times (P_{01})_{S1} = 1.87$ bar.

The temperature at the exit of the compressor is $(T_{03})_{S3} = (T_{01})_{S1} + 3\Delta T_{0S} = 351$ K. With $C_3 = C_1$

$$(T_3)_{S3} = (T_{03})_{S3} - \frac{C_3^2}{2Cp} = 343.836 \text{ K}$$

$$(P_3)_{S3} = \frac{(P_{03})_{S3}}{\left[(T_{03}/T_3)^{\gamma/(\gamma-1)}\right]_{S3}} = 1.74 \text{ bar}$$

$$\rho_3 = \left(\frac{P_3}{RT_3}\right)_{S3} = 1.763 \text{ kg/m}^3$$

Since $\dot{m} = (\rho_3 C_3 A_3)_{S3} = [\rho_3 C_3 (2\pi r_m h_3)]_{S3}$,

$$(h_3)_{S3} = 3.2 \, \text{cm}$$

7. The degree of reaction is expressed as

$$\Lambda = 1 - \frac{C_{u1} + C_{u2}}{2U}$$

$$= 1 - \frac{C_{u2}}{2U} = 1 - \frac{112}{2 \times 188.5} = 0.7$$

13.4 BASIC DESIGN PARAMETERS

The pressure ratio per stage is expressed by the relation

$$\pi_s = \left[1 + \eta_s \frac{\lambda U C_a}{C p T_{01}} (\tan \beta_1 - \tan \beta_2) \right]^{\gamma/(\gamma-1)}$$

To obtain high pressure ratio per stage it is needed to have

1. High blade speed (U)
2. High axial velocity (C_a)
3. High fluid deflection $(\beta_1 - \beta_2)$ in the rotor blade

However, the increase in rotational speed is limited by the maximum acceptable tensile stress at the rotor hub. Furthermore, the increase in the rotational and axial velocities (U and C_a) is limited by the tip Mach number. Compressors of the 1960s had a maximum tip Mach number less than 0.8. This limit was increased for the present transonic compressors to reach some value of 1.4 [4]. The fluid deflection has upper limits determined by the diffusion or de Haller parameters.

13.4.1 CENTRIFUGAL STRESS

Centrifugal stress depends on the rotational speed, ω, blade material density (ρ_b), and the height of the blade. The maximum centrifugal stress occurs in the root. As shown in Figure 13.16, the centrifugal force arising from any element of blade is evaluated as

$$dF_c = \omega^2 r \delta m$$

where (ω) and (r) are the rotational speed and radius of any blade element having a mass of (δm) and length of (dr).

$$\delta m = \rho_b A \, dr$$

Then, the maximum centrifugal stress at the root is the summation of the centrifugal forces of all the elements of the blade from the hub to the tip divided by the blade cross-sectional area at the

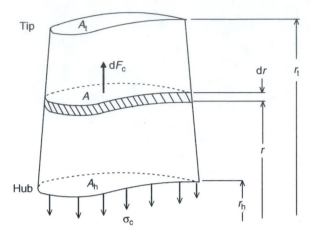

FIGURE 13.16   Nomenclature for centrifugal stress on a rotor blade.

root, or

$$(\sigma_c)_{max} = \frac{\omega^2}{A_{root}} \int_{r_h}^{r_t} r \, \delta m$$

$$(\sigma_c)_{max} = \frac{\rho_b \omega^2}{A_{root}} \int_{r_h}^{r_t} rA \, dr \qquad\qquad (13.13)$$

Special cases:

1. If the blade has a constant area (A) along the blade height, then the maximum stress will be

$$(\sigma_c)_{max} = \frac{\rho_b}{2} (2\pi N)^2 \left(r_t^2 - r_r^2 \right)$$

$$\sigma_{max} = 2\pi N^2 \rho_b A$$

where $A = \pi \left(r_t^2 - r_r^2 \right) =$ annulus area

$$U_t = \omega r_t = 2\pi N r_t$$

$$(\sigma_c)_{max} = \frac{\rho_b U_t^2}{2} \left[1 - \left(\frac{r_r}{r_t} \right)^2 \right]$$

or $$(\sigma_c)_{max} = \frac{\rho_b U_t^2}{2} \left[1 - \zeta^2 \right] \qquad\qquad (13.14)$$

where $\zeta = r_r / r_t$.

2. For linear variation of cross-sectional area with radius

$$(\sigma_{ct})_{max} = \frac{\rho_b U_t^2}{2} \left(1 - \zeta^2 \right) K \qquad\qquad (13.15)$$

$$K = 1 - \frac{(1-d)\left(2-\zeta-\zeta^2\right)}{3\left(1-\zeta^2\right)}$$

where $d = A_{\text{tip}}/A_{\text{root}}$ and $K = 0.55\text{–}0.65$ for tapered blades.

3. If the area is inversely proportional (or $A = \text{constant}/r$), then $r_r A_r = r_t A_t = rA = K$

$$\frac{A_t}{A_r} = \frac{r_r}{r_t} = \zeta$$

$$\therefore \sigma_{c_{\max}} = \frac{\rho_b \omega^2}{A_r} \int_{r_r}^{r_h} rA\,dr = \frac{\rho_b \omega^2 K}{A_r} \int_{r_r}^{r_t} dr$$

$$\sigma_{c_{\max}} = \frac{\left(\rho_b \omega^2\right) K}{A_r} \left[r_t - r_r\right]$$

$$\sigma_{c_{\max}} = \frac{\rho_b \omega^2}{A_r} \left[r_t^2 A_t - r_r^2 A_r\right] = \rho_b U_t^2 \left[\frac{A_t}{A_r} - \zeta^2\right]$$

$$\sigma_{c_{\max}} = \rho_b U_t^2 \left(\zeta - \zeta^2\right) = \rho_b U_t^2 \zeta \left(1 - \zeta\right) \tag{13.16}$$

The same equations are applied for both fans and compressors. If the tip speed of a fan in a high-bypass ratio turbofan is 450 m/s and the blades are long (low hub-to-tip ratios), then the design of the disc becomes critical.

13.4.2 Tip Mach Number

The maximum Mach number is found at the rotor tip of the first stage as it has the maximum radius for any layout of axial compressor constant-tip, constant-mean, or constant-hub radius. Thus, the rotational speed (U) attains a maximum value. For the first stage, the axial velocity is equal to the inlet absolute velocity and has a constant value along the annulus. The corresponding maximum Mach number is

$$M_{\text{tip}} = M_{1\max} = \frac{W_{1\max}}{a} = \frac{\sqrt{C_1^2 + U_{\text{tip}}^2}}{\sqrt{\gamma R T_1}} = \sqrt{\frac{C_1^2 + U_{\text{tip}}^2}{\gamma R \left(T_{01} - \left(C_1^2/2Cp\right)\right)}}$$

$$M_{1\max} = U_{\text{tip}} \sqrt{\frac{\left(C_1/U_{\text{tip}}\right)^2 + 1}{\gamma R \left(T_{01} - \left(C_1^2/2Cp\right)\right)}} \tag{13.17}$$

The relation of maximum relative Mach number (at rotor tip) is plotted versus the tip rotational speed (U_t) for different axial velocities (C_1) (Figure 13.17). Thus for a specified tip Mach number and rotational speed, the axial velocity at inlet is determined.

For a transonic compressor, an acceptable value for the maximum Mach number at the tip is about 1.1 $\left(M_{1\text{tip}} \cong 1.1\right)$, while for the fan in turbofan engines much higher Mach number is expected and allowed, thus $M_{1\text{tip}} = 1.5 - 1.7$. A challenging design is to increase the rotational speed U and keep the Mach number within the acceptable limits. Since, increasing the rotational speed automatically increases the maximum relative speed ($W_{1\max}$), IGVs (an additional ring of stator vanes) are installed upstream of the first stage of the compressor as shown in Figure 13.18 to add a swirl component to the inlet speed (C_1) and thus reduce the relative speed (W_1). Note that the passages of the IGVs are nozzles to increase the absolute speed ($C_1 > C_0$), thus keeping a constant axial velocity C_a.

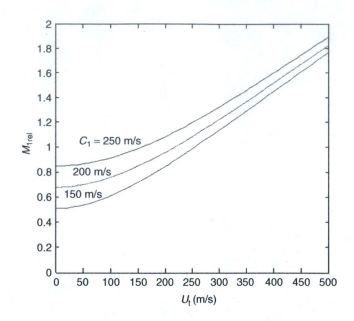

FIGURE 13.17   Relation between tip Mach number and tip rotational speed.

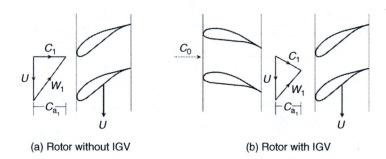

(a) Rotor without IGV   (b) Rotor with IGV

FIGURE 13.18   Inlet velocity triangle with and without IGV.

The maximum relative Mach number at rotor tip is

$$(M_{1\text{rel}})_{\text{max}} = \frac{W_{1\,\text{max}}}{\sqrt{\gamma R T_1}} = \sqrt{\frac{(U - C_1 \sin \alpha_1)^2 + (C_1 \cos \alpha_1)^2}{\gamma R \left(T_{01} - \left(C_1^2/2Cp\right)\right)}}$$

where α_1 is the angle between the absolute velocity and axial direction. Thus, the maximum relative Mach number can be rewritten as

$$(M_{1\text{rel}})_{\text{max}} = \sqrt{\frac{U^2 + C_1^2 - 2UC_1 \sin \alpha_1}{\gamma R \left(T_{01} - \left(C_1^2/2Cp\right)\right)}} \tag{13.18}$$

13.4.3 FLUID DEFLECTION

Fluid deflection is the difference between the outlet and inlet angles $(\beta_2 - \beta_1)$ for the rotor and $(\alpha_3 - \alpha_2)$ for a stator. Excessive deflection, which means a high rate of diffusion, will lead to the

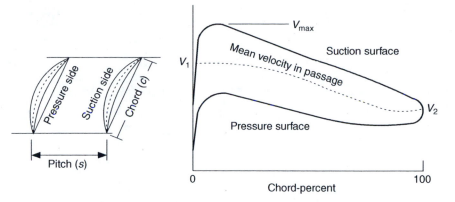

FIGURE 13.19   Velocity distribution in the passage of a cascade.

blade stall. To maximize the specific work, the difference between the flow angles in the rotor must attain its maximum allowable value. If the inlet relative velocity maintains a constant angle, β_1, then the outlet relative angle, β_2, will control either a good design or a stall condition. Since a constant axial velocity is assumed, the relative angle is a measure of the magnitude of the relative velocities (W_1, W_2). An earlier method for assessing the diffusion was the *de Haller number*, defined as V_2/V_1, for which a limit of ($V_2/V_1 \not< 0.72$) was set. Because of its simplicity, it is used until now in preliminary design. However, for final design calculations, the criterion known as the diffusion factor (DF) developed by NACA is preferred [5]. The velocity distribution over the suction and pressure sides of a typical passage of cascade is shown in Figure 13.19. The velocities V_1, V_2, and V_{max} are, respectively, the velocities at the inlet and outlet of the cascade as well as the maximum velocity over the suction surface. DF is defined as

$$\text{DF} \approx \frac{V_{max} - V_2}{V_1} \approx 1 - \frac{V_2}{V_1} + \frac{V_{u1} - V_{u2}}{2(c/s)V_1} = 1 - \frac{V_2}{V_1} + \frac{\Delta C_u}{2\sigma V_1} \qquad (13.19a)$$

where $\sigma = c/s$ is the solidity.

A more precise expression when the mean radii at the inlet and outlet of the blade are not equal is expressed as [6]

$$\text{DF} = 1 - \frac{V_2}{V_1} + \frac{\Delta (rV_u)}{2\sigma rV_1} \qquad (13.19b)$$

Values of DF in excess of 0.6 are thought to indicate blade stall and a value of 0.45 might be taken as a typical design choice [5].

13.5   DESIGN PARAMETERS

The following three design parameters are frequently used in the parametric study of a repeating axial stage:

1. Flow coefficient ϕ
2. Stage loading ψ
3. Degree of reaction Λ

The flow coefficient is defined as the ratio between the axial and the rotational speeds, or,

$$\phi = \frac{C_a}{U} \tag{13.20}$$

The stage loading is defined as the ratio between the total enthalpy rise per stage to the square of the rotational speed, or

$$\psi = \frac{\Delta h_0}{U^2} = \frac{W_s}{U^2} = \frac{\Delta C_u}{U} \tag{13.21}$$

13.5.1 Degree of Reaction

The degree of reaction of a compressor stage has several definitions. For an incompressible flow, the degree of reaction is defined as the ratio of the static pressure rise in the rotor to the static pressure rise in the stage [7], or

$$\Lambda = \frac{p_2 - p_1}{p_3 - p_1}$$

For a compressible flow, the degree of reaction is defined as the ratio of the static enthalpy (or temperature) rise in the rotor to the enthalpy (or temperature) rise in the whole stage, or

$$\Lambda = \frac{h_2 - h_1}{h_3 - h_1} = \frac{\Delta T_R}{\Delta T_R + \Delta T_S} \tag{13.22}$$

A third definition [8] states that the degree of reaction is equal to the change of the static enthalpy rise across the rotor as a fraction of the change in the absolute enthalpy rise in a stage,

$$\Lambda = \frac{h_2 - h_1}{h_{03} - h_{01}} = \frac{h_2 - h_1}{h_{02} - h_{01}}$$

The expression given in Equation 13.22 will be employed in this text.

The following general relation will be derived for the degree of reaction assuming that the axial velocities are not equal ($C_{a2} \neq C_{a1}$). Since the absolute velocities at the inlet and outlet of the stage are assumed equal, that is, $C_1 = C_3$, the static and total temperature rise in a stage are also equal, or $\Delta T = \Delta T_0$. The temperature rise in the stage is

$$\Delta T \equiv \Delta T_0 = \frac{U(C_{u2} - C_{u1})}{C_p}$$

Since the rothalpy (I) is equal at rotor inlet and outlet,

$$I_2 = I_1,$$

where $I = h_{0rel} - (1/2)U^2$.

Since equal mean radii at rotor inlet

$$h_{0rel} = h + \frac{1}{2}w^2$$

and inlet is assumed, that is, $U_2 = U_1$, and $h_{01rel} = h_{02rel}$, or

$$h_1 + \frac{1}{2}W_1^2 = h_2 + W_2^2.$$

The temperature rise within the rotor is

$$\Delta T_R = T_2 - T_1 = \frac{W_1^2 - W_2^2}{2Cp}$$

with $W^2 = C_a^2 + (U - C_u)^2$

$$\therefore \Delta T_R = \frac{C_{a1}^2 + (U - C_{u1})^2 - C_{a2}^2 + (U - C_{u2})^2}{2Cp}$$

$$\Delta T_R = \frac{C_{a1}^2 - C_{a2}^2 + 2U(C_{u2} - C_{u1}) - C_{u2}^2 + C_{u1}^2}{2Cp}$$

The degree of reaction is then

$$\Lambda = 1 + \frac{C_{a2}^2 - C_{a1}^2}{2U(C_{u2} - C_{u1})} - \frac{C_{u1} + C_{u2}}{2U} \tag{13.23a}$$

When the axial velocities at rotor inlet and outlet are assumed to be equal, $C_{a2} = C_{a1}$, then Equation 13.23a is reduced to the following relation:

$$\Lambda = 1 - \frac{C_{u1} + C_{u2}}{2U} \tag{13.23b}$$

In terms of the flow angles, Equation 13.23b may be rewritten as

$$\Lambda = 1 - \frac{C_a}{2U}(\tan \alpha_1 + \tan \alpha_2) \tag{13.23c}$$

From Equation 13.1c with some manipulation, we arrive at the following expressions for the degree of reaction:

$$\Lambda = \frac{1}{2} - \frac{C_a}{2U}(\tan \alpha_1 - \tan \beta_2) \tag{13.23d}$$

$$\Lambda = \frac{C_a}{2U}(\tan \beta_1 + \tan \beta_2) \tag{13.23e}$$

From Equation 13.23e, another expression for the degree of reaction is obtained.

$$\Lambda = \frac{1}{U}\left(\frac{C_a \tan \beta_1 + C_a \tan \beta_2}{2}\right)$$

$$\Lambda = \frac{1}{U}\left(\frac{W_{u1} + W_{u2}}{2}\right)$$

$$\Lambda = \frac{W_{u\,mean}}{U} = \frac{C_a \tan \beta_m}{U}. \tag{13.23f}$$

Three special cases are found here:

1. $\Lambda = 0\%$
2. $\Lambda = 50\%$
3. $\Lambda = 100\%$

The 50% reaction stage provides the best efficiency. If the degree of reaction is much different from 50%, then one of the blade rows must tolerate greater pressure rise than the other. This means that its boundary layer will be more highly loaded and more likely to stall or at least create higher losses. Now, the velocity triangles of this case are examined. From Equation 13.23d,

$$\therefore \tan \alpha_1 = \tan \beta_2$$

$$\therefore \alpha_1 = \beta_2$$

From Equation 13.1,

$$\frac{U}{C_a} = \tan \alpha_1 + \tan \beta_1 = \tan \alpha_2 + \tan \beta_2$$

$$\therefore \alpha_2 = \beta_1$$

This means that these velocity triangles are symmetrical. It is interesting to note that if the degree of reaction is greater than 50%, then $\beta_2 > \alpha_1$ and the velocity triangles will be skewed in the direction of the rotational speed [9]. The static enthalpy rise across the rotor exceeds that of the stator (this is also true for the static pressure). If the degree of reaction is less than 50%, then $\beta_2 < \alpha_1$ and the velocity triangles will be skewed in the opposite direction of the rotational speed. The stator enthalpy rise (and pressure) exceeds that in the rotor.

The case of zero degree of reaction is known as impulse blading. From Equation 13.23e, the magnitude of the relative angles at inlet and outlet of the rotor are equal but have opposite signs, $\beta_2 = -\beta_1$. There is no contribution for the rotor row here. All the enthalpy (or pressure) rises in the stator blade row. The rotor just redirects the incoming flow, such that the outlet angle is equal in magnitude to the inlet angle with opposite sign.

The case of 100% degree of reaction necessitates that $\alpha_2 = -\alpha_1$. All the enthalpy (and static pressure) rises in the rotor while the stator has no contribution.

The three cases described above are illustrated in Figure 13.20.

13.6 THREE-DIMENSIONAL FLOW

In present day turbomachinery, 3D effects are hardly negligible and their incorporation into the design or analysis is essential for the accurate prediction of the performance or improved design of either compressors or turbines [10]. Some of the 3D effects are due to

1. Compressibility, and radial density and pressure gradients
2. Radial variation in blade thickness and geometry
3. Presence of finite hub and annulus walls, annulus area changes, flaring, curvature, and rotation
4. Radially varying work input or output
5. Presence of two phase flow (water injection, rain, sand, ice) and coolant injection
6. Radial component of blade force and effects of blade skew, sweep, lean, and twist
7. Leakage flow due to tip clearance and axial gaps
8. Nonuniform inlet flow and presence of upstream and downstream blade rows

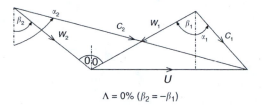

$$\Lambda = 0\% \ (\beta_2 = -\beta_1)$$

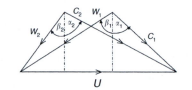

$$\Lambda = 50\% \ (\alpha_1 = \beta_2, \alpha_2 = \beta_1)$$

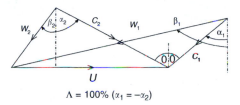

$$\Lambda = 100\% \ (\alpha_1 = -\alpha_2)$$

FIGURE 13.20 Velocity triangles for different degrees of reaction.

9. Mixed-flow (subsonic, supersonic, and transonic flow) regions along the blade with shock-boundary-layer interaction
10. Secondary flow caused by inlet velocity/stagnation pressure gradient and flow turning
11. Air bleed in some intermediate stages of a multistage compressor for surge control
12. Injection of bulky foreign objects such as stones, birds, and humans

Most of these are caused by inviscid effects, which can be treated by the use of inviscid equations of motion. The secondary flow is caused primarily by the presence of viscous layers on the wall.

In the previous sections, two-dimensional flow was assumed in the sense that radial (i.e., spanwise) velocities did not exist. It was assumed that the radial variations in the flow parameters through the compressors annulus are negligible. For a compressor in which the blade height is small relative to the mean diameter, that is, those with high hub/tip ration (of the order of 0.8 or more), this assumption is reasonable. But for blades with low hub-to-tip ratio (e.g., 0.4, as in the case of the fan of a turbofan engine), radial variations are important.

13.6.1 Axisymmetric Flow

The equations of motion governing the turbulent flow through turbomachinery are highly nonlinear; hence, most of the analytical solutions available are for simple flows. These solutions involve several assumptions depending on the type of machinery and the geometry of the blade-row and flow parameters. Numerical solutions for the governing equations started in the late 1960s [11]. The 3D flow in turbomachines is solved by techniques classified as either axisymmetric solutions or nonaxisymmetric solutions. The axisymmetric solutions [10] include

1. Simplified radial equilibrium (SRE) analysis
2. Actuator disc theories (ADT)
3. Passage averaged equations (PAE) and their solutions

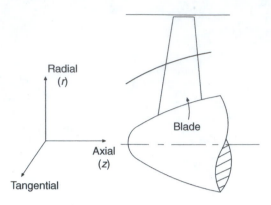

FIGURE 13.21   Axisymmetric flow notations.

The nonaxisymmetric solutions can generally be classified as [10]

1. Lifting line and lifting surface approach
2. Quasi-3D methods
3. Numerical solutions of exact equations (potential, Euler, and Navier Stokes)

The axisymmetric solution is used to predict radial or spanwise variation of properties of flow far upstream and downstream of blade rows. Once the local values of flow parameters are known, the cascade theories can be used to predict the blade-to-blade variation of flow properties.

The equations governing the axisymmetric inviscid and steady flow motion of a fluid particle subjected to a body force having radial, tangential, and axial components (F_r, F_u, F_a) in the cylindrical coordinate system (Figure 13.21) are expressed as

$$\text{Continuity equation}\quad \frac{\partial}{\partial r}(\rho r C_r) + \frac{\partial}{\partial z}(\rho r C_z) = 0 \tag{13.24}$$

$$\text{Radial momentum}\quad \frac{\partial C_r}{\partial t} + C_r\frac{\partial C_r}{\partial r} + C_z\frac{\partial C_r}{\partial z} - \frac{C_u^2}{r} = -\frac{1}{\rho}\frac{\partial p}{\partial r} + F_r \tag{13.25}$$

$$\text{Tangential momentum}\quad \frac{\partial C_u}{\partial t} + C_r\frac{\partial C_u}{\partial r} + C_z\frac{\partial C_u}{\partial z} + \frac{C_u C_r}{r} = F_u \tag{13.26}$$

$$\text{Axial momentum}\quad \frac{\partial C_z}{\partial t} + C_r\frac{\partial C_z}{\partial r} + C_z\frac{\partial C_z}{\partial z} = -\frac{1}{\rho}\frac{\partial p}{\partial z} + F_z \tag{13.27}$$

13.6.2 SIMPLIFIED RADIAL EQUILIBRIUM EQUATION

It is assumed that all the radial motion of the particle takes place within the blade row passages, while in the axial spacing between the successive blade rows (rotor/stator combination) radial equilibrium is assumed. Thus, the axisymmetrical stream surface in the spacing between the blade rows now has a cylindrical stream surface shape (Figure 13.22). Thus, the radial component of velocity is neglected in the spaces between blade rows as

$$C_r \ll C_a, C_r \ll C_a, \quad \text{thus} \quad C_r \approx 0$$

For steady flow, when Equations 13.24–13.27 are solved over these cylindrical surfaces, they are simplified considerably; the resulting radial momentum equation is called the (SRE) equation.

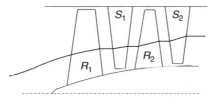

FIGURE 13.22 Axisymmetric and cylindrical surfaces in a one-stage compressor.

These equations are strictly valid away from the blade row. Only the radial equilibrium equation is relevant because the flow does not vary in either the tangential or the axial direction. For a steady inviscid flow in the cylindrical surface without any body force, the variables, C_u, C_a, h, and h_0, are functions of the radius (r). Therefore, the radial component of the angular momentum equation is given by

$$\frac{1}{\rho}\frac{dp}{dr} = \frac{C_u^2}{r} \tag{13.28}$$

Equation 13.28 is the radial equilibrium equation.

Now, the stagnation enthalpy is written as

$$h_0 = h + \frac{C^2}{2} = h + \frac{C_a^2}{2} + \frac{C_u^2}{2} \qquad (C_r \approx 0)$$

$$\frac{dh_0}{dr} = \frac{dh}{dr} + C_a\frac{dC_a}{dr} + C_u\frac{dC_u}{dr}$$

The thermodynamic relation $T ds = dh - (dp/\rho)$

$$dh = T\,ds + \frac{dp}{\rho}$$

$$\frac{dh}{dr} = T\frac{ds}{dr} + \frac{dT}{dr}ds + \frac{1}{\rho}\frac{dp}{dr}$$

Neglecting second-order terms and substituting from Equation 13.28,

$$\frac{dh}{dr} = T\frac{ds}{dr} + \frac{1}{\rho}\frac{dp}{dr} = T\frac{ds}{dr} + \frac{C_u^2}{r}$$

$$\frac{dh_0}{dr} = T\frac{ds}{dr} + \frac{C_u^2}{r} + C_a\frac{dC_a}{dr} + C_u\frac{dC_u}{dr}$$

The term $T\,(ds/dr)$ represents radial variations across the annulus, and may be significant only in detailed design and when the flow is supersonic and shock losses occur. Neglect entropy gradient.

$$\frac{dh_0}{dr} = C_a\frac{dC_a}{dr} + C_u\frac{dC_u}{dr} + \frac{C_u^2}{r} \tag{13.29}$$

This equation is denoted as the vortex energy equation, which is also applicable (as the radial equilibrium equation) only in the axial space between a pair of blade rows. Equation 13.29 is an ordinary differential equation and can be easily solved when all the variables, with the exception of one, are known. For example, for isentropic flow, if C_u and h_0 are known or prescribed, then Equation 13.29 can be solved for C_a.

13.6.3 FREE VORTEX METHOD

Free vortex method is one of the simplest design methods in axial compressors. It is based on the general equation 13.29 with the following simplifications:

1. Assuming constant specific work at all radii, $(dh_0/dr) = 0$

$$\therefore C_a \frac{dC_a}{dr} + C_u \frac{dC_u}{dr} + \frac{C_u^2}{r} = 0$$

2. Assuming constant axial velocity at all radii, $(dC_a/dr) = 0$

$$\therefore C_u \frac{dC_u}{dr} + \frac{C_u^2}{r} = 0$$

$$\frac{dC_u}{C_u} = -\frac{dr}{r}$$

Integrating, we get

$$rC_u = \text{constant} \tag{13.30}$$

Thus, the whirl velocity component of the flow varies inversely with radius, which is known as free vortex.

It is clear that if the following three variables (h_0, C_a, rC_u) are not varying in the radial direction, then the radial equilibrium equation $(1/\rho)(dp/dr) = c_u^2/r$ is satisfied. The resulting blading is denoted free vortex blading. Although it represents a simple mathematical treatment for blade twist, it has two main disadvantages, namely, the large variation of the degree of reaction and large blade twist along blade span (or height).

The degree of reaction from Equation 13.23c is expressed as

$$\Lambda = 1 - \frac{C_a}{2U}(\tan\alpha_1 + \tan\alpha_2) = 1 - \frac{C_{u1} + C_{u2}}{2U}$$

$$\Lambda = 1 - \frac{rC_{u1} + rC_{u2}}{2rU}$$

but

$$\frac{U}{U_m} = \frac{r}{r_m}$$

$$\Lambda = 1 - \frac{\text{constant}}{2U_m r/r_m}$$

$$\Lambda = 1 - \frac{\text{constant}}{r^2} \tag{13.31}$$

A great variation in the degree of reaction from the blade hub to tip is depicted in Equation 13.35. Moreover, some suspect that at the blade root where its radius is a minimum, a negative reaction is possible. One note remaining here is that the free vortex relation Equation 13.30 is applied twice at stations (1) and (2). The following example illustrates the above points.

Example 2   Consider a free vortex design for a compressor stage having a degree of reaction at the mean section equal to 0.5, hub-to-tip ratio of 0.6, and the flow angles $\alpha_{1m} = 30°, \beta_{1m} = 60°$. Calculate the absolute and relative angles at the hub (α_h, β_h) and tip (α_t, β_t) for both stations (1) (upstream of the rotor) and (2) (downstream of the rotor). Also, calculate the degree of reaction at both hub and tip.

Solution: Since the degree of reaction is 50% at the mean section, then $\alpha_{1m} = \beta_{2m}$ and $\alpha_{2m} = \beta_{1m}$.
Since the hub-to-tip ratio is equal to 0.6, then at both stations (1) and (2),

$$r_t = \frac{5}{4}r_m \qquad\qquad r_h = \frac{3}{4}r_m$$

$$U_t = U_m \frac{r_t}{r_m} \qquad\qquad U_t = \frac{5}{4}U_m$$

$$U_h = U_m \frac{r_h}{r_m} \qquad\qquad U_h = \frac{3}{4}U_m.$$

For a free vortex design, $rC_u = $ Constant at stations 1 and 2

$$r_h C_{uh} = r_m C_{um}, \qquad C_{uh} = \frac{r_m}{r_h} C_{um}$$

$$r_t C_{ut} = r_m C_{um}, \qquad C_{ut} = \frac{r_m}{r_t} C_{um}$$

The swirl velocity components are

$$C_{uh} = \frac{4}{3}C_{um} \qquad C_{ut} = \frac{4}{5}C_{um}$$

Then, for both stations 1 and 2

$$C_{u1h} = \frac{4}{3}C_{u1m} \qquad C_{u2h} = \frac{4}{3}C_{u2m} \qquad U_h = \frac{3}{4}U_m$$

$$C_{u1t} = \frac{4}{5}C_{u1m} \qquad C_{u2t} = \frac{4}{5}C_{u2m} \qquad U_t = \frac{5}{4}U_m$$

$$\tan \alpha_h = \frac{C_{uh}}{C_a} = \frac{4}{3}\frac{C_{um}}{C_a} = \frac{4}{3}\tan \alpha_m$$

$$\tan \alpha_t = \frac{C_{ut}}{C_a} = \frac{4}{5}\frac{C_{um}}{C_a} = \frac{4}{5}\tan \alpha_m$$

$$\tan \beta_h = \frac{U_h - C_{uh}}{C_a} = \frac{3/4 U_m - 4/3 C_{um}}{C_a}$$

$$\text{but} \quad \frac{U_m}{C_a} = \tan \alpha_m + \tan \beta_m \quad \text{and} \quad \frac{C_{um}}{C_a} = \tan \alpha_m$$

$$\therefore \tan \beta_h = \frac{3}{4}\tan \beta_m - \frac{7}{12}\tan \alpha_m$$

TABLE 13.2

Characteristics of a Single-Stage Axial Compressor Based on Free Vortex Design Method (All Angles in Degrees)

	α_1	β_1	α_2	β_2	Λ	$\beta_1 - \beta_2$
Hub	37.6	43.9	66.6	−30	0.11	73.9
Mean	30	60	60	30	0.5	30
Tip	24.8	67.5	54.2	56.3	0.68	11.3

Similarly

$$\text{but } \frac{U_m}{C_a} = \tan \alpha_m + \tan \beta_m \text{ and } \frac{C_{um}}{C_a} = \tan \alpha_m$$

$$\tan \beta_t = \frac{U_t - C_{ut}}{C_a} = \frac{5/4 U_m - 4/5 C_{um}}{C_a}$$

$$\therefore \tan \beta_t = \frac{9}{20} \tan \alpha_m + \frac{5}{4} \tan \beta_m$$

All the calculated angles (α_1, α_2, β_1, β_2) at the hub, mean, and tip sections are listed in Table 13.2. From Equations 13.1a and 13.23e, the degree of reaction at any radius may be expressed as

$$\Lambda = \frac{\tan \beta_1 + \tan \beta_2}{2 (\tan \alpha_1 + \tan \beta_1)}$$

The difference in the angles ($\beta_1 - \beta_2$) at different stations along the blade span represents the blade twist. The values of the degree of reaction and blade twist are listed also in Table 13.2. Thus, adopting free vortex design yields a highly twisted blade and a great variation in the degree of reaction from hub to tip. The degree of reaction is only 0.11 at the hub. As explained in Reference 5, because of the lower blade speed at the root section, more fluid deflection is required for a given work input, that is, a great rate of diffusion is required at the root section. It is, therefore, undesirable to have a low degree of reaction in this region, and the problem is aggravated as the hub-to-tip ratio is reduced.

The velocity triangles and the blade twist are shown in Figure 13.23.

Example 3 For an axial compressor stage, using the vortex energy equation

$$\frac{1}{r^2} \frac{d}{dr} \left(r^2 C_u^2 \right) + \frac{d}{dr} \left(C_a^2 \right) = 0$$

Prove that, if the flow is represented by a free vortex, $rC_u = $ constant, then

1. $C_a = $ constant
2. $\tan \alpha_{1h} = \left(\dfrac{1+\zeta}{2\zeta} \right) \tan \alpha_{1m}$
3. $\tan \beta_{1h} = \left(\dfrac{2\zeta}{1+\zeta} \right) \tan \beta_{1h} + \dfrac{(3\zeta + 1)(\zeta - 1)}{2\zeta (1 + \zeta)} \tan \alpha_{1m}$

where ζ is the hub-to-tip ratio, α_1 is the angle between the inlet absolute velocity and the axial direction, β_1 is the angle between the inlet relative velocity and the axial direction, and subscripts (h) and (m) refer to the hub and mean sections, respectively.

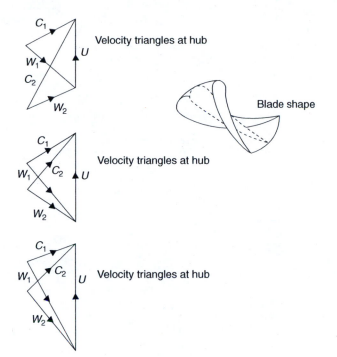

FIGURE 13.23 Free vortex velocity triangle and blade shape.

Solution:

1. For rC_u = constant

$$\therefore \frac{d}{dr}(rC_u) = 0$$

Then, from the vortex energy equation

$$\frac{1}{r^2}\frac{d}{dr}\left(r^2 C_u^2\right) + \frac{d}{dr}\left(C_a^2\right) = \frac{d}{dr}\left(C_a^2\right) = 0.$$

Thus C_a^2 = constant

$$\therefore C_a = \text{constant}$$

A typical velocity triangle is illustrated in Figure 13.24.

2. Since $\tan\alpha_{1h} = C_{u1h}/C_a$, $\tan\alpha_{1m} = C_{u1m}/C_a$, and C_a = constant,

$$\frac{\tan\alpha_{1h}}{\tan\alpha_{1m}} = \frac{C_{u1h}}{C_{u1m}}$$

$$\tan\alpha_{1h} = \frac{C_{u1h}}{C_{u1m}}\tan\alpha_{1m} \qquad (1)$$

but rC_u = constant, or $r_h C_{uh} = r_m C_{um}$

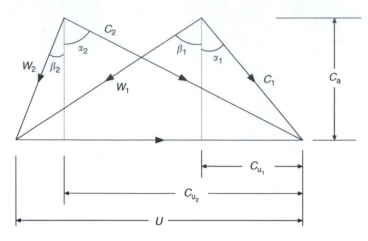

FIGURE 13.24   Velocity triangle.

Then at state (1)

$$r_h C_{u1h} = r_m C_{u1m}$$

$$\frac{C_{u1h}}{C_{u1m}} = \frac{r_m}{r_h}$$

Then from Equation (1)

$$\therefore \tan \alpha_{1h} = \frac{r_m}{r_h} \tan \alpha_{1m}$$

$$\frac{r_m}{r_h} = \frac{r_h + r_t}{2 r_h} = \frac{(r_h/r_t) + 1}{2 (r_h/r_t)} = \frac{1 + \zeta}{2\zeta}$$

$$\tan \alpha_{1h} = \frac{1 + \zeta}{2\zeta} \tan \alpha_{1m}$$

3. From the kinematical relations

$$\tan \alpha_{1h} + \tan \beta_{1h} = \frac{U_h}{C_a}$$

$$\tan \alpha_{1m} + \tan \beta_{1m} = \frac{U_m}{C_a}$$

$$\therefore \frac{\tan \alpha_{1h} + \tan \beta_{1h}}{\tan \alpha_{1m} + \tan \beta_{1m}} = \frac{U_h}{U_m}$$

$$\tan \alpha_{1h} + \tan \beta_{1h} = \frac{U_h}{U_m} (\tan \alpha_{1m} + \tan \beta_{1m})$$

$$\tan \beta_{1h} = \frac{U_h}{U_m} (\tan \alpha_{1m} + \tan \beta_{1m}) - \tan \alpha_{1h}$$

Since

$$\frac{U_h}{U_m} = \frac{r_h}{r_m} = \frac{2\zeta}{\zeta + 1} \quad \text{and} \quad \tan \alpha_{1h} = \frac{1 + \zeta}{2\zeta} \tan \alpha_{1m}$$

$$\therefore \tan \beta_{1h} = \frac{2\zeta}{\zeta + 1} (\tan \alpha_{1m} + \tan \beta_{1m}) - \frac{1+\zeta}{2\zeta} \tan \alpha_{1m}$$

$$\tan \beta_{1h} = \frac{2\zeta}{1+\zeta} \tan \beta_{1m} + \left[\frac{2\zeta}{\zeta + 1} - \frac{1+\zeta}{2\zeta} \right] \tan \alpha_{1m}$$

$$\tan \beta_{1h} = \frac{2\zeta}{\zeta + 1} \tan \beta_{1m} + \frac{4\zeta^2 - (1+\zeta)^2}{2\zeta(1+\zeta)} \tan \alpha_{1m}$$

$$\tan \beta_{1h} = \frac{2\zeta}{\zeta + 1} \tan \beta_{1m} + \frac{(3\zeta+1)(\zeta-1)}{2\zeta(1+\zeta)} \tan \alpha_{1m}$$

13.6.4 GENERAL DESIGN PROCEDURE

The disadvantages of the free vortex method mentioned above motivate searching for other design methods. Thus, if the design is based on arbitrarily chosen radial distribution of any two variables of the three (h_0, C_a, C_u), then the variation of the other is determined from Equation 13.29. However, it is always preferred to employ a constant specific work at all radii along the blade together with an arbitrary whirl velocity distribute for C_{u1}, C_{u2}, that is, compatible with the constant specific work condition. Next, the variation of the axial velocity C_a and the degree of reaction Λ with the radius r are calculated. Equation 13.29 is to be solved twice at both stations (1) and (2), to obtain the radial variation of the axial velocities C_{a1}, C_{a2}.

The designer usually prescribes the whirl velocity somewhere between a *forced-vortex* and a *free-vortex* design according to the relation:

$$C_{u1} = aR^n - \frac{b}{R} \qquad C_{u2} = aR^n + \frac{b}{R} \qquad (13.32)$$

where $R = (r/r_m) = (U/U_m)$.

The value a is determined from the degree of reaction

$$\Lambda = 1 - \frac{C_{u1} + C_{u2}}{2U}$$

At the mean section, the degree of reaction will be

$$\Lambda_m = 1 - \frac{C_{u1m} + C_{u2m}}{2U_m}.$$

Substitute the values of the whirl velocities at the mean section from Equation 13.32, then

$$\Lambda_m = 1 - \frac{(aR_m^n + (b/R_m)) + (aR_m^n - (b/R_m))}{2U_m}, \quad R_m = 1$$

$$\therefore \Lambda_m = 1 - \frac{a}{U_m}$$

$$a = U_m (1 - \Lambda_m) \qquad (13.33)$$

The constant b is calculated from the condition of constant specific work

$$\Delta h_0 = \text{constant} \equiv h_{02} - h_{01}$$

$$W_s \equiv \Delta h_0 = U\left(C_{u2} - C_{u1}\right)$$

$$\Delta h_0 = U_m R\left(aR^n + \frac{b}{R} - aR^n + \frac{b}{R}\right)$$

$$\Delta h_0 = 2bU_m$$

$$b = \frac{\Delta h_0}{2U_m} \tag{13.34}$$

The values of (a) and (b) are constants and independent from the radius (r) as verified by Equations 13.33 and 13.34.

Consider the following three special cases for n, namely, ($n = -1, 1,$ and 0)

$$
\begin{array}{llll}
n = -1 & \text{Free vortex} & rC_u = A \\
n = 0 & \text{Exponential} & rC_u = Ar + B \\
n = 1 & \text{Free power} & rC_u = Ar^2 + B
\end{array}
$$

Rewriting Equation 13.29,

$$\frac{dh_0}{dr} = C_a \frac{dC_a}{dr} + C_u \frac{dC_u}{dr} + \frac{C_u^2}{r}$$

The total enthalpy upstream of the stage (h_{01}) is assumed constant and independent from the radius (r). Since the specific work, $W_s \equiv h_{02} - h_{01}$, is assumed constant, the downstream enthalpy (h_{02}) is also constant. Now Equation 13.29 is to be solved both at states (1) and (2). In both cases, the left-hand side of Equation 13.29 is zero. The equation to be solved is now reduced to

$$C_a \frac{dC_a}{dr} + C_u \frac{dC_u}{dr} + \frac{C_u^2}{r} = 0$$

This equation is to be integrated at station (1), for example, from the mean radius to any station (r). At the mean radius

$$R_m = 1, \quad C_{u1m} = a - b, \quad \text{and} \quad C_{u2m} = a + b$$

Thus, the following equation is obtained:

$$\frac{1}{2}\left(C_{a1}^2 - C_{am}^2\right) = -\frac{1}{2}\left(C_{u1}^2 - C_{u1m}^2\right) - \int_1^R \left(aR^n - \frac{b}{R}\right)^2 \frac{dR}{R}$$

$$= -\frac{1}{2}\left[\left(aR^n - \frac{b}{R}\right)^2 - (a - b)^2\right] - \int_1^R \left(a^2 R^{2n-1} - 2abR^{n-2} + \frac{b^2}{R^3}\right) dR$$

$$\frac{1}{2}\left(C_{a1}^2 - C_{a1m}^2\right) = \frac{1}{2}\left[\left(a^2 R^{2n} - 2abR^{n-1} + \frac{b^2}{R^2}\right) - \left(a^2 - 2ab + b^2\right)\right] + \text{Integral}$$

$$\text{Integral} \equiv \int_1^R \left(a^2 R^{2n-1} - 2abR^{n-2} + \frac{b^2}{R^3}\right) dR$$

$$\text{Integral} = \left[\left(\frac{a^2 R^{2n}}{2n} - \frac{2abR^{n-1}}{n-1} - \frac{b^2}{2R^2}\right) - \left(\frac{a^2}{2n} - \frac{2ab}{n-1} - \frac{b^2}{2}\right)\right]$$

This integral represents a general solution for any value of (n). However, the general solution cannot be applied for the cases of n equals (0) or (1). A similar relation is also obtained for state (2) at the outlet of the rotor stage. Thus, the value of n must be substituted first and the integral is next evaluated. A similar procedure is followed for state (2) downstream of the rotor.

The resulting values for the axial velocities at any radius in stations (1) and (2) are calculated. These results are assembled in Table 13.3 together with the values of the whirl speeds and the degree of reaction.

As stated in Reference 10, for exponential method, the loading as well as the blade twists are moderate. The new blade design tend is toward low aspect ratio blading. The velocity triangles for any type of blading except $n = -1$ is plotted in Figure 13.25. The variation of axial velocities at states (1) and (2), C_{a1} and C_{a2}, are plotted along the blade height and illustrated in Figure 13.26.

Example 4 If the whirl velocities in an axial stage are expressed as

$$C_{u1} = aR^n - \frac{b}{R}, \qquad C_{u2} = aR^n + \frac{b}{R}$$

TABLE 13.3
Governing Equations for Different Design Methods

$$C_{u1} = aR^n - \frac{b}{R}$$
$$a = U_m(1 - \Lambda_m)$$

$$a = U_m(1 - \Lambda_m)$$
$$b = C_p \Delta T_{0s}/2U_m$$

Blading	C_a	Λ	C_u
Free vortex $n = -1$	$C_{a1} = C_{a1m}$ $C_{a2} = C_{a2m}$ $C_{a1m} = C_{a2m}$	$\Lambda = 1 - \dfrac{a}{U_m R^2}$ $\Lambda = 1 - \dfrac{(1-\Lambda_m)}{R^2}$	$C_{u1} = \dfrac{a-b}{R}$ $C_{u2} = \dfrac{a+b}{R}$
Exponential $n = 0$	$C_{a1}^2 - C_{a1m}^2 = -2\left[a^2 \ell nR + \dfrac{ab}{R} - ab\right]$ $C_{a2}^2 - C_{a2m}^2 = -2\left[a^2 \ell nR - \dfrac{ab}{R} + ab\right]$ $C_{a1m} = C_{a2m}$	$\Lambda = 1 + \dfrac{a}{U_m} - \dfrac{2a}{U_m R}$ $\Lambda = 1 + \left(1 - \dfrac{2}{R}\right)(1 - \Lambda_m)$	$C_{u1} = a - \dfrac{b}{R}$ $C_{u2} = a + \dfrac{b}{R}$
First power $n = 1$	$C_{a2}^2 - C_{a1}^2 = 4ab\left(\dfrac{1}{R} - 1\right)$ $C_{a1}^2 - C_{a1m}^2 = -2\left[a^2\left(R^2 - 1\right) - 2ab\ell nR\right]$ $C_{a2}^2 - C_{a2m}^2 = -2\left[a^2\left(R^2 - 1\right) + 2ab\ell nR\right]$ $C_{a2}^2 - C_{a1}^2 = -8ab\ell nR$ $C_{a1m} = C_{a2m}$	$\Lambda = 1 + \dfrac{2a\ell nR}{U_m} - \dfrac{a}{U_m}$ $\Lambda = 1 + (2\ell nR - 1)(1 - \Lambda_m)$	$C_{u1} = aR - \dfrac{b}{R}$ $C_{u2} = aR + \dfrac{b}{R}$

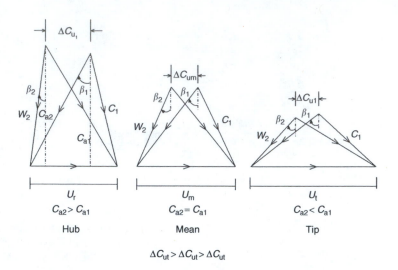

FIGURE 13.25   Velocity triangles at hub, mean, and tip stations for $n \neq -1$.

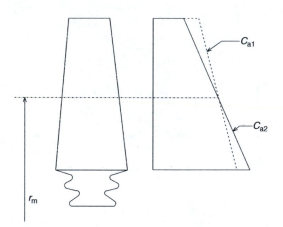

FIGURE 13.26   Variation of axial velocities C_{a1}, C_{a2} along the blade height for $n \neq -1$.

Prove that if a constant specific work is assumed and the exponent, $n = 2$, then

$$C_{a1}^2 - C_{a2}^2 = 12ab\,(R - 1)$$

Solution: The vortex energy equation is expressed by

$$\frac{dh_o}{dr} = C_a \frac{dC_a}{dr} + C_u \frac{dC_u}{dr} + \frac{C_u^2}{r}$$

Since the specific work is constant in the radial direction (or $dh_0/dr = 0$), then the vortex energy equation is reduced to

$$- C_a \frac{dC_a}{dr} = C_u \frac{dC_u}{dr} + \frac{C_u^2}{r}$$

$$- C_a dC_a = C_u dC_u + \frac{C_u^2}{r} dr$$

$$R = \frac{r}{r_{\mathrm{m}}}$$

$$-\int_{C_{\mathrm{am}}}^{C_{\mathrm{a}}} C_{\mathrm{a}} dC_{\mathrm{a}} = \int_{C_{\mathrm{um}}}^{C_{\mathrm{u}}} C_{\mathrm{u}} dC_{\mathrm{u}} + \int_{1}^{R} C_{\mathrm{u}}^2 \frac{dR}{R}$$

$$-\frac{1}{2} \left[C_{\mathrm{a}}^2 - C_{\mathrm{am}}^2 \right] = \frac{1}{2} \left[C_{\mathrm{u}}^2 - C_{\mathrm{um}}^2 \right] + I \tag{1}$$

where $I = \int\limits_{1}^{R} C_{\mathrm{u}}^2 \frac{dR}{R}$

$$n = 2$$

$$C_{\mathrm{u}1} = aR^2 - \frac{b}{R}$$

$$C_{\mathrm{u}2} = aR^2 + \frac{b}{R}$$

At station (1), then equation (1) has the form

$$-\frac{1}{2} \left[C_{\mathrm{a}1}^2 - C_{\mathrm{a}1m}^2 \right] = \frac{1}{2} \left[C_{\mathrm{u}1}^2 - C_{\mathrm{u}1m}^2 \right] + \int_{1}^{R} \left(aR^2 - \frac{b}{R} \right)^2 \frac{dR}{R}$$

$$-\frac{1}{2} \left[C_{\mathrm{a}1}^2 - C_{\mathrm{a}1m}^2 \right] = \frac{1}{2} \left[\left(aR^2 - \frac{b}{R} \right)^2 - (a-b)^2 \right] + \int_{1}^{R} \left(a^2 R^3 - 2ab + \frac{b^2}{R^3} \right) dR$$

$$-\frac{1}{2} \left[C_{\mathrm{a}1}^2 - C_{\mathrm{a}1m}^2 \right] = \frac{1}{2} \left[\left(aR^2 - \frac{b}{R} \right)^2 - (a-b)^2 \right] + \left[\frac{a^2 R^4}{4} - 2abR - \frac{b^2}{2R^2} \right]_{1}^{R}$$

$$-\frac{1}{2} \left[C_{\mathrm{a}1}^2 - C_{\mathrm{a}1m}^2 \right] = \frac{1}{2} \left[\left(a^2 R^4 - 2abR + \frac{b^2}{R^2} \right) - \left(a^2 - 2ab + b^2 \right) \right]$$

$$+ \left[\left(\frac{a^2 R^4}{4} - 2abR - \frac{b^2}{2R^2} \right) - \left(\frac{a^2}{4} - 2ab - \frac{b^2}{2} \right) \right]$$

$$C_{\mathrm{a}1}^2 - C_{\mathrm{a}1m}^2 = \left(\frac{3}{2}a^2 - 6ab \right) - \left(\frac{3}{2}a^2 R^4 - 6abR \right)$$

Similarly

$$C_{\mathrm{a}2}^2 - C_{\mathrm{a}2m}^2 = \left(\tfrac{3}{2}a^2 + 6ab \right) - \left(\tfrac{3}{2}a^2 R^4 + 6abR \right)$$

Since $C_{\mathrm{a}1m} = C_{\mathrm{a}2m}$

$$C_{\mathrm{a}1}^2 - C_{\mathrm{a}2}^2 = 12ab \, (R - 1)$$

From the above equation, the following conclusions are deduced:

$R < 1$	$R = 1$	$R > 1$
$C_{a1} < C_{a2}$	$C_{a1m} = C_{a2m}$	$C_{a1} > C_{a2}$

Example 5   Consider a free vortex design for a compressor stage having a degree of reaction at the mean section equal to 0.6, hub-to-tip ratio of 0.6, and the flow angles $\alpha_{1m} = 30°, \beta_{1m} = 60°$. Calculate the absolute and relative angles at the hub (α_h, β_h) and tip (α_t, β_t) for both stations (1) (upstream of the rotor) and (2) (downstream of the rotor). Also, calculate the degree of reaction at both hub and tip.

If the above compressor stage is to be designed by *exponential method design* instead of free vortex, then recalculate the absolute and relative angles at the hub (α_h, β_h) and tip (α_t, β_t) for both stations 1 (upstream of the rotor) and 2 (downstream of the rotor) and also the degree of reaction at both hub and tip. Additional data are given as follows:

$$a = 100 \text{ m/s}, \quad b = 40 \text{ m/s}$$

Solution: Given data: $\alpha_{1m} = 30°$, $\beta_{1m} = 60°$, and $\zeta = 0.6$.

1. *Free vortex design:*

   The data here are the same as in Example 2, except for the degree of reaction. Thus, the following same results are obtained for the inlet flow along the blade:

   $$\alpha_{1h} = 37.6°, \quad \alpha_{1t} = 24.8°, \quad \beta_{1h} = 43.9°, \quad \beta_{1t} = 67.5°$$

   It was proved in Example 2 that the degree of reaction may be expressed by the relation

   $$\Lambda = \frac{\tan \beta_1 + \tan \beta_2}{2 (\tan \alpha_1 + \tan \beta_1)}$$

   Applying this equation at the mean section, then

   $$\tan \beta_{2m} = 2\Lambda_m (\tan \alpha_{1m} + \tan \beta_{1m}) - \tan \beta_{1m}$$

   $$\tan \beta_{2m} = 2 \times 0.6 \times 2.3094 - 1.732 = 1.0392$$

   $$\beta_{2m} = 46.1°$$

   $$\tan \alpha_{2m} = \frac{U_m}{C_a} - \tan \beta_{2m} = 1.2702$$

   $$\alpha_{2m} = 51.78°$$

   $$\Lambda = \frac{\tan \beta_1 + \tan \beta_2}{2 (\tan \alpha_1 + \tan \beta_1)}$$

   $$\Lambda_h = \frac{\tan \beta_{1h} + \tan \beta_{2h}}{2 (\tan \alpha_{1h} + \tan \beta_{1h})} = 0.2905$$

   $$\Lambda_t = \frac{\tan \beta_{1t} + \tan \beta_{2t}}{2 (\tan \alpha_{1t} + \tan \beta_{1t})} = 0.7438$$

The variation of the angles from hub to tip is plotted in Figure 13.27.

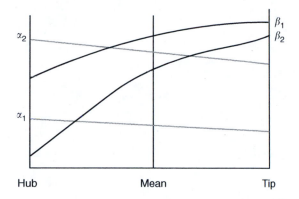

FIGURE 13.27 Variation of angles from hub to tip.

2. *Exponential design method:*
   Since $a = U_m(1 - \Lambda_m) = 100$, $\Lambda_m = 0.6$,

$$U_m = \frac{100}{0.4} = 250 \text{ m/s}$$

Moreover, with $b = 40$ m/s

$$\Delta T_0 = \frac{2bU_m}{Cp} = \frac{2 \times 40 \times 250}{1005} = 19.9 \text{ K}$$

$$C_{u1} = a - \frac{b}{R}$$

Since $R_h = 0.75$, $R_t = 1.25$, then

$$C_{u1h} = 46.7 \text{ m/s}, \quad C_{u1t} = 68 \text{ m/s}$$

Since

$$\frac{U_m}{C_{am}} = \tan\alpha_{1m} + \tan\beta_{1m}$$

$$C_{am} = \frac{250}{\tan 30 + \tan 60} = 108.25 \text{ m/s}$$

$$C_{a1h}^2 = C_{am}^2 - 2\left[a^2 \ln R + \frac{ab}{R} - ab\right]$$

$$C_{a1h}^2 = (108.25)^2 - 2\left[(100)^2 \ln \frac{3}{4} + \frac{100 \times 40}{(3/40)} - 100 \times 40\right]$$

$C_{a1h} = 121.7$ m/s, similarly $C_{a1t} = 94.1$ m/s
   From Table 13.3

$$C_{a2}^2 = C_{a2m}^2 - 2\left[a^2 \ln R - \frac{ab}{R} + ab\right]$$

$$C_{a2h} = 141.9 \text{ m/s and } C_{u2h} = a + (b/R_h) = 153.3 \text{ m/s}$$

$$C_{u2t} = 132 \text{ m/s}$$

$$\tan \alpha_{1h} = \left(\frac{C_{u1}}{C_{a1}}\right)_h = 0.384, \quad \alpha_{1h} = 21°$$

$$\tan \alpha_{2h} = \left(\frac{C_{u2}}{C_{a2}}\right)_h = 0.93, \quad \alpha_{2h} = 42.9$$

$$\tan \alpha_{1t} = \left(\frac{C_{u1}}{C_{a1}}\right)_t = 0.722, \quad \alpha_{1t} = 35.85°$$

$$\tan \alpha_{2t} = \left(\frac{C_{u2}}{C_{a2}}\right)_t = 1.755, \quad \alpha_{2t} = 60.32°$$

$$U_h = \frac{3}{4} U_m = 187.5 \text{ m/s} \quad \text{and} \quad U_t = \frac{5}{4} U_m = 312 \text{ m/s}$$

$$\left(\frac{U}{C_{a1}}\right)_h = \tan \alpha_{1h} + \tan \beta_{1h}$$

$$\tan \beta_{1h} = \frac{187.5}{121.7} - \tan 21 = 1.1568$$

$$\beta_{1h} = 49.15°, \text{ similarly } \beta_{1t} = 68.95°$$

$$\tan \beta_{2h} = \left(\frac{U}{C_{a2}}\right)_h - \tan \alpha_{2h} = 0.392$$

$\beta_{2h} = 21.4°$, similarly $\beta_{2t} = 67.388°$.

From both Tables 13.4 and 13.5 the following conclusions may be deduced:

1. The degree of reaction at hub for exponential design is much higher than the case of free vortex.
2. The rotor twist $(\beta_1 - \beta_2)$ for exponential design is less and much better than free vortex design.

TABLE 13.4

Summary of Results for Free Vortex Design

	$\frac{U}{C_a}$	α_1	β_1	α_2	β_2	Λ	$(\beta_1 - \beta_2)$
Hub	1.725	37.6	43.9	59.25	2.53	0.2905	41.37
Mean	2.3	30	60	51.78	46.1	0.6	13.9
Tip	2.875	24.8	67.5	45.25	61.8	0.7438	5.7

TABLE 13.5
Summary of Results for Exponential Method

	U (m/s)	C_{a1}	C_{a2}	α_1	β_1	α_2	β_2	Λ	$(\beta_1 - \beta_2)$
Hub	187.5	121.7	141.9	21	49.2	42.9	21.4	0.59	27.8
Mean	250	108.25	108.25	30	60	51.78	46.1	0.6	13.9
Tip	312.5	94.1	75.2	38.9	69	60.32	67.4	0.734	1.6

13.7 COMPLETE DESIGN PROCESS FOR COMPRESSOR

Compressors are designed by the *indirect method* [6]. That is, first one defines the flow path (casing and hub) radii and decides the thermodynamic and aerodynamic properties, as well as work input that air must have at each axial location (station) along the flow path. An iterative analysis is conducted in which the continuity, momentum, energy, and state equations are satisfied at each station. Although there are no blades or blade shape in the analysis, their presence is recognized by work input, turning ability, and loss. Then, one designs the blade and vanes to achieve the desired thermodynamic and aerodynamic properties. The designer is constrained to achieve the overall pressure ratio and mass flow rate. The full design details are given in Reference 12.

A concise design procedure is presented in Reference 6. The design steps for this design are

1. Choice of the compressor rotational speed (rpm) and annulus dimensions
2. Determination of the number of stages (assuming efficiency)
3. Calculation of air angles for each stage at the mean section
4. Determination of the variation of air angles from root to tip based on the type of blading (free vortex-exponential–first power)
5. Selection of compressor blading using experimentally cascade data
6. Efficiency check (previously assumed) using cascade data
7. Investigation of compressibility effects
8. Estimation of off-design performance
9. Rig testing

The details of these steps will be described here.

13.8 ROTATIONAL SPEED (RPM) AND ANNULUS DIMENSIONS

From past experience of subsonic compressor design [6], the maximum tip speed, axial velocity, as well as the hub-to-tip ratio of the first stage may have the following typical values:

$$U_t \approx 350 \, \text{m/s}$$

$$C_a \approx 150 - 200 \, \text{m/s}$$

$$\zeta \equiv \frac{r_h}{r_t} \approx 0.4 - 0.6$$

Another design procedure for subsonic compressor is to rely on Mach number rather than specific values of speeds. Thus, the following values are suggested:

Axial Mach number $M_a \approx 0.5 \, (< 0.7)$

Hub-to-tip ratio at inlet $\zeta > 0.3$ and at outlet $\zeta < 0.9$

Relative Mach number: at tip $M_{1tip} < 1.2$ for the high-pressure compressor (HPC), for fan $M_{1tip} < 1.8$

Relative Mach number: at hub $M_{1hub} < 0.9$

However, if this compressor is a part of an aero engine powering a supersonic or future hypersonic aircraft, the total temperature at the inlet to compressor approaches or exceeds 1000 K. In this case, the static temperature is very close to the total values, which leads to very high values for sonic speeds. Thus, the Mach number values suggested above develop unaffordable stresses in the blades. Consequently, these values are drastically reduced, even to half its suggested values. Alternatively, specific values for the mean blade and axial speeds are to be employed.

If this compressor is connected to a turbine, then the rotational speed of the compressor must be the same as the rotational speed of the turbine. This will be a design constraint. Moreover, for the same compressor/turbine spool, the outer dimensions (tip diameter) must be equal or comparable. This will be another constraint.

For the first stage, the flow is axial, or $C_1 = C_{a1}$, and the axial Mach number is calculated as

$$C_1 = M_1 \sqrt{\gamma R T_1}$$

$$C_1 = M_1 \sqrt{\frac{\gamma R T_{01}}{1 + ((\gamma - 1)/2) M^2}} \tag{13.35}$$

Similarly, the relative speed at the blade tip is

$$W_1 = M_{1\mathrm{rel}} \sqrt{\frac{\gamma R T_{01}}{1 + ((\gamma - 1)/2) M^2}} \tag{13.36}$$

The rotational speed is then

$$U_t^2 = W_{1\mathrm{tip}}^2 - C_1^2$$

$$U_t^2 = \left(M_{1\mathrm{tip}}^2 - M_1^2\right) \left(\frac{\gamma R T_{01}}{1 + ((\gamma - 1)/2) M^2}\right) \tag{13.37}$$

The outer diameter and the rotational speed are related by the equation

$$r_t = \frac{U_t}{2\pi N}$$

If either the tip speed or the rotational speed is specified (or have some constraint), then the hub-to-tip ratio is calculated from the continuity equation as follows:

$$\dot{m} = \rho_1 C_{a1} A = \rho_1 C_{a1} \pi \left(r_t^2 - r_h^2\right)$$

$$\zeta^2 = 1 - \frac{\dot{m}}{\rho_1 C_{a1} \pi r_t^2}$$

The density is calculated as follows:

$$C_1 = C_{a1}, \quad T_1 = T_{01} - \frac{C_1^2}{2Cp}, \quad \text{and} \quad P_1 = P_{01} \left(\frac{T_1}{T_{01}}\right)^{\gamma/(\gamma-1)}$$

$$\therefore \rho_1 = \frac{P_1}{R T_1}$$

A check is requested here for the hub-to-tip ratio to be within the specified range.

Example 6   A preliminary design of the low-pressure axial compressor, which is connected to an axial turbine of a small turbojet engine, is requested. The compressor has an inlet total temperature and pressure of 322.2 K and 141 kPa, respectively. The overall pressure ratio of the compressor is 3 and the polytropic efficiency is 0.88. The mass flow rate is 50 kg/s. The turbine rotational speed is 180 rev/s and has a tip diameter of 0.285 m. Assume the flow to be axial at the compressor inlet.

Solution: On the basis of the suggested inlet Mach numbers, assume that the relative tip Mach number $M_{1\text{tip}} = 1.1$ and the axial Mach number $M_a = 0.5$. The procedure is to calculate the static temperature and pressure and then the density. Next, the axial, relative, and rotational speeds are calculated. Finally, the tip radius is calculated, from which the hub-to-tip ratio is calculated. If the tip diameter is far from the turbine outer diameter, then the axial Mach number is to be changed within its design range. The results for the example is arranged in Table 13.6 where the axial Mach number varies from 0.5 to 0.6.
   Since from turbine calculations

$$N = 180 \,\text{rps} \quad \text{and} \quad (r_t)_{\text{turbine}} = 0.285 \,\text{ms}$$

   Choose $M_1 = 0.6$, where $\zeta = 0.5$ and $r_t = 0.283$ m, which is close to the tip radius of the turbine. The corresponding blade rotational speed at tip is $U_t = 320 \,\text{m/s}$.
   The mean radius is $r_m = \left((1 + \zeta)/2\right) r_t = 0.213$ m. The hub radius is 0.143 m. The mean rotational speed U_m, or simply U, is 241 m/s. Here, a decision has to be made concerning the type of variation of the compressor passage at its successive stages. One has to decide whether the compressor will have a constant mean, tip, or hub radius through all stages.
   Now, to determine the outlet dimensions of the compressor, assume that the stage efficiency (or the polytropic efficiency) of the compressor is η_s; the outlet temperature of the compressor is calculated as

$$\frac{T_{02}}{T_{01}} = \left(\frac{P_{02}}{P_{01}}\right)^{(n-1)/n},$$

where $(n - 1)/n = (1/\eta_s)\left((\gamma - 1)/\gamma\right)$.
   Assume that air leaves the stator of the last stage of the compressor axially. Thus $C_2 = C_a = C_1$, which was previously calculated based on the assumed axial Mach number. The static pressure and

TABLE 13.6

Results for a Preliminary Design of an Axial Compressor

M_1	0.5	0.55	0.6
T_1 (K)	307	304	301
P_1 (kPa)	119	115	111
$\rho_1 \left(\text{kg/m}^3\right)$	1.349	1.317	1.28
C_1 (m/s)	175.6	192.2	209
W_1 (m/s)	386.3	386	382
U_t (m/s)	344.1	335	320
r_t	0.304	0.296	0.283
ζ	0.522	0.531	0.506

temperature will be

$$T_2 = T_{02} - \frac{C_2^2}{2Cp}, \quad P_2 = P_{02}\left(\frac{T_2}{T_{02}}\right)^{\gamma/(\gamma-1)}$$

The density and the annulus area are $\rho_2 = P_2/RT_2$, $A_2 = \dot{m}/\rho_2 C_{a2}$.

If the constant mean radius option is selected, then the mean radius is already known and the blade height at outlet is calculated as follows: $h = (A_2/2\pi r_m)$, the tip and root radii at outlet are calculated from the relations

$$r_t = r_m + \frac{h}{2}, \quad r_r = r_m - \frac{h}{2}$$

The results at the compressor outlet are

$$T_{02} = 460.3\,\text{K}, \quad P_{02} = 423\,\text{kPa}, \quad T_2 = 423.6\,\text{K}, \quad P_2 = 357.22\,\text{kPa},$$

$$\rho_2 = 2.8378\,\text{kg/m}^3, \quad r_{t2} = 0.244\,\text{m}, \quad r_{h2} = 0.182\,\text{m}, \quad h = 0.062\,\text{m}$$

13.9 DETERMINE NUMBER OF STAGES (ASSUMING STAGE EFFICIENCY)

To determine the number of stages, the overall temperature rise within the compressor is first determined as $\Delta T_{comp} = T_{02} - T_{01}$. The temperature T_{02} is temperature at the outlet of the compressor which is to be determined through the steps described in Example 6. Second, the stage temperature rise is determined. Divide both values to obtain the number of stages. This figure will be rounded to the nearest integer.

The stage temperature rise will be obtained from Equation 13.9 as follows:

$$\Delta T_0 = \frac{UC_a}{Cp}(\tan\beta_1 - \tan\beta_2)$$

Since the rotor de Haller number is $\text{DH}_r = W_2/W_1$, then for constant axial velocity

$$\text{DH}_r = \frac{\cos\beta_1}{\cos\beta_2} = \frac{\sec\beta_2}{\sec\beta_1} = \sqrt{\frac{1 + \tan^2\beta_2}{1 + \tan^2\beta_1}}$$

$$\therefore \tan\beta_2 = \sqrt{(\text{DH}_r)^2\left(1 + \tan^2\beta_1\right) - 1} \tag{13.38}$$

$$\Delta T_0 = \frac{UC_a}{Cp}\left(\tan\beta_1 - \sqrt{(\text{DH}_r)^2\left(1 + \tan^2\beta_1\right) - 1}\right) \tag{13.39}$$

As a guide, from past experience [5], the temperature rise per stage for the available compressor has the following range:

$$\Delta T_{0s} = 10\text{–}30\,\text{K} \quad \text{subsonic stages}$$

$$\Delta T_{0s} = 45\text{–}55\,\text{K} \quad \text{transonic stage}$$

Then the number of stages is determined as

$$n = \frac{(\Delta T_0)_{\text{compressor}}}{(\Delta T_0)_{\text{stage}}}$$

Again, the above values for temperature rise per stage are only a guide for the present day compressors. Aero engine manufacturers are working hard to increase these values to reach 70° or 80° for transonic compressors.

The polytropic efficiency is related to the compressor efficiency by the relation

$$\eta_c = \frac{\pi_c^{(\gamma-1)/\gamma} - 1}{\pi_c^{(\gamma-1)/\gamma \eta_s} - 1}$$

13.10 CALCULATION OF AIR ANGLES FOR EACH STAGE AT THE MEAN SECTION

For a constant mean diameter and known axial and mean blade speeds (C_a, U_m), the angles $\alpha_1, \beta_1, \alpha_2,$ and β_2 will be calculated for each stage. The degree of reaction at the mean section will also be calculated and checked for reasonable values.

13.10.1 FIRST STAGE

The first stage is characterized by an axial inlet flow ($\alpha_1 = 0$); thus the relative inlet velocity and angle are calculated from the relations

$$W_1 = \sqrt{U_m^2 + C_a^2}$$

$$\tan \beta_1 = \frac{U_m}{C_a}$$

For simplicity, U_m will be written as U. The outlet relative angle is to be calculated from Equation 13.38, where the rotor de Haller number is selected equal or greater than 0.72.

The absolute angle at the rotor outlet is calculated from Equation 13.1b as follows:

$$\tan \alpha_2 = \frac{U}{C_a} - \tan \beta_2$$

which is the inlet angle to the stator, and the outlet angle to the stator (α_3) may be calculated from de Haller number of the stator defined as

$$\text{DH}_s = \frac{C_3}{C_2} = \frac{\cos \alpha_2}{\cos \alpha_3}$$

The value of de Haller number of stator has to be carefully selected; otherwise, it may yield an imaginary value for α_3 (if $\cos \alpha_3 > 1.0$). Thus, it must satisfy the condition

$$\text{DH}_s > \cos \alpha_2 \qquad (13.40)$$

Thus

$$\cos \alpha_3 = \frac{\cos \alpha_2}{\text{DH}_s} \qquad (13.41)$$

A better value for the stage temperature rise can now be calculated by incorporating the work done factor (λ) as follows:

$$\Delta T_{0s} = \frac{\lambda U_m C_a (\tan \beta_1 - \tan \beta_2)}{Cp} \qquad (13.9b)$$

The values of the work done factor for different stages are given in the following table:

Stage Number	First	Second	Third	Fourth (Last)
Λ	0.98	0.93	0.88	0.83

The pressure ratio is calculated from the relation

$$\pi_s = \left(\frac{P_{03}}{P_{01}}\right)_s = \left(1 + \frac{\eta_s \Delta T_{01}}{T_{01}}\right)^{\gamma/(\gamma-1)} \qquad (13.42)$$

The degree of reaction is calculated as follows:

$$\Lambda = \frac{C_a}{2U} (\tan \beta_1 + \tan \beta_2) \qquad (13.23e)$$

13.10.2 Stages from (2) to (n − 1)

The procedure is straightforward as follows:

1. The inlet total conditions to any stage (i) are

$$(P_{01})_{stage(i)} = (P_{01})_{stage(i-1)} \left(\frac{P_{03}}{P_{01}}\right)_{stage(i-1)}$$

$$(T_{01})_{stage(i)} = (T_{01})_{stage(i-1)} + (\Delta T_0)_{stage(i-1)}$$

2. The flow is no longer axial, but the outlet angle α_3 for any stage will be equal to the inlet absolute angle α_1 of the next angle. In general,

$$(\alpha_1)_{stage(i)} = (\alpha_3)_{stage(i-1)}$$

3. The inlet relative angle β_1 is obtained as

$$\tan \beta_1 = \frac{U}{C_a} - \tan \alpha_1$$

4. The outlet relative angle β_2 is obtained from Equation 13.38 with a de Haller number equal to 0.72.
5. The absolute outlet angle α_2 is calculated from the relation

$$\tan \alpha_2 = \frac{U}{C_a} - \tan \beta_2$$

6. The stage temperature rise ΔT_{0s} is calculated from Equation 13.9b using appropriate work done factor.

7. The stage pressure ratio π_s is calculated from the relation 13.42.
8. The degree of reaction Λ is calculated from Equation 13.23e.
9. The appropriate de Haller number DH_s for the stator is calculated from Equation 13.40.
10. The stator outlet angle α_3 is calculated from Equation 13.41.

The above procedure is repeated for all stages up to the last stage.

13.10.3 LAST STAGE

1. The pressure ratio of the stage is evaluated from the overall pressure ratio of the compressor and the product of the stage pressure ratios of all the previous $(n-1)$ stages is as follows:

$$\pi_{\text{last}} = \frac{\pi_{\text{c}}}{\Pi \pi_i}$$

2. The corresponding temperature rise is obtained from the relation

$$(\Delta T_0)_{\text{last}} = \frac{(T_{01})_{\text{last}}}{\eta_s} \left(\pi_{\text{last}}^{(\gamma-1)/\gamma} - 1 \right)$$

3. The inlet relative angle β_1 is calculated as in Section 13.10.2—Steps 2 and 3.
4. The relative outlet blade angle is obtained from the relation

$$\tan \beta_2 = \tan \beta_1 - \frac{Cp \Delta T_0}{\lambda U C_{\text{a}}}$$

   Check for the rotor de Haller number.
5. The outlet absolute angle α_2 is calculated as in Step 4.
6. The degree of reaction Λ is calculated from Equation 13.23.
7. The outlet absolute angle α_3 is assumed to be zero and check the stator de Haller number.

Example 7   A seven-stage axial compressor has the following data:

Overall pressure ratio	4.15
Mean rotational speed	266 m/s
Axial velocity	150 m/s
Stage efficiency	0.9

Inlet temperature and pressure are 288 K, 101 kPa.
   Calculate the following variables for each stage:

$$\pi_s, \ \Delta T_{0s}, \ \alpha_1, \ \beta_1, \ \alpha_2, \ \beta_2, \ \alpha_3, \ \Lambda, \ DH_s, \ DH_r$$

Solution: The total temperature rise for the compressor is calculated following the procedure described in Example 6 and its value is 164.5 K.
   The above procedure is followed and the results are arranged in Table 13.7.
   The summation of total temperature rise for seven stages is 165.3, which is in good comparison to the value calculated above (164.5).

TABLE 13.7

Variations of the Aerodynamic and Thermodynamic Properties of the Seven-Stage Compressor

Stage	1	2	3	4	5	6	7
T_{01} (K)	288	315.4	340.4	364.1	386.6	409.2	431.8
ΔT_{0s} (K)	27.4	25	23.7	22.5	22.6	22.6	21.5
π_s	1.333	1.273	1.238	1.208	1.196	1.185	1.165
α_1	0	11.86	11.73	9.48	9.05	9.05	9.05
β_1	60.6	57.4	57.5	58.05	58.05	58.05	58.05
α_2	35.16	41.59	41.51	40.5	40.5	40.5	39.56
β_2	47	41.65	42.69	42.7	42.7	42.7	43.57
α_3	11.86	11.73	9.84	9.05	9.05	9.05	0
Λ	0.804	0.69	0.692	0.71	0.71	0.71	0.72
DH_r	0.72	0.72	0.72	0.72	0.72	0.72	0.73
DH_s	0.72	0.764	0.76	0.77	0.77	0.77	0.77

13.11 VARIATION OF AIR ANGLES FROM ROOT TO TIP BASED ON THE TYPE OF BLADING (FREE VORTEX–EXPONENTIAL–FIRST POWER)

The three methods that estimate the variations of air properties in the radial direction from blade hub to tip are applied here. The procedure is as follows:

1. Calculate the dimensions at the rotor inlet and outlet for each stage.
2. Calculate the values of the constants (a) and (b) from Equations 13.33 and 13.34.
3. Use Table 13.3 to calculate the variations in axial and whirl velocities as well as the degree of reaction.

Now, to calculate the dimensions upstream and downstream of the rotor, states (1) and (2), the dimensions at state (3) are first calculated from the continuity equation. The air density is calculated from the perfect gas law, and then the annulus area is calculated as the axial speed (as mean section is already known).

$$\rho_3 = \frac{P_3}{RT_3}, \quad A_3 = \frac{\dot{m}}{\rho_3 C_a}$$

This annulus area is used to calculate the blade height (h), $h_3 = A/2\pi r_m$, from which, and assuming constant mean radius, the root and tip radii are calculated as

$$r_{r3} = r_{m3} - \frac{h_3}{2}, \qquad r_{t3} = r_{m3} + \frac{h_3}{2}$$

The corresponding hub and tip radii at the rotor outlet are the mean value of those at states (1) and (3), as shown in Figure 13.28. The dimensions at the rotor outlet are calculated from the following relations:

$$r_{r2} = \frac{r_{r1} + r_{r3}}{2}, \qquad r_{t2} = \frac{r_{t1} + r_{t3}}{2}$$

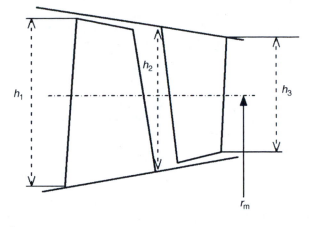

FIGURE 13.28   A single-stage geometry.

It remains to calculate the corresponding nondimensional radii $\left(R = r/r_{\mathrm{m}}\right)$ at the root and tip at states (1) and (2). After calculating the constants (a) and (b), the whirl velocities ($C_{\mathrm{u}1}$ and $C_{\mathrm{u}2}$) are calculated for any design methods from the known values at the mean radius. Moreover, the axial velocities $C_{\mathrm{a}1}$ and $C_{\mathrm{a}2}$ are calculated for any design method other than the free vortex. The degree of reaction is also calculated. It is important to have as high a degree of reaction at the hub as possible.

13.12   BLADE DESIGN

13.12.1   CASCADE MEASUREMENTS

To obtain information with respect to the effect of different blade designs on air flow angles, pressure losses, and energy transfer across blade rows, one must resort to cascade wind tunnels and cascade theory. Experimentation is performed to ensure that the blade row satisfies its objectives. The first objective is to turn air through the required angles $(\beta_1 - \beta_2)$ for rotor and $(\alpha_2 - \alpha_3)$ for stator, with the angle $(\beta_1 - \beta_2)$ as maximum as possible to maximize the stage pressure ratio. The second objective is to achieve the diffusing process with optimum efficiency, that is, with minimum loss of stagnation pressure. Experiments are generally performed in a straight wind tunnel (Figure 13.29) rather than in the annular form of tunnel. The word cascade denotes a row of identical (geometrically similar) blades equally spaced and parallel to each other aligned to the flow direction as shown in Figure 13.30. The height and length of cascade in a cascade wind tunnel, made as large as the available air supply, will allow eliminating the interference effects due to the tunnel walls. Boundary layer suction on the walls is frequently applied to prevent contraction of air stream. Cascade is mounted on a turn-table so that its angular direction with respect to the inflow duct (α_1) can be set in any desired value and thus the incidence angle (i) may be varied. Vertical traverses over two planes usually a distance of one blade chord upstream and downstream of the cascade are provided with pitot tubes and yaw meters to measure the pressure and airflow angles.

Pressure and velocity measurements are made by the usual L-shaped pitot-static tubes. Air direction is found by either claw or cylindrical yaw meters.

Before proceeding further into cascade testing and blade design, it is necessary to define various important geometries and angles relevant to the design. Figure 13.30a and b represent a cascade that has the nomenclature listed in Table 13.8.

The measured velocities and angles above have different meaning for rotor and stator rows as defined in Table 13.9.

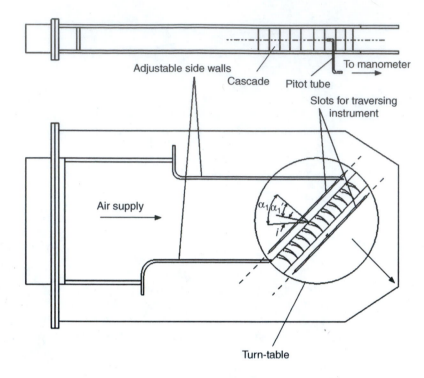

FIGURE 13.29   Cascade wind tunnel.

Other definitions here are

$$\text{Solidity } \sigma = \text{chord/spacing } (c/s)$$

$$\text{Aspect ratio} = \text{height/chord } (h/c)$$

$$\text{Stagnation pressure loss } (P_{01} - P_{02})/(1/2)\,\rho V_1^2 = w/(1/2)\,\rho V_1^2$$

When the pressure and flow direction measuring instruments are traversed along the blade row in the upstream and downstream positions, the results are plotted in Figure 13.31. An average for this stagnation pressure loss [expressed hereafter as $\bar{w}/(1/2)\rho V_1^2$ and an average for the deflection angle $(\bar{\varepsilon})$ are evaluated at this particular incidence angle. The same measurements and averaging process are repeated for different incidence angles. The results are plotted in Figure 13.32. The deflection increases with angle of incidence up to a maximum angle, ε_s. This is the stall point where separation occurs on the suction surface of the blade. Since this angle may not be well defined, it is taken as the angle of incidence where the mean pressure loss is twice the minimum. The practice has been to select a deflection that corresponds to a definite proportion of the stalling deflection. This proportion is found to be most suitable as 80%, thus the selected or nominal deflection, $\varepsilon^* = 0.8\varepsilon_s$.

It was found that the nominal deflection angle is mainly dependent on pitch/chord (s/c) ratio and air outlet angle (α_2),

$$\varepsilon^* = f\left(\frac{s}{c}, \alpha_2\right)$$

Thus, a large number of tests covering different forms of cascade were performed. The results are shown in Figure 13.33. For known flow angles, the deflection angle is calculated. Next, from this deviation angle and the flow outlet angle, the ratio of (s/c) can be obtained. The blade height is

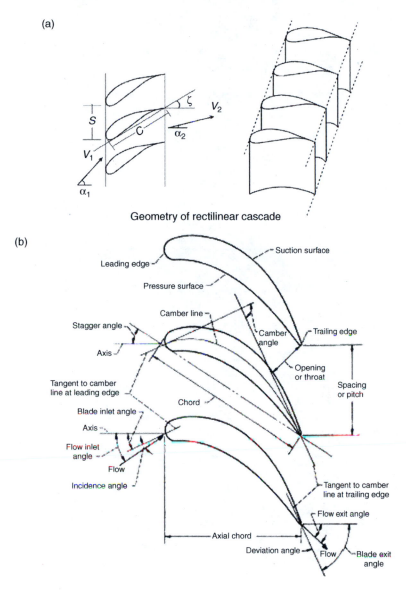

Geometry of rectilinear cascade

FIGURE 13.30   (a) Three-dimensional illustration for cascade and (b) cascade nomenclature.

TABLE 13.8
Cascade Nomenclature

α_1'	Blade (metal) inlet angle	α_1	Air inlet angle
α_2'	Blade (metal) outlet angle	α_2	Air outlet angle
θ	Blade camber angle $= \alpha_1' - \alpha_2'$	i	Incidence angle $= \alpha_1 - \alpha_1'$
ζ	Setting or stagger angle	δ	Deviation angle $= \alpha_2 - \alpha_2'$
s	Pitch (or space)	c	Chord
ε	Air deflection $= \alpha_1 - \alpha_2$	a	Location of maximum camber
V_1	Air inlet velocity	V_2	Air outlet velocity

TABLE 13.9

Proper Values of the Measured Quantities in Stator and Rotor Blade Rows

Cascade	Rotor	Stator
V_1	W_1	C_2
V_2	W_2	C_3
α_1	β_1	α_2
α_2	β_2	α_3

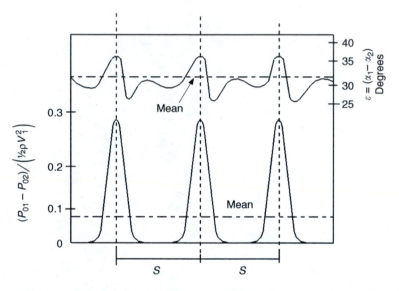

FIGURE 13.31 Variation of stagnation pressure loss and deflection for cascade at a certain incidence angle. (From H.I.H. Saravanamuttoo et al., *Gas Turbine Theory*, 5th edn., 2001, p. 191. With permission.)

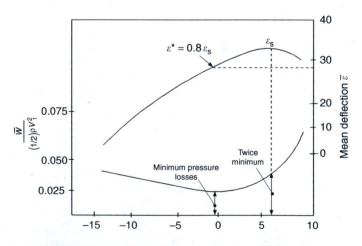

FIGURE 13.32 Mean deflection and mean stagnation pressure loss curves.

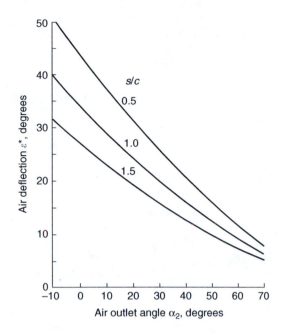

FIGURE 13.33 Cascade nominal deflection versus outlet angle. (From H.I.H. Saravanamuttoo et al., *Gas Turbine Theory*, 5th edn., 2001, p. 191. With permission.)

known from the previous sections and the blade height/chord is a known ratio ($h/c = 3$ for typical subsonic compressor); hence from this known chord, the spacing (s) is calculated. The number of blades is then calculated as

$$n = \frac{2\pi\, r_{\mathrm{m}}}{s}$$

It is recommended to avoid numbers with common multiples for the blades in successive rows (rotor/stator or stator/rotor) to avoid introducing resonant forcing frequencies.

The blade inlet angle (α_1') could be determined from the air inlet angle, assuming zero incidence angle $\alpha_1' = \alpha_1$; ($i = 0$). However, the blade outlet angle (α_2') cannot be determined from flow outlet angle α_2, until the deviation angle δ; ($\alpha_2 - \alpha_2'$) has been determined. The deflection angle is determined from the empirical relation

$$\delta = m\theta\sqrt{\frac{s}{c}} \tag{13.42a}$$

where

$$m = 0.23\left(\frac{2a}{c}\right)^2 + 0.1\left(\frac{\alpha_2^\circ}{50}\right) \tag{13.42b}$$

and $\theta = \alpha_1' - \alpha_2'$.

a is the distance of point of maximum camber from the leading edge of the blade and the angle α_2 is in degrees. For circular arc, $2a/c = 1$, then

$$m = 0.23 + 0.1\left(\frac{\alpha_2^\circ}{50}\right) \tag{13.42c}$$

For IGVs

$$\delta = 0.19\theta\,(s/c) \tag{13.42d}$$

For a circular arc blade camber, the blade setting angle ζ (Figure 13.30a) is given by the relation

$$\zeta = \alpha_1' - \frac{\theta}{2} \tag{13.43}$$

The axial chord of any blade row is then $c_a = c\cos\zeta$. The chord of blade in a subsonic compressor is one-third of the blade height, $c = h/3$. The gap between two successive blade rows is $w = c_a/4$. Thus, the axial length of one stage (including a rotor-gap-stator-gap) will be equal to $l = (5/4)\{(c_a)_r + (c_a)_s\}$.

The same procedure can be followed for any other blade ratios. The length of the compressor is then the sum of axial lengths of all compressor stages.

13.12.2 Choosing the Type of Airfoil

The blading can be designed after the radial distribution of air angles and velocities are determined. First, incidence and deviation angles are selected so that the slope of the airfoil mean line (camber line) at the leading and trailing edges can be established [6]. Incidence angles are selected in the low loss region of its performance envelop. Generally, incidence angle (i) is in the range $-5° \le i \le 5°$. The deviation angle is calculated from Equation 13.42, which is also in the range $5° \le \delta \le 20°$. Mean line shape and thickness are selected to achieve the desired airfoil loadings. The pressure increase is achieved by the diffusion process. The amount of diffusion is monitored by computing the diffusion factor, Equation 13.19a. Values of diffusion factor greater than 0.65 mean flow separation is imminent.

The nature and type of blading depend on the application and Mach number. Subsonic blading usually consists of circular arcs, parabolic arcs, or combinations thereof. Many countries developed their own profiles (e.g., United States, Britain, Japan, West Germany, etc.) for subsonic flows. One of the famous American profiles is the NACA 65 series blades. The famous British profile is the C-series. The NACA 65 series designation is as follows: NACA 65 (x) y, or $65 - x - y$, where x is 10 times the design lift coefficient of an isolated airfoil and y is the maximum thickness in percent of chord. An example for the C-series is 10C4/20P40, where blade is 10% thick C4 profile with a 20° camber angle using a parabolic camber line with maximum camber at 40% of the camber. A 10C4/20C40 blade would be similar, but with a circular arc. Another British series is The RAF 27 profiles. The NACA 65 series and similar British profiles are used in low subsonic flows [12].

13.12.3 Stage Performance

The performance, particularly the efficiency, for a given work input will govern the final pressure ratio. The efficiency is dependent on the total drag coefficient for each of the blade rows comprising the stage. Consider a control volume around a single blade of cascade (Figure 13.34). Two forces are generated, namely, axial and tangential forces (F_a, F_t). These two forces are decomposed into the lift and drag forces (L, D), from which the lift and profile drag coefficients can be obtained. The static pressure rise across the blade can be written as

$$\Delta P = P_2 - P_1 = \left(P_{02} - \frac{1}{2}\rho_2 V_2^2\right) - \left(P_{01} - \frac{1}{2}\rho_1 V_1^2\right). \tag{13.44}$$

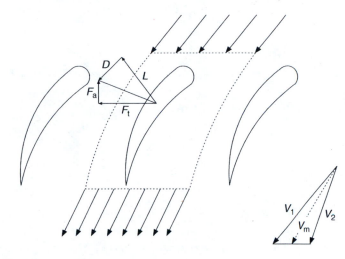

FIGURE 13.34 Forces on cascade.

If the changes in density are negligible, then

$$\Delta P = \frac{1}{2}\rho \left(V_1^2 - V_2^2 \right) - \overline{w} = \frac{1}{2}\rho V_a^2 \left(\sec^2 \alpha_1 - \sec^2 \alpha_2 \right) - \overline{w}$$

$$\Delta P = \frac{1}{2}\rho V_a^2 \left(\tan^2 \alpha_1 - \tan^2 \alpha_2 \right) - \overline{w}$$

Then the axial force per unit length is $F_a = s\Delta P$

$$F_a = \frac{1}{2}\rho V_a^2 s \left(\tan^2 \alpha_1 - \tan^2 \alpha_2 \right) - s\overline{w} \qquad (13.45)$$

The tangential force per unit length is $F_t = \dot{m}\Delta V_u$

$$F_t = \dot{m} \times \text{ (change in swirl velocity component along cascade)}$$
$$F_t = s\rho V_a \times (V_a \tan \alpha_1 - V_a \tan \alpha_2)$$
$$F_t = s\rho V_a^2 (\tan \alpha_1 - \tan \alpha_2) \qquad (13.46)$$

The lift and profile drag coefficients C_L, C_{Dp} depend on the mean angle (α_m). The drag and lift force can be obtained from the axial and tangential forces as follows:

$$D = F_t \sin \alpha_m - F_a \cos \alpha_m$$
$$L = F_t \cos \alpha_m + F_a \sin \alpha_m$$

The lift and drag forces are also expressed as

$$D = \frac{1}{2}\rho V_m^2 c C_{Dp}$$

$$L = \frac{1}{2}\rho V_m^2 c C_L$$

Now with $D = F_t \sin \alpha_m - s\Delta P \cos \alpha_m$

$$D = s\rho V_a^2 (\tan \alpha_1 - \tan \alpha_2) \sin \alpha_m - \frac{1}{2}\rho V_a^2 s \left(\tan^2 \alpha_1 - \tan^2 \alpha_2\right) \cos \alpha_m + s\overline{w} \cos \alpha_m$$

$$\tan^2 \alpha_1 - \tan^2 \alpha_2 = (\tan \alpha_1 - \tan \alpha_2)(\tan \alpha_1 + \tan \alpha_2)$$

with

$$\tan \alpha_m = \frac{1}{2}(\tan \alpha_1 + \tan \alpha_2)$$

$$\tan^2 \alpha_1 - \tan^2 \alpha_2 = (\tan \alpha_1 - \tan \alpha_2)(\tan \alpha_1 + \tan \alpha_2) = 2(\tan \alpha_1 - \tan \alpha_2)\tan \alpha_m$$

$$\therefore D = s\overline{w} \cos \alpha_m$$

but

$$D = \frac{1}{2}\rho V_m^2 c C_{Dp}$$

$$\therefore C_{Dp} = \left(\frac{s}{c}\right)\left(\frac{\overline{w}}{(1/2)\,\rho V_1^2}\right)\left(\frac{\cos^3 \alpha_m}{\cos^2 \alpha_1}\right) \tag{13.47}$$

$$\therefore C_L = 2\left(\frac{s}{c}\right)(\tan \alpha_1 - \tan \alpha_2)\cos \alpha_m - C_{Dp}\tan \alpha_m \tag{13.48a}$$

The lift and drag coefficients depend on the angles α_1, α_2, and α_m, which can be determined as follows:

$$\alpha_1 = \alpha_1' + i, \ \alpha_2 = \alpha_1 - \varepsilon^*, \text{ and } \alpha_m = \tan^{-1}\left[\frac{1}{2}(\tan \alpha_1 + \tan \alpha_2)\right]$$

Using the values of $\overline{w}/((1/2)\,\rho V_1^2)$ and the known value of (s/c), calculate the lift and drag coefficients, C_L and C_{Dp}. Since the $C_{Dp}\tan\alpha_m$ is small, we can write the conventional theoretical values of C_L as

$$C_L = 2\left(\frac{s}{c}\right)(\tan \alpha_1 - \tan \alpha_2)\cos \alpha_m \tag{13.48b}$$

Figure 13.35 illustrates the coefficients of lift and profile drag versus the incidence angle for certain value of s/c. Figure 13.36 illustrates the lift coefficient versus the air outlet angle for different (s/c) values.

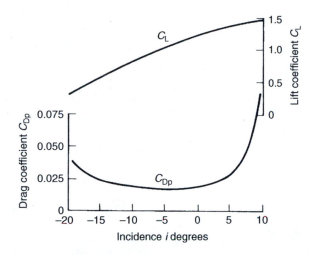

FIGURE 13.35   Variation of the lift and profile drag coefficients with incidence angle. (From H.I.H. Saravanamuttoo et al., *Gas Turbine Theory*, Prentice Hall, 5th edn., 2001, p. 191. With permission.)

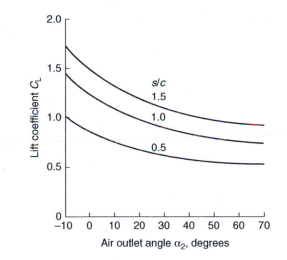

FIGURE 13.36   Variation of the lift coefficient with air outlet angle. (From H.I.H. Saravanamuttoo et al., *Gas Turbine Theory*, Prentice Hall, 5th edn., 2001, p. 191. With permission.)

There are two additional factors affecting the drag:

1. Drag due to the walls of compressor (annulus drag)
2. Drag due to secondary flow

Annulus drag is developed due to boundary layer over the hub and casing. These boundary layers together with the developed boundary layers over the suction and pressure sides of the blades are illustrated in Figure 13.37.

The annulus drag is expressed by the relation

$$C_{DA} = 0.020 \left(s/h \right) \tag{13.49}$$

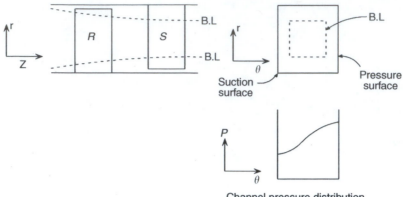

FIGURE 13.37   Annulus drag.

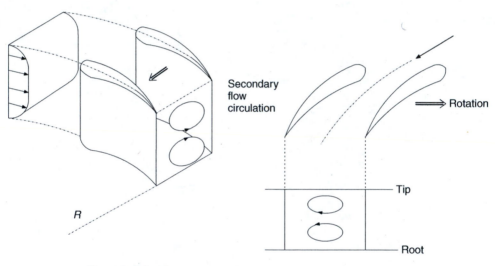

FIGURE 13.38   Secondary flow in a cascade.

Secondary flow is generated due to the curved blade shape, leading to the generation of two opposite circulations as shown in Figure 13.38. Tip clearance will influence the secondary flow. The secondary loss is of the same order of magnitude of the profile drag and may be represented by the relation

$$C_{DS} = 0.018 C_L^2 \tag{13.50}$$

The total drag is then equal to the sum of the three types of drag described above, or

$$\text{Total drag} = \text{profile drag} + \text{secondary flowdrag} + \text{annulus drag}$$
$$C_D = C_{Dp} + C_{DS} + C_{DA}$$

Thus, the loss coefficient can be determined from the relation

$$\frac{\overline{w}}{(1/2)\,\rho V_1^2} = \frac{C_D}{(s/c)\left(\cos^3 \alpha_m / \cos^2 \alpha_1\right)} \tag{13.51}$$

Now to determine the stage efficiency from the cascade measurements, with cascade data, efficiency is evaluated from pressure rise.

13.12.3.1 Blade Efficiency and Stage Efficiency

Since the static pressure rise ΔP is given by the relation

$$\Delta P = \frac{1}{2}\rho V_a^2 \left(\tan^2 \alpha_1 - \tan^2 \alpha_2\right) - \overline{w}$$

the ideal or theoretical pressure rise will be

$$\Delta P_{th} = \frac{1}{2}\rho V_a^2 \left(\tan^2 \alpha_1 - \tan^2 \alpha_2\right)$$

The blade row efficiency is defined as

$$\eta_b = \frac{\text{Actual pressure rise in compressor blade row}}{\text{Theoretical pressure rise in blade row}}$$

$$\eta_b = \frac{\Delta P}{\Delta P_{th}} = \frac{\Delta P_{th} - \overline{w}}{\Delta P_{th}} = 1 - \frac{\overline{w}}{\Delta P_{th}} = 1 - \frac{\overline{w}/\left(\rho V_1^2 / 2\right)}{\Delta P_{th}/\left(\rho V_1^2 / 2\right)}$$

$$\eta_b = 1 - \frac{\overline{w}/\left(\rho V_1^2 / 2\right)}{\Delta P_{th}/\left(\rho V_1^2 / 2\right)} \tag{13.52}$$

$$\frac{\Delta P_{th}}{\left(\rho V_1^2 / 2\right)} = \frac{\left(\rho V_a^2 / 2\right)\left(\tan^2 \alpha_1 - \tan^2 \alpha_2\right)}{\left(\rho V_a^2 / 2\right)\sec^2 \alpha_1} = \frac{\sec^2 \alpha_1 - \sec^2 \alpha_2}{\sec^2 \alpha_1}$$

$$\frac{\Delta P_{th}}{\left(\rho V_1^2 / 2\right)} = 1 - \frac{\sec^2 \alpha_2}{\sec^2 \alpha_1} \tag{13.53}$$

From Equations 13.51 through 13.53, the blade row efficiency can be obtained. For the normal stage, $C_1 = C_3$, the stage isentropic efficiency can be approximated to

$$\eta_s = \frac{T_{3ss} - T_3}{T_3 - T_1}$$

For $\Lambda = 0.5$, half temperature rise occurs in rotor and half at stator, or $T_3 - T_2 = T_2 - T_1$

$$\eta_b \approx \eta_s \tag{13.54a}$$

If the degree of reaction is far from 50%

$$\eta_s = \Lambda \eta_{rotor} + (1 - \Lambda)\,\eta_{stator} \tag{13.54b}$$

FIGURE 13.39 Boeing 737 aircraft.

Example 8 Boeing 737–200 airplanes have the ability to take off and land from unpaved runways (Figure 13.39). Moreover, their engines are close to ground thus, more susceptible to ingestion of solid particles into its core. Solid particles will erode the compressor blades. An examination of one of the compressor blades revealed that erosion led to

1. Shortening of the chord by 10%
2. Increasing the incidence angle (i) from $4°$ to $8°$

Other data of the compressor stage are

Stage total temperature at entry to the stage is $T_{01} = 320\,\mathrm{K}$
Stage total temperature rise $= \Delta T_{0s} = 25\,\mathrm{K}$

For the new blade,

$$\frac{s}{c_{\mathrm{new}}} = 1.0 \quad \text{and} \quad \alpha_{1\mathrm{new}} = 50°$$

For both new and eroded blade, $\alpha_2 = 30°$ and $s/h = 0.25$.
  Calculate

1. Stage efficiency for both new and eroded blade
2. Stage pressure ratio for new and eroded blade
3. Comment!

Solution: Calculations will be arranged in Table 13.10.
  Comments:

1. The stage efficiency has been decreased by 6%
2. The stage pressure ratio has been decreased by 1.4%

TABLE 13.10
Properties of New and Eroded Compressor Stage

	New Blade	Eroded Blade
i	4°	8°
α_1	50°	54°
α_2	30°	30°
$\alpha_m = \tan^{-1}\left[\frac{1}{2}(\tan\alpha_1 + \tan\alpha_2)\right]$	41.5°	44.33°
s/c	1.0	$\frac{1}{0.9} = 1.11$
s/h	0.25	0.25
C_{Dp} (from Figure 13.35)	0.025	0.075
C_L (from Figure 13.35)	0.9	0.94
$C_{DS} = 0.018 C_L^2$	0.01458	0.0159
$C_{DA} = 0.02\frac{s}{h}$	0.005	0.005
$C_D = C_{Dp} + C_{DA} + C_{DS}$	0.04458	0.0959
$\frac{\Delta P_{th}}{(1/2)\rho v_1^2} = 1 - \frac{\cos^2\alpha_1}{\cos^2\alpha_2}$	0.449	0.5393
$\frac{\bar{w}}{(1/2)\rho V_1^2} = \frac{C_D}{(s/c)\left(\cos^3\alpha_m / \cos^2\alpha_1\right)}$	0.04384	0.08154
$\eta_b = 1 - \frac{\bar{w}/\left((1/2)\rho V_1^2\right)}{\Delta P_{th}/\left((1/2)\rho V_1^2\right)}$	0.90236	0.84878
$\eta_b = \eta_s$	0.9024	0.8488
$\pi_s = \left(1 + \eta_s \frac{\Delta T_{0s}}{T_1}\right)^{(\gamma/\gamma-1)}$	1.2693	1.2519

13.13 COMPRESSIBILITY EFFECTS

High-performance gas turbines and aero engines use large airflow rates per unit frontal area, in which the blade tip speed is large, leading to high velocity and Mach numbers in compressors. Early compressors are designed with subsonic velocity throughout. In the present-day compressors, the axial and absolute Mach numbers at inlet are always subsonic. However, since the blade speed (Ωr) is higher at higher radii, the relative Mach numbers may reach supersonic levels near the outer sections of the rotor, while the relative flow near the inner radii remains subsonic. A *transonic* compressor is one where the relative flow remains subsonic at inner radii and supersonic at outer radii, with a transonic region in the middle. This is especially true for high aspect ratio or low hub-to-tip ratio compressors [10]. In some low aspect ratio and compact compressors, there may be cases where the entire flow, hub to tip, has supersonic relative flow. Such a compressor is known as *supersonic compressor*. Supersonic compressors have sharp leading edges and thin blades. The flow pattern in such a supersonic compressor is shown in Figure 13.40. The oblique shock wave is followed by a normal shock wave initiated at the leading edge and propagated toward the suction side of the opposite blade in the blade passage (Figure 13.40). This normal shock wave leads to a jump in both static and total pressures. Nowadays, the most successful designs are the custom-tailored airfoils, which are designed based on the controlled diffusion airfoil and shock-free airfoil design concepts [6]. Thus, the efficiency of supersonic compressor (a single stage), which was as low as 40% during the 1940s, has become nearly 90%. The highest relative Mach number has increased to 2.4 in experimental compressor stage [10]. A historical account of transonic and supersonic compressor development can be found in References 13 and 14.

The velocity triangles for a high-turning supersonic impulse cascade are shown in Figure 13.41 and the velocity triangles for a low-turning supersonic compressor cascade are shown in Figure 13.42.

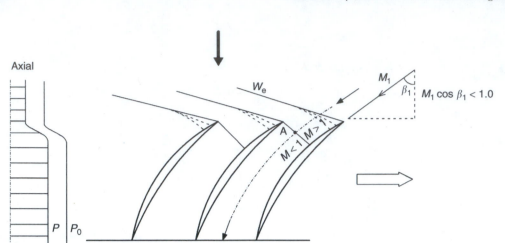

FIGURE 13.40   Flow pattern in a supersonic rotor.

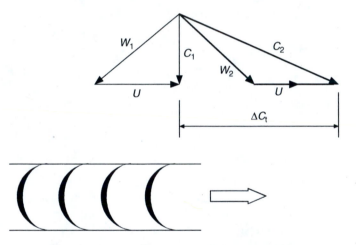

FIGURE 13.41   Velocity triangles of high-turning supersonic impulse cascade.

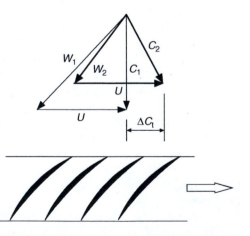

FIGURE 13.42   Velocity triangles low-turning supersonic compressor cascade.

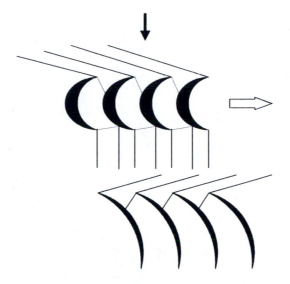

FIGURE 13.43   Shock-in-stator.

The different flow patterns for a single stage in a supersonic compressor may be classified as

1. Shock-in-stator
2. Shock-in-rotor
3. Shock-in-rotor/shock-in-stator

In the first impulse type (Figure 13.41), the flow in both rotor and stator are supersonic; thus, shocks are found in both blade rows (Figure 13.43). In general [15], if the rotor has a reaction type of blading rather than impulse one and has a large turning angle, then the flow in the rotor may be subsonic but the inlet absolute velocity to the stator is supersonic. Thus, the rotor is subsonic while the stator is supersonic and normal shock is generated in the stator. Concerning the second case of reaction blading (Figure 13.42), there are two cases, either the flow in the rotor is supersonic while the flow in the stator is purely subsonic ($M_{1rel} > 1$, $M_2 < 1$) (Figure 13.44), which is identified as "shock-in-rotor" configuration, or the flow in both cases is supersonic ($M_{1rel} > 1$, $M_2 > 1$) (Figure 13.45) and the flow is identified as shock-in-rotor/shock-in-stator. More details are found in Reference 10.

For subsonic flow, the flow is completely specified if the upstream Mach number and the incidence angle are set. For supersonic flow, it is also necessary to specify the downstream pressure, in ratio to that upstream, in order to locate shock position. The loss coefficient is now defined as

$$\overline{\omega} = \frac{P_{01} - \overline{P_{02}}}{P_{01} - P_1} \tag{13.55}$$

where $\overline{P_{02}}$ is the mass average total pressure over the flow downstream cascade. The effect of inlet Mach number on losses is experimentally measured for different blading [16]. Figure 13.46 illustrates the losses for two types of compressor blading, namely, C4 and double circular arc (DCA). Compressibility effect will be most important at the front compressor where the temperature and sonic speed are the lowest. The Mach number corresponding to the velocity relative to the tip of the rotor is the highest. The stator Mach number is highest at the hub. A simplified model for estimating the shock losses [5] is explained with reference to Figure 13.46. A pair of DCA is shown, with a supersonic velocity (M_1), entering in a direction aligned with the leading edge. The supersonic expansion along the uncovered portion of the suction side can be analyzed by Prandtl–Meyer expansion relations or

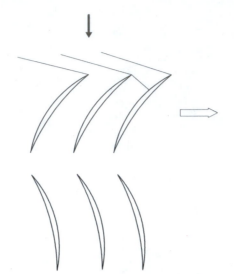

FIGURE 13.44   Shock-in-rotor.

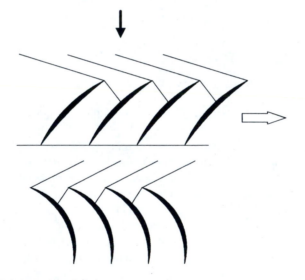

FIGURE 13.45   Shock-in-rotor/shock-in-stator.

tables. The Mach number will increase as the flow proceeds along the suction surface. A normal shock wave stands near the entrance to the passage, striking the suction surface at B. The Mach number at point B, M_2, before the normal shock, is evaluated from Prandtl–Meyer tables once the deflection angle $\Delta\theta$ is known. This angle may be roughly assumed as a portion of the deflection angle $(\beta_1 - \beta_2)$ depending on the (s/c) ratio, say half or less. The static to total pressure ratio is also evaluated from the Prandtl–Meyer tables, (P_1/P_{01}). Then, a normal shock wave is assumed with inlet Mach number equal to that just obtained. From normal shock tables, the outlet Mach number and the pressure ratio (P_{02}/P_1) are determined. Assuming that the downstream pressure is nearly equal to the averaged total pressure, $P_{02} \approx \overline{P}_{02}$, the supersonic loss can be calculated from Equation 13.55.

As an example, if the Mach number at the tip of a rotor is $M_1 = 1.2$, then the air inlet and outlet angles are $\beta_1 = 67°$ and $\beta_2 = 65°$. The total turning angle is $2°$. The turning angle from inlet to the normal shock wave location, $\Delta\theta$, depends on the ratio of the spacing to the chord (s/c). A reasonable

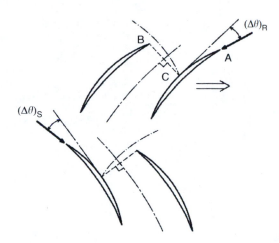

FIGURE 13.46 Shock loss model.

assumption for this angle is $1°$. From the Prandtl–Meyer tables, Mach number just before the normal shock wave is 1.24. The pressure ratio is $P_1/P_{01} = 0.39123$. From normal shock wave tables with Mach number 1.24, the pressure ratio is $P_{02}/P_1 = 2.52629$. Substituting in Equation 13.55 will give the losses $\bar{\omega} = 0.04887$.

Example 9   Figure 13.47 illustrates the air flow pattern through an axial fan. The fan blading and velocity triangles at the tip of a transonic fan rotor are shown in Figure 13.47a, b, d and e while the oblique shock pattern is shown in Figure 13.47c. The inlet and outlet angles of flow relative to the blades are $\beta_1 = 68.8°$ and $\beta_2 = 65°$, respectively. The flow approaches the rotor in the axial direction with an absolute velocity $C_1 = 200$ m/s. The inlet stagnation temperature and pressure are 300 K and 1.15 bar, respectively (radially uniform). The hub-to-tip ratio is 0.4 and the rotor tip diameter is 2.0 m. Inside the rotor blade passage two oblique shocks are generated. Experimental measurements of the static pressure over the blade-to-blade surface indicated that the turning angles of both oblique shocks are, respectively, $\delta_1 = 10°$ and $\delta_2 = 10°$.

At the outlet to the rotor, the relative Mach number equals that in region 3. The absolute inlet speed (C_1) and the specific work (W_s) are assumed constant at all radii.

1. Calculate the static and total pressure ratios across the rotor.
2. Calculate the absolute and relative velocities at the outlet to the rotor, then draw the velocity triangles.
3. Define which portion of the blade is subjected to supersonic relative inlet speed, then draw the velocity triangles at its minimum radius.

Solution: The given data are $T_{01} = 300$ K, $P_{01} = 1.15$ bar, and $C_1 = 200$ m/s. The sonic speed is then $a_1 = \sqrt{\gamma R T_1} = 335$ m/s.

The static conditions are then

$$T_1 = T_{01} - \frac{C_1^2}{2Cp} = 300 - \frac{200^2}{2 \times 1005} = 280 \text{ K}$$

$$P_1 = P_{01} \left(\frac{T_1}{T_{01}}\right)^{3.5} = 1.15 \, (0.933)^{3.5} = 0.90319 \text{ bar}$$

$$\rho_1 = \frac{P_1}{RT_1} = 1.1239 \text{ kg/m}^3$$

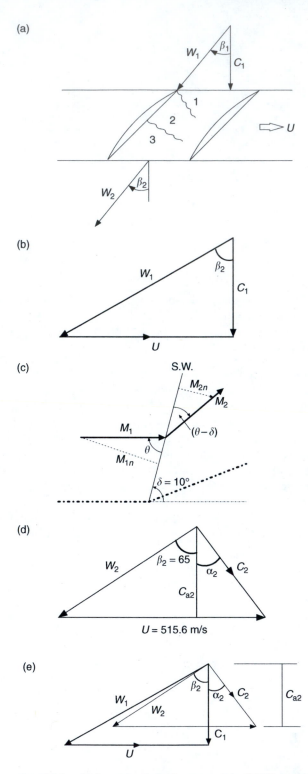

FIGURE 13.47 Transonic axial fan. (a) Layout and shock pattern at tip. (b) Inlet velocity triangle. (c) Oblique shock wave pattern. (d) Outlet velocity triangle. (e) Inlet and outlet velocity triangles in a common apex pattern. (f) Subsonic and supersonic domains on the blade.

(f)

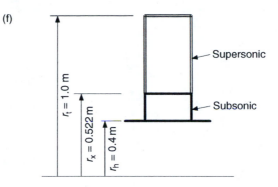

FIGURE 13.47   (Continued)

1. *The first shock wave:*
   The inlet velocity triangle is shown in Figure 13.47b

$$W_1 = \frac{C_1}{\cos \beta_1} = \frac{200}{\cos 68.8} = 553 \text{ m/s}$$

$$M_{W1} = \frac{W_1}{a} = 1.65$$

since $M_{W1} = 1.65$ and $\delta = 10°$, $\theta = 50°$.
   The normal component of the inlet relative Mach number is then

$$M_{W1n} = M_{W1} \sin \theta = 1.65 \ \sin 50 = 1.2639$$

From normal shock wave tables, $M_{W2n} = 0.8052$. As illustrated in Figure 13.47c

$$M_{W2} = \frac{M_{W2n}}{\sin(\theta - \delta)} = \frac{0.8052}{\sin 40°} = 1.2526$$

From normal shock wave tables also

$$\frac{P_2}{P_1} = 1.6976, \quad \frac{\rho_2}{\rho_1} = 1.4528, \quad \frac{T_2}{T_1} = 1.1682, \quad \frac{P_{02}}{P_{01}} = 0.9849$$

2. *The second shock wave:*
Since $M_{W2} = 1.2526$ and $\delta = 10°$, then from shock table [43], $\theta = 57°$. The normal component of the Mach number $M_{W2n} = 1.2526 \sin 57 = 1.0513$. From normal shock wave tables, $M_{W3n} = 0.952$. The Mach number is then

$$M_{W3} = \frac{M_{W3n}}{\sin (\theta - \delta)} = \frac{0.952}{\sin(57 - 2)} = 1.1621$$

From normal shock wave tables

$$\frac{P_3}{P_2} = 1.1229, \quad \frac{\rho_3}{\rho_2} = 1.08619$$

$$\frac{T_3}{T_2} = 1.03365, \quad \frac{P_{03}}{P_{02}} = 0.9985$$

$$P_{03} = \frac{P_{03}}{P_{02}}\frac{P_{02}}{P_{01}}P_{01} = (0.9985)(0.984918)(1.15) = 1.1325 \text{ bar}$$

The static conditions at state 3 are

$$\frac{P_3}{P_1} = \frac{P_3}{P_2}\frac{P_2}{P_1} = 1.906$$

$$P_3 = 1.7217 \text{ bar}$$

$$T_3 = \frac{T_3}{T_2}\frac{T_2}{T_1}T_1 = 338 \text{ K}$$

$$\rho_3 = \frac{\rho_3}{\rho_2}\frac{\rho_2}{\rho_1}\rho_1 = 1.7738 \text{ kg/m}^3$$

The sonic speed is $a_3 = 368.5$ m/s.

If at the outlet to rotor the Mach number is $M_{W3} = 1.1621$, then the rotor outlet relative speed is

$$W_2 = 1.1621 \times 368.5 = 428 \text{ m/s}$$

$$C_{a2} = W_2 \cos 65 = 428 \times 0.4226 = 180.8 \text{ m/s}$$

$$\frac{U}{C_{a2}} = \tan \beta_2 + \tan \alpha_2$$

$$\tan \alpha_2 = \frac{515.6}{180.8} - \tan 65 = 0.7073$$

$$\alpha_2 = 35.27°$$

$$C_2 = \frac{C_{a2}}{\cos \alpha_2} = 221.4 \text{ m/s}$$

$$M_{C2} = \frac{221.4}{368.5} = 0.6$$

The outlet velocity triangle at tip is plotted in Figure 13.47d, while the inlet and outlet velocity triangles having common apex are drawn in Figure 13.47e.

Then, to determine which part of the blade is subjected to supersonic flow, the location at which the inlet relative speed equals the sonic speed $\left(W_1 = a_1 = \sqrt{\gamma R T_1} = 335 \text{ m/s}\right)$ is first determined. Denoting the relative speed at this case with $W_{1s} = 335$ m/s, then

$$\cos \beta_{1s} = \frac{C_1}{W_{1s}} = \frac{200}{335} \quad \text{or} \quad \beta_{1s} = 53.3°$$

From the velocity triangle; similar to that drawn in Figure 13.47b, the rotational speed

$$U_s = C_1 \tan \beta_{1s} = 200 \times 1.3416 = 268.32 \text{ m/s}$$

Since the ratio between rotational speeds at two positions is related to their radii ratio,

$$\frac{U_s}{U_t} = \frac{r_s}{r_{tip}}$$

$$r_s = r_{tip}\frac{U_s}{U_{tip}} = (1.0)\left(\frac{268.3}{515.6}\right) = 0.52\,\text{m}$$

The hub radius is

$$r_h = \left(\frac{r_h}{r_t}\right) r_t = (0.4)(1.0) = 0.4\,\text{m}$$

As shown in Figure 13.47f, the part of the blades extending from hub (radius equals 0.4 m) to a radius of (0.544 m) represents the subsonic part of the blade. The outer portion is then a supersonic part. The fan is thus a transonic fan.

13.14 PERFORMANCE

13.14.1 SINGLE STAGE

Compressor balding has a limited range for incidence before losses became excessive and low efficiency is encountered. At incidence angles away from design point, blade stalling may occur and may lead to surging of the whole compressor, which may lead to engine damage and aircraft hazard. Equation 13.10a may be rewritten again as

$$\frac{Cp\Delta T_{0s}}{U^2} = 1 - \frac{C_a}{U}(\tan\alpha_1 + \tan\beta_2) \tag{13.10a}$$

where α_1 is the absolute inlet angle to the rotor or the outlet angle from the previous stator for a multistage compressor, and β_2 is the air outlet angle from the rotor. Generally, α_1 and β_2 remain unchanged for a fixed geometry compressor. Now defining the *loading coefficient* ψ as

$$\psi = \frac{Cp\Delta T_{0s}}{U^2} = \frac{\Delta C_u}{U}$$

and the *flow coefficient* ϕ as

$$\phi = \frac{C_a}{U}$$

Then

$$\psi = 1 - K\phi \tag{13.56a}$$

where

$$K = \tan\alpha_1 + \tan\beta_2 \tag{13.56b}$$

Plotting relation 13.56a, it is shown that the loading coefficient ψ is a linear function of the flow coefficient (or mass flow parameter) ϕ if α_1 and β_2 are constants (Figure 13.48). The actual performance drops rapidly at flow coefficients above and below the design operating condition.

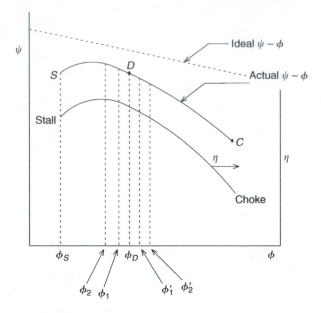

FIGURE 13.48 Actual and ideal performance of a compressor stage.

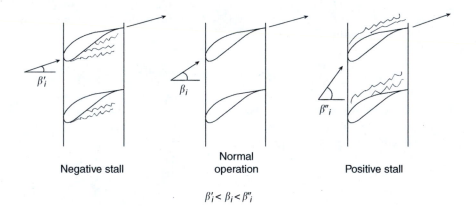

FIGURE 13.49 Positive and negative stall.

The efficiency for the actual performance is also plotted in Figure 13.48. The critical points in this figure are the stall and choking points. The velocity triangles corresponding to both points are also drawn in.

As shown in Figure 13.48, for the ideal case the temperature coefficient decreases as the flow coefficient increases. Moreover, if $\phi = 0$, then $\psi = 1$ or $\Delta C_u = U$, which provides a very large diffusion factor (D) and a drastic drop in efficiency. For satisfactory operation, the temperature coefficient must be in the range $\psi = 0.3 - 0.4$. The real performance indicates that for a flow coefficient ϕ smaller than its design value, that is, $C_a < (C_a)_d$, the relative inlet angle (β_1) and an incidence angle will increase, which will possibly lead to a positive stall. On the other hand, with an increase in the flow coefficient ϕ, that is, $C_a > (C_a)_d$, a decrease in the relative inlet angle (β_1) and a negative incidence results, which may lead to a negative stall (Figure 13.49).

It was emphasized previously that the stage performance is a function of the temperature coefficient, flow coefficient, and efficiency, or

$$\text{Stage performance} = f\left(\phi, \psi, \eta_s\right)$$

Designers, sometimes, employ a pressure coefficient instead of the temperature coefficient in estimating stage performance. The pressure coefficient is defined as

$$\psi_p = \frac{\Delta P_{0s}}{\rho_{01} U^2}$$

$$\frac{P_{02}}{P_{01}} = 1 + \frac{\Delta P_{0s}}{P_{01}} = \left(1 + \frac{\eta_s \Delta T_{0s}}{T_{01}}\right)^{\gamma/(\gamma-1)}$$

since $\Delta T_{0s}/T_{01} \ll 1$

$$\therefore \frac{\Delta P_{0s}}{P_{01}} \cong \frac{\gamma}{\gamma - 1} \frac{\eta_s \Delta T_{0s}}{T_{01}} = \frac{\eta_s Cp \Delta T_{0s}}{RT_{01}}$$

$$\frac{\Delta P_{0s}}{\rho_{01}} = \eta_s Cp \Delta T_{0s}$$

$$\therefore \psi_p \equiv \frac{\Delta P_{0s}}{\rho_{01} U^2} = \frac{\eta_s Cp \Delta T_{0s}}{U^2} = \eta_s \psi$$

$$\psi_p = f(\eta_s, \psi)$$

Thus, the stage performance can be expressed as

$$\text{Stage performance} = G\left(\phi, \psi_p, Cp\right)$$

13.14.2 Multistage Compressor

So far, we have considered an isolated single-stage performance over a range of off-design conditions. For a multistage compressor, the simplest method for examining its performance is to assume that the compressor consists of stages having identical characteristics, which is known as *stage stacking*. It is assumed that all stages operate at the design point ϕ_d. A small departure from the design point at the first stage will lead to a progressively increasing departure from design conditions from the first stage onward. Consider the following two cases:

1. If at constant speed, the compressor is subjected to a slight decease in mass flow rate; this slightly decreases the flow coefficient of the first stage to $\phi_1 < \phi_d$ and slightly increases the specific work or the pressure ratio, $\psi_1 > \psi_d$. Thus, the density of air entering the second stage will increase and the flow coefficient will decrease due to two reasons. First, the decrease in mass flow rate and the extra increase in the density of the air leaving the first stage. Then, the axial velocity of air entering the second stage will decrease ($\phi_2 < \phi_1$). This effect is propagated through the compressor and eventually some of the last stages will stall at decreased ϕ_n (positive stall).

2. Increasing the mass flow in the first stage from design point $\phi_1' > \phi_d$ decreases the temperature rise and pressure ratio at the inlet to the second stage, $\psi_1' < \psi_d$; similarly this will lead to another increase in the axial velocity and flow coefficient of air entering the second stage $\phi_2' > \phi_1'$. Thus, the pressure and temperature coefficients will decrease, or

$\psi_2' < \psi_1'$, due to the increase in air flow rate and decrease in air density. This continuous increase in axial-flow velocity is propagated through the compressor and some stage will be choked.

13.14.3 COMPRESSOR MAP

Compressor performance is usually presented in the form of a compressor map as described in the previous chapter. The dimensionless parameters influencing the compressor performance are derived as

$$\frac{P_{02}}{P_{01}}, \quad \eta_C = f\left(\frac{\dot{m}\sqrt{\gamma R T_{01}}}{P_{01}D^2}, \frac{ND}{\sqrt{RT_{01}}}\right)$$

which for a certain working fluid and a specific compressor are reduced to either

$$\frac{P_{02}}{P_{01}}, \quad \eta_C = f\left(\frac{\dot{m}\sqrt{T_{01}}}{P_{01}}, \frac{N}{\sqrt{T_{01}}}\right) \quad \text{or} \quad \frac{P_{02}}{P_{01}}, \quad \eta_C = f\left(\frac{\dot{m}\sqrt{\theta}}{\delta}, \frac{N}{\sqrt{\theta}}\right)$$

One point in this map represents the design point. However, compressor has to perform efficiently in other operating conditions, frequently identified as *off-design conditions*, namely, during engine starting, idling, reduced power, maximum power, acceleration, and deceleration.

Consequently, compressor has to operate satisfactorily over a wide range of rotational speed (rpm) and inlet conditions. However, if the annulus area and compressor blading are chosen to satisfy the design point condition; then at other conditions, they will not function correctly [5].

The compressor map of an axial compressor is shown in Figure 13.50. It is noticed that for a fixed value of $N/\sqrt{T_{01}}$, the range of mass flow rate is narrower than that in a centrifugal compressor. At high rotational speed, the lines become very steep and ultimately may be vertical.

Surge point occurs before the curves reach a maximum value. Stage characteristics are similar to overall characteristics with much lower pressure ratio. Mass flow is limited by choking in various stages. Before examining the compressor map, it is worth mentioning that the annulus area of the

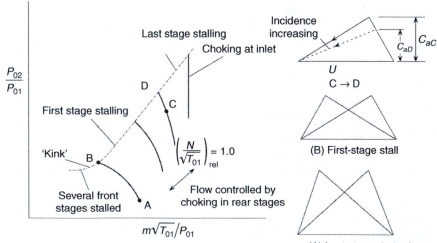

FIGURE 13.50 Compressor map with different operating conditions.

compressor decreases from the front to the rear stages. The reason is that the density increases in the rearward direction (due to pressure increase) and the axial velocity is maintained at a constant.

The following remarks concerning the compressor map can be described:

1. At rotational speeds lower than the design speed, $N/\sqrt{T_{01}} < \left(N/\sqrt{T_{01}}\right)_{design}$, choking will occur at the last stages. This low speed will provide smaller temperature and pressure rise compared to the design values. Thus, the air density will decrease and this reduction in density will lead to an increase in the axial velocity in the rear stages. Thus, choking will occur in the last stages, which will limit the mass flow. This limited mass flow, when coupled to the original density at inlet, will lead to smaller axial velocities at the first stage and a positive stall will follow. Thus, lower speeds result in choking of the rear stages and stalling of the first stage.
2. When the rotational speed increases, the air density is increased to the design value and the rear stages will pass all the flow provided by first stages. Then, choking will occur at the inlet of compressor. Thus, the vertical line is due to inlet choking.
3. At the design speed, the speed line is very steep, thus any decrease in the mass flow rate will bring a reasonable increase in pressure. This will also increase the density and decrease the axial velocity. This effect will propagate in the rearward direction and magnified to the limit of causing stall of the rear stages.

13.14.4 Stall and Surge

Stall is a situation of abnormal airflow through a single stage or multiple stages of the *compressor*, while the whole compressor stall, known as compressor surge, results in a loss of engine power. This power failure may only be momentary or may shut down the engine completely causing a *flameout*. The appropriate response to compressor stalls varies according to the engine type and situation, but usually consists of immediately and steadily decreasing thrust on the affected engine. The most probable cause of a compressor stall is a sudden change in the pressure differential between the intake and combustion chamber. The following factors can induce compressor stall:

- Engine overspeed
- Engine operation outside specified engineering parameters
- Turbulent or disrupted airflow to the engine intake
- Contaminated or damaged engine components

Jet aircraft pilots must be careful when dropping airspeed or increasing throttle.

Physically, rotating stall and surge phenomena in axial flow compressors are the most serious problems in the compressor. Severe performance and reliability problems arise if the compressor is unable to recover from stall. Surge and rotating stall can be described as follows [6]. Consider the performance map in Figure 13.51, plotted on the pressure coefficient and flow coefficient variables $\left(\psi_p - \phi\right)$. As the mass flow is decreased (throttle closed), pressure increases and the compressor will be stable and operate along the unstalled characteristics from points A to B. Further reduction in mass flow below point B will generate rotating stall and the operating point will suddenly drop to a new stable condition in rotating stall, point C. If the mass flow is increased (throttle opened), the compressor will not jump back to point B, but operate along the stalled characteristics to point D. Further increase of mass flow beyond D causes the operating point to jump to point E on the unstalled characteristics. Region B, C, D, E, B is the stall *hysteresis* region. Now, the *rotating stall* [6] is defined as a breakdown at one or more compressor blades. A rotating stall is a stagnated region of air, which moves in the opposite circumferential direction of rotor rotation, but at 50%–70% of the blade speed [10]. It may be described here. If the throttle is closed or the back pressure is increased, the blade loading or the incidence is increased. Then, the adverse pressure gradient on the

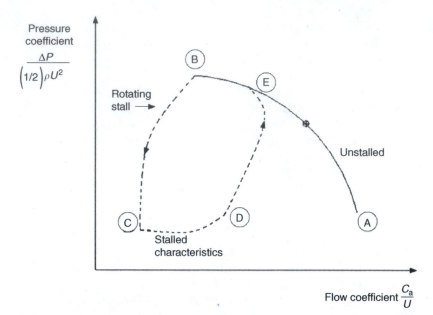

FIGURE 13.51 Performance characteristics for unstalled and stalled operation. (From D.C. Wisler, *Advanced Compressor and Fan Systems*. GE Aircraft Engines, Cincinnati, OH, USA, 1988 General Electric Co. USA, All Rights Reserved. With permission.)

suction surface increases the boundary layer growth, which leads to separation and stall. Not all the blade passages will stall at the same time. Stalling of a blade passage will lead to an increase in the incidence angle of the next passage in the opposite direction of rotation.

Thus, it will stall (Figure 13.52). There may be several stall cells in a blade row rotating at the same time (Figure 13.53). As described in Reference 17, part span is usually encountered in the front stages of the compressor with large aspect ratio blades and low speeds. It is usually encountered during takeoff or start up. Full span stall occurs in the medium speed range and its effect is much more damaging. In brief, rotating stall does not move axially in either direction although it may cause pressure waves to move upstream (compression waves) or downstream (rarefaction waves). Moreover, such standing rotating pockets prevent fresh air from the intake to pass over the stalled rotor blades leading to overheated blades and, consequently, engine wear and possible damage. The compressor surge occurs when there is complete breakdown of the flow field in the entire engine (not just blading). Upon surge, a compression component will unload by permitting the compressed fluid in downstream stages to expand in the upstream direction forming waves strong enough to lead to flow reversal. The compressor can recover and begin again to pump flow. However, if the cause of surge is not removed, then the compressor will surge again and will continue the surge/recovery cycle until some relief is provided. Otherwise, surge will result in violent oscillations in pressure, propagation of pressure waves, and the failure of the entire compression system.

To be cautious against surge led to defining what is called surge margin (very similar to the factor of safety in mechanical system design). It is defined as

$$\text{Stall margin} = \left\{ \frac{(P_{02}/P_{01})_{\text{stall}}}{(P_{02}/P_{01})_{\text{design point}}} \times \frac{\left(\dot{m}\sqrt{T_{01}/P_{01}}\right)_{\text{design point}}}{\left(\dot{m}\sqrt{T_{01}/P_{01}}\right)_{\text{stall}}} - 1 \right\} \%$$

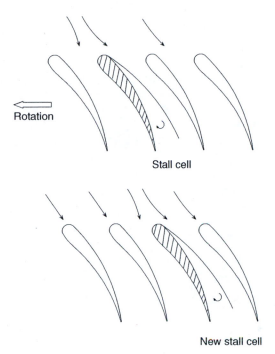

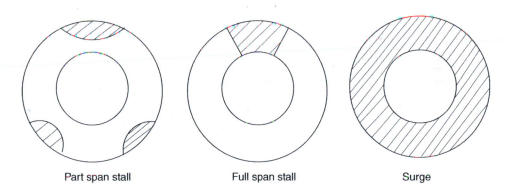

FIGURE 13.52   The propagation of rotating stall.

FIGURE 13.53   Rotating stall and surge.

When surge occurs in an operating engine, it causes at the least a stoppage of the airflow through the compressor [18], which leads to a straightforward expulsion of air out of the intake. In civil transports, this is evidenced by a loud bang emanating from the engine as the combustion process "backfires," which is unsettling to the passengers. This may be accompanied by an increased exhaust gas temperature and *yawing* of the aircraft in the direction of the affected engine. If repeated a number of times, it can do structural damage to the engine inlet or perhaps to the fan (in turbofan engines) or the fan frame. Violent compressor surge might completely destroy the engine and set it on fire. However, if such structural damage is absent, then the engine will usually recover and continue to run. In supersonic aircraft, the consequences can be more serious, because the blockage due to surge can generate a shock wave in the inlet and that may produce overpressures large enough to damage the structure.

13.14.5 Surge Control Methods

As described earlier, surge causes serious problems to compressor, engine, and aircraft. Though it is unavoidable, its recovery is affordable. Since rotating stall represents the onset of surge, stall could be avoided so that surge will not be seen. However, this is an impossible task due to different operating conditions for the compressors installed in aero engines powering airplanes during any trip or mission. For compressors in industrial power plants too, different conditions are found (e.g., during starting, idling, acceleration, deceleration, etc). Foreign object debris (FOD) is another serious circumstance that influences the fluid mass flow rate. Compressor designers proposed and designed some methods for surge control. In the design phase, the single-spool engine is replaced by two-spool engine. Each spool (and consequently compressor) rotates at a different speed. During operation, two alternatives are found, namely, the variable geometry compressor and air bleed. The later techniques are activated by the engine control unit (ECU), which represents a computerized control system that has its inputs from sensors within the engine.

13.14.5.1 Multispool Compressor

Since reduction in compressor speed from design value will cause an increase of incidence in the first stage and a decrease of incidence in the last stage, the incidence could be maintained at the design value by decreasing the speed of the first stage and increasing the speed of the last stage. These conflicting requirements can be met by splitting the compressor into two (or more) sections, each driven by a separate turbine (low-pressure turbine drives the low-pressure compressor and high-pressure turbine drives the high-pressure compressor). Examples for the two-spool engines are CF6, CFM56, GE90, V2500, and JT9 turbofan engines. Examples for three spools are RB211 and the Trent family. Typical compressor maps for the low- and high-pressure spools are illustrated in Figures 13.54 and 13.55 [19].

13.14.5.2 Variable Vanes

Variable stators may be used to unstall those stages precipitating general compressor stall. Since it is the front-end and/or the back-end stages of a compressor that are most likely to encounter stall, the use of variable geometry is normally confined to these stages. Furthermore, with the tendency

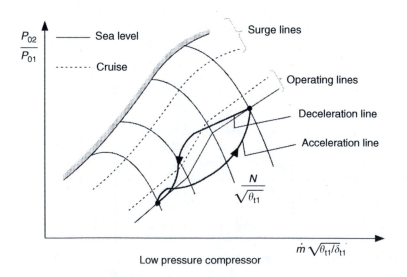

FIGURE 13.54   Map for low-pressure compressor.

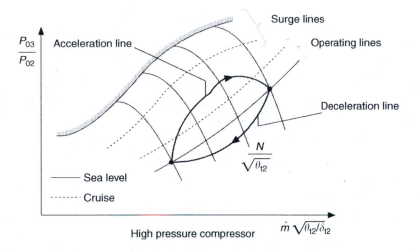

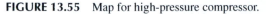

FIGURE 13.55   Map for high-pressure compressor.

of the back-end stages to be more lightly loaded (due to lower Mach numbers) the stall problem is not as severe as in the back stages. As a result, variable stators are normally confined to the front stages and their principal use is to provide acceptable starting and/or low-speed acceleration of the engine. Positive stall is encountered when the inlet relative angle, β_1, and consequently the incidence angle are increased. To reduce this value, two methods are employed. In the first method, the IGV or the stators of the first stages are rotated away from the axial direction to increase the absolute angle at inlet α_1 and thus reduce the relative inlet angle β_1. As an alternative, the rear part of the blade is rotated, thus becoming over cambered, to do the same job as the completely rotated stators (Figure 13.56).

13.14.5.3   Air Bleed

Bleeding of air downstream of a stalled stage or stages will allow the flow to increase ahead of the bleed location. This increase in airflow will in turn result in a reduction of the rotor incidence and unstall the blade (Figure 13.57). The amount of the bleed air will exceed the increase in front flow by a small amount. The flow to the back stages will be thereby reduced. This reduction of flow reduces the axial velocity component to the rear stages and will increase their work output, compensating to a certain degree the loss of output in the front stages (refer to Equation 13.56a).

Example 10   The compressor in a single-spool turbojet engine is composed of a four-stage axial compressor and a single-stage centrifugal compressor. The rotational speed is 290 rev/s. The detailed data of both compressors are given as

Axial compressor (A):

Temperature rise per stage	25 K
Polytropic efficiency η_{pc}	0.9

Centrifugal compressor (B):

Impeller tip diameter	0.5 m
Power input factor ψ	1.04
Slip factor σ	0.9
Efficiency η_c	0.9

The inlet total conditions to the compressor section are

$$T_{01} = 298\,\text{K}, \quad P_{01} = 1.0\,\text{bar}$$

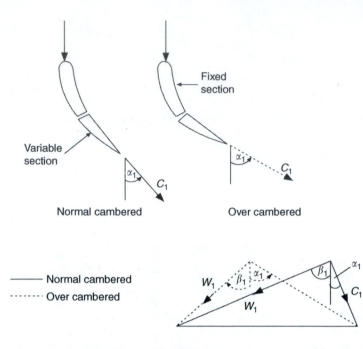

FIGURE 13.56   Variable stator blades.

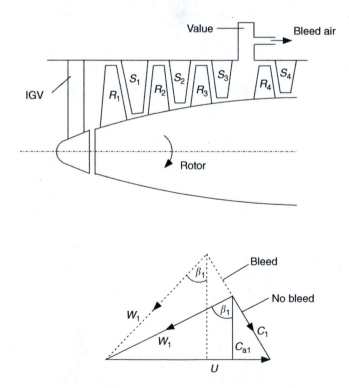

FIGURE 13.57   Bleed as a surge controller.

   As shown in Figure 13.58a, two arrangements are investigated, namely, the axial compressor (A) is installed upstream of the centrifugal compressor (c) (arrangement 1) and the reverse (arrangement 2).

(a) Compare the overall stagnation pressure ratio and temperature rise in both cases.
(b) Plot these operating conditions on the maps of both compressors ($m° = 20$ kg/s).
(c) Calculate the overall efficiency for both arrangements.
(d) Same as in (a) and (c) but with a reduced value for the efficiency of the centrifugal compressor ($\eta_c = 0.8$).
(e) Compare between both layouts from aerodynamics, performance, and mechanical points of view.

Solution:

(a) The temperature difference across these compressors is given by

$$(T_{0e} - T_{0i})_{\text{axial}} = N\Delta T_{0s}$$

$$(T_{0e} - T_{0i})_{\text{centrifugal}} = \frac{\Psi \sigma U_{\text{tip}}^2}{Cp}$$

   Both are constants for the two arrangements.

Arrangement 1
The pressure and temperature at the inlets and outlets of compressor in arrangement (1) is shown in Figure 13.58b.

$$\Delta T_{0\text{axial}} = N\Delta T_{0s} = 4 \times 25 = 100 \text{ K}$$

$$T_{01} = 298 \text{ K}, \quad T_{02} = 398 \text{ K}$$

$$\frac{n}{n-1} = \eta_{PC}\frac{\gamma}{\gamma - 1} = 0.9\frac{1.4}{0.4} = 3.15$$

$$\pi_{\text{axial}} = \left(1 + \frac{\Delta T_0}{T_{01}}\right)^{n/(n-1)}$$

$$\pi_{\text{axial}} = \left(1 + \frac{100}{298}\right)^{n/(n-1)} = (1.3356)^{3.15} = 2.488$$

For the centrifugal compressor

$$U_{\text{tip}} = \pi d_t N = \pi \times 0.5 \times 290 = 455.5 \text{ m/s}$$

$$(\Delta T_0)_{\text{centrifugal}} = \frac{\Psi \sigma U_{\text{tip}}^2}{Cp}$$

$$(\Delta T_0)_{\text{centrifugal}} = \frac{1.04 \times 0.9 \times (455.5)^2}{1005} = 193 \text{ K}$$

$$T_{03} = T_{02} + (\Delta T_0)_{\text{centrifugal}} = 591 \text{ K}$$

$$\pi_{\text{centrifugal}} = \left(1 + \eta_c \frac{\Delta T_0}{T_{02}}\right)^{\gamma/(\gamma-1)} = \left(1 + 0.9\frac{193}{398}\right)^{3.5} = 3.5522$$

$$\pi_{\text{total}} = \pi_{\text{axial}} \times \pi_{\text{centrifugal}} = 8.838$$

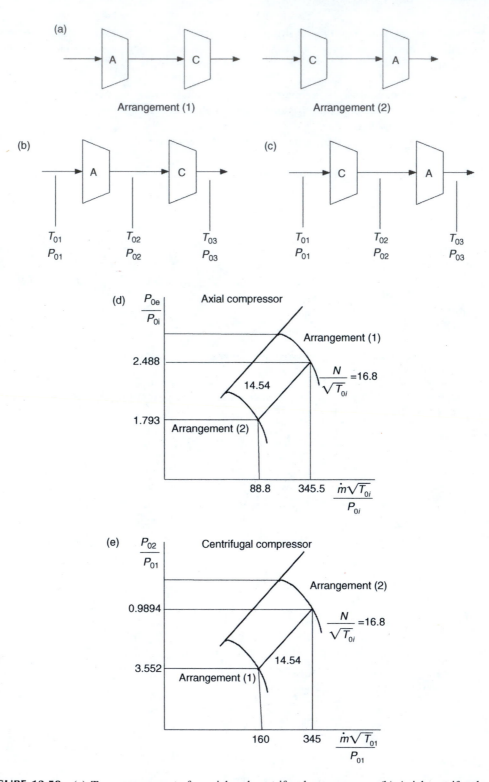

FIGURE 13.58 (a) Two arrangements for axial and centrifugal compressors. (b) Axial-centrifugal compressor (arrangement 1). (c) Centrifugal-axial layout (arrangement 2). (d) Compressor map for axial compressor. (e) Compressor map for centrifugal compressor. (f) Length and diameter pattern for the combined compressor in the two arrangements.

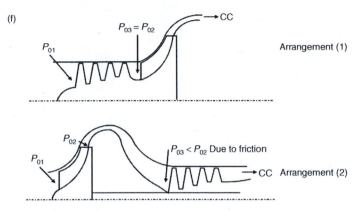

FIGURE 13.58 (Continued)

TABLE 13.11
Output Results for an Axial Compressor

Arrangement	T_{0inlet} K	T_{0exit} K	P_{0i} bar	P_{0e} bar	π	$\dot{m}\sqrt{T_{0i}}/P_{0i}$	$N/\sqrt{T_{0inlet}}$
1	298	398	1.0	2.488	2.488	345.5	16.8
2	491	591	4.989	8.9465	1.793	88.83	13.09

TABLE 13.12
Output Results for a Centrifugal Compressor

Arrangement	T_{0inlet} K	T_{0exit} K	P_{0i} bar	P_{0e} bar	π	$\dot{m}\sqrt{T_{0i}}/P_{0i}$	$N/\sqrt{T_{0inlet}}$
1	398	591	2.488	8.838	3.5522	160.4	14.91
2	298	491	1.0	4.9897	4.9897	345.2	13.09

The overall temperature rise across the two compressors is constant in both arrangements and equal to 293 K.

Arrangement 2

(a) The pressure and temperature at the inlets and outlets of compressor in arrangement (2) is shown in Figure 13.58c.

$$\pi_{centrifugal} = \left(1 + 0.9\frac{193}{298}\right)^{3.5} = (1.5828859)^{3.5} = 4.98969$$

$$\pi_{axial} = \left(1 + \frac{100}{491}\right)^{3.15} = 1.793$$

$$\pi_{total} = 8.9465$$

(b) The results for both arrangements are summarized in Tables 13.11 and 13.12 for axial and centrifugal compressors, respectively. The compressor map for axial compressor is shown in Figure 13.58d, while Figure 13.58e illustrates the compressor map for centrifugal compressor.

(c) The overall efficiency ($\eta_a = \eta_c = 0.9$)

$$\eta = \frac{T_{02'} - T_{01}}{T_{02} - T_{01}} = \frac{T_{01}\left\{(p_{02}/p_{01})^{(\gamma-1)/\gamma} - 1\right\}}{T_{02} - T_{01}}$$

$$\eta_{\text{arrangement}(1)} = \frac{298\left\{(8.839)^{0.4/1.4} - 1\right\}}{293} = \frac{298\,(1.8638 - 1)}{293} = 0.8785$$

$$\eta_{\text{arrangement}(2)} = \frac{298\left\{(8.9465)^{0.4/1.4} - 1\right\}}{293} = \frac{298\,(1.8702 - 1)}{293} = 0.8851$$

(d) The efficiency of the axial compressor is unchanged while the efficiency of the centrifugal compressor drops, or $\eta_a = 0.9$, $\eta_c = 0.8$.

Arrangement 1

$$T_{01} = 298 \text{ K}$$

The compression ratio for the axial compressor will not change.

$$\therefore \pi_{\text{axial}} = 2.488, \quad T_{02} = 398 \text{ K}$$

$$\pi_{\text{centrifugal}} = \left(1 + 0.8\frac{193}{398}\right)^{3.5} = (1.3879396)^{3.5} = 3.1496$$

$$T_{03} = 591 \text{ K}$$

$$\pi_{\text{total}} = 7.836$$

Arrangement 2

$$T_{01} = 298 \text{ K}$$

$$\pi_{\text{centrifugal}} = \left(1 + 0.8\frac{193}{298}\right)^{3.5} = 4.3099$$

$$T_{02} = 491 \text{ K}$$

$$\pi_{\text{axial}} = \left(1 + \frac{100}{491}\right)^{3.15} = 1.793$$

$$\pi_{\text{total}} = 7.7276$$

The overall efficiency (Table 13.13)

$$\eta_{\text{arrangement}(1)} = \frac{298\left\{(7.836)^{0.4/1.4} - 1\right\}}{293} = \frac{298\,(1.8007 - 1)}{293} = 0.81436$$

$$\eta_{\text{arrangement}(2)} = \frac{298\left[(7.7276)^{0.4/1.4} - 1\right]}{293} = \frac{298\,(1.7936 - 1)}{293} = 0.8071$$

TABLE 13.13
Summary for Both Arrangements

	(1)		(2)	
Arrangement	$\eta_a = \eta_c = 0.9$	$\eta_a = 0.9, \eta_c = 0.8$	$\eta_a = \eta_c = 0.9$	$\eta_a = 0.9, \eta_c = 0.8$
Overall pressure ratio	8.838	7.836	8.9465	7.7276
Overall efficiency	0.8785	0.81436	0.8851	0.8071

(e) From the above calculations, which is summarized in Table 13.13, it can be deduced that

1. The overall pressure ratio is slightly higher for arrangement (2) if both compressors have the same efficiency, but if the efficiency of the centrifugal compressor is dropped then arrangement (1) gives a higher pressure ratio.
2. When the efficiency of centrifugal compressor is less than that of the axial one, then the overall efficiency of the compressor in arrangement (1) is larger than the overall efficiency in arrangement (2).
3. Arrangement (1) is appropriate for the installation of combustion chamber while arrangement (2) needs some piping system between both compressors, which adds some mechanical complications to the compressor section. This piping increases the skin friction and thus reduces the total pressure upstream of the axial compressor. The external drag force in arrangement (2) is larger than that in arrangement (1) as shown in Figure 13.58f.

13.15 CASE STUDY

A complete design for a seven-stage axial compressor is presented here. The compressor pressure ratio is 4.15 and the air mass flow rate is 20 kg/s. The ambient temperature and pressure are 288 K and 101 kPa. The rotational speed is 15,000 rpm. The mean flow calculations are given in Table 13.14. The airflow properties along the mean radius of all stages will be given hereafter. Only the second stage will be examined in detail where the variations of different aerodynamic and thermodynamic properties from hub to tip will be given using the three design method. The reader can follow similar steps for the other stages. Finally, the seven-stage compressor will be constructed.

13.15.1 MEAN SECTION DATA

The temperature rise, pressure ratio, flow angles, hub and tip radii, hub-to-tip ratio, and degree of reaction at the mean section of the seven stages are given in Table 13.14. The same procedure described in Sections 13.8 through 13.10 is followed here.

13.15.2 VARIATIONS FROM HUB TO TIP

The free vortex, exponential, and first-power methods are employed for calculating the variations in the flow angles.

Tables 13.15 and 13.16 give the variation of the flow angles and degree of reaction at hub and tip using free vortex method.

Tables 13.17 and 13.18 give the variation of the flow angles and degree of reaction at hub and tip using exponential method.

TABLE 13.14
Mean Radius Characteristics

Stage	1	2	3	4	5	6	7
ΔT_0 (K)	27.35	25.24	23.86	22.50	22.50	22.50	19.85
π	1.3325	1.2756	1.2386	1.2084	1.1955	1.1841	1.1524
α_1	0	8.91	9.21	9.22	9.22	9.22	9.22
α_2	35.06	40.0	40.16	40.17	40.17	40.17	37.37
β_1	60.62	58.30	58.22	58.21	58.21	58.21	58.21
β_2	47.04	43.13	43.0	43.0	43.0	43.0	45.35
α_3	8.91	9.21	9.22	9.22	9.22	9.22	0
Λ	0.8024	0.7196	0.7168	0.7167	0.7166	0.7166	0.7393
ζ	0.5000	0.5722	0.6275	0.6714	0.7066	0.7368	0.7629
r_h (m)	0.1131	0.1234	0.1308	0.1363	0.1404	0.1439	0.1468
r_t (m)	0.2261	0.2158	0.2084	0.2029	0.1988	0.1953	0.1924

TABLE 13.15
Results at Hub Section Using Free Vortex Method

Stage	1	2	3	4	5	6	7
α_1	0.6033	13.9411	14.7497	15.3312	14.8955	14.5530	13.8522
α_2	46.1905	40.0043	46.1618	44.3795	43.5140	42.8187	39.4638
β_1	49.5638	58.3016	47.8898	49.0587	50.3051	51.2802	52.2299
β_2	8.0584	43.1313	18.1698	24.1482	27.5327	30.1270	35.5230
α_3	9.6806	9.2082	9.6782	9.5291	9.4608	9.4077	0
Λ	0.5553	0.7196	0.5237	0.5610	0.5868	0.6064	0.6520

TABLE 13.16
Results at Tip Section Using Free Vortex Method

Stage	1	2	3	4	5	6	7
α_1	0.3017	8.0837	9.3814	10.4300	10.6450	10.8290	10.6542
α_2	27.5295	32.6567	33.1656	33.3068	33.8570	34.3237	32.1321
β_1	67.0605	64.7190	63.6313	62.7436	62.1591	61.6569	61.3031
β_2	61.5658	58.2889	56.8135	55.7395	54.6639	53.7210	54.2055
α_3	8.5645	8.7875	8.8125	8.8196	8.8475	8.8716	0
Λ	0.8888	0.8267	0.8124	0.8021	0.7937	0.7863	0.7975

TABLE 13.17
Results at Hub Section Using Exponential Method

Stage	1	2	3	4	5	6	7
α_1	−9.4648	3.4821	6.3719	8.4192	9.0155	9.4839	9.7826
α_2	38.1211	39.5878	38.9146	38.2140	38.1791	38.1577	35.6433
β_1	53.7386	50.3613	50.8903	51.3491	52.1309	52.7709	53.3723
β_2	16.0130	17.6198	23.3691	27.7422	30.2142	32.1828	36.8703
α_3	9.0875	9.1807	9.1372	9.0932	9.0911	9.0897	0
Λ	0.6047	0.5099	0.5486	0.5780	0.5990	0.6154	0.6583

TABLE 13.18
Results at Tip Section Using Exponential Method

Stage	1	2	3	4	5	6	7
α_1	5.3419	14.3196	14.8613	15.2608	14.9480	14.6812	13.8377
α_2	33.1587	39.6580	39.1979	38.5404	38.4783	38.4300	35.4744
β_1	66.3833	64.1605	63.1053	62.2515	61.6808	61.1976	60.8916
β_2	62.0826	59.1892	57.4469	56.1676	54.9707	53.9351	54.2632
α_3	8.8122	9.1853	9.1553	9.1135	9.1096	9.1066	0
Λ	0.9012	0.8396	0.8223	0.8098	0.7998	0.7912	0.8011

TABLE 13.19
Results at Hub Section Using Exponential Method

Stage	1	2	3	4	5	6	7
α_1	−16.2068	−4.1368	−0.0633	2.9039	4.1835	5.2170	6.2860
α_2	33.0091	33.6940	33.6545	33.5219	33.9764	34.3799	32.4509
β_1	56.5061	53.3604	53.2427	53.2377	53.6923	54.0817	54.3961
β_2	20.6067	22.4275	26.9203	30.4135	32.3290	33.8832	38.0480
α_3	8.8391	8.8048	8.8371	8.8304	8.8536	8.8746	0
Λ	0.6421	0.5415	0.5695	0.5926	0.6098	0.6235	0.6640

TABLE 13.20
Results at Tip Section Using First-Power Method

Stage	1	2	3	4	5	6	7
α_1	12.1762	22.6717	21.8930	21.2576	20.1468	19.2322	17.5194
α_2	41.5821	50.2022	47.6596	45.5026	44.3990	43.5324	39.4774
β_1	65.7243	63.8804	62.7671	61.8879	61.2868	60.7944	60.5044
β_2	63.5145	61.8772	59.2382	57.3591	55.8235	54.5476	54.5261
α_3	9.3171	10.0627	9.8127	9.6217	9.5307	9.4622	0
Λ	0.9161	0.8546	0.8335	0.8184	0.8065	0.7966	0.8051

Tables 13.19 and 13.20 give the variation of the flow angles and degree of reaction at hub and tip using first-power method.

13.15.3 DETAILS OF FLOW IN STAGE NUMBER 2

The variations of the aerodynamics and thermodynamic properties from hub to tip of the flowing air through the fourth stage are calculated and plotted here.

(a) Degree of reaction is illustrated in Figure 13.59.
(b) The angle $(\beta_1 - \beta_2)$ is shown in Figure 13.60 for first power method.
(c) Blade deflection angle $(\beta_1 - \beta_2)$ shown in Figure 13.61.

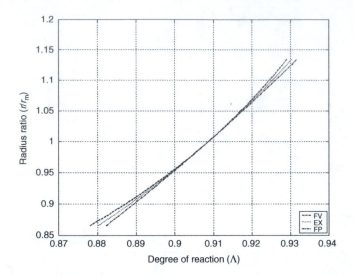

FIGURE 13.59   Variation of degree for reaction from hub to tip.

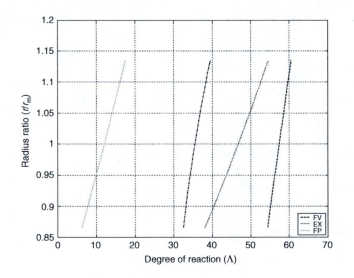

FIGURE 13.60   Variation of stage angles from hub to tip.

(d) The absolute and relative velocities at inlet and outlet C_1, C_2, W_1, W_2 are shown in Figure 13.62 while the tangential and axial components of the absolute velocities $(C_{u1}, C_{u2}, C_{a1}, C_{a2})$ are shown in Figure 13.63.

(e) Flow angles $\beta_1, \beta_2, \alpha_1, \alpha_2$ are illustrated in Figure 13.64.

(f) Variation of inlet condition from hub to tip T_1, P_1, ρ_1 are given in Figure 13.65.

13.15.4   NUMBER OF BLADES AND STRESSES OF THE SEVEN STAGES

Table 13.21 gives the number of blades for the stator and rotor blade rows for all stages. Moreover, the maximum stress and the factor of safety for the rotors in each stage are given. It is assumed that first three stages are made up of titanium alloy while the remaining four are made up of steel.

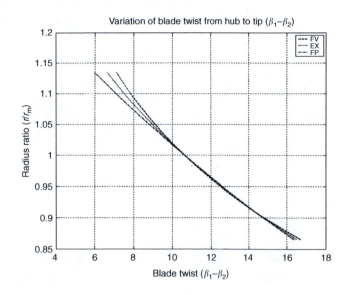

FIGURE 13.61   Variation of blade deflection from hub to tip.

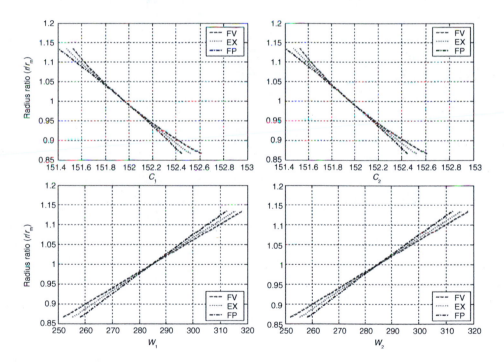

FIGURE 13.62   Variation of velocities from hub to tip C_1, C_2, W_1, W_2.

13.15.5   COMPRESSOR LAYOUT

A complete layout for the seven stages is drawn in Figure 13.66. The number of blades increases in the rearward direction while the blade height as well as their chord at root increases in the forward direction.

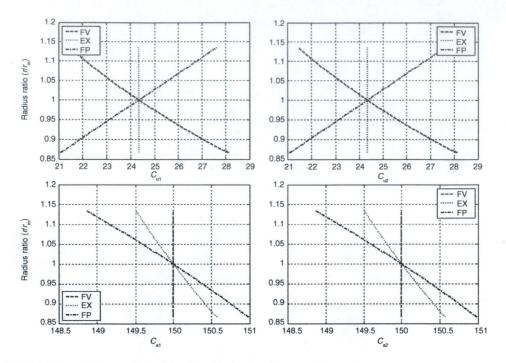

FIGURE 13.63   Variation of velocities from hub to tip ($C_{u1}, C_{u2}, C_{a1}, C_{a2}$).

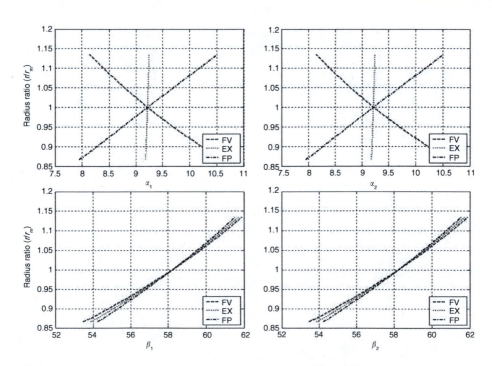

FIGURE 13.64   Variation of absolute and relative angles from hub to tip $\beta_1, \beta_2, \alpha_1, \alpha_2$.

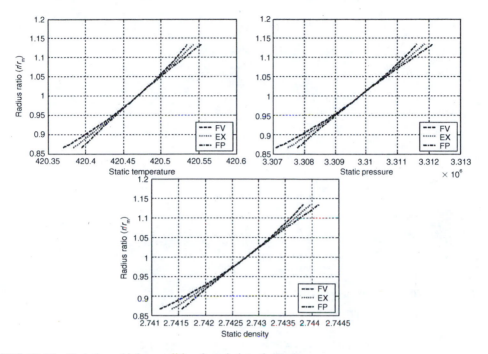

FIGURE 13.65 Variation of inlet condition from hub to tip T_1, P_1, ρ_1.

TABLE 13.21
Number of Blades and Maximum Stresses for the Rotors of the Seven Stages

Stage	1	2	3	4	5	6	7
No. of blade (rotor)	28	34	42	48	56	62	56
No. of blade (stator)	32	39	47	55	63	71	67
σ_{max} (MPa)	158.0	137.0	121.0	190.7	172.3	156.3	142.3
Material	Titanium	Titanium	Titanium	Steel	Steel	Steel	Steel
Factor of safety	4.8	5.5	6.29	2.62	2.90	3.2	3.52

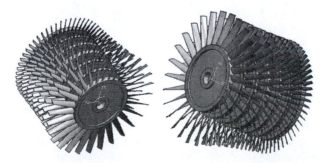

FIGURE 13.66 Three-dimensional illustration for seven-stage axial compressor.

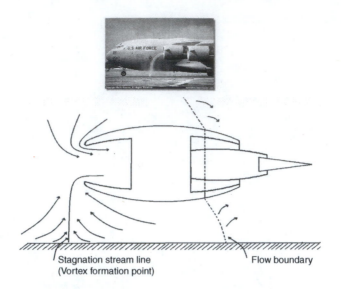

Stagnation stream line   Flow boundary
(Vortex formation point)

FIGURE 13.67   Particle ingestion during aircraft landing.

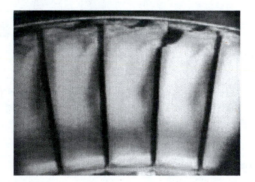

FIGURE 13.68   Erosion of axial compressor.

13.16 EROSION

As described in Chapter 12, the flowing air into gas turbine/aero engines is frequently laden with solid particles such as sand, dust, fly ashes, ice, and volcanic ashes. These particles are ingested into the engine from outside. Owing to their high inertia, they do not follow the carrier phase streamlines and tend to impact the internal surfaces. In case of aero engines, sand particles are ingested into the core during all flight phases. However, the critical flight phases are takeoff and landing ones. Particles of different (but large diameter) sizes are sucked by the intakes during takeoff and landing. Moreover, operation of thrust reversal leads to reingestion of the hot gases and debris during landing (Figure 13.67).

Particle–surface interactions lead to pitting and cutting of the blade's leading and trailing edges, increasing the tip clearance due to changes in the blade geometry and increasing the blade surface roughness (Figure 13.68). This will cause changes in the aerodynamic performance of the fan/compressor modules, which influences their stability, efficiency, and operating range and results in a progressive and rapid degradation of the engine performance. The damage to the blades is caused mainly by direct material impingement of large particles, and even by recirculation of fine particles

800 h, 2000 cycles						8700 h, 5500 cycles

FIGURE 13.69   Erosion of ninth-stage HPC of JT9D.

due to secondary flow through the blade passage [20]. Particles as small as 1–5 μm lead to severe erosion in axial-flow turbomachinery [21,22].

Extensive studies by Pratt and Whitney Aircraft Group proved that erosion was one of the four sources of the JT9D engine performance deterioration [23]. This study was based on historical records and data obtained from five airlines covering the period from early 1973 through December 31, 1976. Erosion of airfoils and outer air seals of compressor resulted in increased roughness and bluntness, loss of camber, loss of blade length, and increased operating clearances. Figure 13.69 shows the effect of flight cycles on the JT9D ninth-stage HPC rotor erosion. The estimated performance losses due to erosion after 3500 flights were about 40% from the total losses.

The strong pressure gradient within the flow passages due to turning of fluid within the rotor/stator blades affects the trajectories of particles and their impact locations [24]. As reported [25], within 6000–8000 h of jet engine operation, compressor blades had blunt leading edges with an increased surface roughness [25]. The inspection of a new CFM56 turbofan after particle ingestion on its high-pressure compressor showed similar effects [26].

Erosion of compressor blades reduces their efficiency. Moreover, the operating line becomes closer to the surge line.

The three-dimensional particle trajectory calculations through axial-flow turbomachines were first reported in Reference 27. Later on, numerous studies were performed for predicting three-dimensional particle trajectories through rotating and stationary turbomachinery blade rows (e.g., refer to [28]). These studies showed many particles impacting the blade leading edges, pressure sides, and trailing edges.

Experimental work was also performed by ingestion of 2.5 kg of sand to examine the effect of erosion on axial compressor performance. These measurements [29] showed that the blade's leading edge and the pressure side were severely eroded with an increase in surface roughness. As a result, the pressure distribution around the blade profile changed. This causes an early boundary layer transition. The blade suction surface remained unaffected for most of the experiments. These tests also showed that the loss in efficiency was related to erosion of the blade's leading edge and increased tip leakage. The stage performance deterioration is linearly related to the erosion rate, and there is an overall shift of the pressure coefficient and efficiency versus flow coefficient. The noticeable conclusions were 3.5% deterioration in stage loading coefficient and 4% in efficiency.

Erosion of axial fans and compressors in aero engines results in blunting their leading edges and reduction of their chords. The aerodynamic performance of a transonic compressor rotor operating with variable chord length blades was examined in Reference 30. The test results indicated that both the surge margin and compressor efficiency are decreased with the decrease in blade chord.

A better analysis of erosion may be developed if the successive changes in blade geometry are continuously employed in the three steps of erosion previously described in Chapter 12. The study may be examined in stepwise time intervals. Thus after a certain time period, say 1000 h, the new shape of the blades is calculated and the gas flow, particle trajectories, and the resulting erosion rates

TABLE 13.22
Parameters of the Particles at Both Takeoff and Cruise Conditions

Particle Parameter	Takeoff	Cruise
Material	Sand	Fly ashes
Density (kg/m^3)	2650	600
Concentration (mass of particles per unit volume of the air, kg/m^3)	0.176×10^{-3}	0.176×10^{-4}
Particle mass flow rate (gm/s)	97.5	8.83
Particle initial velocity (m/s)	128.6	133.5
Initial injection location	Intake inlet	Intake inlet

may be calculated again and used as initial data for the next 1000 time interval. This procedure was used to examine erosion of turbomachines by dividing the 12,000 flight hours into four segments each 3000 h [31,32].

Erosion of an axial fan exposed to particulate flow [33] caused loss in the fan performance due to tip clearance increase and changing in stalling incidence caused by changes in the blade profile, particularly near the leading edge tip region.

The effect of erosion (due to particulated flow) on the performance of the axial transonic fan of a high-bypass ratio turbofan engine (similar to GE-CF6) was investigated [34]. The dynamic behavior of the ingested particles through the intake and fan were investigated in both takeoff and cruise conditions. The effect of the particle size on its trajectory, impact location, and predicted erosion rate was also examined. Moreover, a Rosin Rambler particle diameter distribution was assumed at takeoff and cruise conditions to simulate the problem in a more realistic manner. The mean particle radius during takeoff is 150 μm and during cruise is 15 μm. Table 13.22 shows different parameters of the particles at both takeoff and cruise conditions.

In general, it is found that the deviation of the particle trajectories from the streamlines increases with an increase in the particle size. Consequently, large particle trajectories are greatly influenced by the centrifugal force and successively impact the walls at high speeds and centrifuge faster after impacts. Small-size particles are accelerated to high velocities and their trajectories are as close as possible to the gas streamlines. These particle trajectories define the impact locations with the intake casing, fan nose, blade pressure and suction sides, and the fan hub and casing.

The erosion rate of the fan blade during takeoff is shown in Figure 13.70. Along the blade span, the erosion rate is concentrated both near the trailing and leading edges of the pressure side and near the leading edge of the suction side. On the suction side, the upper half of the blade is exposed to erosion rate ranging from 6.6×10^{-6} to 1.3×10^{-5} mg/g. Other regions on the suction side are less affected by erosion. On the pressure side, the region of maximum erosion rate appears at the corner of the blade between the tip and trailing edge. The value of the erosion rate reaches its maximum in this region, about 3.34×10^{-5} mg/gm. This is because it is exposed to high frequency and high erosion parameter. Moreover, the region on the blade pressure side at about the upper third of the blade height and about 20% chord from the trailing edge shows an increased erosion rate as well. Also, we can see that, near the leading edge of the pressure side, the erosion rate contours are condensed and indicate a moderate value (about 10^{-5}) along the span.

The results of erosion during cruise conditions are shown in Figure 13.71. One can observe high erosion and penetration rate values at the pressure side leading edge. The pressure side trailing edge is exposed to moderate values of erosion and penetration rates. On the blade suction side, the erosion and penetration rates are limited to the blade leading edge at the upper half span.

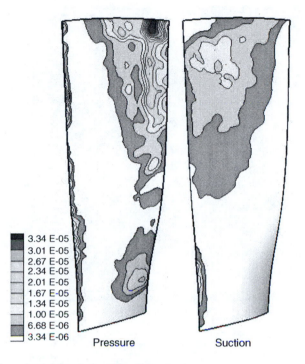

■	3.34 E-05
	3.01 E-05
	2.67 E-05
	2.34 E-05
	2.01 E-05
	1.67 E-05
	1.34 E-05
	1.00 E-05
	6.68 E-06
	3.34 E-06

Pressure   Suction

FIGURE 13.70   Erosion of axial fan during takeoff (mg/gm).

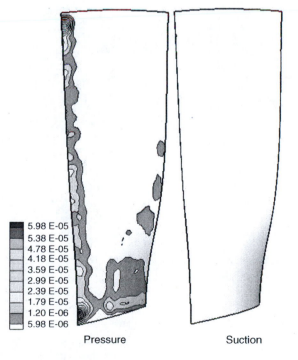

■	5.98 E-05
	5.38 E-05
	4.78 E-05
	4.18 E-05
	3.59 E-05
	2.99 E-05
	2.39 E-05
	1.79 E-05
	1.20 E-06
	5.98 E-06

Pressure   Suction

FIGURE 13.71   Erosion of axial fan during cruise (mg/gm).

FIGURE 13.72   Erosion of the fan of a high-bypass ratio turbofan engine.

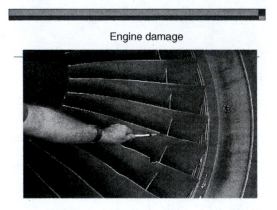

FIGURE 13.73   Foreign object damage of an axial fan of a turbofan engine.

Figure 13.72 illustrates the real shape of eroded fan blades due to solid particle ingestion [44]. Close similarity exists between the calculated and real erosion patterns. Figures 13.73 and 13.74 illustrate the damage in fan blades after exposure to different foreign objects, known as FOD. This abbreviation equally stands for foreign object damage or foreign object debris. These causes of FOD could be wildlife, soft object, ice/hard objects, as well as several unknown sources [35].

13.17 FOULING

Fouling is a main physical problem facing both industrial and aircraft gas turbine engines, especially for those operating in the desert or polluted environments. Fouling is defined as the accumulation of deposits on the blade surfaces causing an increase in surface roughness. This is caused by the adherence of particles to airfoils and annulus surfaces due to the presence of oil or water mists. The result is a build up of material that causes increased surface roughness and to some degree changes the shape of the airfoil. Industrial gas turbine engines are more prone to fouling due to their continued operation in polluted environments. Foulants in the parts per million (ppm) range can cause deposits on balding, resulting in severe performance deterioration [36]. Normal operation of an engine, even if it is equipped with a good filter, results in the accumulation of particles such as dirt, dust, pollen,

FIGURE 13.74   Excessive damage of an axial fan of a turbofan.

insects, and tree sap on the compressor airfoils and gas path surfaces, as well as blocking the inlet filters [37]. If the solid particles are large and hard, their impact may cause erosion damage. It was noted that fouling in compressor blades is mainly caused by dust particles of about 2 μm and less in diameter, which are not removed by filters [38]. Moreover, particles up to 10 μm cause fouling but not erosion [36].

Compressor fouling is a common problem faced by gas turbine operators and has a more serious effect on the engine performance. The effects of compressor fouling are the drop in airflow, pressure ratio, and compressor efficiency. This results in a rematching of the gas turbine and compressor and a drop in power output and thermal efficiency and an increase in the heat rate. The effect of compressor performance deterioration is to force the engine to operate at a higher turbine inlet temperature for a given power, or to limit the available power for a given limiting temperature. Fouling also reduces the compressor surge margin and in extreme cases, it can result in surge problems, as it tends to move the compressor surge line toward the operating line [39]. Typically, about 70%–85% of all gas turbine engine performance loss accumulated during operation is attributed to compressor fouling [40]. All compressors are susceptible to fouling. The degree of fouling, the rate of fouling, and the effect on performance depend on the compressor design, compressor airfoil loading, airfoil incidences, airfoil surface smoothness/coating, type and condition of the airborne contaminates, the site environment, and the climate conditions (high humidity increases the rate of fouling) [41]. It was observed that fouling progress up to 40%–50% in the compressor, primarily affecting the front stages of the compressor [42]. For engines operating in an essentially base-load condition, the increase in fuel burned can be a major economic penalty. The effect of fouling can be mitigated by frequent washing of the compressor. If the period between washes is too long, then it is found that the performance restoration is limited. Fouling is much more serious for a small engine than on a big one, because of smaller flow passage. Both early and late stages may be affected, and sticky deposits may be backed on at the high temperature end of the compressor.

Some information available from open literature indicates that fouling, which causes a 5% reduction in inlet flow, will also reduce the compressor efficiency by about 2.5%, and the resulting decrease in engine power will be about 10% [41]. Usually, the fouling trend is assumed to have linear characteristics with time. Experimental tests on the compressor of a small turboprop gas turbine engine indicated that the prime effect of fouling is on the inlet air flow rather than on the compressor efficiency, and that the reduction in flow varies with the operating speed (being higher at the design speed) [40]. Site data obtained in a large industrial gas turbine indicated that compressor fouling results in a 5% reduction in inlet mass flow rate and about 1.8% reduction in compressor efficiency. This amount of fouling would reduce the engine output by about 7% and increase the heat rate by about 2.5% [40].

PROBLEMS

13.1. For the turboshaft engine shown in Figure 13.p.1, compare between axial and centrifugal compressors.

13.2. (a) For a multistage axial-flow compressor if the pressure ratio in two successive stages $(i-1)$, (i) are equal (i.e., $\pi_{i-1} = \pi_i$), then prove that the ratio between the total temperature rise in the two stages ΔT_i, ΔT_{i-1}, is given by

$$\frac{\Delta T_i}{\Delta T_{i-1}} = \frac{\eta_{i-1}T_i}{\eta_i T_{i-1}}$$

where η_i, η_{i-1} is the stage efficiencies, respectively.

(b) The following data refer to the rotor inlet of the first stage axial compressor (Figure 13.p.2b):

Axial velocity C_a	150 m/s
Mass flow rate	20 kg/s
Mean rotational speed U_m	267 m/s
Mean radius	0.1697 m
Tip radius	0.2262 m
Outlet total temperature T_{03}	310 K
Outlet total pressure P_{03}	1.25 bar
$C_{u2m} = 77\,\text{m/s}$	$\alpha_{3m} = 11°$

Using the free vortex condition $rC_u = $ constant

1. Calculate the angles β_1, α_2, β_2 and the degree of reaction Λ at the root, mean, and tip sections.
2. Draw the velocity triangles for the root, mean, and tip sections.
3. Draw the blade shape.

13.3. One of the intermediate stages of an axial compressor has the following characteristics:

Flow coefficient φ	0.5
Mean rotational speed U	250 m/s
Inlet flow angle α_1	30°
Rotor outlet angle β_1	35°

If these parameters are kept constants, plot the ideal relation $\psi - \varphi$.

Next, if the blade inlet angle is constant and equal to 50°, calculate the incidence angle.

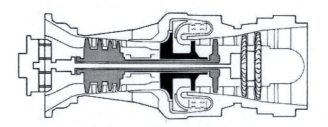

Figure 13.p.1 Turboshaft engine. (Reproduced with the permission of Rolls-Royce plc, copyright © Rolls-Royce plc 2007.)

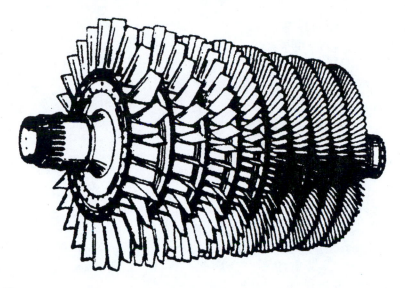

Figure 13.p.2b

Now, if negative and positive stall occur when the incidence angles are (15°) and (−6°), respectively, then calculate the corresponding values of C_a, ψ, and ΔT_{0s}.

Draw the velocity triangles in the original and the two stall cases as well as the flow pattern around the airfoil.

Note that in all cases C_a and U are constants.

13.4. (a) An axial-flow compressor made of AISI 1043 heat-treated steel with yield strength $S_y = 1520\,\text{MPa}$. Prove that for constant blade cross-section from root to tip the maximum centrifugal tensile stress at the blade root is given by

$$\sigma_{max} = \frac{\rho_b}{2} U_t^2 \left(1 - \zeta^2\right)$$

Calculate the factor of safety $\left(S_y / \sigma_{max}\right)$ in the following case:

ρ_b = the blade material density (8160 kg/m³)
U_t = the tip speed (350 m/s)
ζ = the hub-to-tip ratio (0.4)

If the blade is tapered with a linear variation of cross-sectional area, then

$$\sigma_{max} = \frac{\rho_b}{2} U_t^2 \left(1 - \zeta^2\right) K$$

where

$$K = 1 - \frac{(1 - d)\left(2 - \zeta - \zeta^2\right)}{3\left(1 - \zeta^2\right)}$$

where d is the ratio of the cross-sectional area at the tip to that at the root.

Calculate the new value for the factor of safety if $d = 0.4$.

(b) For supersonic compressors it is required to draw figures to illustrate

- Velocity triangles for high-turning impulse cascade
- Velocity triangles for low-turning compressor cascade
- Wave pattern of shock-in-stator configuration

13.5 A multistage axial-flow compressor has the following characteristics:

Inlet total temperature	$T_{01} = 288$ K
Inlet total pressure	$P_{01} = 1$ bar
Hub-to-tip ratio at inlet	$\zeta_{in} = 0.5$
Axial velocity	$C_a = 150$ m/s
Overall pressure ratio	$\pi_c = 4.0$
Compressor isentropic efficiency	$\eta_c = 0.9$
Mass flow rate	$\dot{m} = 20$ kg/s
Tip speed	$U_t = 300$ m/s

Assuming that the flow enters and leaves axially, calculate
(a) The root and tip radii at inlet
(b) The geometry of the compressor at outlet in the following three cases:

  (i) Constant mean diameter for all stages
  (ii) Constant tip diameter for all stages
  (iii) Constant root diameter for all stages

(c) The compressor rotational speed
(d) The number of stages

13.6. The table shown gives some data for the design parameters of the rotor in an axial-flow compressor stage using free vortex blading:

Variable	Root Section	Mean Section	Tip Section
C_{u1} [m/s]	32	24	?
C_{u2} [m/s]	?	150	?
$R[r/r_m]$	?	?	?
U_1 [m/s]	?	?	?
U_2 [m/s]	?	?	?
Λ	?	?	?
C_{a1} [m/s]	109	?	?
C_{a2} [m/s]	?	?	?
ΔT_0 [K]	?	?	?
$U\Delta C_u$ [J/kg]	?	?	?
α_1 [°]	?	?	?
α_2 [°]	?	?	?
β_1 [°]	?	?	?
β_2 [°]	?	?	?
Chord [cm]	6.0	5.5	5.0

(a) Complete the above table.
(b) Draw the angles α_1, β_1, α_2, β_2 against the nondimensional radius (R).
(c) Plot the degree of reaction (Λ) versus R.
(d) Draw a plan view of the blade.

13.7. Centrifugal stresses on the rotor of an axial compressor depend on rotational speed (ω). The maximum centrifugal stress that occurs in the root (for constant rotor cross-sectional area along the blade height) is given by $\sigma_{Cmax} = \rho_b U_1^2 (1 - \zeta^2)/2$.

If the rotor cross-sectional area is inversely proportional to the radius (i.e., $A = \text{constant}/r$), then $\sigma_{Cmax} = \rho_b U_1^2 ((A_t/A_r - \xi^2))$,
where A_t, A_r are the root and tip blade cross-sectional areas and U_1 is the tip rotational speed.

13.8. For a three-dimensional (3D) design of an axial-flow compressor stage, first-power method may be used where

$$C_{u1} = \frac{aR - b}{R} \quad \text{and} \quad C_{u2} = \frac{aR + b}{R}.$$

For the following data:

$$\zeta = 0.6, \quad C_{a1m} = C_{a2m} = 150\,\text{m/s}$$
$$C_{u1m} = 60\,\text{m/s}, \quad C_{u2m} = 150\,\text{m/s}$$

If the specific work $= 21.6$ kJ/kg, then it is required to
1. Calculate $\Delta T, \Lambda_m, a, b$.
2. Calculate Λ_r, Λ_t.
3. Calculate C_{a1}, C_{a2} at the root and tip.
4. Calculate the angles $\alpha_1, \alpha_2, \beta_1,$ *and* β_2 at root, mean, and tip sections.

13.9. (a) Describe three methods for avoiding surge of axial compressors.
(b) What are the types of drag in axial compressor?
(c) Define design and off-design conditions of axial compressor.
(d) Explain three different methods for drawing the compressor map.

13.10. Prove that for an axial compressor stage the pressure ratio is given by

$$\frac{P_{03}}{P_{01}} = \left[1 + \eta_c \frac{M_u^2(\gamma - 1)}{1 + ((\gamma - 1)/2)\,M_1^2} \left\{ 1 - \frac{M_{a1}}{M_u} \left(\tan \alpha_1 + \frac{C_{a2}}{C_{a1}} \tan \beta2 \right) \right\} \right]^{\gamma/(\gamma-1)}$$

where

$$\eta_c = \text{compressor efficiency}$$

$$M_u = U/\sqrt{\gamma R T_1}$$

$$M_{a1} = C_{a1}/\sqrt{\gamma R T_1} \quad \text{and} \quad C_{a1} \neq C_{a2}$$

13.11. The rotor of an intermediate stage is designed using a *free vortex design*. The results for root, mean, and tip sections are given in the following table:

	Root	Mean	Tip
β_1	40.0°	50.9°	60.6°
β_2	−10.0°	28°	5 5°

The number of rotor blades is 49
Mean radius is 0.17 m
Blade height is 0.08 m
It is required to
1. Draw the blade shape at tip, mean, and root sections using an airfoil with

$$\left(\frac{t}{c} \right)_{max} = 0.1$$

2. Draw a plan view of the blade by assembling the above sections employing a stacking axis at the quarter of chord.

13.12. For the velocity triangles shown at the mean section of an axial-flow compressor calculate Λ and ψ. Draw the velocity triangles on common base and common apex configurations (Figure 13.p.12).

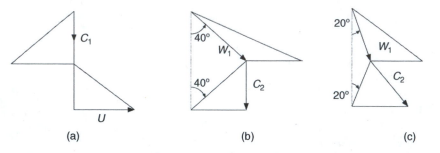

(a)   (b)   (c)

Figure 13.p.12

13.13. The degree of reaction at the mean section of an axial compressor stage is 0.5. The radial variation of the air flow from hub to tip is considered using the three methods: free vortex, exponential, and first power. The minimum degree of reaction is to be chosen as zero. The hub radius is 0.3 m. It is required in the three design methods to

(a) Calculate

   (i)   Hub-to-tip ratio
   (ii) The tip radius
   (iii) Blade height
   (iv) Degree of reaction
   (v)   Nondimensional radius at tip (R_t).

(b) Plot the relation between Λ and R.

13.14a. (a) Define the degree of reaction of an axial compressor stage. PROVE that
   (i)   If $C_1 \neq C_3$ and $C_{a1} \neq C_{a2}$, then

$$\Lambda = \frac{2U\,(C_{u2} - C_{u1}) - \left(C_2^2 - C_1^2\right)}{2U\,(C_{u2} - C_{u1}) - \left(C_3^2 - C_1^2\right)}$$

   (ii) If $C_{a1} = C_{a2}$ and $C_1 \neq C_3$, then

$$\Lambda = \frac{(C_{u2} - C_{u1})\,\{2U - (C_{u1} - C_{u2})\}}{2U\,(C_{u2} - C_{u1}) - \left(C_3^2 - C_1^2\right)}$$

What will be the value of the degree of reaction if $C_1 = C_3$ and $C_{a1} = C_{a2}$?

13.15a. (a) At a particular operating condition an axial-flow compressor has a degree of reaction of 60%, a flow coefficient of 0.5, and a stage loading of 0.35. If the flow exit angles of both the stator and rotor blade rows may be assumed to be remaining unchanged, when the mass flow is throttled, DETERMINE the degree of reaction and stage loading when the mass of air flow is reduced by 10%. Sketch the velocity triangles for the two cases and write your comments. If at a positive stall the angle $\beta_1 = 65°$ and at choking conditions $\beta_1 = 45°$, calculate the ratios of the mass flow in these two cases to the original mass flow rate (assume α_1 and β_2 are constants).

13.16. (a) For the shown high pressure compressor (HPC) of the turbofan engine CFM56-5C (Figure 13.p.16a)

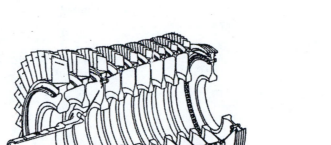

Figure 13.p.16a

(i)   How many stages are at the compressor?

(ii) Compare between the rotor of the first and last stages.

(iii) Plot the variation of the static temperature and pressure, total temperature and pressure, absolute and relative velocities along the compressor stages.

(b) Figure 13.p.16b illustrates one of the rotor blades of the HPC of a high-bypass ratio turbofan engine. It is manufactured from a titanium alloy. The concave side of the upper portion of the blade (extending from radius r_1 to r_t) is coated with tungsten carbide coating, which is an erosion-resistant alloy. The average density of this blade portion is ρ_1 while the inner portion of the blade extending from hub to r_1 is ρ_2. The yield strength of the blade root is S_y. If the requested factor of safety is (FS),

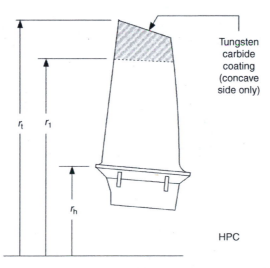

Figure 13.p.16b

Prove that:

For a constant blade cross-sectional area from hub to tip, the factor of safety, FS is given by

$$FS = \frac{2S_y}{U_1^2 \left[(\rho_2 - \rho_1) \zeta_1^2 + \rho_1 - \rho_2 \zeta^2 \right]}$$

where U_1 is the tip rotational speed

$$\zeta_1 = r_1/r_t \quad \text{and} \quad \zeta = r_h/r_1$$

13.17. (a) For a free vortex design, prove that:

$$\Lambda = 1 - \frac{(1 - \Lambda_m)}{R_2}$$

(b) The figures shown illustrate C_{u1} versus R, where $R = r/r_m$.
If $C_{u1} = aR^n - (b/R)$, then identify the value of (n) for each case.
Next, with $C_{u2} = aR^n + (b/R)$, PLOT the curves of C_{u2} on the same figures for different (n) (Figure 13.p.17a).

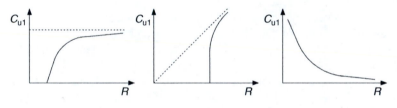

Figure 13.p.17a

(c) The velocity triangles at the inlet of an axial compressor stage at the hub, mean, and tip sections are shown in Figure 13.p.17b. Other data for this stage are

$$U_h = 200 \text{ m/s}, \quad \zeta = 0.5, \quad \dot{m} = 20 \text{ kg/s}$$
$$T_{01} = 290 \text{ K}, \quad P_{01} = 1.05 \text{ bar}$$

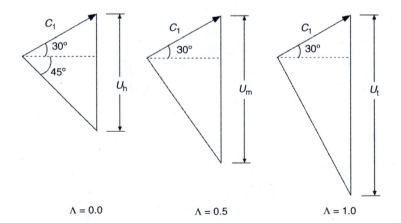

$\Lambda = 0.0$ $\Lambda = 0.5$ $\Lambda = 1.0$

Figure 13.p.17b

It is required to
(i)   Draw the velocity triangles at outlet.
(ii) DO these velocity triangles represent a free vortex design?

13.18. The three-dimensional behavior of an axial compressor stage is to be designed based on the exponential method
The stage has the following data:

$$C_{a1h} = 170 \text{m/s}, \quad C_{a2h} = 190 \text{m/s}, \quad R_h = 0.5$$
$$\alpha_{1h} = 6°, \quad U_h = 150 \text{m/s},$$

It is required to
(i)   Calculate the constants (a) and (b).
(ii)   Do these constants vary from one design method to another?
(iii) Calculate the temperature rise in the stage.
(iv)   Calculate the degree of reaction at the mean section.
(v)   Complete the following table using both first-power and exponential design methods.

	Exponential method	First-power method
C_{a1h} (m/s)		
C_{a2h} (m/s)		
C_{a1t} (m/s)		
C_{a2t} (m/s)		
Λ_h		
Λ_t		

13.19. (a) Figure 13.p.19a illustrates a compressor map. The design point has the following values: $\dot{m} = 50 \text{ kg/s}$, $T_{01} = 288 \text{ K}$,

$$P_{01} = 101.3 \text{ kPa}, \quad N = 6000 \text{ rpm}$$

1. Calculate the values of $\dot{m}$, N at point (A) in the two cases:
    (a) The inlet total pressure and temperature are listed above.
    (b) When the aircraft is flying at an altitude where $T_{01} = 225$ K and $P_{01} = 15$ kPa

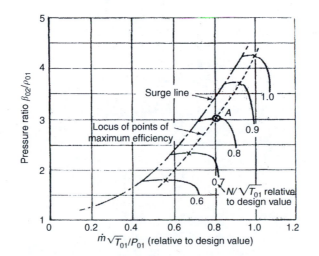

Figure 13.p.19a

2. Calculate the inlet total temperature and pressure if relative to the design point $N/\sqrt{T_{01}} = 0.9$ and $\dot{m}\sqrt{T_{01}}/P_{01} = 0.9$ while the values of N and $\dot{m}$ are equal to the design values.

    (b) The triangle shown illustrates the flow conditions at the rotor inlet of a stage of a multistage axial compressor (Figure 13.p.19b). It is required to *calculate* the following parameters: α_2, β_2, W_2, C_2, temperature rise per stage ΔT_{0s}, and pressure rise per stage if the degree of reaction has the following three values: 0, 0.5, and 1.0.

    Take the work done factor $\lambda = 0.95$ and the stage efficiency $\eta_s = 0.9$.

    Give a clear explanation for your answers.

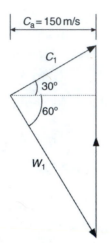

Figure 13.p.19b

13.20. The following data describe the characteristics of an axial compressor stage:

$$W_s = 2.1 \times 10\ \text{m}^2/\text{s}^2, \quad C_a = 150\ \text{m/s}$$
$$U_m = 300\ \text{m/s}, \quad \zeta = 0.5$$
$$\Lambda_m = 0.5, \quad r_m = 0.1697\ \text{m}$$

(a) Calculate the mean flow angles.
(b) If $R_h = 2/3$, using the exponential method for blade design, calculate the parameters: C_{a1}, C_{a2}, C_{u1}, C_{u2} and draw the velocity triangles at the root section.

13.21. Figure 13.p.21a illustrates the civil transport Boeing 747 airplane powered by the turbofan engine CF6 (see Figure 13.p.21b) fitted with axial compressors.

(a) For an axial compressor define the symbols ϕ and ψ
      Prove that for the first stage (where $C_1 = C_{a1}$):
      $\tan \beta_1 = 1/\phi$, $\tan \alpha_2 = \varphi\psi/\phi$, and $\tan \beta_2 = (1 - \psi)/\phi$

(b) In a two-spool turbofan engine, the low-pressure spool is composed of a fan and low-pressure compressor driven by the low-pressure turbine at speed N_1. The fan has the following characteristics at takeoff:

Rotational Speed N_1	3445 rpm
Mass Flow Rate $\dot{m}$	593 kg/s
Fan tip diameter	2.18 m
Fan hub-to-tip (ζ)	0.4

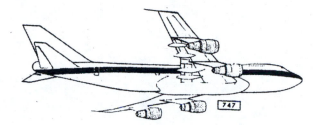

Figure 13.p.21a Two-spool turbofan engine. (Courtesy of General Electric.)

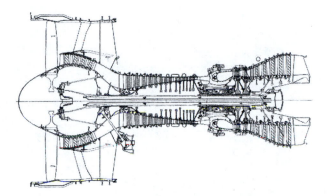

Figure 13.p.21b High bypass ratio turbofan engine. (Courtesy of General Electric.)

Ambient air temperature	15°C
Ambient air density	1.225 kg/m^3
De Haller number at tip radius	0.95
De Haller number at mean radius	0.85

It is required to

(i)   Define the portion of the fan blade subjected to supersonic flow.
(ii) Calculate the angles α_1, β_1, α_2, and β_2 at mean section use Figure (13.p.21c where $\Delta\theta^o$ is the difference between flow inlet and outlet angles, ω is the shock losses coefficient, M_1 and M_2 are the Mach number upstream and downstream the shock wave.

13.22. The high-pressure axial compressor in a turbofan engines is composed of 14 stages. The first six stator stages have variable-angle vanes. The last stage has the following data:
Axial velocity = 160 m/s
Tip diameter 0.64 m, $\alpha_1 = 24°$
Speed N_2 8000 rpm, $\Lambda_m = 0.50$
De Haller number at tip = 0.85
Using Figure (13.p.23c, it is required to

(i)   Calculate the degree of reaction at the tip radius
(ii) Calculate the hub-to-tip ratio (ζ)
(iii) Calculate and plot the values of α_1, α_2, β_1, β_2 and Ω at the hub, mean, and tip sections of blade if free vortex design is assumed
(iv) Describe two methods for improving the surge margin of axial compressors

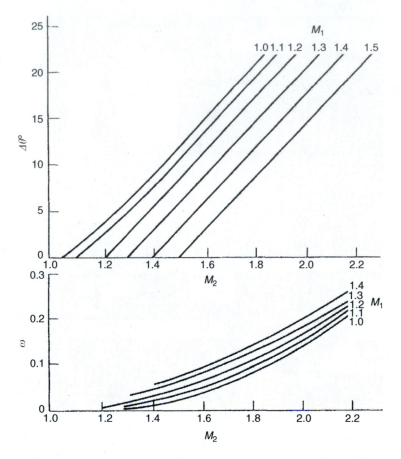

Figure 13.p.21c Shock wave losses. (From H.I.H. Saravanumuttoo et al., *Gas Turbine Theory*, Prentice Hall, 5th edn, 2001, p. 191. With permission.)

13.23. An axial compressor composed of four stages is to be designed using the exponential blading method where

$$C_{u1} = a - \frac{b}{R} \quad \text{and} \quad C_{u2} = a + \frac{b}{R}$$

One of these stages has the following data:

	C_{a1} (m/s)	C_{a2} (m/s)	R_{inlet}	R_{exit}
Root	168	186	0.754	0.79
Mean	150	150	1.0	1.0
Tip	132	116	1.246	1.21

$$r_{mean} = 0.4 \text{ m}, \quad \alpha_{1mean} = \beta_{2mean}, \quad \lambda = 0.88, \quad \eta_s = 0.92$$

The inlet temperature to compressor= 300 K, ΔT_{0stage} = constant for the four stages
It required to
(a) Draw the velocity triangles for hub, mean, and tip sections.

(b) Prove that $h_{02} = $ constant along the blade height.
(c) Calculate the pressure ratio across the compressor.
(d) Explain why one cannot use the following relation (from Table 13.3) for the difference between axial velocities

$$C_{a2}^2 - C_{a1}^2 = -8\,ab\ln R$$

Write down the correct relation

$$C_{a2}^2 - C_{a1}^2 = ?$$

REFERENCES

1. K. Steffens, *Advanced Compressor Technology—Key Success Factor for Competitiveness in Modern Aero Engines*, MTU Aero Engines Internal Report and 15th ISABE, September 2001.
2. K. Hunecke, *Jet Engines, Fundamentals of Theory, Design and Operation*, Motor Books International Publishers & Wholesalers, 1st edn., 6th impression 2003.
3. R.T.C. Harman, *Gas Turbine Engineering, Applications, Cycles and Characteristics*, John Wiley & Sons, 1981, p. 78.
4. N.A. Cumpsty, *Compressor Aerodynamics*, Longman Scientific & Technical, 1989, p. 48.
5. H.I.H. Saravanamuttoo, G.F.C. Rogers, and H. Cohen, *Gas Turbine Theory*, Prentice Hall, 5th edn., 2001, pp. 191.
6. D.C. Wisler, *Advanced Compressor and Fan Systems*. GE Aircraft Engines, Cincinnati, OH, USA, 1988 General Electric Co. USA, All Rights Reserved.
7. R.N. Brown, *Compressors-Selection and Sizing*, Gulf Publication Co., 1981, p. 219.
8. A. Glassman (ed.), *Turbine Design*, NASA SP-290, Vol. I, p. 32.
9. S.L. Dixon, *Fluid Mechanics and Thermodynamics of Turbomachinery*, 4th ed., SI/Metric Units, Butterworth/Heinemann, 1998, p. 144.
10. B. Lakshminarayana, *Fluid Dynamics and Heat Transfer of Turbomachinery*, John Wiley & Sons, Inc., 1996, p. 261.
11. H. Mars, *A Computer Program for the Through Flow Fluid Mechanics in an Arbitrary Turbomachine Using a Matrix Method*, British ARC R&M 3509.
12. I.A. Johnsen and R.O. Bullock (eds.), *Aerodynamic Design of Axial Flow Compressors*, NASA SP-36, 1965.
13. G.K. Serovy, Axial flow compressor aerodynamics, In *Aerothermodynamics of Aircraft Gas Turbine Engines*, Gordon C. Oates (ed.), U.S. Air Force Report AFA APL R-78-52, 1978, chap. 17.
14. A. Wennerstrom, *Low Aspect Ratio Axial Flow Compressors, Why and What It Means*, 1986 SAE, Cliff Garrett Turbomachinery Award Lecture, 3rd, Long Beach, CA, 14 October 1986, 14 pp.
15. J.H. Horlock, *Axial Flow Compressors, Fluid Mechanics and Thermodynamics*, Krieger Publishing Co., 1973, p. 164.
16. D.E. Hobbs and H.D. Weingold, *Development of Controlled Diffusion Airfoils for Multistage Compressor Applications*, American Society of Mechanical Engineers, International Gas Turbine Conference and Exhibit, 28th, Phoenix, AZ; 27–31 March 1983, 11 pp.
17. J. Day, *Stall and Surge in Axial Flow Compressors*, VKI Lecture Series 1992-02, January 1992.
18. J.L. Kerreebrock, *Aircraft Engines and Gas Turbines*, 2nd ed., The MIT Press, 1992, p. 264.
19. United Technologies Pratt Whitney, *The Aircraft Gas Turbine Engine and Its Operation*, P&W Operation Instruction 200, 1988, pp. 3–81.
20. J.H. Neilson and A. Gilchrist, Erosion by a stream of solid particles, *Wear*, 11(2), 111–122, 1968.
21. A.F. El-Sayed, R. Lasser, and W.T. Rouleau, Three dimensional viscous particulate flow in a typical turboexpander, *International Journal of Energy Systems*, 5, 1985.
22. A.F. El-Sayed, R. Lasser, and W.T. Rouleau, Effect of secondary flow on particle motion and erosion in a stationary cascade, *International Journal of Heat and Fluid Flow*, 7(2), 146–154, 1986.

23. G.P. Sallee, *Performance Deterioration Based On Existing (Historical) Data JT9D Jet Engine Diagnostics Program*, United Technologies Corporation, Pratt & Whitney Aircraft Group Commercial Products Division, NASA CR-135448, PWA-5512-21, June 23, 1980.

24. D. Japikse, Review-progress in numerical turbo-machinery analysis, *ASME Transaction Series I— Journal of Fluids Engineering*, 98, 592–606, 1976.

25. R.H. Wulf, W.H. Kramer, and J.E. Paas, *CF6-6D Jet Engine Performance Deterioration*, NASA/CR-159786, NASA, 1980.

26. R.C. Peterson, *Design Features for Performance Retention in the CFM56 Engine*, Turbo-machinery Performance Deterioration FED: Vol. 37 the AIAA/ASME 4th Joint Fluid Mechanics Plasma Dynamics and Lasers Conference, Atlanta, Georgia.

27. M.F. Hussein and W. Tabakoff, Dynamic behavior of solid particles suspended by polluted flow in a turbine stage, *Journal of Aircraft*, 10, 334–340, 1973.

28. M.F. Hussein and W. Tabakoff, Computation and plotting of solid particle flow in rotating cascades, *Computers and Fluids*, 2, 1–15, 1974.

29. W. Tabakoff and C. Balan, Effect of sand erosion on the performance deterioration of a single stage axial flow compressor, *AIAA paper* 83–7053, 1983.

30. W.B. Roberts, A. Armin, G. Kassaseya, K.L. Suder, S.A. Thorp, and A.J. Strazisar, The effect of variable chord length on transonic axial rotor performance, *Journal of Turbomachinery*, 124(3), 351–357, 2002.

31. A.F. El-Sayed and A. Brown, *Computer Prediction of Erosion Damage in Gas Turbine*, ASME Paper 87-GT-127, 32nd ASME International Gas Turbine Conference, 1987, California, USA.

32. A.F. El-Sayed and A. Hegazy, *An Iterative Procedure for Estimating the Effect of Erosion on Turbine Blades Life*, Paper 95-YOKOHAMA-IGTC-58, 1995 Yokohama International Gas Turbine Congress, October 22–27, 1995.

33. Ghenaiet, R.L. Elder, and S.C. Tan, *Particles Trajectories Through an Axial Fan and Performance Degradation Due to Sand Ingestion*, ASME TURBO EXPO, 2001-GT-0497 June 2001.

34. H.H. Zoheir, *Solid Particulate Flow in an Axial Transonic Fan in a High Bypass Ratio Turbofan Engine*, MSc Thesis, Mechanical Power Engineering Department, Zagazig University, Zagazig, 2006.

35. G. Chaplin, et al., *Make it FOD Free- the Ultimate FOD Prevention Program Manual*, FOD, The FOD Corporation, 2004.

36. Cyrus, B. Meher-Homji, and A. Bromley, Gas turbine axial compressor fouling and washing, *Proceedings of the Thirty-Third Turbomachinery Symposium*, Turbolab, Texas A&M university, 2004.

37. J.M. Thames, J.W. Stegmaier, and J.J. Ford, *On-line Washing Practices and Benefits*, ASME paper No. 89-GT-91, June 1989.

38. I.S. Diakunchak, Performance deterioration in industrial gas turbines, *ASME Journal of Gas Turbine and Power*, 114, 161–168, 1992.

39. A.D. Mezheritsky and A.V. Sudarev, *The Mechanism of Fouling and the Cleaning Technique in Application to Flow Parts of the Power Generation Plant Compressors*, ASME paper No. 90-GT-103, 1990.

40. I.S. Diakunchak, Performance deterioration in industrial gas turbines, *ASME Journal of Engineering for Gas Turbines and Power*, 114, 161–168, 1992.

41. H.I.H. Saravanamuttoo and A.N. Lakshminerasimha, *A Preliminary Assessment of Compressor Fouling*, ASME paper No. 85-GT-153, 1985.

42. G.F. Aker and H.I.H. Saravanamuttoo, Predicting gas turbine performance degradation due to compressor fouling using computer simulation techniques, *Journal of Engineering for Gas Turbines and Power*, 111, 343–350, 1989.

43. M.J. Zucrow and J.D. Hoffman, *Gas Dynamics*, John Wiley & Sons, New York, 1975, pp 744-749.

44. http://www3.sympatico.ca/john.otley/AirChina2.jpg

14 Axial Turbines

14.1 INTRODUCTION

Turbines may be defined as turbo machines that extract energy from the fluid and convert it into mechanical/electrical energy. A classification of turbines is shown in Figure 14.1. Turbines may be classified based on the surrounding fluid, whether it is extended or enclosed. Example for extended turbines is the wind turbines, which may be horizontal axis wind turbines (HAWT) or vertical axis wind turbines (VAWT). Enclosed turbines may be classified based on the working fluid as either incompressible or compressible machines. Hydraulic turbines deal with incompressible fluids (mostly water turbines). Compressible turbines may be either steam or gas turbines. Gas turbines may operate as subsonic or supersonic turbines. Turbines may also be classified based on the gas-flow direction within its passage as axial, mixed, or radial turbines. In axial turbines, the flow moves essentially in the axial direction through the rotor. In the radial type, the gas motion is mostly radial as seen in the auxiliary power units (APU). In the mixed-flow turbines, the gases have a combined radial and axial motion. Another classification is related to the role of turbine rotor in power extraction from the gas flowing through its passages. Thus, it may be either impulse or reaction turbines.

Historically, *water turbines* were first built, followed by steam turbines, and finally gas turbines. A water or hydraulic turbine is used to drive electric generators in hydroelectric power stations. The first such station was built in Wisconsin in 1882. The three most common types of hydraulic turbine are the Pelton wheel, the Francis turbine, and the Kaplan turbine.

Two engineers, Charles A. Parsons of Great Britain and Carl G. P. de Laval of Sweden, were pioneers in building *steam turbines* toward the end of the nineteenth century. Continual improvements of their basic machines have caused steam turbines to become the principal power sources used to drive most large electric generators and the propellers of most large ships. Most steam turbines are multistage engines. At the inlet of the turbine, high-pressure steam enters from a boiler and moves through the turbine parallel to the shaft, first striking a row of stationary vanes that directs the steam against the first bladed disc at an optimum speed and angle. The steam then passes through the remaining stages, forcing the discs and the shaft to rotate. At the end of the turbine, the shaft sticks out and can be attached to the machinery.

The *gas turbine* is a relative newcomer. The term gas turbine applies to a turbine with hot gases released from the combustion system as their working fluid. Axial turbines are handled in this chapter while radial turbines will be analyzed in the next chapter. Virtually, all turbines in aircraft engines are of the axial type, regardless of the type of compressor used. Axial turbine is similar to axial compressor operating in reverse. The turbine is composed of a few stages; each has two successive stators (called nozzle or nozzle guide vanes) and rotor blade rows of small airfoil-shaped blades (Figure 14.2). The stators keep the flow from spiraling around the axis by bringing the flow back parallel to the axis. Depending on the engine type, there may be multiple turbine stages present in the engine. Usually, *turbofan* and *turboprop* engines employ a separate turbine and shaft to power the fan and gearbox, respectively, and the engine is termed a two-spool engine. Three-spool configurations exist for some high-performance engines where an additional turbine and shaft power separate parts of the compressor. Nearly, three-fourths of the energy extracted by the turbine is used to drive the connected module (fan/compressor). When energy is extracted from the flow, pressure drops across the turbine. The pressure gradient helps keep the flow attached to the turbine blades; therefore, turbines, unlike, compressors are easier to design. The pressure drop across a single turbine stage can be much greater than the pressure increase across a corresponding compressor stage without

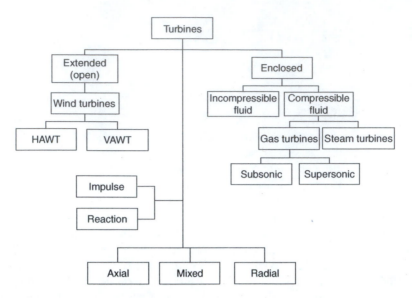

FIGURE 14.1   Classification of turbines.

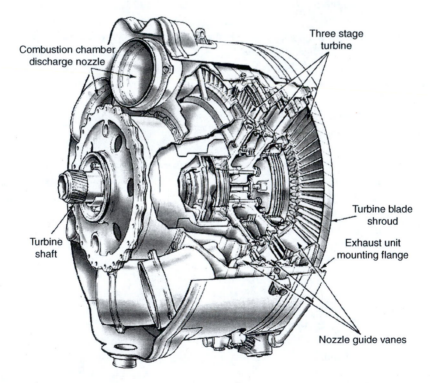

FIGURE 14.2   A triple stage turbine with single shaft. (Reproduced from Rolls-Royce plc, copyright © Rolls-Royce plc 2007. With permission.)

the danger of separating the flow. A single turbine stage can be used to drive multiple compressor stages. Therefore, higher efficiencies can be achieved in a turbine stage. To keep the flow from leaking around the edges of the turbine blades, because of the higher pressure gradient, the blade is shrouded at the tip. The shrouded blades form a band around the perimeter of the turbine, which

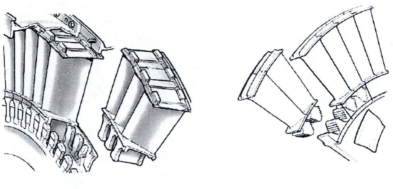

Nozzle guide vanes   Rotor blades

FIGURE 14.3 Shrouded blades of both the nozzle and rotor. (Reproduced from Rolls-Royce plc, copyright © Rolls-Royce plc 2007. With permission.)

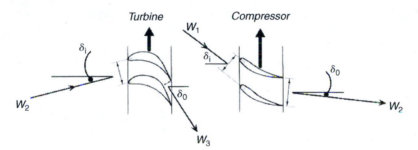

Typical forms of turbine and compressor rotor blades

FIGURE 14.4 The rotor passage for axial turbine and axial compressor.

serves to cut down gas leakage around the tips of the turbine blades, and thus improve the airflow characteristics and increase the efficiency of the turbine (Figure 14.3). The shrouds also serve to reduce blade vibrations.

The turbine blades exist in a much more hostile environment than compressor blades. Sitting just downstream of the burner, the blades are exposed to an entry temperature of 850–1700°C [1]. Therefore, turbine blades, using the advancement in metallurgy and manufacturing processes, are either made up of special metals that can withstand the heat or active cooling. The blade is hollow and cool air, which is bled off the compressor, is pumped through the blade and out through the small holes on the surface to keep the surface cool. Moreover for efficient operation, the turbine rotates at a very high speed (over 40,000 rpm for smaller engines to 8,000 rpm for larger engines), which makes the turbine wheel one of the most highly stressed parts in the engine.

14.2 COMPARISON OF AXIAL FLOW COMPRESSORS AND TURBINES

A single stage of axial turbine drawn side by side with an axial compressor stage is shown in Figure 14.4. The relative speeds in both cases are drawn. Through the turbine the flow is accelerated while it is decelerated for the compressor. From this figure, it is shown that the blade-to-blade passage is convergent for turbine and divergent for compressor. The meridional plane (axial–radial

TABLE 14.1

Comparison between Axial Flow Compressor and Axial Flow Turbine

	Compressor	Turbine
Flow	Decelerating	Accelerating
Meridional passage	Divergent	Convergent
Blade-to-blade passage	Diffuser	Nozzle
Maximum blade thickness	Small	Large
Direction of rotation	From suction to pressure surfaces	From pressure to suction surfaces
Number of stages	3–20	1–4
Hub-to-tip ratio (first stage)	0.4–0.6	> 0.7
Blade height	Large	Small
Annulus area in a multistage	Decreasing	Increasing
Number of blades	Small	Large
$(\Delta T_0)_{stage}$	Positive	Negative
	Small	Large
$P_0, h_0,$ and T_0	Increase	Decrease
Operating temperature [2]	500–1000°R	1000–3500°R
Specific work	$U(C_{u2} - C_{u1})$	$U(C_{u2} + C_{u3})$
Stage arrangement	Rotor–stator	Stator–rotor
$\psi - \phi$ relation	ψ is proportional to $(-\phi)$	ψ is proportional to (ϕ)
Maximum stage efficiency	0.8–0.9	>0.9
Erosion	Owing to foreign objects such as sand, polluted air, and ice	Owing to particles in combustion products
Flow deflection $(\delta_i - \delta_0)$	20–35°	50–180°

directions) is convergent for a compressor and divergent for a turbine. Other comparisons are listed in Table 14.1.

14.3 AERODYNAMICS AND THERMODYNAMICS FOR A TWO-DIMENSIONAL FLOW

14.3.1 VELOCITY TRIANGLES

Single stage of axial turbine is shown in Figure 14.5. The gases leaving the combustion chamber approach the stator (or nozzle) with an absolute velocity (C_1), normally in an axial direction and thus, the absolute angle to the axial direction ($\alpha_1 = 0$). Since the passage represents a nozzle, with either a convergent or convergent-divergent passage, the flow is accelerated leading to an increase in the absolute velocity. The flow leaves the stator passage at a speed (C_2), where ($C_2 > C_1$). The static pressure decreases as usual as the gas passes through the nozzle. Moreover, total pressure decrease due to skin friction and other sources of losses will be described later. Thus, $P_1 > P_2, P_{01} > P_{02}$. In addition, the static enthalpy and total enthalpy drop at the end of the nozzle. The flow leaves the stator with an absolute speed of (C_2) and an angle (α_2) when combined with the rotational speed (U); the relative velocity will be (W_2), inclined at an angle (β_2) to the axial. The velocity triangles at the exit of the nozzle and exit of the rotor are shown in Figure 14.6. The rotor blades extract work from the gases. As the rotor blade-to-blade passages also resemble nozzle shape, the relative velocity (W_3) of the gas will increase ($W_3 > W_2$) while its absolute velocity C_3 decreases

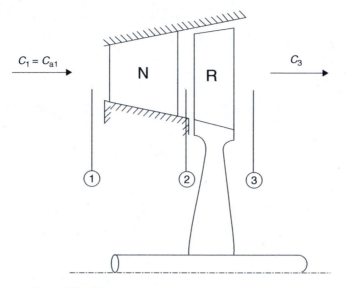

FIGURE 14.5   Layout of an axial turbine stage.

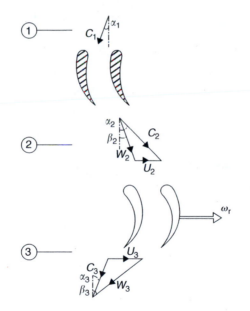

FIGURE 14.6   Velocity triangles.

($C_3 < C_2$) and the flow exit angle (β_3) will be greater than the flow inlet angle ($\beta_3 > \beta_2$). The static and total pressures also drop in the rotor passage ($P_3 < P_2, P_{02} < P_{03}$). The static enthalpy and total enthalpy also drop at the end of the rotor. If there is no static pressure and static enthalpy drop in the rotor, the turbine is called an *impulse turbine*. As the static pressure and temperature decrease in the turbine, the density decreases. The annulus area is increased to avoid excessive high velocities.

The analysis of a turbine stage is similar to that of a compressor stage as will be emphasized here. The same kinematical relation described in centrifugal and axial compressors is valid:

$$\vec{C} = \vec{W} + \vec{U}$$

The three-dimensional velocity field can be further expressed in terms of the axial, radial, and tangential components as

$$\begin{pmatrix} C_a \\ C_r \\ C_u \end{pmatrix} = \begin{pmatrix} W_a \\ W_r \\ W_u \end{pmatrix} + \begin{pmatrix} 0 \\ 0 \\ \pm U \end{pmatrix}$$

In a manner similar to the axial compressor, the following kinematical relations are derived from Figure 14.6:

$$\frac{U_2}{C_{a2}} = \tan \alpha_2 - \tan \beta_2 \tag{14.1a}$$

$$\frac{U_3}{C_{a3}} = \tan \beta_3 - \tan \alpha_3 \tag{14.1b}$$

If the axial velocity is constant at inlet and outlet of the stage, then $C_{a2} = C_{a3} = C_a$ and $U_2 = U_3 = U$.

The kinematical relations in Equations 14.1a and b will be

$$\frac{U}{C_a} = \tan \alpha_2 - \tan \beta_2 = \tan \beta_3 - \tan \alpha_3 \tag{14.1c}$$

Thus, the following relation is reached:

$$\tan \alpha_2 + \tan \alpha_3 = \tan \beta_2 + \tan \beta_3 \tag{14.2}$$

Moreover, assuming a constant rotational speed ($U_2 = U_3 = U$), the following kinematical relation between the absolute and relative velocities at the rotor inlet and outlet is simply arrived:

$$C_{u2} + C_{u3} = W_{u2} + W_{u3} \tag{14.3}$$

The velocity triangles shown in Figure 14.6 may be combined in three different ways. These form the common base (Figure 14.7a) and the common apex velocity triangles (Figure 14.7b). Alternatively, velocity triangles having parallel rotational speed vectors [3] are shown in Figure 14.7c.

14.3.2 EULER'S EQUATION

Euler's equation in turbo machinery represents the conservation of angular momentum, from which the specific work is expressed by the relation

$$W_s = (\vec{U} \times \vec{C}_u)_3 - (\vec{U} \times \vec{C}_u)_2. \tag{14.4a}$$

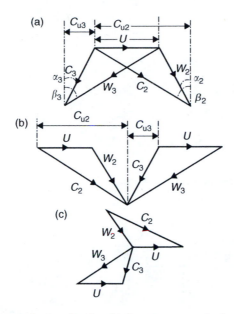

(a)

(b)

(c)

FIGURE 14.7 (a) Common base velocity triangles. (b) Common apex velocity triangles. (c) Parallel rotational speed velocity triangles.

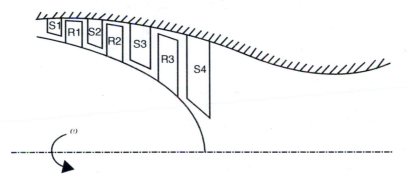

FIGURE 14.8 Layout of an axial turbine having variable mean radius.

Since the swirl velocities $\vec{C}_{u3}$ and $\vec{C}_{u2}$ are in opposite directions,

$$W_s = (UC_u)_3 + (UC_u)_2 \tag{14.4b}$$

In some cases, the turbine has a continuously varying mean radius as shown in Figure 14.8. The corresponding velocity triangles shown in Figure 14.9 represent the most general case where neither the rotational speed nor the axial speed is equal at the inlet and outlet of the rotor.

The specific work will be expressed in the general form

$$W_s = U_2 C_{a_2} \tan \alpha_2 + U_3 C_{a3} \tan \alpha_3 \tag{14.5a}$$

If the axial velocities are unequal, but the mean radius and consequently the rotational speed are equal ($U_2 = U_3 = U$), then

$$W_s = U(C_{a2} \tan \alpha_2 + C_{a3} \tan \alpha_3) = U(C_{a2} \tan \beta_2 + C_{a3} \tan \beta_3) \tag{14.5b}$$

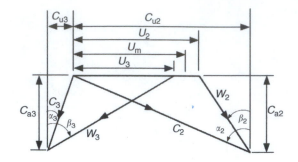

FIGURE 14.9 Velocity triangles for the general case.

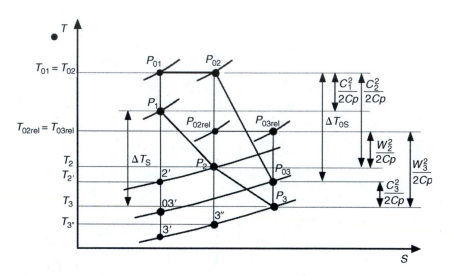

FIGURE 14.10 T–S diagram for a reaction turbine stage having equal mean radii at inlet and outlet of rotor.

If in addition the axial speeds are equal ($C_{a_2} = C_{a3} = C_a$), then Equation 14.5b is reduced to

$$W_s = UC_a(\tan \alpha_2 + \tan \alpha_3) \tag{14.5c}$$

Then from Equation 14.2, the specific work is

$$W_s = UC_a(\tan \beta_2 + \tan \beta_3) \tag{14.5d}$$

The working fluid in turbine is hot gases and has the following properties:

$$\gamma = 1.33 \quad C_p = 1.148(\approx 1.15\,\text{kJ/kg} \cdot \text{K}) \quad R = 287\,\text{J/kg} \cdot \text{K}$$

14.3.3 EFFICIENCY, LOSSES, AND PRESSURE RATIO

The expansion processes in the stator and rotor of an axial turbine are drawn on the T–S diagram shown in Figure 4.10. State 1 represents the inlet to stator, while states 2 and 3 are the inlet and outlet to the rotor, respectively. Across the *nozzle* (stator), no work or heat is added to or rejected from the flow; thus

$$Q_{1-2} = W_{1-2} = 0$$

Accordingly, the inlet and outlet stagnation enthalpies and temperature are equal, or

$$h_{01} = h_{02} \quad \text{and} \quad T_{01} = T_{02} \tag{14.6}$$

Thus, $h_1 - h_2 = \left(C_2^2 - C_1^2\right)/2$.

Across the *rotor*, no heat is exchanged with the surroundings (outside a typical control volume surrounding the stage); thus

$$Q_{2-3} = 0$$

The rotor, however, extracts energy from the hot, thus the enthalpy drops. The corresponding specific work is

$$W_{2-3} = h_{02} - h_{03} = C_p(T_{02} - T_{03})$$

Now, across the *stage*, no heat is transferred, thus

$$Q_s = 0 \tag{14.7}$$

The specific work is then

$$W_{1-3} = h_{01} - h_{03} = (h_{01} - h_{02}) + (h_{02} - h_{03}) = (h_{02} - h_{03}) = Cp(T_{02} - T_{03})$$
$$W_s = h_{02} - h_{03} = C_p(T_{02} - T_{03}) \equiv C_p(T_{01} - T_{03}) = C_p \Delta T_{0s} \tag{14.8}$$

The temperature drop in the stage from Equations 14.5a and 14.8 will be expressed as

$$\Delta T_{0s} = \frac{U_2(C_{a2} \tan \alpha_2 + U_3 C_{a3} \tan \alpha_3)}{C_p} \tag{14.9a}$$

For a constant mean diameter, it will be

$$\Delta T_{0s} = \frac{U(C_{a2} \tan \alpha_2 + C_{a3} \tan \alpha_3)}{C_p} \tag{14.9b}$$

For constant axial speed in addition, it will be

$$\Delta T_{0s} = \frac{U C_a(\tan \alpha_2 + \tan \alpha_3)}{C_p} = \frac{U C_a(\tan \beta_2 + \tan \beta_3)}{C_p} \tag{14.9c}$$

Across the rotor, the *rothalpy* (I), *rot*ational enthalpy, is frequently employed. It is defined as

$$I = h_{0r} - U^2/2 \tag{14.10}$$

Across the rotor, the rothalpy is constant, or

$$I_2 = I_3$$

$$h_{02r} - U_2^2/2 = h_{03r} - U_3^2/2 \tag{14.11a}$$

When the rotor has equal mean diameter and thus equal rotational speeds, the relative total enthalpy is

$$h_{02r} = h_{03r}$$
$$T_{02r} = T_{03r} \tag{14.11b}$$

which is shown on the T–S diagram (Figure 4.10).

Since the relative total enthalpy is expressed by

$$h_{0r} = h + W^2/2$$

for equal relative total relative total enthalpy,

$$h_2 - h_3 = \left(W_3^2 - W_2^2\right)/2$$

And so

$$T_2 - T_3 = \frac{\left(W_3^2 - W_2^2\right)}{2C_p} \tag{14.12}$$

The *stage efficiency* (total-to-total) is defined as

$$\eta_{stage} = \frac{T_{01} - T_{03}}{T_{01} - T_{03}'} = \frac{\Delta T_{0stage}}{T_{01}\left[1 - (P_{03}/P_{01})^{(\gamma-1)/\gamma}\right]}$$

$$\Delta T_{0stage} = \eta_{stage} T_{01}\left[1 - \left(\frac{P_{03}}{P_{01}}\right)^{(\gamma-1)/\gamma}\right]$$

$$\Delta T_{0s} = \eta_s T_{01}\left[1 - \left(\frac{1}{\pi_s}\right)^{(\gamma-1)/\gamma}\right] \tag{14.13}$$

The pressure ratio of a stage, π_s is given by the relation

$$\pi_s = \frac{P_{01}}{P_{03}} = \frac{1}{(1 - (\Delta T_{0s}/\eta_s T_{01}))^{\gamma/(\gamma-1)}} \tag{14.14}$$

The axial velocity ratio (AVR) is expressed as

$$\text{AVR} = \frac{C_{a3}}{C_{a2}} \tag{14.15}$$

From Equations 14.9b and 14.15, then

$$\frac{\Delta T_{0s}}{T_{01}} = \frac{(\gamma - 1)U^2 C_{a2}}{\gamma R T_{01} U}(\tan \alpha_2 + \text{AVR} \tan \alpha_3)$$

Moreover, $T_{01} = T_1(1 + ((\gamma - 1)/2)M_1^2)$, and the rotational and axial Mach numbers are $M_u = U/\sqrt{\gamma RT_1}$ and $M_{a2} = C_{a2}/\sqrt{\gamma RT_1}$, respectively; thus, the temperature rise in the stage as a ratio of the inlet total temperature becomes

$$\frac{\Delta T_{0s}}{T_{01}} = \frac{(\gamma - 1) M_u^2}{\left(1 + (\gamma - 1)/2\right) M_1^2} \frac{M_{a2}}{M_u} (\tan \alpha_2 + \text{AVR} \tan \alpha_3) \tag{14.16a}$$

The pressure ratio of the stage may be expressed in terms of Mach numbers [2] as

$$\frac{P_{03}}{P_{01}} = \left[1 - \frac{(\gamma - 1) M_u^2}{\eta_s \left(1 + ((\gamma - 1)/2) M_1^2\right)} \frac{M_{a2}}{M_u} (\tan \alpha_2 + \text{AVR} \tan \alpha_3)\right]^{\gamma/(\gamma-1)} \tag{14.16b}$$

The turbine efficiency η_t is related to the stage efficiency η_s by the relation

$$\eta_t = \frac{1 - \pi_t^{(\gamma-1)/\gamma}}{1 - \pi_t^{\eta_s(\gamma-1)/\gamma}} \tag{14.17}$$

where the turbine pressure ratio is

$$\pi_t = \frac{(P_0)_{\text{turbine inlet}}}{(P_0)_{\text{turbine outlet}}}$$

Loss coefficient in nozzle: It is expressed either as an enthalpy loss coefficient (λ_N) or as a pressure loss coefficient (Y_N) [4], where

$$\lambda_N = \frac{T_2 - T_2'}{(C_2^2/2) Cp} \tag{14.18}$$

and

$$Y_N = \frac{P_{01} - P_{02}}{P_{02} - P_2} \tag{14.19}$$

A third loss coefficient, namely, velocity loss coefficient [5] is defined as

$$\varphi_N = \frac{C_2}{C_{2_s}} \tag{14.20}$$

which is the ratio between the actual and ideal exit, velocities for the stator.
Loss coefficient in rotor: The enthalpy loss in the rotor

$$\lambda_R = \frac{T_3 - T_3''}{W_3^2/2Cp} \tag{14.21}$$

The pressure loss in the rotor

$$Y_R = \frac{P_{02\text{rel}} - P_{03\text{rel}}}{P_{03\text{rel}} - P_3} \tag{14.22}$$

The velocity loss in the rotor is

$$\varphi_R = \frac{W_3}{W_{3s}} \tag{14.23}$$

which is also the ratio between the actual and ideal exit velocities for the rotor.

The pressure loss may be measured in a cascade tunnel, while the enthalpy loss is used in the blade row design. Both are nearly equal, or

$$Y \approx \lambda \quad \text{or} \quad Y_R \approx \lambda_R, Y_N \approx \lambda_N$$

The total-to-total isentropic efficiency after a lengthy mathematical manipulation [4] can be written as

$$\eta_s \cong \frac{1}{1 + (C_a/2U)\left[\left(\lambda_R \sec^2 \beta_3 + (T_3/T_2)\lambda_N \sec^2 \alpha_2\right)/(\tan \beta_3 + \tan \alpha_2 - (U/c_a))\right]} \tag{14.24}$$

Another expression correlating the stage efficiency and the nozzle and rotor loss coefficients [5] is given by

$$\eta_s = \frac{1}{1 + \left[\left(\lambda_N C_2^2 + \lambda_R W_3^2\right)/2\,(h_{01} - h_{03})\right)]} \tag{14.25}$$

14.3.4 Nondimensional Quantities

1. Blade-loading or temperature drop coefficient
2. Flow coefficient
3. Degree of reaction

The temperature drop or blade loading is expressed as

$$\psi = \frac{Cp\Delta T_{0s}}{U^2}. \tag{14.26a}$$

Most texts and reference books adopt this expression [2,3,5]. The National Gas Turbine Establishment, the United Kingdom, defines this parameter as

$$\psi_{NGTE} = \frac{Cp\Delta T_{0s}}{1/2U^2} = \frac{2Cp\Delta T_{0s}}{U^2} \tag{14.26b}$$

This form is adopted in some texts [4].

Thus,

$$\psi = \frac{\psi_{NGTE}}{2}$$

The flow coefficient has the same definition as in the axial compressor, namely,

$$\phi = \frac{C_{a2}}{U} \tag{14.27a}$$

The temperature drop can be expressed as

$$\psi = \phi \left(\tan \alpha_2 + \text{AVR} \tan \alpha_3 \right) \tag{14.28a}$$

If $C_{a2} = C_{a3} = C_a$, then

$$\phi = \frac{C_a}{U} \tag{14.27b}$$

The corresponding temperature drop is now

$$\psi = \phi \left(\tan \alpha_2 + \tan \alpha_3 \right) = \phi \left(\tan \beta_2 + \tan \beta_3 \right) \tag{14.28b}$$

The degree of reaction is defined in terms of static enthalpy change as

$$\Lambda = \frac{T_2 - T_3}{T_1 - T_3} \equiv \left. \frac{\text{static enthalpy drop in rotor}}{\text{static enthalpy drop in stage}} \right\} \tag{14.29}$$

If equal absolute speeds are assumed in the inlet and outlet of the stage, $C_1 = C_3$, then from Figure 14.10 and Equation 14.9c

$$\Delta T \equiv T_1 - T_3 = \Delta T_0 = \frac{U C_a \left(\tan \beta_2 + \tan \beta_3 \right)}{Cp} \tag{14.30}$$

Moreover, since the relative total enthalpy (and temperature) at the inlet and outlet of the rotor are equal, $h_{02\text{rel}} = h_{03\text{rel}}$

$$h_2 + \frac{W_2^2}{2} = h_3 + \frac{W_3^2}{2}$$

$$T_2 - T_3 = \frac{1}{2Cp} \left(W_3^2 - W_2^2 \right) = \frac{C_a^2}{2Cp} \left(\sec^2 \beta_3 - \sec^2 \beta_2 \right) = \frac{C_a^2}{2Cp} \left(\tan^2 \beta_3 - \tan^2 \beta_2 \right) \tag{14.31}$$

From Equations 14.30 and 14.31, the degree of reaction is

$$\Lambda \equiv \frac{T_2 - T_3}{T_1 - T_3} = \left(\frac{C_a}{2U} \right) \left(\frac{\left(\tan^2 \beta_3 - \tan^2 \beta_2 \right)}{\left(\tan \beta_2 + \tan \beta_3 \right)} \right)$$

$$\Lambda = \frac{C_a}{2U} \left(\tan \beta_3 - \tan \beta_2 \right) \tag{14.32a}$$

Or in terms of the flow coefficient,

$$\Lambda = \frac{\phi}{2} \left(\tan \beta_3 - \tan \beta_2 \right)$$

From Equation 14.27a, for an impulse turbine $\Lambda = 0$, the rotor relative angles at inlet and outlet are equal, or $\beta_2 = \beta_3$.

From Equation 14.1c, the following expressions for the degree of reaction are obtained:

$$\Lambda = \frac{1}{2} - \frac{C_a}{2U} \left(\tan \alpha_2 - \tan \beta_3 \right) \tag{14.32b}$$

TABLE 14.2

Variations of Blading, Velocity Triangles, and Thermodynamics Based on the Degree of Reaction

Λ	Blading	Characteristics	Velocity Triangle	Thermodynamics
0	Impulse	$\beta_2 = \beta_3$	Skewed	$T_2 = T_3$
0.5	Reaction	$\alpha_2 = \beta_3, \alpha_3 = \beta_2$	Symmetrical	$T_2 = (T_1 + T_3)/2$
1.0	Reaction	$\alpha_2 = \alpha_3$	Skewed	$T_1 = T_2$

Or

$$\Lambda = \frac{1}{2} - \frac{\phi}{2} (\tan \alpha_2 - \tan \beta_3)$$

$$\Lambda = 1 - \frac{C_a}{2U} (\tan \beta_3 - \tan \alpha_3) \tag{14.32c}$$

Or

$$\Lambda = 1 - \frac{\phi}{2} (\tan \beta_3 - \tan \alpha_3)$$

Equations 14.32a–c may be used to construct Table 14.2.

Three nondimensional variables may be coupled to give the following relations: If $C_{a2} = C_{a3}$, then

$$\psi = \phi (\tan \alpha_2 + \tan \alpha_3) = \phi (\tan \beta_2 + \tan \beta_3)$$

$$\Lambda = \frac{\phi}{2} (\tan \beta_3 - \tan \beta_2)$$

$$\tan \beta_2 = \frac{1}{2\phi} (\psi - 2\Lambda) \tag{14.33}$$

$$\tan \beta_3 = \frac{1}{2\phi} (\psi + 2\Lambda) \tag{14.34}$$

If

$$\frac{1}{\phi} = \tan \beta_3 - \tan \alpha_3 = \tan \alpha_2 - \tan \beta_2$$

Then

$$\tan \alpha_2 = \tan \beta_2 + \frac{1}{\phi} = \frac{1}{2\phi} (\psi - 2\Lambda + 2) \tag{14.35}$$

and

$$\tan \alpha_3 = \tan \beta_3 - \frac{1}{\phi} = \frac{1}{2\phi} (\psi + 2\Lambda - 2) \tag{14.36}$$

Thus, the temperature loading will be expressed by the following relations:

$$\psi = 2\phi \tan \alpha_2 + 2\Lambda - 2 \qquad (14.37a)$$

$$\psi = 2\phi \tan \alpha_3 - 2\Lambda + 2 \qquad (14.37b)$$

$$\psi = 2\phi \tan \beta_2 + 2\Lambda \qquad (14.37c)$$

$$\psi = 2\phi \tan \beta_3 - 2\Lambda \qquad (14.37d)$$

Figure 14.11 illustrates relations (14.37a) and (14.37b) for a degree of reaction 0.4 for different absolute angles ($\alpha_2 = 50°, 60°, 70°$, and $80°$ and $\alpha_3 = -10°, 0°, 10°, 20°$). A plot for relations (14.37c) and (14.37b) for different values of the degree of reaction (0.4, 0.5, and 0.6) and relative angles ($\beta_2 = 10°$ and $\beta_3 = 70°$) is illustrated in Figure 14.12.

For the special case of

$$\tan \beta_3 = \frac{1}{2\phi} (\psi + 1),$$

$$(14.38)$$

Thus

$$\psi = 2\phi \tan \beta_3 - 1 = 2\phi \tan \alpha_2 - 1 \qquad (14.38a)$$

$$\psi = 2\phi \tan \beta_2 + 1 = 2\phi \tan \alpha_3 + 1 \qquad (14.38b)$$

Figure 14.13 illustrates the relations (14.38a) and (14.38b) for $\beta_2 = 0°, 15°$, and $30°$ and $\beta_3 = 50°, 65°$, and $80°$ temperature loading versus flow coefficient for 50% reaction.

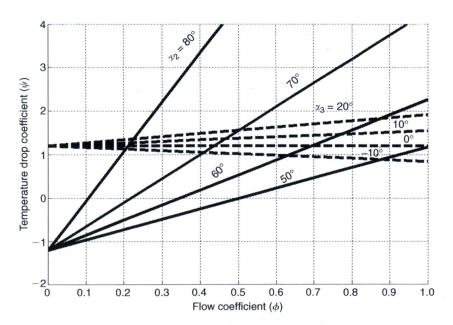

FIGURE 14.11 Temperature loading versus flow coefficient for a 0.4° of reaction.

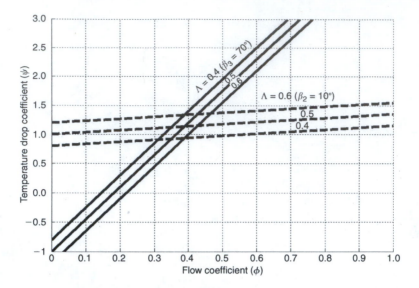

FIGURE 14.12 Temperature loading versus flow coefficient for different degree of reactions.

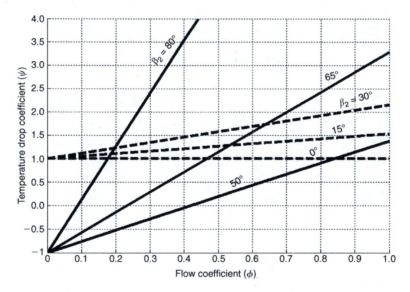

FIGURE 14.13 Temperature loading versus flow coefficient for $\Lambda = 0.5$.

For an impulse turbine ($\Lambda = 0$), then from equation 14.37

$$\psi = 2\phi \tan \beta_2 \tag{14.39}$$

$$\psi = 2\phi \tan \alpha_3 + 2 \tag{14.40}$$

Example 1   Consider a three-stage axial flow turbine where the total inlet temperature is 1100 K, the flow coefficient is 0.8, and the mean rotational speed is constant for all stages and equal to 340 m/s.

Other data are given in the following table:

Stage	Angles	Degree of Reaction (Λ_m)	Efficiency (η_s)	Inlet Temperature (K)	Temperature Drop [ΔT_0 (K)]
1	$\alpha_1 = 0$	0.5	0.938	1100	145
2	$\alpha_1 = \beta_2$	0.5	0.938	?	145
3	$\alpha_3 = 0$	?	0.94	?	120

It is required to

1. Sketch the turbine shape.
2. Draw the velocity triangles at the mean sections for each stage (assume constant mean radius for all stages).
3. Calculate the degree of reaction at the mean section for the third stage.
4. Calculate the pressure ratio of each stage.
5. Prove that for equal pressure ratios of the three stages, the efficiencies must have the following relations:

$$\frac{\eta_{s2}}{\eta_{s1}} = \frac{T_{01} \times \Delta T_{02}}{\Delta T_{01} \times (T_{01} - \Delta T_{01})}$$

$$\frac{\eta_{s3}}{\eta_{s1}} = \frac{T_{01} \Delta T_{03}}{\Delta T_{01} \times (T_{01} - \Delta T_{01} - \Delta T_{02})}$$

Solution

1. The sketch (Figure 14.14) illustrates the three-stage axial turbine having a constant mean radius.
2. To draw the velocity triangles, the angles ($\alpha_2, \alpha_3, \beta_2, \beta_3$) must be calculated first. The appropriate governing equations are the temperature drop and degree of reaction.

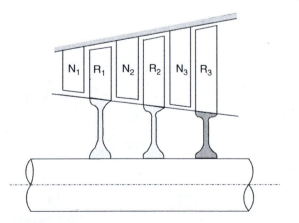

FIGURE 14.14 Layout of the three-stage axial turbine.

From Equations 14.9c and 14.32a,

$$\tan \beta_3 + \tan \beta_2 = \frac{C_p \Delta T_0}{U C_a} \tag{1}$$

$$\tan \beta_3 - \tan \beta_2 = \frac{2\Lambda}{\phi} \tag{2}$$

Now solving these two equations simultaneously results in the angles β_2 and β_3.

For Stage (1)

Since $T_{01} = 1100\,\text{K}$ and $\Delta T_{0s} = 145\,\text{K}$, $T_{03} = 955\,\text{K}$.

Also, as $\Lambda_m = \frac{1}{2}, \alpha_2 = \beta_3, \alpha_3 = \beta_2$

$$\tan \beta_3 + \tan \beta_2 = \frac{1148 \times 145}{340 \times 272} = 1.8 \tag{a}$$

$$\tan \beta_3 - \tan \beta_2 = \frac{2 \times 0.5}{0.8} = 1.25 \tag{b}$$

Then $\beta_2 = \alpha_3 = 15.376°$ and $\beta_3 = \alpha_2 = 56.746°$.

Stage (2)

The same results were obtained for stage (2) as both have the same temperature drop per stage, axial flow coefficient, and axial and rotational speeds.

$$\beta_2 = \alpha_3 = 15.376°$$

$$\beta_3 = \alpha_2 = 56.746°$$

Stage (3)

From Equation 14.9c,

$$\tan \alpha_3 + \tan \alpha_2 = \frac{C_p \Delta T_{0s}}{U C_a}.$$

With

$$\alpha_3 = 0° \quad \text{and} \quad \Delta T_{0s} = 120\,\text{K}$$

$$\therefore \tan \alpha_2 = \frac{C_p \Delta T_{0s}}{U C_a}$$

$$\therefore \alpha_2 = 56.125°$$

To calculate the angles β_2 and β_3, we again use Equations 1 and 2

$$\tan \beta_3 + \tan \beta_2 = \frac{1148 \times 120}{340 \times 272} = 1.4896$$

Also

$$\tan \beta_3 - \tan \beta_2 = \frac{2 \times 0.4042}{0.8} = 1.0105$$

Solving both equations, we get $\beta_2 = 13.471°$ and $\beta_3 = 51.341°$.

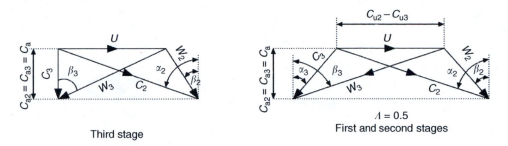

FIGURE 14.15 Velocity triangles for the three stages.

Now, the velocity triangles can be drawn. Stages (1) and (2) have the same velocity triangle, while the third stage has zero exit swirl as shown in Figure 14.15.

3. The degree of reaction for the third stage can be calculated from Equation 14.32c, namely

$$\Lambda = 1 - \frac{\phi}{2}(\tan \alpha_3 - \tan \alpha_2)$$

With $\alpha_3 = 0$ and $\alpha_2 = 56.125°$

$$\Lambda = 0.4042$$

4. The pressure ratio of any stage is calculated from Equation 14.14a.

$$\left(\frac{P_{03}}{P_{01}}\right)_{stage} = \left[1 - \frac{\Delta T_{0s}}{\eta_s T_{01}}\right]^{\gamma/(\gamma-1)}$$

Stage (1)
With $T_{01} = 1100\,K$ and $\Delta T_{0s} = 145\,K$, then $T_{03} = 955\,K$.

$$\left(\frac{P_{03}}{P_{01}}\right) = \left[1 - \frac{145}{(0.938) \times (1100)}\right]^{1.33/0.33} = 0.5432$$

Stage (2)
With $T_{01} = 955\,K$ and $\Delta T_0 = 145\,K$, then $T_{03} = 810\,K$.

$$\left(\frac{P_{03}}{P_{01}}\right) = 0.4935$$

Stage (3)
With $T_{01} = 810\,K$ and $\Delta T_0 = 120\,K$, then

$$\left(\frac{P_{03}}{P_{01}}\right) = 0.50097$$

TABLE 14.3
Summary of results

Stages Parameters	Stage 1	Stage 2	Stage 3
T_{01}	1100	955	810
α_2 (degrees)	56.746	56.746	56.126
β_3 (degrees)	56.746	56.746	51.341
α_3 (degrees)	15.376	15.376	0
β_2 (degrees)	15.376	15.376	13.471
Λ_m	0.5	0.5	0.4042
P_{03}/P_{01}	0.5432	0.4935	0.50097

5. Now to correlate the efficiencies of the first two stages to the temperature drop per stage, since

$$\left(\frac{P_{03}}{P_{01}}\right)_1 = \left(\frac{P_{03}}{P_{01}}\right)_2$$

$$\left(1 - \frac{\Delta T_{01}}{\eta_{s1} T_{01}}\right)^{\gamma/(\gamma-1)} = \left(1 - \frac{\Delta T_{02}}{\eta_{s2}(T_{01} - \Delta T_{01})}\right)^{\gamma/(\gamma-1)}$$

$$\frac{\Delta T_{01}}{\eta_{s1} T_{01}} = \frac{\Delta T_{02}}{\eta_{s2}(T_{01} - \Delta T_{01})}$$

$$\frac{\eta_{s2}}{\eta_{s1}} = \frac{T_{01} \Delta T_{02}}{\Delta T_{01}(T_{01} - \Delta T_{01})}$$

Similarly, assuming equal pressure ratios for all stages:

$$\therefore \left(\frac{P_{03}}{P_{01}}\right)_1 = \left(\frac{P_{03}}{P_{01}}\right)_3$$

$$\left(1 - \frac{\Delta T_{01}}{\eta_{s1} T_{01}}\right)^{\gamma/(\gamma-1)} = \left[1 - \frac{\Delta T_{03}}{\eta_{s3}(T_{01} - \Delta T_{01} - \Delta T_{02})}\right]^{\gamma/(\gamma-1)}$$

$$\therefore \frac{\Delta T_{01}}{\eta_{s1} T_{01}} = \frac{\Delta T_{03}}{\eta_{s3}(T_{01} - \Delta T_{01} - \Delta T_{02})}$$

$$\frac{\eta_{s3}}{\eta_{s1}} = \frac{T_{01} \Delta T_{03}}{\Delta T_{01}(T_{01} - \Delta T_{01} - \Delta T_{02})}.$$

14.3.5 Several Remarks

1. The mass flow parameter can be calculated at any of the three states of the stage. Conservation of mass gives the following relation:

$$\dot{m} = \rho_1 C_{a1} A_1 = \rho_2 C_{a2} A_2 = \rho_3 C_{a3} A_3 \tag{14.41}$$

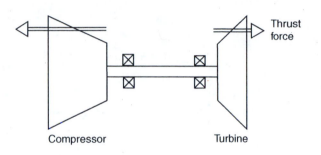

FIGURE 14.16 Thrust force on a spool.

The thrust force on the turbine is a combination of pressure thrust and axial thrust. Thus, the thrust force (T) is equal to

$$\text{Thrust} = \text{Pressure thrust} \pm \text{Axial thrust}$$

$$\text{Pressure thrust} = (P_2A_2 - P_3A_3)$$

$$\text{Axial Thrust} = \dot{m}(Ca_2 - Ca_3)$$

The thrust force (T) is then expressed as

$$T = (P_2A_2 - P_3A_3) + \dot{m}(Ca_2 - Ca_3) \tag{14.42}$$

A pure impulse stage, with no pressure drop and constant axial velocity component across the rotor, will have zero net axial loads. Thrust force generated by compressor either partially or fully balances the thrust force generated on turbine as shown in Figure 14.16. If the axial velocity is constant through the rotor, then the thrust forces arise only from the pressure difference.

2. Aero engines need an axial turbine having a high blade loading, or temperature drop coefficient (ψ), and a high flow coefficient (ϕ). High blade loading means high temperature drop per stage and, consequently, few stages for certain power output. Thus, the weight and size of an aero engine are minimum, which satisfies flight needs. Moreover, high flow coefficient implies high axial velocity and consequently small annulus area. This leads to a small frontal area and a low drag.

On the contrary, the axial turbines in industrial gas turbines are characterized by low blade loading and low axial velocity. These conditions are associated with the best stage efficiency [4]. High efficiency provides the best fuel economy or the lowest specific fuel consumption (SFC). Although low blade loading will lead to a large number of stages and low flow coefficient also leads to a large frontal area, both are of minor importance for ground engines.

Smith [6] examined the data of some 70 turbine tests having a degree of reaction between 0.2 and 0.6, aspect ratio (blade height/chord) in the region 3–4. He plotted these data as ψ–ϕ curves for constant efficiencies. Although these data were published in 1965 for specific data, the present turbine stages even with different geometry and data have the same trends. For this reason, the work of Smith is still used for the preliminary design of turbines. From the above discussion and Smith's work, the following conclusions may be deduced. For aero engines, the optimum value for ψ is from 1.5 to 2.5 and for ϕ from 0.8 to 1.2. Efficiency ranges from 86% to 91%. For industrial gas turbines, the optimum value for ψ ranges from 1.0 to 1.5 while ϕ ranges from 0.4 to 0.8. The corresponding efficiency ranges from 93% to 95%.

3. Stator blade rows have nozzle shapes and for this reason are frequently identified as nozzle vanes. The stator passage may be of the convergent or the convergent–divergent shape. The latter type is not recommended for two reasons: the first reason is that convergent–divergent nozzles are inefficient at part loads or generally at any pressure ratios other than the design value. The second

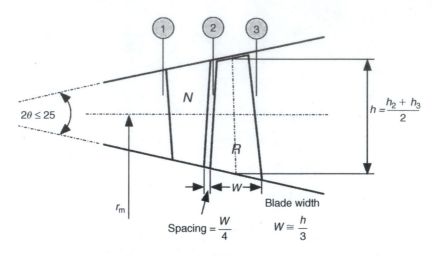

FIGURE 14.17   Suggested geometry for an axial turbine stage.

reason is that the outlet speed from the nozzle C_2 will be high and, consequently, the rotor inlet relative speed W_2 will also be high enough to form shock waves within the rotor passage.

The convergent nozzle must be checked for choking. Thus if $(P_{01}/P_2 < P_{01}/P_c)$, then the nozzle is unchoked. Each ratio can be calculated as follows:

$$\frac{P_{01}}{P_2} = \left(\frac{T_{01}}{T_2'}\right)^{\gamma/(\gamma-1)}$$

where $T_2' = T_2 - \lambda_N C_2^2/2Cp$ and

$$\frac{P_{01}}{P_c} = \left(\frac{\gamma+1}{2}\right)^{\gamma/(\gamma-1)} = \left(\frac{2.33}{2}\right)^{1.33/0.33} = 1.853$$

For choked nozzle, the absolute velocity is equal to the sonic speed, as is illustrated later on. The annulus of turbine stages is generally flared (Figure 14.17). The annulus of turbine stages is generally flared. The flare angle is calculated when the blade heights and widths are calculated first. The blade heights are calculated as follows. The axial velocities at states (1), (2), and (3) are determined first. Next, the absolute velocities at the three states are determined. The static temperature and pressure are calculated as follows: $T = T_0 - (C^2/2C_p)$ and $P = P_0(T/T_0)^{\gamma/(\gamma-1)}$; thus, the density is $\rho = P/RT$. The blade height is then obtained for a constant mean radius from the relation $h = \dot{m}/\pi\rho C_a D_m$. The blade mean height is then calculated as follows: $h_{nozzle} = (h_1 + h_2)/2$ and $h_{rotor} = (h_2 + h_3)/2$. The blade chord ranges from $(h/4)$ to $(h/2)$. The spacing between the blade rows must be greater than 20% of the chord. The flare can then be calculated.

If the included angle of divergence of the walls is large, then a flow separation from the inner wall may be encountered. A safe value for the included angle is to be less than or equal to 25° [7]. Both the flare angles, height-to-width ratio of the stator/rotor blades, and the spacing between the stator and rotor blade rows are design parameters. They depend on aerodynamics and mechanical stresses of the stator and rotor blades. Rotor blades are subjected to vibration stresses as well as mechanical stresses. Vibration stresses arise as the rotor blades pass through the wake of the nozzle blades. Excessive vibration stresses arise when the space between the blade rows is less than 0.2 of the blade width. A value of nearly 0.5 is often used, which will reduce both the vibration stresses and the annulus flare [4].

Example 2   A single-stage axial flow gas turbine has the following data:

- Turbine inlet total temperature $= 1100$ K
- Turbine inlet total pressure $= 3.4$ bar
- Stage temperature drop $\Delta T_0 = 144$ K
- Isentropic efficiency $\eta = 0.9$
- Mean blade speed $U = 298$ m/s
- Mass flow rate $m = 18.75$ kg/s
- Flow coefficient $\phi = 0.95$
- Loss coefficient for nozzle blade $\lambda_N = 0.05$
- Rotational speed $= 200$ rps
- The convergent nozzle is chocked

Calculate

1. Blade-loading coefficient (Ψ)
2. Pressure ratio of the stage
3. The flow angles α_2, α_3, β_2, and β_3

Solution: The gas properties are

$$C_P = 1148 \text{ J/kg} \cdot \text{K}, \quad R = 287 \text{ J/kg} \cdot \text{K}, \quad \gamma = \tfrac{4}{3}.$$

1. The blade loading is

$$\psi = \frac{C_P \Delta T_0}{U^2} = \frac{1148 \times 144}{298^2}$$
$$\psi = 1.8615$$

2. $T_{02} = T_{01} = 1100$ K

$$T_{03} = T_{01} - \Delta T_0 = 1100 - 144 = 956 \text{ K}$$

$$\frac{P_{03}}{P_{01}} = \left[1 - \frac{\Delta T_0}{\eta_t T_{01}} \right]^{\gamma/(\gamma-1)} = \left[1 - \frac{144}{0.9 \times 1100} \right]^4 = 0.533$$

The pressure ratio in the turbine is

$$\frac{P_{01}}{P_{03}} = 1.875, \quad \text{or} \quad P_{03} = 1.813 \text{ bar}$$

3. Since the nozzle is chocked,

(i) $C_2 = \sqrt{\gamma R T_2}$

(ii) $\dfrac{T_{02}}{T_2} = \dfrac{\gamma + 1}{2} = 1.165$

The static temperature at the nozzle outlet is then $T_2 = 944.2 \, \text{K}$.
The absolute speed of the gases leaving the choked nozzle is $C_2 = \sqrt{\gamma R T_2} = 600.3 \, \text{m/s}$.
Since the nozzle losses are expressed by the relation

$$T_2 - T_2' = \lambda_\text{N} \frac{C_2^2}{2C_\text{P}} = \frac{0.05 \times 600^2}{2 \times 1148} = 7.839 \, \text{K}$$

$$T_2' = 936.4 \, \text{K}$$

The pressure ratio is

$$\frac{P_{01}}{P_2} = \left(\frac{T_{01}}{T_2'} \right)^{\gamma/(\gamma-1)} = \left(\frac{1100}{936.4} \right)^4 = 1.904$$

$$P_2 = 1.785 \, \text{bar}$$

$$C_{a2} = U\phi = 298 \times 0.95 = 283 \, \text{m/s}$$

$$\cos \alpha_2 = \frac{C_{a2}}{C_2} = \frac{283}{600} = 0.4716$$

$$\alpha_2 = 62°$$

Since

$$\frac{U}{C_a} = \tan \alpha_2 - \tan \beta_2 = \frac{1}{\phi}$$

$$\tan \beta_2 = \tan \alpha_2 - \frac{1}{\phi} = 0.828$$

$$\beta_2 = 39.6°$$

The specified work W_s is expressed as

$$W_\text{s} = C_\text{P}\Delta T_0 = U(C_{u2} + C_{u3}) = UC_a(\tan \alpha_2 + \tan \alpha_3)$$

$$\tan \alpha_3 = \frac{C_\text{P}\Delta T_0}{UC_a} - \tan \alpha_2 = \frac{1148 \times 144}{298 \times 283} - 1.8807 = 0.0793$$

$$\alpha_3 = 4.54°$$

$$\frac{U}{C_a} = \tan \beta_3 - \tan \alpha_3$$

$$\tan \beta_3 = \frac{1}{\phi} + \tan \alpha_3 = 1.0526 + 0.0794 = 1.132$$

$$\beta_3 = 48.54°$$

Example 3 A single-stage axial flow turbine has a mean radius of 30 cm and a blade height at the stator inlet of 6 cm. The hot gases enter the turbine stage at 1900 kPa and 1200 K and the absolute velocity leaving the stator (C_2) is 600 m/s and inclined at an angle 65° to the axial direction. The relative angles at the inlet and outlet of the rotor blade are 25° and 60°, respectively. The stage efficiency is 0.88. Calculate

1. The absolute angle α_3
2. The rotor rotational speed in rpm
3. The stage pressure ratio
4. Flow coefficient, blade-loading coefficient, and degree of reaction
5. The mass flow rate
6. The power delivered by the turbine (Take $\gamma = 1.33$, $R = 290\,J/kg \cdot K$), and Ca/U is constant through the stage.

Solution:
1. The given angles are $\alpha_2 = 65°$, $\beta_2 = 25°$, $\beta_3 = 60°$.
   From Equation 14.1c,

$$\tan \alpha_3 = \tan \beta_3 - \tan \alpha_2 + \tan \beta_2 = 0.0539$$

Then

$$\alpha_3 = 3.08°.$$

2. Since $C_a = C_2 \cos \alpha_2 = 600 \cos 65 = 253.57\,m/s$
   From Equation 14.1c

$$U = C_a(\tan \alpha_2 - \tan \beta_2) = 253.57 \times (\tan 65 - \tan 25) = 425.5\,m/s$$

But since $U = \pi ND/60$, then

$$N = 60 \times 425.5/\pi \times 0.6 = 13{,}54576\,rpm$$

3. From Equation 14.9c, with $\beta_2 = 25°$, $\beta = 60°$, and

$$Cp = \gamma R/(\gamma - 1) = 1168.8\,J/kg \cdot K$$

$$\Delta T_{0\,stage} = \frac{425.5 \times 253.37}{1168.8}(\tan 25 + \tan 60) = 202.77\,K$$

From Equation 14.14a, the stage pressure ratio is

$$\pi_{stage} = \frac{P_{01}}{P_{03}} = \frac{1}{(1 - (\Delta T_{0s}/\eta_s T_{01}))^{\gamma/(\gamma-1)}} = \frac{1}{(1 - (202.77/(0.88 \times 1200)))^4} = 2.346$$

4. The flow coefficient $\phi = C_a/U = 0.6$.
   The blade-loading coefficient ψ is given by the relation

$$\psi = \frac{C_p \Delta T_{0stage}}{U^2} = \frac{1168.8 \times 202.77}{(425.5)^2} = 1.309$$

The degree of reaction (Equation 14.32a)

$$\Lambda = \phi(\tan \beta_3 - \tan \beta_2)/2 = 0.41$$

5. From Equation 14.1c, $\tan \beta_3 - \tan \alpha_3 = U/C_a$

$$\tan \alpha_3 = \tan 60 - \frac{425.5}{253.57} = 3.1°$$

$$C_3 = \frac{C_a}{\cos \alpha_3} = \frac{253.57}{\cos 3.1} = 253.94 \, \text{m/s}$$

Assuming that $C_1 = C_3$, the static conditions at the stage inlet are calculated as follows:

$$T_1 = T_{01} - \frac{C_1^2}{2C_p} = 1200 - \frac{(253.94)^2}{2 \times 1168.8} = 1172.4 \, \text{K}$$

$$P_1 = P_{01} \left(\frac{T_1}{T_{01}}\right)^{\gamma/(\gamma-1)} = 1900 \times 10^3 \left(\frac{1172.4}{1200}\right)^4 = 1731 \times 10^3 \, \text{Pa}$$

$$\rho_1 = \frac{P_1}{RT_1} = \frac{1731 \times 10^3}{290 \times 1172.4} = 5.09 \, \text{kg/m}^3$$

The annulus area at inlet is $A_1 = \pi D_m h = \pi \times 0.6 \times 0.06 = 0.1131 \, \text{m}^2$.
The mass flow rate is $\dot{m} = \rho_1 C_a A_1 = 5.09 \times 253.57 \times 0.1131 = 145.97 \, \text{kg/m}^3$.
6. Power $= \dot{m} C_p \Delta T_{0\text{stage}} = 145.97 \times 1.1688 \times 202.77 = 34600 \, \text{kW}$
Power $= 34.6 \, \text{MW}$.

14.4 THREE DIMENSIONAL

The procedure similar to that followed in Chapter 13 will be followed here. It is assumed that all the radial motion of the particle takes place within the blade row passages, while radial equilibrium is assumed in the axial spacing between the successive blade rows. Thus, the stream surface in the spacing between the blade rows has a cylindrical stream surface shape.

$$C_r \ll C_a, \quad C_r \ll C_u, \quad \text{thus} \quad C_r \approx 0$$

For steady flow, the *simplified radial equilibrium equation* (SRE) (Equation 13.28) is valid.

$$\frac{1}{\rho} \frac{dp}{dr} = \frac{C_u^2}{r} \tag{14.43}$$

Substituting the radial equilibrium equation into the stagnation enthalpy, neglecting second-order terms and entropy changes in the radial direction, the vortex energy (Equation 13.29) is also derived:

$$\frac{dh_0}{dr} = C_a \frac{dc_a}{dr} + C_u \frac{dC_u}{dr} + \frac{C_u^2}{r} \tag{14.44}$$

Solution of Equation 13.29 provides a simplified three-dimensional analysis to the flow. The three-dimensional nature of the flow affects the variation of the gas angles with radius (from root to tip) as U increases with radius. Also, the whirl component in the flow outlet from the nozzles causes the static pressure and temperature to vary across the annulus. Twisted blading designed to account for the changing gas angles is called vortex blading.
The following three methods will be examined:

1. Free vortex
2. Constant nozzle angle, which is subdivided into three cases:
   (a) General case

(b) Zero exit swirl
(c) Free vortex at rotor outlet
3. General N-case, where the swirl velocity is proportional to the radius raised to the power N

14.4.1 Free Vortex Design

As pointed out in Reference 3, for design purposes, it is usually assumed that the total enthalpy (h_{01}) and entropy are constant at entry to stage, or $dh_{01}/dr = 0$. A *free vortex stage* will be obtained if $dh_{01}/dr = 0$, and the whirl velocity components at rotor inlet and outlet satisfy the following conditions:

$$rC_{u2} = \text{constant} \quad \text{and} \quad rC_{u3} = \text{constant}$$

This can be justified as follows. Since no work is done by the gases in the nozzle, the total enthalpy at the nozzle outlet (h_{02}) must also be constant over the annulus at outlet. At the nozzle outlet, if the whirl velocity component of the flow (C_u) is inversely proportional to radius ($rC_{u2} = \text{constant}$), then it follows from the radial equilibrium equation 13.28 that $dC_{a2}/dr = 0$, and that the axial velocity (C_{a2}) is constant across the annulus. Assuming that the whirl velocity at the rotor inlet is inversely proportional to radius, or $rC_{u3} = \text{constant}$, then the specific work is

$$h_{02} - h_{03} = U(C_{u2} + C_{u3}) = \omega(rC_{u2} + rC_{u3}) = \text{constant}$$

Thus, the total enthalpy at the rotor outlet (h_{03}) is constant over the annulus. With $rC_{u3} = \text{constant}$, the axial velocity ($C_{a3}$) is also constant.

To evaluate the angles at any radius, since axial velocity is constant, $rC_{u2} = rC_{a2} \tan \alpha_2 = \text{constant}$ or $r \tan \alpha_2 = \text{constant}$.

From Equation 14.1a, the relative angle β_2 may be obtained. Same procedure may be followed for the rotor outlet.

As a summary for the free vortex design:

State (1): All the properties are constant along the annulus

State (2): $rC_{u2} = \text{constant}, \quad C_{a2} = \text{constant}$

$$\tan \alpha_2 = \left(\frac{r_m}{r}\right)_2 \tan \alpha_{2m} \tag{14.45a}$$

$$\tan \beta_2 = \left(\frac{r_m}{r}\right)_2 \tan \alpha_{2m} - \left(\frac{r}{r_m}\right)_2 \frac{U_m}{C_{a2}} \tag{14.46a}$$

State (3): $rC_{u3} = \text{constant} \quad C_{a3} = C_{a2} = \text{constant}$

$$\tan \alpha_3 = \left(\frac{r_m}{r}\right)_3 \tan \alpha_{3m} \tag{14.45b}$$

$$\tan \beta_3 = \left(\frac{r_m}{r}\right)_3 \tan \alpha_{3m} + \left(\frac{r}{r_m}\right)_3 \frac{U_m}{C_{a3}} \tag{14.46b}$$

14.4.2 Constant Nozzle Angle Design (α_2)

An alternative design procedure to free vortex flow is the constant nozzle angle. The appropriate conditions that also provide radial equilibrium are

1. Uniform flow in the annulus space between the nozzles and rotor blades, which is satisfied when the outlet flow to nozzle is uniform ($dh_{02}/dr = 0$).

2. Constant nozzle outlet angle (α_2) that avoids manufacturing nozzles of varying outlet angle. Since

$$\cot \alpha_2 = \frac{C_{a2}}{C_{u2}} = \text{constant} \tag{14.47}$$

$$C_{a2} = C_{u2} \cot \alpha_2 \quad \text{and} \quad \frac{dC_{a2}}{dr} = \frac{dC_{u2}}{dr} \cot \alpha_2$$

From the vortex energy Equations 13.29 and 14.47

$$C_{u2} \cot^2 \alpha_2 \frac{dC_{u2}}{dr} + C_{u2} \frac{dC_{u2}}{dr} + \frac{C_{u2}^2}{r} = 0$$

$$C_{u2} \left(1 + \cot^2 \alpha_2 \right) \frac{dC_{u2}}{dr} + \frac{C_{u2}^2}{r} = 0$$

$$\frac{dC_{u2}}{C_{u2}} = -\sin^2 \alpha_2 \frac{dr}{r}$$

Integrating to get

$$C_{u2} r^{(\sin^2 \alpha_2)} = \text{constant} \tag{14.48a}$$

Or

$$C_{u2} = C_{u2m} \left(\frac{r_m}{r} \right)^{\sin^2 \alpha_2} \tag{14.48b}$$

Similarly

$$C_{a_2} r^{\left(\sin^2 \alpha_2 \right)} = \text{constant} \tag{14.49a}$$

Or

$$C_{a2} = C_{a2m} \left(\frac{r_m}{r} \right)^{\sin^2 \alpha_2} \tag{14.49b}$$

Also, the absolute velocity will have the expression

$$C_2 = C_{2m} \left(\frac{r_m}{r} \right)^{\sin^2 \alpha_2} \tag{14.50}$$

Normally, nozzle angles are greater than $60°$ and thus a quite good approximate solution is obtained by designing with a constant nozzle angle and constant angular momentum, or

$$\alpha_2 = \text{constant} \quad \text{and} \quad rC_{u2} = \text{constant}$$

Or in other words,

$$\alpha_2 = \text{constant} \quad \text{and} \quad \frac{C_{u2}}{C_{u2m}} = \frac{C_{a2}}{C_{a2m}} = \frac{C_2}{C_{2m}} = \frac{r_m}{r} \tag{14.51}$$

So far, the conditions at the rotor inlet are properly defined. The conditions at the rotor exit or at state (3) remain to be defined. Three cases are available for rotor outlet, namely,

1. Constant total conditions at outlet
2. Free vortex at outlet
3. Zero whirl at outlet

Case (1): Constant total enthalpy at outlet

The total enthalpy is then constant at all the states of turbine stages, which means that the specific work is also constant.

Since

$$U(C_{u2} + C_{u3}) = \Delta h_0,$$

$$C_{u3} = \frac{\Delta h_0}{U} - C_{u2} = \frac{K}{r} - C_{u2} = \frac{K}{r} - C_{a2} \tan \alpha_2$$

where $K = \Delta h_0 / \omega$.

Or

$$C_{u3} = \frac{K}{r} - C_{a2m} \left(\frac{r_m}{r}\right)^{\sin^2 \alpha_2} \tan \alpha_2 \qquad (14.52)$$

The axial velocity at station 3 is obtained from the vortex energy equation and has the following expression [8]:

$$C_{a3}^2 = C_{a2m}^2 \left(\frac{r_m}{r}\right)^{2\sin^2 \alpha_2} - \frac{\sin(2\alpha_2)}{1 + \sin^2 \alpha_2} \frac{C_{a2m}K}{r_m} \left[\left(\frac{r_m}{r}\right)^{1+\sin^2 \alpha_2} - 1\right] + C_{a3m}^2 - C_{a2m}^2 \qquad (14.53)$$

Case (2): Zero whirl angle ($\alpha_3 = 0$)

Since the flow leaves the rotor axially, $C_{u3} = 0$, the vortex energy equation 13.29 becomes

$$\frac{dh_{03}}{dr} = C_{a3} \frac{dC_{a3}}{dr} \qquad (14.54)$$

Also,

$$h_{03} = h_{01} - UC_{u2} = h_{01} - UC_{u2m} \left(\frac{r_m}{r}\right)^{\sin^2 \alpha_2}$$

$$\frac{dh_{03}}{dr} = -\frac{d}{dr} \left[UC_{u2m} \left(\frac{r_m}{r}\right)^{\sin^2 \alpha_2}\right]. \qquad (14.55)$$

From Equations 14.54 and 14.55,

$$C_{a3}^2 = C_{a3m}^2 + 2U_m C_{u2m} \left(1 - \left(\frac{r}{r_m}\right)^{\cos^2 \alpha_2}\right) \qquad (14.56)$$

Case (3): Free vortex at rotor outlet (rC_{u3} = constant)
The whirl component at different radii is expressed as

$$C_{u3} = \frac{r_m}{r} C_{u3m}$$

Following the same procedure as described earlier, the axial velocity will have the same expression in Equation 14.56.

14.4.3 GENERAL CASE

If the total enthalpy is radially constant, the vortex energy equation 13.29 is expressed by

$$C_a \frac{dC_a}{dr} + C_u \frac{dC_u}{dr} + \frac{C_u^2}{r} = 0 \tag{14.57}$$

Consider a general variation for swirl velocity (C_u) with radius [9] as

$$C_u = Kr^N. \tag{14.58a}$$

Or in terms of mean-section conditions

$$\frac{C_u}{C_{um}} = \left(\frac{r}{r_m}\right)^N \tag{14.58b}$$

Substituting Equation 14.58a into a vortex energy equation and then integrating between the limits of r_m and r

$$\frac{C_a}{C_{am}} = \left\{ 1 - \tan^2 \alpha_m \left(\frac{N+1}{N}\right) \left[\left(\frac{r}{r_m}\right)^{2N} - 1 \right] \right\}^{1/2}, \quad N \neq 0 \tag{14.59}$$

This equation is not valid for the special case of $N = 0$ (constant C_u). For this special case, Equation 14.57 will be reduced to

$$\frac{C_u^2}{r} + \frac{1}{2} \frac{d}{dr} \left(C_a^2\right) = 0$$

Integrating, we get

$$\frac{C_a}{C_{am}} = \left[1 - 2\tan^2 \alpha_m \ell n \frac{r}{r_m} \right]^{1/2}, \quad N = 0 \tag{14.60}$$

In the general case, Equation 14.58 is valid for any value of $N \neq 0$. The case of free vortex, $N = -1$, provides constant axial velocity in the radial direction (or $C_a/C_{am} = 1$) as was proved earlier. The radial variation in the axial velocity is largely dependent on the specified swirl velocity variation (value of N). Two important notes have to be stated here [9]:

1. Axial velocities and flow angles (α_2) associated with certain values of N cannot be obtained with all blade length.
2. The range of N that can be used for design purposes becomes larger as the blades become shorter (values of r_h/r_m and r_t/r_m closer to 1).

As described in Reference 10, free vortex is commonly used so that all other designs are often classified under the common heading "nonfree vortex." The nonfree vortex designs are used to alleviate some of the potential disadvantages associated with the free vortex design, namely, large variation of the degree of reaction across the blade and large blade twist. Thus, a turbine blade has the potential to a low degree of reaction at root, nearly an impulse blading at blade root and a large degree of reaction at the tip. The latter can also be troublesome because it increases the leakage across the blade tip clearance space. The following cases are distanced:

1. The case $N = -2$ is called super vortex
2. The case $N = -1$ is the free vortex
3. The case $N = 0$ is the constant swirl velocity
4. The case $N = 1$ is the wheel flow or solid rotation

14.4.4 Constant Specific Mass Flow Stage

Another design procedure that satisfies the radial equilibrium equation is the constant specific mass flow or the constant mass flow per unit area at all radii. Consider an infinitesimal annulus area between the hub and the casing of an axial turbine (Figure 14.18).

$$d\dot{m} = \rho C_a dA$$

$$\frac{d\dot{m}}{dA} = \rho C_a = \text{constant} \equiv K_1$$

$$\rho^2 = \frac{K_1^2}{C^2 \cos^2 \alpha} \tag{14.61a}$$

$$\cos \alpha = \frac{\rho_m C_m}{\rho C} \cos \alpha_m \tag{14.61b}$$

But in the radial equilibrium equation 14.43

$$\frac{1}{\rho}\frac{dp}{dr} = \frac{1}{\rho}\left(\frac{dp}{d\rho}\right)_s \frac{d\rho}{dr} = \frac{a^2}{\rho}\frac{d\rho}{dr} = \frac{C_u^2}{r} = \frac{C^2 \sin^2 \alpha}{r} \tag{14.62}$$

where $a = \sqrt{\gamma R T}$ is the speed of sound.

The density temperature relation is

$$\frac{T}{\rho^{\gamma-1}} = K_2 = \text{constant} \tag{14.63}$$

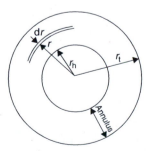

FIGURE 14.18   Flow through infinitesimal annulus in axial turbine.

From Equations 14.62 and 14.63

$$\frac{1}{\rho}(K_2 \gamma R \rho^{\gamma-1})\frac{d\rho}{dr} = \frac{C^2 \sin^2 \alpha}{r}$$

$$\gamma R K_2 \frac{\rho^{\gamma}}{\rho^2}\frac{d\rho}{dr} = \frac{C^2 \sin^2 \alpha}{r}$$

Substituting ρ^2 from Equation 14.62

$$\therefore \frac{\gamma R K_2 \rho^{\gamma}}{(K_1^2/C^2 \cos^2 \alpha)}\frac{d\rho}{dr} = \frac{C^2 \sin^2 \alpha}{r}$$

$$\frac{K_2}{K_1^2}\gamma R \int_{\rho_m}^{\rho} \rho^{\gamma} d\rho = \int_{r_m}^{r} \tan^2 \alpha \frac{dr}{r}$$

$$\left(\frac{K_2}{K_1^2}\right)\left(\frac{\gamma R}{\gamma+1}\right)\left(\rho^{\gamma+1} - \rho_m^{\gamma+1}\right) = \int_{r_m}^{r} \tan^2 \alpha \frac{dr}{r}$$

$$\rho = \left[\rho_m^{\gamma+1} + \left(\frac{K_1^2}{K_2}\right)\left(\frac{\gamma+1}{\gamma R}\right)\int_{r_m}^{r} \tan^2 \alpha \frac{dr}{r}\right]^{1/(\gamma+1)} \tag{14.64}$$

To derive a relation for the absolute velocity variation across the annulus, from the total enthalpy relation $(h = h_0 - (C^2/2))$ and entropy enthalpy relation $(Tds = dh - (dp/\rho))$, assuming that the total enthalpy and entropy are constants along the radial direction,

$$\frac{dh}{dr} = \frac{dh_0}{dr} - \frac{1}{2}\frac{d}{dr}(C^2)$$

$$T\frac{ds}{dr} = \frac{dh}{dr} - \frac{1}{\rho}\frac{dp}{dr} = \frac{dh_0}{dr} - \frac{1}{2}\frac{d}{dr}(C^2) - \frac{1}{\rho}\frac{dp}{dr}$$

$$\frac{1}{\rho}\frac{dp}{dr} = -C\frac{dC}{dr} \tag{14.65a}$$

From the radial equilibrium equation

$$\frac{1}{\rho}\frac{dp}{dr} = \frac{C_u^2}{r} = \frac{C^2 \sin^2 \alpha}{r}. \tag{14.65b}$$

From Equations 14.65a and b

$$\frac{C^2 \sin^2 \alpha}{r} = -C\frac{dC}{dr}$$

$$\int_{C_m}^{C} \frac{dC}{C} = -\int_{r_m}^{r} \sin^2 \alpha \frac{dr}{r} = \int_{r}^{r_m} \sin^2 \alpha \frac{dr}{r}$$

$$\ln \frac{C}{C_m} = \int_r^{r_m} \sin^2 \alpha \, \frac{dr}{r}$$

$$C = C_m \exp \left(\int_r^{r_m} \sin^2 \alpha \, \frac{dr}{r} \right) \tag{14.66}$$

Thus, Equations 14.65, 14.64, and 14.61 are used to determine the variations in density (ρ), velocity (C), and angle (α), respectively, along the radial direction. The calculations can be performed by adopting a step-by-step method for infinitesimal lengths Δr along the blade height. The value of (α) may be assumed constant over the small span Δr [9].

Example 4 An axial turbine has the following data:

$$\zeta = 0.8, \quad \alpha_{2m} = 60, \quad \alpha_3 = 0, \quad \Lambda_m = 0.5 \quad \text{and} \quad C_{am} = 200 \, \text{m/s}$$

Three design methods are to be examined. The governing equations for each are as follows:
Method (1)

$$C_{u2m} = C_{u2h}$$

Method (2)

$$\frac{C_{a2t}}{C_{a2h}} = \zeta^{\sin^2 \alpha_2}$$

Method (3)

$$\frac{C_{u2t}}{C_{u2h}} = \zeta$$

It is required to

1. Identify the design method.
2. Calculate the values of the angles $\alpha_2, \alpha_3, \beta_2, \beta_3$ at the hub, mean, tip sections.
3. Draw the velocity triangles at hub and tip.

Solution: Given data

$$\zeta = 0.8, \quad \alpha_3 = 0, \quad \alpha_{2m} = 60$$
$$\Lambda_m = 0.5 \quad C_{am} = 200 \, \text{m/s}$$

At any radius along the blade

$$\alpha_2 = \tan^{-1}\left(\frac{C_{u2}}{C_{a2}}\right), \quad \beta_2 = \tan^{-1}\left(\frac{C_{u2}-U}{C_{a2}}\right)$$

$$\frac{r_h}{r_m} = \frac{2\zeta}{1+\zeta} = \frac{1.6}{1.8} = 0.889$$

$$\frac{r_t}{r_m} = \frac{2}{1+\zeta} = \frac{2}{1.8} = 1.111$$

At mean section

$$C_{a2} = 200\,\text{m/s}$$

$$C_{u2} = C_{a2} \times \tan\alpha_2 = 346.41\,\text{m/s}$$

$$C_{u3} = 0$$

Moreover, since $\Lambda = 0.5$, then

$$\alpha_2 = \beta_3 = 60°, \quad \alpha_3 = \beta_2 = 0$$

$$U = C_{u2} = 346.41\,\text{m/s}$$

The rotational speed at any radius is

$$U_i = \frac{r_i}{r_m} \times U_m, \quad \text{thus}$$

$$U_h = 307.92\,\text{m/s} \quad \text{and} \quad U_t = 384.9\,\text{m/s}$$

Case (1)
Since $C_{u2} = $ constant along the blade height, from Equation 14.57b this design method is the *general power case with N = 0*.

The axial speed (C_{a2}) is calculated from Equation 14.59, which is expressed in state (2) as

$$C_{a2} = C_{am}\left[1 - 2\tan^2\alpha_m \ln\frac{r}{r_m}\right]^{1/2}$$

Since $\alpha_3 = 0$, from the same Equation 14.59 but applied at state (2), $C_a3 = $ constant from blade hub to tip. Moreover, since $\alpha_3 = 0$, $C_{u3} = 0$ along the blade height. The angles β_2 and β_3 are calculated from Equation 14.1c. The results are summarized in Table 14.4a.

A plot for the results is shown in Figure 14.19.

The velocity triangles are shown in Figure 14.20.

TABLE 14.4a
Variations of Speeds and Flow Angles along the Blade Height for Case (1)

	C_{a2} (m/s)	C_{a3} (m/s)	C_{u2} (m/s)	C_{u3} (m/s)	α_2	α_3	β_2	β_3
Hub	261.3	200	346.4	0	52.97	0	8.4	56.99
Mean	200	200	346.4	0	60	0	0	60
Tip	121.3	200	346.4	0	70.7	0	−17	62.54

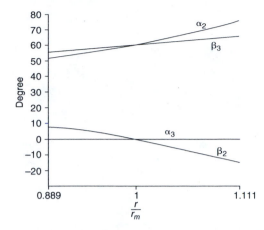

FIGURE 14.19   Variation of flow angles for case (1).

Case 1

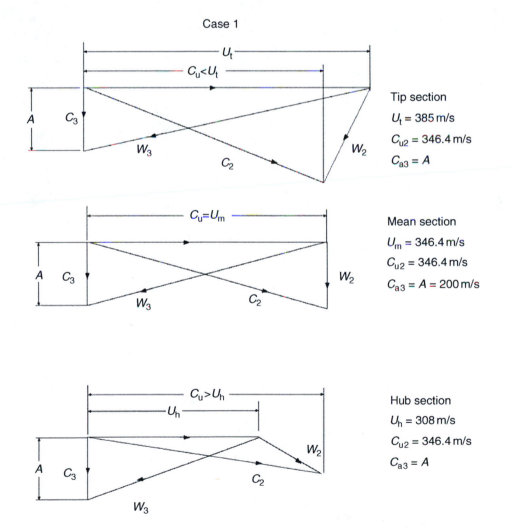

Tip section

$U_t = 385\,\text{m/s}$

$C_{u2} = 346.4\,\text{m/s}$

$C_{a3} = A$

Mean section

$U_m = 346.4\,\text{m/s}$

$C_{u2} = 346.4\,\text{m/s}$

$C_{a3} = A = 200\,\text{m/s}$

Hub section

$U_h = 308\,\text{m/s}$

$C_{u2} = 346.4\,\text{m/s}$

$C_{a3} = A$

FIGURE 14.20   Velocity triangles for case (1).

Case (2)

Since $\dfrac{C_{a2t}}{C_{a2h}} = \zeta^{\sin^2 \alpha_2}$,

$$\frac{C_{a2t}}{C_{a2h}} = \left(\frac{r_h}{r_t}\right)^{\sin^2 \alpha_2} = \left(\frac{r_m}{r_t}\right)^{\sin^2 \alpha_2} \left(\frac{r_h}{r_m}\right)^{\sin^2 \alpha_2}$$

and

$$\frac{C_{a2t}}{C_{a2h}} = \frac{C_{a2t}}{C_{a2m}} = \frac{C_{a2m}}{C_{a2h}}$$

This design method is the *constant nozzle angle* at state (2), or

$$C_{a2} = C_{a2m} \left(\frac{r_m}{r}\right)^{\sin^2 \alpha_2}, \quad C_{u2} = C_{u2m} \left(\frac{r_m}{r}\right)^{\sin^2 \alpha_2}, \quad C_2 = C_{2m} \left(\frac{r_m}{r}\right)^{\sin^2 \alpha_2}$$

At state (3), $\alpha_3 = 0$, $\quad C_{u3} = 0$

$$C_{a3}^2 = C_{a3m}^2 + 2U_m C_{u2m} \left[1 - \left(\frac{r}{r_m}\right)^{\cos^2 \alpha_2}\right]$$

It is deduced here that all the velocity components C_{u2}, C_{a2}, and C_{a3} are variables. The final results are given in Table 14.4b.

A plot for the results is shown in Figure 14.21.

TABLE 14.4b

Variations of speeds and flow angles along the blade height for case (2)

	C_{a2}	C_{a3}	C_{u2}	C_{u3}	α_2	α_3	β_2
Hub	218.47	216.71	378.4	0	60	0	17.885
Mean	200	200	346.41	0	60	0	0
Tip	184.8	183.29	320.09	0	60	0	−19.35

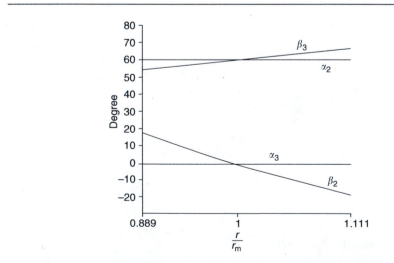

FIGURE 14.21   Variation of flow angles for case (2).

The velocity triangles are shown in Figure 14.22.
Case (3)
Since

$$\frac{C_{u2t}}{C_{u2h}} = \zeta = \frac{r_h}{r_t}$$

or $rC_{u2} = $ constant, this design method is called *free vortex*.

At the rotor outlet, the flow is also a free vortex, or $rC_{u3} = $ constant. The axial velocity is then constant at both states (2) and (3), or

$$C_{a2} = C_{a2m} = C_{a3} = C_{a3m} = \text{constant}$$

The results are summarized in Table 14.4c.

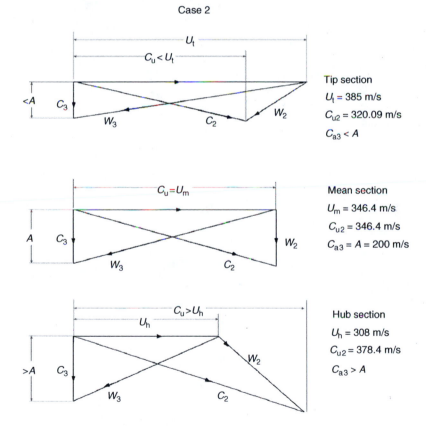

Case 2

Tip section
$U_t = 385$ m/s
$C_{u2} = 320.09$ m/s
$C_{a3} < A$

Mean section
$U_m = 346.4$ m/s
$C_{u2} = 346.4$ m/s
$C_{a3} = A = 200$ m/s

Hub section
$U_h = 308$ m/s
$C_{u2} = 378.4$ m/s
$C_{a3} > A$

FIGURE 14.22 Velocity triangles for case (2).

TABLE 14.4c
Variations of speeds and flow angles along the blade height for case (3)

	C_{a2}	C_{a3}	C_{u2}	C_{u3}	α_2	α_3	β_2	β_3
Hub	200	200	389.7	0	62.833	0	22.247	56.994
Mean	200	200	346.41	0	60	0	0	60
Tip	200	200	311.77	0	57.32	0	−20.85	62.549

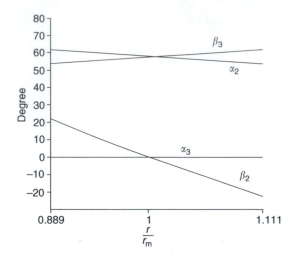

FIGURE 14.23 Variation of flow angles for case (3).

Case 3

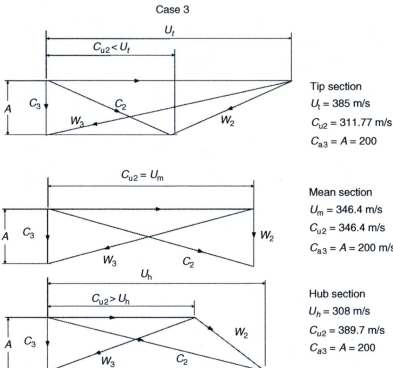

Tip section
$U_t = 385$ m/s
$C_{u2} = 311.77$ m/s
$C_{a3} = A = 200$

Mean section
$U_m = 346.4$ m/s
$C_{u2} = 346.4$ m/s
$C_{a3} = A = 200$ m/s

Hub section
$U_h = 308$ m/s
$C_{u2} = 389.7$ m/s
$C_{a3} = A = 200$

FIGURE 14.24 Velocity triangles for case (3).

A plot for the results is shown in Figure 14.23, while the velocity triangles are shown in Figure 14.24.

Example 5 (A) Two design methods are used in an axial turbine design. The swirl velocity at radius (r) in the first method is (C_u) and swirl velocity at the same radius (r) in the second method is (C_{u}'). If the ratio between two swirl velocities is given by the ratio

$$\frac{C_u}{C_u^\backslash} = \lambda r^2$$

1. What will be the value of the constant (λ) and what may be those design methods?
2. Deduce a relation for the ratio between the axial velocities (C_a/C_a') at radius (r) for both design methods.
(b) A constant nozzle design method is used for the design of an axial turbine stage with a free vortex design at state (3); use the following data to obtain the variations from hub-to-tip in the angles and speeds:

$$\alpha_{2m} = 58.5, \quad \beta_{2m} = 20.5, \quad \alpha_{3m} = 10, \quad \beta_{3m} = 55$$

$$\left(\frac{r_m}{r_r}\right)_2 = 1.16, \quad \left(\frac{r_m}{r_t}\right)_2 = 0.88, \quad \left(\frac{r_m}{r_r}\right)_3 = 1.22, \quad \left(\frac{r_m}{r_t}\right)_3 = 0.85$$

$$\frac{U_m}{C_{a2}} = \frac{U_m}{C_{a3}} = 1.25, \quad U_m = 340 \, \text{m/s}$$

Solution (a)
(1) For general power with $N = 1$ at radius (r), the whirl velocity is C_U.

$$C_u = C_{um} \left(\frac{r}{r_m}\right)^N$$

For $N = 1$

$$C_u = C_{um} \left(\frac{r}{r_m}\right)$$

(2) For free vortex design at radius (r), the whirl velocity is C_u', where

$$r C_u' = \text{constant} \quad \text{or} \quad r C_u' = r_m C_{um}$$

$$C_u' = \frac{r_m}{r} C_{um}$$

Then

$$\frac{C_u}{C_u'} = \frac{r}{r_m} \times \frac{r}{r_m} = \frac{r^2}{r_m^2}$$

(3) The value of the constant $\lambda = 1/r_m^2$.
(4) The relation between the axial velocities.

(i) For general power with $N = 1$ at radius (r), the whirl velocity is C_a

$$\frac{C_a}{C_{am}} = \left\{ 1 - \tan^2 \alpha_m \left(\frac{N+1}{N}\right) \left[\left(\frac{r}{r_m}\right)^{2N} - 1 \right] \right\}^{1/2}$$

For $N = 1$

$$\frac{C_a}{C_{am}} = \left\{1 - 2\tan^2\alpha_m\left[\left(\frac{r}{r_m}\right)^2 - 1\right]\right\}^{1/2}.$$

(ii) For free vortex design at radius (r), the whirl velocity is C'_a

$$C'_a = C_{am} \quad \text{or} \quad \frac{C'_a}{C_{am}} = 1$$

Then

$$\frac{C_a}{C'_a} = \left\{1 - 2\tan^2\alpha_m\left[\left(\frac{r}{r_m}\right)^2 - 1\right]\right\}^{1/2}$$

B

(i) *At the mean* (Figure 14.25)

$$U_m = 340\,\text{m/s} \quad \text{and} \quad \frac{U_m}{C_{a2}} = 1.25, \text{ then}$$

$$\frac{340}{C_{a2}} = 1.25,$$

$$C_{a2m} = C_{a3m} = 272\,\text{m/s}$$

$$C_{u2} = C_{a2}\tan\alpha_2$$

$$C_{u2} = 272\tan(58.5), \quad C_{u2m} = 443.86\,\text{m/s}$$

$$C_{u3} = C_{a3}\tan\alpha_3$$

$$C_{u3} = 272\tan(10), \quad C_{u3m} = 47.96\,\text{m/s}$$

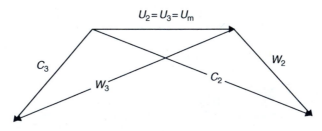

FIGURE 14.25   Mean section velocity triangles.

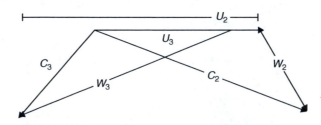

FIGURE 14.26   Velocity triangles at root section.

(ii) *At the root* (Figure 14.26)

$$\alpha_{2r} = \alpha_{2m} = 58.5°$$

$$\frac{U_m}{U_{r2}} = \left(\frac{r_m}{r_r}\right)_2$$

$$\frac{340}{U_{r2}} = 1.16, \quad U_{r2} = 293.1 \, \text{m/s}$$

$$\frac{U_m}{U_{r3}} = \left(\frac{r_m}{r_r}\right)_3$$

$$\frac{340}{U_{r3}} = 1.22, \quad U_{r3} = 278.69 \, \text{m/s}$$

$$C_{a2r} = C_{a2m} \left(\frac{r_m}{r_r}\right)_2^{\sin^2 \alpha_2}$$

$$C_{a2r} = 272(1.16)^{\sin^2 (58.5)} = 302.99 \, \text{m/s}$$

$$C_{u2r} = C_{u2m} \left(\frac{r_m}{r_r}\right)_2^{\sin^2 \alpha_2}$$

$$C_{u2r} = 443.86(1.16)^{\sin^2 (58.5)} = 494.43 \, \text{m/s}$$

$$\tan \alpha_2 - \tan \beta_{2r} = \frac{U_{2r}}{C_{a2r}}$$

$$\tan(58.5) - \tan \beta_{2r} = \frac{293.1}{302.99}, \quad \beta_{2r} = 33.6$$

$$C_{a3r}^2 = C_{a3m}^2 + 2U_m C_{u2m} \left[1 - \left(\frac{r_r}{r_m}\right)_3^{\cos^2 \alpha_2}\right]$$

$$C_{a3r}^2 = (272)^2 + 2 \times 340 \times 443.86 \left[1 - \left(\frac{1}{1.22}\right)^{\cos^2 (58.5)}\right], \quad C_{a3r} = 299.89 \, \text{m/s}$$

$$(rC_u)_3 = \text{constant}$$

$$(r_r C_{ur})_3 = (r_m C_{um})_3$$

$$C_{u3r} = C_{u3m} \left(\frac{r_m}{r_r}\right)_3$$

$$C_{u3r} = 47.96 \times 1.22 = 58.511 \, \text{m/s}$$

$$\tan \alpha_3 = \frac{C_{u3}}{C_{a3}} = \frac{58.11}{299.89}, \quad \alpha_3 = 11.04°$$

$$\tan \beta_3 - \tan \alpha_3 = \frac{U_3}{C_{a3}}$$

$$\tan \beta_3 - \tan(11.04) = \frac{278.69}{299.89}, \quad \beta_3 = 48.35°$$

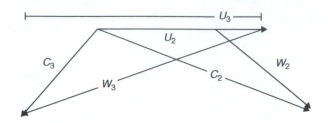

FIGURE 14.27   Velocity triangles at tip section.

(iii) At the tip (Figure 14.27)

$$\alpha_{2t} = \alpha_{2m} = 58.5°$$

$$\frac{U_m}{U_{t2}} = \left(\frac{r_m}{r_t}\right)_2$$

$$\frac{340}{U_{t2}} = 0.88, \quad U_{t2} = 386.36\,\text{m/s}$$

$$\frac{U_m}{U_{t3}} = \left(\frac{r_m}{r_t}\right)_3$$

$$\frac{340}{U_{t3}} = 0.85, \quad U_{t3} = 400\,\text{m/s}$$

$$C_{a2t} = C_{a2m}\left(\frac{r_m}{r_t}\right)_2^{\sin^2\alpha_2}$$

$$C_{a2t} = 272(0.88)^{\sin^2(58.5)} = 247.86\,\text{m/s}$$

$$C_{u2t} = C_{u2m}\left(\frac{r_m}{r_t}\right)_2^{\sin^2\alpha_2}$$

$$C_{u2t} = 443.86(0.88)^{\sin^2(58.5)} = 404.46\,\text{m/s}$$

$$\tan\alpha_{2t} - \tan\beta_{2t} = \frac{U_{2t}}{C_{a2t}}$$

$$\tan(58.5) - \tan\beta_{2t} = \frac{386.36}{247.86}, \quad \beta_{2t} = 3.87°$$

$$C_{a3t}^2 = C_{a3m}^2 + 2U_m C_{u2m}\left[1 - \left(\frac{r_t}{r_m}\right)_3^{\cos^2\alpha_2}\right]$$

$$C_{a3t}^2 = (272)^2 + 2\times340\times443.86\left[1 - \left(\frac{1}{0.85}\right)^{\cos^2(58.5)}\right]$$

$$C_{a3t} = 245.54\,\text{m/s}$$

$$(rC_u)_3 = \text{constant}$$

$$(r_t C_{ut})_3 = (r_m C_{um})_3$$

$$C_{u3t} = C_{u3m}\left(\frac{r_m}{r_t}\right)_3$$

$$C_{u3t} = 47.96 \times 0.85 = 40.766 \, \text{m/s}$$

$$\tan \alpha_{3t} = \frac{C_{u3t}}{C_{a3t}} = \frac{40.766}{245.54}, \qquad \alpha_{3t} = 9.43$$

$$\tan \beta_{3t} - \tan \alpha_{3t} = \frac{U_{3t}}{C_{a3t}}$$

$$\tan \beta_{3t} - \tan(9.43) = \frac{400}{245.54}, \qquad \beta_{3t} = 60.88°$$

Example 6 An axial turbine stage is to be designed by a constant nozzle design. It has the following data:

$$\Delta T_0 = 150 \, \text{K}, \quad U_{2h} = 300 \, \text{m/s}, \quad U_{2t} = 400 \, \text{m/s}, \quad \alpha_2 = 60, \quad \alpha_3 = 0, \quad \text{and} \quad \zeta_3 = 0.75$$

(a) Draw the velocity triangles at the hub, mean, and tip sections.
(b) Use the approximate solution for the constant nozzle blading, namely, $\alpha_2 = $ constant and $rC_{u2} = $ constant to calculate the axial and tangential velocities at the hub, mean, and tip sections at stations 2 and 3.
(c) Using free vortex design calculate the axial and tangential velocity as above.

Solution
(*a*) *Constant nozzle angle design*
The axial turbine stage has a constant nozzle outlet angle and zero exit swirl. Since $\zeta = 0.75$,

$$\frac{r_m}{r_t} = \frac{U_m}{U_t} = \frac{1+\zeta}{2} = 0.875 \qquad\qquad\qquad (a)$$

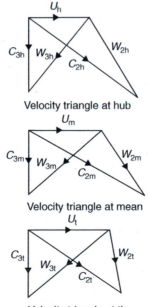

Velocity triangle at hub

Velocity triangle at mean

Velocity triangle at tip

FIGURE 14.28 Velocity triangles at hub, mean, and tip sections.

and

$$\frac{r_m}{r_h} = \frac{U_m}{U_h} = \frac{1+\zeta}{2\zeta} = 1.166$$

The velocity triangle are

$$\frac{C_2}{C_{2m}} = \frac{C_{a2}}{C_{a2m}} = \frac{C_{u2}}{C_{u2m}} = \left(\frac{r_m}{r}\right)^{(\sin \alpha_2)^2}$$

$$(C_{a3})^2 = (C_{a3m})^2 + 2U_m \times C_{u2m} \left(1 - \left(\frac{r}{r_m}\right)^{(\cos \alpha_2)^2}\right)$$

The specific work is given by the relation

$$Cp\Delta T_0 = U_m \Delta C_{um} = U_m C_{u2m}$$

$$C_{u2m} = \frac{Cp\Delta T_0}{U_m} = 492 \,\text{m/s}$$

$$C_{a2m} = C_{u2m} \cot \alpha_2$$

$$C_{a2m} = C_{a3m} = 284 \,\text{m/s}$$

$$\frac{C_{a2}}{C_{a2m}} = \frac{C_{u2}}{C_{u2m}} = \left(\frac{r_m}{r}\right)^{(\sin \alpha_2)^2}$$

At hub section

$$C_{a2h} = C_{a2m} \left(\frac{r_m}{r_h}\right)^{(\sin \alpha_2)^2} = 318.8 \,\text{m/s}$$

$$C_{u2h} = C_{u2m} \left(\frac{r_m}{r_h}\right)^{(\sin \alpha_2)^2} = 552.2 \,\text{m/s}$$

At the tip section

$$C_{a2t} = C_{a2m} \left(\frac{r_m}{r_t}\right)^{(\sin \alpha_2)^2} = 256.93 \,\text{m/s}$$

$$C_{u2t} = C_{u2m} \left(\frac{r_m}{r_t}\right)^{(\sin \alpha_2)^2} = 446.6 \,\text{m/s}$$

At the rotor outlet

$$(C_{a3})^2 - (C_{a3m})^2 = 2U_m \times C_{u2m} \left(1 - \left(\frac{r}{r_m}\right)^{(\cos \alpha_2)^2}\right).$$

$$(C_{a3t}) = \sqrt{(C_{a3m})^2 + 2U_m \times C_{u2m} \left(1 - \left(\frac{r_t}{r_m}\right)^{(\cos \alpha_2)^2}\right)} = 262.1 \,\text{m/s}$$

$$(C_{a3h}) = \sqrt{(C_{a3m})^2 + 2U_m \times C_{u2m}\left(1 - \left(\frac{r_h}{r_m}\right)^{(\cos\alpha_2)^2}\right)} = 306 \text{ m/s}$$

(b) *Approximates constant nozzle* (α_2 = constant, $r\,C_{u2}$ = constant)
    The governing equations are listed here:

$$r_m C_{u2m} = r_h C_{u2h} = r_t C_{u2t}$$

$$C_{u2h} = \frac{r_m C_{u2m}}{r_h}$$

$$C_{a2h} = C_{u2h} \cot\alpha_2$$

$$C_{u2t} = \frac{r_m C_{u2m}}{r_t}$$

$$C_{a2t} = C_{u2t} \cot\alpha_2$$

$$C_{u3h} = C_{u3t} = C_{u3m} = 0$$

$$C_{a3}^2 = C_{a3m}^2 + 2U_m C_{u2m}\left[1 - \left(\frac{r}{r_m}\right)^{\cos^2\alpha_2}\right]$$

$$(C_{a3t}) = \sqrt{(C_{a3m})^2 + 2U_m \times C_{u2m}\left(1 - \left(\frac{r_t}{r_m}\right)^{(\cos\alpha_2)^2}\right)}$$

$$(C_{a3h}) = \sqrt{(C_{a3m})^2 + 2U_m \times C_{u2m}\left(1 - \left(\frac{r_h}{r_m}\right)^{(\cos\alpha_2)^2}\right)}$$

(c) *Free vortex design*

$$(rC_u)_2 = \text{constant}$$

$$C_a = \text{constant at inlet and outlet to rotor}$$

(d) Summary of the velocity variations using different design methods

	Constant Nozzle (m/s)	Approximate Constant Nozzle (m/s)	Free Vortex (m/s)
C_{a2h}	318.8	331.2	284
C_{a2m}	284	284	284
C_{a2t}	256.9	248.5	284
C_{u2h}	552.2	573.7	573.7
C_{u2m}	492	492	492
C_{u2t}	446.6	430.5	430
C_{a3h}	306	331.2	284
C_{a3m}	284	284	284
C_{a3t}	262.6	248.5	284

14.5 PRELIMINARY DESIGN

Usually, design procedure is started by *need recognition*. Here, market research defines specific customer requirements. The design itself begins with preliminary studies such as cycle selection and analysis as described in Chapters 4 through 8, type of turbomachinery used (axial/mixed/radial), and suggested layout. This preliminary procedure defines the inlet total conditions, expansion ratio, power required, mass flow rate, and shaft speed.

If axial turbine is selected, then the design process starts by preliminary design, which includes aerodynamic, mechanical, and thermal steps that are next followed by sophisticated analyses using computational fluid dynamics (CFD) for simulating the real flow to the best available accuracy. This last step is followed by numerous experimental testing for scaled and prototype models. Design procedure is known as an iterative procedure. Thus, several checks were carried out during the above steps and modifications are adopted when there is a necessity.

14.5.1 MAIN DESIGN STEPS

The following steps are used in preliminary analysis:

1. The number of stages (n) is first determined by assuming the temperature drop per stage. For aero engine it is between 140 and 200 K, thus

$$n = \frac{(\Delta T_0)_{\text{turbine}}}{(\Delta T_0)_{\text{stage}}} \qquad (14.67)$$

2. *Aerodynamic design*
    It may be subdivided into mean line design and variations along the blade span.
3. *Blade profile selection*
    The blade profile and the number of blades for both stator and rotor are determined.
4. *Structural analysis*
    It includes mechanical design for blades and discs, rotor dynamic analysis, and modal analysis. Normally, both aerodynamic and mechanical designs are closely connected and there is considerable iteration between them.
5. *Cooling*
    The different methods of cooling are examined. In most cases, combinations of these methods are adopted to satisfy an adequate lifetime. Structural and cooling analyses determine the material of both stator and rotor blades.
6. Check stage efficiency.
7. Off-design.
8. Rig testing.

14.5.2 AERODYNAMIC DESIGN

It may be subdivided into mean blade design and variations along the blade span.
(a) *Mean line design analysis*
It starts by preliminary choices for the stage loading, flow coefficient, and degree of reaction. For aero engines, the following design parameters may be used as a guide:

*   Stage loading ranges from 1.5 to 2.5
*   Flow coefficient ranges from 0.8 to 1.0
*   Stage efficiency ranges from 0.89 to 0.94

- Absolute outlet angle ranges from $0°$ to $20°$
- Stator and rotor blade loss coefficients $\lambda_N \approx \lambda_R \approx 0.05$

Next, the following procedure can be followed:

— The stage expansion ratio is determined from Equation 14.14a.
— The rotational speed, from Equation 14.26a, $U_m = \sqrt{Cp\Delta T_{0s}/\psi}$.
— If the turbine drives a compressor, the compressor design may require a certain speed or speed range to hold a stable operation at high efficiency. The stable range of operation of a compressor and the region of high efficiency of a compressor is much limited in comparison with the turbine, so that usually the compressor determines the turbine speed rather than vice versa [11].

The mean radius is next obtained from the relation $r_m = U_m/2\pi N$.

— The angles $\beta_2, \beta_3, \alpha_2, \alpha_3$ are determined from Equations 14.33 through 14.36.
— The axial speed, from Equation 14.27b, $C_{a1} = \phi U$.

Assuming that if the flow is axial at the stage inlet, then

$$C_1 = C_{a1} = C_3$$

— Thus, the velocity triangles for the stage can be drawn and the speeds (C_2, W_2, W_3) are calculated.
— The thermodynamic properties at the inlet and outlet of both stator and rotor blade rows are calculated (T_1, T_2, T_3), (P_1, P_2, P_3), and (ρ_1, ρ_2, ρ_3).
— The blade heights (h_1, h_2, h_3), blade widths (W_1, W_2), and spacing between the nozzle and rotor blade rows are calculated.

Some checks are performed here:

(i) Flare angle is recommended to be less than $25°$.
(ii) Low swirl angle at the stage exit $(\alpha_3 < 20°)$; otherwise, exit guide vanes may be needed to minimize the losses in the jet pipe and propelling nozzle.
(iii) The pressure ratio in the nozzle must be less than the critical pressure ratio, or $(P_{01}/P_2) < (P_{01}/P_c)$, if the convergent nozzle is chosen.

(b) Three-dimensional flow

The variation of properties in the radial direction is considered using different methods: free vortex, constant nozzle angle, and general power N and constant specific mass flow methods are employed. Though early turbine designs relied on free vortex method, it is still used in a large number of applications for which it proved its suitability [11]. In several applications, there is a need to vary the blade angle in an arbitrary function of radius. Thus, non–free vortex methods may be employed. Radial equilibrium equation must still apply and implies a certain, not necessarily constant, distribution of work across the span.

Proper selection of the design method may be based on which method satisfies the minimum losses and highest degree of reaction at root.

14.5.3 BLADE PROFILE SELECTION

Originally, airfoil sections were laid out manually using a set of circular arcs and straight lines according to various empirical rules based on the experience of successful designs. Turbine airfoil sections are thicker than those of compressor to withstand very high temperatures and to allow for cooling holes. Special series are used for turbine airfoils such as NACA 4412 and 4415 as well as the British series T6, C7, and RAF 27. Gas turbine airfoil sections were originally derived from steam turbine airfoils with some modifications [4], which include an increase in maximum thickness and radius of the trailing edge. As defined in Reference 12, the airfoil is described by a set of 11 parameters including the airfoil radius, axial and tangential chords, inlet and exit blade angles, leading and trailing edge radii, together with a number of blades and throat. Modern airfoil section generator will typically define the blade pressure and suction surfaces using Bezier polynomials that provide the designer with a great flexibility while maintaining a high degree of continuity of the curve and derivatives.

Usually, a minimum of three sections at hub, mean, and tip radii are used. This suffices only for the simplest blade geometries. For complex or three-dimensional shapes, 10 or more sections may be used. Once the blade sections have been generated, cascade testing or a simple computational two-dimensional blade-to-blade analysis may be used. Thus, the velocity or Mach number distribution over the suction and pressure surfaces is determined. The next step is to stack the airfoil section to create the three-dimensional blade shape. A stacking line is defined, which determines the relative position of each of the sections with one another. The stacking line need not be straight or radial, but it can be curved or leaned to produce complex blade shapes [11]. Sections may be stacked on the center area of each section, or the leading, the trailing edge, or some other arbitrary line.

The blade profile and the number of blades for both stator and rotor are determined from Figure 14.29. From known inlet and efflux angles for stator (α_1, α_2) and rotor (β_2, β_3) of the mean line flow, the optimum pitch/chord ratio (s/c) is determined. The blade aspect ratio (h/c) ranges from 3 to 4. The mean blade heights for nozzle and rotor are determined as

$$h_N = (h_1 + h_2)/2 \quad \text{and} \quad h_R = (h_2 + h_3)/2$$

The corresponding chords for nozzle and rotor (c_N, c_R) are determined and thus the spacing for the nozzle and rotor blade rows (s_N, s_R) are determined. Now, the number of blades is determined

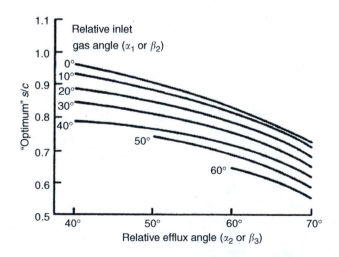

FIGURE 14.29 Optimum pitch/chord ratio. (Courtesy H.I.H. Saravanamuttoo, et al., *Gas Turbine Theory*, Prentice Hall, 5th edn., 2001, p. 313. With permission.)

from the relation

$$n_N = 2\pi r_m/s_N \quad \text{and} \quad n_R = 2\pi r_m/s_R$$

It is usual to avoid numbers with common multiples to reduce the possibilities of resonant frequencies. It is a common practice to use an even number for the nozzle and a prime number for the rotor blades.

14.5.4 Mechanical and Structural Designs

Mechanical and structural designs have a direct impact on aerodynamic design. Both designs are interrelated to achieve high aerodynamic performance, and adequate life and structural integrity. Vibration analysis is very important for rotor blades to avoid resonance and the possibility of blade breakage. An assessment of stress level is considered here.

There are two types of stresses that must be considered, namely, steady and unsteady stresses.

Steady-state stresses arise from the centrifugal, pressure loading, and thermal sources. Unsteady-state stresses arise from the interaction of the rotating blade with stationary features in its vicinity (similar to any two successive blade rows), thermal cycling for turbine downstream combustion chambers as the products of combustion are not completely uniform in the tangential direction [11]. There will also be some mechanical effects caused by residual imbalance and rub out of bearings. Failure in both cases has different features. Failure due to steady stresses causes the ductile blade material to creep or elongate and fail ultimately in plastic deformation. Failure due to unsteady stresses causes the blade to fail through fatigue due to crack initiation and growth.

Turbine blades in aero engines are designed to avoid creep. The allowable stress in turbines depends strongly on the operating temperature and specified temperature, because that stress will not exceed a creep extension of say 1% for 100,000 h at the specified temperature since 100,000 h (11.4 years) is a long period for testing [13]. It is common to compare materials to confirm whether the stress causes rupture after 100 h. The rupture stress divided by blade material density is plotted versus the working temperature. An increase in the working temperature is associated with a decrease in the 100 h rupture stress/density (σ/ρ) ratio. Ceramics have superior strength–temperature characteristics. However, they tend to be brittle and unreliable. Adopting ceramic composites, for example, by including high-strength tin whiskers, may provide a solution to the problem where high strength and reliability could be achieved. For example, Si_3N_4 ceramic may have a 100 rupture stress/density ratio of 35 ($kPa/kg/m^3$) at as high working temperature as 1200°C.

In general, there are three types of mechanical stresses for turbomachines. For turbines, there is an additional thermal stress due to high temperature gradient within the blades. The mechanical stresses are

1. Centrifugal stresses
2. Gas bending stress
3. Centrifugal bending stress

Centrifugal stress (which is a tensile stress) is important for both blades and the discs that support them. It is the largest, but not necessarily the most important, since it is a steady stress [4]. Bending stresses arise in the blades from the aerodynamic forces acting on them due to steady or unsteady conditions. Thermal stresses arise due to the temperature gradient within the blade, which may be caused by the surrounding high-temperature gases and the cold air flowing into the cooling passages.

14.5.4.1 Centrifugal Stresses

For a given hub/tip ratio, centrifugal stress at blade root depends on

1. Blade density (ρ_b)

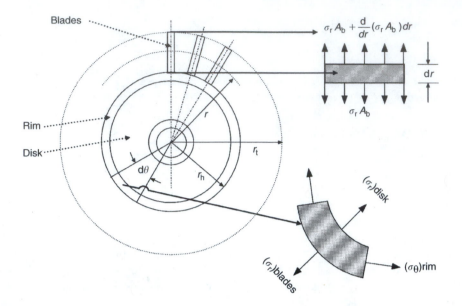

FIGURE 14.30   Stresses in blades.

2. Hub-to-tip radius ratio (ζ)
3. Blade-tip speed (U_t)

It increases with blade twist due to centrifugal loading. It can be reduced by tapering the blade cross section.

14.5.4.2 Centrifugal Stresses on Blades

Consider the layout of a rotor blade row shown in Figure 14.30. Consider an element on the blade at a radius (r) having a cross-sectional area, A_b, and thickness, dr. The element is subjected to a radial stress (σ_r) and a centrifugal force due to its rotation with an angular speed Ω. Force balance of this element yields the following relation:

$$\frac{d}{dr}(\sigma_r A_b) = -(\rho_b A_b)\,\Omega^2 r$$

where ρ_b is the blade material density. If the cross-sectional area is assumed to be constant, then this relation is reduced to

$$\frac{d\sigma_r}{dr} = -\rho_b \Omega^2 r$$

Integrating from any radius to the tip radius and noting that the centrifugal stress is zero at blade tip, the stress at any radius will be

$$\sigma_r = \rho_b \frac{\Omega^2}{2}\left[r_t^2 - r^2\right] = \rho_b \frac{\Omega^2 r_t^2}{2}\left[1 - \left(\frac{r}{r_t}\right)^2\right] = \rho_b \frac{U_t^2}{2}\left[1 - \left(\frac{r}{r_t}\right)^2\right]$$

The maximum centrifugal stress occurs at the blade hub radius ($r = r_h$) or

$$\sigma_b = \rho_b \frac{U_t^2}{2}\left[1 - \zeta^2\right] = 2\rho_b \pi N^2 A \tag{14.68}$$

where $\zeta = r_h/r_t$ is the hub-to-tip ratio, N is the rotational speed in rev/s, and A is the annulus area.

From Equation 14.67, the maximum blade stress factor, $\sigma_b/\rho_b U_t^2$, has a maximum value of 0.5 (at a fictitious hub-to-tip ratio of zero) and decreases rapidly as the hub-to-tip ratio increases.

Since turbine blades are tapered blades both in chord and thickness, the ratio between tip and root areas (a_t/a_h) is between 1/4 and 1/3; thus, it is accurate enough to assume that tapering of the blade reduces the stress to 2/3 of its value [4] as given in Equation 14.68, or

$$(\sigma_b)_{tapered} \approx \rho_b \frac{U_t^2}{3}\left[1 - \zeta^2\right] \approx \frac{4}{3}\rho_b \pi N^2 A \tag{14.69}$$

For linear tapering, the exact expression

$$(\sigma_b)_{tapered} = \frac{\rho_b \omega^2 A}{4\pi}\left[2 - \frac{2}{3}\left(1 - \frac{a_t}{a_h}\right)\left(1 + \frac{1}{1+\zeta}\right)\right] \tag{14.70}$$

14.5.4.3 Centrifugal Stresses on Discs

The centrifugal stress is checked in three locations, namely, the outer rim, inner rim, and in the disc material between both rims (Figure 14.31). The following assumptions are considered:

1. Centrifugal force applied by the blades to the rims can be considered equivalent to a radially uniformly distributed load.
2. The disc is tapered such that its circumferential and radial stresses are constant everywhere and equal to (σ).
3. Rims are thin enough to have uniform tangential stress equal to (σ).

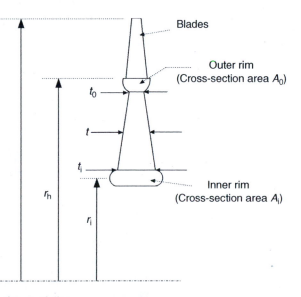

FIGURE 14.31 Nomenclature of disc.

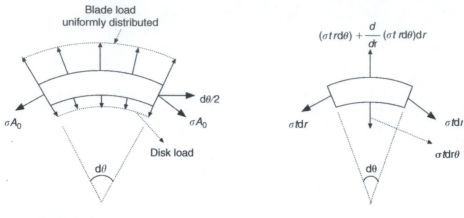

FIGURE 14.32 Stresses on outer and inner rims of disc.

1. *Outer rim*: Applying Newton's second law of motion over a small element on the outer rim (Figure 14.32), the sum of force (on the left-hand side) will be equal to the mass multiplied by acceleration (right-hand side).

$$2\sigma A_0 \frac{d\theta}{2} - N_b \sigma_b A_b \left(\frac{d\theta}{2\pi}\right) + \sigma\,(r_h d\theta)\,t_0 = \rho_b\,(A_0 r_h d\theta)\,r_h \Omega^2$$

Dividing by $A_0 d\theta \rho_b U_t^2$ gives

$$\frac{\sigma}{\rho_b U_t^2} = \frac{(N_b/2\pi)\,(A_b/A_0)\,\left(\sigma_b/\rho_b U_t^2\right) + \zeta^2}{1 + (r_h t_0/A_0)} \tag{14.71}$$

2. *Disc segment*: Also, apply Newton's second law to a segment on the disc. Referring to Figure 14.31, a segment of the disc has a width of (dr), arc length of $(rd\theta)$, and thickness (t). Force balance will give the relation

$$(2\sigma t dr)\,\frac{d\theta}{2} - \frac{d}{dr}\,(\sigma t r d\theta)\,dr = \rho_b\,(tr d\theta dr)\,\Omega^2 r$$

Since σ is assumed to be constant, simplifying to get

$$t - \frac{d}{dr}\,(tr) = -r\frac{dt}{dr} = \frac{\rho_b \Omega^2 r^2 t}{\sigma}$$

$$\frac{dt}{t} = -\frac{\rho \Omega^2}{\sigma} r dr$$

Integrate between the radii of the inner and outer rims to get

$$\ln\frac{t_0}{t_i} = -\frac{\rho \Omega^2}{2\sigma}\left(r_h^2 - r_i^2\right)$$

The stress in the disc can also be rewritten as a stress factor in the following form:

$$\frac{\sigma}{\rho_b U_t^2} = \frac{\zeta^2 \left[1 - (r_i/r_h)^2\right]}{2 \ln (t_i/t_0)} \tag{14.72}$$

3. *Inner rim*: Force balance also gives

$$2\sigma A_i \frac{\mathrm{d}\theta}{2} - \sigma \left(r_i \mathrm{d}\theta\right) t_i = \Omega^2 r_i \left(\rho_b A_i r_i \mathrm{d}\theta\right)$$

which gives

$$\frac{\sigma}{\rho_b U_t^2} = \frac{\zeta^2 (r_i/r_h)^2}{1 - (t_i r_i/A_i)} \tag{14.73}$$

From Equations 14.71 through 14.73, the stresses at any point in the disc can be calculated and compared with the blade root stress using Equation 14.69. Disc stresses can exceed the blade root stress at high hub-to-tip ratios.

14.5.4.4 Gas Bending Stress

The gas flow through the rotor passages generate two forces, namely, tangential and axial forces (F_t, F_a). For stator blade row only axial force is generated. The tangential force also develops a gas bending moment (M_t) about the axial direction, while the axial force develops another gas bending moment (M_a) about the tangential direction (Figure 14.33). The neutral axes are inclined at an angle ϕ to the original axial and tangential axes.

Resolve M_a and M_t in the new X and Y directions to get

$$M_X = M_t \cos\phi + M_a \sin\phi, \qquad M_Y = M_a \cos\phi - M_t \sin\phi$$

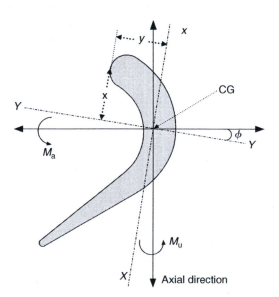

FIGURE 14.33   Gas bending stress.

The corresponding gas bending stress is

$$\sigma_{gb} = \frac{-M_x Y}{I_{xx}} + \frac{M_Y X}{I_{YY}}$$

which is tensile at the blade leading edge and compression at the blade trailing edge. Now to evaluate the order of magnitude of each term, the axial force is evaluated first. The axial force is developed from two sources:

- Momentum change which is the case when there is a change in axial velocity ($C_{a2} \neq C_{a3}$). The axial force per blade is then

$$\frac{|F_a| = \dot{m} |C_{a2} - C_{a3}|}{n},$$

  where n is the number of blades.
- Pressure difference, thus the force *per unit blade height* is

$$F_a = (P_2 - P_3) \left(\frac{2\pi r}{n} \right).$$

The tangential force is determined from the power, which is correlated to the specific work w_s by the relation

$$P = \dot{m} w_s = \dot{m} U (C_{u2m} + C_{u3m})$$

However, the power is also expressed as $P = F_t \times U$.
Then, the tangential force is $F_t = \dot{m} (C_{um} + C_{u3m})$
The tangential force per blade is then

$$F_{t1} = \frac{\dot{m} (C_{u2} + C_{u3})}{n}$$

The tangential moment per blade is

$$M_t \cong F_{t1} \frac{h}{2} = \frac{\dot{m} (C_{u2m} + C_{u3m})}{n} \frac{h}{2}$$

If the tangential momentum is far greater than the axial momentum and the angle ϕ is small, then the gas bending moment is approximated as

$$\left(\sigma_{gb} \right)_{max} \approx -\frac{M_t y}{I_{xx}}$$

$$\left| \left(\sigma_{gb} \right)_{max} \right| = \frac{\dot{m} (C_{u2m} + C_{u3m})}{n} \times \frac{h}{2} \times \frac{y}{I_{xx}} \tag{14.74}$$

where $I_{xx}/y = S = zc^3$ is the smallest value of section modulus.
As reported in Reference 4, the section modulus is expressed as

$$S = \frac{I_{xx}}{y_{max}} = \frac{1}{B} \left(10\frac{t}{c} \right)^n c^3 \tag{14.75}$$

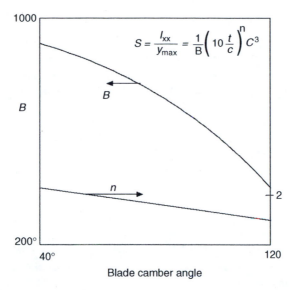

$$S = \frac{I_{xx}}{y_{max}} = \frac{1}{B}\left(10\frac{t}{c}\right)^{n}c^3$$

Blade camber angle

FIGURE 14.34 Approximate rule for section module. (Courtesy H.I.H. Saravanamuttoo, et al., *Gas Turbine Theory*, Prentice Hall, 5th edn., 2001, p. 313.)

The variables (B, n) are plotted as a function of the blade camber angle in Figure 14.34. As depicted from Equations 14.74 and 14.75, the gas bending stress is inversely proportional to the cube of the blade chord. By doubling the chord, the stress will be reduced by nearly one order of magnitude. Thus, it is always possible to keep the gas bending stress smaller than the centrifugal one.

Equations 14.74 and 14.75 represent the gas bending stress for the case of steady aerodynamics. However, as each blade row passes through the wake shed by the blades of the preceding blade row, it experiences a small aerodynamic force fluctuation that could be serious if one of the critical vibration frequencies of the blade was to coincide with the wake passing frequency. During such resonance, destructive strain amplitudes can develop even if the force amplitude is small. These stresses are not serious in turbines compared with compressors. It is much less than centrifugal stresses.

14.5.4.5 Centrifugal Bending Stress

Centrifugal bending stresses arise if the centroids of the blade cross sections at different radii do not lie on a radial line. However, the torsion stress arising from this source is small enough to be neglected [4].

14.5.4.6 Thermal Stress

The blade temperature is not uniform both along the blade height and across its cross section. The blade is surrounded by hot gases and through its cold air flowing through the internal passages used for cooling. At each cross section, there is an average temperature depending on its position along the blade height. For any point at this section, there is a temperature difference ΔT above or below this average temperature. Thus a local thermal stress is developed, which is proportional to the product $(E\alpha\Delta T)$ where E is the modulus of elasticity of the blade material and α is the coefficient of thermal expansion. For average values of $E \approx 2 \times 10^5$ MPa, $\alpha \approx 7 \times 10^{-7}/°C$, $\Delta T \approx 50$ degrees, the thermal stress will be 70 MPa, which has a serious effect on turbine blades particularly if the cooling passages are small.

14.5.5 TURBINE COOLING

The designer of gas turbines and aircraft engines aims at increasing the specific power output (or specific thrust for an aero engine), increasing the thermal efficiency, reducing the SFC, and reducing the engine weight. One way to achieve these objectives is to increase the turbine inlet temperature (TIT). TIT has increased considerably over the past year and will continue to do so. This trend has been made possible by advancement in materials and technology and the use of advanced turbine blade cooling techniques. Improvement in material has allowed slow increase in TIT, while cooling allowed faster rates for the increase of TITs. Since 1950, advancement in material and cooling technology has led to an annual increase of about 10° C [14]. Cooling allows the turbine operational temperature to exceed the material's melting point without affecting the blade and vane integrity. Coolant is either air, which is bled from compressor, or liquid, which is impractical in aero engines. When air is used as a coolant, 1.5%–2% of air mass flow rate (bled from the compressor) is used for cooling each turbine blade row. Then the blade temperature can be reduced by 200–300° C. On practically all large aero engines, the first stage nozzle and rotor are cooled. Cooling is accomplished by directing compressor bleed air through passages inside the engine to the turbine area where the air is led to longitudinal holes, tubes, or cavities in the vanes and the blades [15]. After entering the passages in the vanes and blades, the air (coolant) in all present gas turbines is distributed through holes at the leading and trailing edges of the vanes and blades. The air impinges along the vane and blade surfaces and then passes out of the engine with the engine exhaust.

The use of current alloys permits the TIT to reach 1650 K. Generally, ceramics- and carbon-based materials promise substantial increase in blade temperature. However, some problems have to be solved to overcome their inherent brittleness [11].

14.5.5.1 Turbine Cooling Techniques

As shown in Figure 14.35, the following methods are used in turbine cooling:

1. Convection
2. Impingement
3. Film cooling
4. Full coverage film cooling
5. Transpiration

1. *Convection:* Convection cooling with a single internal passage was the only available cooling technique in the 1960s (Figure 14.36). Cooling air was injected through the airfoil attachment and to the inside of airfoil. The cold air was discharged at the blade tip. Development has led to multipass internal cooling. The blades are either cast, using cores to from the cooling passages, or forged with holes of any size and shape that are produced by electrochemical drilling. The effectiveness of convection cooling is limited by

- Size of the internal passages
- Quantity of cooling air available

The main disadvantage of early convection technique is its failure to cool the thin trailing edges of blades, which is serious as trailing edge is very thin. Moreover, it fails to cool the leading edge, which is subjected to the highest temperature. However, in the 1970s, the evolution in convection cooling led to the cooling of trailing edge by discharging the cold air through the trailing edges (Figure 14.35).

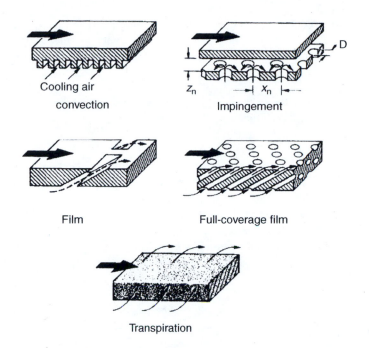

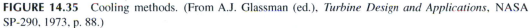

FIGURE 14.35 Cooling methods. (From A.J. Glassman (ed.), *Turbine Design and Applications*, NASA SP-290, 1973, p. 88.)

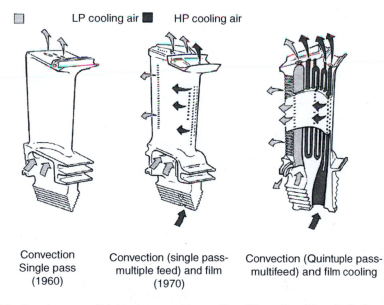

FIGURE 14.36 Development of high-pressure turbine cooling. (Reproduced from Rolls-Royce plc, copyright © Rolls-Royce plc 2007. With permission.)

2. *Impingement cooling*: It is a form of convection; the main difference between this method and conventional cooling method is that air flows radially in convection cooling; however, in impingement method, air is brought radially to one or more center sections, then turned normal to the radial direction and passed through a series of holes so that it impinges on the inside of blade, usually just opposite to the stagnation point of the blade. It is mostly used in stators and may be used in rotor if sufficient space is available to include the required hardware inside the blade.

3. *Film cooling*: It involves the injection of a secondary fluid (cold air) into the boundary layer of the primary fluid (hot gases). This is an effective method to protect the surface from the hot gases as the cooling air acts as an insulating layer to maintain a lower blade material temperature. *Disadvantages*:

   (a) Causes turbine losses due to injection into the boundary layer.
   (b) If too much air is used or if ejected at high speeds it could penetrate the boundary layer.
   (c) If close holes are used, they cause stress concentration.

However, it is more effective than normal convection or impingement methods. The cooling air absorbs energy as it passes inside the blade and through the holes, then it further reduces the blade temperature by reducing the amount of energy transferred from the hot gases to the blade. It must be noted that the cold air used must be at high pressure, as this is not always satisfied, especially for the leading edge of the first stator (nozzle).

Current-cooled airfoil uses a combination of sophisticated internal convection and film cooling.

4. *Full-coverage film cooling*: It involves the injection of cold air from an array of discrete holes. Thus it represents an attempt to draw on some advantages of transpiration cooling without paying its penalties.

5. *Transpiration cooling*: Transpiration cooling of a porous blade wall is the most efficient cooling technique and requires the least cooling air in comparison with the other techniques. It involves the use of a porous material through which the cooling air is forced into boundary layer to form a relatively cooling, insulating film or layer. However, it has some significant drawbacks. For efficient transpiration cooling, the pores should be small, which leads to problems of blockage due to oxidation and foreign contaminate.

Generally, cooling has the following detrimental effects [16]:

   (a) Added cost of producing turbine blades
   (b) Turbine blade reliability
   (c) Loss of turbine work due to the cooling air bypassing one or two of the turbine stages
   (d) Loss due to the cooling air being mixed with hot gas steam

14.5.5.2 Mathematical Modeling

In this section, the temperature of the turbine blade material will be calculated in three different cases, namely, assuming no cooling at all, film, and convection cooling.

1. *No cooling*: The rotor blade of the first stage represents the critical section from the mechanical point of view. It is subjected to the maximum mechanical stress. If for some reasons there is no cooling for this part, then this section represents the worst section in the turbine [13]. Now, let us assume that there is

- No blade cooling
- No conduction of heat away from the blade root into turbine disc

We assume that, in this case, the blade temperature is nearly equal to the relative stagnation temperature of the hot gasses ($T_s = T_{02rel}$).

At the hub section

$$T_{2h} = T_{02} - \frac{C_{2h}^2}{2Cp} = T_{02} - \frac{\gamma - 1}{2\gamma R}\left(C_{u2h}^2 + C_{a2h}^2\right)$$

where subscript (u) and (a) refer to tangential and axial as usual.

Dividing by the rotational speed at tip (U_t) and noting that $\zeta = (r_h/r_t) = (U_h/U_t)$, then

$$\frac{T_{2h}}{T_{02}} = 1 - \frac{\gamma - 1}{2}\left(\frac{U_t}{a_0}\right)^2 \zeta^2 \left[\left(\frac{C_{u2}}{U}\right)_h^2 + \left(\frac{C_{a2}}{U}\right)_h^2\right] \qquad (14.76)$$

Subscript (h) in Equation 14.76 refer to the hub section. The relative stagnation temperature is

$$T_{02rel} = T_{2h} + \frac{W_{2h}^2}{2Cp}$$

With no blade cooling or conduction into the disc, the blade would take a temperature somewhat less than T_{02rel}, or

$$T_s = T_{2h} + r_f \frac{W_{2h}^2}{2Cp} \qquad (14.77)$$

where r_f is the recovery factor, which reflects the fact that the deceleration process in the surrounding boundary layer is not adiabatic. Some heat transfer takes place between the streamlines and the surrounding moving boundaries. The recovery factor has the following range, $r_f = 0.9 - 0.95$. Since

$$\frac{W_{2h}^2}{2CpT_0} = \frac{\gamma - 1}{2\gamma RT_0}\left[W_{u2h}^2 + W_{a2h}^2\right] = \frac{\gamma - 1}{2}\left(\frac{U_t}{a_0}\right)^2 \zeta^2 \left[\left(\frac{C_{u2}}{U} - 1\right)_h^2 + \left(\frac{C_{a2}}{U}\right)_h^2\right]$$

Then from the above equation and Equation 14.76, the surface temperature at hub is

$$\left(\frac{T_{s\,hub}}{T_{01}}\right) = 1 - \frac{\gamma - 1}{2}\left(\frac{U_t}{a_0}\right)^2 \zeta^2 \left[\left(\frac{C_{u2}}{U}\right)_h^2 + \left(\frac{C_{a2}}{U}\right)_h^2 - r_f\left\{\left(\frac{C_{u2}}{U} - 1\right)_h^2 + \left(\frac{C_{a2}}{U}\right)_h^2\right\}\right]$$
$$(14.78)$$

It will be found in the succeeding examples that even without cooling, the blade temperature is less than the stagnation (absolute total temperature) at the inlet to turbine. With blade cooling, the blade root temperature is 300–400 K below the absolute stagnation temperature of gas entering the turbine [13]. This necessitates extracting a cold air from both the low- and high-pressure compressors to cool the turbine blades as shown in Figure 14.37. The disc, stator, and rotor blades need less than 10% of the compressor exit air flow for their cooling.

In the following sections, the cases of blade cooling using either film or convection methods will be analyzed.

2. *Film cooling*: From Eckert relation, the heat flux from the hot gas to the walls q is given by

$$q = h_0 (T_{aw} - T_w), \qquad (14.79)$$

where h_0 is the film coefficient for convective heat transfer to the wall, T_w is the wall temperature, and T_{aw} is the adiabatic wall temperature.

Air is bled from the compressor at a certain location where its temperature is T_i and ejected into the turbine disc where its temperature becomes T_c. Thus, the heat added to the bled air up to the point of injection

$$\dot{Q}_a = \dot{m}_a Cp_a (T_c - T_i) \qquad (14.80)$$

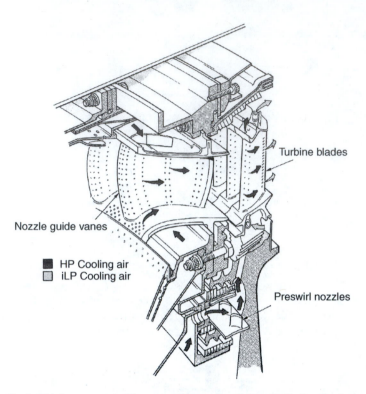

FIGURE 14.37   Cooled high-pressure turbine stage. (Reproduced from Rolls-Royce plc, copyright © Rolls-Royce plc 2007. With permission.)

If heat conduction from blade to disc is small, then this heat $\dot{Q}_a$ will be balanced by the heat transfer from the hot gases to the blade surface $\dot{Q}_g$. Using Eckert law, $\dot{Q}_g$ will be expressed as

$$\dot{Q}_g = h_0 \left(T_{aw} - T_b \right) A_s \tag{14.81}$$

where A_s is the surface area of blade and T_b is the blade temperature.
From Equations 14.80 and 14.81, the air mass flow rate is

$$\dot{m}_a = \frac{h_0 \left(T_{aw} - T_b \right) A_s}{C p_a \left(T_c - T_i \right)}$$

Stanton number is defined as

$$st_0 = \frac{h_0}{\rho_g C p_g W_g},$$

where $(\rho_g, C p_g, W_g)$ are the gas density, specific heat at constant pressure, and inlet speed. With the gas mass flow rate defined as $\dot{m}_g = \rho_g W_g A_g$, where A_g is the hot gas flow area or the area perpendicular to the gas speed, then Stanton number will be expressed as

$$st_0 = \frac{h_0 A_g}{C p_g \dot{m}_g}$$

The ratio between the mass flow rates of the bled air and hot gases is then

$$\frac{\dot{m}_a}{\dot{m}_g} = st_0 \frac{A_s}{A_g} \frac{Cp_g}{Cp_a} \left(\frac{T_{aw} - T_b}{T_c - T_i} \right).$$

Now if the cooling effectiveness η_f is defined as

$$\eta_f = \frac{T_g - T_{aw}}{T_g - T_c},$$

then the adiabatic wall temperature will be $T_{aw} = T_g - \eta_f (T_g - T_c)$.

The ratio between the bled air and gas flow rates is then

$$\frac{\dot{m}_a}{\dot{m}_g} = st_0 \frac{A_s}{A_g} \frac{Cp_g}{Cp_a} \left(\frac{T_g - T_b}{T_c - T_i} - \eta_f \frac{T_g - T_c}{T_c - T_i} \right). \tag{14.82}$$

In Equation 14.82, four temperatures are listed, which are again summarized as follows:

1. T_g gas temperature
2. T_b blade temperature
3. T_i cold air temperature as it bled from compressor
4. T_c coolant ejection temperature

Stanton number is dependent on the gas Reynolds number and ranges from 0.002 to 0.003 for the prevailing values of Reynolds numbers. Typical value for the average film cooling effectiveness is 0.2.

3. *Convection cooling:* Figure 14.38 illustrates a turbine blade cooled by a forced convection method [4]. Cold air bled from compressor is pumped inside internal cooling holes of turbine blade. This cold air is heated as it passes within the blade and becomes less effective in cooling. For this reason together with the low thermal conductivity of the blade, it may be assumed that the spanwise conduction is negligible.

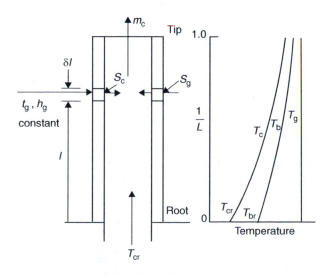

FIGURE 14.38 Forced convection of a turbine blade.

Consider a blade element of length (dl) at a distance (l) from the root. The heat balance between the outer gases and the blade outer surface from one side and through cold air and the inner surface of blade from the other side results in

$$h_g S_g \left(T_g - T_b\right) = h_c S_c \left(T_b - T_c\right) \tag{14.83a}$$

or

$$T_c = T_b - \frac{h_g S_g}{h_c S_c} \left(T_g - T_b\right) \tag{14.83b}$$

Since the gas temperature is constant,

$$\frac{dT_c}{dl} = \left[1 + \frac{h_g S_g}{h_c S_c}\right] \frac{dT_b}{dl} \tag{14.83c}$$

where h_g and h_c are the convection heat transfer coefficients for the hot gases/blade and cold air/blade, respectively. Moreover, S_g and S_c are the wetted perimeters of the outer blade surface and the combined coolant passages, respectively. The coolant temperature gradient in the radial direction will be given by the relation

$$\dot{m}_c C_{pc} \frac{dT_c}{dl} = h_c S_c \left(T_b - T_c\right) \tag{14.84}$$

From Equations 14.83a and 14.84,

$$\frac{dT_c}{dl} = \frac{h_c S_c}{\dot{m}_c C_{pc}} \left(T_b - T_c\right) = \frac{h_g S_g}{\dot{m}_c C_{pc}} \left(T_g - T_b\right)$$

From Equation 14.83c,

$$\frac{dT_c}{dl} = \left[1 + \frac{h_g S_g}{h_c S_c}\right] \frac{dT_b}{dl} = \frac{h_g S_g}{\dot{m}_c C_{pc}} \left(T_g - T_b\right)$$

With

$$\frac{dT_b}{dl} = -\frac{d\left(T_g - T_b\right)}{dl},$$

$$-\frac{d\left(T_g - T_b\right)}{dl} \left[1 + \frac{h_g S_g}{h_c S_c}\right] = \frac{h_g S_g}{\dot{m}_c C_{pc}} \left(T_g - T_b\right)$$

Integrating from root to any distance along the blade gives

$$\int_{T_{br}}^{l} \frac{d\left(T_g - T_b\right)}{\left(T_g - T_b\right)} = K \int_{0}^{l} dl$$

$$T_g - T_b = \left(T_g - T_{br}\right) e^{-Kl/L} \tag{14.85a}$$

where

$$\frac{K}{L} = \frac{h_g S_g}{\dot{m}_c C_{pc} \left[1 + (h_g S_g / h_c S_c)\right]}.$$ (14.85b)

From Equation 14.83b by adding T_g to both sides,

$$T_g - T_c = (T_g - T_b) \left[1 + \frac{h_g S_g}{h_c h_c}\right]$$

From Equation 14.85

$$T_g - T_c = (T_g - T_{br}) \left[1 + \frac{h_g S_g}{h_c S_c}\right] \times e^{-Kl/L}$$ (14.86)

At root $l = 0$, the coolant temperature is $T_c = T_{cr}$

$$T_g - T_{cr} = (T_g - T_{br}) \left[1 + \frac{h_g S_g}{h_c S_c}\right]$$ (14.87)

Combining Equations 14.86 and 14.87 we get

$$T_g - T_c = (T_g - T_{cr}) e^{-Kl/L}$$ (14.88)

From Equations 14.8 and 14.87, one gets

$$T_b - T_{cr} = (T_g - T_{br}) \left[1 + \frac{h_g S_g}{h_c S_c} - e^{-Kl/L}\right]$$ (14.89)

The relative blade temperature at any radius is determined from the relation

$$\frac{T_b - T_{cr}}{T_g - T_{cr}} = 1 - \frac{e^{-Kl/L}}{\left[1 + (h_g S_g / h_c S_c)\right]}$$ (14.90)

Equation 14.90 allows the calculation of the blade temperature at any radius. This temperature depends on the ratio between the mass flow rates of both coolant and gas $(\dot{m}_c / \dot{m}_g)$. Increase in this ratio decreases the blade temperature. Moreover, the heat transfer coefficients (h_c, h_g) depend on Nusselt numbers for coolant and gas main stream, N_{uc}, N_{ug}.

Finally, it is important to assess the cooling performance of a blade using the effectiveness parameter (ϕ) defined as

$$\phi = \frac{T_G - T_M}{T_G - T_C}$$ (14.91)

where subscripts C, G, and M refer to the coolant, gas, and metal, respectively. The value of $\phi = 0$ represents no cooling while the value $\phi = 1$ is the ideal case of perfect cooling where the blade metal and coolant temperatures are equal. Figure 14.39 illustrates the effectiveness of typical cooling methods for different cooling airflow ratios.

The cooling effectiveness is a function of many variables such as the blade design, the arrangement of cooling air passages, the way in which the coolant air is ejected, and the percentage of the coolant air as a ratio of the hot gas main stream.

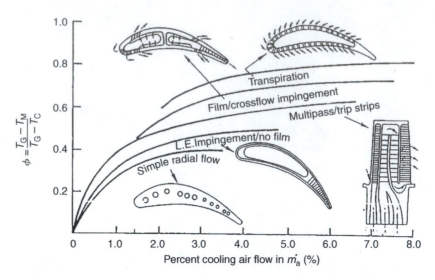

FIGURE 14.39 Effectiveness of different cooling methods as a function of coolant airflow ratio. (Courtesy from Moustapha *et al.* [12].)

14.5.6 Losses and Efficiency

Losses are dependent on several parameters including the blade geometry, incidence angle, Reynolds number, the ratios $s/c, h/c.t_{max}/c$, Mach number, and turbulence level. There are three major components of loss, namely, the profile loss, annulus and secondary flow losses, and tip clearance loss. The overall blade loss coefficient is identified as either Y or λ, which is equal to the sum of these three types.

14.5.6.1 Profile Loss (Y_p)

The profile loss is the loss due to skin friction on the area of the blade surface. It depends on several factors including the area of blade in contact with fluid, the surface finish, and the Reynolds and Mach numbers of the flow through the passage. These factors are directly influencing the boundary layer over the blade surface and any possible boundary layer separation. The surface finish itself is in turn influenced by both erosion and fouling. The profile loss is measured directly in cascade wind tunnels similar to the compressors.

One of the commonly used approaches in defining the profile loss coefficient is from a set of cascade test results [7]. The profile loss coefficient (Figure 14.40) is expressed in terms of pitch/chord ratio and cascade exit angle (β_3) for two special cases of (β_2), namely, axial approach of flow ($\beta_2 = 0$) and impulse blading ($\beta_2 = \beta_3$).

For any other combination of angles, these graphs are interpolated by means of the following relation:

$$Y_p = \left\{ Y_{p(\beta_2=0)} + \left(\frac{\beta_2}{\beta_3}\right)^2 \left[Y_{p(\beta_2=\beta_3)} - Y_{p(\beta_2=0)} \right] \left(\frac{t/c}{0.2}\right)^{\beta_2/\beta_3} \right\} \qquad (14.92a)$$

When Equation 14.92 is used for a stator blade row, the angle α_1 is used instead of β_2 and α_2 is used instead of β_3.

The main assumptions for the validity of Equation 14.92 are

- Subsonic flow turbine stage
- The trailing edge thickness to the pitch ratio: $t_e/s \approx 0.02$

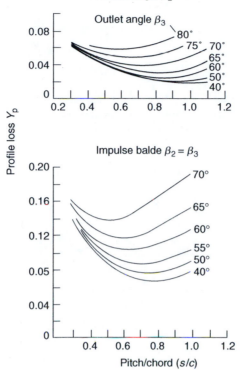

FIGURE 14.40 Profile loss coefficient for conventional blading with $t/c = 0.2$. (From Saravanamuttoo, et al. *Gas Turbine Theory*, Prentice Hall, 5th edn., 2001, p. 313. With permission.)

- Zero incidence angle: $i = 0$
- Reynolds number $R_e = 2 \times 10^5$

Since cascade data and other loss correlation are limited to subsonic conditions, it was suggested in Reference 16 that a correction Equation 14.92 to account for supersonic Mach numbers at either of the outlets of rotor or stator blade row is as follows:

$$Y_p = \left(Y_p\right)_{\text{Equation 14.92a}} \times \left[1 + 60 \left(M - 1\right)^2\right] \tag{14.92b}$$

where M is the absolute Mach number at the nozzle outlet and the relative Mach number at the rotor outlet.

14.5.6.2 Annulus Loss

The profile loss is caused by friction and associated with the boundary layer growth over the inner and outer walls of the annulus. Annulus losses are similar to profile losses as both are caused by friction. However, a fresh boundary layer grows from the leading edge of blade whereas the annulus boundary layer may have its origin some way upstream of the leading edge depending on the details of the annulus itself.

14.5.6.3 Secondary Flow Loss

Secondary flows are contrarotating vortices that occur due to curvature of the passage and boundary layers. Secondary flows tend to scrub both the end wall and blade boundary layers and redistribute low momentum fluid through the passage [11]. Secondary flow losses represent the major source of losses.

Both annulus loss and secondary flow loss cannot be separated and they are accounted for by a secondary loss coefficient (Y_s).

14.5.6.4 Tip Clearance Loss (Y_k)

Tip clearance loss occurs in the rotors. Some fluid leaks in the gap between the blade tip and the shroud, and therefore contributes little or no expansion work.

The combined secondary loss and tip leakage are expressed by the relation

$$Y_s + Y_k = \left[\lambda + B\left(\frac{k}{h}\right)\right]\left(\frac{C_L}{s/c}\right)^2\left[\frac{\cos^2 \beta_3}{\cos^3 \beta_m}\right] \tag{14.93}$$

where

$$\frac{C_L}{(s/c)} = 2\left(\tan \beta_2 + \tan \beta_3\right)\cos \beta_m$$

and

$$\beta_m = \tan^{-1}\left[(\tan \beta_3 - \tan \beta_2)/2\right]$$

The parameter B depends on whether the blade is shrouded or not. For unshrouded $B = 0.5$ while for shrouded $B = 0.25$. Moreover, the parameter λ is a function of the parameter

$$\lambda = f\left[\frac{(A_3 \cos \beta_3/A_2 \cos \beta_2)^2}{(1 + (r_r/r_t))}\right]$$

and is evaluated from Figure 14.41.

Then, the total loss for the nozzle is

$$Y_N = \left(Y_p\right)_N + \left(Y_s + Y_k\right)_N$$

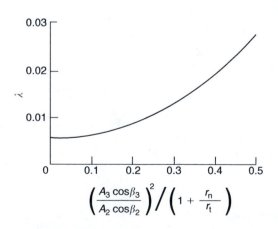

FIGURE 14.41 Secondary loss parameter. (From H.I.H. Saravanamuttoo, et al., *Gas Turbine Theory*, Prentice Hall, 5th edn., 2001, p. 313. With permission.)

Similarly, the total loss for the rotor is

$$Y_R = (Y_p)_R + (Y_s + Y_k)_R$$

The equivalent loss coefficients defined in temperature are given hereafter for the nozzle and rotor as

$$\lambda_N = \frac{Y_N}{T_{02}/T_2'}$$

and

$$\lambda_R = \frac{Y_R}{T_{03rel}/T_3''}$$

The stage efficiency is then calculated as

$$\eta_s = \frac{1}{1 + \left[\lambda_R \left(W_3^2/2C_p\right) + (T_3/T_2) \lambda_N \left(C_2^2/2C_p\right)\right]/(T_{01} - T_{03})} \tag{14.94}$$

The value calculated from Equation 14.94 is checked with the assumed value. If not satisfactory, some changes are made in the previously assumed degree of reaction of the stage-loading coefficient and flow coefficient.

14.6 TURBINE MAP

The overall turbine performance is defined in the same way as for the compressor in terms of similar nondimensional parameters, namely, mass flow parameter $(\dot{m}\sqrt{T_{0i}}/P_{0i})$, speed parameter $(N/\sqrt{T_{0i}})$, pressure ratio, and thermal efficiency. If the turbine is composed of more than one stage then the efficiency previously calculated by Equation 14.94 is considered as a polytrophic efficiency and the turbine.

Efficiency is calculated from the relation

$$\eta_t = \frac{1 - (P_{0e}/P_{0i})^A}{1 - (P_{0e}/P_{0i})^B}$$

where

$$A = \frac{\eta_s (\gamma - 1)}{\gamma}, \quad B = \frac{(\gamma - 1)}{\gamma}$$

The maximum mass flow rate is called choking conditions, which is constant for a certain range of turbine pressure ratio as shown in Figure 14.42a. The efficiency is also plotted versus the turbine pressure ratio (Figure 14.42b). It is noticeable here that the variations in the mass flow parameter and efficiency are not sensitive to the speed parameter as in the case of the compressor.

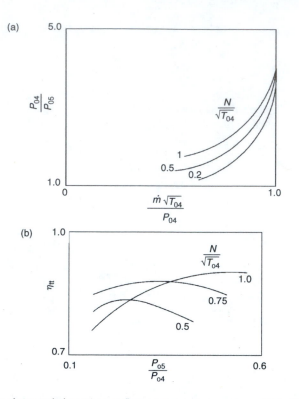

FIGURE 14.42 Turbine characteristics: (a) mass flow versus pressure ratio and (b) efficiency versus pressure ratio.

Example 7 An axial flow turbine is to be cooled by forced convection. The temperature of hot gases is 1200 K and the temperature of the coolant air extracted from the compressor is 350 K. The gas mass flow rate is 50 kg/s and the coolant air is 2% of the hot gases. The heat transfer coefficients of the hot gases and coolant are, respectively, 1200 and 1000 W/m^2 · K. The perimeter of the turbine blade section and the cooling passages are 0.12 and 0.06 m, respectively. The blade height is 0.1 m. It is required to calculate the temperatures of the coolant and blade height along the blade.
Solution: The above data can be rewritten as

$$T_g = 1200 \text{ K}, \quad T_{cr} = 350 \text{ K}, \quad \dot{m}_g = 50 \text{ kg/s}, \quad \dot{m}_c = 1 \text{ kg/s},$$

$$h_g = 1200 \text{ W/m}^2/\text{K}, \quad h_c = 1000 \text{ W/m}^2/\text{K}, \quad S_g = 0.12 \text{ m}, \quad S_c = 0.06 \text{ m}$$

$$L = 0.1 \text{ m}$$

From Equation 14.85a

$$\frac{K}{L} = \frac{h_g S_g}{\dot{m}_c C_{PC} \left[1 + \left(h_g S_g / h_c S_c\right)\right]} = \frac{1200 \times 0.12}{1 \times 1005 \left[1 + \dfrac{1200 \times 0.12}{1000 \times 0.06}\right]} = 0.04214$$

(i) *The temperature of the coolant* (T_c) *is obtained from Equation 14.88, namely,*

$$\frac{T_g - T_c}{T_g - T_{cr}} = e^{-Kl/L}$$

$$\frac{1200 - T_c}{1200 - 350} = e^{-0.042141}$$

Thus, an expression for the coolant temperature along the blade height is expressed as

$$T_C = 1200 - 850e^{-0.04214l}$$

(ii) *The Temperature of the blade* (T_b) is obtained from Equation 14.90, or

$$\frac{T_b - T_{cr}}{T_g - T_{cr}} = 1 - \frac{e^{(-K/L)l}}{\left[1 + \left(h_g S_g / h_c S_c\right)\right]}$$

$$\frac{T_b - 350}{1200 - 350} = 1 - \frac{e^{-0.04214\,l}}{\left[1 + \dfrac{1200 \times 0.12}{1000 \times 0.06}\right]}$$

Thus, the blade temperature at any point along its height is listed in Table 14.5.

Example 8 For a forced convection cooling of an axial flow turbine blade, the following data are given:

Gas temperature	$T_g = 1500$ K
Coolant temperature at inlet	$T_{cr} = 320$ K
Coolant mass flow rate	$\dot{m}_C = 10$ kg/s
Gas flow rate	$\dot{m}_g = 33$ kg/s
Blade height	$L = 0.1$ m
Gas heat transfer coefficient	$h_g = 1500$ W/m^2 K
Wetted perimeter of blade profile	$S_g = 0.15$ m
Parameter	$\left(h_g S_g / h_c S_c\right) = 2$

It is required to

(a) Calculate and plot the temperature of the blade material and coolant along the blade height.
(b) Calculate the parameter $\phi = T_G - T_M / T_G - T_C$ from Figure 14.39, where T_M is the maximum blade temperature at tip as calculated above, and T_C is the coolant temperature at the compressor outlet; here it is 695 K.
(c) What is its value for forced convection?

TABLE 14.5
Variations of the Temperatures of Coolant, Blade, and Gas Along the Blade Height

l	l/L	T_C	T_b	T_g
0.0	0.0	350	950	1200
0.01	0.1	350.36	950.1	1200
0.02	0.2	350.716	950.21	1200
0.03	0.3	351.074	950.316	1200
0.04	0.4	351.432	950.421	1200
0.05	0.5	351.789	950.526	1200
0.06	0.6	352.146	950.631	1200
0.07	0.7	352.504	950.736	1200
0.08	0.8	352.86	950.841	1200
0.09	0.9	353.218	950.946	1200
0.1	1	353.574	951.05	1200

(d) Next use the same bleed mass flow rate ratio $\dot{m}_c/\dot{m}_g$ to calculate the maximum blade temperature T_m at the blade tip for other cooling techniques shown in Figure 14.43.

Solution

1. The blade temperature is expressed by the relation (14.90), from which

$$T_b = T_{cr} + (T_g - T_{cr})\left[1 - \frac{e^{-Kl/L}}{1 + h_gS_g/h_cS_c}\right] = 320 + 1180\left[1 - \frac{e^{-Kl/L}}{3}\right]$$

$$\frac{K}{L} = \frac{h_gS_g}{\dot{m}_cC_{pc}\left[1 + (h_gS_g/h_cS_c)\right]} = \frac{1500 \times 0.15}{1.0 \times 1005(1+2)} = 0.0746$$

$$\therefore T_b = 320 + 1180\left[1 - \frac{e^{-0.0746\,l}}{3}\right] \tag{1}$$

The coolant temperature is to be calculated from the relation

$$T_C = T_g - (T_g - T_b)\left[1 + (h_gS_g/h_cS_c)\right]$$
$$T_C = 1500 - 3(1500 - T_b) \quad \rightarrow (2)$$

From Equations 1 and 2, T_b and T_g are calculated at different blade heights

l(m)	T_b(K)	T_C(K)	
0	1106.6	319.8	Root
0.025	1107.4	322.2	
0.05	1108.1	324.3	Mean
0.075	1108.8	326.4	
0.1	1109.9	329.7	Tip

The blade and coolant temperature distribution along the blade height are plotted in Figure 14.43. Variation of blade and coolant temperatures along the blade height.

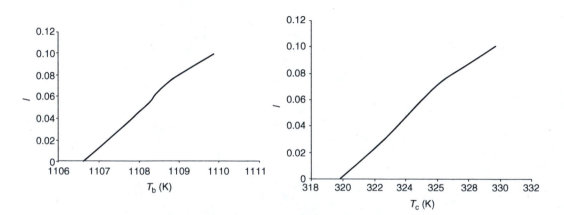

FIGURE 14.43 Blade and coolant temperature distribution along the blade height.

2. Calculate the parameter (ϕ) and (T_M)

$$\text{Since } \phi = \frac{T_G - T_M}{T_G - T_C}$$

$$T_M = T_G - \phi\,(T_G - T_C)$$

Since $\dot{m}_c/\dot{m}_g = 3\%$, from Figure 14.39, the value of ϕ is determined and the maximum blade temperature for different cooling techniques is calculated and listed in Table 14.6.

As seen from the table, transpiration method gives the minimum value of the maximum temperature, which assures that it is the best cooling technique.

14.7 CASE STUDY

The following section represents a complete preliminary design of an axial turbine. The steps followed in the design are given below.

Design steps
1. Calculation of flow properties at mean sections: flow angles, absolute and relative velocities, as well as total and static pressures and temperatures
2. Calculation of annulus dimensions
3. Calculation of the values of flow properties at different sections along the blade height from hub to tip, including airflow angles, absolute and relative velocities, together with the static pressures and temperatures
4. Estimation of the number of blades from data at mean section for nozzles and rotors
5. Calculation of the chord length at any section along blade height for nozzles and rotors
6. Selection of blade material that resists thermal and mechanical stresses
7. Calculation of stresses on rotor blades
8. Estimation of design point performance
9. Calculation of turbine efficiency
10. Turbine cooling

14.7.1 DESIGN POINT

The turbine to be designed is a part or a module of either an aero engine or a gas turbine. The following data represent the values of the design variables determined from the performance analysis of the complete aero engine or gas turbine.

TABLE 14.6

Variation of the Maximum Blade Temperature Based on Different Cooling Techniques

	Cooling Technique	ϕ	$T_m(K)$
1	Simple radial flow	0.395	1182
2	L.E Impingement/no film	0.466	1124.9
3	Multipass/trip strips	0.536	1068.5
4	Film/cross flow impingement	0.611	1008
5	Transpiration	0.69	944.6

Parameter	Value
$\dot{m}$ (kg/s)	110
Inlet pressure (bar)	10.803
Outlet pressure (bar)	6.584
Inlet temperature (K)	1600
Outlet temperature (K)	1430.5
Temperature drop (ΔT_{0s}) (K)	169.5
Isentropic efficiency (η_t)	0.91
Suggested number of stages	1

Only one stage is needed for this temperature drop. The rotational speed of the turbine is determined from the compressor coupled to the turbine as 150 rev/s. Moreover, the mean radius is assumed to be 0.4 m. The following values of some parameters are suggested.

Λ_m	ϕ_m	α_{1m} (deg.)	λ_N
0.45	0.8	0	0.05

The specific heat ratio and specific heat are assumed to be $\gamma = 1.333$ and $C_p = 1148$ J/kg · K. From the temperature drop $\Delta T_{0s} = 169.5$ K, the stage-loading coefficient is $\Psi_m = 1.5717$. The design assumes a constant mean radius along the nozzle (stator) and rotor blade rows.

Mean line flow

The governing equations for mean flow calculations are described previously. The results of the mean flow analysis are given in Table 14.7 for the three stations upstream of the stator and rotor and downstream of the rotor.

TABLE 14.7

Summary of the Geometry, Aerodynamic, and Thermodynamic Properties of a Turbine Stage

	Station (1)	Station (2)	Station (3)
α_m°	0	59.0841	16.4269
β_m°	—	22.7740	57.0841
U_m (m/s)	—	351.8584	351.8584
C_{am} (m/s)	293.4656	281.4867	281.4867
C_{um} (m/s)	—	470.0339	82.9897
W_m (m/s)	—	305.2871	518.0035
T_0 (K)	1600	1600	1430.5
T_m (K)	1562.5	1469.3	1393
P_0 (bar)	10.803	10.612	6.5847
P_m (bar)	9.8250	7.5460	5.9208
ρ_m (kg/m³)	2.1910	1.7895	1.4810
A(m²)	0.1711	0.2184	0.2639
h(m)	0.0681	0.0869	0.1050
r_t(m)	0.4340	0.4434	0.4525
r_m(m)	0.4	0.4	0.4
r_h(m)	0.3660	0.3566	0.3475

TABLE 14.8

Variations of the Annulus Area, Blade Height, and Radii at the Stage Three Stations

	Station (1)	Station (2)	Station (3)
A_m (m^2)	0.1711	0.2184	0.2639
h(m)	0.0681	0.0869	0.1050
r_t (m)	0.4340	0.4434	0.4525
r_m (m)	0.4	0.4	0.4
r_h (m)	0.3660	0.3566	0.3475

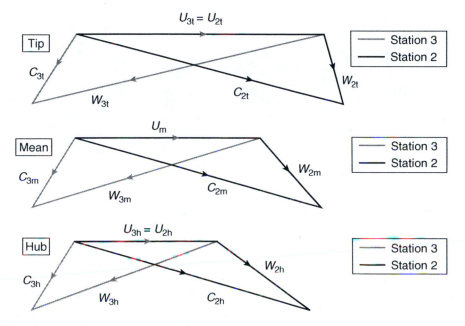

FIGURE 14.44 Variation of velocity triangles.

The annulus areas, blade heights, tip and hub radii at the three stations are given in Table 14.8.

Three-dimensional variations

Variation of the flow parameters from hub to tip is calculated at three sections only (hub, mean, and tip) using the different design methods. The results of the free vortex method are given in Table 14.9. The flow is assumed to enter the stage axially $\alpha_1 = 0$.

The following points may be concluded from Table 14.9:

1. The flow leaves the rotor at a nonzero absolute angles
2. The static temperature and pressure increase from hub to tip at states 2 and 3
3. The maximum relative Mach number occurs at the hub for rotor inlet and tip for rotor outlet. Figure 14.44 illustrates the three velocity triangles at hub, mean and tip sections.

It is worth mentioning here that the ordinate of any line in Figures 14.45 through 14.50 is proportional to the annulus height at stations 1, 2, and 3.

TABLE 14.9

Three-Dimensional Variations of Aerodynamic and Thermodynamic Properties Using Free Vortex

	Hub	Mean	Tip
α_1°	0	0	0
α_2°	61.9057	59.0841	56.4195
α_3°	18.7453	16.4269	14.6076
β_2°	37.2001	22.7740	6.8695
β_3°	54.9466	57.0841	59.1572
U_2 (m/s)	313.6432	351.8584	390.0735
U_3 (m/s)	305.6816	351.8584	398.0352
C_{a1} (m/s)	293.4656	293.4656	293.4656
C_{a2} (m/s)	281.4867	281.4867	281.4867
C_{a3} (m/s)	281.4867	281.4867	281.4867
C_{u2} (m/s)	527.3041	470.0339	423.9851
C_{u3} (m/s)	95.5263	82.9897	73.3619
C_1 (m/s)	293.4656	293.4656	293.4656
C_2 (m/s)	597.7327	547.8747	508.9186
C_3 (m/s)	297.2542	293.4656	290.8896
W_2 (m/s)	353.3918	305.2871	283.5221
W_3 (m/s)	490.1046	518.0035	549.0446
T_{01} (K)	1600	1600	1600
T_{02} (K)	1600	1600	1600
T_{03} (K)	1430.5	1430.5	1430.5
T_1 (K)	1562.5	1562.5	1562.5
T_2 (K)	1444.4	1469.3	1487.2
T_3 (K)	1392	1393	1393.6
P_{01} (bar)	10.803	10.803	10.803
P_{02} (bar)	10.612	10.612	10.612
P_{03} (bar)	6.5847	6.5847	6.5847
P_1 (bar)	9.8250	9.8250	9.8250
P_2 (bar)	7.1746	7.6818	8.0638
P_3 (bar)	5.9042	5.9208	5.9319
M_{w2}	0.4753	0.4071	0.3758
M_{w3}	0.6715	0.7095	0.7518

Number of blades for nozzle and rotor

From calculations at mean section, the blade heights at stations 1, 2, and 3 are determined. Next, assuming that the aspect ratio (h/c) is in the range (2–4), the blade chord at mean section (c) is calculated.

From Figure 14.29, we can get the optimum pitch/chord ratio (s/c) as the flow angles at the mean section. Then the number of blades is estimated from the relation

$$\text{Number of blades} = \frac{2\pi r_m}{s}$$

The results for calculations and the number of blades for the nozzle and rotor after rounding are given in Table 14.10.

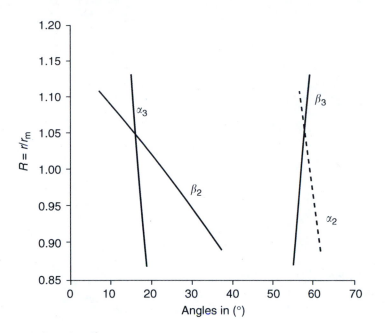

FIGURE 14.45   Variation of angles.

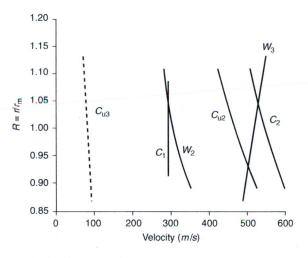

FIGURE 14.46   Variation of velocities.

Chord length at any section along blade height for nozzle and rotor
After calculating the number of blades and pitch/chord at any station, the chord at any station will
be recalculated as follows:

$$\text{The spacing or pitch is } s = \frac{2\pi r_m}{\text{number of blades}}$$

$$c = \frac{s}{(s/c)}$$

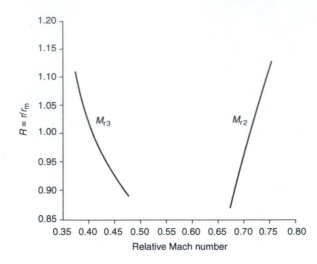

FIGURE 14.47   Variation of relative Mach number.

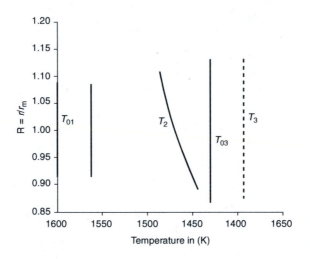

FIGURE 14.48   Variation of temperatures.

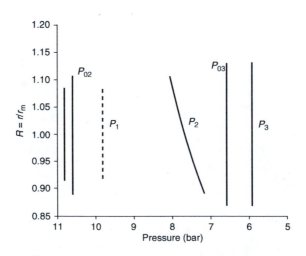

FIGURE 14.49   Variation of pressures.

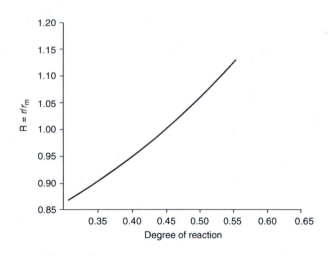

FIGURE 14.50   Variation of degree of reaction.

TABLE 14.10
Blade Geometry and Number of Blades for Nozzle and Rotor Blades

	Nozzle	Rotor
h/c	2	2
s/c	0.8490	0.8015
No. of blades	78	67

Blade material selection
The nozzle and rotor blade materials are selected to resist very high thermal and mechanical stresses. A suggested material for the present turbine blades is (SiC/SiC Ceramic Matrix Composites) as it can resist the very high temperature (up to 1900 K); besides it has a light weight ($\rho_b = 3200 (kg/m^3)$) to reduce stresses on blades.

Stresses on rotor blades
On the basis of the rotational speed of the rotor as well as its dimensions and material, the centrifugal stress is calculated from Equation 14.69 while the gas bending stress is calculated from Equations 14.74 and 14.75. The following values of stresses are obtained.

$(\sigma_{ct})_{max}$ (MPa)	$(\sigma_{gb})_{max}$ (MPa)
95	118.69

The chord length at different locations is tabulated in Table 14.11.

Losses calculations
The procedure described in Section 14.5.6 is followed here. Equations 14.92 and 14.93 are used to evaluate the loss coefficients for nozzle and rotor. The results are given in Table 14.12.

TABLE 14.11
Chord Length at Root, Mean and Tip Sections for Nozzle and Rotor

	Hub	Mean	Tip
Chord of nozzle (c_N) (m)	0.0357	0.0380	0.0403
Chord of rotor (c_R) (m)	0.0443	0.0468	0.0498

TABLE 14.12
Total and Equivalent Loss Coefficients

	Nozzle	Rotor
Total loss coefficient (Y)	0.1002	0.1508
Equivalent loss coefficient (λ)	0.0916	0.1381

Turbine efficiency

From the above values of loss coefficients and Equation 14.94, the value of the efficiency of the turbine is $\eta_t = 0.8904$. The assumed value was 0.91. The deviation between the assumed and calculated value is nearly 2%.

SUMMARY

The following table summarizes the different methods that handle the variations of air properties from the root to tip sections.

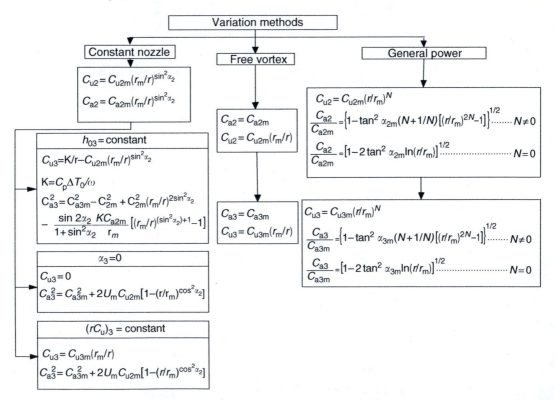

PROBLEMS

14.1 A single-stage axial flow gas turbine has the following data:

Inlet stagnation temperature, T_{01}	1200 K
Stator outlet angle, α_2	60°
Rotor blade inlet angle, β_2	20°
Rotor blade outlet angle, β_3	60°

Calculate
(a) The flow coefficient
(b) The stage-loading coefficient
(c) The degree of reaction
(d) If the blade speed $U = 360$ m/s, then calculate

   (i)   The stage total temperature drop
   (ii) The stage pressure ratio if its efficiency is 0.9

14.2 An axial flow gas turbine stage has a flow coefficient of 0.65, a constant axial velocity, and the gas leaves the stator blades at an angle of 65° to axial direction. Calculate
(a) The blade-loading coefficient
(b) The rotor blade relative flow angle
(c) The degree of reaction

14.3 A single-stage axial flow gas turbine has a mean radius of 30 cm. The hot gases enter the turbine stage at 1900 kPa and 1200 K and the absolute velocity leaving the stator (C_2) is 600 m/s making 65° with the axial direction (α_2). The rotor inlet and outlet blade angles (β_2 and β_3) are 25° and 60° , respectively. Draw the velocity triangles and calculate
(a) The rotor rotational speed (rpm)
(b) The stage pressure ratio if its efficiency is 0.88
(c) The flow coefficient, loading coefficient, and degree of reaction
(d) The power delivered by the turbine if mass flow rate is 50 kg/s

14.4 An axial flow turbine stage is designed for zero exit swirl ($\alpha_3 = 0$). Sketch the velocity triangles for the speed-to-work ratios of $(\lambda = U/\Delta C_u)$ 0.33, 0.5, and 1. Compare the three designs with respect to the work obtained and the degree of reaction.

14.5 A single-stage axial flow turbine, with axial leaving velocity, has the following data:
Static temperature at stator inlet $= 50$ K
Gas velocity at stator inlet $= 165$ m/s
Blade mean velocity $= 300$ m/s
Degree of reaction at mean radius $= 0.5$
Plot the velocity triangles, determine
(a) The fluid angle (α_2)
(b) The stage-loading coefficient
(c) The flow coefficient
(d) The stage total pressure ratio if $\eta_s = 0.9$

14.6 In a zero reaction gas turbine, the blade speed at the mean diameter is 290 m/s; gas leaves the nozzle ring at an angle 65° to the axial direction while the inlet stage temperature is 1100 K. The following pressures were measured at various locations:

At nozzle entry, stagnation pressure	400 kPa
At nozzle exit, stagnation pressure	390 kPa
At nozzle exit, static pressure	200 kPa
At rotor exit, static pressure	188 kPa

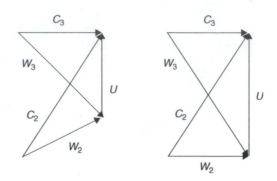

FIGURE 14.51   Velocity triangles for (a) impulse turbine and (b) 50% reaction turbine.

Assuming the magnitude and the direction of velocities at entry and at exit of the stage are the same, determine the total-to-total efficiency of the stage.

14.7 The velocity diagrams for *impulse* (zero reaction) and 50% reaction turbine are shown in Figure 14.51. In both cases, the rotor absolute inlet velocity C_2 is 400 m/s at an angle $\alpha_2 = 70°$, and the absolute exit angle C_3 is axial. Complete the following table:

Quantity	Impulse	50% Reaction
U		
Φ		
Ψ		
β_1		
β_2		
$\Delta T_{0\,stage}$		
Work		

14.8 A small axial flow gas turbine with equal stage inlet and outlet velocities has the following design data based on the mean diameter:

Inlet Temperature T_{01}	1100 K	Inlet Pressure P_{01}	350 kPa
Axial velocity, C_a	260 m/s	Blade speed U	350 m/s
Nozzle efflux angle, α_2	60°	Stage exit angle α_3	12°
Mass flow rate	18 kg/s		

Calculate
(a) The rotor blade angles
(b) The blade-loading coefficient and degree of reaction
(c) The power output
(d) The nozzle throat area required if the throat is situated at the nozzle outlet and nozzle losses coefficient is 0.05

14.9 A single-stage axial flow gas turbine with convergent nozzles has the following data:

Inlet stagnation temperature, T_{01}	1150 K	Mass flow rate	25 kg/s
Inlet stagnation pressure, P_{01}	420 kPa	Mean blade speed, U	340 m/s
Stage stagnation temperature drop	150 K	Rotational speed, N	15000 rpm
Pressure ratio, P_{01}/P_{03}	1.9	Flow coefficient, Φ	0.75
Angle of gas leaving stage, α_3	12°	Nozzle coefficient, λ_N	0.05

If the axial velocity remains constant and the gas velocities at inlet and outlet are the same determine at the mean radius:

(a) The blade angles
(b) The stage efficiency
(c) The blade-loading coefficient and degree of reaction
(d) The required nozzle inlet area
(e) The annulus area at rotor inlet and outlet
(f) The blade height at the above locations

14.10 The following data apply to a single-stage axial flow gas turbine

Inlet stagnation temperature, T_{01}	1200 K	Mass flow rate	36 kg/s
Inlet stagnation pressure, P_{01}	8.0 bar	Mean blade speed, U	320 m/s
Stage stagnation temperature drop	150 K	Rotational speed, N	250 rev/s
Isentropic stage efficiency	0.9	Outlet velocity, C_3	400 m/s

Assuming that the outlet velocity is axial, calculate

(a) The stage specific work and the stage power
(b) Draw the velocity triangle and determine the blade and flow angles
(c) The stage pressure ratio
(d) The inlet rotor and exit Mach numbers
(e) The blade height at rotor inlet and exit if $\lambda_N = 0.05$

14.11 A single-stage axial flow gas turbine with convergent nozzles has the following data:

Inlet stagnation temperature, T_{01}	1100 K	Mass flow rate	18.75 kg/s
Inlet stagnation pressure, P_{01}	3.4 bar	Mean blade speed, U	298 m/s
Stage stagnation temperature drop	144 K	Rotational speed, N	200 rps
Nozzle losses coefficient, λ_N	0.05	Flow coefficient, Φ	0.95
Isentropic stage efficiency	0.9		

The convergent nozzle is chocked, calculate

(a) The pressure ratio in the turbine
(b) The flow angles $\alpha_2, \beta_2, \alpha_3$, and β_3
(c) The blade-loading coefficient
(d) The degree of reaction
(e) The blade height and annulus radius ratio at stations 1, 2, and 3

14.12 The following particulars relate to a single-stage turbine of a free vortex design:

Inlet stagnation temperature, T_{01}	1050 K	Outlet velocity, C_3	275 m/s
Inlet stagnation pressure, P_{01}	3.8 bar	Root blade speed, U	300 m/s
Pressure ratio, P_{01}/P_{03}	2.0	Isentropic efficiency	0.88

The turbine is designed for zero reaction ($\Lambda = 0$) at the root radius and the velocities at inlet and outlet (C_1 & C_3) are both equal and axial. Calculate

(a) The nozzle efflux angle α_2 and the blade inlet angle β_2 at the root radius.
(b) If the tip-to-root radius ratio of the annulus at exit from the nozzle blades is 1.4, determine the nozzle efflux angle and degree of reaction at the tip radius.
(c) Assuming a nozzle blade losses coefficient of 0.05, calculate the static pressure at inlet and outlet of the rotor blades at root radius.

14.13 A single-stage axial flow gas turbine has a degree of reaction at mean radius $\Lambda_m = 0.5$. If the free vortex design is considered, what is the minimum hub-to-tip ratio (ζ) such that the degree of reaction remains positive along the blade height?

14.14 The data below apply to a single-stage axial flow gas turbine designed on free vortex theory. The outlet velocity is axial and the turbine is designed with a constant annulus area through the stage.

Inlet stagnation temperature, T_{01}	1225 K	Mass flow rate	30 kg/s
Inlet stagnation pressure, P_{01}	800 kPa	Mean blade speed, U	330 m/s
Stage stagnation temperature drop	160 K	Rotational speed, N	15000 rpm
Isentropic stage efficiency	0.9	Outlet velocity, C_3	390 m/s
		Nozzle coefficient, λ_N	0.05

Calculate
(a) The stage pressure ratio
(b) The flow and blade angles at the mean radius
(c) The flow coefficient, loading stage coefficient, and degree of reaction
(d) The blade height at inlet and exit from the rotor
(e) The flow and blade angles at root and tip radius using free vortex design
(f) Recalculate the above requirements if the angle of gases leaving the stage α_3 is $12°$.

14.15 A single-stage turbine receives a supply of gases at a temperature of 1400 K and a pressure of 3.2 bar. The inlet and exit flows for the stage are both axial. The rotational speed of the turbine is 10,000 rpm and it provides a power output of 4 MW. The mean radius of the blades is 0.25 m. The stage-loading coefficient is 1.4 and the turbine operates with a flow coefficient of 0.5. It is required to perform a preliminary mean radius design for this stage, specify.
(i) Calculate the axial velocity and mass flow rate.
(ii) Calculate the blade height at the inlet and outlet of stator as well as at the rotor outlet.
(iii) Calculate the number of stator and rotor blades.
(iv) Calculate the relative flow angles at the rotor entry and exit.
(v) Sketch the velocity triangles.
(vi) Estimate the stage efficiency.
The gas may be taken as $C_p = 1148$ J/kg · K, $\gamma = 4/3$.

14.16 (a) Compare between the degree of reaction in an axial turbine and axial compressor.
(b) For an axial turbine, compare the values of ψ and ϕ in the case of
• Aircraft engines
• Industrial gas turbine

14.17 A single spool turbojet engine has a seven-stage axial flow compressor driven by a single axial turbine. The engine's components have the following data:
1. Compressor

Inlet total conditions	$T_{01} = 288$ K, $P_{01} = 1.01$ bar
Mean blade speed	$U = 250$ m/s
Flow coefficient	$\phi = 0.6$
Bleed from compressor's outlet	$b = 8\%$
Mass flow rate	$\dot{m} = 20$ kg/s

Stage	Flow angles (degree)	Rotor's de Haller no.	Work-done factor (λ)	Degree of reaction (Λ)
1	$\beta_2 = 42°$ $\alpha_1' = 28.5°$	0.8	0.98	
2		0.75	0.93	
3		0.74	0.88	0.5
4		0.73	0.83	0.5
5		0.73	0.83	0.5
6		0.73	0.83	0.5
7		0.73	0.83	0.5

2. Combustion chamber

Pressure drop	$\Delta P_{occ} = 2.25\%$
Fuel-to-air ratio	$f = 1.7\%$

3. Turbine

Turbine inlet temperature	1100° K
Isentropic efficiency	$\eta_t = 0.9$
Mechanical efficiency	$\eta_m = 0.95$
Mean blade speed	$U = 298$ m/s
Flow coefficient	$\phi = 0.95$
Loss coefficient for the nozzle blades	$\lambda_N = 0.05$
Rotational speed	200 rps

The convergent nozzle is chocked.

Calculate

(a) The temperature rise in the compressor
(b) The pressure ratio of the compressor
(c) Blade-loading coefficient in the turbine (Ψ)
(d) The pressure ratio in the turbine
(e) The flow angles in the turbine ($\beta_3, \beta_2, \alpha_2, \alpha_3$)
(f) The degree of reaction of the turbine
(g) The blade height and the annulus radius ratio at stations 1, 2, and 3 of the turbine

14.18 At mid radius of an axial turbine stage is designed at 50% reaction, zero outlet swirl, and axial velocity component of 200 m/s (constant through the stage). The absolute flow angle at outlet from the stator is 65° from the axial direction.

(a) At the design point, what is the work done by the fluid at mid radius on the rotor (kJ/kg)?
(b) For no change from turbine design flow conditions, but with a 20% drop from design rotor rpm, estimate the work done (kJ/kg) by the rotor at mid radius.
(c) Calculate the flow coefficient for the design point.
(d) Calculate the work coefficient for the design point.
(e) If the air enters the stator at 1200° K, calculate the stagnation temperature and pressure ratios ($\eta_s = 0.94$) for the design point.
(f) If the tip diameter is 0.8 m, the hub-to-tip ratio is 0.9, and the blade material density is 8000 kg/m^3, calculate the maximum centrifugal stress and specify where it occurs (for the design point).
(g) If the blade's maximum, average, and minimum temperatures are 1300, 1000, and 900 K, respectively, then calculate the maximum and minimum thermal stresses for the design point ($E = 2 \times 10^5$MPa, $\alpha = 7 \times 10^{-6}$/K)

14.19 The surface temperature of the hub section of the rotor of an axial turbine stage is given by the relation

$$\frac{T_{Shub}}{T_{01}} = 1 - \left(\frac{\gamma - 1}{2}\right)\left(\frac{U_t}{a_{01}}\right)^2 \zeta^2 \left[\left(\frac{C_{u2}}{U}\right)_h^2 + \left(\frac{C_{a2}}{U}\right)_h^2 - r_f\left\{\left(\frac{C_{u2}}{U} - 1\right)_h^2 + \left(\frac{C_{u2}}{U}\right)_h^2\right\}\right]$$

An axial stage turbine has the following data:

$$\left(\frac{C_{a2}}{U}\right)_h = 0.6, \quad U_t = 500 \text{ m/s}, \quad \zeta = 0.9$$

$r_f = 0.9, \gamma = 1.33, T_{01} = 1000$ K, $T_{Shub} = 850$ K, employing free vortex design,

(a) Calculate $(C_{u2}/U)_h$.

(b) If $\Lambda_m = 0.5$, then draw the velocity triangles at the hub (assume $C_{a2} = C_{a3}$).

(c) Calculate ΔT_{0s} and ψ.

14.20 Prove that if the nozzle of an axial turbine is chocked, then the velocity for gases leaving the nozzle (C_2) is expressed by

$$C_2 = \sqrt{\left(\frac{2\gamma R}{\gamma + 1}\right) T_{02}}$$

A single-stage axial flow turbine has the following data:

Inlet temperature	T_{01}	1100 K
Temperature drop	$T_{01}-T_{03}$	145 K
Inlet pressure	P_{01}	4 bar
Mean blade speed	U	340 m/s
Swirl angle at exit	α_3	10°
Flow coefficient	ψ	0.8
Nozzle loss coefficient	λ_N	0.05

The nozzle is chocked and the gases have the properties: $C_p = 1.248$ kJ/kg, $\gamma = 1.33$

It is required to

(a) Calculate the temperature drop coefficient.

(b) Calculate the degree of reaction.

(c) Calculate the velocities C_1, C_2.

(d) Draw the velocity triangles.

(e) Draw the stator and rotor blades at mean section.

14.21 In an axial turbine, the variation of the swirl velocity in the radial direction is given by

$$C_u = kr^N$$

The variation of axial velocity is given by

$$\frac{C_a}{C_{am}} = \left\{1 - \tan^2 \alpha_m \left(\frac{N+1}{N}\right)\left[\left(\frac{r}{r_m}\right)^{2N} - 1\right]\right\}^{1/2}$$

Provided that $N \neq 0$

Consider an axial stage having the following data at mean section

$\beta_2 = 0$, $\Lambda_m = 0.5$, $U_m = 360$ m/s, $\Phi_m = 0.8$, $N = 1$ and $\zeta = 0.8$.

Prove that at any station $C_{u2} = U$.

Draw the velocity triangles at the hub and tip sections. Draw also the rotor shape.

14.22. Complete the following statements:

(i) When $P_{02} = P_{03}$, then = ... zero   and .... = Zero

(ii) When $T2 = T_3$, then .... = zero   and .... = ...

(iii) When $T_1 = T_2$, then .... =   and .... = 1.0

(iv) When $\alpha_1 = \alpha_2 = \alpha_3$, then the stator blades ....   and .... = Zero

For a forced convection cooling of an axial flow turbine blade, the following data are given:

Gas temperature $T = 1500$ K
Coolant temperature at inlet $T_{cr} = 400$ K
Coolant mass flow rate $\dot{m}_c = 1.5$ kg/s
Gas flow rate $\dot{m}_g = 50$ kg/s
Blade height $L = 0.1$ m
Gas heat-transfer coefficient $h = 1500$ W/m^2 K
Wetted perimeter of blade profile $S = 0.15$ m
Parameter $(h_g S_g / h_c S_c) = 2$

(i) Calculate and plot the temperature variation of the blade material and coolant along the blade height.

(ii) Calculate the parameter

$$\Phi = \frac{T_g - T_m}{T_g - T_c}$$

where T_m is the maximum blade temperature at tip as calculated above and T_c is the coolant temperature at the compressor outlet (here it is 710 K).

Next use the same bleed mass flow rate ratio $(\dot{m}_c / \dot{m}_g)$ to calculate the maximum blade temperature T_m at the blade tip for other cooling techniques as shown in Figure 14.39.

REFERENCES

1. *The Jet Engine*, Rolls Royce plc, 5th edn., reprinted 1996, p. 45.
2. B. Lakshminarayana, *Fluid Dynamics and Heat Transfer of Turbomachinery*, John Wiley & sons, Inc., 1996, p. 145.
3. J.H. Horlock, *Axial Flow Turbine—Fluid Mechanics and Thermodynamics*, Robert E. Krieger Publishing Co., Inc, 1982, p. 51.
4. H.I.H. Saravanamuttoo, G.F.C. Rogers, and H. Cohen, *Gas Turbine Theory*, Prentice Hall, 5th edn., 2001, p. 313.
5. H. Moustapha, *Gas Turbine Design*, Pratt & Whitney Canada Inc., Lecture Notes, p. 81.
6. S.F. Smith, A Simple Correlation of Turbine Efficiency, *Journal Royal Aeronautical Society* 69, 1965, pp. 467–470.
7. D.G. Ainley and G.C.R. Mathieson, *An Examination of the Flow and Pressure Losses in Blade Row of Axial Flow Turbines*, Aeronautical Research Council R&M 2891 (HMSO 1955).
8. M.H. Vavra, *Aero-Thermodynamics and Flow in Turbomachines*, Robert E. Krieger Publishing Co., Inc, 1974, p. 456.
9. S.M. Yahya, *Turbines Compressors and Fans*, TATA McGraw-Hill, 1983, p. 334.
10. A.J. Glassman (ed.), *Turbine Design and Applications*, NASA SP-290, 1973, p. 88.
11. H. Moustapha, M.F. Zelessky, N.C. Baines, and D. Japikse, *Axial and Radial Turbines*, Concepts NREC, 2003, p. 31.
12. L.J. Pritchar, *An Eleven Parameter Axial Turbine Airfoil Geometry Model*, ASME P 85-GT-219, 1985.
13. P. Hill and C. Peterson, *Mechanics and Thermodynamics of Propulsion*, 2nd edn., Addison Wesley Publication Company, Inc., 1992, p. 387.
14. M.P. Boyce, *Gas Turbine Engineering Handbook*, (2nd edn.), Butterworth-Heinemann, 2002, p. 351.
15. W.W. Bathie, *Fundamentals of Gas Turbines*, 2nd edn., John Wiley& Sons, Inc., 1996, p. 325.
16. J. Dunham and P.M. Came, *Improvements to the Ainley-Mathieson Method of Turbine Performance Prediction*, ASME Paper 70-GT-2, 1970.

15 Radial Inflow Turbines

15.1 INTRODUCTION

Radial inflow turbines have a long history and predate axial machines by many years. The first truly effective radial turbines were water turbines. In this chapter, only radial inflow turbines working with compressible fluids are discussed. Radial inflow turbines are not as common as radial flow (centrifugal) compressors. Like the latter most radial inflow turbines are relatively low specific speed machines suitable for applications with low flow rates and large pressure drops. They are also simpler and cheaper than axial turbines and so are used when weight and cost are more important than efficiency. The inward-flow radial (IFR) turbine covers tremendous ranges of power, rates of mass flow, and rotational speeds, from very large Francis turbines used in hydroelectric power generation and developing hundreds of megawatts down to tiny closed cycle gas turbines employed in space power generation of a few kilowatts [1].

Most applications are at a relatively small scale representing compact power sources such as turbochargers for cars, buses and trucks, railway locomotives and diesel power generators, cryogenic and process expanders, small gas turbines such as the auxiliary power unit (APU) in most present airliners, helicopter engines, space power systems, rocket engine turbopumps, and other systems where compact power sources are required.

Radial inflow turbines (similar to centrifugal compressors) are found in one of the following two layouts: either as a part of a small gas generator, or in other words, coupled to a compressor and situated downstream of a combustion chamber. In this case, the turbine is coupled to a centrifugal compressor and normally has a back-to-back configuration. Moreover, the in-between combustion chamber is frequently of the reverse type. Thus, this small gas turbine will have a minimum length, which is advantageous in many applications, particularly space ones. In the second layout, the turbine is a part of a module, say, in a cryogenic expander where the first element of the module is an inlet scroll or volute. The flow enters from a pipe and is dispersed around the annulus of the turbine in the volute, which is very similar to the centrifugal compressor and its outward scroll.

The desirable characteristics of radial turbines are

1. High efficiency
2. Ease of manufacture
3. Sturdy construction
4. High reliability

Radial turbines are similar to centrifugal compressors, as both have mixed flow directions. Fluid enters and leaves in two perpendicular directions. For radial turbines, the flow enters radially and leaves axially close to the axis of rotation. This turning of the flow takes place in the rotor passage, which is relatively long and narrow. Figure 15.1 illustrates a radial inflow turbine.

Radial inflow turbine is composed of four elements, namely, volute (or scroll), nozzle vanes, rotor, and a discharge diffuser. The flow enters the turbine via a scroll, which provides some swirl velocity before distributing in the stator blades. Small turbines used in turbochargers do not have stator blades and rely on the scroll to provide all the angular momentum entering the rotor. The ring of nozzle upstream of the rotor accelerates the flow. Flow enters the diffuser at a large radius and leaves it at a smaller radius. At the rotor inlet where the flow velocity relative to the rotor has little or no tangential component, the rotor blades are usually straight and radial. This straight section of

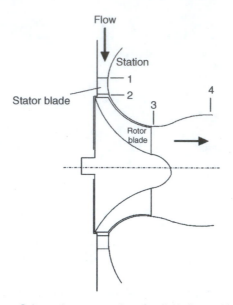

Schematic cross section of radial-inflow turbine

FIGURE 15.1   Layout of a radial inflow turbine

the rotor blade is generally is rather highly loaded, since angular momentums $(r \times C_u)$ (where (r) is the radius and (C_u) is the tangential component of absolute velocity) varies with the square of the radius. (Since $W_u \approx 0$, $C_u = u \propto r$; therefore $rC_u \propto r^2$). Usually downstream of the rotor, there is a diffuser to reduce the exhaust velocity to a negligible value.

15.2 THERMODYNAMICS

The expansion process in a radial turbine differs appreciably from that in an axial turbine because of the radius change in the rotor. The relative total temperature and pressure decrease with decreasing radius; this is a distinct advantage for the radial turbine because it permits the use of a lower velocity level for a given overall expansion. This can be seen from the temperature–entropy diagram; see Figure 15.2.

For the radial turbine, as shown in Figure 15.2, the turbine exists either downstream a volute or a combustion chamber. The first element considered in this analysis is the nozzle. The flow may experience a slight total pressure drop, but the total temperature is unchanged.

$$T_{02} = T_{01}, \quad P_{02} = P_{01} - \Delta P_{0\,\text{nozzle}} \tag{15.1}$$

For preliminary design, it may be satisfactory to assume that $\Delta P_{0\,\text{nozzle}} = 0$. For the rotor, the outlet total pressure (P_{03}) line is farther removed from the (P_{02}) line. This difference is due to both the rotor losses and the change in radius.

A radial turbine velocity diagram is shown in Figure 15.3 for a turbine with prewhirl in the inlet volute and a mean diameter ratio (exit-mean to inlet) of about 0.5.

The absolute inlet velocity to the rotor C_2 has two components in the radial and tangential directions (C_{r2}, C_{u2}). The radial component is compatible with the mass flow rate:

$$\dot{m} = \rho_2 A_2 (1 - B_2) C_{r2} \tag{15.2}$$

where A_2 is the annulus area at the rotor tip and B_2 is the allowance for boundary layer blockage [2].

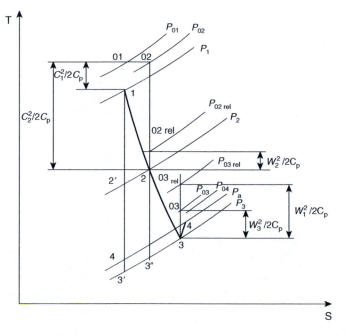

FIGURE 15.2   Temperature–entropy diagram.

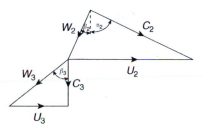

FIGURE 15.3   Velocity triangles in a radial inflow turbine.

The specific work output is given by Euler's equation for turbomachinery

$$W = U_2 C_{u2} - U_3 C_{u3} \tag{15.3}$$

which is expressed in terms of enthalpy drop as

$$W = Cp\,(T_{01} - T_{03}) = h_{01} - h_{03}$$

If the flow at the rotor inlet is radial then the relative angle β_2 is zero (thus $U_2 = C_{u2}$). The flow leaves the rotor axially, thus C_{u3} is zero.

$$W = Cp\,(T_{01} - T_{03}) = U_2 C_{u2} = U_2^2 \tag{15.4}$$

If the flow is assumed ideal isentropic through the turbine with a perfect diffuser, then define the specific work output as

$$W' = Cp \, (T_{01} - T_4) = \frac{C_0^2}{2} \tag{15.5a}$$

$$C_0^2 = 2CpT_{01} \left[1 - \left(\frac{P_4}{P_{01}} \right)^{(\gamma-1/\gamma)} \right] \tag{15.5b}$$

where C_0 is the spouting velocity defined as the velocity equivalent of the isentropic enthalpy drop. From Equations 15.4 and 15.5a

$$U_2^2 = \frac{C_0^2}{2} \tag{15.6}$$

This velocity ratio is identified as $\nu = U_2/C_o$

For ideal case, $\nu \equiv U_2/C_o = 0.707$. For nonisentropic flow [3], this speed ratio becomes $U_2/C_0 \approx 0.68 - 0.71$.

Three efficiencies are defined here: the isentropic efficiency of the turbine alone and the isentropic efficiency for the turbine and diffuser. The isentropic efficiency of the turbine is either expressed as total-to-total or as total-to-static efficiency. The turbine total-to-static efficiency (η_{ts}) is defined as

$$\eta_{ts} = \frac{T_{01} - T_{03}}{T_{01} - T_3'} \tag{15.7}$$

The total-to-total efficiency (η_{tt}) is defined as

$$\eta_{tt} = \frac{T_{01} - T_{03}}{T_{01} - T_{03s}} = \frac{\Delta T_0}{T_{01} - T_{03s}} \tag{15.8}$$

The combined turbine and diffuser efficiency is

$$\eta_0 = \frac{T_{01} - T_{03}}{T_{01} - T_4'} \tag{15.9}$$

Nozzle loss coefficient is also defined as

$$\lambda_N = \frac{T_2 - T_2'}{(C_2^2/2Cp)} \tag{15.10}$$

Rotor loss coefficient is given by

$$\lambda_R = \frac{T_3 - T_3''}{(W_3^2/2Cp)} \tag{15.11}$$

For constant pressure process $Cp\delta T = T\delta s$, then the nozzle loss coefficient can be expressed as:

$$\lambda_N = \frac{T_3'' - T_3'}{(C_2^2/2Cp)} \frac{T_2'}{T_3'}$$

The turbine isentropic efficiency can be expressed in terms of the nozzle and rotor loss coefficients

$$\eta_{ts} = \left[1 + \frac{1}{2Cp\,(T_{01} - T_{03})} \left\{ C_3^2 + \lambda_R W_3^2 + \lambda_N \frac{T_3'}{T_2'} C_2^2 \right\} \right]^{-1} \tag{15.12a}$$

Since $C_3 = U_2(r_3/r_2)\cot\beta_3$, $W_3 = U_2(r_3/r_2)\mathrm{cosec}\beta_3$, and if $\beta_2 = 0$, then $C_2 = U_2\mathrm{cosec}\alpha_2$, and applying Equation 15.4, then Equation 15.12a can be rewritten as

$$\eta_{ts} = \left[1 + \frac{1}{2}\left\{\left(\frac{r_3}{r_2}\right)^2 (\cot^2\beta_3 + \lambda_R\mathrm{cosec}^2\beta_3) + \lambda_N\frac{T_3'}{T_2'}\mathrm{cosec}^2\alpha_2\right\}\right]^{-1} \tag{15.12b}$$

The term T_3'/T_2' can be calculated from the relation

$$\frac{T_3'}{T_2'} = 1 - \frac{U_2^2}{2CpT_2}\left[1 + \left(\frac{r_3}{r_2}\right)^2 \left\{(1+\lambda_R)\,\mathrm{cosec}^2\beta_3 - 1\right\} - \cot^2\alpha_2\right] \tag{15.13}$$

where $T_2 = T_{01} - \left(U_2^2/2Cp\right)\mathrm{cosec}^2\alpha_2$ \hfill (15.14)

Example 1 A small radial inflow turbine develops 60 kW power when its rotor is running at 60,000 rpm. The pressure ratio P_{01}/P_3 is 2.0. The inlet total temperature is 1200 K. The rotor has an inlet tip diameter of 12 cm and rotor exit tip diameter of 7.5 cm. The hub–tip ratio at exit is 0.3. The mass flow rate of gases is 0.35 kg/s. They enter the rotor radially and leave axially and have the following inlet and outlet angles $\alpha_2 = 70°$ and $\beta_3 = 40°$. The nozzle loss coefficient $(\lambda_N) = 0.07$. It is required to calculate

1. The isentropic efficiency (η_{ts})
2. The rotor loss coefficient (λ_R)

Solution: The rotor tip rotational speed is $U_2 = \pi \times D_2 \times N = 377$ m/s.
From the inlet velocity triangle; Figure 15.3 with $\beta_2 = 0$, then

$$\sin\alpha_2 = \frac{U_2}{C_2}$$

The absolute velocity at inlet is $C_2 = U_2 \times \mathrm{cosec}\,\alpha_2 = 401.185$ m/s.
The static temperature at inlet is $T_2 = T_{02} - \left(C_2^2/2Cp\right) = 1130$ K

$$T_{01} - T_3' = T_{01}\left[1 - \frac{T_3'}{T_{01}}\right] = T_{01}\left[1 - \left(\frac{P_3}{P_{01}}\right)^{(\gamma-1/\gamma)}\right] = 190.92 \text{ K}$$

The turbine power is $\mathscr{P} = \dot{m} \times Cp \times (T_{01} - T_{03})$

$$T_{o1} - T_{o3} = \frac{60,000}{0.35 \times 1,148} = 149.328 \text{ K}$$

Since $\eta_{ts} = (T_{01} - T_{03})/(T_{01} - T_3') = 0.782$

$$\frac{r_3}{r_2} = \frac{D_{3h} + D_{3s}}{2D_2} = \frac{\xi \times D_{3s} + D_{3s}}{2 \times D_2} = 0.40625$$

From Equation 15.13

$$\frac{T_3'}{T_2'} = 1 - \frac{U_2^2}{2 \times Cp \times T_2}\left[1 + \left(\frac{r_3}{r_2}\right)^2 \left\{(1+\lambda_R)\,\text{cosec}^2\beta_3 - 1\right\} - \cot^2\alpha_2\right]$$

$$\frac{T_3'}{T_2'} = 0.9396 - 0.02187 \times \lambda_R$$

Substituting the value of T_3'/T_2' in Equation 15.12b

$$\eta_{ts} = \left[1 + 0.5\left\{\left(\frac{r_3}{r_2}\right)^2\left(\cot^2\beta_3 + \lambda_R\cos ec^2\beta_3\right) + \lambda_N\frac{T_3'}{T_2'}\cos ec^2\alpha_2\right\}\right]^{-1}$$

$$(0.782)^{-1} = 1.1544 + 0.2005 \times \lambda_R$$

$$\lambda_R = 0.62$$

15.3 DIMENSIONLESS PARAMETERS

The following dimensionless parameters are frequently used in preliminary design of radial inflow turbines:

1. *Stage loading:* The stage-loading coefficient can be expressed using the Euler turbomachinery equation as:

$$\psi = \frac{C_{u2}U_2 - C_{u3}U_3}{U_2^2}$$

$$\psi = \frac{C_{u2}}{U_2} - \frac{r_3}{r_2}\frac{C_{u3}}{U_3} \tag{15.15a}$$

   where r_3/r_2 is the rotor radius ratio. The exit swirl is normally fairly small, so that the second term on the right-hand side is small in comparison with the first, and the loading coefficient can be approximated as

$$\psi \approx \frac{C_{u2}}{U_2} \tag{15.15b}$$

2. *Flow coefficient:* The flow coefficient is defined in terms of the exit meridional velocity, also nondimensionalized by the inlet blade speeds:

$$\phi = \frac{C_{a3}}{U_2} \tag{15.16}$$

3. *Rotor meridional velocity ratio:* The rotor meridional velocity ratio ξ

$$\xi = \frac{C_{r2}}{C_{a3}} \approx 1.0 \tag{15.17}$$

4. *Specific speed:* Specific speed is defined as

$$N_s = \frac{N\sqrt{Q}}{(\Delta h_{0s})^{3/4}} \qquad (15.18)$$

where N is the rotational speed and Q is the volume rate. Substituting for Δh_{0s} ($\equiv C_0^2/2$) and considering the volume flow rate at station 3, then the specific speed has the form

$$N_s = 0.18\sqrt{\frac{Q_3}{ND_2^3}}$$

or $\Omega_s = 1.131\sqrt{Q_3/ND_2^3}$ (rad).

Specific speed is a function only of the operating conditions and contains no parameters relating directly to the turbine geometry [2]. Performance measurements of a large number of radial inflow turbines outline that the best efficiencies occur at specific speeds in the range of about 0.4–0.8.

15.4 PRELIMINARY DESIGN

Appropriate values for the following variables are to be selected such that an optimum efficiency is attained.

- Stage loading (ψ), in the range 0.8–1.0
- Flow coefficient (ϕ), in the range 0.3–0.4
- Total-to-total or total-to-static efficiency, equal to or greater than 0.7
- Meridional velocity ratio (ξ), in most cases selected as unity
- The ratio between exit hub to the inlet radii (r_{3h}/r_2), in the range of 0.25–0.35

Generally some data will be available for the turbine to be designed. These include

- The power
- Rotational speed
- Working fluid
- Mass flow rate
- Inlet total conditions

The design steps are as follows:

1. Calculate the total enthalpy (or temperature) drop from the relation

$$\mathscr{P} = \dot{m} \times Cp \times (T_{01} - T_{03})$$

$$\Delta T_0 \equiv (T_{01} - T_{03}) = \frac{P}{\dot{m} \times Cp}$$

2. Calculate the rotor tip speed from

$$U_2 = \sqrt{\Delta h} = \sqrt{Cp\,(T_{01} - T_{03})}$$

3. Calculate the axial speed at exit from the relation

$$C_{a3} = \phi\, U_2$$

The axial velocity at rotor outlet is nearly equal to the absolute velocity at rotor outlet or

$$C_3 \cong C_{a3}$$

4. Calculate the rotor inlet radial velocity component from the relation

$$C_{r2} = \xi\phi\, U_2 = \xi C_{a3}$$

5. Calculate the rotor inlet swirl speed from the relation

$$C_{u2} = \psi\, U_2$$

6. Calculate the rotor inlet absolute flow angle from the relation

$$\alpha_2 = \tan^{-1} \frac{C_{u2}}{C_{r2}}$$

7. Calculate the rotor inlet relative angle

$$\beta_2 = \tan^{-1} \frac{U_2 - C_{u2}}{C_{r2}}$$

8. Calculate the inlet absolute velocity

$$C_2 = (C_{r2} + C_{u2})^{1/2}$$

9. Calculate the rotor inlet static temperature

$$T_2 = T_{02} - \frac{C_2^2}{2C_p}$$

10. Calculate the rotor inlet static pressure

$$P_2 = P_{02} \left(\frac{T_2}{T_{02}}\right)^{\gamma/\gamma - 1}$$

For preliminary calculations, it may be first assumed that $P_{02} = P_{01}$.

11. Calculate the rotor inlet area

$$A_2 = \frac{\dot{m} R T_2}{P_2 C_{r2}}$$

12. Calculate the rotor inlet radius

$$r_2 = \frac{U_2}{2\pi N}$$

13. Calculate the rotor inlet width

$$b_2 = \frac{A_2}{2\pi r_2}$$

14. Calculate the rotor outlet total temperature

$$T_{03} = T_{01} - \Delta T_0$$

15. Calculate the static temperature at rotor outlet

$$T_3 = T_{03} - \frac{C_3^2}{2Cp}$$

16. Calculate the rotor total outlet pressure

$$P_{03} = P_{01} \left(1 - \frac{\Delta T_0}{T_{01} \eta_{tt}} \right)^{\gamma/\gamma-1}$$

17. Calculate the static pressure at rotor outlet

$$P_3 = P_{03} \left(\frac{T_3}{T_{03}} \right)^{\gamma/\gamma-1}$$

Alternatively, if the total-to-static efficiency is given, then the static pressure is obtained from the relation

$$\frac{P_3}{P_{01}} = \left[1 - \frac{1}{\eta_{ts}} \left(1 - \frac{T_{03}}{T_{01}} \right) \right]^{\gamma/\gamma-1}$$

The total pressure at rotor outlet is next calculated

$$P_{03} = P_3 \left(\frac{T_{03}}{T_3} \right)^{\gamma/\gamma-1}$$

18. The rotor outlet area is calculated

$$A_3 = \frac{\dot{m} R T_3}{C_3 P_3}$$

19. The hub radius at rotor outlet is calculated from the known ratio (r_{3h}/r_2) as

$$r_{3_h} = \left(\frac{r_{3h}}{r_2} \right) \times r_2$$

20. The tip radius as

$$r_{3t} = \sqrt{\frac{A}{\pi} - r_{3h}^2}$$

21. The mean rotational speed at rotor outlet

$$U_3 \, (\text{RMS}) = \pi \, (r_{3h} + r_{3t}) \, N/2$$

22. For the design followed here, zero swirl is assumed at outlet; thus the flow leaves the rotor axially or $C_{u3} = 0$. The outlet relative angle is then

$$\beta_3 = \tan^{-1} \frac{U_3}{C_3}$$

23. The outlet relative speed is

$$W_3 = U_3 \text{ (RMS) cosec } \beta_3$$

Some notes may be added here:

- In the above procedure, no exit swirl is assumed. The effect of exit swirl is discussed in detail [4]. However, for most practical purposes designers usually add exit swirl only when it is necessary to increase or decrease the power output without changing the speed of rotation. This is clear from Equation 15.3. It must be noted that if the exit swirl angle exceeds about 15–20° diffuser performance falls into conventional diffuser cases. This constraint is similar to that in nozzles following axial turbines.

- Rotor blade inlet and exit angles are determined from the flow inlet and outlet angles with known incidence and deviation angles.

$$\beta_{b2} = \beta_2 - i$$
$$\beta_{b3} = \beta_3 + \delta_3$$

As regards incidence angle, it must be consistent with the general experience from which it is noticed that the best efficiency occurs at some value in the range 20–30° in magnitude.

15.5 BREAKDOWN OF LOSSES

Figure 15.4 illustrates a division of the losses in a radial turbine stage and emphasizes the large contribution made to the stage by the rotor and rotor-related losses [5]; namely,

- Nozzle blade row boundary layer
- Rotor passage boundary layer
- Rotor blade tip clearance
- Disk windage (or the back surface of rotor)
- Kinetic energy loss at exit

From Figure 15.4 it may be concluded that for low values of specific speed, the stator and rotor viscous losses are very large because of the high ratio of wall area to flow area. Also, the clearance loss is large because the blade-to-shroud clearance is a relatively large fraction of the passage height. The windage loss, which depends on primarily the diameter and rotational speed, is also large at low specific speed because of the low flow rate. As specific speed increases, the stator and rotor losses, clearance loss, and windage loss all decrease because of the increased flow and area. The exit kinetic energy loss becomes predominant at high values of specific speed.

The effects of geometry and velocity diagram characteristics were examined by calculating the previously mentioned losses for a large number of combinations of the following parameters:

- Nozzle exit angle α_2
- Rotor diameter ratio D_{3av}/D_2
- Rotor blade entry height to exit diameter ratio b_2/D_{3av}

The static efficiency is plotted against specific speed; see Figure 15.5. For the range of values used in the study, all the calculated points fell in the shaded areas in Figure 15.5. Stator-exit angle is seen to be a prime determinant of efficiency, which falls into a small region for each stator-exit flow angle. The dashed curve is the envelope of all the computed static efficiencies and the solid curve above it represents the corresponding total efficiencies.

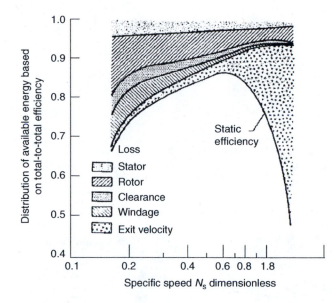

FIGURE 15.4 Breakdown of losses of radial turbine. (H. E. Rohlik, *Analytical Determination of Radial Inflow Turbine Design Geometry for Maximum Efficiency*, NASA TN D-4384. With permission.)

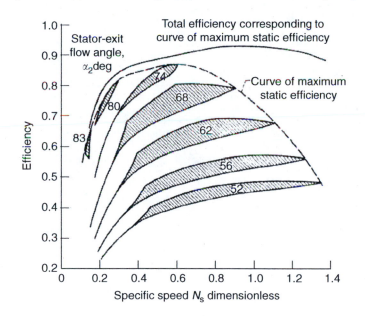

FIGURE 15.5 Effect of design speed on total-to-static efficiency. (A. J. Glassman, *Turbine Design and Application*, NASA SP-290, p. 279. With permission.)

Example 2 A radial inflow turbine has the following characteristics and operating conditions:

Inlet total temperature	1400 K
Inlet total pressure	3.0 bar
Rotor tip rotational speed	500 m/s
Rotational speed	40,000 rpm
Specific speed	$\Omega_S \equiv 1.131\sqrt{\dfrac{Q_3}{N D_2^3}} = 0.58$

$$\beta_{3av} = 63°, \qquad \frac{D_{3s}}{D_2} = 0.7, \qquad \alpha_2 = 74°$$

$$\frac{D_{3h}}{D_{3s}} = 0.4, \qquad \frac{U_2}{C_0} = 0.7, \quad \text{and} \quad \frac{b}{D_2} = 0.0873$$

Using Figure 15.5, it is required to find

(a) The turbine total-to-static efficiency
(b) Geometrical dimensions of turbine $D_2, b_2, D_3, D_{3h}, D_{3av}$
(c) The outlet total and static temperature and outlet static pressure (T_{03}, T_3, P_3)
(d) Rate of mass flow $(\dot{m})$

Solution

(a) *Total-to-static efficiency*

From Figure 15.5, with specific speed of $\Omega_S = 0.58$ rad and $\alpha_2 = 74°$, then $\eta_{ts} = 0.85$.

(b) *Geometrical dimensions*

- $D_2 = \frac{60 U_2}{\pi N} = \frac{60 \times 500}{\pi \times 40,000} = 0.2387$ m

- $\frac{b_2}{D_2} = 0.0873$

- $b_2 = 0.0873 \times 0.2387$ m

- $\frac{D_{3s}}{D_2} = 0.7$

- $D_{3s} = 0.1671$ m

- $\frac{D_{3h}}{D_{3s}} = 0.4$

- $D_{3h} = 0.06684$ m

- $D_{3av} = \frac{D_{3h} + D_{3s}}{2} = \frac{0.06684 + 0.1671}{2}$

- $D_{3av} = 0.11697$ m

(c) *The outlet conditions*

- From Equation 15.4, the specific work is

$$W = Cp \, (T_{01} - T_{03}) = U_2 \, C_{u2} = U_2^2$$

Then $1148 \times (1400 - T_{03}) = (500)^2$

$$T_{03} = 1182.23 \text{K}$$

- The specific speed $\Omega_S = 1.131 \sqrt{\dfrac{Q_3}{N \, D_2^3}}$

$$0.58 = 1.131 \sqrt{\frac{Q_3}{(40,000/60) \, (0.2387)^3}}$$

Then $Q_3 = 2.3845 \, \text{m}^3/\text{s}$.

Since

$$Q_3 = C_{a3} \left(\pi/4 \right) \left(D_{3S}^2 - D_{3h}^2 \right)$$

Then

$$2.3845 = C_{a3} \left(\pi/4 \right) \left((0.1671)^2 - (0.06684)^2 \right)$$

$$C_{a3} = 129.44 \ \text{m/s} = C_3$$

$$T_3 = T_{03} - \frac{C_3^2}{2 \, Cp} = 1182.23 - \frac{(129.44)^2}{2 \times 1148}$$

$$T_3 = 1174.93 \ \text{K}$$

- The static pressure is correlated to the total-to-static efficiency by the relation

$$\frac{P_3}{P_{01}} = \left[1 - \frac{1}{\eta_{ts}} \left(1 - \frac{T_{03}}{T_{01}} \right) \right]^{(\gamma/\gamma-1)} = \left[1 - \frac{1}{0.85} \left(1 - \frac{1182.23}{1400} \right) \right]^{(\gamma/\gamma-1)} = 0.4455$$

$$P_3 = 1.337 \ \text{bar}$$

(d) *Mass flow rate*

$$\dot{m} = \rho_3 \, Q_3 = \frac{P_3}{R \, T_3} \times Q_3$$

$$\dot{m} = \frac{1.337 \times 10^5}{287 \times 1174.93} \times 2.3845 = 0.9454 \ \text{kg/s}$$

15.6 DESIGN FOR OPTIMUM EFFICIENCY

A general one-dimensional design procedure for the radial inflow turbines was presented in Reference 7. Starting from the kinematical relation for velocities at rotor inlet

$$U_2 - C_{u2} = C_{r2} \tan \beta_2 = C_{u2} \left(\frac{\tan \beta_2}{\tan \alpha_2} \right)$$

The following relation is reached:

$$\frac{U_2 C_{u2}}{C_{r2}^2} - \frac{C_{u2}^2}{C_{r2}^2} - \tan \alpha_2 \tan \beta_2 = 0$$

Solving the above relation gives for positive values of (α_2), the following relation is obtained

$$\tan \alpha_2 = \frac{\sin \alpha_2}{1 - \cos \alpha_2} \tag{15.19a}$$

The above relation yields the following simple numerical relation between the absolute and relative angles at inlet

$$\alpha_2 = 90 - \frac{\beta_2}{2} \tag{15.19b}$$

From Equations 15.19a and b, the following relation can be obtained

$$\frac{C_{u2}}{U_2} = \cos\beta_2 \tag{15.20}$$

The number of blades is obtained from the following relation known as Whitefield equation

$$n = \frac{1}{\cos^2\alpha_2} \tag{15.21}$$

Thus, for a given number of rotor blades, the absolute flow angle at rotor inlet is calculated from Equation 5.21. The relative flow angle at rotor inlet is next determined from Equation 15.19b.

The ratio between the swirl velocity and rotational speed at rotor inlet is determined from Equation 15.20. From Equation 15.4 and for known specific power, both the rotational speed and swirl velocity are determined. The next step is to determine the rotor inlet diameter from the rotational speed and the blade speed (rev/sec). Other dimensions at rotor outlet are determined from known ratios between outlet mean radius and rotor inlet radius and the hub to shroud ratio at outlet. The rotor inlet width may be calculated from the conservation of mass.

Here are some other expressions that define the number of blades. Jamieson [8] developed the following relation:

$$n_{min} = 2\pi \tan\alpha_2 \tag{15.22}$$

Glassman [9] developed the relation

$$n = \frac{\pi}{30}\left(110° - \alpha_2\right)\tan\alpha_2 \tag{15.23}$$

Example 3 An IFR turbine with 12 vanes is required to develop 229.6 KW from a supply of hot gases available at stagnation temperature of 1000 K and a flow rate of 1 kg/s. It has the following data:

$$\frac{C_{a3}}{U_2} = 0.25, \quad \zeta = 0.4, \quad \frac{r_{3s}}{r_2} = 0.7, \quad \text{and} \quad \frac{W_{3m}}{W_2} = 2.0$$

Total-to-static efficiency 0.85
Static pressure at rotor exist is 100 kPa.
Nozzle enthalpy loss coefficient is $\lambda_N = 0.06$.
Use the optimum efficiency design method.

Determine

1. Absolute and relative flow angles at rotor inlet (α_1, β_2)
2. Overall pressure ratio P_{01}/P_3
3. Rotor tip speed (U_2)
4. Ratio of relative velocity W_{3s}/W_2 at shroud
5. Diameter and width of rotor, and its speed of rotation

Solution: Whitefield equation

$$\cos^2 \alpha_2 = \frac{1}{n} = \frac{1}{12} = 0.08333$$

$$\alpha_2 = 73.22°$$

Also

$$\beta_2 = 2\,(90 - \alpha_2) = 2\,(90 - 73.22) = 33.56°$$

$$\Delta T_0 = \frac{W}{Cp} = \frac{229.6}{1.148} = 200 \text{ K}$$

$$\frac{P_{01}}{P_3} = \frac{1}{\left(1 - (\Delta T_0 / \eta_{ts} T_{01})\right)^{(\gamma/\gamma - 1)}} = \left(1 - \frac{0.2}{0.81}\right)^{-4} = 2.924$$

From the specific power relation, with

$$\frac{C_{u2}}{U_2} = \cos \beta_2$$

$$W = U_2 C_{u2} = U_2^2 \cos \beta_2$$

$$U_2 = 524.9 \text{ m/s}$$

$$\frac{r_{3m}}{r_{3s}} = \frac{r_{3s} + r_{3h}}{2} \frac{1}{r_{3s}} = \frac{1 + \zeta}{2} = 0.7$$

$$\frac{r_{3m}}{r_2} = \frac{r_{3m}}{r_{3s}} \frac{r_{3s}}{r_2} = 0.7 \times 0.7 = 0.49$$

$$\cos \beta_{3m} = \frac{C_{a3}}{U_2} \frac{r_2}{r_{3m}} = \frac{0.25}{0.49} = 0.5102$$

$$\beta_{3m} = 59.32°$$

$$\cos \beta_{3s} = \frac{C_{a3}}{U_2} \frac{r_2}{r_{3s}} = \frac{0.25}{0.7} = 0.3571$$

$$\beta_{3s} = 69.077°$$

$$\frac{W_{3s}}{W_2} = \frac{W_{3s}}{W_{3m}} \frac{W_{3m}}{W_2} = \frac{\sec \beta_{3s}}{\sec \beta_{3m}} \times 2.0 = 2.702$$

$$T_{03} = T_{01} - \Delta T_0 = 1000 - 200 = 800 \text{ K}$$

$$T_3 = T_{03} - \frac{C_3^2}{2 \times Cp} = T_{03} - \left(\frac{C_{a3}}{U_2}\right)^2 \frac{U_2^2}{2 \times Cp} = 800 - (0.25)^2 \frac{(524.9)^2}{2 \times 1148} = 792.5 \text{ K}$$

$$\dot{m} = \rho_3 C_{a3} A_3 = \left(\frac{P_3}{RT_3}\right) \left(\frac{C_{a3}}{U_2}\right) U_2 \pi \left(\frac{r_{3s}}{r_2}\right)^2 \left(1 - \zeta^2\right) r_2^2$$

$$1.0 = \left(\frac{10^5}{287 \times 792.5}\right)(0.25) \times 524.9 \times \pi \times 0.7^2 \times \left(1 - 0.4^2\right) r_2^2$$

$$r_2^2 = 0.0134$$

$$r_2 = 0.1158 \text{ m}$$

$$D_2 = 0.2316 \text{ m}$$

$$\omega = \frac{U_2}{r_2} = 4532.8 \text{ rad/s} \qquad (N = 43285.2 \text{ rpm})$$

Also since

$$\dot{m} = \rho_2 C_{r2} A_2 = \rho_2 C_{r_2} \pi D_2 b_2$$

$$\frac{b_2}{D_2} = \frac{\dot{m}}{4\pi \times \rho_2 \times C_{r2} \times r_2^2}$$

To obtain the density ρ_2, the following steps have to be followed:

$$C_{u2} = U_2 \cos \beta_2 = 437.4 \text{ m/s}$$

$$C_{r2} = \frac{C_{u2}}{\tan \alpha_2} = \frac{437.4}{3.3163} = 131.89 \text{ m/s}$$

$$C_2 = \frac{C_{u2}}{\sin \alpha_2} = \frac{437.4}{0.9574} = 456.8 \text{ m/s}$$

$$T_2 = T_{02} - \frac{C_2^2}{2Cp} = 1000 - \frac{(456.8)^2}{2 \times 1148} = 909.1 \text{ K}$$

$$T_2' = T_2 - \frac{\lambda_N C_2^2}{2Cp} = 909.1 - \frac{0.06 \times (456.8)^2}{2 \times 1148} = 903.64 \text{ K}$$

$$P_2 = P_{01} \left(\frac{T_2'}{T_{01}}\right)^{(\gamma/\gamma-1)} = 2.924 \times 10^5 \times \left(\frac{903.64}{1000}\right)^4 = 1.9497 \times 10^5 \text{ Pa}$$

$$\frac{b_2}{D_2} = \frac{\dot{m}}{4\pi \times (P_2/RT_2) \times C_{r2} \times r_2^2} = \frac{1.0}{4\pi \times ((1.9496 \times 10^5)/(287 \times 954.5)) \times 131.89 \times 0.0134}$$

$$= 0.06322$$

$$b_2 = 0.0146 \text{ m}$$

The geometry of radial turbine is shown in Figure 15.6.

Example 4   Compare between the specific power of axial and radial turbines in the following case:
Axial flow turbine having the following angles $\alpha_2 = \beta_3 = 60$ and $\alpha_3 = \beta_2 = 0$.
Radial inflow turbine with $\alpha_2 = 60$, $\alpha_3 = \beta_2 = 0$, and $\beta_3 = 0$.
The rotational speed U_2 is equal in both turbines.

Solution: The velocity triangles for both axial and radial turbines are shown in Figure 15.7.

(a) *Axial flow turbine:* Since $\alpha_2 = \beta_3 = 60$ and $\alpha_3 = \beta_2 = 0$, and the specific work is

$$W_{\text{axial}} = U_2 C_{u2} + U_3 C_{u3}$$

$$W_{\text{axial}} = U_2^2$$

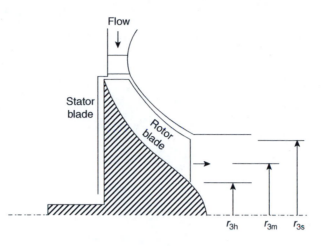

FIGURE 15.6   Geometry of radial turbine.

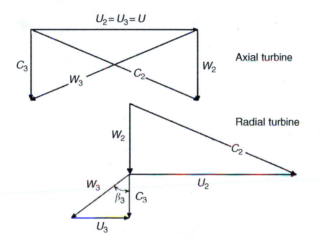

FIGURE 15.7   Velocity triangles at inlet and outlet for both axial and radial turbines.

(b)   *Radial flow turbine:* Since $\alpha_2 = 60°$, $\alpha_3 = \beta_2 = 0$, and $\beta_3 = 45°$ and the specific work is

$$W_{\text{radial}} = U_2 C_{u2} - U_3 C_{u3}$$

$$W_{\text{radial}} = U_2 U_2 - U_3 \times 0$$

$$W_{\text{radial}} = U_2^2$$

$$\frac{W_{\text{axial}}}{W_{\text{radial}}} = \frac{U_2^2}{U_2^2} = 1$$

Thus, in this special case the specific powers of both turbines are equal.

15.7   COOLING

Increasing the turbine entry temperature has been the preoccupation of gas turbine designers for many years. As described in Reference 10, cooling of the rotor nozzle and rotor is performed using

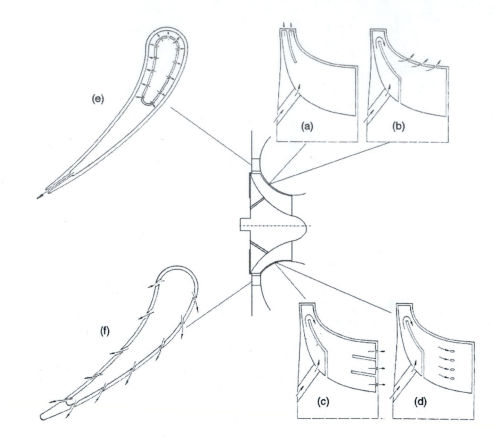

FIGURE 15.8 Cooling techniques for both the rotor and stator of a radial inflow turbine.

different techniques. Convection, impingement, and film cooling are used for the rotor. Figure 15.8
illustrates different cooling methods for both of the rotor and stator. In Figure 15.8A convection
cooling with leading edge ejection is employed, which provides good cooling in the hottest part of
the blade. Figure 15.8B illustrates impingement cooling of the leading edge together with tip surface
ejection. Figure 15.8C illustrates trailing edge ejection together with leading edge impingement.
Figure 15.8D illustrates film cooling through different holes in the blade surface together with
leading edge impingement also.

Nozzle cooling may be either impingement plus trailing edge ejection (Figure 15.8E) or surface
film ejection (Figure 15.8F).

PROBLEMS

15.1 A radial inflow turbine for small application like APU in aircraft and has the following data:

Pressure ratio:	$\dfrac{P_{01}}{P_3}$	2.1
Inlet temperature:	T_{01}	1100 K
Output work		43 kW
Rotational speed		800 rps

Gas mass flow rate $\dot{m}$		0.35 kg/s
Rotor inlet tip diameter		12.0 cm
Rotor exit tip diameter		7.5 cm
Hub–tip ratio at exit		0.3
Nozzle efflux angle	α_2	70°
Rotor vane outlet angle	β_3	40°
Nozzle loss coefficient	λ_N	0.07

Calculate

    (a) The isentropic efficiency η_{ts}.

    (b) The rotor loss coefficient λ_R.

15.2 (a)   Prove that the specific work for any turbomachinery is expressed by Euler's equation:

$$W = \frac{1}{2}\left[\left(U_2^2 - U_3^2\right) - \left(W_2^2 - W_3^2\right) + \left(C_2^2 - C_3^2\right)\right]$$

    (b)   Compare between axial and radial inflow turbines.

    (c)   Plot on T–S diagram the expansion process in a radial inflow turbine and an axial flow turbine.

    (d)   Describe the different methods of cooling in a radial inflow turbine.

15.3 A radial inflow turbine has the following characteristics and operating conditions:

Inlet total temperature	1,000 K
Inlet total pressure	2.0 bar
Rotor tip rotational speed	500 m/s
Rotational speed	140,000 rpm

$$\beta_{3av} = 63°, \quad \frac{D_{3s}}{D_2} = 0.7, \quad \alpha_2 = 62°, \quad \text{and} \quad \frac{b_2}{D_2} = 0.0873$$

The turbine operates at the maximum total-to-static efficiency.
Using Figure 15.5, find

    (a)   The specific speed

    (b)   Geometrical dimensions of turbine $D_2, b_2, D_{3s}, D_{3h}, D_{3av}$

    (c)   The outlet total and static temperature and outlet static pressure $(T_{03}, T_3, P)_3$

    (d)   Rate of mass flow $(\dot{m})$

    (e)   Rotor inlet static temperature and pressure (T_2, P_2)

15.4 A small inward flow radial gas turbine comprising a radial vaned impeller, a ring of nozzle blades, and an axial diffuser operates at its design point with an overall total-to-total efficiency of 0.91. At turbine entry the stagnation pressure and temperature of the gas are 410 kPa and 1100 K, respectively. The flow leaving the turbine is diffused to a pressure of 101 kPa and has negligible final velocity. Given that the flow is just choked at nozzle exit, determine the impeller peripheral speed and the flow outlet angle from the nozzle. The working fluid is hot gases having $\gamma = 1.333$ and $R = 287$ J/kg°C.

15.5 A radial turbine is proposed as the gas expansion element in a chemical plant. The pressure and temperature conditions through the stage at the design point are to be as follows:

Upstream of the nozzle, $P_{01} = 700$ kPa, $T_{01} = 1150$ K
Nozzle exit, $P_2 = 530$ kPa, $T_2 = 1030$ K
Rotor exit, $PP_3 = 385$ kPa, $T_3 = 915$ K, $\quad T_{03} = 925$ K

The ratio of rotor exit mean diameter to rotor inlet tip diameter is chosen as 0.45 and the rotational speed is 25,000 rpm. Assume the relative flow at rotor inlet is radial and the absolute flow at rotor exit is axial. The gases have a molecular weight of 40 and a ratio of specific heats of 1.6. Determine the following:

(a)   The total-to-static efficiency of the turbine
(b)   The rotor diameter
(c)   The losses factors λ_N, λ_R
(d)   The angles α_2, β_2

15.6   A small inward flow radial gas turbine fitted with an axial diffuser operates at its design point with a total-to-total efficiency of 0.9. At turbine entry the stagnation pressure and temperature of the gas are 410 kPa and 1100 K, respectively. The flow leaving the turbine is diffused to a pressure of 101 kPa and has negligible final velocity. Given that the flow is just choked at nozzle exit, determine the impeller peripheral speed and the flow outlet angle from the nozzle. The working fluid is hot gases having $\gamma = 1.333$ and $R = 287$ J/kg°C.

15.7   Following the design procedure described in Section 15.4 it is required to design a radial inflow turbine having the following specifications:

Mass flow rate	0.145 kg/s
Speed	130,000 rpm
Power	20.0 kW
Inlet total temperature	1080 K
Inlet total pressure	2.1 bar
Stage loading ψ	0.9
Flow coefficient ϕ	0.3
Meridional velocity ratio C_{r2}/C_{a3}	1.0
Total-to-total efficiency	0.8
r_{3h}/r_2	0.3

Calculate

(a)   The velocities $U_2, C_2, W_2, U_3, C_3, W_3$
(b)   The dimensions at inlet r_2, b_2
(c)   The dimensions at outlet r_{3h}, r_{3t}

REFERENCES

1. S. L. Dixon, *Fluid Mechanics and Thermodynamics of Turbomachinery*, Oxford, Butterworth-Heinemann,
    4th edn., Oxford, 1998, p. 236.
2. Hany Moustapha, Mark F. Zelessky, Nicholas C. Baines, and David Japikse, *Axial and Radial Turbines*, Concepts NREC, 2003, p. 209.
3. H. I. H. Saravanamuttoo, G. F. C. Rogers, and H. Cohen, *Gas Turbine Theory*, Prentice Hall, 5th edn., Harlow, Essex, England, 2001, p. 367.
4. H. Chen and N. C. Baines, Analytical optimization design of radial and mixed flow turbines, *Proc. Institution of Mechanical Engineers*, 206, 177–187.
5. H. E. Rohlik, *Analytical Determination of Radial Inflow Turbine Design Geometry for Maximum Efficiency*, NASA TN D-4384, NASA, Washington DC, 1968.
6. A. J. Glassman, *Turbine Design and Application*, NASA SP-290, NASA, Washington DC, 1975, p. 279.
7. A. Whitefield, The preliminary design of radial inflow turbines, *J. Turbomachinery, Trans. Am. Soc. Mech. Engr.*, 112, 50–57.
8. A. W. Jamieson, The radial turbine, In *Gas Turbine Principles and Practice*, Sir H. Roxbee-Cox (Ed.), Newnes, 1955 (Chapter 9), pp. 236–271.
9. A. J. Glassman, *Computer Program for Design and Analysis of Radial Inflow Turbines*, NASA TN 8164, NASA, Washington DC, 1976.
10. A. Whitefield and N. C. Baines, *Design of Radial Turbomachines*, Longmans Scientific and Technical, Essex, England, 1990, p. 357.

16 Module Matching

16.1 INTRODUCTION

Module matching means the interplay of the engine geometry and engine parameters. It is not only a matching between a turbine and the driven compressor, but is also the global synchronization of the different modules including the diffuser, combustion chamber, nozzle and fan (for turbofan engines), afterburner (for afterburning engines), propeller (for turboprop engines), and load (for gas turbines and turbo shaft engines). Turbine is coupled either to a compressor or to a fan, thus forming an engine spool as described in Chapters 4 through 8. However, the performance matching between these two modules is influenced by their front and back modules. Choked nozzle influences the performance of its preceding elements. The diffuser, on the contrary, controls the mass flow and total pressure/temperatures of all the succeeding modules. If the diffuser incorporates a water injection system, the mass flow in the succeeding elements is influenced. Fuel addition and pressure drop in combustion chamber influences the performance of the following turbine and nozzle. Afterburner influences the mass flow rate and inlet pressure to the succeeding nozzle.

As described in the preceding four chapters, each compressor/turbine has a certain design point where the rotational speed (N), pressure ratio (π_c), and mass flow ($\dot{m}$) are specified. However, it is requested for these modules to operate efficiently over the complete range of speed and power output for aero engine or power plant. These conditions are identified as the *off-design conditions*. These include starting, idling, reduced flow, maximum power, part load, acceleration, and declaration. For aero engines, performance is more complex as the inlet conditions are further changed due to flying at different flight speeds and altitudes.

No matter what type of engine is being considered, the steady state performance of an engine at each speed (N) is determined by two conditions: conservation of mass (or continuity of flow) and conservation of energy (or power balance), which lead to the "match point." However, for aero engines a third condition is added, namely, conservation of momentum, which develops the necessary, thrust for propelling aircraft at different flight conditions.

16.2 OFF-DESIGN OPERATION OF A SINGLE-SHAFT GAS TURBINE DRIVING A LOAD

A single spool composed of a turbine driving both the compressor and load is shown in Figure 16.1. The load could be a generator, marine screw, and propeller of a turboprop engine. It is assumed that the losses in the inlet and exhaust modules are ignored. Thus the inlet total pressure and temperature of the compressor (T_{01}, P_{01}) are equal to the ambient conditions (T_a, P_a). Moreover, the gases expand in the turbine to the ambient pressure.

The cycle of this layout is illustrated in Figure 16.2. From a conservation of mass through the spool, the mass flow rate through the turbine ($\dot{m}_3$) is equal to the mass flow rate through the compressor ($\dot{m}_1$) less any bleeds ($\dot{m}_b$), but supplemented by the fuel flow ($\dot{m}_f$) and any water injected if water injection ($\dot{m}_{w.i}$) is added. The mass of fuel added is approximately equal to the bled air. The fuel to air ratio (f) and the bleed ratio (b) may be given instead.

The compressor and turbine maps are known together with other data such as the ambient conditions (T_a, P_a), fuel flow rate (f) and bleed ratio (b), pressure drop in the combustion chamber (ΔP_{cc}), as well as load power. The turbine pressure is then related to the compressor outlet pressure

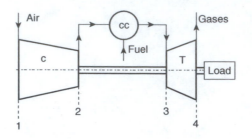

FIGURE 16.1   Single-shaft gas turbine.

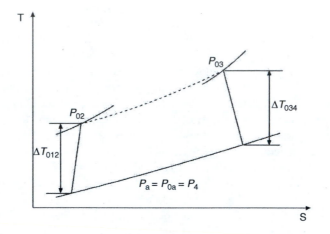

FIGURE 16.2   T–S diagram for single shaft gas turbine.

by the relation

$$P_{03} = P_{02}(1 - \Delta P_{cc}) \tag{16.1}$$

16.2.1 MATCHING PROCEDURE

The following steps summarize the matching procedure:

1. Select a rotational speed (N): From the known ambient conditions determine ($N/\sqrt{T_{01}}$), which defines the operating line on the compressor map.
2. Assume turbine inlet temperature (T_{03}): Calculate ($N/\sqrt{T_{03}}$), which determines the turbine operating line.
3. Assume the compressor pressure ratio (P_{02}/P_{01}): From the compressor map, the operating point is completely defined (i.e., the mass flow parameter $\dot{m}_1\sqrt{T_{01}}/P_{01}$ and compressor efficiency η_c); see Figure 16.3.
4. The compressor mass flow rate is calculated from the relation

$$\dot{m}_1 = \frac{\dot{m}_1\sqrt{T_{01}}}{P_{01}}\frac{P_{01}}{\sqrt{T_{01}}} \tag{16.2}$$

and the turbine mass flow rate is then obtained from the relation

$$\dot{m}_3 = \dot{m}_1(1 + f - b) \tag{16.3}$$

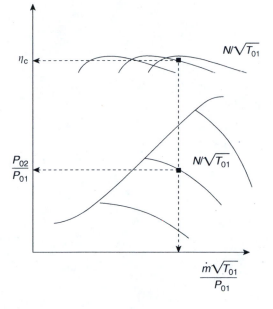

FIGURE 16.3   Operating point on compressor map.

5. Calculate the compressor specific work from the relation

$$w_c = \frac{Cp_c T_{01}}{\eta_c} \left[\left(\frac{P_{02}}{P_{01}} \right)^{(\gamma_c - 1)/\gamma_c} - 1 \right] \tag{16.4}$$

As seen above the variations in specific heat and specific heat ratio are taken into account in the different gas turbine modules.

6. The turbine mass flow parameter is determined as follows:

$$\frac{\dot{m}_3 \sqrt{T_{03}}}{P_{03}} = \frac{\dot{m}_1 \sqrt{T_{01}}}{P_{01}} \frac{P_{01}}{P_{02}} \frac{P_{02}}{P_{03}} \frac{\dot{m}_3}{\dot{m}_1} \sqrt{\frac{T_{03}}{T_{01}}} \tag{16.5}$$

From this mass parameter and the operating speed line, the turbine efficiency can be determined (Figure 16.4). It is worthy to say again here that the turbine pressure ratio P_{03}/P_{04} is equal to the compressor pressure ratio (P_{02}/P_{01}).

7. The turbine work can be determined then from the relation

$$w_t = \eta_t Cp_t T_{03} \left[1 - \frac{1}{(P_{03}/P_{04})^{(\gamma_t - 1)/\gamma_t}} \right] \tag{16.6}$$

8. The net power (P) is then evaluated from the relation

$$P = \dot{m}_3 w_t - \frac{\dot{m}_1 w_c}{\eta_m} \tag{16.7}$$

where η_m is the mechanical efficiency for the spool. This net power is then checked with the load power, if they are unequal, then a new value for the compressor pressure ratio is selected, and the above procedure is repeated until convergence.

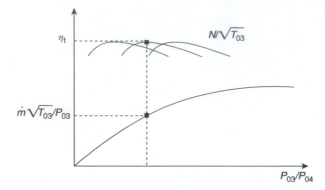

FIGURE 16.4 Operating point of turbine map.

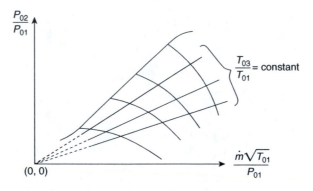

FIGURE 16.5 Constant turbine inlet temperature on the compressor map.

Two points are discussed here. The first is the case of choked nozzle. If the nozzle is choked, then the left-hand side of Equation 16.5 will be constant, or

$$\frac{\dot{m}_3\sqrt{T_{03}}}{P_{03}} = \frac{\dot{m}_1\sqrt{T_{01}}}{P_{01}}\frac{P_{01}}{P_{02}}\frac{P_{02}}{P_{03}}\frac{\dot{m}_3}{\dot{m}_1}\sqrt{\frac{T_{03}}{T_{01}}} = \text{const.}$$

For constant pressure ratio within the combustion chamber ($P_{03}/P_{02} = $ const.) and constant mass flow rate in both of the compressor and turbine ($\dot{m}_3 = \dot{m}_1$), then

$$\frac{P_{02}}{P_{01}} = K_1\frac{\dot{m}_1\sqrt{T_{01}}}{P_{01}}\sqrt{\frac{T_{03}}{T_{01}}}$$

where k_1, is constant.

If $(T_{03}/T_{01}) = $ const., then the compressor pressure ratio will be expressed as

$$\frac{P_{02}}{P_{01}} = K\frac{\dot{m}_1\sqrt{T_{01}}}{P_{01}} \tag{16.8}$$

where k is also a constant.

This linear relation when plotted on the compressor map will have the shape illustrated in Figure 16.5. However, this relation shows that the pressure ratio is zero when the mass flow rate is zero, which is impossible. Since Equation 16.8 assumes that the turbine is choked, which never occurs at low mass flow rate, it is then logical to assume that these lines are used for mass flow rates away from origin [1–3].

Example 1 The following data refer to a single shaft of a turboprop engine during ground run. The gas turbine is running at its design point.

Compressor characteristics			Turbine characteristics		
P_{02}/P_{01}	$\dot{m}_1\sqrt{T_{01}}/P_{01}$	η_C	$\dot{m}_3\sqrt{T_{03}}/P_{03}$	P_{03}/P_{04}	η_t
4.0	350	0.76	130	4.2	0.88
4.5	330	0.8	144	4.6	0.9
5.0	310	0.84	144	5.0	0.91

The compressor delivery pressure and the ambient conditions are 1.01 bar and 288 K. The nondimensional flows are based on $\dot{m}$ in kg/s, P in bar, and T in K. Neglect mechanical losses. The pressure losses in the combustion chamber are 3%. Assume also that the mass flow rate in both compressor and turbine are equal. Calculate the power consumed in the propeller when the turbine inlet temperature is 1200 K.

Solution: Flow compatibility is expressed by the relation (16.5):

$$\frac{\dot{m}_3\sqrt{T_{03}}}{P_{03}} = \frac{\dot{m}_1\sqrt{T_{01}}}{P_{01}}\frac{P_{01}}{P_{02}}\frac{P_{02}}{P_{03}}\frac{\dot{m}_3}{\dot{m}_1}\sqrt{\frac{T_{03}}{T_{01}}}$$

For equal mass flow rate through the compressor and turbine, 3% pressure loss in the combustion chamber, and equality of compressor inlet pressure and turbine outlet pressure ($P_{01} = P_{04}$), then

$$\frac{\dot{m}\sqrt{T_{03}}}{P_{03}} = \frac{\dot{m}\sqrt{T_{01}}}{P_{01}}\frac{P_{01}}{P_{03}}\sqrt{\frac{1200}{288}} = 2.041\frac{\dot{m}\sqrt{T_{01}}}{P_{01}}\frac{P_{04}}{P_{03}} \tag{1}$$

The mass flow parameter is calculated from the Equation 1 for different turbine pressure ratios and tabulated along with the values given in the data above in the following table:

P_{03}/P_{04}	$\dot{m}_3\sqrt{T_{03}}/P_{03}$ From turbine map	$\dot{m}_3\sqrt{T_{03}}/P_{03}$ From Equation 1
4.2	130	170.1
4.6	144	146.42
5.0	144	126.54

The two sets of turbine mass flow parameter are plotted in Figure 16.6.

From Figure 16.6, equilibrium between compressor and turbine occurs at $P_{03}/P_{04} = 4.64$, $\dot{m}_3\sqrt{T_{03}}/P_{03} = 144$. Plotting the turbine efficiency on the same figure gives the corresponding turbine efficiency as $\eta_t = 0.9012$.

The corresponding compressor pressure ratio is

$$P_{02}/P_{01} = \frac{P_{03}/P_{04}}{0.97} = \frac{4.64}{0.97} = 4.7835$$

The corresponding values of the compressor mass flow rate and efficiency are determined from Figure 16.7, which are $\dot{m}_1\sqrt{T_{01}}/P_{01} = 319$ and $\eta_C = 0.82$, respectively.

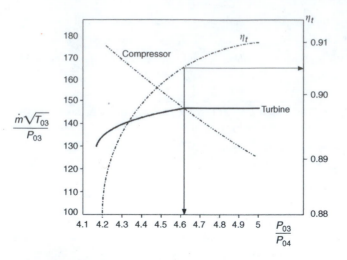

FIGURE 16.6   Equilibrium point for compressor and turbine.

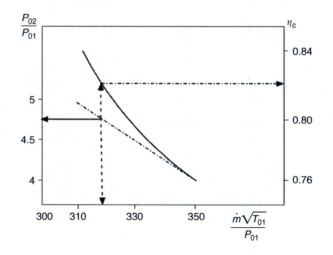

FIGURE 16.7   Matching point on compressor map.

From Equation 16.2, the mass flow rate is

$$\dot{m}_1 = \frac{\dot{m}_1\sqrt{T_{01}}}{P_{01}}\frac{P_{01}}{\sqrt{T_{01}}} = \frac{319 \times 1.01}{\sqrt{288}} = 18.98 \text{ kg/s}$$

The available power is evaluated from Equation 16.7.

$$P = \dot{m}_3 w_t - \dot{m}_1 w_c = w_t = \dot{m}\left\{\eta_t Cp_t T_{03}\left[1 - \frac{1}{(P_{03}/P_{04})^{(\gamma_t-1)/\gamma_t}}\right] - \frac{Cp_c T_{01}}{\eta_c}\left[\left(\frac{P_{02}}{P_{01}}\right)^{(\gamma_c)/\gamma_c} - 1\right]\right\}$$

$$= 19.98\left\{0.9012 \times 1.148 \times 1200\left[1 - \left(\frac{1}{4.64}\right)^{0.25}\right] - \frac{1.005}{0.82}\left[(4.7835)^{0.286} - 1\right]\right\}$$

$$= 3922.3 \text{ kW}$$

16.2.2 Different Loads

The load could be either a propeller or an electric generator. The propeller could be of the constant or variable speed type. For a fixed pitch propeller, the propeller power is proportional to the cube of its rotational speed. This speed is equal to the turbine speed times the gear reduction ratio. Thus if this ratio and the mechanical efficiency of the gearbox is known, then the propeller power can be plotted versus the turbine speed. Thus for each speed line on the compressor map, only a single point will give the required power [2], which is determined by trial and error after taking several points on the compressor map and finding the available power. Repeating this procedure for different speed lines a series of points are obtained, which when joined together form the equilibrium running line. Variable pitch propellers are commonly used in aircraft, which run at constant rotational speed. The load in this case varies by changing the pitch. Thus, the equilibrium lines for these loads would correspond to a particular line of constant nondimensional speed. Each point on this line corresponds to a different value of fuel flow, turbine inlet temperature, and power outpour.

The case of electrical generator is similar to the variable pitch propeller.

16.3 OFF DESIGN OF FREE TURBINE ENGINE

Figure 16.8 illustrates the gas turbine incorporating a free (power) turbine. The corresponding temperature–entropy diagram is shown in Figure 16.9.

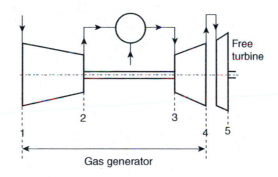

FIGURE 16.8   Gas turbine with a free turbine.

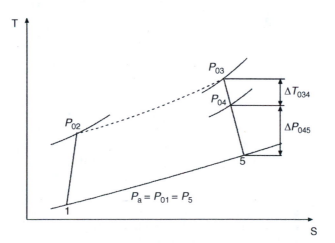

FIGURE 16.9   Temperature–entropy diagram for a gas turbine with a free turbine.

The gas generator is in equilibrium, or in other words, the power of the gas generator turbine is exactly equal to the compressor power. The remaining power in the gases leaving the first (generator) turbine is absorbed by the free turbine and used to drive the load. The analysis to be followed here is divided into two steps:

1. Matching between the compressor and turbine of the gas generator
2. Off-design analysis of the free turbine

16.3.1 GAS GENERATOR

The procedure is an alternative to that described in Section 16.2.

Procedure
The following parameters are known in advance:
Inlet conditions (T_a, P_a), fuel to air ratio (f) and bleed ratio (b), pressure drop in the combustion chamber (ΔP_{cc})

1. Select a constant speed line on compressor map and choose any point on this line; thus, the parameters $N/\sqrt{T_{01}}$, P_{02}/P_{01} and $\dot{m}\sqrt{T_{01}}/P_{01}$ are determined.
2. The compressor-specific work is determined from Equation 16.4.
3. Guess a value of the turbine pressure ratio (P_{03}/P_{04}), thus, the value of the turbine mass flow parameter is determined. It is assumed here that the turbine nondimensional flow is independent of the nondimensional speed and that it is a function in the turbine pressure ratio. The turbine mass flow rate is determined from Equation 16.3.
4. Calculate the turbine temperature ratio (T_{03}/T_{01}) from Equation 16.5 as follows:

$$\left(\frac{T_{03}}{T_{01}}\right)_A = \left(\frac{\left(\dot{m}_3\sqrt{T_{03}}/P_{03}\right)}{\left(\dot{m}_1\sqrt{T_{01}}/P_{01}\right)}\frac{P_{02}}{P_{01}}\frac{P_{03}}{P_{02}}\frac{\dot{m}_1}{\dot{m}_3}\right)^2 \tag{16.9}$$

Thus, the turbine inlet temperature is defined from which $N/\sqrt{T_{03}}$ and turbine efficiency η_t are determined.

5. Calculate another value for the turbine inlet temperature ratio from the energy balance for of the gas generator ($W_t = W_c$). The following relation is obtained:

$$\left(\frac{T_{03}}{T_{01}}\right)_B = \frac{\dot{m}_1}{\dot{m}_3}\frac{Cp_c}{\eta_c\eta_tCp_t}\frac{\left[(P_{02}/P_{01})^{(\gamma_c-1)/\gamma_c}-1\right]}{\left[1-\frac{1}{(P_{03}/P_{04})^{(\gamma_t-1)/\gamma_t}}\right]} \tag{16.10}$$

In general the values of $(T_{03}/T_{01})_A$ and $(T_{03}/T_{01})_B$ obtained from Equations 16.9 and 16.10 are not equal, thus the guessed value of turbine pressure ratio (P_{03}/P_{04}) is not valid for an operating point on the equilibrium running line. Thus, another value for the turbine pressure ratio is assumed and the above procedure is repeated until convergence is satisfied.

Remarks
In the matching procedure, it was assumed that the turbine nondimensional mass flow parameter is independent of the speed parameter. Thus, only one line exists for the unchoked part of the mass flow parameter. If the mass flow parameter is dependent on the nondimensional speed parameter, another matching procedure is to be followed.

Steps 3 and 4 above are to be modified as follows:

(3) Assume a turbine temperature ratio (T_{03}/T_{01}); thus the turbine speed parameter $N/\sqrt{T_{03}}$ and the turbine mass flow parameter $\dot{m}_3\sqrt{T_{03}}/P_{03}$ are obtained from Equation 16.5. From both parameters, the turbine pressure ratio and efficiency η_t are determined.

(4) From work compatibility, the value of turbine temperature ratio (T_{03}/T_{01}) is obtained from Equation 16.10. This value is compared with the assumed value. If unequal, another turbine temperature ratio is assumed and the above procedure is repeated until convergence.

The steps described above are arranged in a flow chart given in Figure 16.10.

16.3.2 FREE POWER TURBINE

The parameters of the power turbine are the mass flow parameter $\dot{m}\sqrt{T_{04}}/P_{04}$, pressure ratio P_{04}/P_a, speed parameter $N/\sqrt{T_{04}}$, and efficiency η_{ft}. The free turbine map is illustrated in Figure 16.11.

Here are the additional compatibility conditions after proper matching of the gas generator modules. The outlet conditions of the gas generator turbine (T_{04}, P_{04}) are known, from which the mass flow parameter and speed parameter can be determined.

The mass flow parameter is

$$\frac{\dot{m}\sqrt{T_{04}}}{P_{04}} = \frac{\dot{m}\sqrt{T_{03}}}{P_{03}} \times \frac{P_{03}}{P_{04}} \times \sqrt{\frac{T_{04}}{T_{03}}} \tag{16.11}$$

The pressure ratio across the free power turbine is determined from the relation

$$\frac{P_{04}}{P_a} = \frac{P_{04}}{P_{03}}\frac{P_{03}}{P_{02}}\frac{P_{02}}{P_{01}}\frac{P_{01}}{P_a} \tag{16.12}$$

It is noted that the outlet pressure of the free turbine is assumed equal to the compressor inlet pressure by neglecting the losses in the inlet diffuser and outlet duct of industrial gas turbines; or

$$P_a = P_5 = P_{01}$$

Matching procedure for a free power turbine is straightforward as given below:

1. Select a constant speed line on the compressor map.
2. Select a point on this speed line on the compressor map.
3. Follow the same procedure described above for matching between this point and a corresponding point on the gas generator turbine.
4. Calculate the pressure ratio of the free power turbine from Equation 16.12.
5. Determine the mass flow parameter from the free turbine map.
6. Compare this value with the value calculated from Equation 16.11.
7. If agreement is not reached, the above procedure is repeated by selecting another point on the same constant speed line of compressor and repeat the procedure until convergence is achieved.
8. Repeat the above procedure for other constant speed lines.
9. The successful points on the compressor map is joined together to provide an operating line for such application as shown in Figure 16.12.

As concluded in Reference 2, the equilibrium running line for the free power turbine is independent of the load and is determined by the swallowing capacity of the free power turbine.

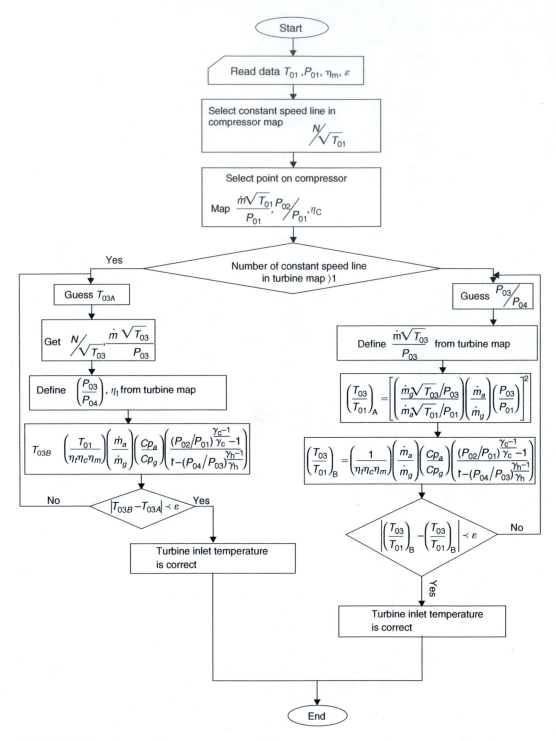

FIGURE 16.10 Flow chart for gas generator matching.

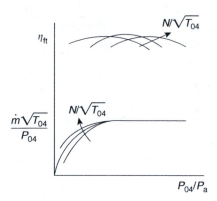

FIGURE 16.11 Map of free turbine.

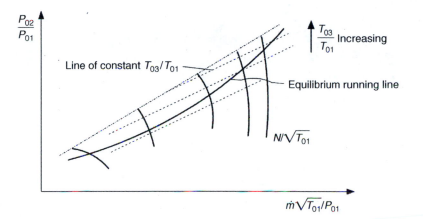

FIGURE 16.12 Operating line for a free power turbine.

The power generated from the free power turbine is

$$\text{Power output} = \dot{m}_4 C p_4 \Delta T_{045} \tag{16.13a}$$

$$\text{Power output} = \dot{m}_4 C p_4 \eta_{ft} T_{04} \left[1 - \left(\frac{1}{P_{04}/P_a} \right)^{(\gamma_t - 1)/\gamma_t} \right] \tag{16.13b}$$

The mass flow rate can be obtained from the known output conditions of the gas generator turbine

$$\dot{m}_4 = \frac{\dot{m}\sqrt{T_{04}}}{P_{04}} \frac{P_{04}}{\sqrt{T_{04}}}$$

The efficiency can be also obtained from the known speed and mass flow parameters.

The gas generator turbine and free power turbines represent two turbines in series. An important rule for them is that if the free turbine is choked, the gas generator turbine will be constrained to operate at a fixed nondimensional point; thus $(\Delta T_{034}/T_{03})$ has a constant value equal to the design value.

Example 2   The following data refer to a gas turbine with a free power turbine, operating at design speed.

	Compressor characteristics			**Gas generator turbine characteristics**	
η_C	$\dot{m}\sqrt{T_{01}}/P_{01}$	P_{02}/P_{01}	η_t	$\dot{m}\sqrt{T_{03}}/P_{03}$	P_{03}/P_{04}
6.0	250	0.83	2.8	100	0.85
5.6	270	0.84	2.5	100	0.85
5.2	290	0.83	2.2	95	0.85

Assuming that the power turbine is chocked, the value of $\dot{m}\sqrt{T_{04}}/P_{04}$ being 220, determine the design values of compressor pressure ratio and turbine inlet temperature. Assume pressure losses in combustion chamber of 3% and assume the mechanical efficiency of the gas generator rotor to be 0.98 and take the ambient temperature as 288 K. The nondimensional flows quoted are based on $\dot{m}$ in kg/s, and P in bar and T in K, all pressures and temperatures being stagnation values.

Solution
For chocked free turbine $\dot{m}\sqrt{T_{04}}/P_{04} = $ constant, or $\dot{m}\sqrt{T_{04}}/P_{04} = 220$. Compatibility of mass flow rate between two turbines

$$\frac{\dot{m}\sqrt{T_{04}}}{P_{04}} = \frac{\dot{m}\sqrt{T_{03}}}{P_{03}} \times \sqrt{\frac{T_{04}}{T_{03}}} \times \frac{P_{03}}{P_{04}}$$

where

$$\frac{T_{04}}{T_{03}} = 1 - \eta_t\left(1 - \frac{1}{\pi_t^{G_t}}\right)$$

and π_t is the pressure ratio of gas generator turbine and $G_t = (\gamma_t - 1)/\gamma_t$.
Define the calculated mass flow parameter of the free power turbine as $\left(\dot{m}\sqrt{T_{04}}/P_{04}\right)_A$, which is calculated from Equation 1.

$$\left(\frac{\dot{m}\sqrt{T_{04}}}{P_{04}}\right)_A = \frac{\dot{m}\sqrt{T_{03}}}{P_{03}} \times \left[1 - \eta_t\left(1 - \frac{1}{\pi_t^{G_t}}\right)\right]^{0.5} \times \pi_t \qquad (1)$$

P_{03}/P_{04}	η_t	$\dot{m}\sqrt{T_{03}}/P_{03}$	$\left(1 - \frac{1}{\pi_t^{G_t}}\right)$	$\left[1 - \eta_t\left(1 - \frac{1}{\pi_t^{G_t}}\right)\right]^{0.5} \times \pi_t$	$\left(\dot{m}\sqrt{T_{04}}/P_{04}\right)_A$
2.8	0.85	100	0.2269	2.5155	251.54
2.5	0.85	100	0.2047	2.2721	227.2
2.2	0.84	95	0.1789	2.0279	192.7

The calculated mass flow parameter of the free turbine $\left(\dot{m}\sqrt{T_{04}}/P_{04}\right)_A$ is plotted together with the given choked value $\dot{m}\sqrt{T_{04}}/P_{04} = 220$ as shown in Figure 16.13.
The intersection point gives the matching point data; namely,

$$\pi_t = \frac{P_{03}}{P_{04}} = 2.438 \quad \text{and} \quad \frac{\dot{m}\sqrt{T_{03}}}{P_{03}} = 98.9667$$

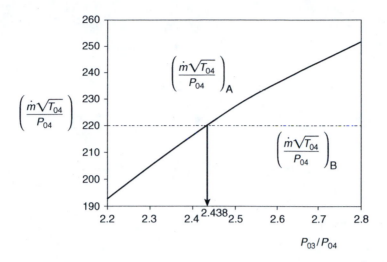

FIGURE 16.13 Matching point.

Next, matching between the compressor and gas generator turbine may be done as follows:
Continuity equation

$$\frac{\dot{m}\sqrt{T_{01}}}{P_{01}} = \frac{\dot{m}\sqrt{T_{03}}}{P_{03}}\sqrt{\frac{T_{01}}{T_{03}}}\frac{P_{03}}{P_{02}}\frac{P_{02}}{P_{01}} = \frac{\dot{m}\sqrt{T_{03}}}{P_{03}}\sqrt{\frac{T_{01}}{T_{03}}}\pi_{b}\pi_{c} = 98.967 \times 0.97\sqrt{\frac{T_{01}}{T_{03}}} \times \pi_{c}$$

$$\frac{\dot{m}\sqrt{T_{01}}}{P_{01}} = 96\sqrt{\frac{T_{01}}{T_{03}}} \times \pi_{c} \tag{2}$$

Power balance

$$\dot{m}C_{P_{c}}T_{01} \times \frac{1}{\eta_{C}}\left[\pi_{c}^{G_{c}} - 1\right] = \dot{m}C_{P_{t}}T_{03}\eta_{m}\eta_{t}\left[1 - \left(\frac{1}{\pi_{t}}\right)^{G_{t}}\right]$$

where $G_{c} = (\gamma_{c} - 1)/\gamma_{c}$, thus

$$\frac{T_{01}}{T_{03}} = \frac{C_{P_{t}}}{C_{P_{c}}}\eta_{t}\eta_{m}\left[1 - \left(\frac{1}{\pi_{t}}\right)^{G_{t}}\right]\frac{\eta_{C}}{\pi_{c}^{G_{c}} - 1}$$

$$\frac{T_{01}}{T_{03}} = 0.19\frac{\eta_{C}}{\pi_{c}^{G_{c}} - 1} \tag{3}$$

Equations (2) and (3) are used to calculate the mass flow parameter of the compressor that is tabulated in following table against the given input values for the compressor map:

π_{c}	η_{c}	$\frac{T_{01}}{T_{03}}$	$\left(\frac{\dot{m}\sqrt{T_{01}}}{P_{01}}\right)_{calculated}$	$\left(\frac{\dot{m}\sqrt{T_{01}}}{P_{01}}\right)_{map}$
6.0	0.83	0.2356	279.6	250
5.6	0.84	0.2506	269.19	270
5.2	0.83	0.2618	255.5	290

Graphical solution of both values $\left(\left(\dot{m}\sqrt{T_{01}}/P_{01} \right)_{\text{calculated}} \right.$ and $\left. \left(\dot{m}\sqrt{T_{01}}/P_{01} \right)_{\text{map}} \right)$ gives the following results for the compressor operating point:

$$\pi_c = 5.6, \quad \eta_c = 0.84, \quad \frac{T_{01}}{T_{03}} = 0.2506 \quad \text{and} \quad \frac{\dot{m}\sqrt{T_{01}}}{P_{01}} = 269.5$$

The turbine inlet temperature is then

$$T_{03} = \frac{T_{01}}{0.2506} = \frac{288}{0.2506} = 1149.24 \, \text{K}$$

16.4 OFF DESIGN OF TURBOJET ENGINE

There is similarity between the flow characteristics of a nozzle and a turbine [2]. The static operation of the turbojet engine is similar to the operation of a free power turbine. The equilibrium running line of the compressor (Figure 16.12) can be obtained by the same method described in the previous section. During flight, the forward speed produces a ram pressure dependent on the flight speed (Mach number) and intake efficiency as follows:

$$\frac{P_{01}}{P_a} = \left(1 + \eta_d \frac{\gamma - 1}{2} M_a^2 \right)^{\gamma/(\gamma-1)} \tag{16.14}$$

The temperature ratio is

$$\frac{T_{01}}{T_a} = \left(1 + \frac{\gamma - 1}{2} M_a^2 \right) \tag{16.15}$$

The ram pressure will increase the compressor delivery pressure, which in turn increases the outlet pressure of the turbine and thus increases the pressure ratio of the nozzle. Once, the nozzle is choked the nondimensional flow will reach its maximum value and will be independent from the nozzle pressure ratio and in turn the forward speed. This in turn fixes the turbine operating point due to compatibility conditions between the turbine and nozzle. The present aero engines operate with choked nozzle for all its flight envelops except during taxiing, approaching, and landing. The nozzle map is illustrated in Figure 16.14.

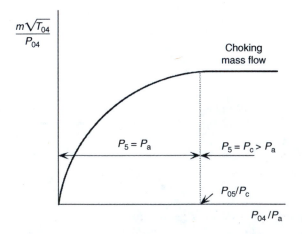

FIGURE 16.14 Nozzle map.

The mass flow parameter of the nozzle is related to the ram pressure by the relation

$$\frac{P_{04}}{P_a} = \frac{P_{04}}{P_{03}} \times \frac{P_{03}}{P_{02}} \times \frac{P_{02}}{P_{01}} \times \frac{P_{01}}{P_a} \qquad (16.16)$$

where P_{01}/P_a is obtained from Equation 16.14.

Referring back to Figure 16.5, the constant temperature ratio (T_{03}/T_{01}) had a straight-line pattern. When these lines are drawn if the gas generator is coupled to a nozzle, each line will have a fan of equilibrium running lines each for a fixed flight Mach number. These lines are found at the region of low compressor pressure ratios and consequently low mass flow rate. As the pressure ratio increases these lines merge into the single running line as before, which represents the choked nozzle operation.

The engine thrust force is then calculated from the relation

$$T = \dot{m}_5 (V_5 - V_f) + A_5 (P_5 - P_a) \qquad (16.17)$$

The flight speed is given by the relation

$$V_f = M_a \sqrt{\gamma R T_a} \qquad (16.18)$$

The exhaust speed (V_5) depends on the nozzle choking.
For a choked nozzle

$$V_5 = \sqrt{\gamma R T_5} = \sqrt{\frac{2\gamma R T_{04}}{\gamma + 1}} \qquad (16.19a)$$

For unchoked nozzle

$$V_5 = \sqrt{2Cp (T_{04} - T_5)} = \sqrt{2CpT_{04}\left[1 - \left(\frac{P_a}{P_{04}}\right)^{(\gamma-1)/\gamma}\right]} \qquad (16.19b)$$

For choked conditions

$$P_5 = P_c = P_{04}\left(1 - \frac{1}{\eta_n}\frac{\gamma - 1}{\gamma + 1}\right)^{\gamma/(\gamma-1)} \qquad (16.20)$$

For unchoked conditions $P_5 = P_a$.

The performance of turbojet engine appears to be dependent on the nondimensional engine speed, but in fact, it is dependent on the actual *mechanical speed* that is controlled by the turbine stresses and governors [2]. The mechanical speed at takeoff represents the maximum speed that may be withstood due to stress limitations. Climb speed will be lower, say 98% and cruise may be less say about 95%. Ambient conditions also have pronounced effects. An increase in ambient temperature will decrease $N/\sqrt{T_a}$ and consequently $N/\sqrt{T_{01}}$. This will move the operating point toward lower values of pressure ratio and mass flow parameter, or be equivalent to lower mechanical speed. This results in an appreciable decrease in the available thrust.

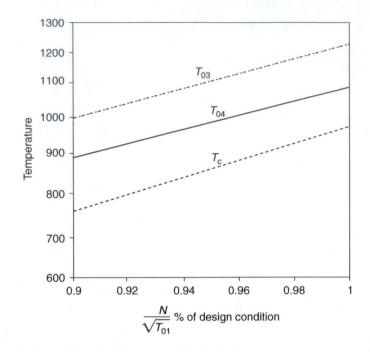

FIGURE 16.15 Turbine inlet and outlet temperature and exhaust temperature versus the engine speed percentage.

Example 3 During a ground test for a turbojet engine at ambient condition of 1 bar and 288 K, the following data are recorded as the engine accelerated from 90% to 100% of the design point:

$N/\sqrt{T_{01}}$% of design	$\dot{m}\sqrt{T_{01}}/P_{01}$	P_{02}/P_{01}	η_c
90%	370	3.6	0.858
95%	421	4.07	0.8585
100%	454	4.6	0.86

The turbine inlet temperature at 90% of design speed is 1000 K. Both the turbine and nozzle are choked all the time and have constant efficiencies of 0.9 and 0.95 respectively.

For choked turbine, the following conditions are to be satisfied:

$P_{04}/P_{03} = $ constant, $T_{04}/T_{03} = $ constant, and N is proportional to $\sqrt{T_{03}}$.

It is required to

- Calculate the turbine inlet and outlet temperature at 95% and 100% of the design speed.
- Calculate the specific thrust of the engine $T/\dot{m}$.
- Plot the parameters $T_{03}, T_{04}, T_e, V_e, T/\dot{m}$ against the nondimensional speed parameters $N/\sqrt{T_{01}}$% of design.

Solution: The ambient conditions are the same as the compressor inlet conditions as the engine is during ground run.

$$T_{01} = 288\,\text{K}, \quad P_{01} = 1\,\text{bar}$$

At 90% of design speed

$$\Delta T_{012} = \frac{T_{01}}{\eta_c}\left[\left(\frac{P_{02}}{P_{01}}\right)^{(\gamma_c - 1)/\gamma_c} - 1\right] = 148.5 \text{ K}$$

$$W_c = W_t$$

$$\Delta T_{034} = \Delta T_{012} \times \frac{Cp_a}{Cp_g} = 130 \text{ K}$$

$$\left(\frac{P_{04}}{P_{03}}\right) = \left[1 - \frac{\Delta T_{034}}{\eta_t T_{03}}\right]^{\gamma_h/(\gamma_h - 1)} = 0.535$$

$$\frac{P_{03}}{P_{04}} = 1.866$$

$$T_{04} = T_{03} - \Delta T_{034} = 870 \text{ K}$$

$$\frac{T_{03}}{T_{04}} = 1.149$$

During acceleration, the pressure and temperature ratio across the turbine is constant for a choked condition. Moreover, the rotational speed is proportional to the square root of the turbine inlet temperature thus,

$$\frac{P_{04}}{P_{03}} = \text{const.}, \quad \frac{T_{04}}{T_{03}} = \text{const.}, \quad \text{and} \quad N\alpha\sqrt{T_{03}}$$

At 95% of design speed

$$\sqrt{\frac{T_{03}}{1000}} = \frac{0.95}{0.9}$$

$$\therefore T_{03} = 1114 \text{ K}$$

At 100% design speed

$$\sqrt{\frac{T_{03}}{1000}} = \frac{1.0}{0.9}$$

$$\therefore T_{03} = 1234.6 \text{ K}$$

The specific thrust at ground run and choked nozzle is given by the relation:

$$\frac{T}{\dot{m}} = V_c + \frac{A_e}{\dot{m}}(P_c - P_a)$$

Here, the governing equations for $\dot{m}$, V_c, P are given:

$$\frac{P_c}{P_{04}} = \left(1 - \frac{1}{\eta_n}\frac{\gamma - 1}{\gamma + 1}\right)^{\gamma/(\gamma - 1)}$$

$$P_C = \frac{P_C}{P_{04}}\frac{P_{04}}{P_{03}}\frac{P_{03}}{P_{02}}\frac{P_{02}}{P_{01}}P_{01}$$

$$T_{04} = T_{03} \times \frac{T_{04}}{T_{03}} = \frac{T_{03}}{1.149}$$

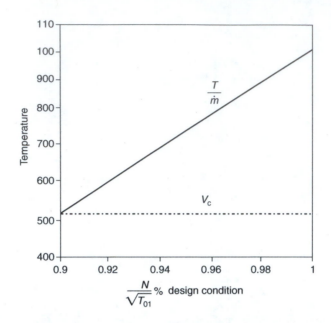

FIGURE 16.16 Specific thrust and exhaust speed versus engine speed percentage.

The exhaust speed is calculated from Equation 16.19a

$$\dot{m} = \left(\frac{\dot{m}\sqrt{T_{01}}}{P_{01}}\right)(P_{01})\left(\frac{1}{\sqrt{T_{01}}}\right) = \left(\frac{\dot{m}\sqrt{T_{01}}}{P_{01}}\right) \times \frac{1}{\sqrt{288}} = \frac{\left(\dot{m}\sqrt{T_{01}}/P_0\right)}{16.97}$$

$$\frac{A_e}{\dot{m}} = \frac{0.5}{\dot{m}}$$

Calculations are arranged in the following table. Plots for the temperature at the inlet and outlet of the turbine as well as at the nozzle outlet against the rotational speed are shown in Figure 16.15. Plots for the exhaust (here critical) speed and the specific thrust against the rotational speed ratio are given in Figure 16.16.

$N/\sqrt{T_{o1}}$% of design value	90%	95%	100%
$\dfrac{P_{02}}{P_{01}} = \dfrac{P_{03}}{P_{01}}$	3.6	4.07	4.6
$\dfrac{P_{04}}{P_{03}}$	0.535	0.535	0.535
$\dfrac{P_{04}}{P_c}$	1.919	1.919	1.919
P_c (bar)	1.0036	1.1346	1.28243
T_{03} (K)	1000	1114	1234.6
T_{04} (K)	870	969.5	1074.4
T_c (K)	745.7	831	920.9
V_c (m/s)	534.18	563.9	593.6
$\dot{m}$ (kg/s)	21.802	24.8	26.752
$\dfrac{A_e}{\dot{m}}$ (m^2 · s/kg)	0.02293	0.02015	0.01869
$\dfrac{T}{\dot{m}}$ (N^2 · s/kg)	542.4	835.119	1121.46

PROBLEMS

16.1 A two-spool gas turbine is fitted with a free power turbine driving a load, and the other spool is a typical gas generator. If the compressor pressure ratio is 2.7. The combustion chamber pressure loss is 3% of the compressor delivery pressure and exhaust pressure loss can be ignored. The turbine characteristics are given below:

Gas generator turbine			Power turbine		
P_{03}/P_{04}	$\dot{m}\sqrt{T_{03}}/P_{03}$	η_t	P_{04}/P_a	$\dot{m}\sqrt{T_{04}}/P_{04}$	
1.35	25	0.85	1.50	65	
1.60	50	0.86	1.70	90	
1.85	65	0.85	1.85	100	

The nondimensional flows quoted are based on $\dot{m}$ in kg/s, and P in bar, and T in K. Calculate the gas generator turbine pressure ratio at this condition.

16.2 The following data refer to a single shaft of a turboprop engine during ground run. The gas turbine is running at its design point.

Compressor characteristics			Turbine characteristics		
P_{02}/P_{01}	$\dot{m}_1\sqrt{T_{01}}/P_{01}$	η_C	$\dot{m}_3\sqrt{T_{03}}/P_{03}$	P_{03}/P_{04}	η_t
3.8	320	0.77	110	4.0	0.88
4.3	300	0.8	135	4.4	0.9
4.8	290	0.83	135	4.8	0.91

The compressor inlet conditions are 1.01 bar and 288 K. The nondimensional flows are based on $\dot{m}$ in kg/s, and P in bar, and T in K. The mechanical efficiency is 0.98. The pressure losses in the combustion chamber are 2%. Assume that the bled air is 3% of the compressor outlet mass flow rate. Calculate the power consumed in the propeller when the turbine inlet temperature is 1200 K.

$N/\sqrt{T_{01}}$ (% design)	$\dot{m}\sqrt{T_{01}}/P_{01}$	P_{02}/P_{01}	η_C
90	380	3.7	0.855
95	430	4.3	0.865
100	460	4.7	0.86

16.3 A turbojet engine when tested at ambient conditions of 1.0 bar and 288 K gives the following results:

The units of the mass flow rate, temperature, and pressure are kg/s, K, and bar.

The turbine inlet temperature at 100% design speed is 1100 K. The turbine efficiency remains constant in all cases at a value of 0.92. Calculate at 100% design speed:

1. The mass flow rate
2. The thrust force

If a similar test is performed in a hot day where the ambient temperature is 303 K and the pressure is also 1.0 bar and the mechanical speed is 100% of the design condition,

Calculate

(a) The new design speed

(b) The new mass flow rate

(c) The thrust force

Assume that the nozzle is choked during all the above tests in cold and hot days and has an exit area of 0.4 m^2. The efficiency of the turbine in the hot day remains 0.92.

16.4 A turbojet engine when tested at ambient conditions of 1.0 bar and 288 K gives the following results:

$N/\sqrt{T_{01}}$% of design	$\dot{m}\sqrt{T_{01}}/P_{01}$	P_{02}/P_{01}	η_c
80	380	3.7	0.855
90	430	4.3	0.865
100	460	4.7	0.86

The units of the mass flow rate, temperature, and pressure are kg/s, K, and bar. The turbine inlet temperature at 100% design speed is 1200 K. The turbine efficiency remains constant in all cases at a value of 0.91.

(a) Calculate at 100% design speed:

1. The mass flow rate

2. The thrust force

(b) If a similar test is performed in a hot day where the ambient temperature is 300 K and the pressure is also 1.0 bar and the mechanical speed is 100% of the design condition, Calculate

1. The new design speed

2. The new mass flow rate

3. The new thrust force

Assume that the nozzle is choked during all the above tests in cold and hot days and has an exit area of 0.45 m^2. The efficiency of the turbine in the hot day remains 0.91.

16.5 The following data refer to a gas turbine with a free power turbine, operating at design speed:

Compressor characteristics			Gas generator turbine characteristics		
P_{02}/P_{01}	$\dot{m}\sqrt{T_{01}}/P_{01}$	η_c	P_{03}/P_{04}	$\dot{m}\sqrt{T_{03}}/P_{03}$	η_t
6.2	250	0.83	2.9	110	0.85
5.8	270	0.84	2.5	110	0.85
5.4	290	0.83	2.1	100	0.85

Assuming that the power turbine is chocked, the value of $\dot{m}\sqrt{T_{04}/P_{04}}$ being 230, determine the design values of compressor pressure ratio and turbine inlet temperature. Assume pressure losses in combustion chamber of 3% and assume the mechanical efficiency of the gas generator rotor to be 0.98 and take the ambient temperature as 288 K. The nondimensional flows quoted are based on $\dot{m}$ in kg/s, and P in bar and T in K, all pressures and temperatures being stagnation values.

16.6 Describe the matching procedure for:

(a) A two-spool turbojet engine

(b) A two-spool unmixed turbofan engine

16.7 Outline the matching procedure for:

(a) A three-spool unmixed turbofan engine

(b) A three-spool mixed turbofan engine

REFERENCES

1. W.W. Bathie, *Fundamentals of Gas Turbines*, 2nd edn., John Wiley & Sons, Inc., 1996, pp. 348–368.
2. H.I.H. Saravanamuttoo, G.F.C. Rogers, and H. Cohen, *Gas Turbine Theory*, Prentice Hall, 5th edn., 2001, pp. 387–410.
3. P. Hill and C. Peterson, *Mechanics and Thermodynamics of Propulsion*, 2nd edn., Addison-Wesley Pub. Co., Inc., pp. 402–406.

Appendix A

GLOSSARY

Absolute pressure   The total pressure measured from absolute zero (i.e., from an absolute vacuum).

Absolute temperature   The temperature of a body with reference to the absolute zero, at which point the volume of an ideal gas theoretically becomes zero. (Fahrenheit scale is $-459.67°$F/Celsius scale is $-273.15°$C).

Adiabatic efficiency   Ratio between measured shaft power and the adiabatic compression power, referring to measured mass flow.

Afterburner   The afterburner is a second combustion chamber; afterburning (or reheat) is a method of augmenting the basic thrust of an engine to improve the aircraft takeoff, climb, and (for military aircraft) combat performance. Afterburning consists of the introduction and burning of fuel between the engine turbine and the jet pipe-propelling nozzle; utilizing the unburned oxygen in the exhaust gas increases the velocity of the jet leaving the propelling nozzle and therefore increases the engine thrust.

ATFI   The abbreviation **ATFI** stands for Advance Technology Fan Integrator, a revolutionary propulsion concept for future aircraft that features a reduction gearbox between the low-pressure turbine and the fan.

Augmenter   Augmenters are afterburners on low-bypass turbofan engines. Core airflow and bypass (fan) airflow are mixed aft of the turbines, in the exhaust. Fuel nozzles supply atomized fuel into the airflow and an igniter ignites the fuel–air mixture. Augmenters are used on low-bypass turbofan engines to increase thrust for short periods during takeoff, climb, and combat flight.

Augmenter exhaust nozzles   Augmenter exhaust nozzles make up the aft end of augmented low-bypass turbofan engines. It has a flame holder, fuel nozzles, an igniter, and a variable exhaust nozzle. The fuel nozzles supply atomized fuel into the exhaust airflow and the igniter makes the fuel–air mixture burn. Augmenter exhaust nozzles are used on low-bypass turbofan engines to increase thrust.

Axial-flow compressor   A compressor through which the air passes mostly in axial direction, that is, along the engine axis; in a centrifugal-flow compressor the air flow direction is transverse to the engine axis. Both variants are still in use today. Their application depends on the engine concept in each case.

Blade load   Blade load is the ratio between the static pressure differential upstream and downstream of the cascade and the dynamic inlet pressure or, more generally, the total of aerodynamic loads acting on the airfoil.

Booster compressor   Machine for compressing air or gas from an initial pressure, which is above atmospheric pressure, to a still higher pressure.

Brake horse power (BHP)   The maximum rate at which an engine can do work as measured by the resistance of an applied brake, expressed in horsepower.

Bypass ratio (BPR)   With commercial engines the ratio between cold and hot airflow is very high. A BPR of 6:1, for example, means that the air volume flowing through the fan and bypassing the core engine is six times the air volume flowing through the core engine.

Can-type combustion chamber   Can-type combustion chambers are particularly suitable for engines with centrifugal-flow compressors as the airflow is already divided by the compressor outlet diffusers. Each flame tube has its own secondary air duct. The separate flame tubes are all interconnected. Ignition problems may occur, particularly at high altitudes. The entire combustion section consists of 8–12 cans that are arranged around the engine. Individual cans are also used as combustion chambers for small engines or auxiliary power units.

Cascade spacing   Distance between the skeleton lines of two adjacent airfoils.

Centrifugal flow compressor   The direction of the flow of compressed air discharges outward at 90° to the spool axis. The centrifugal compressor consists of a rating impellor, a fixed diffuser, and a manifold that collects and turns the compressed air.

Centrifugal flow turbojet   A turbojet engine with a centrifugal compressor rather than an axial flow compressor.

Cogeneration   (1) Any of several processes that either use waste heat produced by electricity generation to satisfy thermal needs, or process waste heat to electricity, or produce mechanical energy or (2) the use of a single prime fuel source in a reciprocating engine or gas turbine to generate both electrical and thermal energy to optimize fuel efficiency. The dominant demand for energy may be either electrical or thermal. Usually it is thermal with excess electrical energy, if any, being transmitted into the local power supply lines.

Combustion chamber   The purpose of the combusting chamber is to provide a stream of hot gas that releases its energy to the turbine and nozzle sections of the engine. There are three types of combustion chambers.

Component efficiency   Component efficiency is the efficiency of individual engine components, such as compressor, combustion chamber, or turbine. One constituent of the component efficiency is the mechanical efficiency, which takes bearing losses and friction between gas and air flows and engine components into account.

Compressor   The compressor is the first component in the engine core. The compressor squeezes the air that enters it into smaller areas, resulting in an increase in the air pressure. This results in an increase in the energy potential of the air. The normal parts of a compressor include compressor front frame, compressor casing with stator vanes, a rotor with rotor blades, and a compressor rear frame.

Core   The core engine module is aft of the fan module and forward of the turbine stator case and is made up of three components; compressor rotor and stator, combustion liner, and stage 1 HPT nozzle. The core is responsible for supplying approximately 20% of the total engine thrust and the torque for operation of all accessories.

Cowling   A removable metal covering placed over and around an airplane's engine(s).

Degree of reaction   The degree of reaction is the ratio between the specific flow in the rotor (static change over rotor) and the specific flow work between inlet and outlet of the stage (static change over stage).

3D Design   Today, 3D methods are used to design the blading of turbomachinery components (compressors and turbines). This means that the airfoils are no longer designed section by section to obtain the necessary deflection of the gas flow but over their entire radial height so that a rather uniform load distribution is achieved and local flow separations and excessive losses are prevented.

Diffuser   A stationary passage surrounding an impeller in which velocity pressure imparted to the flow medium by the impeller is converted into static pressure.

Engine 3E   Engine 3E is a German Government–sponsored research program where 3E stands for environment, efficiency, and economy. Under this program, Motoren- und Turbinen-Union (MTU) is maturing technologies to reduce noise levels by 10%, carbon dioxide emissions by 20%, and NOx emissions by 85% by the year 2010.

Engine price   The price of a production engine for the Airbus A320 aircraft family, for example, amounts to approximately US $5.5 million and that of a jumbo jet engine is in the region of US $10 million.

Exhaust   The exhaust section is located behind the turbine section and at the rear of the engine. It is made up of either a fixed or variable nozzle assembly, depending on the aircraft application. The exhaust section directs the exhaust gases aft and further accelerates the exhaust gases to produce forward thrust. Variable nozzles are usually found on military engines while fixed ones are typically associated with commercial turbofans.

Fan   The fan is the first component on the engine. The spinning fan sucks in large quantities of air. Most blades of the fan are made of titanium. It then speeds this air up and splits it into two parts. One part continues through the "core" or center of the engine, where it is acted upon by the other engine components. The fan module typically supplies approximately 80% of the engine thrust. The second part "bypasses" the core of the engine. It goes through a duct that surrounds the core to the back of the engine where it produces much of the force that propels the airplane forward. This cooler air helps quiet the engine as well as adding thrust to the engine.

Fuel consumption   It is the fuel consumed in liters for traveling 100 km per passenger seat. For a medium range aircraft it is 5–6 liters while for a long range—like Airbus A340—it is 3.7 liters/seat.

Gas generator   The combination of the compressor, combustion chamber, and the turbine.

Gas turbine   The turbine converts gas energy into mechanical work to drive the compressor by a rigid shaft.

Geared fan   Normally, the fan, the low-pressure compressor, and the low-pressure turbine are fitted on one shaft. In contrast, a geared fan is "uncoupled" from the low-pressure system by means of a reduction gearbox. Thus, the low-pressure turbine and the low-pressure compressor can be operated at their respective optimum high speeds whereas the fan rotates at a much lower speed (ratio approx. 3:1). Advantages of the geared fan concept include a markedly improved overall engine efficiency and signification noise reduction.

High-pressure turbine (HPT)   The HPT module is aft of the compressor rear frame and forward of the LPT stator case. The HPT module is made up of the HPT rotor and HPT stator and it removes energy from the combustion gases to turn the high-pressure compressor and accessory gearbox.

Horsepower (HP)   A unit of power equal to 33,000 ft-lb of work per minute.

Hydrocarbons   Chemicals containing carbon and hydrogen.

Hypersonic   Very fast speed of flight, 3500–7000 mph (or Mach 5–10). The space shuttle travels this fast, once it is in space.

IGV   Inlet guide-vane valve: a valve assembly at the air inlet of a "blower" (single stage, low pressure, centrifugal air compressor) usually advised to be mounted in *very close proximity* to the "blower" impeller. Provides "pre-swirl" of airflow in same rotational direction as "blower" impeller. Proven to improve efficiency (reduced BHP) during *throttled-down modulation* of "blowers." Effectiveness, when used with multistage centrifugal air compressors, degrades rapidly.

Impeller   The part of the rotating element of a dynamic compressor that imparts energy to the flowing medium by means of centrifugal force. It consists of a number of blades mounted so as to rotate with the shaft.

Impingement cooling   A high-velocity cold air jet is directed from a hole vertically onto the component surface to be cooled. As the cooling air jet hits the surface, it is diverted in all directions parallel to the impingement surface. The cooling effect is high, but decreases continuously as the distance from the point of impingement increases.

Industrial gas turbine   The operating principle of an individual gas turbine is essentially the same as that of an aero engine. However, a so-called power turbine is used in place of the low-pressure turbine of an aero engine that drives the fan. This power turbine delivers the necessary power—directly or via a gearbox—to a generator or pump, and so forth. Nearly all industrial gas turbines of the lower and intermediate power classes are euro engine derivatives.

Jet engine   Ant of a class of reaction engines that propel aircraft by means of the rearward discharge of a jet of fluid, usually hot exhaust gases generated by burning fuel with air drawn in from the atmosphere. The aircraft engine provides a constant source of thrust to give the airplane forward movement.

Low-pressure compressor   To achieve pressure ratios of over 30 in present-day engines, two different types of compressors are used: low-pressure compressors and high-pressure compressors.

These compressors that rotate at different speeds are driven by their counterparts in the turbine via concentric shafts.

Low-pressure turbine   The LPT module is in the rear of the engine, aft of the HPT stator case. LPT components include the LPT rotor, LPT nozzle stator case, and turbine rear frame. The LPT removes energy from the combustion gases to drive the low-pressure compressor (N1) rotor assembly.

Mixer   A mixer is a sheet-metal cone at the rear end of some engines which, owing to its special wavy form, ensures intensive mixing of the high velocity and, thus, loud exhaust gases exiting the low-pressure turbine with the slowly flowing bypass air. This results in a marked reduction of noise levels.

Multistage centrifugal compressor   A machine having two or more impellers operating in series on a single shaft and in a single casing.

Multistage compressor   A machine employing two or more stages.

Normal turbine stage   The normal turbine stage consists of a stator and a rotator, with the rotor being arranged downstream of the stator.

Nozzle   The nozzle is the exhaust duct of the engine. This is the engine part that actually produces the thrust for the plane. The energy depleted airflow that passed the turbine, in addition to the colder air that bypassed the engine core, produces a force when exiting the nozzle that acts to propel the engine, and therefore the airplane, forward. The combination of the hot air and cold air is expelled and produces an exhaust, which causes a forward thrust.

Pressure ratio   The pressure ratio indicates the ratio between the pressure of the air entering the engine and the pressure of the air leaving the compressor. A pressure ratio of 30:1 (1 being the pressure of the inlet air) or just 30 thus means that the air pressure at the compressor exit (just upstream of the combustion chamber inlet) is 30 times the pressure value at the compressor inlet (first compressor stage or, in most engines, fan).

Principle of energy transfer   To allow the transfer of energy in a turbomachine the flow of the working fluid must be deflected in bladed rotors. This means that the generation of energy (compressor) and absorption of energy (turbine) take place in the rotors only, with the absolute flow being unsteady.

Propeller spinner   A cone-shaped piece of an airplane, mounted on a propeller, which reduces air resistance or drag.

Prop fan   The prop fan features a huge fan propeller that is driven by a turbine and allows bypass ratios of 15:1 or more to be achieved. Fuel consumption, noise, and emissions are significantly reduced.

Propulsion   (As a field of study in relation to aeronautics.) Is the study of how to design an engine that will provide the thrust that is needed for a plane to take off and fly through the air.

Propulsion efficiency   The external efficiency—also called propulsion efficiency—is a measure of the quality of an engine. It depends on the outlet velocity/flight velocity ratio, that is, the smaller this ratio the higher is the propulsion efficiency.

Pusher engine   A turboprop engine with a propeller mounted behind the engine and thus the generated thrust forces pushes the airplane. The engine is then identified as pusher, while if the propeller is in front of the engine it is called puller as it pulls the airplane.

Radial-flow machine   The flow direction is predominantly radial. Radial-flow machines feature centrifugal stages (flow from inside to outside) and centripetal stages (flow from outside to inside). With a view to ensuring the necessary energy transfer, the centrifugal principle is used in the compressor and the centripetal principle in the turbine.

Ram   The amount of pressure buildup above ambient pressure at the engine's compressor inlet, due to forward motion of the engine through the air—air's initial momentum.

Ramjet   A jet engine with no mechanical compressor, consisting of specially shaped tubes or ducts open at both ends. The air necessary for combustion is shoved into the duct and compressed by the forward motion of the engine.

Ram ratio   The ability of an engine's air inlet duct to take advantage of ram pressure.

Regimes of flight   The ranges of speed at which airplanes fly. Subsonic: 100–350 mph; transonic: 350–750 mph; supersonic: 760–3500 mph; hypersonic: 3500–7000 mph.

Reheaters   Heat exchangers for raising the temperature of compressed air to increase its volume.

Reverse-flow annular combustion chamber   The advantages and disadvantages of reverse–flow annular combustion chambers are the same as with annular combustion chambers. They reduce the length of the engine, however, at the expense of a larger cross-section. Reverse-flow annular combustion chambers are used mainly in engines whose last compressor stage is a centrifugal-flow compressor.

Single-crystal turbine blades   Single-crystal turbine blades have been developed for use in the high-temperature environment of advanced aero engines. With single-crystal blades the cooling process is controlled such that a single crystal is produced.

Slip   The internal leakage within a rotary compressor. It represents gas at least partially compressed but not delivered. It is experimentally determined and expressed in CFM to be deducted from the displacement to obtain capacity.

Solar-powered aircraft   Solar-powered aircraft, such as the pathfinder; use photovoltaic cells to convert energy from the sun into electricity to power electric motors that drive the aircraft.

SPC   Specific power consumption.

Specific fuel consumption   The ratio of fuel consumption to compressor capacity.

Specific work of a stage   The mechanical work of a stage per 1 kg mass flow is the specific work of a stage.

Speed of sound   When a plane travels faster than at 760 mph, a sound barrier forms in front of the plane. If a plane is going at the speed of sound it is traveling at Mach 1.

Stage   A stage is the combination of a stator and a rotor.

Stall   A flight condition wherein the airflow separates from the airfoil surface, or the airflow around the airfoil becomes turbulent, causing the airfoil to lose lift. It is usually a result of insufficient airspeed or excessive angle of attack.

Straight cascade   Blades/vanes arranged at regular intervals along a straight line, that is, the cascade axis (two-dimensional geometry).

Surge   The reversal of flow within a dynamic compressor that takes place when the capacity being handled is reduced to a point where insufficient pressure is being generated to maintain flow (also known as pumping).

Thermal barrier coatings   During operation, engine components are subjected to a combination of mechanical, thermal, and chemical loads. They must therefore be protected by high-temperature-resistant coatings capable of withstanding temperatures as high as 2500°C. Thermal barrier coatings are mainly used in combustion chambers and on turbine blades and vanes.

Thermal efficiency   The thermal or internal efficiency of an aero engine is the ratio between the amount of heat converted to work and the total amount of heat supplied. The thermal efficiency is a measure of the quality of an engine.

Thrust   The forward force that pushes the engine and, therefore, the airplane forward. Sir Isaac Newton discovered that for "every action there is an equal and opposite reaction." Aircraft engine uses this principle.

Thrust class   Turbojet engines are grouped into three thrust classes: engines with a thrust of up to 20,000 lbs, engine with a thrust between 20,000 lbs and ~ 50,000 lbs, and engines with a thrust in excess of 50,000 lbs.

Thrust, gross   The thrust developed by the engine, not taking into account any presence of initial-air–mass momentum.

Thrust, net   The effective thrust developed by the engine during the flight, taking into consideration the initial momentum of the aircraft speed mass before entering the influence of the engine.

Thrust reverser   Thrust reversers serve as an aircraft's main brakes on landing. There are three types of thrust reversers: translating cowl, clamshell, and turboprop reverse pitch. All three literally reverse the engine's thrust by closing in when deployed by the pilot pushing the air out the front of the engine rather than the back. This motion decreases the speed of the aircraft and is the loud noise you hear when landing.

Thrust, specific fuel consumption   The fuel that the engine must burn to generates 1 lb of thrust.

Thrust, static   Thrust developed by engine, without any initial air—mass momentum due to engine's static state.

Thrust-to-weight ratio   The thrust-to-weight ratio is a characteristic value indicating the technical standard of an engine. The thrust-to-weight ratio of present-day military engines, for example, is 10:1.

Turbine   The high-energy gas flow coming out of the combustor goes into the turbine, causing the turbine blades to rotate. This rotation extracts some energy from the high-energy flow that is used to drive the fan and the compressor. The gases produced in the combustion chamber move through the turbine and spin its blades. The task of a turbine is to convert gas energy into mechanical work to drive the compressor.

Turbofan engine   In turbofan engines only a small part of the thrust is generated in the core engine. The major part is produced by the bypass flow through the fan. The fan feeds large air volumes downstream that surround the engine like an envelope and thus reduce the engine noise. With increasing bypass ratio, that is, ratio between internal and external airflow, operating cost, emissions, and noise are reduced. Bypass ratios of turbofans for modern airlines are now about 6:1, that is, the bypass flow through the fan and around the engine is six times the airflow through the core engine.

Turbojet engine   Turbojet engines, the first jet engine generation, are single-flow engines in which the total volume of air ingested flows through the compressor, the combustion chamber, and the turbine. The thrust is then produced by the exhaust gases exiting the engine at high velocities. Because of the low efficiency and the high noise levels, turbojet engines are no longer produced today.

Turboprop   A turboprop is a jet engine attached to a propeller. The turbine at the back is turned by the hot gases generated by the engine, and this turns a shaft that drives the propeller. A variety of smaller aircraft are powered by turboprops. Like the turbojet, the turboprop engine consists of a compressor, combustion chamber, and turbine, and the air and gas pressure is used to run the turbine, which then creates power to drive the compressor. Compared with a turbojet engine, the turboprop has better propulsion efficiency at flight speeds below about 500 mph. Modern turboprop engines are equipped with propellers that have a smaller diameter but a larger number of blades for efficient operation at much higher flight speeds. To accommodate the higher flight speeds, the blades are scimitar-shaped with swept-back leading edges at the blade tips. Engines featuring such propellers are called propfans.

Turboshaft engine   Turboshaft engines are mainly used to power helicopters. They may include an additional free power turbine with a gearbox to convert the thrust into driving power for the rotor.

Two-stage compressor   Machines in which air or gas is compressed from initial pressure to an intermediate pressure in one or more cylinders or casings.

Variable-pitch propeller   An engine-driven device, designed to drive an airplane forward, whose efficiency can be improved by turning its blades in midflight.

Velocity triangle   The velocity triangle is the geometric representation of the kinematics condition.

Zeppelin   Named after its German inventor, this airship craft is controllable, powered, and lighter than air, with a rigid structure.

Appendix B

TURBOFAN

High-Thrust Class Turbofan Engines (>200 kN) [1]

	GE-90	CF6-50C2	CF6-80C2
Company	General Electric (USA)	General Electric (USA)	General Electric (USA)
In use since	September 1995	1978	October 1985
First flew on	Airbus A-340 & B-777	KC-10 (Military)	A-300/310, 747/767
Description	High bypass TF	Two-shaft high BPR TF	Two-shaft high BPR TF
Weight (Dry)	—	3960 kg	4144 kg
Overall Length	4775 mm	4394 mm	4087 mm
Intake/fan diameter	3124 mm	2195 mm	2362 mm
Pressure ratio	39.3	29.13	30.4
Bypass ratio	8.4	5.7	5.05
Thrust at TO	388.8 kN	233.5 kN	276 kN
Thrust during Cruise	70 kN	50.3 kN	50.4 kN
S.F.C. (SLS)	8.30 mg/N · s	10.51 mg/N · s	9.32 mg/N · s
Air mass flow rate	1350 kg/s	591 kg/s	802 kg/s
Presence of FADEC*	Yes	No	Yes
Other information	33% lower NO_x emission. Less noise than other TFs in its class (due to low fan tip speed)	TET of LPT is 1144 K	Lower fuel burn (s.f.c.) than other engines, long life, high reliability

* FADEC: Full Authority Digital Electronic Control.

	RB-211-524G/H	Trent-882	JT-9D-7R4
Company	Rolls Royce (UK)	Rolls Royce (UK)	Pratt & Whitney (USA)
In use since	February 1990	August 1994 (Cert.)	February 1969 (first)
First flew on	747-400 and 767-300	Boeing 777	Boeing 747/767, A310
Description	Three-shaft axial TF	Three shaft TF	Twin-spool TF
Weight (Dry)	4479 kg	5447 kg	4029 kg
Overall length	3175 mm	4369 mm	3371 mm
Intake/fan diameter	2192 mm	2794 mm	2463 mm
Pressure ratio	33	33+	22
Bypass ratio	4.3	4.3+	5
Thrust at TO	269.4 kN	366.1 kN	202.3 kN
Thrust during cruise	52.1 kN	72.2 kN	176.3 kN
S.F.C.	15.95 mg/N · s (cruise)	15.66 mg/N · s (cruise)	10.06 mg/N · s

(Continued)

(Continued)

	RB-211-524G/H	Trent-882	JT-9D-7R4
Air mass flow rate	728 kg/s	728+ kg/s	687 kg/s
FADEC (Y/N)	No	Yes	No
Other information		Most powerful conventional a/c engine in contract (till September 1995) in the world (Trent 772)	

[1] http://www.anirudh.net/seminar/ge90.pdf

Low-Thrust Class Turbofan Engines (<200 kN) [1]

	CFM56-5C2	JT-8D-17R	V 2500-A1
Company	CFM International (France) & GE (USA)	Pratt & Whitney (USA)	Intl. Aero Engines (USA)
In use since	Late 1992	February 1970	July 1988
First flew on	Airbus A-340	Boeing 727/737 & DC-9	Airbus A-320
Description	Two-shaft subsonic TF	Axial flow twin-spool TF	Twin-spool subsonic TF
Weight (Dry)	2492 kg (Bare Engine) 3856 kg (approx.)	1585 kg	2242 kg (Bare Engine) 3311 kg (with powerplant)
Overall length	2616 mm	3137 mm	3200 mm
Intake/fan diameter	1836 mm	1080 mm	1600 mm
Pressure ratio	37.4	17.3	29.4
Bypass ratio	6.6	1.00	5.42
Thrust at TO	138.8 kN	72.9 kN	111.25 kN
Thrust during cruise	30.78 kN	18.9 kN	21.6 kN
S.F.C.	16.06 mg/N · s	23.37 mg/N · s	16.29 mg/N · s
Air mass flow rate	466 kg/s	148 kg/s	355 kg/s
FADEC (Y/N)	Yes	No	Yes
Other information			

[1] http://www.anirudh.net/seminar/ge90.pdf

Appendix C

Samples of Gas Turbines (Representative Manufacturers)

Model	Year	Rating (kW)	Pressure Ratio	Mass Flow (lb)	Turbine Speed (rpm)	Turbine Inlet Temperature	Exhaust Temperature	Approximate Dimensions L × W × H
ABB Alstom Power								
GT35	1968	17,000	12.0	203.0 lb	3000 3600	—	374°C	14.1 × 4.0 × 3.7 m
GT 11 N2	1993	113,700	15.1	842.0 lb	3000	—	524°C	8.5 × 5.5 × 10.0 m
GT 26	1994	265,000	30.0	1238.0 lb	3000	—	640°C	12.3 × 5.0 × 5.5 m
AlliedSignal								
ASE40	1978	3284	8.8	28.1 lb	15,400	2022°F	1116°F	4.3 × 2.5 × 3.0 ft
ASE120	2000	9784	20.5	69.2 lb	7910	2321°F	970°F	8.5 × 5.9 × 4.0 ft
Bharat Heavy Electronics								
PG6561(B)	2000	42,100	12.2	311.0 lb	5163	—	1011°F	11.6 × 5.0 × 3.8 m
PG9351(FA)	1996	255.600	15.4	1422.0 lb	3000	—	1129°F	22.6 × 5.0 × 5.4 m
Dresser-Rand								
DR60G	1990	13,958	21.5	103.0 lb	7000	1358°F	909°F	42.0 × 11.5 × 10.5 ft
DR61GP	1998	28,429	21.5	186.3 lb	3600	1490°F	941°F	48.0 × 11.5 × 18.0 ft
Ebara								
PW-12E	1990	1180	7.3	14.1 lb	1500/ 1800	—	598°C	5.5 × 3.2 × 2.2 m
FT8 Twin	1990	51,500	20.2	371.0 lb	3000/ 3600	—	458°C	39.0 × 12.0 × 9.0 m
GE Energy Products, Europe								
PG6101(FA)	1993	70,140	15.0	437.0 lb	5235	—	1107°F	120.0 × 20.0 × 34.0 ft
PG9351(FA)	1996	255,600	15.4	1375.0 lb	3000	—	1129°F	112.0 × 25.0 × 50.0 ft

Continued

(Continued)

Model	Year	Rating (kW)	Pressure Ratio	Mass Flow (lb)	Turbine Speed (rpm)	Turbine Inlet Temperature	Exhaust Temperature	Approximate Dimensions $L \times W \times H$
GE Industrial Aeroderivative Gas Turbine								
LM1600-PB STIG	1991	16,900	25.1	116.0	7000	1355°F	878°F	15.0 × 8.0 × 7.0 ft
LM6000-PC	1997	44,090	29.4	280.0	3600	1540°F	842°F	16.3 × 7.0 × 7.0 ft
GE Power Systems								
PG9351(FA)	1996	255,600	15.4	1475.0	3000	—	1129°F	112.0 × 25.0 × 50.0 ft
Hitachi								
PG6101(FA)	1993	70,140	15.0	433.0	5247	—	1107°F	120.0 × 20.0 × 34.0 ft
PG9331(FA)	1995	243,000	14.8	1422.0	3000	—	1106°F	112.0 × 25.0 × 50.0 ft
Ishikawajima-Harima Heavy Industries								
LM1600PA	1988	13,900	22.3	103.6	7000	1406	914°F	4.3 × 2.6 × 2.2 m
STIG-IM5000	1986	51,160	30.6	344.0	3600	1450°F	727°F	—
Kawasaki Heavy Industries								
M1A-11	1989	1235	9.3	18.0	1500/1800	—	463°C	2.5 × 1.6 × 2.0 m
M7A-02	1997	6958	15.9	59.8	1500/1800	—	525°C	3.7 × 1.5 × 1.7 m
Mitsubishi Heavy Industries								
ASE40	1996	2600	8.4	27.1	15,400	—	1042°F	1.3 × 0.9 × 1.1 m
M501G	1997	254,000	20.0	1250.0	3600	—	1105°F	15.2 × 4.6 × 5.0 m
M701g	1997	334,000	21.0	1625.0	3000	—	1089°F	18.2 × 6.2 × 6.2 m
Mitsubishi Engineering and Shipbuilding								
MSC100	1989	10,690	17.1	91.8	1500/1800	—	488°C	14.5 × 2.8 × 3.3 m
MTU Motoren- und Turbinen-Union Friedrichshafen GmbH								
LM6000PC	1997	20,070	29.4	280.0	3600/1500	1540°F	842°F	16.3 × 7.0 × 7.0 ft
Nuovo Pignone-Turbotecnica								
MS9001FA	1991	255,600	15.4	1406.5	3000	—	1129°F	—
Pratt & Whitney Canada								
ST6L-813	1978	848	8.5	8.6	300000	—	1051°F	4.2 × 1.4 × 1.6 ft
ST40 Water Inj.	1999	4547	19.0	32.9	14875	—	1010°F	5.4 × 2.2 × 3.2 ft
Rolls-Royce								
Avon	1964	14,580	8.8	173.0	5500	—	827°F	50.0 × 13.5 × 13.5 ft

Model	Year	Rating (kW)	Pressure Ratio	Mass Flow (lb)	Turbine Speed (rpm)	Turbine Inlet Temperature	Exhaust Temperature	Approximate Dimensions *L* × *W* × *H*
Siemens, Siemens Westinghouse								
RB211-6562DLE	1992	27,516	20.8	202.2	4800	—	932°F	71.0 × 13.5 × 13.5 ft
Trent	1996	51,190	35.0	351.0	3600	—	800°F	65.9 × 14.1 × 14.3 ft
V94.2	1981	157,000	11.1	1122.0	3000	2150°F	999°F	45.9 × 41.0 × 27.6 ft
V94.3A	1995	258,000	17.0	1398.0	3000	2400°F	1054°F	41.0 × 20.0 × 24.6 ft
Solar Turbines								
Mercury 50	1998	4072	9.1	37.1	14170	—	663°F	35.1 × 10.1 × 13.7 ft
Titan 130	1998	13,500	16.0	110.1	11200	—	913°F	48.0 × 10.4 × 13.4 ft
Sulzer Turbo								
R7	1970	10,600	7.7	140.0	6400	925°C	342°C	38.0 × 12.0 × 13.0 ft
Toshiba								
PG7231FA	1994	171,700	15.5	925.0	3600	—	1116°F	180.0 × 75.0 × 31.0 ft
PG9351FA	1996	255,600	15.4	1375.0	3000	—	1129°F	112.0 × 25.0 × 50.0 ft
Turbo Power								
FT8 Twin Pac	1990	51,290	20.2	376.0	3000/ 3600	—	851°F	130.0 × 40.0 × 30.0 ft
Volvo Aero								
VT4400	1999	4,400	12.4	44.5	7200	1052°C	481°C	1.5 × 1.0 × 1.0 m

Index